DOE/RL-97-1047

Hanford Site Historic District

History of the Plutonium Production Facilities 1943-1990

Manhattan Project 1943 - 1946

Cold War Era 1947 - 1990

Hanford Cultural and Historic Resources Program
U.S. Department of Energy

Battelle Press
Columbus • Richland

This document was prepared by the Pacific Northwest National Laboratory for the
U.S. Department of Energy, Richland Operations Office
Hanford Cultural and Historical Resources Program
formerly under the direction of Dee Lloyd, currently under the direction of Annabelle L. Rodriguez.
Contract DE-AC06-76RL01830

Chris Willingham — **Release Approval** 4-18-02 — **Date**

Library of Congress Cataloging-in-Publication Data

Hanford Site historic district : history of the plutonium production facilities, 1943-1990 / Hanford Cultural and Historic Resources Program, U.S. Department of Energy.
p. cm.
Includes bibliographical references and index.
ISBN 1-57477-133-7 (pbk. : alk. paper)
1. Nuclear facilities—Washington (State)—Richland—History. 2. Hanford Site (Wash.)—History. 3. Nuclear weapons—Materials. 4. Plutonium. I. Hanford Cultural and Historic Resources Program.
TK9025 .H36 2003
623.4'5119'0979751—dc21 2002034473

Printed in the United States of America

Battelle Press
505 King Avenue
Columbus, Ohio 43201, USA
614-424-6393 or 1-800-451-3543
Fax: 1-614-424-3819
Internet: press@battelle.org
Web Bookstore: www.battelle.org/bookstore

Collage l-r: B Reactor, 1940s Construction Workers at Lunch, N Reactor

This book is printed on 30% post-consumer recycled, acid free, and ground-wood free paper.

AUTHORS

THOMAS E. MARCEAU – Bechtel Hanford, Inc.
Supervisor, Environmental Restoration Contractor Cultural Resources Program, 1994-Present
Chair, Hanford Site Historic Buildings Task Group, 1994-1997
Author   Chapter 1 - Historic Overview
          Chapter 4 - Recommendations for Future Uses of Structures

DAVID W. HARVEY – Pacific Northwest National Laboratory
Research Scientist/Historian, Hanford Cultural and Historic Resources Program, 1993-Present
Key Contributor, Hanford Site Historic Buildings Mitigation Project, 1994-Present
Author   Section 1 - Construction History
          Section 2 - Fuel Manufacturing
          Section 9 - Military Operations
          Appendix B - Metal Fuels Fabrication Building
Co-Author   Section 8 - Site Security
          Section 11 - Transportation, Communication, and General Site Support
          Appendix B - Development Laboratory Complex
          Appendix B - Plutonium Concentration Facility
          Appendix B - Waste Incinerator Facility

DARBY C. STAPP – Pacific Northwest National Laboratory
Manager, Hanford Site Historic Buildings Mitigation Project, 1999-Present
Director, Hanford Cultural Resources Laboratory, 1998-Present
Author   Section 3 - Reactor Operations
          Section 12 - History of Workers
Co-Author   Section 7 - Research and Development

* * * * * * * *

SANDRA D. CANNON – Pacific Northwest National Laboratory
Communications Manager, Hanford Site Historic Buildings Mitigation Project, 1996-Present
Author   Chapter 3 - Guide to Resources

CHARLES A. CONWAY – Radio-Chemical Process Consultant
Consultant to the Hanford Site Historic Buildings Mitigation Project, 1999
Co-Author   Section 4 - Chemical Separations

DENNIS H. DEFORD – Hanford Search
President and Senior Researcher, Hanford Search
Retired Senior Scientist CH2M HILL Hanford, Inc. and Westinghouse Hanford Company
Retired Senior Management Analyst, U.S. Department of Energy
Author   Section 6 - Waste Management

BRIAN J. FREER – CH2M HILL Hanford, Inc.
Researcher, Historical Buildings Mitigation Project, 1998-1999
Author   Section 10 - Worker Health and Safety
Co-Author   Section 4 - Chemical Separations

MICHELE S. GERBER – Fluor Hanford, Inc.
Historian, Hanford Site
Author   Section 5 - Plutonium Finishing
Co-Author   Appendix B - Development Laboratory Complex
          Appendix B - Plutonium Concentration Facility
          Appendix B - Waste Incinerator Facility

JOY K. KEATING – Pacific Northwest National Laboratory
Researcher, Hanford Site Historic Buildings Mitigation Project, 1998-2000
Co-Author   Section 8 - Site Security
          Section 11 - Transportation, Communication, and General Site Support

CHRISTINE F. NOONAN – Pacific Northwest National Laboratory
Researcher, Hanford Site Historic Buildings Mitigation Project, 1999-2000
Co-Author   Section 7 - Research and Development

GENE WEISSKOPF - B Reactor Museum Association
Project Coordinator and Lead Author for the B Reactor Museum Association
Author   Appendix B - B Reactor
          Appendix B - T Plant

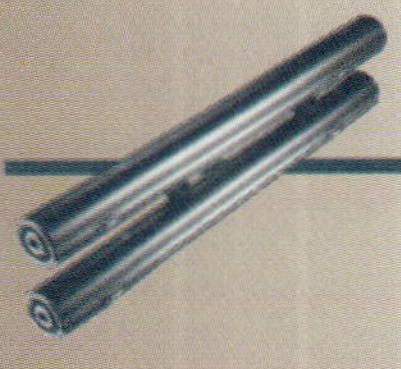

CONTENTS

ACKNOWLEDGEMENTS

Numerous people shared their knowledge of facilities and processes at the Hanford Site with us during the course of writing this book. Numerous others reviewed and made recommendations on early drafts. We specifically requested experts in fields applicable to this history of facilities to review an early draft, but many other reviewers read the book and made recommendations out of shear interest in helping us achieve as accurate a portrayal of the facilities as possible. To all who have helped improve this book, we offer our sincere thanks!

CONTRIBUTORS

Paul Beardsley, retired Hanford Patrol – Site Security

Bruce Cameron, Hanford Patrol Training Instructor – Site Security

Roy Gephart, Waste Management Technical Expert – Waste Management

Shirley Gydesen, Classification Officer – Site Security

Don Sorenson, Plutonium Finishing Technologist – Plutonium Finishing

Jim Spracklen, Safeguards and Security Director – Site Security

Judy Tokarz-Hames, Emergency Preparedness – Emergency Preparedness subsection in Site Security

Al Walker, Security Operations Program Manager – Site Security

Everett Weakley, Senior Principal Engineer – Fuel Manufacturing

TECHNICAL PEER REVIEWERS

Marie Boutte, Medical Anthropologist, University of Nevada at Reno – Health and Safety

Rodney Carlisle, Professor of History, Rutgers University – Fuel Manufacturing, Reactor Operations, Plutonium Finishing

Cathryn Carson, Assistant Professor of History of Science, University of California at Berkeley – Reactor Operations

Sharon Ghamari-Tabrizi, Senior Research Fellow in Cold War Science – Military Operations, Overview of Book

Hugh Gusterson, Associate Professor of Anthropology & Science, Massachusetts Institute of Technology – Research and Development

Louise Kaplan, Assistant Professor of Nursing, Pacific Lutheran University – Health and Safety

Ronald Kathren, Professor Emeritus in Biokinetics and Radiological Health, Washington State University – Waste Management and Health and Safety

Kevin Kennedy, Lecturer of Environmental Sciences, University of California at Berkeley – Chemical Separations

Scott Kirsch, Assistant Professor, Department of Geography, University of North Carolina – Site Security

Russell Olwell, Lecturer of Natural Sciences, University of Michigan – Site Security and History of Workers

Joshua Silverman, Doctoral Candidate in History, Carnegie Mellon University – Waste Management

John K. Smith, Associate Professor of History, Lehigh University – Chemical Separations

Andrea Szasz, Associate Professor of Sociology, University of California at Santa Cruz – Waste Management and History of Workers

Bryan Taylor, Associate Professor of Communications, University of Colorado at Boulder – Construction History, Overview of Book

Harry Thayer, Author of *Management of the Hanford Engineer Works in World War II: How the Corps, Du Pont and the Metallurgical Laboratory Fast Tracked the Original Plutonium Works* – Transportation, Communications, and Site Support

PUBLICATION TEAM

Communications Manager – Sandra D. Cannon

Editor – Rosalind E. Schrempf

Reference and Bibliography Editor – Janet K. Tarantino

Book Designer – Joan K. Slavens

Text Designers – Elaine C. Schneider, Cindi L. Gregg

Indexer – Kristin L. Manke

Webmaster – Debora A. Schulz, Jodi C. Melland

PREFACE

The U.S. Department of Energy Richland Operations Office and the Washington State Historic Preservation Office designated the Manhattan Project and Cold War facilities of the Hanford Site a historic district eligible for listing in the National Register of Historic Places in 1996. The purpose of this book is to preserve in words, diagrams, and photographs the structures at the Hanford Site during the Manhattan Project and Cold War Era because the majority of them have been or will be demolished. Because of the national and international significance of the Hanford Site, we hope certain properties will be left in place to illustrate the scope and scale of the various missions that developed the atomic energy program and supported national security. Those facilities that must be demolished because of safety concerns and lack of a future use will at least live on in this book.

This book is the result of a regulatory agreement between the U.S. Department of Energy, the Washington State Historic Preservation Office, and the Advisory Council on Historic Preservation. They agreed that a historic narrative (this book) would be written to record

- Information about the roles played by different property types and/or processes at the Hanford Site
- Numbers of buildings/structures within the different property types and their locations
- Changes in technology, design, and use of the different property types over time
- Photographs, plans, and cross sections of representative samples that document the form or manifestation of different property types

The U.S. Department of Energy's present mission at the Hanford Site is cleanup: restore the Columbia River Corridor, transition the Central Plateau to long-term waste storage and treatment operations, and prepare for the future. None of the accomplishments fulfilling the U.S. Department of Energy's present mission are described in this book. The time period covered in this book is 1943-1990—the years spanning the Manhattan Project and Cold War era. The book is the result of research into the facilities and processes that took place within those facilities at the Hanford Site during that time period. This book is not a definitive history nor an investigation or interpretation of issues nor a prognosis on the clean up of the Hanford Site or activities beyond 1990. That we leave to future researchers, who we hope will use the resources in this book as a springboard for their work.

Part of the agreement with the Washington State Historic Preservation Office and the Advisory Council on Historic Preservation is that the book will be published as a hard copy. This has limited the amount of space we can devote to each topic. Even with that limitation, the book is voluminous. We are publishing the book on the Internet to make it accessible to more readers. The appendices will only be published on the Internet, not in the hard copy.

CHAPTER 1

HISTORIC OVERVIEW

T.E. MARCEAU
BECHTEL HANFORD, INC.

Contents

Graphic: Uncle Sam by R.S. Sloan, Rockwell Hanford Operations.

CHAPTER 1 - HISTORIC OVERVIEW

For nearly 50 years (1943-1990), the history of the Manhattan Project and the Hanford Site's role in it and the subsequent Cold War have been largely invisible. For national security reasons, the government labeled these facilities "classified" and deliberately kept the public in the dark about what was going on behind the fences. With the end of the Cold War, this history no longer needs to be kept secret. In recognition of the global significance of events that occurred on an isolated stretch of the Columbia River in south-central Washington State, the Richland Operations Office of the U.S. Department of Energy (DOE) conducted a 4-year project to document the plutonium production facilities at the Hanford Site (Figure 1.1). Through this project, the roles these facilities played in World War II and the Cold War Era, together with the perspective of the workers who staffed them, are captured in this book supplemented with historic photographs and oral histories – many of which appear in print for the first time.

PURPOSE AND OBJECTIVES

The dual purpose of this book is to formally identify the Hanford Site Manhattan Project and Cold War Era Historic District and to preserve in words, diagrams, and photographs its significant structures because the majority of these structures have been or will be demolished. The designation of the Manhattan Project and Cold War Era facilities at the Hanford Site as a Historic District came about through an agreement between the U.S. Department of Energy and the Washington State Historic Preservation Office (SHPO).

The National Historic Preservation Act (NHPA) was passed in 1966. Among the reasons for its passage, Congress noted that "historic properties significant to the Nation's heritage are being lost or substantially altered, often inadvertently, with increasing frequency" (16 USC 470(b)(3)). From its inception, NHPA has emphasized the identification and protection of discrete sites, buildings, structures, and objects tied to a specific geographic location.

The principal tool for identification, evaluation, and protection of significant properties is the National Register of Historic Places – the "nation's inventory of historic places and the national repository of documentation on the variety of historic property types, significance, abundance, condition, ownership, needs, and other information" (NPS 1990, p. 1). Section 106 of NHPA prescribes for each undertaking that Federal agencies "take into account the effect of the undertaking on any district, site, building, structure, or object that is included in or eligible for inclusion in the National Register" (16 USC 470f).

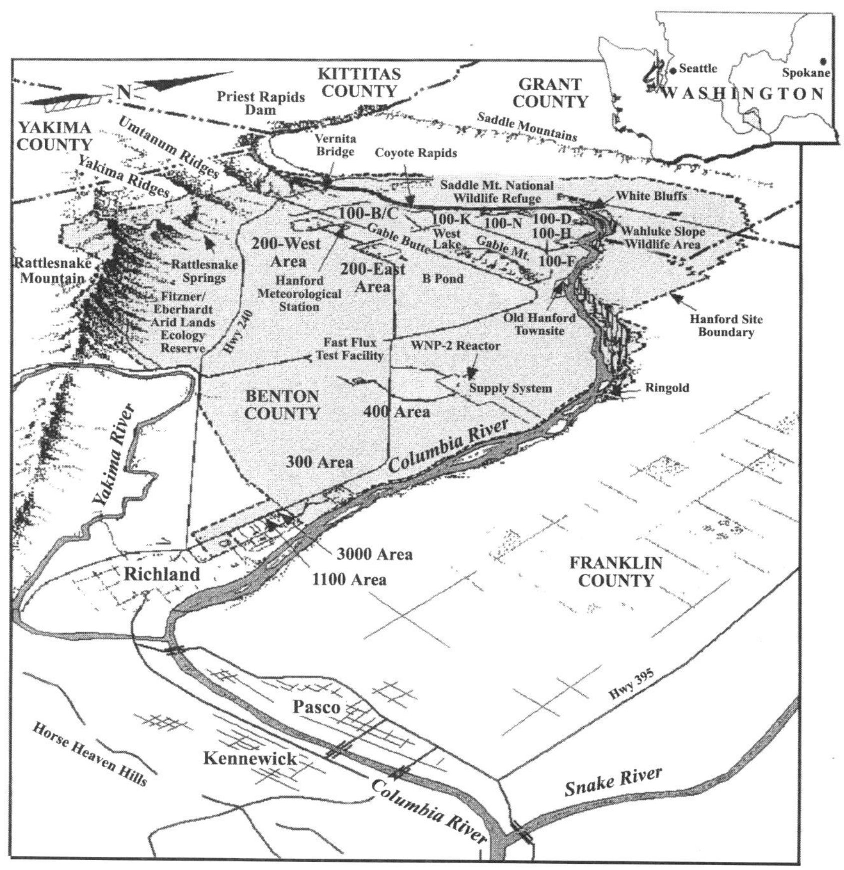

Figure 1.1. *Hanford Site (Gray Area) Located in Washington State along the Columbia River*

Section 110 of NHPA requires that agencies "initiate measures to assure [sic] that where...an historic property is to be substantially altered or demolished, timely steps are taken to make or have made appropriate records...for future use and reference" (16 USC 470h-2(b)). With the change in mission from plutonium production to environmental restoration and the demolition of surplus properties, DOE resolved to satisfy its responsibilities under NHPA by initiating an innovative management strategy for those properties related to the Manhattan Project and Cold War.

In 1996, DOE, SHPO, and the Advisory Council on Historic Preservation signed a Programmatic Agreement that modified compliance with Sections 106 and 110 of NHPA with respect to historic buildings on the Hanford Site (DOE 1996a). In deliberations leading to the Programmatic Agreement, DOE and SHPO first determined that the Hanford Site was a designed industrial landscape whose buildings, grouped by function within designated geographic complexes, were united historically and thematically by the production of plutonium for national defense. Given this finding, DOE and SHPO agreed that the Hanford Site met the requirements for an Historic District, as defined by the National Park Service, because it possessed a "significant concentration, linkage, or continuity of sites, buildings, structures, or objects united historically...by plan or physical development" (NPS 1990, p. 5). By identifying the Hanford Site Manhattan Project and Cold War Era Historic District in the Programmatic Agreement, DOE and SHPO were able to replace the documenting and mitigating of each building at the Hanford Site with a systematic treatment of the Hanford Site itself.

Key to this strategy was to develop property types and identify the buildings that best represented each type. DOE selected the primary functions of Fuel Manufacturing, Reactor Operations, Chemical Separations, and Plutonium Finishing as well as the support functions of Waste Management, Research and Development, Site Security, Military Operations, Health and Safety, and Infrastructure as categories in which the buildings would be classified. Using this classification matrix, DOE assigned 1100 buildings to specific property types and evaluated each building for its eligibility for listing in the National Register of Historic Places as a contributing or non-contributing property within the Historic District. Of the 527 buildings determined to be contributing properties, DOE and SHPO ultimately selected 190 (initially 187) properties for individual documentation using Historic American Engineering Records (HAER), expanded Historic Property Inventory Forms (ExHPIF), or standard Historic Property Inventory Forms (HPIF). These key properties collectively represented the events and activities that took place on the Hanford Site from 1943-1990. The *Hanford Site Manhattan Project and Cold War Era Historic District Treatment Plan* (Marceau 1998) shows the original classification matrix of 1100 buildings and the properties recommended for individual documentation.

DOE's current mission of environmental restoration, which includes the demolition of surplus properties, will adversely affect the historic properties that comprise the Hanford Site Manhattan Project and Cold War Era Historic District. However, SHPO agreed that recording the key events that occurred at the Hanford Site from 1943-1990 in this book and documenting each of the 190 representative properties would address the effects of decommissioning and environmental restoration. This book, *History of the Plutonium Production Facilities of the Hanford Site Historic District, 1943-1990,* satisfies Stipulation VI of the Programmatic Agreement (DOE 1996a) under which DOE agreed to write a "synthetic, integrated Hanford Site historic narrative" that would include:

- Contextual information about the different property types and processes associated with them – see Chapter 2 of this book and Appendix B on the Internet
- Numbers and locations of buildings within property types – see Table A.5 in the *Hanford Site Manhattan Project and Cold War Historic District Treatment Plan* at http://www.hanford.gov/doe/culres/historic/index.htm (Marceau 1998) and Appendix B on the Internet
- Descriptions of changes in technology, design, and use of property types over time – see Appendix B on the Internet
- Photographs, plans, and cross sections of representative examples of the different property types – see Appendix B on the Internet

The narrative was further defined in the *Hanford Site Manhattan Project and Cold War Era Historic District Treatment Plan* (Marceau 1998), which was written in compliance with Stipulation IV of the Programmatic Agreement. In scope, the historic narrative would be a "report which will chronicle the history of the Hanford Site, its technology, and the people who worked here" (DOE 1998a, p. 1). Selection of the word "chronicle" was deliberate since, according to *Webster's New Collegiate Dictionary*, it defined the end state of the report as an "historical account of events arranged in order of time *without analysis or interpretation*" (emphasis added). From the outset, this book was designed to reflect the emphasis of the National Register program and the Secretary of the Interior's Standards and Guidelines for Historical Documentation.

> ***"[Historical] Documentation is a detailed record, in the form of a report or other written document, of the historic context(s) and significance of a property. Historical research to create documentation uses archival materials, oral history techniques, ethnohistories, prior research contained in secondary sources and other sources to make a detailed record of previously identified values or to investigate particular questions about the established significance of a property or properties....Documentation generally results in both greater factual knowledge about the specific property and its values, and in better understanding of the property in its historical context...Documentation should incorporate rather than duplicate the findings of previous research." - 48 FR 44716, p. 44729***

In keeping with this guidance (see sidebar box), the authors of this book used existing information to the fullest extent possible to recount the events that occurred at the Hanford Site from 1943 through 1990 to highlight the significance of the buildings and structures in which those events occurred. The purpose of this book is not an academically oriented history of the Hanford Site. The authors of this book did not use a uniform or classic historic approach, nor did they interpret or explain the events that took place at the Hanford Site over the past 50 years. Rather, they sought to present these events like a reporter who informs the audience and allows for independent conclusions.

The authors of this book were selected because of their backgrounds in the particular topics to be researched. The authors' biographies are in Appendix A on the Internet (http://www.hanford.gov/docs/rl-97-1047/biographies/index.htm). The variety of topics covered required authors from various disciplines with the result that each author took a slightly different approach to the method of research and manner of recording it. The sections in Chapter 2, therefore, are not always parallel in the type or depth of focus. While the reader may question the extent of the research underlying each section, how critically the authors examined their sources, the explicit or implicit filters they applied in selecting pertinent passages, and how completely they synthesized their data, the authors neither present, nor intend that this book be taken as, a definitive history of the Hanford Site or an investigation or interpretation of specific issues, such as the long-term effects of operations on groundwater. That undertaking we leave to others, who we hope will use the resources in this book as a springboard for their work.

This book provides a brief chronicle of the facilities at the Hanford Site organized by the processes that define their reason for existence, mainly Fuel Manufacturing, Reactor Operations, Chemical Separations, Plutonium Finishing, and related activities. It complies with the requirements of the National Historic Preservation Act to document the facets of the properties that qualify them for listing in the National Register and their role in the Manhattan Project and subsequent Cold War. It also corresponds with the intent of the National Register program to recognize physical properties and document their appearance and importance.

SCOPE

This book covers the time period 1943-1990 – the years spanning the Manhattan Project and the Cold War. Chapter 1 is a historic overview summarizing some of the events that accounted for and influenced the creation and operation of the Hanford Site as well as some of the activities undertaken on the Hanford Site in response to these events.

Chapter 2 consists of twelve sections with details, descriptions, and illustrations of the primary production operations (manufacturing the fuel, irradiating it in the reactors, chemically separating it to isolate plutonium, and finishing the plutonium into the form needed for a bomb), support activities (construction, waste management, research and development, security, military operations, worker health and safety, and infrastructure), and most importantly a perspective from the workers who made it all happen.

Chapter 3 is a guide to resources and information for future researchers. Ideas generated or omissions noted within this text may be investigated more fully by the interested reader using these resources.

Chapter 4 looks to the future and provides recommendations for future uses for some of the facilities still standing on the Hanford Site. It envisions opportunities for economic development and heritage tourism as a way to extend vestiges of the past into the 21st century.

Appendix A contains the biographies of those who contributed directly or indirectly to the book. Appendix B contains the Historic American Engineering Records and Historic Property Inventory Forms for the focal properties designated in the Programmatic Agreement for individual documentation.

The following topics are not discussed because they are either treated more extensively or appropriately in other documents or they fall outside the scope of this book:

Waste Products/Cleanup: Each step in the plutonium production cycle had an associated waste stream. This book discusses only that wastes were produced during the fuel manufacturing, irradiation, separation, and concentration processes and were introduced into the environment in pits, drains, cribs, trenches, basins, and tanks. With the exception of personal and environmental health, no other topic in this book will likely generate as much concern among readers – particularly those within proximity of the Hanford Site. However, this book will not explore or debate the consequences or effects of the waste products generated and stored on the Hanford Site nor the cleanup of the Hanford Site which began after the Cold War Era.

Health Effects: This book discusses the programs established to monitor worker health and safety and the environment. The long-term effectiveness of these programs and the effects of working with or living near radiation and chemicals are briefly discussed. These topics are covered more appropriately and explicitly in other sources including the results reported from the Hanford Environmental Dose Reconstruction (HEDR) Project (HEDR 1994a), the Screening Assessment and Requirements for a Comprehensive Assessment, Columbia River Comprehensive Impact Assessment (DOE 1998a), as well as risk assessments contained in Environmental Impact Statements, CERCLA documents, and Safety Assessment Reports.

The Decision to Use the Bomb: The decision to use atomic weapons was not made lightly. The question was debated vigorously on moral, political, technical, and military grounds. President Truman's reasoning, the views of individual members of the Interim Committee, and the split within the Scientific Panel culminating in a petition Leo Szilard drafted and 155 of the scientists developing the bomb signed arguing that it not be used are all well documented. Readers interested in this topic are directed to sources such as The Making of the Atomic Bomb (Rhodes 1986) or "Hiroshima: World Culture as the Event War" (Lanouette 1999).

The agreement with SHPO and the Advisory Council on Historic Preservation requires publishing this book as a hard copy, which limits the amount of space we can devote to each topic. Each of the chapters and sections in Chapter 2 has been restricted to roughly fifty pages including figures. Because of cost constraints inherent in publication, the appendices have been published only on the Internet at http://www.hanford.gov/docs/rl-97-1047/index.htm. If funding permits, a compact disk containing the complete book may be available in the future.

This book gives the historical measurements, which are usually rounded off. For exact measurements, the reader should refer to the original documents.

The References Cited Section lists the resources used in researching this book. The Bibliography lists additional resources not used directly but which may be of interest to the reader in pursuing additional research.

SETTING THE STAGE

The Hanford Site was established for one reason – to produce plutonium for use in weapons. With a production record of 54.5 metric tons (quantity published to date, which is 60 percent of the total amount of weapons-grade plutonium produced by the United States), the Hanford Site fulfilled its Manhattan Project and Cold War mission (DOE 1996b, Figures 6 and 7, Table 2). At its inception, the Hanford Site was largely an experiment, the production-scale culmination of two key events – the demonstration of a controlled, sustainable nuclear chain reaction capable of transforming uranium into plutonium, and the discovery of a process to separate plutonium from uranium following irradiation.

The theme that the Hanford Site was an experiment runs throughout the Manhattan Project and the Cold War Era. Creating plutonium was a new science. Producing plutonium on an industrial scale was a major challenge. The time pressures of war added to the uncertainties that accompanied each decision. Yet critical decisions had to be made within very short time frames despite the uncertainties. Such a situation helps explain the mistakes, the risk taken, the resourcefulness of the workers, and the achievements that made the Hanford Site a Historic District today.

Throughout the book we read of decisions that had to be made without full knowledge of the consequences and the workers who met these challenges. In describing the need to proceed with the Manhattan Project and construct the Hanford Site, Lieutenant General Leslie R. Groves as commander of the Manhattan Project relates "We would go ahead with their design and construction as fast as possible, even though we would have to base our work on the most meager laboratory data...we could not afford to wait" (Groves 1983, p. 72). We read of how one of the production line crew solved the problem of aluminum crumpling within 24 hours when manufacturing the fuel elements. We also learn of the problems of isolating the plutonium once it had been created: "The lack of actual irradiated uranium for experimentation, ignorance of the chemical properties of plutonium, and engineering inexperience with high radiation levels, complicated the development effort [to extract and purify the plutonium] tremendously" (DOE 1997e, p. 171). Describing the challenge of chemically separating out plutonium, one of Du Pont's chemical engineers, Raymond Genereaux, illustrates the theme of the Hanford Site as an experiment: "We realized immediately we had to design flexibility into it if we were going to move ahead and get any designing done before we knew for sure what we were doing, before we knew what the process would be" (Genereaux, quoted in Sanger 1989, p. 40).

Creating plutonium was a challenge because it is an artificially made element. It exists only rarely in nature. Plutonium is a product of uranium fissioning and is created by the absorption of neutrons within the nuclei of uranium atoms that have been subjected to neutron bombardment (see Figure 1.2). Plutonium is not a daughter product of uranium. It does not form directly from uranium. Rather, the transmutation is a two-stage process in which an intermediary element is formed. Under neutron bombardment within a nuclear reactor, uranium-238 (one of the two main isotopes of uranium, element 92) first changes to uranium-239, a highly unstable isotope marked by the addition of an extra neutron. Within 20 minutes, on average, uranium-239 spontaneously transforms to neptunium-239 (element 93). With a half-life of only 2.3 days, neptunium is also highly unstable, and when undergoing radioactive decay, spontaneously changes to plutonium-239 (element 94), a radioactive element with a half-life of approximately 24,000 years (Rhodes 1995).

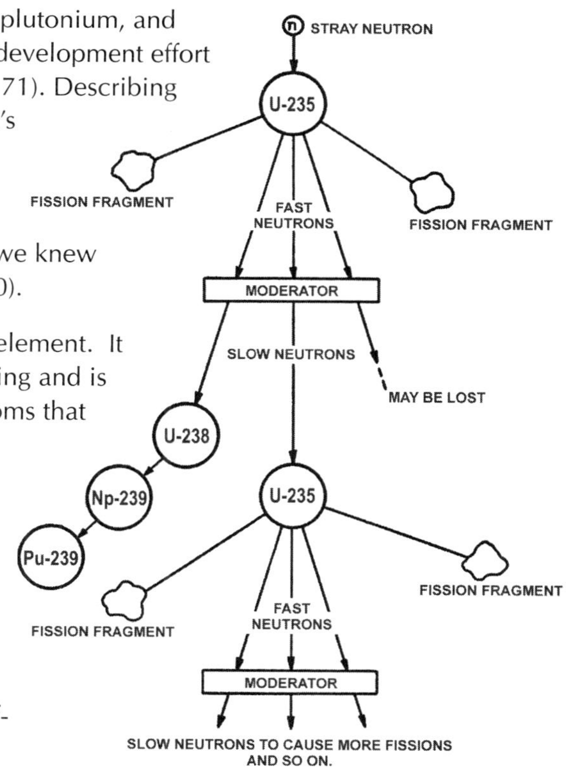

Figure 1.2. *Creation of Plutonium—a Product of Uranium Fissioning*

Plutonium is highly fissile. This means its nuclei can be readily split, releasing large amounts of energy either under controlled conditions as a nuclear fuel or under uncontrolled conditions as a nuclear explosion. Physicists have devoted years to intense research to arrive at this understanding. In fact, theoretical physicists predicted the transmutation of uranium to plutonium several years before they actually observed it. The discovery of plutonium and its chemistry was intimately tied to the investigation of the fissile properties of uranium.

NUCLEAR FISSION

> ***"The discovery of fission came just as Germany was girding itself to abandon expansion by intimidation and resort to armed conquests...World War II erupted at a moment when the promise of atomic energy had progressed from being possible to being probable. It was not clear whether this energy could be released explosively however."* - Sublette 1999, p.1**

The nuclear arms race began with the race to develop the atomic bomb. The United States was a late entrant whose participation came largely at the prodding of emigrant scientists including Leo Szilard, Edward Teller, and Eugene Wigner who "regarded it as their responsibility to alert Americans to the possibility that German scientists might win the race to build an atomic bomb and to warn that Hitler would be more than willing to resort to such a weapon" (Gosling 1999, p. vii). Great Britain, Japan, and the Soviet Union also participated in this race; however, "only the Americans... protected by oceans on both sides, managed to take the discovery of fission from the laboratory to the battlefield and gain a short-lived atomic monopoly" (Gosling 1999, p.vii).

The history of high-energy physics and nuclear fission is extensive. Its retelling is well beyond the scope of this book. However, a recounting of key events is instrumental in establishing a framework for the existence of the Hanford Site and the establishment of the Historic District. Many dates and events could be chosen as the starting point, but this much abbreviated discussion begins in 1932 when James Chadwick demonstrated the existence of the neutron, or non-charged particle, within the nucleus of an atom. Later that year, J. D. Cockroft and E. T. S. Walton successfully split lithium atoms in a particle accelerator (Gosling 1999, Sublette 1997). In 1933, Szilard first envisioned the dual potential of the atom when he surmised that the collision of neutrons within a chain reaction would release energy and speculated on the use of this energy in making bombs. Szilard formalized these thoughts in a patent application on July 4, 1934 that described how explosions could be induced through chain reactions and introduced the concept of critical mass. It would be 2 years, however, before the British Admiralty accepted Szilard's offer of his patents (Sublette 1997).

In contrast to prevailing studies that required a high energy source to accelerate positive-charged proton beams and alpha particles, Enrico Fermi experimentally bombarded 63 elements with neutrons, reasoning that little resistance would be encountered by uncharged particles entering the nucleus. One of the experimental elements was uranium (Gosling 1999). Over the next several years, the international scientific community focused on neutron bombardment as a more promising technology for splitting atoms and uranium as a key element.

> ***In July 1937, Japan invaded China and in November joined Germany and Italy in the Axis Alliance.***

Otto Hahn and Fritz Strassmann, working in Berlin, discovered that uranium nuclei split into radioactive barium and fragments of uranium. Hahn published the experimental results in *Naturwissenschaften* on December 21, 1938 (Sublette 1997). On January 13, 1939, Otto Frisch substantiated these results and, together with Lise Meitner, an Austrian émigré living in Sweden at the time, calculated the unprecedented amount of energy released. Although she did not receive the credit she deserved, Meitner is recognized as the first to theorize the fission process (Meitner and Frisch 1939, Simes 1996). Based on their calculations, Frisch applied the term "fission," from biological cell division, to name this process (Sublette 1997, Gosling 1999).

Niels Bohr announced the results of the Hahn-Strassmann uranium experiments and the Meitner-Frisch calculations to the American scientific community on January 26, 1939. In March of that year, a Columbia University team consisting of Fermi, Szilard, Walter H. Zinn, and Herbert Anderson discovered that "at least two secondary neutrons emerge from each neutron-induced fission" of uranium (Rhodes 1986, p. 294). On August 31, Bohr and John A. Wheeler, working at Princeton University, published their theory that the isotope uranium-235, present in trace quantities within uranium-238, was more fissile than uranium-238 and should become the focus of uranium research. In this publication, they also postulated that a then unnamed, unobserved transuranic element (referred to simply as 94^{239} or, more descriptively, as "high octane") produced during fissioning of uranium-238 would be highly fissionable (Sublette 1997). With these findings, the theoretical foundation for atomic energy was substantially set.

On September 1, 1939, Germany initiated World War II by invading Poland.

In response to this aggressive act by Germany, U.S. uranium research objectives began to alter on October 11, 1939 when Alexander Sachs delivered a letter to President Franklin D. Roosevelt drafted by Szilard and signed by Albert Einstein. Szilard was among the most vocal of those advocating a program to develop bombs based on recent findings in nuclear physics and chemistry. The letter warned the President of the potential for and consequences of atomic weapons and prevailed upon him to take immediate action to counter the work being done in Germany at the Kaiser Wilhelm Institute in Berlin. Ten days later the Presidential Advisory Committee on Uranium met for the first time in Washington D.C. President Roosevelt had acted quickly, but the Committee, under the direction of Lyman Briggs, initially would take a cautious approach by continuing to focus primarily on uranium research and not applications (Sublette 1997).

On April 9, 1940, Germany invaded Denmark and Norway and on May 10 invaded Holland, Belgium, and France.

If German scientists were moving ahead as quickly in developing an atomic bomb as the German army was in subjugating Western Europe, the United States needed to accelerate its uranium program. The National Defense Research Committee replaced the Presidential Advisory Committee on Uranium in June 1940 and afforded the American scientific community a presence within the executive branch. Vannevar Bush, president of the Carnegie Foundation, was appointed head of the new committee. By eliminating the former committee's military members, Bush concentrated power with the scientists. He also introduced security and secrecy into nuclear research by "barr[ing] foreign-born scientists from committee membership and block[ing] the further publication of articles on uranium research" (Gosling 1999, p. 7). These themes of secrecy and exclusion would become hallmarks of the nuclear weapons program. Using $40,000 awarded by the National Defense Research Committee, in November 1940 Fermi and Szilard began constructing a sub-critical, graphite-moderated, uranium oxide "pile" (the original name for reactors) at Columbia University to further investigate chain reactions (Sublette 1997).

On June 28, 1941, the nuclear scientific community assumed an even stronger presence within the government when President Roosevelt established the Office of Scientific Research and Development by executive order. Vannevar Bush was named its director and now reported directly to the President. President Roosevelt created the Office of Scientific Research and Development within a week of Germany invading the Soviet Union and less than a month following a presentation Tokutaro Haiwara made at the University of Kyoto in which he speculated about the potential for a fusion explosion using a fission ignition – the basis for thermonuclear weapons (Sublette 1997, Gosling 1999).

The British counterpart to the Office of Scientific Research and Development, the MAUD Committee, issued its final report on July 15, 1941, in which they described the technical details of atomic bombs together with proposals and cost estimates for their development (Sublette 1997). "Americans had been in touch with the MAUD Committee since fall

1940, but it was the July 1941 MAUD report that helped the American bomb effort turn the corner. Here were specific plans for producing a bomb, produced by a distinguished group of scientists with high credibility in the United States." (Gosling 1999, p. 9).

Meeting with President Roosevelt and Vice President Henry A. Wallace on October 9, Bush relayed the findings in the MAUD report and received instructions from the President to quicken the pace of research to determine the feasibility of a bomb but not to proceed to production without explicit authorization (Gosling 1999). On November 6, the National Academy of Sciences, under Arthur Compton, issued a report that verified the conclusions of the MAUD report, although less optimistically. Compton estimated that "a critical mass of between two and 100 kilograms of uranium-235 would produce a powerful fission bomb and that for $50-100 million isotope separation in sufficient quantities could be accomplished" (Gosling 1999, p. 10). Bush delivered the report to President Roosevelt on November 27. Building on this report and acting on the President's instructions, Bush organized an accelerated research project on December 6 that would investigate gaseous diffusion, electromagnetic separation, centrifuge separation, chain reactions, heavy water production, and plutonium production. At the same time, he began laying the foundation for engineering studies and the construction of pilot plants (Gosling 1999, Sublette 1997).

On December 7, 1941, Japan attacked Pearl Harbor.

America's entry into the war ended the bureaucratic struggling that had characterized fission study by focusing investigations on the production of atomic weapons (Sublette 1999). The National Defense Research Committee yielded all "broad policy decisions relating to uranium" to the Top Policy Group consisting of Vice President Wallace, Secretary of War Henry L. Stimson, Army Chief of Staff George C. Marshall, Vannevar Bush, and James Conant (Gosling 1999, p. 10). America's full commitment to the development of fission weapons came on December 18, 1941, when Arthur Compton called the first meeting of the newly created S-1 Committee (Section One, Uranium, of the Office of Scientific Research and Development). The Committee awarded $400,000 to Ernest O. Lawrence to continue his work on electromagnetic separation of uranium-235 at the University of California, Berkley (Gosling 1999). "With the United States now at war and with the fear that the American bomb effort was behind Nazi Germany's, a sense of urgency permeated the federal government's science enterprise...By spring 1942...the situation had changed from one of too little money and no deadlines to one of a clear goal, plenty of money, but too little time" (Gosling 1999, p. 10).

The accelerated research program Bush established at the end of 1941 and President Roosevelt approved on January 19, 1942 meant several lines of research were being investigated concurrently so an atomic bomb could be delivered to the President by 1944. For the development of the Hanford Site, the most pertinent research was that devoted to plutonium production through a sustained chain reaction using uranium as a fuel. In January 1942, Compton moved "pile research" (reactor research) from Columbia University to the University of Chicago where he established the Metallurgical Laboratory. Over the next several months, Fermi's experiments there brought theory ever closer to reality. In April, his efforts shifted from "demonstrating feasibility to securing graphite and uranium of adequate purity and in sufficient quantity to build the [Chicago Pile 1 or CP-1] reactor" (Sublette 1997). By mid-August, Fermi had evinced that a chain reaction was a certainty. On December 2, 1942, at 3:49 p.m., Fermi and Samuel K. Allison achieved the world's first controlled, self-sustained nuclear chain reaction in an experimental reactor of natural uranium and graphite constructed under the squash courts at Stagg Field (Gosling 1999, Sublette 1997). The stage had been set for reactor operations at the Hanford Site.

ISOTOPE SEPARATION

Concurrent with the search for technologies to produce the isotopes uranium-235 and plutonium-239 was the search for a method to separate these isotopes once they were produced. Vastly different techniques would be required.

Early in 1939, Niels Bohr surmised that uranium-235 and uranium-238 possessed differing fission characteristics and concluded that the observed slow fission in uranium was attributable to uranium-235. Together with John A. Wheeler, Bohr published this theory on August 31. By March 1940, John R. Dunning of Columbia University experimentally confirmed their finding in collaboration with Alfred O. Nier of the University of Minnesota (Gosling 1999). However, to determine whether uranium-235 would support an explosive nuclear chain reaction, concentrated samples of this isotope, sufficient in size to form a critical mass, would need to be obtained. Separation of this isotope from uranium could not be accomplished through chemical means because isotopes of the same element, such as uranium-235 and uranium-238, are chemically indistinguishable. Physical separation, based solely on their differing atomic weights, would have to be used. Electromagnetic separation, centrifuge separation, and gaseous diffusion were candidate techniques.

Each of these techniques had champions, and each had problems to overcome in moving from an experimental to a production stage. Entirely new technologies would have to be designed and built with many of their components individually manufactured to exacting tolerances not found in standard industrial applications. Therefore, on May 23, 1942, given the uncertainties of success for any one of these separation methods to concentrate uranuim-235, or at this point the likelihood of sustaining a nuclear reaction that would transform uranium into plutonium, Conant urged the S-1 Committee to proceed with all options simultaneously "regardless of cost." He argued convincingly that "redundant development" would reduce the time required to produce enriched uranium or plutonium (Sublette 1997). The decision to push all options forward as fast as possible "reflected the inability of the committee to distinguish a front-runner and its consequent unwillingness to abandon any method. With funds readily available and the outcome of the war conceivably in the balance, the S-1 leadership recommended that [gaseous diffusion, centrifuge and electromagnetic separation, and plutonium production] proceed to the pilot plant stage and to full production planning" (Gosling 1999, p. 11).

To ensure timely success, the United States "simultaneously pursued plutonium and highly enriched uranium as fissile materials for atomic weapons" (DOE 1997e, p. 136). Implementation of this strategy with respect to uranium enrichment led to the establishment of the Clinton Engineer Works, renamed the Oak Ridge Reservation after the war. Research that resulted first in the discovery of plutonium and then to a chemical means by which it could be separated from uranium led to the establishment of the Hanford Engineer Works, later renamed the Hanford Site.

In their August 1939 publication, Bohr and Wheeler had theorized that a uranium fission byproduct – assigned the nondescript elemental placeholder 94^{239} – would be highly fissionable. Following this lead, Edwin McMillian and Philip Abelson announced on May 27, 1940 that they had discovered a new element – element 93^{239} (neptunium) – produced by neutron bombardment of uranium-238 within the cyclotron at the University of California, Berkeley. Demonstration of this short-lived transuranic element was significant because neptunium, if the Bohr-Wheeler theory of transmutation were correct, would become 94^{239} following an additional beta-decay. Glenn T. Seaborg and Arthur Wahl, also of the University of California, Berkeley, brought closure to the theory when they successfully demonstrated the existence of element 94^{239} on February 26, 1941 (Sublette 1997). In 1942, they named this element plutonium, after the protocol of naming new elements after the outer planets, that is, uranium (Uranus), neptunium (Neptune), and plutonium (Pluto) (Rhodes 1986).

Having demonstrated that plutonium-239 could be created as a fission product from uranium-238, Seaborg, together with Emillio Segrè, began to investigate the chemical properties of plutonium. On May 18, 1941, fewer than 3 months after its discovery, they concluded that plutonium-239 was 1.7 times as fissionable as uranium-235, thereby "proving it to be an even better prospective nuclear explosive" (Sublette 1997). Seaborg subsequently moved to the Metallurgical Laboratory in Chicago at Compton's request where he began work on developing a "large-scale, remote-controlled" chemical separation and purification process to extract and concentrate plutonium (Rhodes 1986, p. 381). On August 20, 1942, Seaborg successfully concluded a sequence of chemical oxidation and reduction cycles using "ultramicrochemical equipment" that produced a microgram (a millionth of a gram) of plutonium (Rhodes 1986, p. 414). While Sublette (1997) correctly notes that Seaborg had identified a methodology that could be used on an industrial scale, actual scale-up of the equipment "had to keep abreast of, and in some cases ahead of, the development of the [separations] process itself" (Groves 1983, p. 85).

MANHATTAN PROJECT: CREATION OF THE HANFORD SITE

> *"...with the agreement of everyone who held any degree of responsibility for the project, I had decided almost at the very beginning that we would have to abandon completely all normal, orderly procedures in the development of the production plants. We would go ahead with their design and construction as fast as possible, even though we would have to base our work on the most meager laboratory data. Nothing like this had ever been attempted before, but with time as the controlling factor, we could not afford to wait to be sure of anything."* **- Groves 1983, p. 72**

Fermi had demonstrated that a controlled, self-sustained nuclear reaction capable of producing plutonium-239 was no longer theoretical. Seaborg had provided a means whereby plutonium could be chemically separated from irradiated uranium. The prerequisites necessary to enable operations at the Hanford Site had been met. Nevertheless, the gap between laboratory demonstrations and full-scale production was formidable. Lieutenant General Leslie R. Groves, who would command the Manhattan Project, both encouraged and demanded risk taking in bridging that gap.

BEGINNINGS OF THE MANHATTAN PROJECT

By mid-1942, government-supported research was concentrated at Columbia University in New York, the University of California in Berkeley, and the University of Chicago Metallurgical Laboratory (DOE 1997e). Although directed towards the development of atomic weapons, the S-1 program was still largely an academic endeavor in which scientists held the leadership role. This changed within a matter of months.

> *"The need for security suggested placing the S-1 program within one of the armed forces, and the construction expertise of the Corps of Engineers made it a logical choice to build the production facilities...With this reorganization in place, the nature of the American atomic bomb effort changed from one dominated by research scientists to one in which scientists played a supporting role in the construction enterprise run by the United States Army Corps of Engineers."* **- Gosling 1999, pp. 11-12**

As early as October 9, 1941, the military presence, which had largely been excluded with the establishment of the National Defense Research Committee in June 1940, began to reemerge when President Roosevelt authorized Bush to "explore construction needs with the Army" (Gosling 1999, p. 9). The S-1 Committee decision on May 23, 1942 to move from laboratory research to full production prompted a central role for the military since the prevailing thought was that weapons production should be "brought under an organization experienced in producing weapons" (Sublette 1999, p. 2). Bush selected the U.S. Army Corps of Engineers.

In June 1942, President Roosevelt approved $85 million to support atomic weapons development, which was a phenomenal increase from the $6,000 supplied for "uranium research" just 2 years earlier (Sublette 1997). The Corps assumed responsibility for process development, procurement, design, and site selection. They received $54 million in funding. The remaining $31 million was reserved for research and pilot plant studies under the direction of the S-1 Committee in the Office of Scientific Research and Development (Gosling 1999). However, this split between the Corps and Office of Scientific Research and Development proved counterproductive, and on June 18, 1942, Brigadier General Wilhelm D. Styer ordered Colonel James C. Marshall to organize an Army Corps of Engineer District devoted exclusively to managing and coordinating atomic weapons development. Initially named the Laboratory for the Development of Substitute Metals, by August 13, 1942, the organization was renamed the Manhattan Engineer District.

Such an industrial scale project under wartime conditions and pressure to produce demanded effective, proven leadership. Just a month after its formation, the Army tapped Lieutenant General (then Colonel) Leslie R. Groves to head the Manhattan Engineer District and its namesake, the "Manhattan Project." Groves was appointed on September 17, 1942 and "seized immediate and decisive control. In just two days he resolved issues that had dragged on for months under Compton" (Sublette 1999):

- Purchasing and shipping 1,250 tons of high quality Belgian Congo uranium to Staten Island
- Purchasing 52,000 acres on the Clinch River in Tennessee
- Raising emergency procurement priority to AAA, the highest allowed by the War Production Board (Gosling 1999, Sublette 1997).

Groves' decisiveness was rewarded on September 23, 1942 with his promotion to Brigadier General. He firmly believed that the exigencies of war required that the focus shift from laboratory research to full-scale development and production in the shortest time possible. He impressed this view on the scientific community. "Groves' pushy, even overbearing, demeanor won him few friends among the scientists on the Manhattan Project (in particular a special enmity developed between Groves and Szilard). Many detested him at the time, considering him a boor and a buffoon. It was only after the war that many scientists began to appreciate how critical his organizational and managerial genius was to the MED" (Sublette 1999, p. 3).

On October 15, 1942, Groves asked J. Robert Oppenheimer to head a laboratory in which research on the physics and design of atomic weapons would be centralized. The S-1 Executive Committee had discussed the need for this laboratory on September 13. On November 16, 1942, Groves and Oppenheimer selected a site on the Los Alamos mesa in New Mexico for the laboratory. The location was code-named "Site Y." In December 1942, Groves initiated development planning for the Clinton Engineer Works, code-named "Site X." This location on the Clinch River in Tennessee was to contain an experimental reactor, a chemical separations plant, and an electromagnetic separation facility (Sublette 1999).

Construction of the Clinton Engineer Works was predicated on the need to accelerate an already compressed schedule. Meeting with the Military Policy Committee on November 12, 1942, Groves and Conant argued that centrifuge separation should be abandoned and that gaseous diffusion, electromagnetic separation, and plutonium production should bypass the pilot plant stage and proceed directly to full-scale operations. The S-1 Executive Committee seconded these decisions on November 14 (Gosling 1999). On December 28, 1942, following a final review of the decisions reached by the Military Policy Committee and S-1 Executive Committee, President Roosevelt approved $500 million for the Manhattan Project (Gosling 1999, Sublette 1997). "No schedule could guarantee that the United States would overtake Germany in the race for the bomb, but by the beginning of 1943 the Manhattan Project had the complete support of President Roosevelt and the military leadership, the services of some of the nation's most distinguished scientists, and a sense of urgency driven by fear" (Gosling 1999, p. 17).

SELECTION OF THE HANFORD SITE

Even before Fermi had achieved a sustained nuclear reaction, competition had begun to design a prototype large-scale production reactor. In June 1942, Thomas V. Moore and his team developed a plan for a helium-cooled pilot reactor to be built at the Argonne Forest Preserve outside Chicago. Because of the spherical segmentation of its outer shell, the design came to be known as the "Mae West." Eugene Wigner and Gale Young proposed a vertical water-cooled design, while Leo Szilard championed a reactor cooled by bismuth, a liquid metal. On October 5, with design frozen for months by indecision, Groves gave Compton and the Metallurgical Laboratory scientists 1 week to reach a decision with the admonishment that "even wrong decisions are better than no decisions" (Gosling 1999, p. 27). Compton decided to build a water-cooled reactor at Argonne, and an air-cooled semiworks at the Clinton Engineer Works. The shift from helium to air reflected an analysis by E. I. Du Pont de Nemours and Company in Wilmington, Delaware of the options proposed by the Metallurgical Laboratory (Gosling 1999). Du Pont soon assumed an influential role in the design, engineering, and construction of plutonium production and separation facilities.

The Manhattan Project was on a very fast track. One example among many serves to illustrate this point. In December 1942, the M. M. Sundt Company began constructing the Los Alamos Laboratory "without plans or blueprints in order to finish as quickly as possible" (Sublette 1997, p. 10). Speed, while paramount, did not completely displace the concern for safety, however. The Manhattan Project was investing in unproven technology, much of it being developed for the first time. Safety consideration as much as the requirement of secrecy ultimately led Groves to locate facilities in isolated areas.

For example, Compton's decision to relocate Fermi's second experimental reactor (CP-2) from the University of Chicago to the Argonne Forest Preserve was based primarily on safety considerations, that is, the potential effect on an urban population should something go wrong with the reactor. This was also the concern underlying Groves' decision to relocate the X-10 pilot plutonium reactor and separation plant from Chicago to Tennessee. This latter decision, however, had an unanticipated consequence. A pilot facility at the Clinton Engineer Works negated the construction of a full-scale production complex there due to siting requirements. There was not enough room for both facilities, and Knoxville was too close (Gosling 1999). Additionally, Groves was concerned that if an accident did occur in an untested, large reactor, the uranium enrichment facilities planned for the Clinton Engineer Works could be destroyed as well. This, together with the loss of life, would compromise the security and viability of the project (Groves 1983). Another location would have to be found for the industrial-scale plutonium production and chemical separations facilities.

On December 14, 1942, Metallurgical Laboratory scientists, Du Pont engineers, and Corps staff established the siting criteria. While he did not attend the meeting, Groves recalls his instructions to the participants and the outcome of the meeting:

> *"I arranged for a meeting at the Du Pont offices in Wilmington to ensure that the Du Pont organization, the scientific people at Chicago, and the MED would all have the same understanding of the then-accepted atomic theories, the known scientific and technical facts, the scientific probabilities, and the construction and operating problems. After reaching a common understanding, they were to arrive at the criteria for the plutonium plant site, with special attention to the limitations imposed by safety."* - Groves 1983, p. 70

Once they agreed on the criteria, Colonel Franklin T. Matthias, one of Groves' top assistants, and Du Pont engineers Gilbert Church and A. E. S. Hall began the scouting work on December 16. Concentrating on six locations in California, Oregon, and Washington, they selected an area around the small farming and ranching communities of White Bluffs, Hanford, and Richland on an isolated stretch of the Columbia River in Washington as the closest match to the siting criteria. Groves viewed the proposed site on January 16, 1943 and authorized the establishment of what they called the "Hanford Engineer Works" (today the Hanford Site) on 670 square miles of land. The location was assigned the code name "Site W" (Gosling 1999). It was the last site selected for the Manhattan Project.

COMMUNITY SACRIFICE

The Columbia River was the focal point of the Hanford Site. Clear, cold water from this source would dissipate the heat generated from nuclear fission within the reactors. Turbines driven by the river's current at Bonneville and Grand Coulee dams would supply electrical power. And, in the eyes of Matthias and Groves at least, the area the river flowed through was sparsely settled, essentially an empty region particularly well suited for a secret government project. To those directly affected by its placement within their communities, it held a different meaning. The land on which the Hanford Site would be built was home to Native Americans with long ancestral ties to the land, and Euro-Americans who had settled in the area beginning in the 1860s. In constructing the Hanford Site, the Army forcefully removed over 1500 people who sacrificed their lands, homes, and communities to a war effort the government cloaked from their view.

Native Americans

Concentrated on the Columbia River, near inland water sources and areas of high topographic relief, the archaeological record of Native American occupation of the Hanford Site stretches back thousands of years. Typical archaeological sites include pit house villages, open campsites, fishing sites, hunting/kill sites, game drive complexes, quarries, and spirit quest sites. Physical features such as rock cairns, rock alignments, or artifacts mark these sites. Other sites relating to subsistence and ceremonial activities, which are not marked by physical remains, are also present but generally unrecognized by non-Native Americans. In 1943, the Wanapum were still resident in the Priest Rapids Valley – the area soon to become the Hanford Site. They called the Columbia River *Ci Wana* (big river) and from it took their name (the river people). They used many of the same locations as their ancestors and engaged in a largely traditional lifestyle of fishing, hunting, and gathering, although white settlement had restricted their access to some areas. Their tie to the land was deeply rooted, but the Manhattan Project needed a secure area.

"Never in their history had the Wanapums failed the government that first subdued and then ignored them. Puck Hyah Toot met with the colonel [Matthias] and heard his story that the government needed the land and that the people could roam at will no longer. He understood but fragments of the colonel's talk, realizing only that it meant another move and that all but the last old village site was lost...The last Wanapums, their hearts wavering, quietly surrendered their ancestral fishing ground and rifted canyon walls at White Bluffs, because the government said the land was needed." **- Relander 1956, p. 257**

When the Wanapum had to abandon their camps at White Bluffs and other locations on the Hanford Site, they concentrated their activities at Priest Rapids. They continually sought to retain entry to the Hanford Site to fish for salmon at their traditional camps since salmon was a primary staple and central to their religion. Shortly after construction began and the security fences had been established, Matthias made inquiries about "restricting the Indian fishing rights in the project area." He was told that "it would be difficult to obtain an out and out restriction, but that it might be possible to deal locally" (Matthias 1945). On September 15, 1943, Matthias entered the following in his diary:

> *"...when the Indians arrive in this area in the fall, they spend two or three weeks, usually in October, fishing in the river opposite and north of White Bluffs. These fish are cut up and dried and provide the principal item of food for the tribe during the winter. A number of proposals were offered to the Indians in which they indicated very little interest, among them was (1) a cash settlement equivalent to the value of the fish, (2) a proposal that we deliver to them the amount of fish they would normally catch in a year, (3) that they would be permitted to fish in the White Bluffs area this fall, but that they would not come back next fall, including a provision for a payment of their fishing privileges. It was finally agreed that we provide the Indians with a truck and a driver who will, at their request, during the fishing season haul the Indians and the fish from White Bluffs to their camp at Priest Rapids once a day. This would permit the Indians to do their fishing under supervision, it would avoid the necessity of their living and sleeping in the area, and would assure them of as much fish as they now get. A cash settlement to be paid annually for their privilege to fish and be in that part of the river was rejected by the Chief. His only interest was that he get fish."* - Matthias 1945, pp. 176-177

This arrangement was superseded by an arrangement made on April 2, 1944 whereby "it was agreed that these Indians would be passed through the reservation on presentation of a pass by one of three people, the Chief and his two assistants, who would be permitted to escort other Indians of the Tribe. Indians without these three people will not be admitted" (Matthias 1945, p. 57). Issuance of these passes was eventually revoked for "security reasons" (Relander 1956).

When the Wanapum were relocated for the Manhattan Project, their possessions were left behind. Frances Riddell described the consequences of relocation at one site adjacent to the Columbia River:

> *"This village was suddenly abandoned at the beginning of World War II when the Federal government took over the area in order to establish a military reservation. As the Indians were not present at the time, one informant told me, all of their houses and belongings remained just as the inhabitants had left them, the Indians were not allowed to return for their belongings. However, the houses were soon broken into by irresponsible whites and the belongings of the Indians taken or strewn about the area."* - Riddell 1948, p. 1

Concern about their material possessions was far overshadowed by their concern for their ancestors as Matthias notes of Chief Johnny Buck's request to "treat Indian Graves which we find with reverence and dispose of the bodies in some reasonable way" (Matthias 1945, p. 57). It would be nearly 15 years before the Wanapum regained access to the Hanford Site to mark their cemeteries. In the fall of 1997, indicative of post-production changes, DOE approved a long-standing request, and Wanapum Elders began to bring Tribal members, especially children and young-adults, to see and learn about their sacred sites and traditional cultural places first-hand on an annual basis.

Euro-Americans

The transformation of the Priest Rapids Valley from an agrarian to an industrial landscape began with the Manhattan Project and continued throughout the Cold War. When Matthias came to view the site in December 1942, the area was still recovering from the effects of the Depression (Sharpe 1999). The principal towns were White Bluffs, Hanford, and Richland. Smaller communities like Vernita, Allard, Wahluke, and Ringold were primarily ferrying stations facilitating crossings of the Columbia River. Active and abandoned farmsteads lay between these centers. From his vantage point, Matthias missed more than a hundred years of history and the roots these settlers had set down in the desert sand.

The Second War Powers Act authorized the Secretary of War to "acquire lands by purchase, as needed for military or other war purposes" (Sharpe 1999, p. 1). On February 8, 1943, Henry Stimson began land acquisition proceedings under directive RE-D 2161 (Gerber et al. 1997, p. 5.6). Ultimately, 670 square miles of land were withdrawn from settlement (Figure 1.3). On March 6, residents received an official notification, known as a "declaration of taking," that informed them the Army was taking their land for a top-secret project. The government provided from 90 days to as little as 48 hours for residents to relocate, forcing many to leave most of their possessions behind (Chatters 1989, Sharpe 1999). "There was no time to harvest crops or say proper goodbyes before being separated from lifelong friends" (Anonymous 1996). Walt Grisham, a White Bluffs resident, was serving in the U.S Army Air Corps stationed in Wormingford, England when he received a letter telling him his home and farm had been condemned and his family moved because of an important project. "I wondered why the postmark was from Hanford, and not White Bluffs. I learned that the families were asked not to mention much to their sons serving in the war" (Sorenson 1999).

For some, the enforced relocation came at a very high price: "It was a pretty hard thing. Some of them ended up at Medical Lake (the location of a state mental hospital), you know, they couldn't quite face it. My mother never really adjusted to it. The fact that the government could come in and take your home away" (Ashton 1999).

The Army used many of the abandoned homes for offices and living quarters and brought in prison labor to harvest the crops. Under condemnation, the government offered only the value for the home as assessed by non-local appraisers and not the value of improvements, equipment, or the crops in the field (Heriford 1995). Many landowners rejected the Army's offer and filed petitions in court to secure a better appraisal. "Matthias adopted a strategy of settling out of court to save time, time being a more important commodity than money to the Manhattan Project" (Gosling 1999, p. 32).

When they were done with construction of the Manhattan Project at the Hanford Site, the Army razed the homes and cut down the orchards. While the people were proud of what they had contributed to the war effort when "the secret" was revealed in 1945, their resentment over how they were treated remains strong even today. "Gone are those peaceful autumns; gone, in fact, is White Bluffs itself. It cannot even be called a ghost town, because the buildings too are gone, except for the remains of a few foundations which stand like teeth whose fillings have dropped out...White Bluffs was a victim of the age of the atom; it is a ghost which marks man's "progress" (Harris 1972, p. 1).

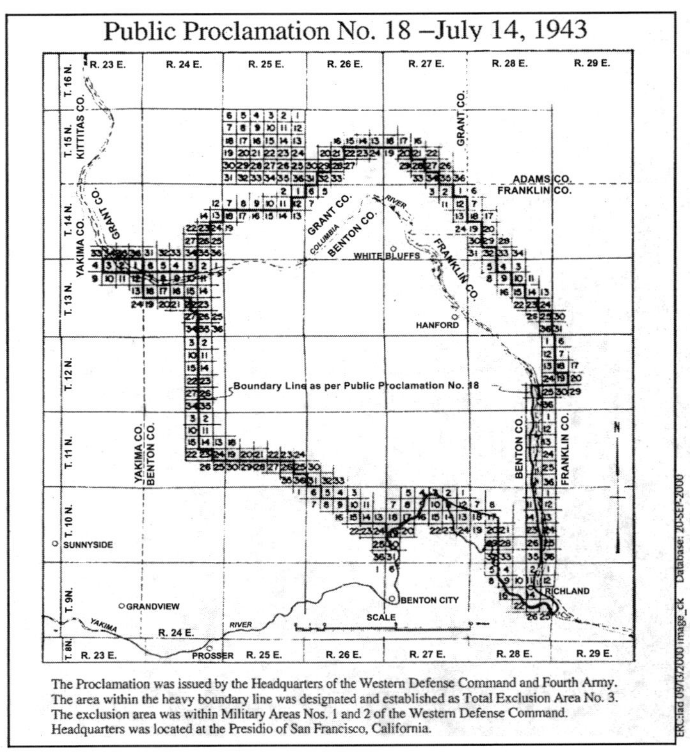

The Proclamation was issued by the Headquarters of the Western Defense Command and Fourth Army. The area within the heavy boundary line was designated and established as Total Exclusion Area No. 3. The exclusion area was within Military Areas Nos. 1 and 2 of the Western Defense Command. Headquarters was located at the Presidio of San Francisco, California.

Figure 1.3. *Initial Map Designating the Area of the Hanford Site, 1943*

Findlay and Hevly (1995, p. 11) offer a less personal but more encompassing summary of the displacement of the local population: "The Army separated almost all people from their homes and lands on the site; only a handful of people, who got work on the project, stayed behind in their homes, and they now paid rent to Du Pont for the privilege. Even the dead were evicted. The government exhumed 177 burials from the White Bluffs cemetery and reinterred them in the nearby town of Prosser."

For the record, the Hanford townsite cemetery was also relocated to Prosser. The Army's actions at White Bluffs and Hanford were similar to those undertaken at Kingston and Clinton, Tennessee. There the Army acquired clear title for immediate use of approximately 56,000 acres between October 1942 and March 1943 through condemnation. Nearly 1000 families were forced to relocate, some within 2 weeks of receiving their eviction notice. Compensation came only after the property had been vacated, and relocation expenses were not compensated. There too, structures not retained for "war-related uses" were demolished (Souza et al. 1996, p. 3-35).

INITIAL CONSTRUCTION

Site Name:	Hanford Engineer Works - December 1943 to December 1946
Site Manager:	Col. Franklin T. Matthias - February 1943 to August 1945
Responsible Agency:	U.S. Army Corps of Engineers - December 1942 to December 1946
Site Contractor:	E. I. Du Pont de Nemours and Co, Inc. - December 1942 to August 1946

Groves succinctly recounts the rules common to all Manhattan Project construction sites: "All design was governed by three rules: 1) safety first against both known and unknown hazards; 2) certainty of operation – every possible chance of failure was guarded against; and 3) the utmost saving of time in achieving full production" (Groves 1983, p. 83).

The initial construction phase for the Hanford Engineer Works (originally named the Gable Project by the War Department and later named the Hanford Site) extended from March 1943 to February 1945 (Gerber 1992b). During that time, the Corps transformed an open agrarian landscape into a closed military-industrial complex described by one former resident as "a vast impersonal government site...with guards stationed at strategic points" (Harris 1972, pp. 1-2).

Chosen as much for its isolation as its resources of water, sand, and gravel, the Corps now had to construct not only the complex itself, but the infrastructure necessary to support it. That they did this under adverse conditions (particularly wind, weather, and a short labor supply) during wartime and under tight security was a major engineering accomplishment: "Once the land was procured, construction proceeded at a nearly unbelievable pace. In just the 30 months between groundbreaking in March 1943, and the end of the war in August 1945, the MED built 554 buildings not dedicated to living requirements" (Gerber 1992b, p. 6).

These buildings ranged from reactors, separation plants, and laboratories to craft shops, warehouses, and electrical substations. Many of these buildings, such as the reactors and chemical separations plants, were first-of-a-kind since nothing of their type or scale had ever existed before (see Sections 3 and 4 in Chapter 2 for expanded discussions). For comparison, Fermi's first "pile" (the Chicago Pile 1 Reactor or CP-1), designed only to demonstrate a sustained chain reaction, was an ellipse that measured 25 feet wide by 20 feet high and used 56 tons of uranium fuel, 400 tons of graphite, had no cooling system, and operated at a maximum power level of 200 watts-thermal. The X-10 Reactor at the Clinton Engineer Works, designed as an experimental reactor to provide research quantities of plutonium, was a 24-foot cube, used milled uranium-metal fuel elements encased in aluminum, was air-cooled, and operated at a maximum power level of 4 million watts-thermal. The Hanford Site Manhattan Project reactors (B, D, and F), designed as plutonium production plants, were rectangular, measured 36 feet long by 28 feet wide by 36 feet high, used 200 tons of uranium-metal fuel, used 1200 tons of graphite, were water-cooled, and operated at an initial power level of 250 million watts-thermal.

While the million-fold scale-up in reactors was remarkable, the Hanford Site chemical separations plants (221-B, 221-T, and 221-U) truly amplified the extant model developed at the Metallurgical Laboratory: "Plutonium production at Hanford depended as much on chemical separation as it did on chain-reacting piles [reactors]. The chemistry was Glenn Seaborg's, spectacularly scaled up a billionfold directly from his team's earlier ultramicrochemical work" (Rhodes 1986, p. 603).

Just as the selection of the Hanford Site had been preconditioned by siting criteria, so too was the design of the complex itself. The Corps and Du Pont used both the available space and the natural landforms to isolate and contain the primary production areas (see Figure 1.4). The reactors, six of which were planned, had to be adjacent to the Columbia River because of their cooling requirements. Starting inland from the western boundary and farthest from the communities of Richland, Kennewick, and Pasco, the reactors would be sited at intervals of at least 1 mile along the river. The reactor complexes were given sequential alphabetic identifications A through F. To increase separation between reactors and their isolation within the Hanford Site, the Corps selected B, D, and F as the initial locations for construction. By using alternating locations, the separation between reactors averaged nearly 6 miles. The reactor complexes were assigned the Area identification codes 100-B, 100-D, and 100-F. Each area (B, D, etc.) contained one reactor and support facilities.

The chemical separations plants, four of which were planned, would be built some 10 miles south of the reactors within the interior of the Hanford Site. Two of these plants would be built in the 200 East Area, while the other two would be built some 4 miles away in the 200 West Area. The basalt landforms of Gable Butte and Gable Mountain lay between the reactors and the separation plants, providing additional safety should something happen at either the irradiation or separation plants. Finally, the fuel manufacturing facilities (300 Area), considered the least threatening of the primary operations, would be located on the river approximately 20 miles southeast of the separation plants and 10 miles north of the administrative area (700 Area) and residential/commercial center (1100 Area) planned for Richland (Gerber 1994a, Gosling 1999, Rhodes 1986).

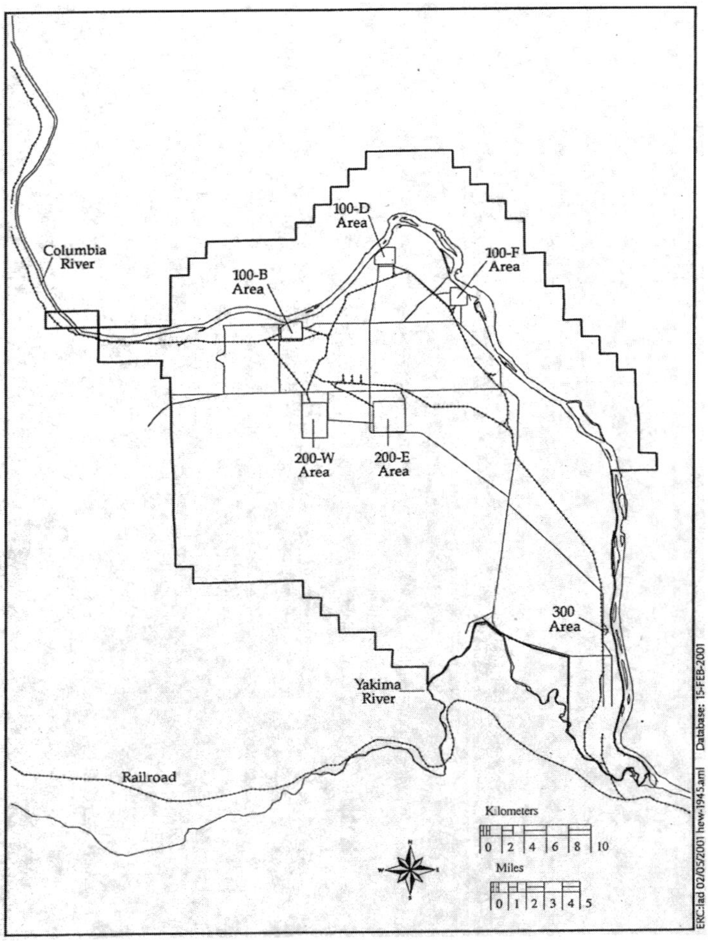

Figure 1.4. *Hanford Site during the Manhattan Project Around 1945 When the Site Was Called the Hanford Engineer Works*

The uncertainties surrounding the emerging technologies and the consequences associated with their failure marked the Hanford Site as one of the greatest scientific experiments of modern time or one of the riskiest. The Corps therefore designed each process area as an independent unit complete with operations facilities, operations support facilities, and administrative, security, health, communication, utility, and waste support facilities. This structural redundancy was a hedge against failure in any one unit since the overriding objective was the uninterrupted production of plutonium. To knit these units together, the Corps established a transportation, utilities, and communications infrastructure.

The single railroad line to the Hanford Site and the existing road system constituted the backbone of the transportation system that would move workers and materials. However, these facilities needed to be upgraded. New construction added "158 miles of railroad and 386 miles of roadway" (Chatters 1989, p. D.85).

Electrical power for the 100 and 200 Areas would be obtained from the Grand Coulee-to-Bonneville grid, while the remainder of the Hanford Site would receive power from the Midway-to-Walla Walla tie-line. To route this power, the Corps constructed 52 miles of transmission lines. These electrical lines, together with the "distribution substations, fence and road lighting, fire alarm systems, and telephones and telephone cables," were a non-geographic area but nonetheless assigned a numeric area identification, the 500 Area, to meet the military need for order and organization (Gerber et al. 1997, p. 5.22).

Similarly, the 800 Area referred to the steam power plants and all "overhead pipe line facilities, specifically all pipe supports, steam lines, air lines, and process lines," while the 900 Area identified all "underground pipe line facilities, specifically the export water lines and valve houses, raw water and fire protection lines, sanitary sewers, process sewers, wells, and pumps" (Gerber et al. 1997, p. 5.23).

One of the most visible and ubiquitous items constructed was fencing. Hundreds of miles of fences were placed around the Hanford Site and its facilities. Fences were erected around key facilities within fenced complexes, and even the administrative center within the residential community was enclosed by fences. See Chapter 2, Sections 1 and 11, for additional discussion on construction and infrastructure.

RECRUITING A WORKFORCE AND SECURING MATERIALS

Recruiters made direct inquiries to Du Pont staff engineers, scientists, and consultants and enticed many to relocate with offers of better pay or more challenging work, although the specific nature of that work could not be explicitly stated (DeJong 1995). Du Pont, working with the War Manpower Commission and the U.S. Employment Service, conducted aggressive advertisement campaigns in 47 of the 48 existing states as well as Alaska and Canada. Tennessee was excluded because of recruitment for the Clinton Engineer Works. The advertisements touted "attractive" wages and extended, steady hours for skilled and unskilled workers. Unskilled workers could make twice the daily rate offered to laborers elsewhere, while skilled labor was offered a daily rate of $15 – 50 percent more than the prevailing rate of $10. Even with these incentives, maintaining a full labor force remained more an objective than an accomplishment throughout initial construction: "The work proceeded slowly, dogged by recruiting problems. The nation at war had moved beyond full employment to severe labor shortages and men and women willing to camp out on godforsaken scrubland far from any major city were hard to find" (Rhodes 1986, p. 499).

Recruiters had to work hard since "they found themselves competing against other employers, including the armed services, which enjoyed a head start in mobilizing for war" (Findlay and Hevly 1995, p. 15). To counter the effects of the draft and voluntary military enlistment, Du Pont reached out to people who were not at the top of the Selective Service's list: "Slightly more than half of all employees were 38 years or older, and therefore above the maximum drafting age, and three-quarters of those between 18 and 26 years old had 4-F ratings" (Findlay and Hevly 1995, p. 17). Women made up nearly 13 percent of the workforce, and African Americans made up an additional 16 percent. Both of these groups were segregated within the Hanford construction camp. While initially reluctant to hire Hispanic Americans, Matthias eventually conceded to the War Manpower Commission but housed those who were hired "near Pasco, off the Hanford Site and away from the other workers" (Findlay and Hevly 1995, p. 18).

Extraordinary measures were taken as necessary to ensure that construction continued at an accelerated schedule: "Later in 1944, when the project required trained pipe-fitters for completing the three piles [reactors], the Manhattan Engineer District (MED) furloughed enlisted men from active duty, enrolled them in reserve units, and employed them as civilian workers at Hanford, with a promise to reinduct them into the Army if their work proved unsatisfactory. Matthias and Du Pont protected other highly skilled personnel, which the project could not afford to lose, by reaching agreements with local draft boards" (Findlay and Hevly 1995, p. 16).

As with all aspects of the Manhattan Project, the need for secrecy hampered Du Pont's ability to hire and maintain workers. Since they could not be told specifically what the purpose of their work was, many employees left because they felt they were "not making a contribution to the war effort." Even those who stayed developed "a transient mentality" because they did not know if the Hanford Site would exist after the war. Union relations were difficult because of the secrecy issue although appeals to patriotism and skilled labor shortages "forced organized labor to become accommodating" (Findlay and Hevly 1995, p. 21). See Chapter 2, Section 12 for additional discussion of this topic.

The issue of secrecy extended to the acquisition of supplies. So as not to tip off the War Production Board, the Hanford Site was constrained by procurement and transportation ratings of less than the optimum AAA. Gerber (1992b, p. 11) notes, for example, materials such as "lower grades of steel, and even of earthenware" sometimes were substituted where "a high grade of Colombian (Austenitic) stainless steel" should have been used. Other materials like "heavy-gauge railroad track, cement, copper, lumber, iron, electrical wire, graphite and aluminum were always in short supply...[but] the Hanford project could acquire an overriding priority when necessary."

Hanford Construction Camp

Colonel Matthias was charged with delivering plutonium to Los Alamos as soon as possible, and Lieutenant General Groves made several visits to the Hanford Site to remind him of the urgency. Yet for most of 1943, Matthias had to postpone construction of the production facilities and concentrate the workforce on constructing the facilities needed to house and support the workers, most of whom lived in tents through the first 6 months of construction (Findlay and Hevly 1995). Du Pont selected the now evacuated Hanford townsite to serve as the genesis of the construction camp because of its proximity to the 100 and 200 Areas and the utilities already existing there. Already numbering 12,500 in August 1943, the construction force increased nearly four-fold to 45,000 at its peak in June 1944 (Thayer 1996). By that time, the "self-contained town" had "131 barracks for 24,892 men and another 64 barracks for 4,357 women; 880 hutments [Quonset huts] for men; and 3,639 trailer lots" (Findlay and Hevly 1995, p. 25). In addition to housing, the camp also included 65 construction shops, a 24-building administrative area, eight mess halls, saloons, stores, a recreation hall and auditorium, a theater, a bowling alley, and a softball diamond among other amenities. Services included eight schools, a day nursery, five fire stations, and a large infirmary and clinic (Gerber 1999, p. 5).

Rhodes (1986, p. 499) reports that meat rationing, common during the war, did not occur at the Hanford construction camp. In fact, the ready availability of meat was used in recruiting the workforce, which was amply fed as evidenced by the information obtained from a posting found on a barrack wall by Artie McDaniels. A copy of the posting is on file at the Library of the U.S. Department of Energy's Hanford Cultural Resources Laboratory in Richland, Washington (see sidebar box).

Various authors have described the Hanford construction camp as a "sea of tents and barracks, where workers had little to do and nowhere to go" (Gosling 1999, p. 32), where "boredom and loneliness" prevailed (Chatters 1989, p. D.85), and even the saloons were built "with windows hinged for easy tear-gas lobbing" (Rhodes 1986, p. 499). Findlay and Hevly (1995, p. 26) seek to put this "notoriety" in perspective by observing that the camp was not dissimilar from "the kind of rough-and-tumble community – like mining, lumber, and other construction camps – that emerged in the American West whenever a hard-working, well-paid, heavily male population was concentrated temporarily in one spot."

Just to be sure, however, the Corps did lock up their women, who lived in segregated barracks behind barbed wire under the watchful protection of "housemothers" and guards. The effectiveness of this enforced separation may be suspect, though, given events at Los Alamos where one women's dormitory came under review for closure until "a determined group of bachelors argued even more persuasively against closing the dorm. It seems that the girls had been doing a flourishing business of requiting the basic needs of our young men, and at a price" even with roving military police (Robert Wilson, quoted in Rhodes 1986, pp. 566-567).

A more typical image of life at the camp is provided by Susanna Brown, who together with her husband, moved to the Hanford Site from Texas in 1944: "Primer and I lived in different barracks with fences and a guard you had to show a pass to, going in and out. I worked the late shift and my husband worked on another so we didn't see much of each other. Primer would visit me in the sitting room – all of the women's barracks had one" (Sorenson 1996, p. 5).

Chapter 2, Section 12 of this book adds substantially to these images of life not only at the Hanford Camp but also throughout the operational life of the Hanford Site. It gives voice to the people who worked here and the issues they faced.

World War II Posting Found on a Barrack Wall

- 1,000,000 meal ticket cards on file
- $3,500,000 worth of meal tickets sold in 8 months through payroll deduction (does not include meal tickets sold for cash)
- Menus prepared 60 days in advance
- Eight mess halls, 2700 workers in each hall at each setting, usually three settings per meal
- 3000 pounds of sausage used for one breakfast
- 2500 pounds of pot roast for one meal
- 18,000 pork chops for one meal for one mess hall
- 11,000 pounds of Swiss steak for one meal for all mess halls
- 250,000 pounds of meat used for all mess halls for 1 week
- 15 tons of potatoes for one mess hall each day
- 5000 heads of lettuce for each meal for one mess hall
- 1200 pounds of onions for one meal for one mess hall
- 900 full pies for one meal for one mess hall
- 600 gallons of ice cream a day
- 250 good cows needed to supply the milk for one breakfast
- 1000 pounds of coffee for one day for one mess hall
- 700 cases of Coca-Cola a day
- 2200 loaves of bread used each day for sandwiches, not counting bread on tables
- Three sandwich machines, each making 360 sandwiches an hour
- 272,000 pounds of processed meat, ready for oven or grill used in 1 week
- 4,000,000 lunch boxes sold from July 1943 to October 1944
- 30,000 doughnuts for one day
- An automatic doughnut machine making 18,000 doughnuts per hour
- 6500 eggs used for Sunday breakfast (only time fried eggs could be served because meals were served over a longer period of time)
- 12,000 turkeys for Thanksgiving (22 tons of turkey, 12 tons of ham)

Other information supplied on this sheet indicates:

- Number of employees quartered in barracks: 40,000
- Number of employees quartered in trailer camp: 8,000
- 10,000 newspapers sold each day by recreation halls
- 16,000 packages of cigarettes sold each day
- 12,000 gallons of beer consumed each week (13 carloads)
- 2000 keys for barracks lost each month by employees
- 1,785,000 sheets washed. If these were tied together they would reach from Hanford to New York City
- Hanford is the largest voting precinct in the United States
- Hanford has the largest general-delivery post office in the world

From the beginning the workweek was long, and, as deadlines neared, it only got longer. It was initially a 6-day, 8-hour shift, then the Corps added an extra hour to each day in September 1943 and finally added floodlights to allow "around-the-clock construction" at the 100-B and 100-D Areas (Gerber 1992b, p. 8). Despite the cold and heat and the winds that blew the freshly released desert sands into and over everyone and everything, the workforce housed at the Hanford construction camp succeeded in building the world's first industrial-scale plutonium production and extraction complex in just under 2 years – a truly remarkable feat considering that today the paperwork alone required to permit construction of a single reactor would take significantly more time. By late February 1945 as the Corps and Du Pont were gearing up for operations, the camp was abandoned. Facilities that could be used elsewhere were moved out. Trailers, for example, were relocated to North Richland in 1946 to house General Electric and Atomic Energy Commission employees and their families, while the larger buildings such as the mess hall, recreation hall, and barracks were either dismantled or relocated to meet a growing postwar need for housing and temporary office space (Chatters 1989, p. D.85). Within a year after the war ended, whatever remained of the camp was removed and the area leveled, leaving only the roadway grid and a few isolated foundations low enough to have escaped complete destruction by the bulldozer's blade.

Richland Village

Though construction began on both the Hanford construction camp and the Richland Village (1100 Area) in 1943, the areas provide a study in contrasts. "Du Pont built the Hanford camp to house the workers during construction of the first three production reactors, two chemical separations canyons, and assorted other facilities. It erected the Richland Village to house the employees who would administer and operate the plant" (Findlay and Hevly 1995, p. 12). The authors go on to list a number of juxtapositions beginning with their designations: Hanford was a camp, Richland was a village. Hanford was meant to last only as long as construction required, Richland would be a permanent settlement. Hanford was hastily built, Richland was planned and revised with changing conditions. Hanford was working class, Richland was middle class. Indeed, Matthias fully intended that Richland Village house supervisors, engineers, operators, and essential office staff and that craftsmen and laborers would be assimilated within the communities of Kennewick and Pasco.

To build the administrative and residential center, the Corps condemned the original Richland townsite and farms adjacent to it. Since the Hanford construction camp held a higher priority in terms of completion, residents of Richland were given until November 15 to evacuate their homes. As with the communities of White Bluffs and Hanford, the Corps identified 150 houses it could use together with some commercial properties and the irrigation system, then destroyed the rest creating an open, dusty landscape where the town once stood. While agreeing in principle that the Village would serve the managers and operators, the Corps and Du Pont disagreed on how well and at what cost. Groves ordered that Richland Village be built "as cheaply as possible" and pressed the same requirement on Los Alamos and the Clinton Engineer Works. However, Du Pont, in effect looking out for its future staff who would be the operators and managers of the Hanford Site, insisted that quality and comfort be included within the design. For example, while Groves proposed "barracks and dormitories," Du Pont held firm for houses and submitted designs commensurate with the employee's status – the alphabet houses that still distinguish the former Richland Village within contemporary Richland. Unlike Los Alamos and the Clinton Engineer Works, Richland Village was not enclosed within a fence, although it was tightly guarded and monitored by the Army. The only fenced area within the Village included the Administrative or 700 Area (Findlay and Hevly 1995, p. 34-39).

> ***"When completed, wartime Richland represented a compromise between the visions of Du Pont and Groves. As a result of the general's efforts to economize, the town wound up with inadequate numbers of sidewalks, garages, stores, and shopping areas, no civic center, and roads too narrow for much auto traffic...Wartime Richland also had its share of temporary housing – 25 dormitories by 1945 – so that it could house both construction and operations personnel when those two phases of Hanford's development overlapped."*** **- Findlay and Hevly 1995, p. 38**

G. Albin Pehrson, the Spokane architectural engineer Du Pont engaged to design Richland Village, developed a "neighborhood-oriented plan" that would provide an "island of refuge" from the military atmosphere surrounding the Hanford Site. By 1945, the Village contained 4329 houses and a population of 16,000. The houses, eight conventional styles (A, B, D, E, F, G, H, and L) and three pre-fabricated styles (A-1, B-1, and C-1), were classified based on the number of bedrooms and total cost (Gerber et al. 1997, pp. 5.24-5.25). For the few who moved from the construction camp to the Village, the difference was described by Nell MacGregor: "We felt unspeakably elegant in rooms with plastered walls, painted woodwork and splinterless floors" (Findlay and Hevly 1995, p. 34).

Richland Village also contained 24 commercial or retail stores including food and drug stores, barber and beauty shops, a hardware and variety store, and an icehouse. Community buildings included three schools, a church, a hospital, a movie theater, and a combined police and fire station. Recreational activities were enhanced by the incorporation of open spaces and green belts within the residential areas and a park along the Columbia River (Gerber et al., 1997, p. 5.25). Interestingly, in an ironic way, Du Pont and the Army found themselves on opposing sides with respect to the operation of the Village:

> *"In building Richland, the Army had wanted to skimp on expenses while Du Pont had wished to spend more; in operating the village, those roles were reversed. W.O. Simon, Du Pont's manager at HEW, called the Army "liberal" in its treatment of residents and pointed out a number of instances where the Army had demonstrated too much initiative in providing services to townspeople. Du Pont's more "conservative" approach, according to Simon, "would avoid any activity or endeavor not necessary to its immediate needs and would follow rather than anticipate public opinion."* - Findlay and Hevly 1995, p. 43

Anytime the Army is called "liberal," a significant difference of opinion must exist. Engaging as this thread may be, the interested reader is directed to continue with the Findlay and Hevly manuscript for a better understanding of the cultural history of the Hanford Site and the Tri-Cities.

Constructing the Operations Facilities

Construction of the operating facilities was a shared responsibility among the Corps, the Metallurgical Laboratory, and Du Pont. The Corps was responsible for funding, scheduling, and decision making. The Metallurgical Laboratory scientists were responsible for the theoretical construct. Du Pont was responsible for design and engineering. As with many multiparty undertakings, the degree to which the parties disparage one another varies inversely with the level of success being achieved. In this case, the scientists were quick and consistent in criticizing Du Pont's "caution, expertise, and management skills," though in hindsight many credit Du Pont's "industrial experience, its engineering orientation, and its cautious policies" with bringing order to chaos and ultimately ensuring "the timely success of Hanford" (Findlay and Hevly 1995, pp. 13-14). Groves singles out Du Pont manager Crawford H. Greenewalt for praise in handling a difficult situation:

> *"[Greenewalt] served as the bridge between the hard-driving, thoroughly competent, industrial-minded, scientific engineers and executives at Wilmington and the highly intelligent and theoretically inclined scientists at Chicago. This means that he had to shuttle back and forth between Chicago and Wilmington and later Hanford, easing tensions and calming tempers and, at the same time, seeing to it that needed scientific decisions were promptly reached at Chicago."* - Groves 1983, p. 79

Gosling (1999, p. 30) maintains that a sense of lessened importance, occasioned by the construction of the Hanford Site and the Clinton Engineer Works, fueled some of the scientists' criticism as "Met Lab research became increasingly unimportant in the race for the bomb and the scientists found themselves serving primarily as consultants for Du Pont." Time, or more specifically the lack of it, was also a factor: "The scientists' complaint about Du Pont, and General Groves' constant pressure on Matthias, stemmed primarily from the extreme urgency of the project from start to finish...Such urgency pushed construction and operations at Hanford at a pace that would have been inconceivable outside of wartime" (Findlay and Hevly 1995, p. 14).

Uncertainty coupled with accelerated schedules contributed even more to the frustration of all parties. Construction would be ongoing while the basic research underlying nuclear reaction and chemical separations was still being developed. However, as Groves later noted, "it is essential to keep in mind the truly pioneering nature of the plutonium development...gigantic steps [were taken]...moving rapidly...from the idea stage to an operating plant" (Groves quoted in Gerber 1992b, p. 11).

Design and construction proceeded simultaneously and more often than not out of step since the exigencies of war allowed little or no time for testing or validating designs. For example, groundbreaking for the cooling system for the first Hanford Site production reactor, 105-B, began on August 27, 1943, barely 8 months after Fermi had started CP-1, a short-term experimental reactor operated with no cooling system, and 3 months before initial operation of the air-cooled X-10 Reactor. Cooling a reactor with water was unprecedented, yet the success of full-scale plutonium production and the Hanford Site's contribution to the Manhattan Project in large part rested on this entirely new and untested technology: "Because of the tremendous energy associated with fission, a great amount of heat is developed. To produce one ounce of plutonium, the reactor must develop a thousand kilowatts of heat power for a period of a month. Where many pounds are involved, a major engineering problem is the removal of this heat" (Compton 1956, p. 161).

On October 4, 1943, Du Pont engineers released the design drawings for the B Reactor allowing construction to begin 6 days later with general site preparation (Sublette 1997). "Starting with the foundations for the pile [reactor] and the deepwater basins behind it where the irradiated slugs [fuel elements] would be collected after discharge, the work crews were well above ground by the end of the year" (Rhodes 1986, pp. 499-500). See the Historic American Engineering Record (HAER) for the B Reactor in Appendix B of this book on the Internet for a detailed discussion of the reactor's construction and operations.

More than half a continent away, Du Pont was also constructing the X-10 complex at the Clinton Engineer Works to test the principles of reactor operation and plutonium separation. Groundbreaking on this pilot reactor and its associated chemical separations plant containing six hot cells had begun during February 1943. Within 9 months of initiating construction, the X-10 Graphite Reactor went critical (condition in which a material undergoes nuclear fission at a self-sustaining rate) on November 4, 1943 and began "producing the first substantial (gram) amounts of plutonium to assist research into its properties" (Sublette 1997). With the world's supply of plutonium at this time standing at 2.5 milligrams, all produced in cyclotrons, the 1.54 milligrams extracted within the separation cells at the Clinton Engineer Works by the end of the year nearly doubled the amount available for research. A secret courier carried the plutonium to the Metallurgical Laboratory in Chicago using a container resembling a penlight (Souza et al. 1996, p. 3-39). Although this air-cooled experimental reactor was not a true prototype, experience gained in its construction assisted Du Pont's efforts with the Hanford Site reactors. The Hanford Site benefited from the X-10 reactor in other ways as well. Before June 1944, Du Pont sent future Hanford Site reactor operators to the Clinton Engineer Works for on-the-job training.

As noted earlier, security and secrecy were hallmarks of the Manhattan Project. This extended even to the blueprints used in constructing critical buildings. As described in the B Reactor HAER in Appendix B on the Internet, Rudy DeJong, a construction foreman at the B Reactor, recalls that the "foremen, engineers, and superintendents were the only ones who ever saw a drawing! We had to go into a vault inside the reactor building, look at the drawings, figure and get your dimensions, make notes, then go out to your crew and tell them what to do and what your dimensions were!" Du Pont also, for security reasons, designated the reactor buildings "classified" and further restricted worker access to those possessing the proper badge. A separate division engineer supervised construction of the reactor buildings. This helped maintain "compartmentalization," which was the heart of security and secrecy. Compartmentalization, as Groves established it, meant "each man should know everything he needed to know to do his job and nothing else" (Groves 1983, p. 140).

Fewer than one percent of the workers were told the true nature of the project. One who was briefed, Kenneth McCreight, a supervisor in charge of over 1800 electricians, remembers some very long workdays extending 18-20 hours and switching assignments or terminating workers when necessary if too many questions were asked. He "hand picked" workers for the reactors and "the ones who were less skilled or couldn't get along with others" were assigned "less sensitive work such as building offices and septic systems" (Gerber 1994b, p. 16).

In addition to the three 100 Area reactor complexes along the Columbia River, Du Pont also constructed three chemical separations complexes within the interior of the Hanford Site. The 200 West Area contained the 221-T and 221-U Plants, while the 221-B Plant was located in the 200 East Area. The second separations plant (221-C) planned for the 200 East Area was cancelled when Groves determined it was unnecessary based on experience and information gained from operations at the Clinton Semiworks (Gerber et al. 1997). Indeed, only the 221-B and 221-T Plants were operated during World War II although Du Pont had constructed all three by early 1945. Gosling (1999, p. 33) observes that "labor shortages and the lack of blueprints" so constrained construction activities in the 200 Areas that Du Pont redirected workers to the 100-B Area "with the result that the 1943 construction progress on chemical separations was limited to digging two huge holes in the ground."

The Hanford Site chemical separations plants (locally called "canyons" because of their architectural design featuring a very long, high-walled, windowless operations area) dwarfed the pilot plant built at the Clinton Engineer Works. The Clinton Semiworks contained six hot cells, while the process plants on the Hanford Site contained forty. Measuring more than 800 feet long, 65 feet wide, and with a surface height of 80 feet (an additional 20 feet lay below grade), the workers also nicknamed the buildings "Queen Marys" after the British ocean liner. Stark, nearly featureless concrete structures on the outside, "the interior had an eerie quality as operators behind thick concrete shielding manipulated remote control equipment by looking through television monitors and periscopes from an upper gallery" (Gosling 1999, p. 34). To protect process workers from intense radiation, Du Pont engineer Raymond P. Genereaux designed each process cell to be encased within concrete walls 7 feet thick and enclosed each cell with a 6-foot-thick cover weighing 35 tons. A traveling overhead crane was used to lift the covers when necessary (Gerber et al. 1997).

"Once the Queen Marys were contaminated with radioactivity no repair crews could enter them. Equipment operators had to be able to maintain them by remote control. The operators trained at Du Pont in Delaware, at Oak Ridge and on mockups at Hanford, but the engineer in charge...sought more authoritative qualifications. And found it: he required his operators, one hundred of whom arrived at Hanford in October 1944, to install the process equipment into the first completed separation building by remote control, pretending the canyon was already radioactive. They did, awkwardly at first but with increasing confidence as practice improved their remote-manipulation skills." - **Rhodes 1986, p. 604**

To receive the high level radioactive liquid wastes (radioisotopes and chemicals) resulting from the chemical separations process, Du Pont constructed 64 single-shell underground storage tanks in four complexes (241-B, 241-C, 241-T, 241-U), known as "tank farms." Each complex consisted of sixteen tanks – twelve 500,000-gallon tanks and four 54,500-gallon tanks (Gerber 1992b). Each of the 500,000-gallon tanks measured 75 feet in diameter and was constructed of "reinforced concrete with a one-fourth inch welded [carbon-] steel plate lining" (Gerber et al. 1997, p. 5.83). The four smaller tanks were 20 feet in diameter. Gerber (1992c, p. 46) reports that by the end of the Manhattan Project "half of these tanks were 100 percent full and the other half were 40 percent full." This is understandable given the fact that nearly 10,000 gallons of high level radioactive waste was generated per day at both 221-T and 221-B Plants (Anderson1990).

Stretching the workforce even further, Du Pont was also busy constructing the fuel manufacturing and testing facilities within the 300 Area. While irradiation and chemical separations were more visible steps in producing weapons-grade plutonium (based on the sheer number of facilities devoted to each of these operations), both were sequential steps that relied on precisely milled and processed uranium metal fuel elements, commonly known as "slugs." The 313 Metal Fabrication Building and 314 Press Building were key facilities that supplied the fuel elements for irradiation in the reactors.

Both the 313 and 314 Buildings were operational well before the reactors were completed because without fuel the reactors could not be charged. Completed in the fall 1943, about the same time that reactor construction began, the 313 Building was reconfigured with eight additions within a 10-month period, each addition responding to "process improvements and changes in the very new, untried, and unique uranium fuel fabrication activities being carried out in the facility" (Gerber 1993b, p. 1). In January 1945, workers at the 314 Building began to extrude (press) uranium billets (uranium metal cylinders approximately 20 inches long and weighing 200 pounds) into rods that machinists shaped into fuel elements in the 313 Building (DOE 1997e, p. 151). Prior to this onsite capability, the Hanford Site received pre-extruded uranium metal rods directly from outside sources beginning October 1943 (see Chapter 2, Section 2 for additional discussion and illustrations).

The 305 Test Pile (reactor), of necessity, was also constructed and operational before the production reactors in the 100 Areas. Beginning in March 1944, this 16-foot graphite cube, air-cooled, natural uranium-powered reactor, the first to operate on the Hanford Site, "compared the performance of material samples under irradiation with samples of known quality and graded each lot" (Gerber 1993b, p. 11). For example, to maximize the movement of neutrons, only very pure graphite could be used in constructing the production reactors. To ensure that suitable graphite was delivered to the 100 Areas, the 305 Test Pile was pushed to 16 hours a day almost as soon as it became operational. By May, it was testing graphite bars 24 hours a day. This schedule was maintained until August when operations were scaled back to a single 8-hour shift. In addition to graphite bars, the 305 Test Pile irradiated and tested uranium billets (called "eggs"), fuel elements, aluminum canning materials, aluminum welding rods, and aluminum process tubes. This reactor also provided a radiation source for developing and calibrating radiation detection and measurement instruments (Gerber1993b, pp. 11-13).

Like the 305 Test Pile, the 321 Separations Laboratory and the 3706 Radiochemistry Laboratory provided onsite research and trouble-shooting capabilities for the chemical separations process still under development as the 200 Area canyon buildings were being designed and constructed. Completed in December 1944, the 321 Building was a pilot-scale plant used "to replicate, study, and develop solutions to problems developed in the early bismuth phosphate chemical separations process used in Hanford's T Plant and B Plant" as well as "equipment corrosion studies and methods of decontamination" (Gerber 1993b, p. 15). The 3706 Building was the first radiochemistry laboratory at the Hanford Site. Completed early in 1945, researchers used the 57 laboratories "to perform small-scale experiments with both low- and high-activity radioactive materials in support of all HEW production activities" (Gerber 1993b, p. 19). The Historic Property Inventory Forms (HPIF) for the 305, 321, and 3706 buildings give more detailed discussions of activities, accomplishments, and architecture (see Appendix B on the Internet).

Construction of the Hanford Site was a massive undertaking. Simultaneously, Du Pont built three reactor complexes, two chemical separations complexes, a fuel manufacturing and research and development center, a construction camp, and an employee village. The B Reactor HAER in Appendix B on the Internet summarizes the comments made by a number of the people directly involved in constructing the Hanford Site (see sidebar box).

> ***"The work that was performed at the Hanford Site was of monumental proportions, rising from bare desert to fully functional industrial complex in a scant two years, while creating a nuclear technology that had only just been discovered. Even under peacetime conditions, the project would have garnered notoriety for its cutting-edge developments and, especially, its complete success. During the incredible push of wartime production and urgency, however, the story takes on legendary proportions." -***
> **B Reactor HAER**

Each production complex was self-contained, complete with administrative, operational, health, security, maintenance, supply, and utility support facilities. This designed redundancy ensured that plutonium production would not be interrupted – an imperative for the Manhattan Project. A sense of the scale of construction, momentarily discounting its engineering complexity, is imparted through a summary listing of numbers alone. Within the 100 Areas, 94 permanent buildings and 69 service facilities were constructed. Additionally, "over 100 small support service TC [temporary construction] structures were built in each

100 Area including warming sheds, privies, check booths, miscellaneous sheds and guard/badge houses" (Gerber et al. 1997, p. 5.16). This number does not include the 50 or more temporary construction structures used for the Division Engineers Office, Government Field Office, Safety Office, Machine-Millwright and Sheetmetal Shop, and structures for other administrative and craft support. The 200 Areas contained 108 permanent buildings and 62 service facilities and an additional 100 temporary construction buildings and structures. The 300 Area contained 34 permanent buildings and 25 service facilities. Only seven temporary construction buildings are accounted for in the 300 Area, which is considerably fewer than the other Areas, "because more of the 300 Area work was done by subcontractors who provided their own facilities" (Gerber et al. 1997, p. 5.20). These numbers are further augmented by the 9 permanent buildings and 14 service facilities in the 200 North Area, the "machine shops, pipe and electrical shops and carpentry shops" constructed for site-wide support at White Bluffs (Gerber 1999, p. 4), the Hot Mix Plant, and the Riverland Classification Railroad Yard among other operational and support facilities outside the designated Areas (Gerber et al. 1997).

"It's Atomic Bombs": Hanford Site Operations 1943-1945

On July 4, 1944, Oppenheimer informed the Los Alamos staff that the "neutron emission of reactor-produced plutonium was too high for the gun assembly to work" (Sublette 1999). The gun assembly required that two subcritical masses of the proper size and shape be forced together at high speed to obtain a supercritical mass that would initiate a chain reaction and subsequent explosion. Predetonation of either or both of the subcritical masses because of "spontaneously emitted neutrons" would render the bomb useless (Gosling 1999, p. 38). Given the rate of emission determined by Emilio Segrè, the planned plutonium gun had to be abandoned.

At this point in 1944, with construction of B Reactor nearly complete, the irony in Oppenheimer's announcement is sadly apparent: "The only workable bomb design at hand, the gun-type weapon, required U-235 which had no proven practical production methods available...Plutonium production had not yet begun, but the production technique appeared to have a high probability of success. However plausible approaches to building a plutonium bomb did not yet exist" (Sublette 1999).

Fortunately, Seth H. Neddermeyer of the ordnance staff at Los Alamos had been investigating an alternative, though highly theoretical, method for explosion with uranium of a purity less than optimal. These emergent studies were now directed to plutonium. Implosion would use high explosives to direct symmetrical shockwaves inward towards a subcritical mass of imbedded plutonium. The force of the explosion would compress the plutonium into a supercritical mass and initiate the chain reaction leading to explosion. Implosion research became a top priority. "With just 12 months to go before expected weapon delivery a new fundamental technology...had to be invented, made reliable, and an enormous array of engineering problems had to be solved" (Sublette 1999). Oppenheimer assigned the task to Neddermeyer and George B. Kistiakowski.

The unplanned change to implosion also had an impact on the Hanford Site. Even before they were operational, production rates would be accelerated to counter the uncertainty of how much plutonium would be required. Given Grove's directive to deliver as much plutonium as possible as quickly as possible, initial activation of the B Reactor did not bode well. First-person accounts of the Hanford Site operators' startup of B Reactor can be found in Sections 3.1-3.3 of the B Reactor HAER in Appendix B on the Internet.

Ten months after the X-10 reactor had gone critical, Enrico Fermi was present to witness the startup of the Hanford Site's B Reactor. On September 13, 1944, Fermi loaded the first fuel element into the reactor: "the Pope," says Rhodes (1986, p. 557), "conferring his blessing as he had on the piles [reactors] at Chicago and Oak Ridge." It took 2 weeks to hand load 1595 process tubes, slightly more than the 1500 calculated by Eugene Wigner in his initial design.

"Tuesday evening, September 26, 1944, the largest atomic pile [reactor] yet assembled on earth was ready...The operators withdrew the control rods in stages just as Fermi had once directed at CP-1...The pile [reactor] went critical a few minutes past midnight; by 2 A.M. it was operating at a higher level of power than any previous chain reaction. For the space of an hour all went well...Early Wednesday evening B pile died...Early Thursday morning the pile came back to life. By 7 A.M. it was running well above critical again. But twelve hours later it began another decline. Princeton theoretician John A. Wheeler...had been "concerned for months about fission product poisons." B pile's heavy breathing convinced him such a poisoning had occurred." **- Rhodes 1986, pp. 557-558**

Wheeler suspected that xenon-135, a daughter product of the reaction-induced isotope iodine-135, was the neutron-absorbing poison that was causing the reactor to fail. Once this short-lived isotope had decayed, the reaction would revive and continue building until the xenon level increased to the point where it again began to absorb the neutrons necessary to sustain the chain reaction.

To determine if this were the case, Crawford Greenewalt, also present for the startup, called Samuel Allison at the Argonne laboratory to test Wheeler's theory using the CP-3 reactor: "Disbelieving, [Walter] Zinn started the 300-kilowatt reactor up again and ran it at full power for twelve hours. It was primarily a research instrument and it had never been run for so long at full power before. He found the xenon effect...Groves received the news acidly. He had ordered Compton to run CP-3 at full power full time to look for just such trouble" (Rhodes 1986, p. 559).

Extra reactivity would have to be added to overcome this effect before production could begin. Fortunately, and providing a concrete example of Du Pont's conservative engineering approach overriding the theoretical calculations of the Metallurgical Laboratory scientists, Wheeler had anticipated this chance of failure: "Wheeler had fretted about fission-product poisoning...[and] had advised the chemical company to increase the count of uranium channels for a margin of safety...That necessitated drilling out the shield blocks, which delayed construction and added millions to the cost. Du Pont had accepted the delay and drilled the extra channels. They were in place now when they were needed, although not yet connected to the water supply" (Rhodes 1986, p. 560).

It would take 2 months to connect the remaining 409 channels, thereby achieving the reactor's full design configuration of 2004 process tubes. In the interim, however, having benefited from Du Pont's earlier engineering decision, the D Reactor went critical on December 17, 1944, "with sufficient reactivity to overcome fission product poisoning effects" (Sublette 1999). By December 28, 1944, all work required at B Reactor was completed. With these two reactors operating in tandem, large-scale plutonium production was under way. When the F Reactor went on-line on February 25, 1945, theoretical production capacity from the combined reactors rose to 21 kilograms per month (Sublette 1997).

Nevertheless, initial production efforts at the Hanford Site remained tied to B Reactor. The B Reactor was not idle during the 2 months it took to supply cooling water to all 2004 process tubes. Between September 30 and December 28, the reactor operated intermittently at steadily increasing power levels while fuel was added and discharged and additional process tubes were connected to the cooling system. As described in the B Reactor HAER (see Appendix B on the Internet), a test run of irradiated fuel elements was discharged from a single process tube on November 6, 1944, "much earlier than would be normal for fuel processing [usually 100 days], but even mildly hot fuel was badly needed to test the various fuel-handling and chemical separations facilities in the 100 and 200 Areas."

The first scheduled discharge of 42 tubes took place between November 24 and 28. These fuel elements were transported by rail to the 200-North Area Lag Storage Buildings (212-N, 212-P, and 212-R) where they were allowed to cool. Within 30 days they were shipped to the T Plant where the world's first production-scale chemical separations processing began on December 26, 1944. Processing was complete on February 2, 1945. Colonel Matthias himself drove the first plutonium shipment from the Hanford Site to Portland, Oregon, then boarded a train for Los Angeles, where he delivered the product to an "agent from Los Alamos" on February 3, 1945 (Findlay and Hevly 1995, p. 50).

The B Plant came on-line on April 13, 1945. With this increased separations capacity, workers prepared a plutonium shipment for rail transportation every 5 days. To speed delivery, heavily guarded truck caravans replaced rail transportation beginning in May. To further accelerate delivery, Du Pont abandoned ground transportation and sent planes directly to Los Alamos beginning in late July. In the face of intense pressure from Oppenheimer in Los Alamos and Groves in Washington, D.C., Matthias continuously pressed Du Pont to produce more plutonium. In response, Du Pont not only "ran the reactors above their rated power level" (Findlay and Hevly 1995, p. 50), but also reduced the radiological cooling period for irradiated fuel elements to potentially unsafe levels to meet immediate demands: "Throughout the spring and early summer of 1945, metal cooling times fell, as HEW rushed to produce plutonium for the Trinity and Nagasaki bombs. Exactly how short the metal decay periods became is unclear, but it is known that they fell below 30 days and to perhaps as low as a few weeks" (Gerber 1994a, p. 1).

Even with accelerated production, Groves never let up and "continued to push for heightened production until the day Japan surrendered" (Findlay and Hevly 1995, p. 51).

PRODUCTION CYCLE

Much as Niels Bohr had surmised in March 1939, the production of nuclear weapons in the United States became "a complex series of integrated manufacturing activities executed at multiple sites across the country" (DOE 1997e, p. 117). The Hanford Site was one of only three Manhattan Project complexes built to develop the atomic bomb. Its mission, to produce plutonium, complimented that of the Oak Ridge Reservation, where enriched uranium was produced. Together, these two "production" sites supplied the scientists at Los Alamos with the nuclear materials necessary to test and fabricate the atomic fuels that powered the Fat Man and Little Boy bombs, respectively. After World War II, however, the Hanford Site, Oak Ridge Reservation, and Los Alamos became part of an expanding nuclear weapons complex that ultimately spanned the United States with facilities in 28 states (Figure 1.5). Most facilities were highly specialized, contributing key services or materials in the national production line. Examples include uranium mining in Wyoming, uranium refinement in Ohio, weapons assembly in Texas, and weapons testing in Nevada. No one facility could do it all, and none existed in isolation. This was the height of compartmentalization. Throughout its operational history, the Hanford Site remained a principal component of the U.S. nuclear weapons complex.

When writing the history of nuclear weapons production, DOE (1997e, p. 117) grouped production activities into eight major processes:

- Mining, milling, and refining of uranium
- Isotope separation of uranium, lithium, boron, and heavy water
- Fuel and target fabrication for production reactors
- Reactor operations to irradiate fuel and targets to produce nuclear materials
- Chemical separations of plutonium, uranium, and tritium
- Component fabrication of both nuclear and nonnuclear components
- Weapon operations, including assembly, maintenance, modification, and development of nuclear weapons
- Research, development, and testing

The Hanford Site played a part in five of these activities – fuel and target fabrication, reactor operations, chemical separations, component fabrication, and research and development. Component fabrication did not begin at the Hanford Site until 1949 with construction of the Plutonium Finishing Plant (234-5Z Building) where plutonium nitrate paste was further refined into plutonium metal (called "buttons") and machined into weapons parts (called "pits").

Figure 1.6 is a stylized presentation of the Hanford Site production cycle as it existed throughout the Manhattan Project. The cycle began with the arrival of uranium billets from offsite sources. These billets were pressed, machined, canned, tested, and stored in the 300 Area. From there, the unirradiated fuel elements ("fresh metal") were transported to the 100 Areas where they were irradiated within the reactors. Irradiated fuel "cooled" (a term for the decay of short-term radioisotopes) in the 200 North Area before being transported to the 200 East or 200 West Areas for separation within the chemical separations plants. Following separation and concentration, the plutonium, in the form of plutonium nitrate

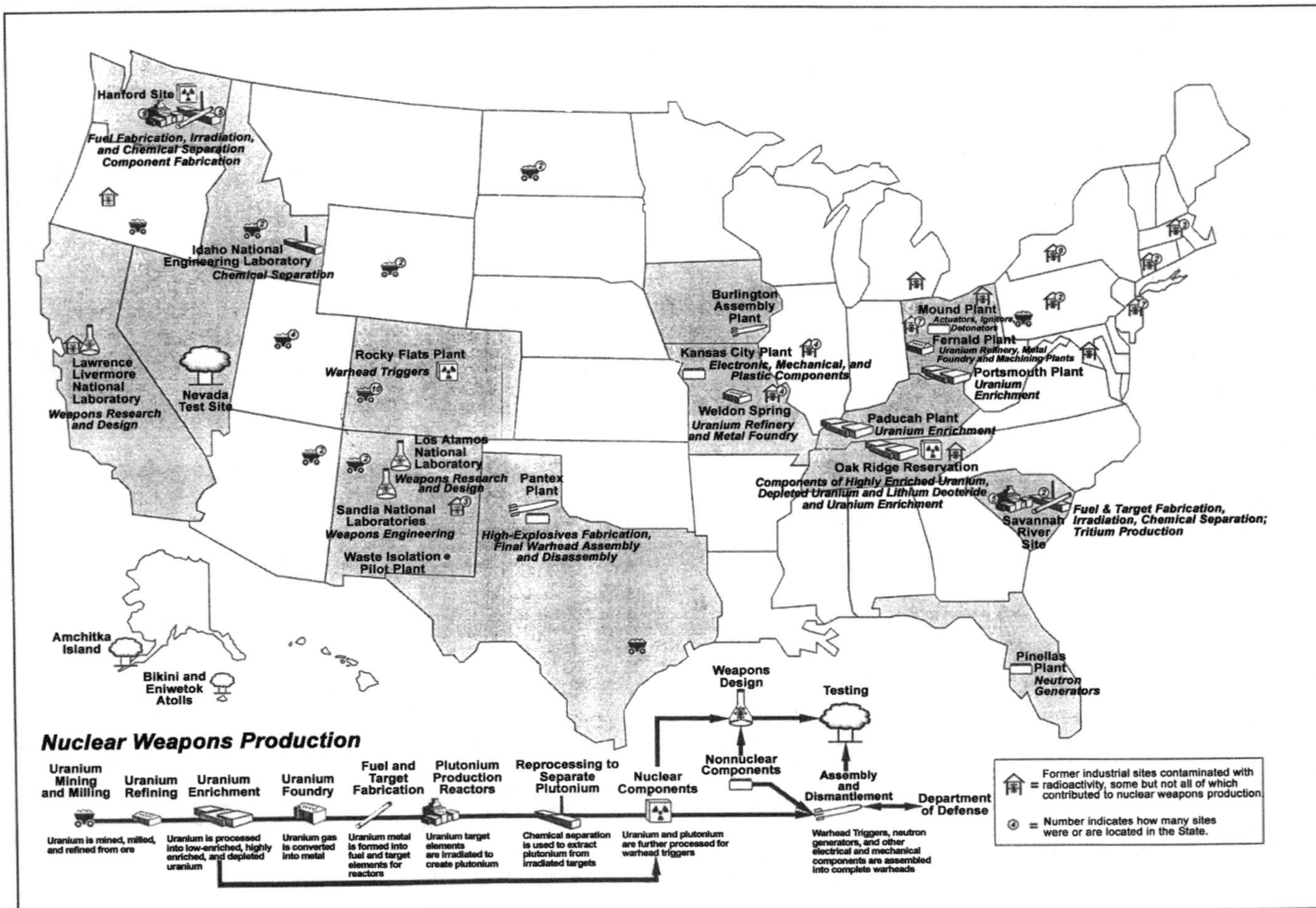

The United States nuclear weapons complex comprised dozens of industrial facilities and laboratories across the country. The weapons production infrastructure originated with the Manhattan Project during World War II and evolved and operated until the late 1980s. It typically employed more than100,000 contractor personnel at any one time. From the Manhattan Project to the present, the United States has spent approximately $300 billion on nuclear weapons research, production, and testing (in 1995 dollars).

Figure 1.5. *U.S. Nuclear Weapons Complex*

paste, was returned to the 200 North Area for storage before being shipped to Los Alamos. Detailed accounts of these operations are presented in Chapter 2, Sections 2 through 4.

Fuel Manufacturing

The first essential step in the plutonium production process was to manufacture uranium fuel. From the time pre-pressed uranium metal rods began to arrive at the Hanford Site in October 1943, workers systematically increased the manufacturing operations on site. Millwrights began to machine extruded uranium rods in December 1943. Canning operations began in March 1944. By September 1944, workers began straightening and "outgassing" (a term for removing hydrogen and other gases) fuel rods on site. Uranium metal billets first arrived in November 1944. Machinists began to extrude fuel rods from these billets in January 1945 (Gerber et al.1997, DOE 1997e).

The following steps in the fuel manufacturing process are a composite, assuming uranium billets were the initial raw material, and all activities were conducted on site. When shipments of fresh metal in the form of billets first arrived at the Hanford Site, workers stored the billets in one of the nine 303 Buildings (303-A, B, C, D, E, F, G, J, and K) within the 300 Area. Inspectors selected samples for testing in the 305 Test Pile (reactor). Lots matching purity requirements were sent to the 314 Building to be heated within a furnace in an inert (argon) atmosphere then forced through a press to form rods from 12-14 feet in length. These rods were straightened and outgassed to ensure distortion did not occur within the

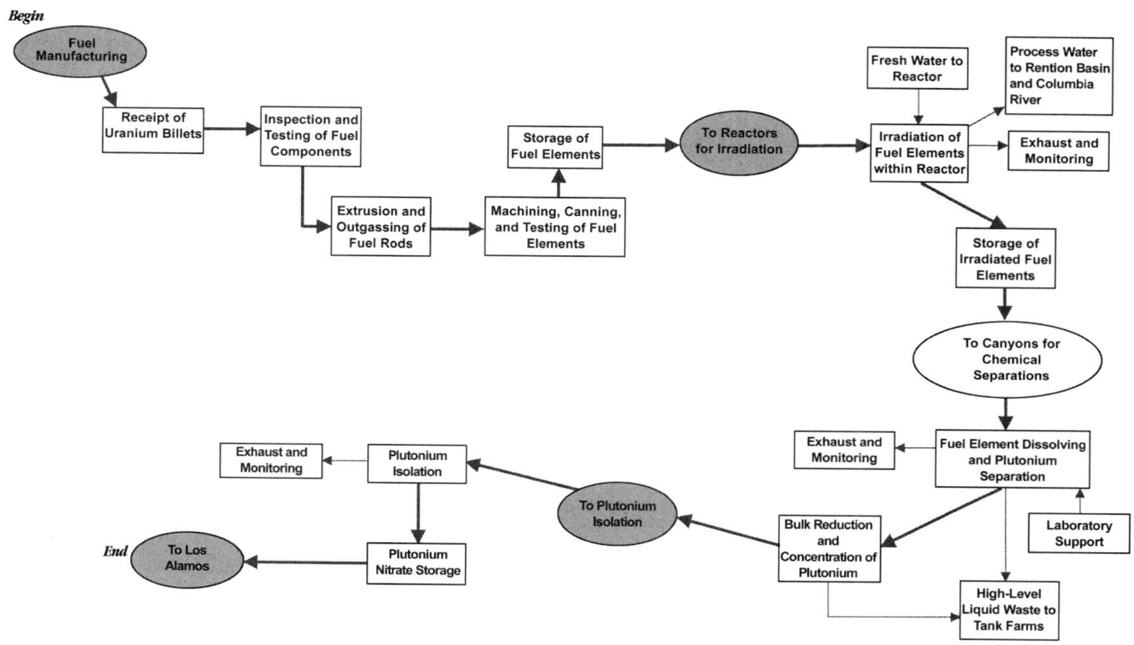

Figure 1.6. *Plutonium Production Cycle as It Existed throughout the Manhattan Project at the Hanford Site*

reactor. The rods were then sent to the 313 Building where they were machined (cut and turned) under very exacting tolerances to the required length and diameter. The cores (machined sections of rods) were "degreased" (dirt and oils removed in solvents) and "pickled" (solvent residue removed with an acid solution). The cleaned cores were then "canned" – a two-step process that involved preparing the core and inserting it within a sealed aluminum container. Short-lived electrical canning of cores began in March 1944.

The primary piece of equipment, an electric heater press nicknamed the "whiz bang," never performed well, and the entire process was replaced by the triple-dip method in August 1944. The fuel elements (canned cores) were subjected to three tests to ensure quality. Two of these, the Frost Test and Etch Test, were performed in the 313 Building. Each was designed to detect bonding or canning failures. The final test, steam autoclaving, was designed to detect leaks that would cause the fuel element to burst within the reactor. This test was conducted in the 314 Building. Every step had accompanying inspection points, and pieces not meeting rigorous standards were sent back for recycling. Chapter 2, Section 2 describes the fuel manufacturing process in detail.

Reactor Operations

Reactor operations included loading the fuel, operating the reactor, removing the fuel, and maintaining the reactor. It took 1 metric ton of uranium metal to produce 250 grams of plutonium. The movement of two key elements (water and fuel) is at the core of this second step in the production cycle.

The B, D, and F Reactors were designed as single-pass, water-cooled reactors. A continuous flow of water would pass through the reactors, absorbing and removing the heat generated by the nuclear chain reaction. In part, the Hanford Site was selected because the Columbia River afforded a near limitless supply of cold water. On the clean side of the reactor, the 108, 181, 182, 183, 185, 189, and 190 Buildings supplied and treated the water essential to operating the reactors. The 107 structures received heated process water flowing out of the reactors before its return to the Columbia River.

Water was withdrawn from the Columbia River through the 181 River Pump Houses located directly on the river. Electric pumps provided the power that pulled water from below the normal low-water level through the 181 Buildings and into the 182 Reservoir and Pump Houses. The 182 Buildings stored 25 million gallons of primary and secondary (backup) water. Primary water was pumped to the 183 Filter/Treatment Plants, and secondary water was held in reserve for emergency situations. Within the 183 Buildings, raw water was purified with chemicals (mixed and supplied via the 108 Chemical Pump Houses), filtered to remove particulate matter, and stored for use in multiple "clearwells" – each with a holding capacity of 5 million gallons.

Throughout World War II, filtered water was then subjected to supplemental treatments that were later discontinued as unnecessary. In the supplemental treatments, dissolved gases, "particularly carbon dioxide and oxygen," were removed within the 185 Deaeration Plants to improve the "heat transfer capacity of the coolant" (Gerber et al. 1997, p. 5.78). An additional step was taken at the 100-D and 100-F Areas. Here, water was next pumped to the 189 Refrigeration Buildings where it was chilled "allowing the reactors to operate at higher power levels and still not heat the process water to the point where it would flash to steam" (Gerber et al. 1997, p. 5.33). Prepared water was then pumped to the Filtered Water Storage Tanks within the 190 Process Pump Houses. Pumps housed in the 190 Buildings then injected water into the reactor cores at a rate of almost 30,000 gallons per minute. Following its single pass through the reactor core, process water was then discharged to the 107 Retention Basins where it cooled from 2-4 hours while short-lived radionuclides picked up in the core decayed. From the retention basins, the water re-entered the Columbia River via outfall structures and underground pipelines emerging at mid-channel (see Chapter 2, Section 3 for an extended discussion of water movement and Section 6 for reactor discharges).

Given the importance of water in cooling the reactor, and the consequences inherent in failing to maintain the reactor operating temperature within prescribed limits, Du Pont designed three independent systems to ensure uninterrupted water flow within the reactor. The primary pumping system operated on electricity. Nearly 75 percent of the electrical demand within the 100 Areas went to pumping and processing water. The secondary system relied on steam-driven turbines housed within the 184 Powerhouses. The tertiary system employed gravity feed from two 300,000-gallon elevated water tanks, the 187 structures. The sole purpose of the backup systems was to provide cooling water to the reactor while it was being shut down under emergency conditions, such as an electrical outage.

The second key element in reactor operations was the uranium fuel. Processed fuel elements were transported from the 300 Area to the 100 Areas where they were stored in the 103 Fresh Metal Buildings (103-B, 103-D, 103-F) before being loaded into the reactors. In reality, except for the initial fuel loading of a process tube, loading (charging) and unloading (discharging) were simultaneous processes. On a conceptual level, raw fuel loaded into the front face of the reactor would push processed fuel out the rear face, a simple concept that was difficult to execute. See the B Reactor HAER in Appendix B on the Internet for an extended discussion of the loading and unloading process.

Because of its configuration, irradiation of fuel elements within the reactor did not take place at a uniform rate. Fuel elements near the center of the reactor had a higher neutron flux than those around the edges. Physicists calculated the point at which irradiation had produced the optimum amount of plutonium-239 based on the position of the fuel elements within the reactor. Under-irradiation decreased the plutonium yield per unit of uranium, while over- irradiation increased the amount of undesirable isotopes such as plutonium-240 and plutonium-241. Tracking the progress of the fuel elements within each of the 2004 process tubes, these scientists called for an outage when approximately 25 percent of the tubes were ready for discharge.

Once discharged, the fuel elements were cooled in the fuel storage basin. Workers then prepared the fuel elements for rail shipment to the 200 North Area. Using a monorail, workers brought the buckets containing irradiated fuel elements to the adjacent transfer bay where they remotely placed the buckets within specifically designed lead casks and sealed the casks for shipment. Using an overhead crane, they pulled the filled casks from the deep-water basin and placed them within water-filled tanks on a flatbed rail car. Throughout the war years, fuel elements remained within the fuel storage basin for a very short period, usually days, before they were packaged for shipment to the 212 Lag Storage Buildings. Further radiological cooling occurred within a deep, water-filled basin in these buildings (Gerber et al. 1997, p. 5.81; see also the B Reactor HAER in Appendix B on the Internet).

Staff operated the reactor from the Control Room. Startup began when the reactor operator began the gradual and sequenced withdrawal of the nine horizontal control rods. The operator used seven of these control rods to advance, retard, or maintain the neutron flux throughout the reactor as required. The remaining two rods were used to fine-tune the reaction.

Under normal operating conditions, the reactor was shut down by fully inserting the horizontal control rods. When required, the reactor could be shut down quickly by inserting 29 vertical safety rods normally suspended above the reactor. Under emergency conditions, the "electromagnetic clutch holding each rod in the out position would be de-energized (de-magnetized), and the rods would free-fall by gravity into channels penetrating the reactor" (Gerber et al. 1997, p. 5.31). A second fail-safe system consisted of a liquid boron solution contained in tanks on top of the reactor that could be fed into the reactor through the vertical safety rod openings. As with other safety configurations, electrical systems were backed by hydraulic systems, which were backed by gravity. Chapter 2, Section 3 and the B Reactor HAER in Appendix B on the Internet describe in detail the reactor operations.

Chemical Separations

Although Du Pont constructed three fully independent chemical separations facilities on the Hanford Site, only the T and B complexes were used throughout World War II. This decision derived from the integrated nature of the Manhattan Project: "Originally eight separation plants were considered necessary, then six, then four. Finally, with the benefit of the operating experience and information obtained from the Clinton semi-works, we decided to build only three, of which two would operate and one [221-U] would serve as a reserve" (Groves 1983, p. 85).

Dissolving irradiated fuel elements, and isolating and concentrating the plutonium they contained was the third step in the production cycle. Leaving as they had arrived on rail cask cars, irradiated fuel elements were transported from the Lag Storage Buildings directly to Cell 3 of the process plants—a unique section of the building "designed to provide a 23-foot cell with adequate shielding to house the railroad tunnel into the building" (Gerber 1994a, p. 7). The chemical separations process began when the bridge crane operator removed the irradiated fuel elements from the casks and placed them into a dissolver tank. The tank was full or "charged" when a load of 1.5 metric tons of fuel elements had been transferred. Through remote manipulation, operators then chemically dissolved the aluminum cans encasing the irradiated fuel and transformed the solid fuel to a liquid solution. Operators, stationed at control panels in the shielded operating gallery, remotely processed this solution through a series of chemical and physical procedures from one cell to the next down the canyon until the 1.5 metric ton batch of irradiated fuel was reduced to 330 gallons of plutonium-bearing solution. The bismuth phosphate process separated plutonium from uranium and other irradiation-induced isotopes "by varying the valent state of the plutonium-239 and then repeatedly dissolving and centrifuging plutonium-bearing solutions" (Gerber 1994a, p. 4). Liquid waste from all process cycles was jetted to the single-shell, high level waste tanks constructed in the 200 Areas.

At this point, processing moved from the 221 Plants to the 224 Bulk Reduction Buildings (224-T, 224-B) where additional dissolving and centrifuging continued to concentrate the plutonium in solution, first using bismuth phosphate as a carrier then lanthanum fluoride. In the final step, the "solid plutonium lanthanum oxide was...dissolved with nitric acid, making plutonium nitrate" (Gerber 1994a, p. 11). The 330-gallon batch of plutonium-bearing solution piped into the 224 Building exited as 8 gallons of concentrated plutonium nitrate. Carefully placed within specially designed cans, teamsters transported the plutonium nitrate to the 231-Z Isolation Building in the 200-West Area. The designation "Z" was applied because the final step in the production cycle (purification) occurred in this building: "...a final product precipitation using sulfates and peroxide took place. The precipitate cake was dissolved with nitric acid, placed in small (10-inch-high) shipping cans, and boiled right in the cans using hot air. It was reduced to a wet nitrate paste (the HEW product)" (Gerber 1994a, p. 11).

These shipping cans, each holding approximately 1 kilogram of plutonium, were then transported to the 213 Structure in the 200-North Area. Excavated into the southeast face of Gable Mountain, this facility contained two magazine rooms (213-J, 213-K Vaults) "lined with concrete shelving interspersed with concrete brick partitions" (Gerber et al. 1997,

p. 5.82). Plutonium was stored in these vaults until it was shipped to Los Alamos. Sections 4 and 5 of Chapter 2 describe the chemical separations and plutonium finishing processes in detail.

Administrative, Facility, and Operations Support

Adhering to the siting requirements Groves established, the production cycle was segmented into discrete steps assigned to specific geographic areas of the Hanford Site, and, with the singular exceptions of fuel manufacturing and plutonium purification, duplicated at independent facilities within those areas. In reality then, the Hanford Site industrial complex consisted of seven production centers, each of which required administrative, facility, and operations support. For additional information on the construction, appearance, function, or history of many of these support facilities, see the Expanded Historic Property Inventory Forms (ExHPIF) or standard HPIF in Appendix B on the Internet.

Within the 100 Areas (100-B, 100-D, 100-F), a diverse workforce provided security, first aid, training, and maintenance services within specific ancillary buildings. For example, before and after each shift, all workers had to pass through the 1701 Guard House. Guards stationed at the entrances verified the identification of each employee and visitor before allowing access to any of the 100 Areas. Those who worked in the exclusion area containing the reactor complex then had to pass an additional identification check in the 1702 Badge House before proceeding to their assigned work area. A Patrol Headquarters (1720 Building), First Aid Station (1719 Building), and Fire Station (1709 Building) also were established within each 100 Area. Du Pont stressed fire awareness and the need for quick fire suppression as early as June 1943 because of the remoteness of the facilities, the vegetation surrounding them, and the number of wooden buildings contained within each Area (Gerber 2000, p. 3).

Working within the 1717 Combined Shop or 1722 Area Shop, craftsmen fabricated or repaired the equipment and materials needed to maintain ongoing operations. For instance, the 1717 Building contained "a machine shop, a carpenter shop, a pipe shop, and a sheet metal shop, an electric shop, a forge shop, a tool room, and six offices" (Gerber et al. 1997, p. 5.54). Vehicles were repaired and serviced within the 1716 Automotive Repair Shop. Craft supplies could be obtained from the 1713 Warehouse or 1715 Storage Building. Workers needing to change into coveralls and other personal protection equipment before working with radioactive materials would go to the 1707 Change House, where they might also shower or have lunch. Administrative staff oversaw operations, processed paper, maintained payrolls, ordered materials, and performed numerous other tasks within the 1704 Supervisor's Office.

Within the 200 Areas (200 East and 200 West), staff provided the same administrative, security, health, maintenance, and utility functions in facilities whose only distinction from those in the 100 Areas was often the number displayed above the entryway. As illustration, the 182 Pump Houses in the 100 Areas were labeled 282 in the 200 Areas, the 184 Power Houses and 1802 Overhead Steam Lines were identified as 284 and 2802, in the 200 Areas, and the 1707 Change Houses in the 100 Areas became 2707 in the 200 Areas. Like the 100 Areas, the 200 Areas also contained their own "coal-fired central power station to provide steam for heating and process equipment needs, a water supply, sanitary system, [and] road and railroad grid" (Gerber et al. 1997, p. 5.18).

The 200 Areas also contained support facilities specific to their function as chemical separations areas. The chemicals necessary to accomplish separations were delivered and stored in the 211 Tank Farms. Fresh chemicals could be supplied directly to the 221 Plant process cells from any of the 20 above-ground storage tanks or to the 271 Chemical Preparation and Service Buildings where they were mixed before being piped to the 221 Plants or 224 Buildings. Scientists in the 222 Sample Preparation Laboratories tested process solution samples withdrawn throughout the separation process. "Since the entire separation process was conducted remotely, the only way to verify that the process was working within specifications was to draw and test samples" (Gerber et al. 1997, p. 5.38). Finally, in this sequencing from start to finish, the 292 Exhaust Gas Laboratories supplied information on the level of chemical and radioactive contaminants exiting the process exhaust stack.

The 300 Area also contained a complete set of administrative, health, safety, and maintenance support facilities. However, operations support facilities within the 300 Area were directed not only toward fuel manufacturing (the area's primary mission) but also toward chemical separations process improvement and essential materials testing (Gerber et al. 1997). An example is the role of the 305 Test Pile (reactor) in materials testing for the Hanford Site production reactors.

Another primary investigative focus within the 300 Area was research and development of the chemical separations process. As another example of the synergetic integration of the Manhattan Project, research conducted at the Clinton Semiworks resulted in changes in the design and mission of the 321 Separations Laboratory at the Hanford Site. Originally designed only for "cold" pilot-scale research using unirradiated sources, a "hot lab" was added just before completion of the 321 Building because scientists at the Clinton Semiworks had demonstrated the importance of understanding and controlling factors affecting chemical separations under actual operating conditions.

To meet accelerating production demands, research staff in the Hanford Site 3706 Radiochemistry Laboratory examined each operation phase to make the bismuth phosphate process faster and more efficient, thereby extracting the greatest amount of plutonium in the least amount of time. Research conducted in the 3706 Building was tested experimentally in the 321 Building "canyon," then applied within the 221 separations plants. Rhodes (1986, p. 604) recounts Glenn Seaborg's proud notation that "yields in the first plant runs...ranged between 60 and 70 per cent...[and] reached 90 per cent early in February 1945." While this research was under way, other scientists in the 3706 Building focused on "metallurgical examination of irradiated fuel elements from the reactors, fuel development for the 313 Building, examination of graphite from the experimental levels of the 100 Area piles [reactors], special sample analyses from the spectroscopy and radiocounting activities, and multifaceted analyses for environmental and personal survey programs" (Gerber et al. 1997, p. 5.39). See Chapter 2, Section 7 for extensive coverage of research and development activities at the Hanford Site.

DISCHARGE AND WASTE: THE RESIDUAL OF PRODUCTION

Leona Marshall, the only woman physicist on Fermi's staff, had occasion to visit the Hanford Site. Some years later, she offered this first-hand remembrance: "When the Queen Marys began to function, dissolving the irradiated slugs [fuel elements] in concentrated nitric acid, great plumes of brown fumes blossomed above the concrete canyons, climbed thousands of feet into the air, and drifted sideways as they cooled, blown by the winds aloft" (Marshall quoted in Rhodes 1986, p. 604).

As disquieting as this mental image may be, we are reminded that hindsight is a stern judge. While the discharge and disposal methods used throughout World War II, indeed through 1970, were primitive by today's standards, "the nuclear weapons industry typically used waste-disposal methods that were considered acceptable at the time" (DOE 1995e, p. 23). At the time, the "imperatives of the nuclear arms race...demanded that weapons production and testing, rather than waste management and the control of environmental contamination, be given the first priority" (DOE 1995e, p. 5). Along these same lines, Findlay and Hevly (1995, p. 50), referring specifically to the Hanford Site, provide a summary less encumbered by government perspective: "Wartime exigencies dictated haste at the expense of safety and environmental concerns. Du Pont and the Army "took risks to get results" because they were in a hurry. They also cut corners with the expectation that such emergency conditions would last only a short while and that more effective safeguards would be introduced once the war ended."

Groves' recollection in the early 1960s confirms Findlay and Hevly's later analysis: "We had always thought that it would be possible to eliminate much of the radioactive problem in the future. We also hoped to recover the uranium remaining in the existing solutions and to reduce the bulk of the radioactive waste materials, thus making them easier to handle" (Groves 1983, p. 91). An abiding concern with the long-term effects of radioactive waste on future generations was also apparent at the time. In an interview given in 1948, Dr. Lyle Borst, then director of the Brookhaven National Laboratory Pile Project and the former supervisor of research at the X-10 Graphite Reactor, observed:

> *"For the first time man is creating something that cannot be destroyed – radioactive materials...If their radiation gets into biological organisms, including human beings, it will destroy them...[present disposal methods, that is, thick concrete tanks, sunk deep into the ground, are adequate]. But since we're thinking in terms of geological*

ages, how are we going to guarantee that some damn-fool archeologist won't go sticking his nose into the stuff ten thousand years from now?" - Lang 1948, p. 203

Plutonium production generated large volumes of radioactive and chemical waste. Hanford workers had to contend with these wastes in three forms: solid, liquid, and gas. The overview in this chapter discusses only methods developed and used during the Manhattan Project for handling and managing radioactive or toxic wastes. Chapter 2, Section 6 and the pertinent ExHPIFs and HPIFs offer more detail on how waste has been managed throughout the Manhattan Project and the Cold War Era.

Solid Waste

Burial of radioactively contaminated solid waste within shallow excavations was the preferred method of disposal ever since this type of waste began to accumulate at the Hanford Site. Low-level solid waste contained small amounts of radioactivity dispersed in large amounts of materials. Examples included protective clothing, tools, equipment, machinery, debris, sludge and other items contaminated with radionuclides. Within the 100 Areas, low-level solid waste burial grounds consisted of simple pit or trench excavations with no lining materials. Generally, one large burial ground was established at each reactor complex to facilitate burial of materials generated within that complex. Smaller pits or trenches were excavated as needed or to accommodate specific items. Low-level solid waste in the 200 Areas, namely protective clothing, obsolete or failed equipment, and other types of materials similar to those generated in the 100 Areas were disposed of in unlined burial grounds near the process facilities. Mixing of wastes (such as hazardous and radioactive, combustible, and non combustible) within the same burial ground was a defining characteristic of early waste disposal methods within both areas.

In addition to low-level solid waste such as protective clothing and process equipment, generated in common with the other areas, workers in the 300 Area also had to dispose of wastes resulting from the manufacture of fuel elements and research and development activities. Uranium scraps, primarily lathe turnings, billet ends, and container residue, were stored in 5-gallon cans in and around the 303 Buildings. Metal oxide from fuel elements that had failed the autoclave test was stored in 30-gallon drums as was uranium sludge (Gerber 1992c, p. 6). Two solid waste burial grounds were constructed within the 300 Area during the Manhattan Project. The first (Burial Ground 618-8) was used from 1943-1944, the second (Burial Ground 618-1) from 1945 to 1951 (Gerber 1993b, p. 59).

Liquid Waste

Within the 100 Areas, low-level liquid waste, primarily reactor process water, was discharged to retention basins and ultimately to the Columbia River. Fuel storage basin overflows, maintenance and decontamination activities, and other activities also generated low-level radioactive liquid wastes. Many of these wastes were often discharged to open pits or cribs (an underground chamber constructed of loosely spaced timbers) adjacent to the reactor buildings. Contaminated water from individual process tubes in which a fuel cladding failure had occurred was generally discharged to a more distant crib where it percolated into the soil. No high-level liquid wastes were generated in the 100 Areas.

It is important to note for the 200 Area that high-level radioactive liquid wastes, generated at a rate of 10,000 gallons per ton of irradiated fuel processed, were always sent directly to underground storage tanks during World War II. Such was not the case, however, with low-level liquid waste. Herbert M. Parker, chief health physicist for the Hanford Site during the Manhattan Project, characterized the earliest low-level liquid waste disposal methods within the 200 Areas as temporary and expedient, devised "to avoid absurd costs on tank storage, [or] evaporation equipment" (Gerber 1992b, p. 45). His reference was to direct discharge of low-level wastes within natural depressions. This practice was abandoned early in 1945 because these resulting swamps produced higher than acceptable radiation levels. Throughout the remainder of the war, indeed until their general abandonment in mid-1947, most low-level liquid wastes were discharged to reverse wells (perforated casings extending to near groundwater depth). Some process wastes, however, were handled differently. For example, piped first to a settling tank where suspended particulates dropped out, process waste from the 224 Building was then mixed with discharged process cooling water from the 221 Plant before being piped to one of two 500,000-gallon retention basins. These basins "overflowed into open earthen drainage ditches that ran far out into the desert" (Gerber 1994a, p. 23).

Low-level liquid wastes from the 300 Area laboratories and fabrication facilities were collected in a centralized process sewer system (3904 Structure) and discharged to a process pond adjacent to the Columbia River. The process sewer system was distinct from and not connected to the sanitary sewer system (3903 Structure). All high-level liquid wastes from the 3706 Building were sent to the process sewer system, while high-level liquid waste from the 321 Building was discharged to four underground tanks, a reverse well, or the process sewer (Gerber 1992c).

Gaseous Waste

In designing the Hanford Site reactors, Du Pont realized, as had Fermi, that reactivity would increase if air were removed from the area within which the reaction occurred. This efficiency would occur because nitrogen gas, the largest component of air, attracts neutrons and therefore acts as a poison. Another inducement to remove air from the reactor arose from the fact that argon gas, a very minor component of air, quickly became radioactive under neutron bombardment. Escaping from the reaction area, this gas would be hazardous.

To address these issues, the design engineers made the reactor air-tight and installed a "circulating helium atmosphere" that "displaced neutron-absorbing air from the reactor core, and removed gases generated by the reactor" (DOE 1997e, p. 164). The 115 Gas Purification Building, adjacent to the reactor building, contained all the equipment necessary to purify and circulate the helium, which was stored in high-pressure storage tanks in the 110 Process Gas Storage Building. An integrated ventilation and exhaust system was installed throughout the reactor building that channeled air from areas of no contamination to areas of high contamination where they were collected and released. Engineers designed this negative pressure gradient for every building where radiation would be present throughout the Hanford Site. For the 100 Area reactors, two supply fans "washed, filtered, and tempered" outside air coming into the building, while two exhaust fans discharged circulated air through an underground concrete duct to the base of the 116 Exhaust Stack. Released without filtration, low-level process gases passed to the atmosphere from the top of the stack at 200 feet (Wahlen 1989, p. 28).

From the beginning, the Metallurgical Laboratory scientists knew that chemical separation would generate highly toxic fumes in the metal dissolving phase. Early operations within the Clinton Semiworks demonstrated that gases generated throughout the separations process contained radioactive fission products. To address these off-gases at the Hanford Site, Du Pont engineered a ventilation and exhaust system in the canyon buildings with the 291 Exhaust Building and Stack as the point of release. Measuring 200 feet high, the stack "functioned to exhaust process gases…and to provide additional 'diluting air' deemed essential to the safe dispersion of process gases to the atmosphere" (Gerber 1994a, p. 16). To assist with their unfiltered dispersion into the atmosphere, Du Pont engineers increased the normal flow of air and gases from 40,000-60,000 cubic feet per minute using three large fans within the stack.

In the 321 Building, nitrous oxide and fission gases from research to improve the chemical separations process were released directly to the atmosphere through an exhaust stack on the building's roof. Laboratory exhausts generated within the 3706 Building simply discharged through ventilation dormers on the roof (Gerber 1992c).

Worker and Environmental Monitoring

In reviewing the formative stages of personal and environmental monitoring programs, Gerber (1992b, p. 11) observed that "the new sciences of health physics and environmental monitoring developed rapidly at the [Hanford] Site, pioneering advances in instrumentation procedures, working conditions, and methods of analytical chemistry to process substances never before sampled." After developing a pilot health physics program at the Clinton Engineer Works X-10 complex, Herbert Parker, a radiological physicist, left Tennessee to return to Washington State. Together with Carl C. Gamertsfelder (an x-ray diffraction scientist), William D. Norwood (a physician), and a team of fifteen others, Parker established the Health Instruments Section for the Hanford Site in July 1944.

The Health Instruments Section "was charged with defining and measuring radiological hazards, establishing procedures to make jobs in plutonium production safe for workers, and developing and calibrating instruments." These tasks were made very challenging "because strict MED security regulations precluded revealing to most employees even the existence of radioactivity" (Gerber 1992b, p. 12). While operating in an information vacuum, workers found that the regulations drafted by the Health Instruments Section were meticulously enforced. For example, each employee wore a film badge and two "pencils" (a small, tubular radiation monitor) throughout their shift. They were required to have the pencils read daily, while the film badges were submitted for reading weekly. Those who worked in proximity to radiation sources wore additional personal monitors and provided samples for urinalysis regularly. All this as protection against a danger about which they could not even be informed. Indeed, discussion of radiation monitoring was so sensitive that even the term could not be used. It was referred to by its code named "health physics."

Groves' selection of the Hanford Site was itself an acknowledgement of the dangers inherent in working with radiation. Even amidst the intense pressures of construction, Groves understood the need to determine where and how contamination might spread, and what effects it might have: "An accident might release radioactivity into the air; that called for thorough meteorological work. The river water needed study; so did the river's valuable salmon, to see how they would take to mild doses of transient radioactivity from pile [reactor] discharge flow" (Rhodes 1986, p. 497).

Beginning in June 1943, a meteorological team began to document "dominant wind patterns, velocities, and variances" at the Hanford Site. Once a baseline had been established, general meteorological studies gave way to "wind dispersion tests with oil fog (SO_2 [sulfur dioxide]), beginning in the partially completed T Plant Stack (291-T) as early as April 1944." By Fall 1944, the C Plant Stack and then the Meteorological Tower (622 Structure), a 400-foot-high metal lattice work erected outside of the 200 West Area, were used in sequence to determine off-gas trajectories because pre-operation activities precluded the use of the 291-T Stack. "So extensive was the weather study and forecasting effort that by the end of 1944, over 36,000 individual readings on wind dilution patterns had been recorded in the 200 West Area" (Gerber 1994a, p. 29). Du Pont would use the information gained in these studies to determine the optimum conditions for atmospheric dispersal of process wastes, particularly the off-gases produced in chemical separations.

Twenty-nine Air Monitoring Stations (614 Buildings) were constructed to track the waste stream and evaluate the dispersion model. These small, wooden structures housed continuous air samplers that technicians periodically analyzed for the presence of plutonium nitrate dust as well as iodine-131 and other fission isotopes. Twelve of these sampling stations were placed in the 200 Areas, fifteen were distributed across the Hanford Site (including Richland Village), and two were located off site. Acting on the data obtained from these stations, Du Pont determined by June 1945 that initial dissolving operations should be limited to the night shift since dispersal characteristics were generally more favorable. Also concurrent with the commencement of chemical separations activities, the Health Instruments Section began regional investigations of the deposition of airborne fission materials, particularly iodine-131, and their accumulation and effect within both animals and vegetation (Gerber 1994a, Gerber et al. 1997).

Before operations started, in fact beginning with design changes to the reactor intake pipes, the Corps considered the effects of plutonium production on the Columbia River and its aquatic resources, particularly salmon. For example, the Corps, on Groves' suggestion, established the Applied Fisheries Laboratory at the University of Washington in 1943 under the direction of Lauren Donaldson (Gerber 1992b). Studies on the effects of radiation on fish were initially conducted on the Seattle campus to maintain what would now be called "plausible deniability" between the research being conducted there and the production of plutonium at the Hanford Site. Exposing fish to x-rays, the radiation source available for research, provided useful but limited data. If questions regarding the effects of reactor effluent on fish were to be answered, then direct exposure to reactor effluent, and the isotopes it contained, had to occur.

Consequently in June 1945, Richard Foster began to conduct onsite research within the 100-F Area. Effluent water of differing strengths and temperatures was discharged to troughs and ponds holding salmon and trout to determine tolerances, diseases, and reproductive viability. Initial results indicated that "[y]oung fish, possibly due to higher metabolic rates, accumulated proportionately more radioactivity in their tissues than did adults. Accumulation of radioactivity occurred more slowly in cooler water, and activity levels differed among various fish tissues" (Gerber 1992b, p. 36). Disclosure of either the conduct of these first-of-a-kind studies or their results was not allowed until after World War II. Dr. Ronald Katherine, a leading expert in radiological health physics, maintains that these two studies mark the beginning of the longest continuously operating environmental monitoring program in the United States. Chapter 2, Section 7 discusses the atmospheric, radiobiological, and radioecological research conducted at the Hanford Site.

SECRECY AND SECURITY

Secrecy enveloped atomic research from the onset of the War in Europe until the flash at Hiroshima made clear to the world that one nation had realized the vision patented by Leo Szilard in 1934. Security measures, many of which violated individual civil liberties, were constants throughout the Manhattan Project. As Findlay and Hevly (1995, p. 39) observed, the Richland Village may have escaped the fence that enclosed both Los Alamos and the Clinton Engineer Works, but the "Army would still police Richland and watch its residents carefully – wartime censors examined each departing letter; security personnel listened in on phone calls; [and] hotel porters acted as 'counter-espionage agents'." In fact, undercover agents were everywhere, assuming roles that provided them with access to the operating facilities, laboratories, and craft shops. Many recruited informants to extend their reach.

Security concerns associated with atomic weapons development, however, were not confined to the Hanford Site and continued after the end of World War II. Discussing line work at the Rocky Flats Plant, an Atomic Energy Commission installation devoted to the manufacture of plutonium pits and other weapon components established in 1951, Jack Weaver recalled: "Nobody talked about what we were doing in those days. We didn't even talk about it amongst ourselves, let alone with our families and friends offsite. On the floor even, we referred to "Y" or "Z" or "U" and not to beryllium or plutonium or enriched uranium" (DOE 1995e, p. 17).

These remarks mirror those described elsewhere by retired Hanford Site employees and demonstrate that Groves' philosophy of "compartmentalization" permeated the national atomic weapons program. As Groves recounted: "We made certain that each member of the project thoroughly understood his part in the total effort – that, and nothing more" (Groves 1983, p. xv).

The Hanford Site Patrol, established under the Plant Protection Program developed by Du Pont and approved by Colonel Matthias in March 1943, provided front-line security for the Hanford Site. The Hanford Site Patrol partnered with and sought assistance from the Corps and the Federal Bureau of Investigation, both of whom established a local presence in conducting background checks, safeguarding information, and preventing theft of materials among other tasks. No new employee was hired without a background investigation commensurate with the position they would fill. Having accepted a position, each employee signed a "Declaration of Secrecy," enforceable under the National Espionage Act or the Federal Sabotage Act with imprisonment or heavy fines if breached. The Corps established classified areas and materials, and the Hanford Site Patrol provided security orientation and education so employees did not transgress boundaries with respect to either facilities or documents to which they did not have authorized access. Employees seeking information not essential to their specific tasks or speculating in rumors detrimental to the workforce were promptly disciplined with reprimand or termination. Some found these measures oppressive or excessive and, as a final irony, were required to resign a Declaration of Secrecy before they could terminate their employment. In fact, all employees renewed this pledge of allegiance before terminating employment under any circumstances.

The Hanford Site Patrol conducted foot, vehicle, and boat patrol of the Hanford Site perimeter as well as the entry points to all operational areas from 1943 until August 1944. At that time, the Corps assumed responsibility for perimeter patrol and also established an Air Patrol to enforce the restricted air space established over the Hanford Site. The Hanford Site Patrol then concentrated on patrolling the operational areas and the facilities within and between them. Ensuring the security of classified information remained one of their primary tasks. The Hanford Site Patrol made daily inspections to verify that filing cabinets containing classified data remained locked. An unlocked cabinet was investigated to determine that its contents were intact. Armed Hanford Site Patrol staff also accompanied every shipment of uranium metal, fuel elements, irradiated fuel, and concentrated plutonium nitrate throughout the production cycle. Chapter 2, Section 8 describes in detail the secrecy and security measures in place at the Hanford Site.

ALAMOGORDO BOMBING RANGE, HIROSHIMA, AND NAGASAKI

By late 1944, the Hanford Site was fully operational, dependably delivering plutonium to Los Alamos for research, testing, and assembly. Though still tightly under wrap, the contribution the Hanford Site workforce made to the Manhattan Project would soon become apparent, pending resolution of one critical problem (see sidebar box).

> ***"Field tests performed with uranium-235 prototypes in late 1944 eased doubts about the artillery method to be employed in the uranium bomb. It was clear that the uranium-235 from Oak Ridge would be used in a gun-type nuclear device to meet the August 1 [1945] deadline Groves had given General Marshall and the Joint Chiefs of Staff. The plutonium produced at such expense and effort at Hanford would not fit into wartime planning unless a breakthrough in implosion technology occurred."*** **- Gosling 1999, p. 42**

The hard-fought breakthrough came in April 1945 when the combined efforts of inveterate chemists and metallurgists at Los Alamos led to precise techniques for producing plutonium metal. This occurred just 3 months ahead of the July 4 test firing Oppenheimer had scheduled for the plutonium bomb he christened "Trinity" (Gosling 1999). When Groves asked many years later why he had chosen that name, Oppenheimer responded that the reason was not clear even to him, but noted that certain thoughts had been in his mind. In particular, he mentioned a devotional poem by John Donne, which opened with the line: "Batter my heart, three person'd God..." "Beyond this," he said, "I have no clue" (Oppenheimer, quoted in Rhodes 1986, pp. 571-572).

Gosling (1999, p. 48) writes that "At precisely 5:30 a.m., on Monday, July 16, 1945 the atomic age began" at an isolated location within the Alamogordo Bombing Range (which later became the White Sands Missile Range) also known by the Spanish name Jornada del Muerto – the Journey of Death. With an explosive force of approximately 18.6 kilotons of power, the Trinity "device" turned darkness into light and the cool, damp desert air into searing heat. Isidor Rabi recounts the moment of detonation:

> *"Suddenly, there was an enormous flash of light, the brightest light I have ever seen or that I think anyone has ever seen. It blasted; it pounced; it bored its way right through you. It was a vision which was seen with more than the eye. It was seen to last forever. You would wish it would stop; altogether it lasted about two seconds. Finally it was over, diminishing, and we looked toward the place where the bomb had been; there was an enormous ball of fire which grew and grew and rolled as it grew; it went up into the air, in yellow flashes and into scarlet and green. It looked menacing. It seemed to come toward one."* - Rabi, quoted in Rhodes 1986, p. 672

No longer was there any doubt that the implosion bomb, fueled by plutonium irradiated at the Hanford Site's B Reactor and separated at T Plant, would work.

President Roosevelt did not live to see the atomic age. He died of a cerebral hemorrhage on April 12, 1945 in Warm Springs, Georgia. Harry S. Truman assumed the Presidency that evening and with it the need to determine national policy in the waning days of World War II. Before he was fully advised of the Manhattan Project, a bombing raid on Tokyo on April 13 destroyed the building containing Japan's gaseous thermal diffusion experiment. "It burned to the ground and took the Japanese atomic bomb project with it" (Rhodes 1986, p. 612). About the same time, American forces had confiscated the Belgian uranium ore stored in Strassfurt, Germany, crippling the German atomic weapons program and thereby eliminating any potential atomic threat from the Axis powers.

Despite Germany's surrender, Japan continued to resist the unconditional surrender demanded by the Allied Forces. The success of the Trinity test and word that both the enriched uranium and plutonium bombs would be ready for use by early August enabled President Truman to avoid extending an offer of surrender to Japan that allowed the Imperial

> ***On May 7, 1945, Germany surrendered unconditionally.***

Emperor to continue to rule. Nuclear weapons might force a surrender and negate the need to invade Japan, which, based on bitter experience gained from military encounters on the Pacific islands, would have resulted in significant loss of lives on both sides. Without an invasion, Russian entry into the Pacific theater would not be necessary, which was equally important given Russia's "designs on Eastern Europe and Germany." Russia could not "expect to share in the postwar administration of Japan" (Gosling 1999, pp. 50-51).

On July 26, 1945, the Potsdam Declaration was issued via radio to Japan. In it, the President of the United States, the President of Nationalist China, and the Prime Minister of Great Britain called on the Japanese government to "proclaim now the unconditional surrender of all Japanese armed forces...The alternative for Japan is prompt and utter destruction" (Rhodes 1986, p. 692). Rejection came quickly on July 29.

At 2:45 a.m. on the morning of August 6, 1945, the *Enola Gay*, code named Dimples Eight Two, began its take-off run down the Tinian airfield with "Little Boy," the enriched uranium gun-style bomb, securely shackled within the bomb bay. Little Boy was the brother of "Fat Man"—the other bomb planned for Japan. Confident that the gun design would work with an enriched uranium core, Oppenheimer never tested a prototype of this bomb as he had done with the Trinity plutonium device. This mission was both test and application: "Little Boy exploded at 8:16:02 Hiroshima time, 43 seconds after it left the Enola Gay, 1,900 feet above the courtyard of Shima Hospital, 550 feet southeast of Thomas Ferebee's aiming point, Aioi Bridge," (Rhodes 1986, p. 711) with a yield equivalent to 15,000 tons of TNT (DOE 2000a).

> ***"Little Boy killed 70,000 people (including about twenty American airmen being held as POWs) and injured another 70,000. By the end of 1945, the Hiroshima death toll rose to 140,000 as radiation sickness deaths mounted. Five years later the total reached 200,000. The bomb caused total devastation for five square miles, with almost all of the buildings in the city either destroyed or damaged." -*** **Gosling 1999, p. 51**

Radio stations all over the United States broadcast the news of Hiroshima within hours of its bombing. Reading from a press release issued by President Truman, announcers informed the public for the first time of the existence of the atomic bomb and of President Truman's resolve to drop additional atomic bombs if Japan did not offer its unconditional surrender. Rhodes (1986, p. 736) opines that the civilian Japanese population viewed the atomic bomb "like a golden opportunity to surrender without shame, but the admirals and generals still despised unconditional surrender and refused to concur." Foreign Minister Togo again reached out to Russia to mediate an agreement. Instead, Ambassador Naotake Sato was informed on August 8 that Russia would declare war on Japan at midnight, August 9 (Rhodes 1986).

> ***"Fat Man exploded 1,650 feet above the slopes of the city with a force of 21,000 tons of TNT. Fat Man killed 40,000 people and injured 60,000 more. Three square miles of the city were destroyed, less than Hiroshima because of the steep hills surrounding Nagasaki. By January 1946, 70,000 people had died in Nagasaki. The total eventually reached 140,000, with a death rate similar to that of Hiroshima." -*** **Gosling 1999, pp. 53-54**

The Fat Man bomb (named in honor of Winston Churchill and fueled by Hanford Site plutonium) was originally scheduled to be dropped August 11, 1945, but the drop date was advanced by 2 days to take advantage of forecasted good weather. At 3:47 a.m. on August 9, *Bock's Car* lifted off from Tinian under the command of Major Charles W. Sweeny rather than the B-29's namesake Frederick Bock. Kokura Arsenal on Kyushu Island was the primary target. The Mitsubishi plant at Nagasaki, where the torpedoes used at Pearl Harbor originated, was the secondary target. Bad weather conditions caused Sweeny to abort the drop over Kokura, despite making three passes over the target area. Running low on fuel because the reserve tank was not operational, Sweeny turned toward Nagasaki where he could make one pass before heading for an emergency landing at Okinawa. A "twenty-second visual run" through an otherwise clouded sky provided the only opportunity to locate the aiming point. The bombardier only had time to fix on a stadium several miles upriver (Rhodes 1986).

Even after the Nagasaki bombing, "civilian and military leaders continued to struggle in deadlocked debate" (Rhodes 1986, p. 743). Taking the initiative, Emperor Hirohito "forced the issue" and delivered an offer of conditional surrender to Washington through Switzerland on August 10. The condition was that he remain as sovereign. Taking advice from Stimson, Byrnes, and Secretary of the Navy James Forrestal, President Truman authorized Byrnes to draft a response that stated that the Emperor's rule would be "subject to the Supreme Commander of the Allied Powers." The reply was both broadcast on August 11 and routed for return through Switzerland.

News of Japan's surrender was celebrated across the United States, but no more so than in the "atomic city" of Richland, whose "victory celebrations were covered in newspapers and on radio programs throughout the nation, and the little city basked in admiration and praise." Groves came to the Hanford Site in October and presented each employee, for their part in the war effort, the Army-Navy "E" Award, "the highest civilian production commendation," and an "A-bomb" pin authorized by the War Department (Gerber 1992b, p. 17).

End Of The Manhattan Project: 1946

The Manhattan Project did not end with the announcement of the Hiroshima bomb. A well-controlled breach in the wall of secrecy allowed the government to indulge public curiosity without revealing sensitive information. *Atomic Energy for Military Purposes*, authored by physicist Henry DeWolf Smyth and issued on August 12, 1946 made information on atomic bomb development available to the general public: "It appalled the British, enlightened the Soviets on which approaches to isotope separation not to pursue and – Groves' intention in releasing it – defined what might be public and what secret about the atomic bomb program, thereby forestalling information leaks" (Rhodes 1986, p. 750).

While the future of atomic weapons and atomic research would be debated somewhat more openly now, public disclosure did not stop the secret production and stockpiling of atomic weapons that continued, albeit at a reduced pace. In September 1945 under a plan approved by Stimson and Marshall, Groves either closed or idled operations at the Clinton Engineer Works. At the Hanford Site, Groves first reduced the operating power at all three reactors, then shut down B Reactor altogether in December 1946. During this slowdown only T Plant was used for chemical separations. Finally, Groves re-assigned weapon assembly responsibility from Los Alamos to the Sandia Base in Albuquerque (Gerber 1992b, Gosling 1999). Manhattan Project operations ratcheted down while the formulation of a national atomic policy moved to the forefront.

As early as spring 1944, Niels Bohr attempted to impress on both President Roosevelt and Prime Minister Churchill the need to freely exchange information on nuclear weapons development both to secure international agreement on control and avoid international proliferation – an arms race. Rhodes (1986, p. 528) summarizes the political implications of Bohr's message to these world leaders: "Tell the Soviet Union soon, before the first bombs were nearly built, that a bomb project was under way, and the confidence might lead to negotiations on postwar arms control. Let the Soviet Union discover the information on its own, build the bombs and drop them, oppose the Soviets at the end of the war with an Anglo-American nuclear monopoly, and the likeliest outcome was a nuclear arms race."

Bohr did not succeed with either man. Churchill chastised him, and while President Roosevelt was more cordial and agreeable, he agreed in the end with Churchill that secrecy would be maintained. Focusing on the immediate situation, both lost sight of Bohr's deeper message, that relationships among nations would be changed fundamentally after the bomb since no one could win a war between nations armed with atomic weapons.

Like Bohr, many of the scientists who developed the atomic bomb expressed misgivings about its use. Leo Szilard, the man who pushed hardest for the production of nuclear weapons in 1939, became one of the most ardent voices in seeking to control their use and spread. He made four attempts to influence policy during the spring and summer of

1945. In March, Szilard drafted a memorandum for President Roosevelt and requested a letter of introduction from Einstein to secure a meeting with the President. Roosevelt died before their meeting could be arranged. Undeterred, Szilard sought to meet with President Truman in May. President Truman instead directed him to James Byrnes in South Carolina. In response to Szilard's concern that "the greatest immediate danger which faces us is the probability that our 'demonstration' of the bomb will precipitate a race in the production of these devices between the United States and Russia," Byrnes responded first that the use of the bomb would satisfy Congress that the two billion dollars spent secretly in its development was justified, and that a demonstration of the power of the bomb might incline Russia to withdraw its troops from eastern Europe after the war (Szilard, quoted in Rhodes 1986, p. 637).

Discouraged but still undeterred, Szilard worked with James Franck, Glenn Seaborg, and other members of the Metallurgical Laboratory's Committee on the Social and Political Implications of the Atomic Bomb to assert their positions to the Interim Committee's Scientific Panel that only international control of atomic power would prevent an arms race and that the bomb should be demonstrated on an uninhabited area before being used against Japan. Theirs was the minority view, as the Scientific Panel met with the Interim Committee on May 31 and opted for maintaining nuclear monopoly and conducting an unannounced strike on Japan (Gosling 1999).

Finally, in July, Szilard drafted and circulated a petition to President Truman in which he urged him not to use the bomb: "The last few years show a marked tendency toward increasing ruthlessness. At present our Air Force, striking at the Japanese cities, are using the same methods of warfare which were condemned by American public opinion only a few years ago when applied by the Germans to the cities of England. Our use of atomic bombs in this war would carry the world a long way further on this path of ruthlessness" (Szilard 1945).

Lanouette (1999, p. 6) notes that 155 scientists from the Metallurgical Laboratory and the Clinton Engineer Works signed drafts of this petition that Oppenheimer banned at Los Alamos. Groves went further than Oppenheimer. Looking to dismiss the petition, he polled the scientists, but "when 72 percent of those responding favored a demonstration of the bomb, the Army bureaucracy suppressed both the petition and the poll."

Two events in late September 1945 framed the debate for international control of atomic weapons. Shortly after Japan's surrender, having seen and keenly felt the destructive power unleashed by the atomic bomb, Oppenheimer brought a letter drafted by the Interim Committee's Scientific Panel to Washington to deliver to Stimson. In the letter, the Panel outlined their concerns that further research would introduce far more powerful bombs, that no military counter measures existed then or would be found later to defend against atomic bombs, and that safety lay not in developing or stockpiling weapons but in "making future wars impossible" (Rhodes 1986, pp. 751-752). Clearly, the Panel's opinion had changed, moved both by the aftermath of the bombing and the earlier admonishments of Bohr and Szilard. Oppenheimer did not see Stimson, who was away when Oppenheimer arrived in Washington. Instead, Oppenheimer met with Vannevar Bush and Stimson's aide George L. Harrison. Neither was very receptive. When the letter reached Byrnes through Harrison, Byrnes instructed Oppenheimer to continue the work initiated under the Manhattan Project "full steam ahead" because of the deteriorating international situation (Rhodes 1986, p. 752).

The second event, the Conference on Atomic Energy Control, was a 2-day meeting organized by Szilard and University of Chicago chancellor Robert M. Hutchins. Following the revelation of the atomic bomb, Szilard felt the time was right to "get thoughtful and influential people to think about what the bomb might mean to the world, and how the world and America could adjust to its existence" (Szilard, quoted in Rhodes 1986, p. 750). Jacob Viner expounded on the concept of "psychological warfare" between equally armed nations wherein fear of the retaliatory effects of the bomb would deter its use. If the "war of nerves" maintained fear at a level appropriate for deterrence, the net effect would be peace. Szilard noted that deterrence might bring a "durable peace," but only "world government" could ensure a "permanent peace" (Rhodes 1986, p. 753).

Unfortunately for those proposing full disclosure and international control, initial congressional legislation, the May-Johnson bill, drew heavily from the draft legislation prepared by the Interim Committee prior to Hiroshima and Nagasaki bombings. Acting on President Truman's charge to formulate a nuclear weapons policy, the Interim Committee had developed draft legislation in July that proposed a "peacetime organization with responsibilities very similar to those of the Manhattan Project." The government would continue to control nuclear research and development through a board of commissioners weighted in favor of the military. Security would be maintained through fines and imprisonment. President Truman initially pushed for passage of the May-Johnson bill on October 3, 1945. Opponents of the bill complained that its provision for continued military control "tolerable during war...was unacceptable during peacetime when free scientific interchange should be resumed." Debate on the bill stalled "through a parliamentary maneuver" (Gosling 1999, p. 57).

On December 20, 1945, Brien McMahon, chairman of the Senate Special Committee on Atomic Energy, introduced an alternative to the May-Johnson bill, which had steadily lost ground, first from scientific opposition, then from President Truman's withdrawal of support. The bill was debated for 5 months. Groves objected to the bill and cited its "weak security provisions, the low military presence, and...[the] provision that atomic weapons be held in civilian rather than military custody." Ultimately, civilian control prevailed. President Truman signed the bill, known as the Atomic Energy Act of 1946, on August 1 and thereby gave control of atomic research and development to the newly created United States Atomic Energy Commission effective January 1, 1947 (Gosling 1999, p. 57).

Two of Groves' last acts as commander of the Manhattan Project demonstrated his acknowledgement of increased civilian interaction and control. In June 1946, nine plutonium bombs constituted the entire U.S. nuclear weapons arsenal (Rhodes 1986). During July of that year, two were detonated under *Operation Crossroads*. With a "large, invited audience of journalists, scientists, military officers, congressmen, and foreign observers" in attendance, Shot Able and Shot Baker, each with an explosive yield of 21,000 tons of TNT, were detonated on Bikini Atoll on July 1 and 15 respectively. These public tests were the last conducted by the Manhattan Project (Gosling 1999, p. 55).

Late in 1946 in response to Oppenheimer's request for increased research and development funding, Groves established three national laboratories – Argonne, Brookhaven, and Oak Ridge (Gerber 1992b). The Argonne and Oak Ridge national laboratories grew out of their military background, and even their civilian focus remained tied to applied atomic research, that is, process improvement. The Brookhaven National Laboratory, on the other hand, was a "civilian institution from its conception, and its program was...oriented to basic research," that is, the discovery or understanding of previously unknown structures or characteristics principally in the fields of physics, engineering, biology, and medicine (Crease 1999, p. 1).

The Manhattan Project lasted 4 years from 1942-1946 and estimates on total expenditures range from $1.8 to $2.2 billion. Today, this would be equivalent to over $20 billion (Bracchini 1997). It succeeded in what it had set out to do – develop an atomic bomb for military use. Whether it had made the world more secure was another question.

COLD WAR: EXPANSION OF THE HANFORD SITE

Site Name:	Hanford Works - January 1947 to December 1974
Site Manager:	Lt. Col. Fredrick J. Clarke - August 1945 to August 1947
	Carlton Shugg - August 1947 to September 1948
Responsible Agency:	U.S. Army Corps of Engineers
	Atomic Energy Commission - January 1947 to December 1974
Site Contractor:	General Electric Company - September 1946 to January 1964

Rhodes (1986, p. 756) points out that the problem following the war "was not lack of support but lack of authority." Congressional authorization was required before funds could be released. The release of funds was tied to the development of policy objectives and legislation. With the creation of the Atomic Energy Commission, initial legislation and congressional authorization were essentially resolved for the moment. Following the Atomic Energy Commission's introduction into the open political arena, however, the development of a national policy would continually shift with world events and presidential elections.

> ***"Secrecy proved to be a blessing in disguise. Although it dictated remote site locations, required subterfuge in obtaining labor and supplies, and served as a constant irritant to the academic scientists on the project, it had one overwhelming advantage: Secrecy made it possible to make decisions with little regard to normal peacetime political considerations."*** **- Gosling 1999, p. 19**

In an interesting analysis of the United States and Russia at the end of the war, Rhodes (1995, p. 180) compares their relative youth as nations, their shared origin through revolution, and their populations and resources. However, despite their similarities, he concludes that there remains "an intractable difference between them. They were opposite experiments in the large organization of people and natural wealth, the one through liberty and competition, the other through terror and centralized control, an open society and a closed – a crystal of quartz and a crystal of onyx, Robert Oppenheimer once contrasted them – and each was convinced that the other side's intentions were malevolent."

The beginning of the Cold War was neither as abrupt nor as visible as that of World War II. Instead, it grew in the course of boasts and retaliatory statements beginning in 1946 with the Soviet Union's "there can be no permanent peaceful coexistence," Prime Minister Churchill's "an iron curtain has descended across the continent," and the Clifford-Elsey report that Soviet foreign policy was "a direct threat to American security." As a signal of U.S. resolve, the Truman Doctrine, delivered on March 12, 1947, assured the world that the United States would "support free peoples who are resisting attempted subjugation by armed minorities or by outside pressures" (Rhodes 1995, p. 295). In large measure, the Truman Doctrine was the military counterpunch to the Marshall Plan, first articulated by Secretary of State George C. Marshall on June 5, 1947. His economic recovery plan proposed that the United States "do whatever it is able to do to assist in the return of normal economic health in the world, without which there can be no political stability and no assured peace" (Marshall, quoted in Rhodes 1995, p. 297).

With international positions hardening, the Atomic Energy Commission visited each of the Manhattan Project sites early in 1947. David Lilienthal, Chairman of the Atomic Energy Commission, remembered their January inspection "as one of the saddest days of my life" when he determined there was only "one [bomb] that had a good chance of being operable" (Lilienthal, quoted in Rhodes 1995, p. 283). "Meeting early in the year after a whirlwind tour of the production sites and laboratories, the AEC's General Advisory Committee assigned its highest priority to weapons research and production. By February, improvement and expansion of the plutonium production units at Hanford topped the list of AEC goals" (Gerber 1992b, p. 18).

In August 1947, the General Electric Company, having assumed management of the Hanford Site from Du Pont on September 1, 1946, announced that two new reactors would be built and research leading to the development of a new chemical separations process would begin. The news was well received in Richland, whose continued prosperity relied on government support for the Hanford Site.

To enhance plutonium production at a time when the existing reactors were aging, General Electric commissioned construction of the DR and H Reactor complexes to meet Atomic Energy Commission objectives. Completed in March 1949, the DR (D Replacement) Reactor was meant to take the place of the D Reactor, whose graphite core was observed to be expanding and distorting. In fact, expansion was occurring in all three reactors. The distortion was caused by graphite's reaction to intense neutron bombardment. Called "Wigner's disease," after the physicist Eugene Wigner who had predicted the effect in 1942, energy acquired under bombardment realigned the carbon atoms within the graphite's crystal lattice. This rearrangement caused the graphite to expand (Rhodes 1995). Operators monitoring the expansion at B Reactor noted that temperature, in addition to the intensity of the neutron flux, was causing disproportional expansion. As described in the B Reactor HAER in Appendix B on the Internet, "Temperature was so crucial to expansion that the cooler edges of the pile [reactor] were actually expanding quite a bit *more* than the graphite in the highly irradiated but thermally hot center portion." As a consequence of expansion, the process tubes were bowing, making charging and discharging more difficult, the alignments of the control rod and safety rod channels were altering, and the neoprene seals along the shielding walls were being stressed. Unchecked distortion would render the reactors inoperable or, more ominously, uncontrollable.

Reactor expansion is the reason Groves shut down the B Reactor and reduced the operating power of the D and F Reactors in March 1946. Until the problem was resolved or new reactors were built, B Reactor would be held in reserve. Groves closed B Reactor to ensure that production of plutonium and polonium-210 could continue if D and F had to be taken offline. His primary concern was maintaining a source of polonium-210, which was used as the initiator or trigger in plutonium bombs. To produce polonium-210, "lead-bismuth alloy targets (called 'B Metal') [were] welded...into unbonded aluminum cans," and irradiated in the production reactors. The first targets were fabricated in 1944 in the 313 Building, and production of polonium would continue until the early 1950s (DOE 1997e, p. 158). Polonium-210 is a short-lived isotope. With a half-life of only 138 days, "if the Hanford piles [reactors] broke down and polonium production ceased, the atomic bombs in the nations small stockpile would become unreliable within a year" (Rhodes 1995, p. 277). However, while construction of the DR Reactor was under way, scientists at the Hanford Site's Pile Technology Division found the solution to graphite swelling. By adding carbon dioxide to the helium atmosphere surrounding the reactor, the temperature of the graphite could be raised from 100° C to 250° C. This thermal increase during neutron bombardment within the reactor, termed nuclear annealing, negated the effects of crystal realignment. General Electric restarted B Reactor in July 1948 and elevated the operating power of all three Manhattan Project reactors, which increased product yield. Given this increased productivity, the Atomic Energy Commission held the DR Reactor in standby. The DR Reactor would not be started up until October 1950.

Advancing on the existing design, General Electric engineered the H Reactor to operate at 400 megawatts-thermal – a 60 percent increase in power relative to the B, D, F, and DR reactors. Starting up October 1949, the H Reactor complex was the first post-Manhattan Project site constructed in the 100 Area. It also marked a deviation from the siting plan worked out by Groves and Du Pont in that it was located not downriver from the 100-F Area, as might be expected given the alphabetic sequence established in 1943, but rather up river between the 100-D and 100-F areas. In addition, the DR Reactor was located within one-half mile of the D Reactor so their later simultaneous operation also violated the 1943 siting requirement of a separation of not less than one mile between production reactors. These initial siting requirements would be further abrogated by future expansions.

Construction of two new reactors essentially improved existing capabilities; however, completion of the Plutonium Finishing Plant (234-5Z Building or Z Plant) in 1949 provided an entirely new capability. As described in the ExHPIF for the 234-5Z Building in Appendix B on the Internet: "Although the nitrate compound that left Hanford had been evaporated into a very thick, dry paste for safety reasons, the managing federal agencies (Manhattan Engineer District and later the Atomic Energy Commission) nonetheless viewed the shipment of non-solid forms of Pu [plutonium] as potentially hazardous, and thus looked for a way to create metallic Pu at the Hanford Site."

The expansion ordered in 1947 provided that opportunity. Using plutonium nitrate from the 231-Z Building as feed, operators at the Plutonium Finishing Plant chemically converted plutonium nitrate, the World War II end-product, into hockey puck-shaped plutonium metal "buttons," which resemble hockey pucks. Machinists then fabricated these buttons into "pits" for use in nuclear weapons. Operations began on July 5, 1949, using the Rubber Glove Line, an assembly of 28 interconnected glove boxes. The Rubber Glove Line performed less than satisfactorily. Airborne contamination releases and corrosion failures caused slowdowns and periodic shutdowns at a time when the Atomic Energy Commission was calling for increasing production (see Chapter 2, Section 5 for an extended discussion).

The 234-5Z Building contained its own analytical laboratory where chemists analyzed samples drawn from stages throughout the finishing process to ensure that safety and isotopic standards were maintained. Another section housed the developmental laboratory where researchers sought ways to refine measurement accuracy, extract more plutonium (thereby reducing wastes), and solve production problems (see Chapter 2, Section 5). Support facilities constructed in association with 234-5Z included the 216-Z-1 and 216-Z-2 Waste Cribs, the 241-Z Waste Storage Building, the 272-Z Shop and Warehouse, the 291-Z Stack, the 2701-Z Badge House, the 2704-Z Office Building, the 2705-Z Temporary Technical Office Building, the 2719-Z First Aid Station, and the 2901-Z Elevated Water Storage Tank.

Scientists and politicians alike considered uranium a very limited resource in the late 1940s, so much so that the United States and Great Britain each tried to control the world supply (Rhodes 1995). Dissatisfied with the fact that uranium was flushed away as waste in the initial stages of the bismuth phosphate process, the Atomic Energy Commission issued "direct and urgent orders" to develop a better separations process early in 1947 (Gerber 1992c, p. 84). Consequently, General Electric decontaminated several laboratories within the 3706 Building and assigned their scientists the task of perfecting an alternative process. The scientists labeled the process they invented reduction-oxidation or simply REDOX. In contrast to the bismuth phosphate batch process, REDOX was a continuous, solvent extraction process that was more efficient (in terms of total plutonium extracted per ton of irradiated fuel elements) and more economic (since uranium could be separated out and recovered) (Gerber 1992c, Gerber 1993b). However, construction of the REDOX Plant (202-S Building), the process canyon named after the pioneering chemical separations process, would not begin until late 1949. Chapter 2, Section 4 describes the REDOX process in detail.

To handle the additional waste stream from increased production within the 200 Area, General Electric built the 241-BX, 241-BY, and 241-TX Tank Farms, constructing an additional 42 single-shell tanks (Gerber 1992b). In 1948, General Electric built the 300 North Cribs (321 Cribs). Because monitoring of the main Process Pond determined that radioactivity levels had risen "precipitously, climbing by three to thirty orders of magnitude in water and mud samples taken from 1945 through 1948" (Gerber 1993b, p. 55), the 300 North Cribs were seperated from the 300 Area by nearly 5 miles. A second, larger process pond (the North Process Pond) was constructed that same year after a break in the dike of the main Process Pond (now called the South Process Pond) released 14.5 million gallons of waste water to the Columbia River in 90 minutes (Gerber 1993b)."

The workers who constructed these new facilities during the postwar expansion resided within the 3000 Area, a newly created construction camp located 5 miles north of the Richland Village. Single men lived in barracks acquired from the former Pasco Naval Air Station, while married men lived in small trailers. Some 12,000 construction workers and 13,000 dependents called the 3000 Area camp home in 1948. Population within the Richland Village increased to nearly 23,000 during this expansion period, fueled by "the highest birth rate in the nation!" (Gerber 1992b, p. 22).

The operations workforce alone at the newly renamed Hanford Works expanded from 4479 to 8628 in the 2 years between 1946-1948 (Findlay and Hevly 1995). In their attempt to make Richland "attractive enough to recruit and retain plant employees," the Atomic Energy Commission instructed General Electric that "consistent with security and other requirements, residents...shall enjoy those facilities, services, and activities which are properly a part of American life." Still, to some Richland Village retained its feeling as a police state since the "company, not the citizens, ran the town, and ran it on behalf of the AEC." To act on some of their concerns, General Electric established the Richland Community Council in 1948. The Council "provided a forum for residents to raise questions or propose changes, but it served only as an advisory body, with no legislative power and only a minimal budget provided by GE" (Findlay and Hevly 1995, pp. 80-88). Self-government would not come for another decade.

SOVIET "GIANT" AWAKES: 1947-1949

"In the winter and spring of 1948," Rhodes (1995, p. 317) writes, "the Soviet Union and the Western allies gave up any remaining pretense of continuing their wartime collaboration." New alliances, or forced allegiances, would be formed in the post-war era. Among the first was the alliance between Britain, France, Belgium, the Netherlands, and Luxembourg established by the Treaty of Brussels on March 17, 1948. The occupation of Czechoslovakia by Russian troops on February 25, 1948 was among the first of the Soviet-enforced allegiances (Rhodes 1995).

On June 24, 1948, the Soviets closed all rail traffic leading to Berlin from West Germany, effectively isolating the city. Rhodes (1995, p. 319), citing Soviet sources, proposes that the "possibility that the West might rearm Germany [by including West Germany in a military alliance with the United States, Britain, Canada, and Western Austria] may well have precipitated the Soviet decision to move against Berlin." The State Department had initiated secret discussions on March 22 on establishing a western military alliance with Britain and Canada to defend Western Europe. By mid-June, with the secret long exposed, the Senate opened the way for "public negotiations towards a North Atlantic Treaty Organization" (Rhodes 1995, p. 322).

President Truman responded to the gauntlet the Soviets had thrown down by approving an airlift operation to supply the American, British, and French sectors of Berlin on June 28. The airlift lasted until the Soviets opened all rail lines in May 1949, 11 months after the crisis began. In one of the first misinformation campaigns of the Cold War, "the government made a point of revealing" that the sixty B-29s the U.S. Air Force flew to East Anglia in support of the airlift in July 1948 "were atomic-capable and hinted that they carried atomic bombs" (Rhodes 1995, p. 326). The allusion was untrue, but the bluff indicated the deterrent effect afforded atomic weapons and foreshadowed the National Security Council's adoption of deterrence as official national policy in November 1948.

Just prior to the Berlin blockade, weapon developers from Los Alamos traveled to Eniwetok atoll to test a new series of bombs, the Mark IV. Norris Bradbury, Oppenheimer's successor at Los Alamos, had directed his staff to develop "small atomic weapons" (Mitchell 1999, p. 21). These small tactical weapons "offered a technological solution that would bring nuclear weapons back to the battlefield, in a sense controlling the awesome destructive capabilities within some acceptable level. The destruction of concentrations of enemy troops, planes, tanks, and ships was morally more acceptable than that of whole cities and their inhabitants" (Mitchell 1999, p. 22).

The Mark IV bombs had new cores, which increased the yield substantially. The *Sandstone* tests of the Mark IV bombs consisted of three explosions: the first of which yielded 37 kilotons, the second 49 kilotons–the largest yield of any atomic weapon yet. The third bomb, tested on May 14, 1948, exploded with a yield of 18 kilotons (DOE 2000a).

Not only were these new cores highly effective, but they were also highly efficient. Using this design, the existing stockpile of nuclear weapons could be upgraded and expanded, and new weapons with higher yields could be added more quickly: "Fabrication of standard nuclear cores stopped immediately so that all fissionable material would henceforth go into new models. The day of tailor-made weapons was fading fast; with Mark 4 would come mass production of components and assembly-line techniques" (Hewlett and Duncan, quoted in Mitchell 1999, p. 22).

At the Conference on Atomic Energy Control held at the University of Chicago in September 1945, Jacob Viner had characterized the United States and the Soviet Union as two giants whose mutual opposition would deter or ignite world conflict (Rhodes 1986, p. 753). Having faced down the Soviets in Berlin and with the West's nuclear capability increasing through expanded production facilities and redesigned weapons, Secretary of State Dean Acheson expressed the administration's general feeling that in July 1949 the "position of the West…[has] grown greatly in strength, and the position of the Soviet Union…has changed from the offensive to the defensive" (Acheson, quoted in Gerber 1992b, p. 23). This feeling was short-lived.

On the isolated, high, short-grass steppes of northeastern Kazakhstan, the Soviets had established a site to detonate their first nuclear device, RDS-1. The test, which the Soviets code-named "First Lightning," was set off at 7:00 a.m. on August 29, 1949. The RDS-1 was a near replica of the Trinity device, whose plans had been acquired through espionage (Rhodes 1995, pp. 167-174). As with Trinity, the test was delayed by light early-morning showers but was successful, exploding with a yield of 20 kilotons. Unlike Trinity, however, the Soviets had constructed single- and multi-story houses, built bridges and tunnels, positioned military hardware, and placed animals both in open pens and within buildings to observe and measure the effects of the explosion and its attendant radiation throughout the test site. Little survived the blast (Rhodes 1995, pp. 364-368). The test marked the culmination of Soviet covert construction operations that had been going on for nearly 4 years.

The Soviets began construction of the F-1 (Physics-1) test reactor in July 1946 based on a design two Soviet scientists, Igor Kurchakov and Igor S. Panasyuk, had proposed in July 1943. However, based on declassified information, physicist Arnold Kramish concluded in 1955 that F-1 "was practically a carbon copy of the American 305 reactor built at Hanford," whose design was being developed at the Metallurgical Laboratory in June 1943 (Kramish, quoted in Rhodes

1995, p. 266). Their function was also identical – testing the purity of graphite and uranium before emplacement within production reactors. It started operation on Christmas day 1946.

Although Stalin did not publicize the success of the "First Lightning" test, western scientists literally pulled the information out of thin air. When a U.S. Air Force "sniffer plane" was on a routine air-monitoring flight over the Kamchatka Peninsula on September 3, 1949, it detected radioactivity "300 percent greater than the level...established as an alert: measurements quickly confirmed that the radioactivity was fission derived" (Rhodes 1995, p. 371). By September 14, experts associated with Los Alamos, the U.S. Navy, and the British monitoring program confirmed these findings. The scientist at Tracerlab concluded that the blast, given the name "Joe 1" by Kramish, had taken place at 6:00 a.m. on August 29 – they were off by one hour (Rhodes 1995). Following a period of denial, skepticism, and guarded acceptance, President Truman made the news public on September 23. The Soviet giant had awakened.

President Truman had made it known on July 14, 1949 before the detonation of RDS-1, that nuclear superiority was paramount to U.S. national security (Mitchell 1999, p. 20). Now that the Soviets had the atomic bomb, his words held a special meaning for Edward Teller, who judged that the time had come to conclude his long-standing push for development of thermonuclear weapons – the "Super." Rhodes (1986, p. 754) explains: "The end sought was a bomb burning about a cubic meter of liquid deuterium. For such a bomb the energy release will be about ten million tons [10 megatons] of TNT." Such a bomb would be 1,000 times more powerful than the one exploded over Hiroshima.

As early as 1943, Teller had lobbied for the development of a hydrogen bomb. Indeed he was "so preoccupied with the idea of the Super during the war-time Manhattan Project that when Oppenheimer assigned Teller work on calculations for Fat Man, the work had to be reassigned" (Mitchell 1999, p. 25). Ernest Rutherford, Marcus Oliphant, and Paul Harteck had discovered the "hydrogen fusion reaction" in 1934 while conducting experiments with the particle accelerator at Cambridge University. When concentrated heavy water (hydrogen-2 or deuterium) was bombarded with deuterium-accelerated nuclei, "the deuterium nuclei had fused together...Neutrons, heat and intense gamma radiation came out of the reaction as...the new nucleus [of helium] adjusted its energy level and stabilized." The reaction required "heating the nuclei until their thermal motion overcame their electrical repulsion" and thereby acquired the name "thermonuclear fusion" (Rhodes 1995, p. 247).

From April 18-20, 1946, the Los Alamos scientist had held a "Super Conference" to review work conducted to date, to debate the feasibility of a hydrogen bomb, and to investigate its design. Teller, summarizing a report he authored with six colleagues, concluded that the Super could be made and that a large-scale development program should be initiated towards that end. Others in attendance were not convinced. The ignition might take place, they argued, but whether or not it would continue to burn was highly problematical. Noting, however, the importance of tritium as a booster for either fission or fusion bombs, Teller recommended its production. This proposal did not generate resistance. The report issued following the Super Conference concluded that development would be costly "using up a fair portion of the national nuclear-weapons budget for some years to come." With the war concluded and with no apparent need for a weapon of this destructive magnitude, research on the Super was restricted to the Theoretical Division and then to only half-time over the next few years (Rhodes 1995, pp. 252-255).

With President Truman's announcement of the Soviet Union's nuclear capability on September 23, 1949 as a catalyst, the Atomic Energy Commission's Joint Committee on Atomic Energy proposed a number of ways to expand the production of atomic weapons including dramatically accelerating work on the hydrogen bomb and building a new generation of reactors so that tritium production would not come at the expense of plutonium production. However, convening the Atomic Energy Commission on October 5, David Lilienthal did not place thermonuclear weapons on the agenda. Incredulous that the hydrogen bomb had been left out of the discussion, Strauss contacted Rear Admiral Sidney Souers, who was to meet with President Truman the following morning. President Truman was briefed on the Super for the first time on October 6, 1949 (Rhodes 1995, pp. 380-381).

On January 31, 1950, President Truman approved development of the Super. National security policy was based on deterrence, and Secretary of Defense Louis Johnson succinctly summarized the implications of that policy with respect to Russia: "We want a military establishment sufficient to deter that aggressor and sufficient to kick the hell out of her if she doesn't stay deterred" (Johnson, quoted in Rhodes 1995, p. 405). Within a few months, two events occurred which made President Truman's decision to add thermonuclear weapons to the U.S. arsenal look prescient. In May, the Communist forces of Mao Tse-Tung expelled Chiang Kai-shek and his Nationalist followers from China. Stalin and Tse-Tung had entered into a mutual assistance agreement on February 14. On June 25, the Communists forces of North Korea invaded the southern Korean peninsula. These events played to the nation's worst fears of unchecked communist expansion.

KOREAN WAR: 1949-1952

Site Name:	Hanford Works - January 1947 to December 1974
Site Manager:	Fred C. Schlemmer - September 1948 to May 1950
	David F. Shaw - June 1950 to June 1955
Responsible Agency:	Atomic Energy Commission - January 1947 to December 1974
Site Contractor:	General Electric Company - September 1946 to January 1964

"Using the Korean War as a club," the Senate persuaded President Truman to increase Atomic Energy Commission funding from $260 million to $1.4 billion between July and October 1950. A National Security Council Special Committee report, issued on October 2, cited the need not only for "more weapons of existing designs, but for new designs in large numbers," and focused attention on their delivery systems as well. By mid-1951, the Joint Committee on Atomic Energy was already pressing for an additional enhancement of 50 to 150 percent in atomic weapons production (Mitchell 1999, p. 31).

Gerber (1992b, p. 25) has characterized the period from late 1949 to 1952 as "the greatest era of expansion in U.S. atomic/nuclear history." During this interval, the Atomic Energy Commission consolidated uranium metal production at the Fernald Feed Materials Production Plant in Ohio; constructed a second plutonium production site, the Savannah River Plant, in South Carolina; and constructed a second uranium enrichment site, the Paducah Gaseous Diffusion Plant, in Kentucky. Two new laboratories, the Reactor Testing Station in Idaho and the Sandia Laboratory, were opened. The Pantex Plant in Texas was established as a weapons assembly site. Finally, the Atomic Energy Commission established the Nevada Test Site and expanded the Pacific Proving Ground at Bikini and Eniwetok atolls (Gerber 1992b). At the same time these new sites were being constructed, the Atomic Energy Commission added a number of facilities to the existing Manhattan Project sites.

Workers broke ground for the C Reactor complex, the sixth production reactor at the Hanford Site, on June 6, 1951. It started up 17 months later on November 18, 1952. Like its predecessors, the C Reactor was a single-pass, water-cooled, graphite-moderated reactor. Unlike its predecessors, however, the design team engineered it to operate at 650 megawatts-thermal, which was a 60 percent increase over the design rating of the H Reactor completed only a year earlier and a 160 percent increase over the original design specifications for the Manhattan Project reactors (Gerber et al. 1997, p. 5.17). To accelerate construction, General Electric located the C Reactor within the 100-B Area and less than a mile from the B Reactor to "take economic advantage of existing utilities, services, and facilities" (Gerber 1993a, p. 106).

Nonetheless, as production capabilities rose at the Hanford Site, the Atomic Energy Commission production requirements expanded even more. To meet ever-spiraling demands, Hanford Site physicists looked for ways to increase power levels first within the Manhattan Project reactors and then all reactors. In April 1949, physicists began incremental testing at D Reactor that would push the operating level from 250 to 330 megawatts-thermal. The D Reactor was chosen because with the DR Reactor nearly complete, the D Reactor was considered expendable should something go wrong. Exceeding expectations, D Reactor was operating at 400 megawatts-thermal by January 1950 (Gerber 1993a).

> ***"To increase plutonium production capacity, Hanford began adding low-enriched uranium fuel slugs [elements] to its reactors as early as 1950. Enriched uranium fuel also allowed Hanford engineers to even out the reactor's temperature and power distribution, reducing problems caused by uneven thermal expansion and radiation-induced swelling of the graphite core. Neutron absorbing "poison" slugs, also made on site, also helped to even out the reactor's power distribution. Most of these enriched uranium slugs were manufactured using the same techniques as the natural uranium slugs. However, some of the fuel elements were made of highly-enriched uranium alloyed with aluminum, which required special fabrication techniques to prevent accidental criticalities."*** **- DOE 1997e, p. 154**

Sensing that operating levels could be pushed further, the Atomic Energy Commission recommended that the Hanford Site scientists "test whether B, D, and F Reactors could operate at 600 megawatts-thermal, without causing undue system failures or increases in the plutonium-240 content of the product." Design reviews necessary to identify what equipment changes would be required to reach this level began shortly thereafter (Gerber 1993a, p. 4). However, the increased operating level was already causing problems as "fuel element ruptures, feared since World War II, became a reality." The first fuel failure occurred in the F Reactor in 1948. Subsequent failures "increased slowly during 1949 to 1950, but expanded dramatically in 1951" (Gerber 1993a, p. 26). Chapter 2, Section 3 discusses in detail the failure of fuel elements that resulted from increasing the power levels.

In constructing the C Reactor and extending the power levels of the existing reactors, General Electric increased plutonium production capabilities at the Hanford Site and supported the Atomic Energy Commission's immediate need to meet escalating military requirements. Augmentation in plutonium production also allowed General Electric to support development of the Super from February 1949 to March 1952 by releasing some of the process tubes in the existing reactors for tritium production. Production of tritium gas, conducted under the code name "P-10," began with the jacketing of lithium-aluminum alloy target elements in unbonded aluminum-silicon cans in the 313 Building. Initially manufactured offsite, machinists began to fabricate these target elements on the Hanford Site in the mid-1950s. The lithium targets, "surrounded by highly enriched uranium 'driver' elements," were irradiated primarily in "H Reactor but sometimes in B Reactor" (Gerber 1993a, p. 44). The irradiated targets were stored in the 104-B-2 Tritium Laboratory to await processing. To process the tritium, General Electric converted the four-story 108-B Chemical Pump House into offices, laboratories, and operations support rooms and installed five process lines contained within exhaust hoods on the third floor. A 300-foot ventilation stack was added to the 108-B Pump House to exhaust process gases. Tritium gas was extracted from the irradiated target elements, bottled in shipping casks, and then stored in the 104-B-1 Tritium Vault. The 1703-B Technology Service Building was also constructed to provide additional offices as well as storage space for process equipment (Gerber 1993a, pp. 44-46).

With plutonium production levels rising and after nearly 5 years of developmental testing at the Hanford Site, the REDOX Plant (202-S Building), the "only continuous solvent extraction plant in the world," began processing irradiated fuel elements in January 1952 (Gerber 1992b, p. 25). Chemists in the 222-S Control Laboratory provided process sampling and operations support not only for plutonium separation work conducted in the REDOX Plant but also for uranium recovery work in the 221-U and 224-U plants and waste reduction operations in the 242-B and 242-T Waste Evaporator buildings as well. With REDOX operational, General Electric shut down B Plant (Gerber 1994a).

Responding to the need to conserve, recover, and recycle uranium (a strategic material believed to be scarce at this time), General Electric modified the 224-U Bulk Reduction Building in 1951 to serve as the primary facility for the Uranium Metal Recovery program. Now known as the Uranium Trioxide Facility or simply UO_3, it began full-scale operations in 1952 to convert liquid uranyl nitrate hexahydrate (UNH) to uranium trioxide powder (UO_3) through a calcination process (Bailey and Gerber 1997, p. 13). The uranium fraction (UNH), isolated during REDOX separations, was piped to the Uranium Trioxide Facility for processing. A second uranium feed for the 224-U Bulk Reduction Building came from an unanticipated source. In 1952, General Electric changed the mission of the U Plant from training operators for B and T Plants to extracting the uranium contained within the bismuth phosphate process waste stored in the single-shell tank farms. General Electric retrofitted the process cells in the U Plant with new equipment to accommodate the tri-butyl phosphate process, developed by Hanford Site chemists, to effect separation (Gerber et al. 1997). The Atomic Energy Commission supplied the uranium trioxide powder to the defense program for use in armor plating, projectile tips, and nuclear submarine fuel among other items.

A program to extract a formerly discarded material – uranium within high-level liquid waste – not only met a strategic need but also indicated the ongoing efforts to manage waste generated at the Hanford Site. The 242-B and 242-T Waste Evaporator Buildings represented another program aimed at waste reduction. General Electric constructed these facilities in 1951 to increase the storage capacity of the high-level liquid waste storage tanks by decreasing the amount of liquid contained in the waste. Operators first pumped liquid waste from the single-shell storage tank to a feed tank within the 242 Building. From the feed tank, the liquid waste passed to one of two preheaters before entering the evaporator tank. Continued heating in the evaporator tank concentrated the waste by boiling off liquid, leaving behind a slurry. This slurry was pumped back into the storage tanks where it solidified as it cooled. The net result was a reduction in waste volume. Low-level liquid condensate was disposed of in cribs or trenches while liquid vapor was scrubbed, filtered, and vented through an exhaust stack (see the 242-T HPIF in Appendix B on the Internet for additional information). Despite these efforts, General Electric constructed an additional eighteen single-shell, high-level liquid waste storage tanks (241-S and 241-TY Tank Farms) because of the increased productivity requirements the Atomic Energy Commission mandated.

At the same time that liquid wastes were being recycled and reduced, Hanford Site scientists were investigating ways to control or limit airborne releases of radionuclides from the process canyon stacks. Increasing production goals and expanding production capabilities placed additional burdens on the existing chemical separations facilities. Before the REDOX Plant opened, B Plant and T Plant were supporting five reactors operating at increased output levels. Not surprisingly, the emission of process gases had risen with each production quota. By late 1947, scientists had documented contaminated vegetation over a wide area of Idaho, Oregon, and Washington that was linked to emissions from the 291-B and 291-T Stacks.

After nearly 2 years of operations, General Electric installed water scrubbers between the exhaust fans and the stack openings and within the exhaust ducts for each process cell to filter the air before it was released. General Electric also replaced the fans, fan casings, and ductwork within the stacks because of radiologically contaminated particulates or "specs" being emitted with the exhaust gases (Gerber 1994a, p. 33). Further modifications were made to the exhaust flow in 1948 when General Electric added large underground sand filters to treat process gases before their release through the stacks. Measuring 110 feet long by 48 feet wide, these filters were designed to strip radiologically charged "droplets" or "acid mists" (formed from condensed process gases) as well as particulates out of the waste stream. However, "these filters were only marginally successful because the sand beds plugged and the resistance (pressure drop) within the unit increased rapidly" (Gerber 1994a, p. 34). Fiberglass filters installed within the dissolver off-gas lines offered little additional improvement. Finally late in 1950, Hanford Site scientists discovered that fiberglass embedded with silver-nitrate ("silver reactor" filters) was 99.9 percent effective in capturing iodine-131 when placed in the ducts between the dissolver cells and the main exhaust lines. However, this success rate was achieved only when the silver reactor filters were replaced or regenerated frequently and irradiated fuel cooling periods were extended to 80-100 days before dissolving (see Gerber 1994a, pp. 33-37 for additional discussion).

Operating conditions in the Soviet Union during this time provide perspective for waste management practices in the United States. In commenting on working conditions within the Soviet Union's chemical separations plant at Chelyabinsk-40, Rhodes observes:

> *"It was impossible to work at the plutonium separation plant without dangerous radiation exposure. Sixty-six percent of B installation workers received subclinical but excessive doses in the first year of operation of up to 100 rem; an unlucky 7 percent received above 100 rem where clinical signs such as vomiting and blood changes begin to appear. (For comparison, the average lifetime dose of workers in the US and British nuclear-*

weapons industry has been estimated at from 3 to 11 rem.).... B installation, by design, discharged its intensely radioactive fission waste directly into the Techa River. By 1951, radioactivity from Chelyabinsk-40 had been measured in river water discharging into the Arctic Ocean more than a thousand miles north." - Rhodes 1995, pp. 350-351

These statements highlight a philosophical difference underlying the two nuclear weapons development programs. The Soviets viewed the general workforce as largely expendable and gave little weight to environmental damage in their design considerations. From its inception, the U.S. program implemented measures to monitor worker health and safety while seeking ways to first understand and then limit environmental degradation. This is not to imply that the U.S. program always succeeded in these objectives or that the consequences of ill-informed decisions are not still with us. It does demonstrate, however, that the United States was aware of at least some of the potential dangers, investigated their effects, and, when not rushed to action, sought to act accordingly. The Hanford Site offers a case in point.

Applied research, as well as production capabilities and process improvement, benefited from increased funding during the first years of the Korean War. The Hanford Site Biology Program had outgrown the facility assigned to it in 1946 – a single Quonset hut in the 100-F Area. Responding to a 1948 budget request from Hanford Site administrators, the Atomic Energy Commission Advisory Committee for Biology and Medicine determined that "it was unwise to set up an additional national laboratory at Hanford as it was primarily a production center and research in biology and medicine there should be directly applicable to local problems. Nevertheless, the Committee believed it highly desirable that adequate facilities should be provided for aquatic biology and for a farm for animals because of specific concern to operations at Hanford" (see the 108-F Building ExHPIF in Appendix B on the Internet). In 1949, the Hanford Site Biology Program acquired the remodeled 108-F Chemical Pump House for office and laboratory space and constructed an animal farm and agricultural plots nearby. New radioecology and aquatic biology laboratories were added in 1951, with additional expansions in 1953 and 1956 devoted to biological research. Scientists concentrated on three major research areas: biological effects of reactor effluent on native fish within the Columbia River, biological effects of radioisotopes (such as cesium-137, iodine-131, and strontium-90) on living tissue within large and small animals, and botanical effects on plants grown in soils containing radionuclides.

The Atomic Energy Commission was not the only agency expanding its presence at the Hanford Site. In response to increased tensions due to the escalating Cold War as expressed by the active conflict in Korea, the U.S. Department of Defense initiated additional measures to address the security of U.S. nuclear facilities. On March 28, 1951, the Army established Camp Hanford in North Richland and assigned troops from Ft. Lewis the task of providing air defense for the Hanford Site. The Army built sixteen anti-aircraft artillery (AAA) batteries, encircling the primary production complex, along what came to be called Army Loop Road (Gerber et al. 1997, p. 5.47). Each AAA site was equipped with 90 mm and 120 mm guns placed in four revetments constructed from sandbags. Each site also contained barracks for the enlisted men and officers and support facilities such as radar facilities, a mess hall, a motor pool, and craft shops.

Camp Hanford, situated in the 3000 Area just south of the North Richland Construction Camp, "consisted of cantonment/barracks areas, an administrative area, commercial district, a trailer park, 'Bremerton' housing (residential units from the Puget Sound Naval Shipyard in Bremerton, Washington), medical facilities, recreation centers, and an industrial area" (Gerber et al. 1997, p. 5.26). The Camp Hanford industrial area, located at the southern terminus of the Camp, contained an electronic and signal maintenance shop (1154 Building), brigade motor pool (1226 Building), ordnance maintenance and crafts fabrication shop (1240 Building), and shipping and receiving warehouses (1250 and 1252 Buildings) among other facilities that provided support for the defense installations (Gerber et al. 1997, p. 5.48). Chapter 2, Section 9 describes Camp Hanford and its 10-year mission.

KOREAN WAR: 1952-1955

With the Korean War continuing, President Truman approved further expansion on January 16, 1952. At $4.9 billion, this authorization was 250 percent higher than the 1950 appropriation and indicated the exponential growth of the Atomic Energy Commission under its new Chairman, Gordon Dean. The "atomic weapons program was now so large that in the fall of 1952 the hitherto insatiable [Atomic Energy Commission] Military Liaison Committee reported that it found no reason for further expansion" (Wells, quoted in Mitchell 1999, p. 33). All installations and improvements carried under this authorization were due for completion by January 1957.

On September 24 and October 18, 1951, the Soviets had detonated their second ("Joe 2") and third ("Joe 3") atomic bombs, which were smaller, lighter, and more powerful than their first (Rhodes 1995). That same fall, with Stanislaw Ulam having found a way to make the Super work and fearful the Soviets might be making equal advances, the Atomic Energy Commission pressed forward with a dedicated thermonuclear weapon program. The Panda Committee, headed not by Teller but by Marshall Holloway, had 1 year to deliver a workable bomb. Teller left Los Alamos in September 1951 in protest. Primarily in response to Teller's unending efforts to establish a second weapons laboratory after leaving Los Alamos, the Atomic Energy Commission opened the Livermore Laboratory in California in July 1952. Although no longer a member of the design team, Teller now had a place to conduct thermonuclear diagnostic research (Rhodes 1995).

> ***"Once the explosion broke through the casing, it expanded in seconds to a blinding white fireball more than three miles across (the Hiroshima fireball had measured little more than one-tenth of a mile) and rose over the horizon like a dark sun; the crews of the task force, thirty miles away, felt a swell of heat as if someone had opened a hot oven, heat that persisted long enough to seem menacing...At its furthest extent, the Mike cloud billowed out above a thirty-mile stem to form a huge canopy more than one hundred miles wide that loomed over the [Eniwetok] atoll. Radioactive mud fell out, followed by heavy rain...The fireball had vaporized the entire [Elugelab] island, leaving behind a circular crater two hundred feet deep and more than a mile across filled with seawater...The explosion vaporized and lifted into the air some eighty million tons of solid material that would fall out around the world."*** **- Rhodes 1995, pp. 508-509**

The nuclear community grew on October 3, 1952 when the British exploded their first atomic bomb ("Hurricane") with a yield of 25 kilotons (Rhodes 1995). Just under a month later, the balance of power shifted dramatically when the United States detonated the world's first thermonuclear bomb. At 7:15 a.m. on November 1, 1952, "Mike" ("The Sausage") exploded with a yield of 10.4 megatons, the equivalent of 20 billion pounds of TNT – twice as much as "all the explosives used during the Second World War" (Rhodes 1995, p. 494).

The transitions of 1952 were not limited to nuclear armaments. Nikita Khrushchev assumed nominal control of the Soviet Union on June 26, 1952 following Stalin's death that March. After losing the New Hampshire primary in the spring of 1952, President Truman withdrew from the presidential race. His successor as the democratic candidate, Adlai Stevenson, lost to Dwight D. Eisenhower that November. Close on the heels of these political transitions, the Soviets detonated their first thermonuclear bomb ("Joe 4") at the Semipalatinsk test site on August 12, 1953. The explosive yield of 400 kilotons, fewer than the Atomic Energy Commission's Ivy series fission bomb, indicated that the Soviets had not discovered or through espionage knew of the Teller-Ulam invention of staging and fission compression (Rhodes 1995). However, because the Soviet device was thermonuclear, it was sufficient to alter again the perception of the balance of power and reinvigorate the arms race.

On January 12, 1954, John Foster Dulles, Secretary of State under President Eisenhower, announced the administration's national policy of "massive retaliation," which anticipated the use of "nuclear weapons, tactical as well as strategic, whenever their use would be desirable from a military standpoint" (Mitchell 1999, p. 36). For the first time, the use of nuclear weapons was placed on a par with that of conventional arms and as an acceptable response to conventional aggression. In part, this policy was foreshadowed in a memorandum from President Eisenhower to Dulles in September 1953. Assessing Soviet aggressiveness, President Eisenhower advised that "our own preparation could no longer be geared to a policy that attempts only to avert disaster during the early 'surprise' stages of a war, and so gain time for full mobilization. Rather, we would have to be constantly ready, on an instantaneous basis, to inflict greater loss upon the enemy than he could reasonably hope to inflict on us" (Eisenhower, quoted in Rhodes 1995, p. 528).

Interestingly, President Eisenhower had told Lewis Strauss, upon appointing him Chairman of the Atomic Energy Commission in May 1953, that his first assignment was to demilitarize atomic energy so the world would no longer have to live with the fear of a nuclear war (Rhodes 1995). President Eisenhower would translate this message into action under his "Atoms for Peace" initiative. However, his defense policy of retaliation had the opposite consequence. It opened the floodgate of nuclear weapons development and deployment. The Atomic Energy Commission budget rose again from $4.9 billion in 1952 to $9 billion in 1955, an amount "exceeding the capital investment of General Motors, Bethlehem and US Steel, Alcoa, Du Pont and Goodyear combined" (Rhodes 1995, p. 561): "More production capacity meant more weapons, which diversified from strategic bombs into tactical and strategic warheads attached to everything from depth charges to atomic cannons to anti-aircraft missiles to ballistic missiles of every range from battlefield to intercontinental" (Rhodes 1995, p. 561).

By 1952, the Joint Chiefs of Staff had identified from 5,000-6,000 nuclear targets, nearly 100 times the 66 cities that had been marked in 1945. Initially, the number of targets was a function of the number of weapons in the nuclear stockpile. But with improved bomb designs extracting greater yields from a lower quantity of fissile material and greatly expanded production capabilities, the number of targets became the driving force and production responded accordingly (Rhodes 1995).

As part of "Project X", General Electric began building the KW and KE reactors – the largest single-pass, water-cooled, graphite-moderated reactors constructed at the Hanford Site – in November 1952 and January 1953, respectively. While the Atomic Energy Commission and General Electric knew by the early 1950s that heavy-water or beryllium would serve as a better moderator, they returned time and again to the graphite-moderated, single-pass, water-cooled reactor design because the nation needed plutonium and the proven design guaranteed results. With the KE and KW Reactors, General Electric carried the design to a second-generation status by dramatically increasing the scale. At 1850 megawatts-thermal, their design was for a power operating level three times that of the C Reactor and nearly nine times higher than the original Manhattan Project reactors. Each core contained 3220 process tubes and more experimental ports than any of the other reactors (Gerber et al. 1997, p. 5.17). The KW Reactor started up in December 1954 and the KE Reactor in February 1955.

Further abrogating the siting requirements Lieutenant General Groves established in 1943, General Electric sited the KW and KE reactors within the same area (100-K Area), less than a mile apart, and approximately halfway between the "two other twin reactor sites, B/C and D/DR." Their decision to co-locate the KW and KE reactors downplayed "the possibility of a simultaneous accident," focusing instead on their "rapid, economical construction" (Findlay and Hevly 1995, pp. 168-169).

At the same time, General Electric continued to increase the power levels of the existing reactors to meet rising production demands. For example, by 1956 "a thorough set of modifications and retrofittings designed to increase coolant flow to the pile [reactor]" had raised the operating level at B Reactor to 800 megawatts-thermal (Gerber 1993a, p. 21). These changes involved all facets of the water treatment, delivery, and circulation system so that 71,000 gallons per minute could be pushed through the reactor (Gerber 1993a). Before the reactors were shut down, their operating levels would be pushed far higher, some to nearly ten times their design rating.

With augmented irradiation capability came the need for increased separations capability. Workers completed construction of the Plutonium-Uranium Extraction (PUREX) Plant (202-A Building) in April 1955. This was the last chemical separations canyon built at the Hanford Site. Like the KE and KW "jumbo" reactors, PUREX would become the "workhorse" separations plant both because of its design capacity and the increased efficiency of its extraction process. PUREX used a continuous solvent extraction process similar to the REDOX process. However, rather than relying on gravity to maintain the countercurrent flow of solutions within the extraction columns, the PUREX Plant used piston-driven pumps to circulate solutions within pulsed extraction columns. Extraction solvents of the two processes also differed. The PUREX process used tri-butyl phosphate rather than hexone. The PUREX process also separated a greater number of products including both weapons-grade and fuel-grade plutonium, depleted and slightly enriched uranium, neptunium, and thorium (see Chapter 2, Section 4 for more information on PUREX). The PUREX Plant went on line in January 1956. Because of the combined efficiency of REDOX and PUREX, T Plant, which by 1953 was reduced to processing only 12 percent of the fuel being irradiated in the 100 Areas, ceased processing altogether on March 20, 1956 and received a new mission (Gerber 1994a).

The REDOX complex also expanded during this period. One of the more significant additions was the Plutonium Concentration Facility (233-S Building) constructed in 1955 "as part of the REDOX Canyon and Service Facility's Phase II capacity increase" (Gerber et al. 1997, p. 5.39). Using an ion-exchange process, operators produced a final solution with a higher plutonium concentration level than had been achieved before. This solution was transported to the 234-5Z Plutonium Finishing Plant for final processing (see the 233-S HAER in Appendix B on the Internet for detailed discussion).

In 1952, operators de-emphasized use of the Rubber Glove Line in the 234-5Z Plutonium Finishing Plant and on March 18 began operations on the world's first large-scale Remote Mechanical Line, known as the RMA Line. The switch from manual to mechanical operations immediately increased production rates by 200 percent. General Electric retired the Rubber Glove Line in May 1953. To accommodate spiraling Atomic Energy Commission production demands, General Electric installed two machining glove boxes on the RMA Line so that two separate weapon pits could be produced simultaneously. By September of that year, Hanford Site plutonium was also being sent directly to the Rocky Flats Plant northwest of Denver for final fabrication. Hanford Site operators met or exceeded all Atomic Energy Commission production quotas for finished pits and non-machined buttons throughout 1954 and 1955 (see Chapter 2, Section 5 for additional details).

Like the uranium recovery process initiated at the UO_3 Facility in 1952, Hanford Site chemists pioneered a process "to recover plutonium from PFP waste streams" in the 234-5Z Building (Gerber et al. 1997, p. 5.40). Beginning in 1951, researchers in the 1706 and 321 Buildings, and later the 325 Building, contributed to the development of the RECUPLEX (RECovery of Uranium and PLutonium by EXtraction) process. RECUPLEX, a dissolver/solvent extraction process conducted in three glove boxes, coaxed plutonium from solid and liquid waste streams derived from the plutonium finishing operations. From July 1955 to May 1962, RECUPLEX operators provided highly concentrated plutonium-nitrate solutions to the RMA Line for conversion to plutonium metal, thereby adding hundreds of kilograms of plutonium to the U.S. nuclear arsenal (see Chapter 2, Section 5).

In 1955, General Electric also added the 224-UA Building, containing "six continuous-action calciners," to the UO_3 Facility (224-U Building) to "improve powder- and waste-handling" capabilities for the Uranium Metal Recovery program (Bailey and Gerber 1997, p. 13). Liquid uranium nitrate, obtained initially from REDOX and subsequently from PUREX as well, was now concentrated in the 224-U Building and converted to uranium oxide powder in the 224-UA Building (Gerber et al. 1997, p. 5.35).

Given increased production efficiency and demands, workers constructed twenty-one additional single-shell, high-level liquid waste storage tanks (241-SX and 241-A Tank Farms) in the 200 Area between 1953 and 1955.

General Electric opened five new research and development laboratories and a second test reactor facility within the 300 Area at the Hanford Site between 1952 and 1955. In Findlay and Hevly's opinion (1995, p. 146), research at the Hanford Site fulfilled two overriding objectives, the "steady development of graphite-moderated reactors" and the discovery of "measures to avoid production losses." "Hanford's first eight reactors were graphite and water-moderated, water-cooled uranium piles [reactors]; they were recognized by the end of World War II as dead ends in terms of the development of nuclear reactor technology. But they remained throughout the Cold War as cost-effective, reliable sources of plutonium, and ironically the U.S. strategy of a high-technology defense mandated that Hanford stick to a relatively low-technology method of producing plutonium" (Findlay and Hevly 1995, p. 141).

The largest of the research and development laboratories was the 325 Radiochemistry Building, dedicated to chemical separations process improvement. As completed in 1953, this facility contained eight hot cells, "each 6 feet by 6 feet by 5.5 feet (deep) surrounded by 2.5-foot-thick concrete walls with stainless steel liners [designed] to safely house and handle multicurie-level chemical development work with high-activity substances" (Gerber 1993b, p. 21). With its completion, much of the research formerly conducted in the 3706 Building transferred to the 325 Building. Scientists in the 325 Building worked to improve the PUREX, REDOX, and UO_3 processes; to lower the activity levels of chemical separations waste; and to develop the RECUPLEX process (Gerber 1992b, 1993b).

The 326 Pile (reactor) Technology Building, completed in 1953, had two primary missions, one relating to reactor physics, the other to metallurgy. "Experimentation with higher power levels, new lattice configurations, and cooling methods and trials with varied materials and designs in process tubes and fuel elements were proceeding quickly and simultaneously" (Gerber 1993a, p. 27). Continuing the research initially conducted in the 189-D Building, scientists in the 326 Building pushed the leading edge in Approach-to-Critical studies in developing lattice configurations that not only increased productivity but also improved safety. Lattice configuration, "the spacing of fuel tubes and moderators within the pile [reactor]," was of critical importance because their arrangement was the primary determiner of reactivity and, therefore, power (see the 305-B ExHPIF in Appendix B on the Internet). In addition to lattice configuration, scientists also examined reactor components and fuel elements to effect improvements using "[s]olid-state nondestructive examination (NDE) methods." Another area of metallurgical research included analyses to improve fuel element jacketing and thereby reduce fuel failures (Gerber 1993a, p. 27).

To accelerate reactor design studies and relieve the burden then being placed on the 305 Test Pile (reactor), General Electric completed the subsurface component of the 305-B Experimental Test Reactor Building in 1954. Two small reactors were housed in this below-grade structure. Both were dedicated to improving lattice configuration. The Physical Constants Test Reactor (PCTR) began operations in October 1955. Physicists used the graphite-moderated PCTR to test experimental lattice designs and measure their efficiency. The second test reactor, the Thermal Test Reactor (TTR), a 5-foot graphite cube, was acquired in 1954 from the Knolls Atomic Power Laboratory but was not placed in service until 1957-1958 (Gerber 1992c, p. 72). Physicists used the TTR "as a neutron source for an external thermal column, as a low power irradiation facility, or for danger co-efficient measurements" in conducting a variety of exponential reactor physics experiments (see the 305-B ExHPIF in Appendix B on the Internet).

While scientists worked primarily with unirradiated materials in the 326 Building, scientists in the 327 Radiometallurgy Building (operational in 1953) worked with "hot" materials to understand and resolve the problems such as fuel element distortion, uranium growth, and process tube corrosion that accompanied increased power levels.

"Irradiated materials, including ruptured or failed fuel rods [elements] containing plutonium and fresh fission products, were examined while they were "green" (i.e., when they had experienced very little decay or stabilization time). Destructive examination of an irradiated fuel rod called for drilling a pinhole from which was collected a fission gas sample. Next the element was sawed open, and a metallurgical sample was ground and polished for analysis. Irradiated process tubes and other reactor components likewise were cut, ground, and polished to produce workable samples." **- Gerber 1993a, p. 31**

Scientists conducted destructive and non-destructive analyses in the 327 Building's eight hot cells and two basins. The 327 Building replaced the 111-B "Test Building" where Hanford Site scientists had first initiated studies "to determine the 'nature and causes of dimensional instability' in fuel elements and the effects of irradiation stresses on pile [reactor] materials" (Gerber 1992c, p. 154).

Concurrent with these studies on materials and process improvement, General Electric built the 329 Biophysics Laboratory in 1953 "to support the pioneering...environmental monitoring and bioassay programs that were developed at the Hanford Site during the 1940s and 1950s" (Gerber 1993a, p. 24). Scientists working in this facility developed new radiation counting procedures as well as the instrumentation necessary to conduct the analyses. They examined air, water, soil, vegetation, and terrestrial and marine wildlife samples from areas within and adjacent to the Hanford Site to as far away as the Pacific Proving Grounds. Hanford Site scientists also played "a major role in monitoring and counting fallout from atmospheric nuclear bomb tests conducted by the United States and the USSR" (Gerber 1993a, p. 24).

To support these laboratories, General Electric built the 328 Mechanical Development Building to fabricate and service the "specialized and intricate apparatus and equipment" that the biologists, chemists, metallurgists, and physicists working in the 325, 326, 327, and 329 buildings used. The only repairs not conducted in the 328 Building were those involving contaminated equipment that had to remain within the "hot" laboratories. The 328 Building, opened in 1952, replaced the Manhattan Project 3717 Instrument Shop and the two 3722 Area Shops (Gerber 1993a, p. 47).

The 618-10 Burial Ground... ***"consisted of trenches and rows of burial caissons known as "pipe fields." The caissons were made of 5 to 6 open-bottomed 55-gallon drums welded together and buried upright. From the mid-1950s until about 1960, solid radioactive wastes were collected from operations buildings in cardboard containers and then stored in lead pans known as "gunk catchers" and transported to 300 North [618-10] in shielded "load luggers." The cardboard waste containers then were dropped from the gunk catchers down the caissons, and the holes were filled with sand and dirt until radiation levels declined to a safe or "tolerance" reading. If radiation levels could not be reduced to tolerance ranges, concrete was poured down the hole until such levels were achieved."*** **- Gerber 1993a, p. 59**

Liquid radioactive wastes from these laboratories were piped to sampling tanks in the new 340 Retention and Neutralization Building through a complex of "single-walled stainless steel pipes" known as the Radioactive Liquid Waste System. Technicians sampled the waste within these tanks and diverted wastes with no detectable radioactivity to one of two newly constructed Liquid Waste Trenches (307 Structures). Waste registering higher than allowable release limits was pumped into holding tanks to await tanker truck transport to 200 Area cribs (Gerber 1993a, p. 57). High-level radiochemical and radiometallurgical solid wastes, primarily from the 325 and 326 buildings, were disposed of first to the 618-2 Burial Ground in 1953 and then to the 618-10 Burial Ground, located more then 4 miles northwest of the 300 Area, from 1954-1964.

Following the expansions of 1949-1955, the Hanford Site produced more than twice as much weapons-grade plutonium per year as the Savannah River Site through 1967 (Figures 6 and 7 in DOE 1996b). It is likely, though unsubstantiated pending release of production amounts from non-U.S. facilities, that the Hanford Site produced more weapons-grade plutonium than any other complex in the world.

PEAK PRODUCTION: 1956-1965

The Killian report, named after James Killian, President of the Massachusetts Institute of Technology, whom President Eisenhower appointed to head the Technical Capabilities Panel in 1954, reached Eisenhower early in 1955. The report recommended "crash programs on intercontinental ballistic missiles (ICBM), intermediate-range ballistic missiles (IRBM) based in Europe and on submarines, new military communications systems, intelligence systems, dispersal of Strategic Air Command (SAC) bombers with some portion on continuous airborne alert, and new antiballistic missiles to destroy (as yet undeveloped) Soviet missiles" (Mitchell 1999, p. 37). In short, the report reflected the growing paranoia of a Soviet surprise attack and the measures necessary to survive and counter it. This fear was amplified in 1956 when Khrushchev warned the West on November 17, "We will bury you."

In the spring of 1957, President Eisenhower asked H. Rowan Gaither to lead a panel of the Science Advisory Committee of the Office of Defense Mobilization in evaluating civil defense needs. Gaither's deputy, Robert C. Sprague, was successful in expanding the scope to include investigation of a "ready and effective second-strike capability" as well. That fall the Soviets placed the first man-made object into outer space with the launch of Sputnik I on October 4, 1957 (Mitchell 1999). "Sputnik, meaning "fellow traveler" in Russian,...was technologically crude and uninteresting to those involved in the U.S. space program, and had no military value, but the nation was horrified, in a state of near panic" (Mitchell 1999, p. 38).

The Soviets now appeared capable of delivering nuclear warheads in minutes from space rather than in the hours required to deliver bombs by conventional aircraft. The United States had not demonstrated this ability. The Gaither report, issued to President Eisenhower on November 7, 1957 and largely authored by Sprague and Paul Nitze, played up the "spectacular progress" made by the Soviets with respect to nuclear weapons and asserted that the Soviets had "probably surpassed" the United States in missile technology. The report recommended that an additional $44.2 billion be added to the defense budget over 5 years to address this "missile gap." In briefing Eisenhower, Sprague did not believe the President understood the gravity of the Soviet threat. However, Mitchell argues that "Eisenhower had the military background to know that wars tend to be initiated during periods of tense international crises, rather than bolt-from-the-blue scenarios, especially one of such magnitude as to involve nuclear weapons." Displeased with President Eisenhower's equivocal response to the report, "some members of the Gaither committee leaked the highly classified report to the press, creating the sensationalism and public outcry that they thought would bring action" (Mitchell 1999, p. 40).

By May 1958, the Soviets had also launched Sputnik II and Sputnik III, and "near-hysterical descriptions of Soviet nuclear capabilities and threats" figured prominently in the national and highly partisan debate leading to the 1960 presidential election (Mitchell 1999, p. 41). Based on high-resolution aerial photographs taken by U-2 spy planes and knowledge that the U.S. military missile program would be operational by 1960, President Eisenhower knew that the claims of Soviet superiority and the charges being leveled against his administration of providing a "second-best defense" were unfounded. However, his inability to release this top-secret information may have contributed to Nixon's defeat. "Ironically," Mitchell (1999, p. 43) points out, "this two-year period (1958-1959) of near-hysterical demand for nuclear weapons...was in fact the two-year period of greatest U.S. stockpile growth in the entire history of the arms race."

Having won the election, John F. Kennedy learned the true nature of the U.S. nuclear weapons force as well as intelligence estimates of the vastly over-extolled Soviet stockpile. However, his campaign had raised public expectations and created a momentum for increased armament. Of more immediate concern, he discovered that the Navy and the Air Force were engaged in a power struggle to shape and control the nuclear weapons program.

With the launch of the U.S.S. George Washington, the world's first submarine both nuclear-powered and armed with sea-launched ballistic missiles in May 1960, Admiral Arleigh Burke recommended that the administration adopt a deterrence-based strategy using "submerged nuclear-powered submarines with free roam of two-thirds of the earth's surface." The Air Force, on the other hand, put forward a war-fighting strategy, developed by RAND analyst William Kaufmann, that relied upon "more accurate land-based missiles and aircraft-delivered bombs" (Mitchell 1999, pp. 45-47). Robert McNamara, Secretary of Defense under President Kennedy, had commissioned a study earlier from his staff to investigate strategic alternatives. This report was delivered shortly after both the Navy and Air Force had briefed McNamara on the strengths of their plans and the weaknesses of the other's. Acting on the recommendations made by his staff, McNamara established "...a new nuclear war plan that would emphasize destroying enemy military forces, avoiding their cities and protecting a ready reserve force aimed at their cities but withholding this force as a bargaining lever with which to coerce the enemy to end the war and come to terms favorable to the United States" (Mitchell 1999, p. 49).

In adopting this policy, the Kennedy administration largely aligned itself with the Air Force's position that the United States could not only fight a nuclear war but win it. McNamara referred to this policy as "flexible response." To succeed in these policy objectives, the United States would have to continue investing heavily in the nuclear stockpile. "Swept up in a political current of their own making, both Kennedy and McNamara approved additional spending on nuclear weapons and delivery systems for what McNamara would later admit was an already over-built stockpile" (Mitchell 1999, p. 51).

Two pivotal events in the early 1960s set the temperature of the Cold War for the next two decades: the erection of the Berlin Wall and the facedown over nuclear missiles in Cuba. In "practicing diplomacy by nuclear threat" (Rhodes 1995, p. 569) throughout the 1950s, the United States and the Soviet Union gained experience in intimidating each other across the globe. The two superpowers rattled missiles and bombers in closing out the Korean War in 1953, at the Suez Canal in 1956, and over Lebanon and the islands of Quemoy and Matsu in 1958. The world had become divided ideologically, and nowhere would this be demonstrated more clearly than in Berlin. Three years after demanding that the forces of the North Atlantic Treaty Organization leave West Berlin and to stem the "hemorrhage of East Germans defecting to the West" (Rhodes 1995, p. 570), Khrushchev erected the most visible and psychological symbol of the Cold War in 1961 – the Berlin Wall. Indeed, the destruction of the Berlin Wall on November 11, 1989 generally marks the end of the Cold War.

The hottest point of the Cold War occurred during the week of October 22-28, 1962. On October 14, a U-2 flight photographed Soviet/Cuban construction of medium-range ballistic missile (MRBM) sites in western Cuba. Khrushchev meant to counter the U.S. placement of fifteen Jupiter MRBM missiles in Turkey by deploying Soviet MRBMs in Cuba. In his own words, Khrushchev maintained, "We had no desire to start a war. On the contrary, our principal aim was only to deter America from starting a war" (Khrushchev, quoted in Rhodes 1995, p. 570). But the war he sought to avoid came closer to occurring that week than at any other point in time.

While Kennedy and Khrushchev "began an exchange of belligerent messages" (Rhodes 1995, p. 572), the U.S. military, by presidential order, moved from Defense Condition (DefCon) 5 to DefCon 3 with President Kennedy's nationally televised announcement of the crisis and his intention to blockade Cuba on October 22. Following that address, the nation was "gripped by dread of a nuclear holocaust. Schoolchildren were practicing duck-and-cover drills" (Thomas 2000, p. 55). By October 24, the military stood at DefCon 2, a state of full alert, and war seemed imminent. The crisis peaked that day as the U.S. Navy stopped two Soviet freighters at the quarantine line. Upon receiving the news that the blockade had held, Secretary of State Dean Rusk issued one of the most oft-quoted phrases of the Cold War, "We are eyeball to eyeball, and I think the other fellow just blinked" (Thomas 2000, p. 56). The crisis concluded on Sunday, October 28, with Khrushchev's announcement to the Soviet Presidium that "In order to save the world, we must retreat" (Thomas 2000, p. 59). "The Soviet Union never went to full nuclear alert in all the years of the Cold War. After the Cuban missile crisis, the United States never did again. Nor did the two nations ever again directly confront each other" (Rhodes 1995, p. 576).

> ***"Security depends upon assuming a worst plausible case, able to absorb the total weight of nuclear attack on our country – on our retaliatory forces, on our command and control apparatus, on our industrial capacity, on our cities, and on our population – and still be capable of damaging the aggressor to the point that his society would be simply no longer viable in the twentieth-century terms. That is what deterrence of nuclear aggression means. It means the certainty of suicide to the aggressor, not merely to his military forces, but to his society as a whole."*** **- McNamara quoted in Mitchell 1999, p. 54**

Lee Harvey Oswald assassinated President Kennedy on November 22, 1963. Lyndon Johnson assumed the presidency that day retaining McNamara as his Secretary of Defense. This continuity was meaningful because McNamara had an opportunity to rethink the war-fighting strategy while reflecting on the events in Berlin and Cuba. McNamara's belief in an ability to inflict significant damage on Soviet nuclear forces sufficient to avert massive U.S. civilian casualties through retaliatory strikes began to wane with the Soviet deployment of submarine-launched missiles and the confirmed existence of their land-based silos in 1962. His resolve diminished further with receipt of mathematician Glenn Kent's systems analysis review of "flexible response" in January 1964. Kent concluded that "offense will always win over defense" in a nuclear war and will do so "at a lower cost." He demonstrated, conclusively to McNamara at least, that "fighting and winning a nuclear war is virtually impossible"(Mitchell 1999, p. 53). Given these realizations, McNamara effected a shift in U.S. policy to a deterrence-based strategy referred to as "assured destruction."

Mitchell (1999, p. 55) argues that this change in philosophy, together with the diversion of defense funds to the emerging war in Vietnam, President Johnson's overall frugality, and his focus on domestic policy led to the "first ever leveling-off and eventual decline in the U.S. nuclear weapon stockpile." In McNamara's view, the United States had far surpassed the number of weapons necessary to ensure the destruction of the Soviet Union. In his State of the Union Address, delivered on January 8, 1964, President Johnson announced a number of cutbacks for the Atomic Energy Commission – the closure of three Hanford Site plutonium production reactors was among them (Mitchell 1999).

> ***"Weapons complex configuration and weapons design and manufacturing processes in the U.S. [had] changed substantially from the Manhattan Project era. Laboratories and production plants developed better technologies to increase their capabilities, output, and efficiency. The weapons themselves [had] evolved considerably, becoming smaller, lighter, more powerful and versatile, safer, and more reliable. The federal government centralized the weapons complex in the early 1950s. By the mid-1960s, stockpiles of some key weapons materials became plentiful enough that the complex ceased producing them."*** **- DOE 1997e, p. 117**

RISE AND FALL OF DEFENSE PRODUCTION

Site Name:	Hanford Works - January 1947 to December 1974
Site Manager:	James E. Travis - June 1955 to June 1965
Responsible Agency:	Atomic Energy Commission - January 1947 to December 1974
Site Contractor:	General Electric Company - September 1946 to January 1964

Essentially, the military used both the "massive retaliation" and "flexible response" strategies "to justify larger and larger procurement budgets, and consequently more and more nuclear weapons" (Mitchell 1999, p. 50). With "assured destruction," McNamara eliminated "damage limitation and counterforce as a military rationale for more weapons" since a "still greater nuclear superiority could not be theoretically justified" (Mitchell 1999, pp. 53, 55). Figure 1.7 shows the effects of these policy decisions. Atomic Energy Commission production of weapons-grade plutonium on the Hanford Site rose dramatically from 1956 to 1960, reached a sustained peak between 1960 and 1965, then dropped off from 1966 through 1969 even more precipitously than it had risen.

In 1961, the year before the standoff over missiles in Cuba, the Hanford Site workforce produced the most weapons-grade plutonium of any year—almost 4500 kilograms (Figure 6 in DOE 1996b). In the Hanford Site's peak production years of 1956-1965, the workforce produced 37,146 kilograms of weapons-grade plutonium—almost 70 percent of the total amount produced between 1945-1989.

General Electric continuously raised the operating levels of the production reactors between 1956 and 1964 to meet the ever increasing military demands for plutonium. By the time the Atomic Energy Commission ordered closure of the original Manhattan Project reactors, they had attained operating levels nearly ten times their initial design rating at 2210 megawatts-thermal (B Reactor), 2165 megawatts-thermal (D Reactor), and 2140 megawatts-thermal (F Reactor). The early post-war reactors were operating at 2140 megawatts-thermal (H Reactor) and 2015 megawatts-thermal (DR Reactor), while the Korean War reactors had attained operating levels of 2500 megawatts-thermal (C Reactor) and 4400 megawatts-thermal (KW and KE reactors) (DOE 1996b). All of these operating levels were several orders of magnitude above their design ratings. The numbers for the "jumbo" reactors might have gone higher still but the Atomic

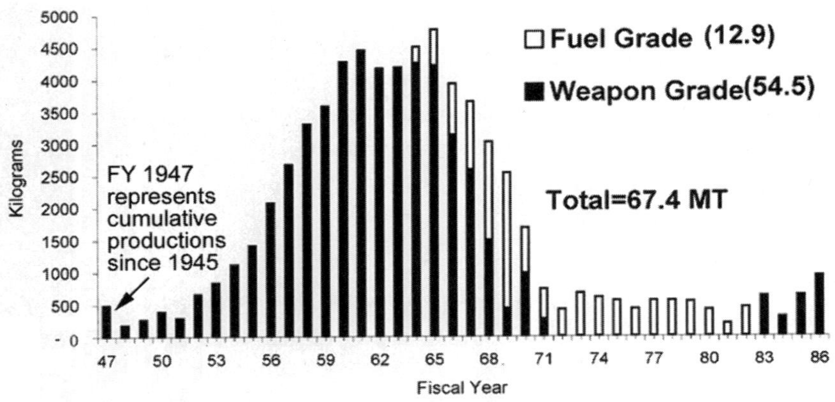

Figure 1.7. *Total Annual Production of Plutonium at the Hanford Site*

Energy Commission imposed "an administrative power limit of 4000 megawatts-thermal at the KE and KW-Reactors" (DOE 1996b, Section 9.1.1). General Electric achieved these operating levels by investing heavily in cooling, electrical, and mechanical improvements to the older reactors and their support facilities as well as modifications in fuel element design and process tube fabrication (Gerber 1993a, pp. 62-63).

Gerber (1993a, p. 62) attributes most of the power increase to a new fuel element design introduced in 1957. In fabricating internally and externally cooled fuel elements (known as I&E fuel), technicians in the 313 Building machined a horizontal hole through the uranium metal to allow water to circulate around and within the fuel element simultaneously as it passed through the reactor. The internally and externally cooled fuel "had a vastly augmented cooling capacity" and "provided less hydraulic resistance to the coolant flow." In a related improvement, General Electric began to use zircaloy-2 to replace the original aluminum process tubes. The higher tensile strength and higher melting point of zirconium made the new process tubes both thinner, thereby increasing heat exchange and water volume, and safer in the event of water loss (Gerber 1993a). Adding to these engineering changes, General Electric also effected an environmental change:

> *"Another innovation that allowed higher operating power levels in the late 1950s and early 1960s, specifically in the hotter summer and autumn months, was artificial cooling of the Hanford Reach of the Columbia River by a controlled spill of cold water from the bottom levels of Lake Roosevelt, the reservoir behind the Grand Coulee Dam. The cooling program began during the summer of 1959, and continued through the summer of 1965. It cooled the Columbia's water in the vicinity of Hanford reactors by an average of 1° C, and by a maximum of 2.3° C during September 1960. The lower inlet water temperature made possible higher power levels in reactor operations, while still not exceeding the maximum bulk outlet temperatures that were allowed."* - Gerber 1993a, p. 62

Pushing the Hanford Site reactors did not come without consequences, however, as the higher operating levels brought more frequent and more serious fuel ruptures, accelerated process tube corrosion and failures, and increased leaks in the effluent pipes among other problems. Hanford Site scientists investigated ways to reduce or eliminate these problems in the 185/189-D Complex and the 300 Area research laboratories. For example as early as January 1952, scientists had initiated "heat transfer and fuel corrosion studies" in the newly converted 185/189-D Building, "a thermal hydraulics and coolant systems development studies facility" (Gerber et al. 1997). According to the 190-D Complex HAER in Appendix B on the Internet, using mock-ups of pipes and tubes loaded with dummy fuel elements, scientists within the 185/189 "Flow Lab" developed "coolant flow information that allowed repeated power level increases in the Hanford reactors."

Exhaust gases also increased in volume as reactor operating levels rose. To address this rise in emissions, General Electric constructed large, primarily below-grade filters in the early 1960s to treat process gases before their release through the reactor stacks. The 117 Air Filter Buildings, consisting of two filter cells divided by an operating gallery, were tied into the existing exhaust ductwork using concrete inlet and outlet tunnels. Engineers designed "an exhaust filtration system that would entrap a small percentage of the pile [reactor] noble gases (kryton-85, argon 39, 41, and 42, and xenon-133), 70 to 95% of the halogens (iodine-131 and bromine-82), and most of the remainder of the particulates and aerosols in reactor gases (including cesium-137, tellurium-129, selenium-79, ruthenium-103/106, and others)" Gerber 1993a, p. 67). Technicians monitored the exhaust gases using instrumentation placed in the 119 Air Sample Buildings. Prior to this time, reactor gases had not been filtered before being dispersed into the atmosphere.

The Atomic Energy Commission placed equal pressure on the Savannah River Site to produce beyond its engineered capacity during this period. With an original design rating of less than 500 megawatts-thermal, "the thermal power level of the reactors was increased to approximately 2500 mw" between 1955 and 1965 (DOE 1996b, Section 9.1.2). Indeed, the Atomic Energy Commission demanded increased productivity throughout its nuclear weapons complex during this interval.

Hanford Site chemical separations operators met the challenge of increased plutonium production from the site's eight enhanced reactors with only two operating canyons (REDOX and PUREX), although the real ability to meet accelerated demand rested primarily with PUREX. Engineers had designed PUREX to operate at a processing capacity of 8.33 metric tons of uranium per day. Within its first year of operations, PUREX staff had pushed operations to 16 metric tons of uranium per day, twice the design rating. By 1958, PUREX accounted for 79 percent of the Hanford Site production total. In the fall of 1960, in just under 5 years of operations, PUREX "reached the point of having processed 22,000 tons of irradiated fuel, thus surpassing the combined total of B and T plants and the REDOX facility for all their years of operations. For short periods of time, PUREX demonstrated the capacity to operate at 3.6 times its design capacity" (Gerber 1996, p. 4-14). This production rate notwithstanding, General Electric re-engineered the process equipment first between 1959 and 1961 and then again in 1965 to bring operations to 33 metric tons of uranium per day (Gerber 1996).

Responding to the need to increase efficiency to further production, General Electric modified the RMA Line in the 234-5Z Building in 1955 so that the "precipitation and purification activities" formerly conducted in the 231-Z Building would now be the first steps undertaken at the Plutonium Finishing Plant (Gerber 1996, p. 5-3). This action concentrated plutonium finishing within one building, thereby increasing efficiency, and eliminated the need for twice handling and transporting plutonium between the chemical separations and metal production phases, thereby increasing safety. Typical of the Hanford Site, the 231-Z Building did not stand idle following this loss of function. General Electric partially restructured the building's interior and established the Plutonium Fabrication Laboratory. Scientists working in this facility conducted "specialized plutonium metallurgy work...associated with research for the development of more sophisticated nuclear weaponry. Research was also conducted in the facility that was associated with plutonium fuels development for the commercial nuclear industry" (Gerber et al. 1997, p. 5.37).

General Electric further accelerated production by installing the Remote Mechanical C (RMC) Line in the Plutonium Finishing Plant in 1960. The RMC Line, "a completely self-contained, remotely operated series of glove boxes," produced both plutonium metal and machined weapons parts. Operating the RMA and RMC Lines in tandem, technicians provided the Atomic Energy Commission with weapons components of "precise, specified dimensions and configuration" (Gerber 1996, pp. 5-2 to 5-4) until December 1965 when the AEC shut down machining operations at the Hanford Site and transferred this task to the Rocky Flats Plant.

General Electric sought to contain or control the wastes generated by plutonium finishing operations. The RECUPLEX process was one of the measures they applied. However, a criticality incident in April 1962 resulted in the permanent shutdown of the facility. Prior to this closure, engineers had been designing a "safer and a more flexible version of the RECUPLEX." Completed in December 1963 and operational by May 1964, operators in the Plutonium Reclamation Facility (236-Z Building) used a continuous solvent extraction process to recover and purify plutonium from plutonium-bearing scrap materials generated at the Plutonium Finishing Plant. The resultant plutonium nitrate was then returned to the PFP for processing (Gerber et al. 1997, p. 5.40). Using industrial solid waste rather than process waste as feed material, operators in the 232-Z Incinerator Building burned and/or leached plutonium-bearing waste materials generated across the Hanford Site to recover usable plutonium, which they returned to PFP beginning in January 1962. General Electric also constructed the 242-Z Waste Treatment Facility to recover plutonium from the liquid waste generated in the Plutonium Finishing Plant and contained in the 241-Z Tanks. Recovery operations began in 1963 (Gerber et al. 1997, p. 5.41).

Given the production rates attained by the PUREX Plant, General Electric built four additional single-shell, high-level liquid waste storage tanks (241-AX Tank Farm) in the 200 East Area. These tanks, the last single-shell tanks constructed at the Hanford Site, were available to receive PUREX process wastes from 1963-1964 (Gerber 1992b). The total number of single-shell tanks containing high-level liquid waste now stood at 149 (see Gephart 1999 for current information on liquid waste tanks at the Hanford Site).

Following 2 years of chemical cleaning that had begun with an initial nitric acid wash on April 1, 1956, General Electric placed T Plant into service as the Hanford Site's central decontamination facility in 1958. Although decontamination operations began that year, cleanup continued through 1969 as workers removed most of the chemical separations processing equipment and buried it within solid waste burial grounds near T Plant. In 1959, General Electric added the 2706-T Decontamination Annex to handle "equipment too large to be moved into T Plant or pieces having lower contamination levels than those decontaminated in T Canyon itself" (Gerber 1994a, p. 39). In addition to cleaning Hanford Site equipment and returning it to service, an activity that netted between $100,000 and $200,000 in monthly savings, decontamination workers also processed offsite materials including "high-exposure, irradiated fuel from the Shippingport (Pennsylvania) power reactor" in the late 1970s and the "zeolite beds...loaded with cesium-137 from the Three Mile Island (Pennsylvania) power reactor" in 1983 (Gerber 1994a, p. 42). Limited Operations status was imposed in January 1987 and remained in effect until both the 221-T Plant and 2706-T Decontamination Annex were inactivated in January 1990 awaiting major upgrades.

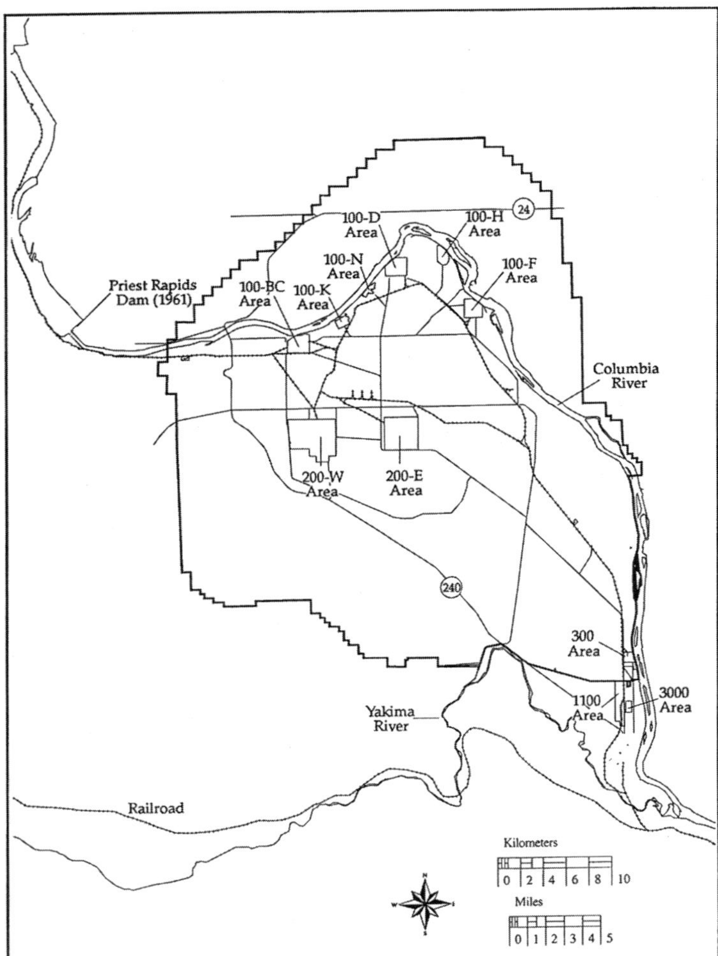

Figure 1.8. *Hanford Site during the Cold War Era Around 1965 When the Site Was Called Hanford Works*

It was during the period of peak plutonium production (the 1960s) that General Electric built the ninth and final graphite-moderated, water-cooled production reactor at the Hanford Site, the N Reactor. Figure 1.8 shows the additional reactor areas, roads, and Priest Rapids Dam added since the time of the Manhattan Project. The Atomic Energy Commission advanced the N Reactor (originally referred to as the New Production Reactor and then just N Reactor) as a model to demonstrate the "technological and economic feasibility of constructing a dual-purpose reactor capable of producing special nuclear materials for weapons and also using the steam for electricity" (Stapp and Marceau 1996, p. 18). Senator Henry Jackson of Washington, a member of the Joint Committee on Atomic Energy, was the driving force behind the Atomic Energy Commission's position. In fact, by arguing in 1956 that there could never be enough atomic bombs, Findlay and Hevly (1995, p. 171) suggest that Jackson used the Senate "to force a reluctant AEC to build a dual-purpose reactor at Hanford." Arguments for and against construction were long and embittered, and in 1958 Congress approved only those funds sufficient to construct a weapons-production reactor, albeit a safer one incorporating new design technology.

Engineers designed another significant difference into the cooling system for the N Reactor. Abandoning the single-pass design for the first time, General Electric incorporated a recirculation configuration. This design change accounted for a number of benefits. First and foremost, "innovation of the light-water recirculation design configuration greatly reduced the direct release of radionuclides to the environment" since the primary cooling water was not directly discharged to the Columbia River. Because recirculation required that the water be pressurized, higher operating levels were attainable with less film deposited in the process tubes due to higher flow rates. Since less water was necessary under pressurized conditions, a smaller water intake system would suffice. Finally, the "recirculation design also enabled the generation of steam to produce electricity" (Stapp and Marceau 1996, p. 23). "By putting the water in the pipes under a pressure of 3206 pounds per square inch, the water could be prevented from turning to steam and superheated water could be run through a heat exchanger to provide steam for a generating plant..." (Findlay and Hevly 1995, p. 170).

> ***"A significant difference from the Hanford [Site]'s earlier reactor was N Reactor's change from a positive-void coefficient design to a far superior negative-void coefficient design. When a steam bubble, or void, develops in any one of the process tubes, the neutron flux increases logarithmically and enhances the possibility of a serious nuclear accident. N Reactor's design incorporated a negative-void coefficient that, in the case of a steam bubble or void in the process tube, tends to shut the reactor down."*** **- Stapp and Marceau 1996, p. 23**

To meet the startup deadline, the reactor operators started up N Reactor on December 31, 1963 and immediately shut it down. Actual startup of operations was in March 1964, and it was several years before it operated in its dual capacity of producing both plutonium and steam for electricity.

The N Reactor, "having benefited from nearly two decades of experience with Hanford's eight older reactors," was fueled by a new type of fuel element fabricated through a new process developed in the 306 Fuel Element Pilot Plant. Constructed in 1956, General Electric used the 306 Building to "pilot process improvements in single-pass reactor fuel fabrication methods" (Gerber 1993b, pp. 7-8). Anticipating the operation of the N Reactor, General Electric expanded the facility in 1960 and assigned its staff the task of developing a fuel element that could withstand increased heat and pressure. Rising to the challenge, Hanford Site metallurgists pioneered a "coextrusion fabrication process" wherein "all the fuel-element components, including the uranium core and all the cladding materials, were cleaned, assembled, and then extruded together. This method provided a better, more uniform bond than the earlier process of jacketing each fuel element separately" (Gerber 1993b, p. 7). Besides the process improvement, N Reactor fuel elements were also composed of two pieces, an inner and outer element, designed as a "tube-in-tube configuration (with a hole running down the entire length of the element)," which "allowed cooling water to circulate optimally around the elements" (Gerber 1993b, pp. 7-8). Machinists began making the fuel elements in the 333 Fuels Manufacturing Building, originally known as the New Fuel Cladding Facility, in 1960 (see Chapter 2, Section 2 for a detailed discussion of the process and facilities).

Because the Hanford Site was the leading producer of weapons-grade plutonium for the Atomic Energy Commission, its air defense remained critical. In 1955, the U.S. Army placed four Nike Ajax missile complexes on the Hanford Site. These guided missiles replaced the AAA gun batteries since the missiles provided "an air defense system with the capacity of engaging high speed aerial targets at greater ranges than conventional antiaircraft artillery." Three systems were emplaced on the North Slope, the area north of the Columbia River and south of Saddle Mountain. The fourth system was placed at the base of and atop Rattlesnake Mountain. By the late 1950s, a more advanced Nike missile, the Hercules, replaced the Ajax series. This replacement was short-lived, however, as "the development of the intercontinental ballistic missiles had rendered Nike missiles obsolete, and the Nike sites were abandoned when Camp Hanford was deactivated in 1960 and closed in 1961" (Gerber et al. 1997, p. 5.47).

Budgetary considerations and an abundance of formerly scarce nuclear materials resulted in a shift from redundant sites to single-mission sites and a shutdown of some sites and materials production facilities in the mid-1960s (DOE 1997e, p. 18). For example, the Atomic Energy Commission eliminated the production of plutonium weapons parts on the Hanford Site in 1965 leaving the Rocky Flats Plant as the sole source of plutonium components (DOE 1997e, p. 26).

More significant than the loss of a function, the Hanford Site began to loose its primary mission when the Atomic Energy Commission began shutting down the production reactors. Following President Johnson's announcement of cutbacks, the Atomic Energy Commission ordered the closure of the DR Reactor, effective December 1964, and closure of H and F reactors effective April and June 1965, respectively. The Atomic Energy Commission would issue additional closure orders in 1967, 1968, and 1969.

Emergence of Non-Defense Applications: Atoms for Peace

In December 1953, President Eisenhower announced his "Atoms for Peace" initiative designed to share "technology and nuclear material with other nations, including nuclear materials for research and power reactor programs." Before this could happen, however, Congress had to amend the Atomic Energy Act of 1946. They did so in 1954. The amendment allowed "civilian peaceful use, though not ownership, of special nuclear materials" and further allowed "U.S. assistance to foreign countries developing peaceful nuclear programs" (DOE 1996b, Section 3.1). With this amendment, President Eisenhower acted on his own directive to Lewis Strauss, delivered on Strauss' appointment as chairman of the Atomic Energy Commission, to develop alternatives other than military applications for atomic energy. In 1957, the International Atomic Energy Agency was formed "to promote peaceful nuclear energy and control nuclear material" (DOE 1996b, Section 3.4). "The mid-to-late 1950s was a very expansive time in United States nuclear history. The belief was widespread that atomic energy would be the fuel of the future adopted to virtually all civilian and industrial power needs" (Gerber 1993b, p. 33).

In 1949, the Atomic Energy Commission had established the National Reactor Testing Station (now known as the Idaho National Engineering and Environmental Laboratory) northwest of Idaho Falls, Idaho. This facility offered an isolated location to design, build, and test prototype nuclear reactors. In the 1950s and 1960s, the Atomic Energy Commission charged the facility staff with developing nuclear power for peacetime use and finding methods to effectively deal with radioactive waste (DOE 1997e). The Atomic Energy Commission had not selected the Hanford Site for the initial Materials Testing Reactor in 1952 because the reactor "was designed to be used by researchers for research; a secondary group and a secondary function at Hanford, as its role was defined by the AEC." The Atomic Energy Commission built the Materials Testing Reactor in Idaho because they viewed the Hanford Site as a "professional workplace dominated by engineers rather than scientists, and, among the scientists, by the most industrially-oriented disciplines" (Findlay and Hevly 1995, pp. 150-151). By 1960, the Hanford Site scientific community was not content to let this assessment persist.

Knowing both the importance of and the opportunities afforded in developing, testing, and fabricating alternate fuels for commercial applications, General Electric completed construction of both the 308 Plutonium Fabrication Pilot Plant (PFPP) and the 309 Plutonium Recycle Test Reactor (PRTR) in the 300 Area in 1960. "Anxious to participate in this new atomic frontier," Hanford Site scientists then "embark[ed] on a large research effort...directed at demonstrating the effectiveness of various plutonium oxide and mixed (plutonium, uranium, and other metallic) oxide fuel blends" (Gerber 1993b, pp. 33, 36). Scientists working on the Plutonium Fuels Utilization Program developed and fabricated plutonium-based fuel elements in the PFPP, first in metallic form and then as "ceramic fuel blends ('cermets')," from 1960 through 1965 (Gerber 1993b, p. 36). They tested these "developmental fuels" in the PRTR, "a heavy water moderated, 100 megawatt experimental reactor" used not only to test alternative fuels but also "process tubes, and the physical parameters of plutonium fuels" (Gerber 1992b, p. 40). Design engineers had located the PRTR partially below grade within the 309 Building containment vessel. In addition to three large, below-grade cells containing reactor process, experimental, and instrumentation equipment, the 309 Building also contained a subsurface Hot Fuel Examination Facility, an irradiated fuel storage pit, and a fresh fuel storage pit (Gerber 1993b, p. 33). Scientists had pre-tested many of the PRTR components using mock-ups in the 185/189-D Thermal Hydraulics Laboratory. Between 1968 and 1972, physicists used the High Temperature Lattice Test Reactor, a 10-foot cubed, liquid nitrogen gas-cooled, graphite-moderated reactor housed in the 318 Building, "to test powdered, pelletized and other experimental fuels at high temperatures" (Gerber 1992b, p. 40).

In 1960, General Electric also completed construction of the 325-A High-Level Radiochemistry Annex to further "innovative, non-defense" research programs. This addition to the 325 Building provided three large hot cells within which scientists conducted "pioneering" isotope separation work for the National Aeronautical and Space Administration and medical researchers. Using high-level liquid waste as source material, chemists "fractionized specific isotopes [strontium-90, cesium-130, curium-244, americium-241, and promethium-147 being most in demand]...by ion exchange, carrier precipitation, solvent extraction, and various combinations of these and other methods" (Gerber 1993b, p. 23). Chemists in the 325 Building complex also investigated methods to encapsulate strontium fluoride and cesium chloride and to vitrify high-level wastes in support of waste treatment and waste management objectives (Gerber 1993b).

The Atomic Energy Commission authorized the 324 Chemical Materials Engineering Laboratory (constructed between 1964 and 1966 and originally named the Fuel Recycle Pilot Plant) to provide "chemical reprocessing and metallurgical examination" services for fuel elements irradiated in the PRTR. A second mission was to "house the Waste Solidification Engineering Project, one of the first high-level waste vitrification demonstration programs in the world" (Gerber 1993b, pp. 48-49). Therefore, design engineers provided both hot cells and laboratories for radiochemical and radiometallurgical research as well as numerous "cold" laboratories, a maintenance and fabrication shop, storage vaults, and waste vaults throughout the multi-level building. The support mission for PRTR ended quickly, however, when PRTR shut down in 1969 following the Atomic Energy Commission's decision to terminate mixed oxide fuels research (Gerber 1993b, p. 36). Several "contamination events" associated with PRTR operations likely contributed to this decision to shut down the test reactor (see Gerber 1993b, pp. 35-36 for discussion). On the other hand, vitrification research, using "[e]xtremely high-level waste, sometimes 'spiked' with extra strontium-90," or, post-1975, "spent commercial reactor fuel," continued until 1980 (Gerber 1993b, p. 51).

Hanford Site scientists began to prepare "plutonium oxides in specified, tailored blends for commercial nuclear experiments and development" within the Plutonium Finishing Plant in September 1964. The European Atomic Energy Community (EURATOM) was the customer. Between 1966 and 1968, non-defense work climbed to nearly 30 percent of the workload at the Plutonium Finishing Plant as "large-scale interaction with the commercial nuclear industry, nuclear research customers and foreign nuclear customers became a reality." After 1968, the RMA Line, closed since 1966, became known as the "oxide line" because of its use in non-defense work (Gerber 1996, pp. 5-4 to 5-5).

COLD WAR ENDS: DIVERSIFICATION AT THE HANFORD SITE

The end of the Cold War was not nearly as simple or as linear as the following discussion may suggest. The geopolitics were very complex. The period from 1960-1990 witnessed the rise of the Chinese as a nuclear power (October 16, 1964) and thermonuclear power (June 17, 1967), the emergence of oil as a defining force in world commerce, numerous proxy wars within the "Third World," the introduction of Soviet troops in Afghanistan, and a host of other domestic and international events that shaped and contorted the times. The interested reader is directed to *The Cold War, A History* by Martin Walker (1995) and the sources the author draws on for a detailed accounting.

The Cold War may have ended with the breakup of the Soviet Union in 1991, but glimpses of its demise can be seen in retrospect over a 30-year period in efforts to control the deployment and use of nuclear weapons. Having walked to the brink of nuclear war and measured its consequences, representatives of the United States, Great Britain, and the Soviet Union met during July 1963 to negotiate a comprehensive test ban treaty. When onsite inspections became an insurmountable obstacle, negotiators turned to a limited test ban that would outlaw nuclear tests in the atmosphere, under water, and in outer space. President Kennedy signed the Limited Test Ban Treaty on July 25, 1963. He addressed the American public the following night and expressed his desire to effect a thaw in the Cold War.

> *"In an age when both sides have come to possess enough nuclear power to destroy the human race several times over, the world of communism and the world of free choice have been caught up in a vicious circle of conflicting ideology and interest. Each increase in tension has produced an increase of arms; each increase of arms has produced an increase of tension...Yesterday, a shaft of light cut into the darkness...For the first time an agreement has been reached on bringing the forces of nuclear destruction under international control – a goal*

first sought in 1946 when Bernard Baruch presented a comprehensive control plan to the United Nations...If we are to open new doorways to peace, if we are to seize this rare opportunity for progress, if we are to be as bold and farsighted in our control of weapons as we have been in their invention, then let us now show the world on this side of the wall and the other that a strong America also stands for peace. There is no cause for complacency." - Kennedy 1963

On July 1, 1968, sixty nations signed the Nuclear Arms Nonproliferation Treaty, and on November 17, 1969, the United States and Soviet Union began Strategic Arms Limitation Talks.

The U.S. Senate ratified the Limited Test Ban Treaty by an 80 to 14 vote on September 24, 1963. The Treaty went into effect on October 11. President Kennedy had come a long way from his campaign rhetoric.

Under President Johnson, McNamara had found a way to reverse defense spending on nuclear weapons. In fact, he approved research and development for only one new weapon system, the Multiple Independently-targetable Reentry Vehicle (MIRV) as a way to "approve military requests for additional weapons without the cost of new delivery systems or missile bases" (Mitchell 1999, p. 55). President Johnson, notes Mitchell (1999), was politically able to propose reductions because of the significant increase in weapons procured during the short Kennedy administration. Ironically, deployment of MIRV under the following Nixon administration, together with a shift in national policy, led to an increase in the nuclear stockpile between 1971 and 1974.

On January 10, 1974, James Schlesinger, Secretary of Defense under Richard M. Nixon, formally set aside assured destruction when he announced "that the old policy...was no longer adequate for deterrence." The United States needed "a set of selective options against different sets of targets," primarily military rather than civilian. Called "flexible targeting," this policy emphasized "controlled escalation" (Mitchell 1999, p. 60). Prior to this policy shift, the military had equipped Minuteman III and Poseidon missiles with MIRV warheads in 1970 and 1971, respectively. Following this announcement, the military again pushed for new weapons, in particular the B-1 bomber and the MX missile. Mitchell (1999, p. 61) points out that the Nixon administration buildup, while increasing the number of weapons in the short term, actually reshaped the nuclear stockpile in the years following 1974. "The number of weapons retired from the stockpile, mostly tactical weapons, far outpaced the number of new weapons entering the stockpile, mostly strategic weapons." He argues that "the reduction in fissile material production, initiated by President Johnson in the mid-1960s, was beginning to have an impact on the number of new weapons deployed."

On May 26, 1972, the United States and the Soviet Union signed SALT I.

Even as he built up the U.S. nuclear arsenal, however, President Nixon signaled a desire to reduce the economic effects caused by the arms race. In his first inaugural address, delivered in January 1969, President Nixon told the world, "After a period of confrontation, we are entering a period of negotiation. Let all nations know that during this administration our lines of communication are open...With those who are willing to join us, let us cooperate to reduce the burden of arms, to strengthen the structure of peace" (Nixon, quoted in Walker 1995, p. 216). President Nixon was speaking of détente. Walker (1995, pp. 216-219) suggests that the "public perception of détente as a method of easing tensions was...an almost deliberate fraud [since] détente was the continuation of the Cold War in other places, and by more subtle means." However, President Nixon used détente as an opportunity to pursue arms control, which had the duel objective of reducing the defense budget and easing the debt created by the Vietnam War. The Anti-Ballistic Missile (ABM) Treaty, which President Nixon signed in 1972, addressed his inaugural message and represented the first attempt at mutual restraint between the United States and the Soviet Union. This treaty, which limited the "development and deployment of missiles designed to nullify each other's deterrent by shooting them down in flight," was a stabilizing measure that Walker (1995, p. 204) characterizes as the "first formal recognition that each superpower had an interest in protecting the strategic arsenal of the other." In the Strategic Arms Limitation Talks (SALT) that followed, the United States and Soviet Union sought to codify the number of nuclear weapons each could possess.

More direct than his predecessors, President Jimmy Carter, addressing the nation and the world in his inaugural speech in January 1977, stated bluntly: "We will move this year a step forward toward our ultimate goal – the elimination of all nuclear weapons from this earth" (Carter, quoted in Mitchell 1999, p. 62). Unfortunately for President Carter, the Soviet buildup initiated years before by Leonid Brezhnev "bore fruit with the introduction of four new models of strategic missiles in 1974...[and] the deployment of the SS-20 missiles [in Eastern Europe] in 1977" (Walker 1995, p. 244). While President Carter acted on his "moral conviction to rid the world of nuclear weapons" in signing SALT II on June 18, 1979 (Mitchell 1999, p. 62), the Soviets appeared to be gaining superiority (see sidebar box).

"U.S. interests appeared to be coming under threat throughout the world, with Soviet arms sales to Iraq and Syria, Soviet support for the PLO [Palestine Liberation Organization], the advance of the Soviet Union's Vietnamese allies throughout South-East Asia, and the new challenges to the pro-American regimes in Central America. The Kremlin's Cuban allies had established military control in Angola. On the maps of the world which American news magazines began to publish, with countries coloured in red for Soviet influence, the West's retreat was made to look graphically clear." **- Walker 1995, p. 246**

On December 27, 1979 Soviet Troops invaded Afghanistan and assassinated resistance leader Hafizullah Amin.

Dissatisfied from the start with his administration, the Committee on Present Danger, a conservative group of cold warriors, portrayed President Carter as ineffective and blamed him for turning the United States into a "weak and fading super-power" (Mitchell 1999, p. 64). On December 4, 1979, on the advice of National Security Advisor Zbigniew Brezinski, President Carter approved deployment of 200 MX missiles, a MIRV missile containing ten independently-targeted nuclear warheads. "The MX decision, and Carter's approval of the new Trident II SLBM [submarine launched ballistic missile], Pershing II IRBM [intermediate-range ballistic missile], and cruise missile...were measures by an administration perceived as weak, to demonstrate U.S. resolve, not only to the Soviets, but also to the American people" (Mitchell 1999, p. 63). In a classic case of bad timing, followers of the Ayatollah Khomeini seized the U.S. Embassy in Tehran on November 4, 1979 and held the diplomats and staff hostage for over a year thereby negating President Carter's decision in the eyes of the American public and much of the world. When the rescue attempt failed on April 24, 1980, the people handed the presidency to Ronald Reagan, who, like President Kennedy, had campaigned hard on a platform to restore "strong and decisive leadership and a strengthened nation" (Mitchell 1999, p. 65).

On September 22, 1980, Lech Walesa organized the Solidarity union in Poland.

Ronald Reagan "inherited a clear mandate for change, backed by a new American nationalism, and did not back away from the task at hand" (Mitchell 1999, p. 67). He increased spending on defense by 50 percent in his first 5 years in office. Twenty percent of that increase went to nuclear weapons, the remainder to conventional weapons.

Ironically, the "dramatic rearmament of the Reagan years was largely envisioned by the Carter Administration" (Walker 1995, p. 250). However, President Reagan seized the initiative. "Reagan's rearmament was an investment in national security. It was also the test of Reagan's private conviction that the United States could afford an arms race, while the Soviet Union could not. The Soviet Union would either have to renounce the arms race, or bankrupt itself into collapse in the vain effort to keep up" (Walker 1995, p. 267).

To compound the buildup, President Reagan proposed a new Strategic Defense Initiative (SDI), better known as "Star Wars," on March 23, 1983. President Reagan's call for a massive research project to develop a land- and space-based anti-missile system alarmed the Soviets, whose newly empowered Secretary General Yuri Andropov had proposed a series of initiatives for arms reductions, including "a 25 per cent cut in the strategic arsenals of both the USA and the Soviet Union, combined with a nuclear freeze on any new deployments" (Walker 1995, p. 272). It frightened the

Western Europeans, who believed they would become the target of choice for Soviet missiles if the United States were hidden behind a defensive shield and appeared to most observers to violate the mandates of the ABM Treaty of 1972. The SDI was a double-edged sword. It is credited by many as the final straw that broke the back of the Soviet economic resolve to continue the arms race. As Mitchell (1999, p. 67) points out: "Within four years the U.S. was spending approximately $6 billion annually on research and development for a Star Wars system. SDI would prove to be an up in the economic ante of the arms race that Gorbachev was unwilling or unable to match." It also delayed agreements on arms reductions by several years because of President Reagan's reluctance to give up the system. The Soviet Union, believing that the Reagan buildup and SDI were a repudiation of the Andropov "peace offensive" offered in 1982, became "convinced by its own propaganda that Reagan's America was capable of a pre-emptive nuclear attack" (Walker 1995, p. 277). Strategic Arms Reduction Talks (START), initiated in Geneva on June 29, 1982, made little progress by 1983. The Soviet Union then suspended the talks in December 1984.

Mikhail Gorbachev, a member of the new generation that Andropov believed "could help haul the country from Brezhnev's stagnant legacy," became Secretary General in March 1985 (Walker 1995, p. 272). Gorbachev's domestic agenda had three defining characteristics: "Sweeping economic reform with the profit motive, moves towards political democratization, and a much greater freedom for the Soviet press and media" (Walker 1995, p. 284). The essence of his reform movement was captured in two words, *perestroika* (restructuring) and *glasnost* (openness). The Solidarity movement in Poland, human rights activists in Czechoslovakia, and the democratic movements within Russia itself, fueled by the ever widening "peace movement" with its call for "No Nukes," emerged from the darkness into the light under Gorbachev. "Gorbachev's "Novoye Myshlenniye" or New Thinking in international affairs was first spelt out at the Geneva summit with President Reagan in October 1985, when they agreed in principle to work towards a Strategic Arms Reduction Treaty [START] to cut their nuclear arsenals in half. It was amplified in January of 1986, with Gorbachev's detailed scenario for nuclear disarmament by the year 2000" (Walker 1995, p. 290).

Meeting again in Reykjavik in October 1986, President Reagan and Secretary General Gorbachev acted on the exchange of ideas they had had on disarmament since their meeting in Geneva. They "agreed in principle" to eliminate medium-range nuclear missiles from Europe, eliminate all ABMs within 10 years, and eliminate tactical nuclear weapons. However, the "proposal to abolish nuclear weapons tumbled at the last fence, at Gorbachev's insistence that Reagan abandon his cherished SDI, or at least confine the research to the laboratory" (Walker 1995, p. 295). What President Reagan could not do, Congress did. With SDI scaled back, President Reagan and Secretary General Gorbachev signed the Intermediate Nuclear Forces (INF) Treaty in December 1987.

On June 3-4, 1989, Chinese forces violently ended the students' democratic movement in Tienanmen Square.

In December 1988, Gorbachev addressed the United Nations and set in motion the breakup of the Soviet Union. "Force or the threat of force," he said, "neither can nor should be instruments of foreign policy...Freedom of choice is a universal principle. It knows no exceptions" (Gorbachev, quoted in Walker 1995, p. 309). Taking him at his word, the people of Eastern Europe began to reclaim their countries, first Hungary, then Lithuania, Estonia, Latvia, and Poland, all within the first 6 months of 1989. East Germany left the Soviet fold in October 1989, destroying the Berlin Wall on November 10 to remove all doubt. Czechoslovakia followed on November 30 (Walker 1995, pp. 309-313). President George H. Bush and Secretary General Mikhail Gorbachev signed START in July 1991. Following a failed coup by communist hard-liners in Moscow in August, eight of the fifteen Soviet Republics declared complete independence. In an address to the nation on September 27, 1991, President Bush announced that he would destroy the nuclear artillery shells and warheads that had armed the tactical weapons based in Europe; withdraw all the tactical nuclear weapons from surface ships, submarines, and land-based naval aircraft; order the stand down of all strategic bombers and all the ICBMs scheduled for deactivation; and cancel those strategic weapons systems then in development (Mitchell 1999). With these actions, the United States and the Soviet Union ended the Cold War.

ECONOMIC DIVERSIFICATION: KEY TO SURVIVAL FOR THE HANFORD SITE

Site Name:	Hanford Works Hanford Reservation - January 1975 to September 1977 Hanford Site - October 1977 to 1990
Site Manager:	Donald G. Williams - June 1965 to November 1971 Thomas A. Nemzek - November 1971 to September 1973
Responsible Agency:	Atomic Energy Commission Energy Research and Development Administration - January 1975 to September 1977 Department of Energy - October 1977 to 1990
Site Contractor:	Douglas United Nuclear - November 1965 to August 1967 Reactor Operations, Fuel Manufacture
	United Nuclear Industries - September 1967 to April 1973 Reactor Operations, Fuel Manufacture UNC Nuclear Industries, Inc. - April 1973 to June 1987 Reactor Operations, Fuel Manufacture
	Isochem, Inc. - January 1966 to September 1967 Chemical Processing, Waste Management
	Atlantic Richfield Hanford Company - September 1967 to June 1977 Chemical Processing, Waste Management, Management and Integration
	Rockwell Hanford Operations - July 1977 to June 1987 Chemical Processing, Waste Management, Management and Integration
	Westinghouse - July 1970 to June 1987 Hanford Engineering and Development Laboratory, Fast Flux Test Facility
	Battelle - January 1965 to 1990 Pacific Northwest National Laboratory
	ITT Federal Support Services, Inc. - March 1966 to August 1971 Site Administrative Services and Infrastructure

Findlay and Hevly (1995, pp. 72, 78) repeatedly make the point that the Hanford Site was, above all else, a production center whose overriding mission was the "manufacture of as much fissionable material for bombs, as quickly and cost-efficiently as possible." Even as commercial applications began to expand, the Atomic Energy Commission "mandated that Hanford favor certain operations over research, and plutonium production over power production." However, this myopia began to change as defense production fell off rapidly after 1965.

In the mid-1960s, with three reactors already shut down, the Atomic Energy Commission, in concert with community leaders, looked for ways to maintain the viability of the industrial complex. If the Hanford Site were to survive, it clearly had to move beyond defense-related plutonium production. In a strong statement of the restructuring that would be necessary to achieve this objective, the Atomic Energy Commission and General Electric jointly announced the withdrawal of General Electric as the Hanford Site contractor on January 21, 1964. General Electric was not let go because of poor management. Rather, the Atomic Energy Commission acted upon the prevailing perception that a single, large contractor countered effective regional economic development by removing the incentive for other businesses to come (Findlay and Hevly 1995). "In an effort to encourage the use of the site's resources and work force in new ways, the Commission increased its commitment to research activities and power production at Hanford, spread responsibility for managing Hanford's activities among several contractors, and required those contractors to invest in the local economy" (Findlay and Hevly 1995, pp. 217-218).

Finding ways to boost the local economy was imperative as the proportion of weapons-related federal support fell from nearly 100 percent in 1964 to only 25 percent in 1975.

POWER GENERATION

Congress did not approve the power plant that would produce and distribute electricity when it authorized construction of N Reactor in 1958. It voted down a proposal to use the built-in capability for conversion again in 1961. However, when the Washington Public Power Supply System (Supply System) agreed to pay all the expenses for the electrical generation and transmission facility, Congress passed the legislation necessary for a non-federal agency to construct and operate the facility on the Hanford Site on September 26, 1962.

Advocates for the facility, known as the Hanford Generating Plant, had waged a long political struggle against strong opposition from private utilities and the coal industry. Taking the long view, "business advocates saw dual-purpose reactors…as the point of entry for private enterprise into the nuclear arena" (Findlay and Hevly 1995, p. 170). Local advocates saw the Hanford Generating Plant as a way to "link Hanford's production mission to the long-delayed hopes for commercial nuclear power' (Findlay and Hevly 1995, p. 204).

President Kennedy had taken an interest in the Hanford Generating Plant, indeed his administration had placed the conditions on the authorizing legislation, particularly private funding, that won Congressional approval. Suitably, then, in a ceremony open to the public for the first time on the Hanford Site, Kennedy both dedicated N Reactor and initiated construction of the Hanford Generating Plant on September 26, 1963. Noting that it was "exactly one year ago today, after a long and bitter legislative battle, that I was able to sign the Hanford steam plant legislation," Kennedy went on to say that to "have permitted this resource to be wasted would have been in conflict with all the principles of resource conservation and utilization to which we are committed" (Kennedy, quoted in Stapp and Marceau 1996, pp. 18-19).

The Hanford Generating Plant first produced electricity on April 8, 1966. With the addition of the generating facility, N Reactor became the nation's first and only dual-purpose reactor. To produce electricity, operators pumped N Reactor primary cooling water to the 109-N Heat Exchange Building. Circulated through ten parallel heat exchangers, the heat from the process water caused water contained in a secondary system to boil and produce steam. This steam was transferred through overhead pipes to the 185-N Export Powerhouse Turbine Building where it drove two large turbines generating approximately 860 megawatts of electricity that was passed through the 155-N Export Power Switchyard to the Bonneville Power Administration power grid (UNI n.d., DOE 1997a).

In 1957, the Shippingport Reactor in Pennsylvania became the nation's first commercial nuclear-powered electrical generating facility. It was the first of many planned and constructed across the nation in the late 1950s and 1960s. Acting on the success of the Hanford Generating Plant, and in partial fulfillment of a bold local plan to establish a "nuclear power park" at the Hanford Site as a model for resolving the nation's "energy crisis" of the late 1970s, the Supply System proposed constructing three "electricity-generating reactors" – the first of up to twenty. Construction of the reactors began in 1980. However, the project was a huge financial drain and the Supply System terminated construction on two of the reactors in 1982. In 1983, Supply System "defaulted on its bonds, the largest default on municipal bonds in financial history." Only one of the Supply System reactors was completed and began producing electricity in 1984 "seven years later than initially planned" (Findlay and Hevly 1995, p. 296). In 1990, this reactor was the only active nuclear reactor at the Hanford Site.

SERVICE TO THE NUCLEAR AND SCIENTIFIC COMMUNITIES

As the Hanford Site production reactors shut down, fuel manufacturing and chemical separations activities decreased so that research and development programs now constituted the major mission. The research and development facilities included numerous laboratories and large test facilities in support of peaceful uses of plutonium, reactor fuels development, liquid metal technology, fast flux support, gas-cooled reactor programs, and life sciences programs.

Construction and operation of the Plutonium Recycle Critical Facility (PRCF) in 1963 indicated the efforts undertaken by the Atomic Energy Commission to promote local economic development and uncouple the Hanford Site from its historic reliance on national defense. Completed as a second addition to the 309 Building, General Electric designed PRCF to support PRTR operations. However, scientists soon began using PRCF to test experimental lattice systems in support of light-water-cooled reactor design. "A commercial license was obtained from the U.S. Nuclear Regulatory Commission along with special permits from the Atomic Energy Commission. These permits allowed the PRCF to operate as a private business venture, housing experiments for both government projects and for private nuclear fuel vendors until 1976" when the Energy Research and Development Agency (ERDA), Atomic Energy Commission's successor, closed the facility (Gerber 1993b, p. 35).

Originally a condition of their selection in 1965 as the Hanford Site contractor for chemical separations, Isochem, Inc. reneged on its promise to build a $9 million plant to recover and encapsulate radioisotopes from the Hanford Site high-level liquid waste stream. Commercial applications for the radioisotopes included food preservation, sterilization of medical supplies, and chemical manufacturing. However, unlike earlier successful isotope separations work conducted in the 325 High-Level Radiochemistry addition, Isochem determined no commercial market was available for the isotopes. Dissatisfied with Isochem's performance, the Atomic Energy Commission turned the chemical separations operation over to the Atlantic Richfield Hanford Company (ARHCO) in 1967 (Findlay and Hevly 1995). Initially, ARHCO did not pursue commercially based applications. Instead it modified the idle B Plant and provided its second mission, the recovery of cesium-137 and strontium-90 from high-level liquid waste. In 1974, ARHCO added the 225-B Waste Encapsulation and Storage Facility (WESF) to the B Plant complex. Here technicians solidified, encapsulated, and stored the recovered cesium and strontium. Operators completed cesium and strontium recovery in 1983 and 1985, respectively. These encapsulated isotopes were sent to the Oak Ridge National Laboratory for use in their isotope program.

In 1969, the Atomic Energy Commission decided to pursue breeder reactors instead of reactors fueled by mixed oxides. In an unanticipated decision, the Atomic Energy Commission selected the Hanford Site rather than the Argonne National Laboratory or the Idaho National Engineering and Environmental Laboratory as the location for the prototype Fast Fuel Test Reactor "even though production-oriented Hanford had no design team in place for such a new reactor" (Findlay and Hevly 1995, p. 243). Many of the facilities originally constructed for mixed oxide research were converted quickly for use in the Liquid Metal Fast Breeder Reactor program. For example, Westinghouse built a replica of a Fast Flux Test Facility (FFTF) operating cell within the 309 Building and used this facility to "train and requalify operators and to check operating procedures." The High Bay, added to the 308 Building in 1971, contained laboratories "for the initial storage, handling, testing, assembly, and instrumentation of FFTF fuel." In the late 1970s, specialists used the newly installed Training Research Isotopes, General Atomics (TRIGA) reactor to test FFTF fuel elements and jacket integrity (Gerber 1993b, p. 36). Metallurgists also conducted non-destructive examinations of FFTF fuel elements in the Shielded Materials Facility within the 324 Chemical Engineering Laboratory between 1968 and 1980.

To consolidate FFTF research and development, Westinghouse established the Hanford Engineering Development Laboratory. In addition to modifying existing facilities, Westinghouse constructed a number of new buildings within the 300 Area between 1968 and 1972 including the 335 Fast Reactor Thermal Engineering Facility, the 336 Core Segment Development Facility, and the 337 High-Temperature Sodium Facility. The 338 Building, originally constructed in the 100-F Area, was moved into the 300 Area in 1971 and served as the Components Mock-Up and Maintenance Facility to support FFTF. "In them, sodium loops tested various properties of sodium as a coolant, measured the behavior of mechanical components in a sodium environment, and fabricated, assembled and decontaminated equipment" (Gerber 1992b, p. 41).

Construction of the FFTF (405 Building) began in 1970 and extended until 1978. Liquid sodium cooling allows neutrons to achieve higher energy levels than those possible or desirable in a water-cooled reactor. The flow of neutrons, or flux, is therefore faster. This characteristic of sodium-cooled reactors is reflected in the name of the facility. Engineers designed FFTF to operate at 400 megawatts-thermal. It went critical in February 1980. "FFTF was designed and constructed for the irradiation testing of fuels, core components and target assemblies for liquid metal fast breeder reactors. Reactor activities were later expanded to include long-term testing and evaluation of reactor components and systems, fusion power materials testing, passive [safety] testing and production [of] medical isotopes, and space power system research" (Gerber et al. 1997, p. 5.22).

As a final example of services the Hanford Site provided to the nuclear community, the Atomic Energy Commission designated the Plutonium Reclamation Facility (236-Z Building) as the Central Scrap Management Organization in 1972. The Plutonium Reclamation Facility would now recover and recycle plutonium from scrap materials shipped from defense and commercial facilities throughout the nation. "New defense grade oxides and metals could be made after Pu [plutonium] was reclaimed from these scraps, and commercial fuels...could be recycled and blended" (Gerber et al. 1997, p. 5-20).

In service to the biological, ecological, and medical communities, in 1970 Battelle constructed the Life Sciences Laboratory (331 Building) as a replacement for the 108-F Building. Between 1972 and 1984, Battelle added several ancillary support facilities including living areas for the test animals, an animal birthing facility, an animal waste treatment facility, a greenhouse, experimental areas, and warehouses. Continuing pioneering research on the effects of radiation on animals and plants, scientists in the 331 complex conducted studies "devoted to dosimetry, isotope preparation, plant physiology, terrestrial ecology, aquatic biology, and biochemistry." They also conducted research in "inhalation toxicology...virology, histology, hematology...histochemistry, pathology, and microbial physiology" (Gerber 1993b, p. 44). One accomplishment, among many, was the development of new techniques for heart surgery.

LAND RELEASES

The shutdown of reactors, Findlay and Hevly (1995, p. 244), assert "presented chances to diversify Hanford's economy in an environmentally sensitive way, an increasingly important consideration in light of the growing environmental movement." The Atomic Energy Commission had released lands before in response to economic pressures. In 1947, the Atomic Energy Commission, through its Advisory Committee on Reactor Safeguards, established a 280,000-acre safety zone around the three production reactors as a protective measure in the event of an accident within one or more of these facilities (see Figure 1.9). This safety zone, extending beyond the boundaries delineating the Hanford Site in 1943, was divided into a Control Zone and a Secondary Zone. The Control Zone, which the Atomic Energy Commission owned outright as part of the original land acquisition, consisted of the land nearest the reactors and adjacent to the Columbia River. While the Atomic Energy Commission did not own the Secondary Zone, which extended to the crest of the Saddle Mountains, they had successfully prevented any development on or use of these 192,000 acres. Herein lay the conflict.

The lands within the Secondary Zone had been slated for settlement and irrigation under the Columbia Basin Project initiated in 1933. In 1952, the Bureau of Reclamation pointedly told the Atomic Energy Commission that continued withdrawal of the Secondary Zone was "costly to the taxpayer and project and to the Pacific Northwest...[and a] loss to the nation" (Van Arsdol 1980, Attachment 1, p.2). Given mounting pressure from the local community, the Atomic Energy Commission announced on January 8, 1953 that it was releasing some 87,000 acres on the extreme east and west ends of the Secondary Zone. "We feel justified in making this determination on the basis of the knowledge and experience gained during eight years of safe and successful operation of the Hanford Works, during which the safety systems of the plant have steadily improved" (Atomic Energy Commission Commissioners quoted in Van Arsdol 1980, Attachment 2, p. 4).

This concession was insufficient since the Atomic Energy Commission also announced that while they would "permit the temporary construction of canals and roadways through the remaining restricted areas" they would continue to "oppose occupation or regular work there" (Findlay and Hevly 1995, p.122). With the hint of an impending lawsuit from the Columbia Basin Commission and following hearings held by Senator Jackson in October 1957, the Atomic Energy Commission released an additional 105,500 acres from the Secondary Zone on December 30, 1958, thereby reducing direct government control to the 560 square miles currently managed as the Hanford Site.

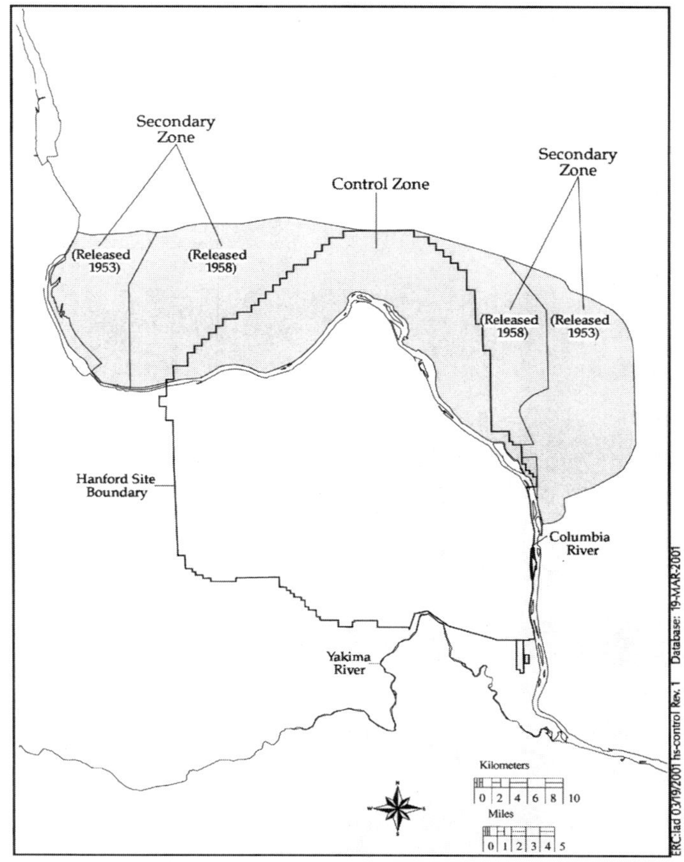

Figure 1.9. Control Zone and Secondary Zones at the Hanford Site

Pressure to release additional lands for economic development continued, but the Atomic Energy Commission found a new ally in the environmental movement. In response to two proposals centering on Rattlesnake Mountain and its adjacent slopes, the Atomic Energy Commission opted for biological and natural resource research rather than grazing and irrigation. On March 29, 1967, Senator Warren Magnuson announced the creation of the Arid Lands Ecology Reserve (now known as the Fitzner/Eberhardt Arid Lands Ecology Reserve) encompassing the southwestern edge of the Hanford Site. The Atomic Energy Commission set aside these 76,800 acres to "expand knowledge about man's relationship with nature, and thereby help protect the environment"(Findlay and Hevly 1995, p. 245). In discussing this decision further, Findlay and Hevly (1995, p. 247) assert that "proposals oriented more towards preservation gained precedence over economic development, largely because preservation dovetailed with a continuing emphasis on security at Hanford."

Continuing this trend, the Atomic Energy Commission established two additional natural resource areas. In 1971, the U.S. Fish and Wildlife Service assumed management of the Saddle Mountain National Wildlife Refuge on 32,000 acres of land within the western portion of the former Control Zone. That same year, the Washington State Department of Fish and Wildlife obtained management of 55,680 acres of the northern and eastern Control Zone designated as the Wahluke Wildlife Recreation Area. Figure 1.10 shows the additions since the Cold War Era of the Wahluke Wildlife Recreation Area, Saddle Mountain National Wildlife Refuge, Arid Lands Ecology Reserve, the 400 Area where the Fast Flux Test Facility is located, and the Washington Public Power Supply System.

REAGAN BUILDUP

Site Manager: Alex G. Fremling - September 1973 to June 1984
Responsible Agency: Department of Energy
Site Contractor: Westinghouse Hanford Company - June 1987 to 1990

The Atomic Energy Commission shut down D Reactor in June 1967. With four reactors now inactive, the Atomic Energy Commission closed REDOX in December 1967. Then, beginning with B Reactor in February 1968, the Atomic Energy

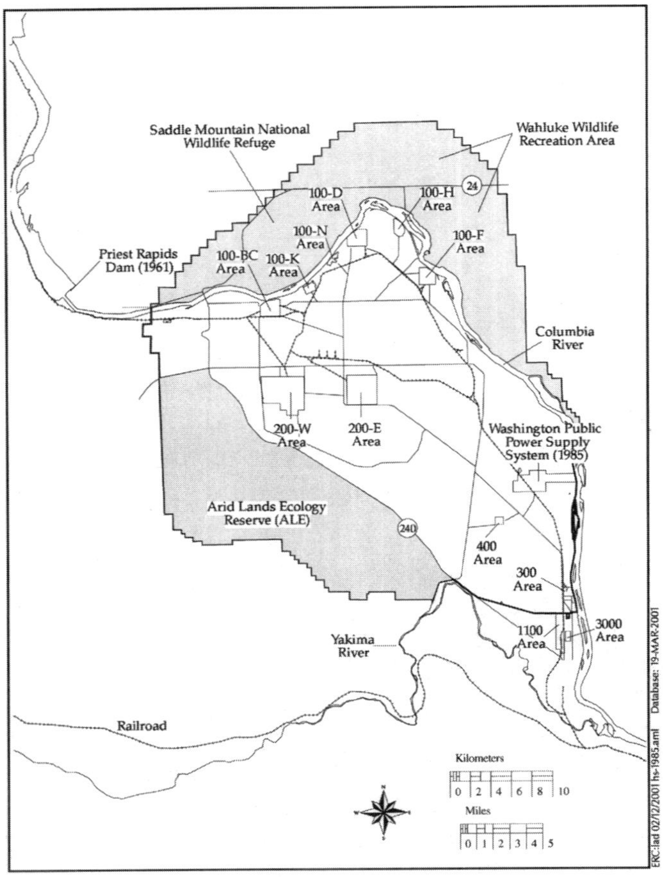

Figure 1.10. *Hanford Site Around 1985 Showing Additions Since the Cold War Era*

Commission ordered the closure of all remaining Hanford Site reactors in relatively quick succession: C Reactor in April 1969, KW Reactor in February 1970, and KE Reactor in January 1971. Only N Reactor expanded its operations to include steam production for the Hanford Generating Plant in 1966 and, for a short period between 1968 and 1970, tritium production. However, the Atomic Energy Commission order, issued for closure of the KE Reactor on January 26, 1971, also required the shutdown of the N Reactor.

Designed and operating now as a dual-purpose reactor, the prospect of loosing a facility that had barely begun fulfilling its commercial energy mission to the Pacific Northwest mobilized opposition against the Atomic Energy Commission decision to shutdown N Reactor. The Bonneville Power Administration, Supply System, Atomic Energy Commission, and the Federal Budget Office negotiated for 4 months, finally reaching an agreement in April 1971 that allowed N Reactor to remain in operation primarily to produce steam to turn the turbines for the Hanford Generating Plant. However, as a consequence of operations, N Reactor would also provide an additional service to the commercial reactor program. It would "pre-produce fuel-grade plutonium for the breeder reactor program" (Stapp and Marceau 1996, p. 24).

As the names indicate, weapons-grade plutonium was produced for weapon use, and fuel-grade plutonium was used as fuel for nuclear power reactors. In contrast to weapons-grade plutonium in which the level of the isotope plutonium-240 is kept below 7.4 percent, fuel-grade plutonium carries a plutonium-240 content of 7.4 percent or more. The N Reactor produced 8.2 metric tons of fuel-grade plutonium (12 percent plutonium-240) between 1973 and 1982. Unfortunately, this fuel-grade plutonium "was (and continues to be) stored in the fuel basins at the 100-K Reactors and was never used by the breeder reactor program as initially intentioned" (Stapp and Marceau 1996, p. 24). Because N Reactor would not be producing weapons-grade plutonium and PUREX was not equipped at the time to process commercial fuel, the Atomic Energy Commission placed PUREX on standby in June 1972.

To support the buildup of nuclear weapons initiated by President Carter and actuated by President Reagan, operators at N Reactor began producing weapons-grade plutonium in 1982. While DOE made the decision in 1980, it took nearly 2 years to upgrade the N Reactor before production could begin. "One fundamental problem was the distortion of the graphite stack, where built-in slip joints could not accommodate all of the local distortion, some block cleavage, and actual separation of blocks had occurred within the central core" (Gerber 1996, p. 2-10). Other issues involved system and facility maintenance on the aging structure as well as the implementation of new safety regulations issued by the U.S Nuclear Regulatory Commission. Westinghouse produced weapons-grade plutonium in N Reactor until 1987 when a number of unrelated events led to its closure.

By the mid-1980s, the international peace movement and its demand for disarmament made major inroads even within the Soviet Union. With the Intermediate Nuclear Forces Treaty nearly signed, and disagreements about strategic arms reduction narrowing, the need for additional plutonium had greatly diminished. Of more immediate concern was the explosion and fire at the Chernobyl nuclear power plant on April 26, 1986. It raised concerns about the safety of N Reactor because of perceived design similarities. "After lengthy study of the plant's flaws and projected costs [for improvement], DOE decided in 1987 and [again in] 1988 not to restart it" (Findlay and Hevly 1995, p. 302). DOE ordered the reactor defueled in 1989. It never operated again.

In 1983, following an 11-year "temporary" shutdown, Westinghouse reactivated PUREX to process N Reactor weapons-grade fuel. During the period from 1972-1983, Westinghouse initiated numerous design changes, including a major conversion project that resulted in the production of plutonium oxide (metal) rather than plutonium nitrate (liquid) as the process end state. This allowed final processing to begin immediately upon receipt of the uranium metal at the Plutonium Finishing Plant. Other changes addressed "environmental concerns that led to providing upgraded filtration systems, seismic safeguards, backup power sources and many other projects" (Gerber 1996, p. 4-15). Additionally, under the Fuels Segregation Program from 1983-1984, PUREX operators also "extracted approximately 425 kilograms of weapon-grade plutonium...[from] some of the fuel assemblies discharged during the N-Reactor fuel grade campaigns" (DOE 1996b, Section 9.1.1). DOE placed PUREX on standby status in October 1990 after having effectively ended operations in December 1988.

Westinghouse operated the Plutonium Reclamation Facility and the RMC Line within the Plutonium Finishing Plant on an as-needed basis to support PUREX during the final defense run. The Plutonium Reclamation Facility reopened between 1984 and December 1987. The Plutonium Finishing Plant operated from 1985 to June 1989 (Gerber 1996, p. 5-5). Processed plutonium metal was stored within the Primary Plutonium Storage Facility (2736-Z Building). The design and configuration of the building's underground storage vaults, constructed between 1972 and 1973, remain classified.

To accept the high-level liquid wastes associated with chemical processing, various contractors built 28 new high-level liquid waste storage tanks between 1968 and 1986. Knowing that a number of the single-shell tanks had developed leaks, engineers developed a new design, the "tank-within-a-tank," referred to simply as a double-shell tank. Each tank had a capacity of 1 million gallons (Gerber et al. 1997, p. 5.83). It was also during this period that the Hanford Site contractors re-evaluated their waste management practices. Among the conclusions reached, they determined in 1970 that solid waste should be labeled and segregated by type before burial, that waste materials should be tracked, and that new burial grounds should be constructed in the 200 Area to receive all future solid wastes. In addressing high-level liquid wastes, contractors added two new Evaporators in 1974 (242-S) and 1976 (242-A) to augment the waste minimization program

TRANSITION TO CLEANUP

Site Manager:	Michael J. Lawrence (June 1984 to July 1990)
	John D. Wagoner (July 1990)
Responsible Agency:	U.S. Department of Energy
Site Contractor:	Westinghouse Hanford Company

Throughout the 1970s and 1980s, the environmental movement grew beyond any expectations held at its birth in the 1960s. "The nation's attitude toward energy in general changed, and its confidence in nuclear power in particular declined" (Findlay and Hevly 1995, p. 275). Popular films such as *The China Syndrome* moved public opinion with its depiction of the "meltdown" of a reactor core. Public reaction to this fictionalized account appeared subdued by contrast when, just a few weeks after the movie's release, the nation witnessed the real thing in 1979 in Pennsylvania at Three Mile Island. The explosion, fire, and subsequent global release of radionuclides from the Chernobyl accident in 1986 galvanized the anti-nuclear movement. "Around the state and around the nation, the growing doubts about nuclear energy and nuclear wastes found political support, and thereby affected Hanford's future. These changes in attitude came at a time when Hanford was particularly vulnerable in government" having lost their powerful Congressional delegates through death or political defeat (Findlay and Hevly 1995, p. 298).

"The production of nuclear weapons had run continuously, beginning during World War II with the startup of the first reactor to produce plutonium for the top-secret Manhattan Project. But growing concerns about safety and environmental problems had caused various parts of the weapons-production complex to be shut down in the 1980s. These shutdowns, at first expected to be temporary, became permanent when the Soviet Union dissolved in 1991. The nuclear arms race of the Cold War came to a halt for the first time since the invention of the atomic bomb." **- DOE 1995e, p. 1**

Along with this lack of confidence in nuclear power, the public also began to change their attitude toward secrecy and security. While the need for secrecy and security arose from an external threat (the race with Germany to develop the atomic bomb and then the race with the Soviet Union to be the world power), the practices related to secrecy and security (classification of information, shredding of documents, surveillance of workers, etc.) also had the effect of preventing American citizens from gaining access to information about potential environmental degradation and health risks emanating from the Hanford Site.

The concept of a nuclear park died with the Supply System bond default. Defense production ended with the closures of N Reactor, PUREX, and the Plutonium Finishing Plant. Hanford Site administrators, with DOE's encouragement and assistance, sought one more time to find a national mission to store and manage the nation's nuclear waste. They saw their opportunity in the Nuclear Waste Policy Act of 1982. Under this legislation, Congress would select an eastern and a western repository among ten candidate sites. DOE spent $300 million examining the basalt formations of the Hanford Site, focusing primarily on Gable Mountain. This attention to the Hanford Site came at the expense of most of the alternative sites because of DOE's "perception that Washingtonians – or at least the state's elected officials – were more receptive to the idea of a repository than the people of other states" (Findlay and Hevly 1995, p. 298).

However in 1987, DOE cancelled all activity on the Basalt Waste Isolation Project on the Hanford Site and turned to Yucca Mountain in Nevada. "...the existence of an immense amount of waste at Hanford, and the record of its management and mismanagement over the years, provoked increasing public criticism. The waste itself, in other words, became a national crisis, so that Hanford was cast in the public mind as a problem rather than a solution to a problem" (Findlay and Hevly 1995, p. 274).

Cleaning up the Hanford Site became the mission. In May 1989, DOE, the U.S. Environmental Protection Agency, and the Washington State Department of Ecology signed the *Hanford Federal Facility Agreement and Consent Order*, better known as the Tri-Party Agreement, which committed DOE to cleaning up the Hanford Site within 30 years. The signatories divided the Hanford Site into 78 Operable Units, each containing a number of "past practice" waste sites, each of which was to be investigated and remediated separately under the Comprehensive Environmental Response, Compensation, and Liability Act of 1980 (CERCLA) program or the Resources Conservation and Recovery Act (RCRA) program.

With that, the Hanford Site's Manhattan Project and Cold War mission to produce weapons-grade plutonium finally ended. With each passing year, the once active industrial complex faded further into obscurity. Facilities for which there were no safety or security concerns, or those that were not needed to house personnel or materials for cleanup, or those with no future use were demolished or simply left unattended while they deteriorated. The 100 Area facilities were particularly hard-hit.

Beginning in 1972, Decontamination and Decommissioning (D&D) concentrated on the clean structures (those not contaminated with radioactivity) within the reactor complexes, such as those for water treatment. Demolitions increased or waned over the years, consistent with funding allocated to the task. During the late 1980s, D&D activity reached a peak as workers took down the majority of buildings within the reactor complexes, in many cases leaving little more than the reactor itself. As demolition and cleanup moved to the forefront, the history associated with a specific building became less important than the history of the wastes it produced. Given the health risk associated with much of that waste, it is critical that DOE understand where and how waste was generated and how to effect remediation.

On the other hand, under the Historic Preservation Program, it is also important that DOE shed light on the history of the facilities themselves, long-hidden behind national security and fences. At first glance, little distinguishes the mundane buildings distributed across the Hanford Site. Outwardly, they appear to comprise just another large industrial complex composed of production and support facilities. It is only after one looks behind the structural facades that their importance emerges.

The world-changing events associated with these buildings have local, state, national, and international significance that cannot be dismissed or denied. The Hanford Site mission both shaped and was shaped by world events in the last half of the 20th century. The buildings and structures remaining on the Hanford Site are the cumulative legacy and physical manifestation of that mission. This book introduces the reader to the science, technology, craftsmanship, and politics that infuse these facilities with meaning. The reader is encouraged to explore the events, processes, accomplishments, and failures that embody the story of the facilities and their role at the Hanford Site.

AREAS FOR FURTHER RESEARCH

The theme that the Hanford Site was an experiment runs throughout the Manhattan Project and the Cold War Era. Creating plutonium was a new science. Producing plutonium on an industrial scale was a major challenge. The time pressures of war and need for secrecy added to the uncertainties that accompanied each decision. Yet critical decisions had to be made within very short time frames despite the uncertainties. Because this book is a history of facilities and not intended to be the definitive history of producing plutonium at the Hanford Site, numerous facets of this theme of the Hanford Site as an experiment could not be explored. Hopefully future researchers will use this book as a springboard for their own studies to investigate such topics as the following:

- How was the geological history of the Hanford area affected by the load it had to bear when the Hanford Site was built?
- The yearly budgets for the Hanford Site have never been released to the public. When they are, future researchers would be able to evaluate the effect of the budgets over the years on the operations at the Hanford Site.
- How did the different corporate cultures of the various contractors, Du Pont and General Electric in particular, affect operations at the Hanford Site?
- Because the production of plutonium on an industrial scale was new and on a wartime schedule, workers had to be extraordinarily creative in making processes with which no one had experience work. A study of their creativity and the results would contribute greatly to understanding how the Hanford Site was able to produce plutonium against numerous odds.
- Developing an electronic, annotated bibliography of all references pertaining to the Hanford Site would facilitate future researchers' ability to find the particular information they seek.

Overview of the Managers and Contractors at the Hanford Site

Date	Contractor	Responsibility
Responsible Agency: U.S. Army Corps of Engineers		
Site Manager: Col.Franklin T. Matthias, 1943-1945 – Lt.Col.Fredrick J. Clarke, 1945-1947		
Site Name: Hanford Engineer Works		
1943-1946	**E. I. Du Pont de Nemours and Company**	**Prime Contractor** responsible for all activities
Responsible Agency: Atomic Energy Commission		
Site Manager: Carlton Shugg, 1947-1948 - Fred C. Schlemmer, 1948-1950 - David F. Shaw, 1950-1954 James E. Travis, 1955-1965 - Donald G. Williams, 1965-1971 – Thomas A. Nemzek, 1971-1973 - Alex G. Fremling, 1973 - 1984		
Site Name: Hanford Works		
1947–1974	**General Electric Company**	**Prime Contractor** responsible for all activities except as indicated
	Giffels & Vallet, Inc.	Engineering
	Charles T. Main	Engineering
	J. A. Jones	Construction Work
	Vitro Engineering, Co.	Architect-Engineer Services
	Douglas-United Nuclear, Inc.	Reactor Operations
	Isochem	Chemical Processing Operations
	Atlantic Richfield Hanford	**Prime Contractor** responsible for all activities except as indicated
	ITT Federal Support Services, Inc.	Support Services
	U.S. Testing Company	Bioassays, film badges, environmental samples
	Battelle Memorial Institute	National Laboratory
	Westinghouse Hanford Company	Fast Flux Test Facility
	Hanford Environmental Health Foundation	Health Services
Responsible Agency: Energy Research and Development Administration		
Site Manager: Alex G. Fremling 1973-1984		
Site Name: Hanford Reservation		
1975-1977	(see contractor list above)	
Responsible Agency: U.S. Department of Energy		
Site Manager: Michael J. Lawrence, 1984-1990, John D. Wagoner, 1990		
Site Name: Hanford Site		
1977-1990	**Rockwell Hanford Operations**	**Prime Contractor** responsible for all activities except as indicated above
	Westinghouse Hanford Company	**Prime Contractor** responsible for all activities except as indicated below
	ICF Kaiser Hanford Company	Architecture and Engineering
	Hanford Environmental Health Foundation	Health Services
	Battelle Memorial Institute	National Laboratory

CHAPTER 2

PLUTONIUM PRODUCTION FACILITIES

CONTENTS

Photo: B Reactor under Construction, December 9, 1943 (Neg. P520, Box D28790)

SECTION 1

CONSTRUCTION HISTORY

D.W. HARVEY
PACIFIC NORTHWEST NATIONAL LABORATORY

Section 1 - Construction History

From the outset, numerous factors influenced the physical characteristics of the Hanford Site's facilities, but functional considerations were the primary determinants of the design features and layout. Once built to meet a specific function, industrial and laboratory facilities at the Hanford Site were subjected to numerous internal and external construction modifications to accommodate technological advances, changing mission and scientific objectives, and expansion of plutonium production as well as non-defense facilities. These changes had a significant effect on the layout of the site and influenced the selection of construction styles, building materials, and industrial architectural designs.

The initial layout and construction also reflected the federal government's desire for cost-effective, wartime mobilization. As with other World War II military installations, speed of design and construction was of the utmost necessity. Stabilizing construction design was difficult because no previous experience or precedent was available from which to draw upon. Because of the wartime emergency situation, the U.S. Army Corps of Engineers only permitted slight design variations were permitted at the Hanford Site.

Before designing and constructing facilities, however, the Corps had the responsibility to select where the Hanford Site should be built. In late 1942 when the Corps contracted with E.I. Du Pont de Nemours and Company for the construction and management of the world's first plutonium production facility, they originally considered locating the site at the Oak Ridge Reservation in Tennessee. The Oak Ridge Reservation, however, did not meet the federal government's siting criteria. The reservation did not have sufficient land for six reactors to be separated from each other by at least 1 mile and three chemical separations plants to be separated from each other and from the reactors by at least 4 miles. Also, an isolated area was not to be artificially created by the evacuation of large numbers of people from the selected site, which would have occurred if the Oak Ridge Reservation had been chosen. Most importantly, the Corps and Du Pont were concerned that premature criticality of plutonium would cause an explosion, devastating the Oak Ridge Reservation and sending a lethal contamination cloud over nearby Knoxville and strategic war industries in the Tennessee Valley Authority region (Groves 1983).

Wartime economic considerations also influenced siting criteria. Arable land was not to be taken. Local supplies of gravel for the vast amount of concrete required and regional suppliers of coal and oil were preferred to reduce transportation costs and conserve precious wartime fuel. Because of the wartime scarcity of metal, the site was to be close to the power source so it would not be necessary to build long transmission lines (USACE 1947, Vol. 3).

Wartime urgencies necessitated the construction of a plutonium production facility in a short time span.

> *"...in order to keep to the tight schedule mandated by the U.S. Army for the Manhattan Project's effort to build an atomic bomb. The project required an immense amount of expertise, labor, and building materials, at a time when each of these commodities was in short supply. It demanded tremendous secrecy, despite its immense size, and it entailed the construction of a permanent set of facilities, including a whole new town, even though it was not clear that these facilities would be needed after the war."*
> - Findlay and Hevly 1995, p. 1

The need for the Manhattan Project to proceed with haste despite the reliance on untested technologies was expressed years later by Lieutenant General Leslie R. Groves, commander of the Manhattan Project, when he wrote "Nothing like this had ever been attempted before, but...we could not afford to wait...." (Groves 1983, p. 72).

Contents

Divider Page Photo: Construction Camp, June 22, 1944 (Neg. P3437, Box 028791)

Because of the hazards inherent in the production and separation of plutonium and the handling and disposal of large quantities of radioactive materials and waste, the design and layout of the world's first plutonium production facility had to satisfy the Corps's safety, location, and natural resource requirements. The site had to have (USACE 1947, Vol. 3):

- Area of at least 12 miles by 16 miles
- Remote setting with no population greater than 1,000 within 20 miles
- Abundant water supply of at least 25,000 gallons per minute to cool the reactors
- Dependable hydroelectric power source to supply at least 100,000 kilowatts of electricity
- Convenient access to railroad and highway facilities
- Relatively flat landscape
- Available fuel and concrete aggregate

The siting criteria narrowed the choice of a location to a few isolated places in the far West. From the beginning, the Pacific Northwest was at the top of the list. The Bonneville Power Administration's excess quantities of electrical power were available year-round without the need to install additional generating equipment. The discovery that a high-voltage power line from Grand Coulee Dam (1941) to the Bonneville Dam (1937) ran through the future Hanford Site made the area even more attractive (HAI 1987).

After a tour of the mid-Columbia region, the Corps concluded that the Hanford area met all siting conditions better than the other West Coast areas being considered. The site had the quantity of land required, was remote from major populated areas, had enough clean water from the nearby Columbia River, ample electrical power from Grand Coulee Dam and associated transmission facilities, a functional railroad, a flat landscape, and coarse glacial-fluvial sediment that provided sand and gravel aggregate for constructing large concrete structures. Additionally, the site was far enough inland from the coast to satisfy War Department officials (Groves 1983). Finally, soil investigations indicated that the land surface was capable of sustaining the considerable stress from the construction of processing buildings. "Although not an original siting requirement, Hanford's arid environment and soil features allowed large amounts of liquid waste to be released to the ground without immediately descending to groundwater" (Gray and Becker 1993, p. 462).

The planners siting the Manhattan Project decided early on to locate production facilities in remote areas, partly for secrecy and military security and partly out of concern for public safety in the event of a catastrophic reactor accident. In choosing isolated areas like Hanford, the Manhattan Project officers dispensed with the usual practice of locating a large industrial facility near adequate housing, services, construction labor, and a skilled work force. Instead, the Manhattan Project had to build and administer whole new communities and draw masses of people from other places to work at the sites (HAI 1987).

Figure 2-1.1. *Hanford Site Construction Camp Located at the Former Hanford Townsite Around 1944*

Hanford was one of the largest procurements of land handled during the war with a total of 428,850 acres. Within a year after the federal government acquired it, they transformed the site from a sparsely populated, arid shrub-steppe landscape into a major military and manufacturing complex that included the establishment of a construction camp at the former Hanford townsite (see Figure 2-1.1) and a new community in Richland, which consisted of over 4300 new residential units to house some 6000 operations workers and 11,000 family members (USACE 1947, Vol. 3).

Site Design and Layout

For safety and security reasons, the planners of the Hanford Site separated the production areas by relatively large distances. Due to the nature of the production processes, they designed the site areas as independent units constructed in widely separated districts because "the possibility of explosions of catastrophic proportions and the possibility of releasing to the atmosphere of intensely radioactive gases would dictate the selection of a site of sufficient area to permit the several manufacturing areas to be separated by distances of several miles...that included six primary manufacturing areas separated by distances of not less than one mile, and four secondary manufacturing areas separated from each other by not less than four miles...." (USACE 1947, Vol. 3, p. 2.1). This decision was made so an accident in any one area would not affect the operation of the remaining units.

Early considerations in the design of the Hanford Site took into account the supplying of adequate electrical power, communication, rail, and highway facilities to all construction, operating, and housing areas. Because of the magnitude of the project and the demand on these facilities, the Corps found it necessary not only to expand the existing facilities but to design additional and reliable power and communication lines, roadways, and railroads (USACE 1947, Vol. 3). Totals included "158 miles of railway, 386 miles of new highway, 52 miles of 230 Kv transmission lines, a water distribution system ranking in magnitude with the large metropolitan areas...." (Matthias n.d., p. 3).

For security, safety, and functional reasons, the Corps designed and constructed all the process areas as semi-autonomous units. Each area contained support buildings and structures that provided administrative, security, health and safety, communications, utility and maintenance, waste treatment and environmental monitoring services.

300 Area

The 300 Area was where the uranium fuel was manufactured before it was sent for irradiation in the reactors in the 100 Areas. As such, the 300 Area housed the first step in the plutonium production process. The Hanford Site fabricated nuclear fuel in the form of pipe-like cylinders (referred to as "fuel elements" or "slugs") from metallic uranium from offsite production facilities. Metallic uranium was extruded into the proper shape and encapsulated in aluminum or zirconium cladding. Railroad cars then transported the fuel cylinders to the reactors in the 100 Areas for irradiation. For more information on the fuel cladding and canning processes, see Section 2, Fuel Manufacturing.

Because the 300 Area's fuel fabrication activities were the least likely of the production processes to experience a serious accident, the Corps and Du Pont considered it safe enough to locate them near a populated area. The 300 Area is approximately 7.5 miles north of the center of Richland and occupies approximately 1.5 square miles in the southeastern portion of the Hanford Site along the west bank of the Columbia River.

The layout of the 300 Area was a grid pattern (see Figure 2-1.2). Originally the facilities were concentrated in the central zone and northwest corner of the 300 Area. Besides the uranium fuel fabrication facilities, the 300 Area was the location of most of the Hanford Site's research and development activities, chemical process laboratories, and test reactors. Over 30 buildings and structures were constructed there during the wartime period.

Figure 2-1.2. *Grid Pattern of 300 Area Looking West, 1979*

The 300 Area was also the location of several test reactors during the Manhattan Project and Cold War periods. In the 305 Building, a small reactor tested samples of the graphite, uranium, and other materials used in essential Hanford Site operations. The Physical Constants Test Reactor and Thermal Test Reactor were located in the 305-B Building and used for lattice configuration experiments. Other 300 Area test reactors included the Plutonium Recycle Test Reactor in the Plutonium Fuels Utilization Program in the 309 Building and the High Temperature Lattice Test Reactor in the 318 Building.

Most of the early buildings and structures were situated in the 300 Area by function and proximity to fuel fabrication/ uranium production and laboratory facilities. Buildings such as the 313 Fuel Manufacturing Support Facility, 314 Press Building, 305 Test Pile, 305 B Engineering Development Lab Annex, and 3706 Radiochemistry Lab were sited in proximity to one another. The 303 Fresh Metal Storage buildings (303-A, B, C, D, E, F, G, J, K) were situated in a relatively east-west linear line perpendicular to the 313 Building due to their functional association with the missions of the facility. But as the size of the 300 Area expanded and its mission diversified, research, laboratory and fuel fabrication facilities were mixed among storage buildings, shops, administrative buildings, health, security and safety facilities, and environmental monitoring structures.

The current layout of the 300 Area consists of three distinct sections. The northern section has experienced minimal building development, consisting mainly of waste sites such as trenches, ponds, and burial grounds. The central section has a congested network of infrastructure systems, Manhattan Project and early/middle Cold War era facilities. The southern section has had minimal infrastructure and building/structural development. The southern section, lacking a grid design, is a sprawling landscape of parking lots, mobile offices/trailers and several buildings/structures that date from the late Cold War era to the post-1990 period.

100 AREAS

Fabricated fuel cylinders were shipped by rail from the 300 Area to the reactors in the 100 Areas for irradiation, the second step in the plutonium production process. The nine plutonium production reactors (typically referred to as piles) and their ancillary/support facilities were designed and constructed along the south shore of the Columbia River. The reactor areas had to be situated close to the river because large quantities of water were required to dissipate the heat generated during reactor operations. For specific information on the design and construction of Hanford Site reactors, see Section 3, Reactor Operations..

Because of the dangers inherent in the irradiation of uranium fuel elements, the planners situated the reactors as far as possible from Richland. The 100 Areas are approximately 30 miles from Richland in the northern portion of the Hanford Site along the south bank of the Columbia River.

Within the 100 Areas, each of the reactors (B, C, D, DR, F, H, KE, KW, and N) was located in an area 1 square mile in size and separated 1-3 miles from one another, a distance assumed adequate enough to prevent operational difficulties in one area from affecting another. The Hanford Site planners thought that this siting requirement would minimize the effects of an explosion or act of sabotage at one reactor from adversely affecting other 100 Areas facilities.

While each of the production facilities in the 100 Areas was designed as a self-contained, functional unit, completely independent of the others, each production reactor was functionally dependent on the two other major processing areas, the 200 and 300 areas.

Each 100 Area was designed to be virtually identical except for100N (see Area Expansions sub-section). In addition to the reactor buildings, each 100 Area had in common a retention basin, pump houses, chemical and gas storage facilities, water purification facilities, reservoirs, filter plants, power houses, water treatment plants, water tanks, electrical substations, waste processing facilities, change houses, warehouses, maintenance shops, oil and gas storage facilities, patrol headquarters and badge houses, first aid stations, fire stations, and water chemistry laboratories.

During the Manhattan Project era, Du Pont and the Corps used temporary construction structures in the 100-B, D, and F areas. Temporary construction facilities included construction offices and shops, storage buildings and yards, electric power distribution facilities, water pumping and transmission facilities, and ground transportation facilities.

Because there had been no experience in the construction and operation of a facility similar to the Hanford Site, it was only natural that the initial design would change. The need for some of the changes was not apparent until the startup of B Reactor. Many of the design changes in the 100 Areas were made as construction expedients rather than because of changes in the process. The greater portion of these changes, mainly piping revisions, occurred in the 100-B Area since it was the first of the three 100 Areas completed (Du Pont 1945a, Vol. 3).

200 AREAS

The 200 Areas with its chemical separations (processing) plants functioned as the third step in the nuclear process at the Hanford Site. The 200 Areas contained all the facilities used to separate, isolate, store, and ship the plutonium. To separate the plutonium from the uranium and fission by-products formed in the irradiation process, the chemical separations plants dissolved irradiated fuel cylinders and then chemically manipulated the resulting plutonium-bearing solution. In addition to the chemical separations facilities in the 200 Areas, tanks were constructed to store the resulting high level radioactive waste. For a description of these tanks, see the Facilities Descriptions sub-section in Section 6, Waste Management.

Because separating out plutonium was the most hazardous of the plutonium production processes, the 200 Areas were situated 20 miles from Richland. The 200 Areas are located on a plateau in the center of the Hanford Site, roughly 5 miles south of the Columbia River. The planners selected the sites because they "lay in the intervening valley between the Rattlesnake Hills, Gable Mountain, and Gable Butte with the latter two providing a natural barricade between the 100 and 200 Process Areas" (Du Pont 1945a, Vol. 3, p. 812).

Another determinant that influenced the siting of the 200 Areas in the central plateau area had to do with its distance above the water table. The water table is approximately 240 feet below the surface of the central plateau, whereas the water table is 50-75 feet below the 100 Areas. While the Corps built the processing facilities in the 200 Areas atop the same permeable gravel as the production reactors, the sediment underlying the 200 Areas was finer, less permeable to liquids, and with less chance of contamination of the groundwater (conversation with R.W. Carpenter, Waste Site Facility researcher, CH2M HILL Hanford, Inc., November 1996).

Within the 200 Areas, the hazardous nature of the separations activities made it undesirable to concentrate the process buildings together so the Corps designated 200 East, West, and North areas. "Certain storage buildings were segregated (in 200 N) and the productive capacity divided (between 200 East and West). The distances required between areas made a large tract necessary" (Du Pont 1945a, Vol. 3, p. 812).

The 200 East and 200 West areas were very similar. The most significant buildings were the separations plants, T Plant and B Plant. The 200 West Area operated initially when T Plant (221-T) dissolved the first irradiated fuel rods received from B Reactor in 1944. In the 200 East Area, B Plant (221-B) began processing irradiated fuel in April 1945 (Gerber 1993c). The planners placed these separations plants approximately 2 miles apart.

Each of the 200 East and West areas contained temporary structures needed for construction such as administrative facilities, warehouses, craft and maintenance shops. Other temporary construction buildings included change houses, warming sheds, privies, and guard houses. In some cases, temporary construction structures were converted to permanent buildings near the end of the wartime period (Gerber 1993c).

The 200 North Area was completely different in design and function from 200 East and 200 West because it was a storage area for the irradiated fuel rods after their removal from the reactors while awaiting chemical dissolution, for the finished product before it was shipped to Los Alamos, and for the empty product storage cans that were returned from Los Alamos awaiting refill. The 200 North Area was the first of the 200 Areas to operate when it began storing irradiated fuel rods from the B Reactor in November 1944 (Gerber 1993c).

The 200 Areas were relatively self-sufficient during the Manhattan Project period. Each area contained a coal-fired central power station to provide steam for heating and for the process equipment. They also had their own water supply, sanitary system, railroads, and administrative and utility buildings.

400 AREA

The 400 Area came into existence when Westinghouse Hanford Company built the Fast Flux Test Facility complex in the 1970s for the U.S. Department of Energy (DOE). The Fast Flux Test Facility complex contains the largest test reactor on the Hanford Site. Westinghouse designed this special nuclear reactor to test fuels and materials for advanced nuclear power plants as well as for the irradiation testing of fuels, core components, and target assemblies for liquid metal fast breeder reactors.

The Fast Flux Test Facility complex includes the reactor, containment structure, and various utilities and ancillary facilities. The complex array of buildings, equipment, and roads are arranged in a grid around the Reactor Containment (405 Building). Being sodium-cooled, the reactor did not need to be located near the river for coolant purposes as did the production reactors. Similar to the 100, 200 and 300 areas, the 400 Area contains support facilities that provided administrative, security, health and safety, utility and maintenance, communications, waste treatment and environmental monitoring services.

500, 800, AND 900 AREAS

These three areas are not geographical areas but consist of electrical, pipeline, power and sewage outlets and facilities, and fences and fire lines found throughout the site.

600 AREA

The 600 Area comprised facilities that served more than one specific area. "Included in the construction of the HEW [Hanford Engineer Works now Hanford Site] are a number of facilities which serve more than one specific area and in cases, such as roads and railroads, serve the entire Project. For this reason they have been designated as 600 Area buildings and facilities, and are not confined to a single location…" (Du Pont 1945a, Vol. 4, p. 1085). The 600 Area designation soon came to refer to all areas of the Hanford Site that were excluded specifically from some other area. Facilities included health, safety, and environmental monitoring properties such as radar and utility/communication towers and structures, site security, military defense, and fire suppression facilities.

The initial step in organizing the 600 Area for construction activities began in 1943. The planners selected the former Hanford townsite as the location for the project construction camp because of its proximity to the 100 and 200 areas, existing infrastructure (for example, roads, buildings, wells, electrical utilities), and the terminus of the Priest Rapids Railway, a branch line of the Chicago, Milwaukee, and St. Paul Railroad. Practically overnight the camp turned into a city, reaching a population of about 45,000 during its peak occupancy in mid-1944, complete with barracks, trailer lots, mess halls, recreation, administrative, and other service facilities.

Du Pont's Wilmington Design Division prepared the layout for the site's intra-area roads and railroads (Du Pont 1945a, Vol. 4). Because of the widespread location of construction areas and the volume and types of materials involved, railroads were an important method of transportation. Existing tracks were upgraded to accommodate the anticipated heavy use during the Manhattan Project.

> *"The isolation of the HEW site [present Hanford Site] and the distances separating the various manufacturing areas made the establishment of a complete transportation network a necessity. An extensive system of roads and railroads was designed and constructed to expedite transfer of the enormous quantities of construction equipment and materials to the various areas as well as to provide adequate transportation facilities in case of an emergency in one or more of the areas during plant operations.... Design of the plant road system was based upon the great peak construction requirements and upon traffic needs during the operating period in the event that mass evacuation of the operating personnel might be necessary."* - USACE 1947, Vol. 3, p. 7.4

Other major 600 Area facilities included the Army's Camp Hanford anti-aircraft artillery sites. Sixteen of them were built in the early 1950s. The most intact anti-aircraft artillery sites are situated along Army Loop Road (see Figure 2-1.3) and on 200 East Hill. Extant resources include the remains of doughnut-shaped artillery placements (defense walls) and other sandbagged/cobblestone structures.

The internal layout of the anti-aircraft artillery sites reflected a standard military arrangement of facilities separated by function. Four semi-circular artillery revetments were arranged in a square separate from the rest of the installation. The more permanent, concrete structures were situated in a rectangular grid that included barracks, latrines, mess halls, recreation halls, motor pools, administrative, and radar facilities. Each anti-aircraft artillery site typically had a small arms range, a water storage cistern, and sanitary/sewage waste facilities. The site facilities were connected by roadways, sidewalks, and parking lots.

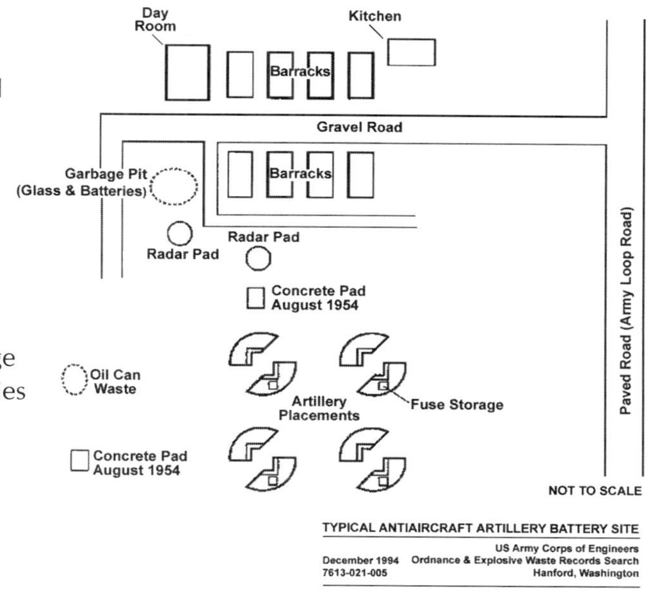

Figure 2-1.3. *Typical Anti-Aircraft Artillery Battery Site*

700 AREA

The 700 Area was the original location of the central administrative functions for the Hanford Site. The 700 Administration Area was contained within the commercial district of downtown Richland. Besides the administrative center of the site, the 700 Area was also comprised of repair and maintenance shops, central supply houses, plant laundries, central supply houses, and central heating plant. Planners placed the administration and shopping district near the center of town so the surrounding lands could be developed exclusively for residential purposes. The 700 Area was bounded by Swift Boulevard on the north, Knight Street on the south, Jadwin Avenue on the east, and Stevens Drive on the west. The area was roughly rectangular in shape with the long portion running east-west.

Although situated within the center of Richland and not containing any buildings of a strictly classified nature, the entire 700 Area was fenced off from the surrounding land, and, within the area, the Administration Building, the Permanent Record Storage facilities and the Employment Building were separated by a fence from the other sections. The 700 Area ceased to be a separate fenced entity after the sale of the property in 1957-1958 to private owners. During that time, the area was incorporated as the City of Richland. Today, only four DOE buildings are extant, all constructed during the Cold War period.

1100 AREA

The 1100 Area is located in North Richland west of Stevens Drive, north of Snyder Avenue, east of the DOE railroad tracks. The 1100 Area had the site support services, such as general stores, shipping, receiving, transportation, maintenance, and general warehouse facilities. The area expanded considerably during the early 1950s in response to the need for a central shipping, receiving, and transportation center due to the accelerated Cold War era growth of the Hanford Site. The industrial and warehouse facilities were strategically situated adjacent to the Hanford Site's main north-south highway and the site railroad.

BUILDING PERIODS

MANHATTAN PROJECT

Du Pont under the direction of the Corps began construction at the Hanford Site in March 1943 and essentially completed it by April 1945. Design and construction of the site included seven process areas (100-B, 100-D, 100-F, 200 East, 200 West, 200 North, and 300), five service areas (500—outside electrical facilities such as substations and transmission lines, 600—general facilities and services, 700—administration, 800—overhead pipe lines, and 900—underground pipe lines), and the 1100 residential and commercial area (Du Pont 1945f).

> ***"The isolation of the site from any existing centers of population presented serious problems with respect to many phases of construction. These problems were related primarily to the procurement, transportation, housing, feeding, health, morale, and retention of a maximum total construction force of about 45,000 persons which number was reached in June, 1944... The urgent need for placing the plant in operation at the earliest possible date precluded the possibility of delaying the start of construction work. In many instances, construction work was in progress during periods when basic research had not been fully developed. Consequently, the burden of construction planning, scheduling, and procurement was extremely great."* - DOW n.d., pp. 4-5**

The construction of the Hanford Site presented a number of unprecedented problems that stemmed both from the project's wartime emergency mission and its immense magnitude. For example, distances were considerable between several of the process plants to be constructed. Wartime scheduling needs allowed no time to test the designs. The secret nature of the project caused coordination difficulties and inflexible military security requirements. The most difficult element, however, was the isolated location of the site.

COLD WAR ERA EXPANSION

Hanford Site's Cold War Mission

International events associated with the intensification of the Cold War during the late 1940s to early 1950s increased allocations to national defense and expansion of America's nuclear weapons program. This translated into a dramatic expansion of facilities at all Atomic Energy Commission sites, including a substantial increase in plutonium production and construction/modification of plutonium production structures at the Hanford Site. Such expansion at the Hanford Site required an enormous amount of resources, resulting in the largest federal peacetime construction project to that date.

In the immediate postwar period, the effort to expand the Hanford Site's production capabilities looked and felt somewhat like the wartime years. Congress decided to expand the nation's plutonium stockpile in response to Soviet actions in eastern Europe. That placed much of the burden of production directly upon the Hanford Site, which was the sole supplier of fissionable plutonium for atomic bombs until the completion of the first reactors at the Savannah River Site in 1953 (Findlay and Hevly 1995).

The Hanford Site's postwar mission underwent two major periods of expansion: first with America's commitment to nuclear weapons as a central part of its postwar defense doctrine and then with the intensification of the Cold War between 1949-1953, especially with the outbreak of the Korean War. Between 1947 and 1955, the Atomic Energy Commission added five new reactors at the Hanford Site while concurrently boosting the output of the three Manhattan Project reactors. Incremental improvements in the basic components of the World War II reactors and a construction program to build reactors that incorporated these changes accounted for the doubling in plutonium output in 1952-1953. New plutonium extraction centers were built to process this output.

Building Expansion at the Hanford Site

Gerber (1992b) characterized the major postwar expansion of the Hanford Site's production facilities between 1947-1963 as follows:

- First Postwar Expansion (1947-1949) resulted in major expansion of the Hanford Site facilities. Two new reactors, DR and H, and ancillary buildings were constructed.
- Korean War Expansion (1950-1952) was the aftermath of the Soviet Union's detonation of its first atomic bomb and a major Communist thrust into South Korea. C Reactor and support facilities were built.
- Eisenhower Expansion (1953-1955) was the result of President Eisenhower's massive retaliation policy and the first Soviet hydrogen bomb detonated in 1953. Two jumbo reactors, KE and KW, were built. The ballistic missile development program was initiated.
- Major Defense Production (1956-1964) was the period of the most intense defense production period at the Hanford Site. N Reactor and other major facilities were constructed.

Early postwar expansion at the Hanford Site saw the construction of numerous buildings and structures that included the Reduction-Oxidation Chemical Processing (REDOX) Plant, the Plutonium-Uranium Extraction (PUREX) Plant, the Plutonium Finishing Plant (Z Plant), "the C Plant (a radiochemical pilot plant), seven complex laboratories in the 300 Area, and 81 additional underground, high-level waste storage tanks" (Gerber 1993d, p. 7). The chronic shortage of tank space led to the decision to build two evaporators, 242-B and 242-T, to concentrate and reduce the volume of high level wastes. Z Plant made possible the conversion of plutonium-nitrate paste to plutonium metal known as buttons. "Work also went forward on the development of the REDOX process, the only continuous solvent extraction plant in the world to save scarce uranium that was wasted in the first precipitation cycle of the old bismuth-phosphate processing. In 1949, Hanford Works [now the Hanford Site] initiated completion of C Plant in the 200 East Area as a pilot plant for the REDOX process" (Gerber 1991, p. 5).

The mid-1950s through 1963 when N Reactor was completed marked the most intense defense production period at the Hanford Site. Additionally, major new capabilities were added at Z Plant that centered on the reclaiming of plutonium from process scraps.

> ***"The most outstanding new facility was N Reactor, an unique plant that combined plutonium production with the steam generation of commercial electric power.... Producing over 65 billion kilowatts in 24 years, N Reactor was the largest electric power producer in the nation in its early years.... Other new facilities constructed during the peak production years...included the small, experimental Plutonium Recycle Test Reactor and Plutonium Fuels Pilot Plant, and the last four single-shell, high-level waste storage tanks."*** **- Gerber 1992b, pp. 31-32**

Non-Defense Programs and Mission Change

Expansion of the Hanford Site facilities during the 1950s also included programs dedicated to the peaceful uses of the atom. President Eisenhower's "Atoms for Peace" program and the "passage of the new Atomic Energy Act of 1954 designed to allow for more private, commercial atomic applications, brought innovative, non-defense programs to the Hanford Atomic Products Operation (HAPO)...HAPO built the 308 Plutonium Fuels Pilot Plant and the 309 Plutonium Recycle Test Reactor" (Gerber 1992b, p. 40) to test developmental fuels for use in commercial reactors.

By the mid-1960s, decreased national demand for special nuclear materials precipitated a down-sizing of the Hanford Site's plutonium production mission. By 1971, all but one reactor were shut down. Plutonium production for national defense use came to a halt between 1972 and 1983 when the site's only fuel processing facility was shut down. Fuel production at the Hanford Site stopped when the N Reactor shut down in 1987 while numerous improvements were made. By 1988, improved relations with the Soviet Union, symbolized by Strategic Arms Reduction Treaty talks, seemingly reduced the need for special nuclear materials, and N Reactor was put in standby mode, never to operate again. The closure of N Reactor brought to an end nearly 45 years of plutonium production at the Hanford Site.

AREA EXPANSIONS

The expansion of buildings and structures in each of the areas was in direct response to the needs of the Manhattan Project and Postwar era.

100 Areas

The General Electric Company completed the DR and H reactors for the Atomic Energy Commission during the first post-World War II expansion (1947-1949) period at the Hanford Site. DR Reactor was completed in March 1949 as a replacement for the D Reactor. "D Reactor was thought to be nearing the end of its effective operational life in the late 1940s due to growth and distortion of its graphite core [see "Graphite Expansion" in the Reactor Operations Section]. It was subsequently determined that the graphite distortion in 105-D could be controlled; both reactors would operate simultaneously" (Carpenter 1993, p. 2-3).

Construction of the H Reactor began in March 1948 and was completed in October 1949. "The reactor initially operated at 400 MW of power. Its operating limits were gradually increased over the years, until 2140 MW was authorized in 1965. Soon thereafter the reactor was shut down and deactivated in April 1965" (DeFord and Einan 1995, p.2-4).

"The 105-C Reactor was the Hanford Site's sixth production reactor. Its construction began in June 1951 and was completed in November 1952. The reactor initially operated at 650 megawatts, but its power level was increased over time to 2360 megawatts. It operated until April 1969 and has remained in deactivated status since then" (Carpenter 1994a, pp. 2-3).

The 100-KE/KW areas contain the KE and KW reactor buildings and their support facilities, including water treatment structures, administrative buildings, laboratories, and maintenance shops. Construction of the 100-K Area reactors began in September 1953 as part of Project X, a large Cold War expansion effort at the Hanford Site (Carpenter 1994b). The construction of the 1850 megawatt reactors was completed in 1954. They were the largest reactors built as of that date. The 100-K Area reactors had more output than the other single-pass reactors because of higher thermal power levels. Their operating limits were gradually increased to a limit of 4400 megawatts by 1961. Operations continued until 1970, when the KW Reactor was deactivated and 1971, when the KE Reactor was deactivated. The reactors have remained in deactivated status since that date. The fuel storage basins and small portions of both reactor buildings, however, remain active for the storage of irradiated fuels from N Reactor.

N Reactor, completed in 1963, was the last of the graphite-moderated reactors. "While basically the same as the previous eight reactors, the N Reactor design differed in several ways to afford greater safety and to enable cogeneration of electricity" (Stapp and Marceau 1996, p. 3). The fabrication process at N Reactor also differed from the other eight reactors. The fuel rods for N Reactor were manufactured in the 333 Fuels Manufacturing Building using the co-extrusion process. (For details on the co-extrusion process, see the Fuel Manufacturing Section.) Testing of tubes made by this process "concluded that the characteristics of zirconium alloy enhanced reactor operations and provided longer life than the aluminum that had been used previously" (Stapp and Marceau 1996, p. 20).

200 Areas

Of the approximately 350 buildings and structures (excluding mobile offices/trailers, subsurface trenches, wells, cribs, towers and storage tanks) remaining today in the 200 East and West areas, 45-50 were built during the Manhattan Project. The rest were constructed during the post-World War II period, reflecting the considerable expansion of the Hanford Site during the Cold War. Likewise, many Manhattan Project era facilities were redesigned over the years to accommodate continually changing missions and technologies at the Hanford Site. Three of the most important buildings constructed were:

The Reduction-Oxidation Plant (REDOX - S Plant), completed in the 200 West Area in 1952, was the first full-scale solvent extraction plant built in the United States for the recovery of plutonium and uranium. The plant employed an advanced organic solvent extraction process that replaced the bismuth phosphate process employed at B and T plants.

The Plutonium-Uranium Extraction Plant (PUREX - A Plant), completed in the 200 East Area in 1955, extracted plutonium and uranium from nuclear fuel. The plutonium was sent in a liquid form to the Plutonium Finishing Plant in the 200 West Area, while the uranium was sent in liquid form to the Uranium Trioxide Plant. The Uranium Trioxide Plant converted the liquid to a solid uranium oxide powder that was shipped to other DOE sites where it was enriched and converted to forms useful in weapons.

The Plutonium Finishing Plant (Z Plant - 234-5 Building), completed in the 200 West Area in 1949, played an important defense role during the Cold War. The Plutonium Finishing Plant, made possible the conversion of plutonium nitrate (a wet paste) to plutonium metal, known as buttons, through the oxalate, oxide, and reduction steps (Gerber 1992b).

300 Area

Indicative of the intensive post-World War II construction at the Hanford Site was the growth of the 300 Area. The postwar expansion of the United States nuclear weapons complex led to a dramatic increase in the construction of research and development laboratories, maintenance and craft shops, and administrative facilities in the 300 Area with approximately 160 buildings and structures (excluding mobile offices/trailers and subsurface tanks/cribs) still extant. "The large defense production expansions that increased the number of reactors, separations facilities and waste tanks at the Hanford Site in the early 1950s also led to growth in the 300 Area. In 1952-1953, five large buildings, along with ancillary service structures, opened in the southern sector of that area.... Soon after these buildings were completed, continuing defense production expansions brought major additions to them" (Gerber 1992b, pp. 39-40).

The construction of research and development facilities accelerated during the early 1950s. The following decade saw the establishment of additional laboratories and expansion of research and development activities to handle the increase in defense and energy research. In the 1970s, additional support and laboratory facilities were constructed in the southern 300 Area for energy research, waste management activities, and biological and environmental sciences research. The additional facilities were needed for scientists to explore the potential danger to human health and the surrounding environment through environmental monitoring and bioassay programs.

Although the number of facilities in the 300 Area increased considerably during the post-World War II era, its original functions as a process improvement and fuel fabrication area remained relatively unchanged through the duration of the Cold War era. During this period, many buildings and structures in the 300 Area had to be retrofitted to accommodate new technologies and changing missions. For example, the 3706 Building, the original radiochemistry laboratory for the Hanford Site, underwent extensive modifications and additions before its completion in 1945. During the 1950s-1960s, the 3706 Building experienced major modifications when many of its laboratories were converted to offices.

The fuel fabrication process in the 300 Area went through numerous changes during and after World War II, especially in the 313 and 314 buildings. Fuel elements were initially fabricated and jacketed in these facilities. The 313 Building even experienced eight additions during its first year (1943-1944) due to process improvements and changes in uranium fuel fabrication activities. A decade later, the 313 Building was extensively remodeled and expanded to accommodate a new fuel canning technology. By 1971, all fuel element preparation activities for the single-pass reactors ceased in the 313 and 314 buildings with the closure of the reactors.

During the early 1950s, General Electric constructed many buildings and facilities for the Atomic Energy Commission in the 300 Area under the Hanford Laboratories Operation Program for research and development activities. The most prominent developmental laboratories and shops included the 324 Chemical Materials Engineering Laboratory, 325 Radiochemistry Building, 326 Pile Technology Building, 327 Radiometallurgy Building, and the 329 Biophysics Laboratory. The 328 Mechanical Development Building was built as a central shop to support the above laboratories.

Facility expansion continued in the 300 Area with the co-extrusion process, a new fuel fabrication method developed in the 1960s to accommodate N Reactor fuel needs. "The unique new process for jacketing or cladding uranium fuel elements for Hanford's N Reactor was initially developed in the 306 Fuel Element Pilot Plant" (Gerber 1993b, p. 7). The pilot plant was completed in 1956 initially to assist 313 Building operations and to pilot process improvements in single-pass reactor fuel fabrication methods. The facility was expanded in 1960 to develop the co-extrusion fabrication process for N Reactor fuel elements. The 333 Building was constructed at the same time to manufacture fuel elements for N Reactor using the new process.

400 Area

The construction of the Fast Flux Test Facility complex commenced in 1970 and was completed in 1978. The reactor reached initial criticality in February 1980 and began operating at full power by 1982. Reactor activities were later expanded to include long-term testing and evaluation of reactor components and systems, fusion power materials testing, passive safety testing, production of medical isotopes, and space power system research (Mayancsik 1988).

During the mid-1980s, DOE constructed the Fuels and Materials Examination Facility in the 400 Area as a major addition to the breeder reactor technology development program at the Hanford Site. Its designed purpose was to inspect irradiated fuel materials from breeder reactors, namely the Fast Flux Test Facility at the Hanford Site and the Clinch River Breeder Reactor Plant in Tennessee. Because the breeder reactor development program was cancelled, the facility was never used for its designed purpose.

600 Area

During the Cold War period, existing roads were widened and new roads built to handle expanded site uses and missions. Some temporary construction structures in the 600 Area, such as the Hot Mix Plant for road materials, survived as permanent buildings, and several miles of temporary construction rail track, roads, and walks were converted for permanent use (Gerber 1993e).

The major expansion in the 600 Area, though, was the construction of anti-aircraft artillery batteries and Nike missile systems that provided air defense of the Hanford Site from 1950 through 1961. Nike missile installations supplanted the anti-aircraft artillery sites during the mid- and late 1950s in the United States. Between 1955 and 1961, the U.S. Army deployed Nike Ajax and Hercules missiles at four locations on the Hanford Site, three on the Wahluke Slope and one on the Fitzner-Eberhardt Arid Lands Ecology Reserve (see Figure 2-1.4). Of the four Nike missile launch sites and radar control sites that replaced the anti-aircraft artillery emplacements in the mid-1950s, only the one on the Arid Lands Ecology Reserve (H-52) remains intact.

The buildings and structures at each Nike site were organized into two separate installations: the launch area and the battery control area. The launch area (H-52 L) contained the underground missile storage magazines, launch equipment, motor pools, generator buildings, acid pits, missile refueling areas, missile assembly and testing buildings, and administrative, housing, recreation, and general maintenance facilities. The main function of the launch area was to maintain missile batteries in a combat-ready posture. The battery control area (H-52 C) contained all the radar, guidance, electronic, and communications equipment needed to identify incoming targets, launch missiles, and direct and guide missiles in flight to intercept enemy aircraft.

The former battery control area is located at the top of Rattlesnake Mountain, while the former associated launching area is situated at the base of the mountain. The location of the battery control areas had to be "between a minimum of one-half mile and a maximum of three miles from the launch area. The minimum distance was determined by the maximum tracking capability in elevation of the missile tracking radar, and the maximum distance by practical considerations of providing communication by cables" (Carlson and Lyon 1996.).

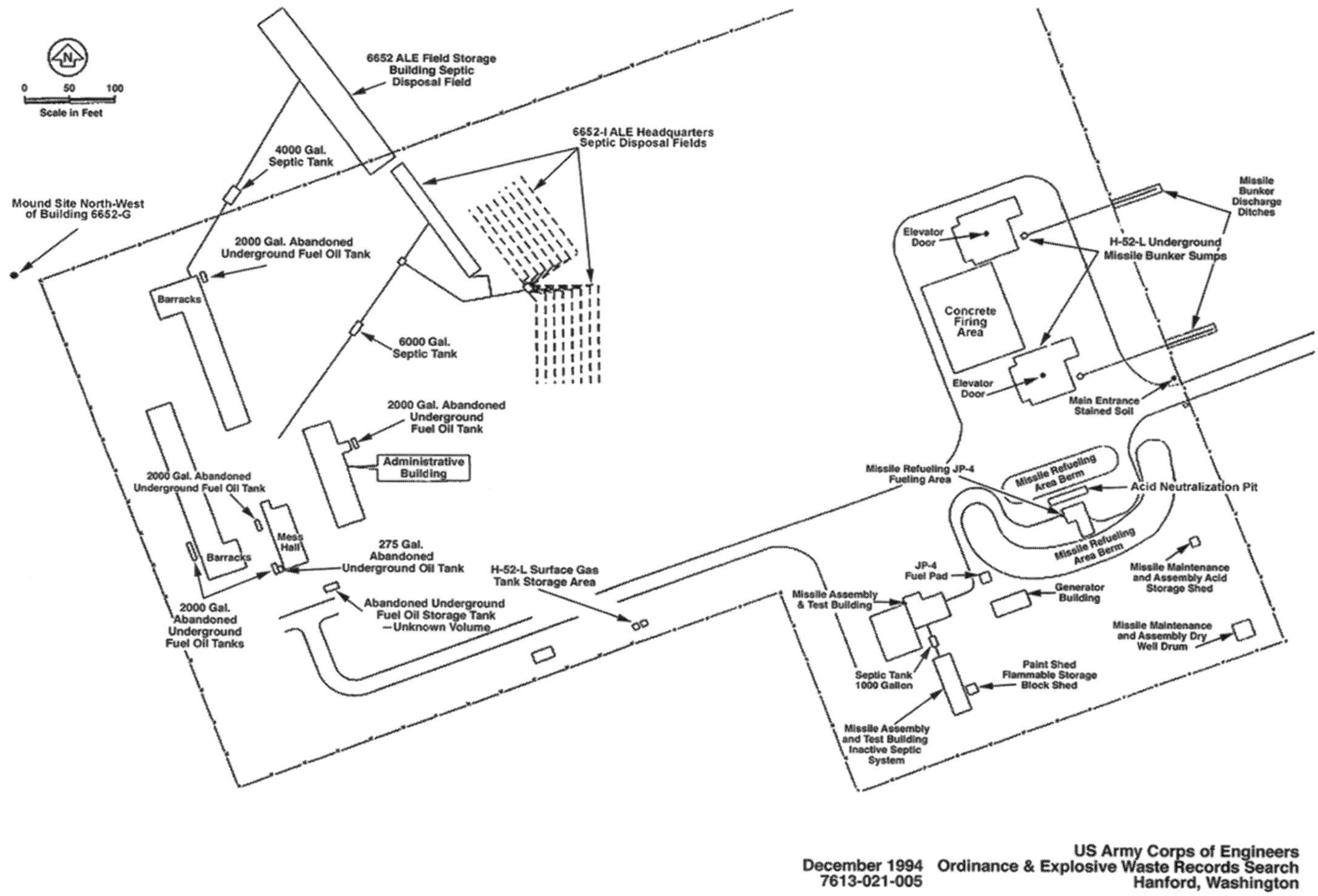

Figure 2-1.4. *Missile Launch Site at the Fitzner-Eberhardt Arid Lands Ecology Reserve*

While the spatial arrangement of the buildings and structures in each launch and battery control area was site-specific with no standard layout plans, the H-52 launch area was divided into two separate zones for functional and safety reasons. One included the administrative, mess hall, residential, and recreational facilities. The other included the underground missile storage, missile refueling, missile assembly and testing, fuel storage, and generator facilities. The H-52 control center combined the administrative and barracks functions in one building with the radar equipment located in adjacent trailers. Some Nike sites nationwide designated separate areas for housing in the operating areas, but the Hanford Site's H-52 installation had barracks in both the launch and battery control areas (McMaster et al. 1984).

1100 Area

The 1100 Area underwent considerable growth during the early 1950s in response to the Cold War induced expansion of the Hanford Site. Representative of the expansion was the construction of the 1167 Warehouse, 1170 Bus Terminal, and 1171 Transportation Maintenance Shop. These facilities and the entire 1100 Area were strategically situated adjacent to rail connections and the main north-south site highway to expediently provide support services to the site.

INDUSTRIAL VERNACULAR ARCHITECTURE

Vernacular architecture is defined as buildings and structures not classified as high style architecture, "...a catch-all term for the study of the kinds of buildings neglected by traditional architectural history...." (Rydell 1985, p. 401). Vernacular architecture can also be categorized as local or regional adaptation of widespread, high style architectural forms.

The design of the Hanford Site's industrial, utilitarian buildings is a vernacular adaptation of the International/Modernist style, an architectural expression of aesthetic functionalism that gained popularity during the post-World War II era. In this industrial environment, "a building was beautiful to the degree that it was functional" (Teague 1940, pp. 15). The Hanford Site's industrial vernacular facilities can be associated with architect Louis Sullivan's famous maxim, "that form ever follow function" (Sullivan 1896, p. 403). Architectural form can reflect industrial function not only in the design of individual buildings but also in the layout of an entire complex like the Hanford Site.

The Hanford Site is an evolving industrial vernacular landscape whose layout and design has been shaped by the variety of functional uses and changing site missions. Function plays a significant role in industrial vernacular landscapes, influencing the type of utilitarian facilities designed and constructed. In the case of the Hanford Site, this includes buildings and structures constructed in the production areas and infrastructure found throughout the site such as roads, railroads, communication and utility/electrical facilities. The Hanford Site's Manhattan Project/Cold War era landscape reflects unembellished industrial architecture, concrete and steel structures that express a sense of architectural balance and functional integrity devoid of non-essential decorative elements and ornamentation.

CONSTRUCTION STYLES AND MATERIALS

The built environment at the Hanford Site reflects industrial and utilitarian functions over aesthetic concerns, not only in the design and layout of the site's production areas but also in the design of individual buildings and construction materials used. Functional, unadorned concrete and steel were the most commonly used materials at the Hanford Site.

Concrete is a name applied to any number of compositions or combination of aggregate, consisting of sand, gravel, crushed stone, or other coarse material, bound together with various kinds of bonding materials, such as lime or cement. Various concrete applications include (Coney n.d.):

- Unreinforced concrete: composite material containing aggregate held together by a cement with water to form a paste. For an example of an unreinforced concrete structure, see Figure 2-1.5. The name derives from its not having iron or steel reinforcing bars.

- Reinforced concrete: concrete strengthened by the inclusion of iron or steel reinforcing bars, which increase the tensile strength of the concrete. For an example of a reinforced concrete structure, see Figure 2-1.6.

- Precast concrete: concrete that is cast and hardened away from the building site and then put in place in the building as a rigid component.

Figure 2-1.5. *Example of Unreinforced Concrete Structure, the 337-B Building in 1971*

Figure 2-1.6. *Example of Reinforced Concrete Structure, the 337 Building in 1970*

- Cast-in-place concrete: concrete poured on-site into a previously erected framework that is removed after the concrete has set.

One of the reasons the Corps chose the Hanford Site as one of the locations for the Manhattan Project was that sufficient aggregate would be available locally to provide enough concrete for site construction needs. Du Pont developed aggregate sources on site for the Corps as one of its first construction activities.

The amount of concrete used during the Manhattan Project was substantial. It was the most extensively used material in the construction of the Hanford Site. Laborers applied more than 780,000 cubic yards of concrete, an amount that equaled approximately 390 miles of concrete highway 20 feet wide by 6 inches thick. About 1,500,000 concrete blocks and 750,000 cement bricks were used in plant construction, enough to build a 1-foot by 6-foot wall over 30 miles long (Du Pont 1945a).

The urgent nature of the Manhattan Project at the Hanford Site dictated an emphasis on speed and functionalism, which translated into a preference for flat roof, concrete, box-like structures over more traditional architectural forms. Exterior walls exhibited minimal non-functional ornamentation. Their steel skeletons allowed the construction of non-loadbearing exterior walls made predominantly of concrete. Cladding consisted of various types of concrete applications, horizontal wood, asphalt/transite shingles, corrugated metal, and galvanized steel panels. Corrugated metal is a lightweight, ribbed cladding that is manufactured by rolling continuous flat sheets of lightweight steel or aluminum into ribbed profiles (*Architecture* 1995).

100 Areas

Three general types of permanent building construction were initially used in the 100 Areas: reinforced concrete, structural steel framing together with concrete block and/or reinforced concrete, and wood framing (Du Pont 1945a, Vol. 3). Common materials and stylistic features used in the 100 Areas during the entire Manhattan Project/Cold War period included concrete foundations and flooring, corrugated and shingled transite siding, concrete block walls, corrugated metal/pressed steel siding and roofing (see Figure 2-1.7 for an example), wood and steel framing, flat precast concrete roofs covered with tar and gravel surfacing or composition shingles, and doors constructed of corrugated, roll up metal or double leaf, single panel wood.

Figure 2-1.7. *Example of Corrugated Metal Siding, 185.5-KW Building in 1995*

Since sand and gravel deposits were found immediately under the surface in the 100 Areas, two aggregate pits were opened in the 100-B Area. However, that aggregate could only be used for road construction and stabilization work. To make concrete, concrete aggregate had to be brought in from the Haven gravel pit located approximately 1/2 mile west of the 100-B Area. "The Haven pit was capable of producing 9,600 tons of concrete aggregate daily. The Hanford Site gravel pit, located just west of Hanford proper, was capable of producing 14,832 tons of concrete aggregate daily" (Du Pont 1945a, Vol. 3, p. 650).

The Hanford Site concrete production plants were in the 100 Areas. "One concrete mixing plant was erected in the immediate vicinity of the 105 [Reactor] Building[s] in each of the 100-B and D areas by the Hanford concrete contractors to furnish ready-mixed concrete for construction. Since a sizable portion of concrete yardage was concentrated at the 105 Building and the 185, 189 and 190 building group, a Pumpcrete method (concrete mix pumped through pipes) of placing concrete from a central pumping plant was used" (Du Pont 1945a, Vol. 3, p. 651).

The availability of prepared concrete materials expanded at the Hanford Site during the post-World War II period. In the 100-H Area, the bulk of the concrete was prepared in the central mixing plant erected within the 100-H Area. Some amounts were also furnished from the White Bluffs plant.

The cladding of the buildings for each of the reactors was consistent. Reinforced concrete shielding walls and corrugated asbestos cement siding were commonly used. Roof construction, except for reinforced concrete slabs over the inner rod room and rear face enclosure, was poured insulated concrete. The ceilings were constructed of laminated gypsum board. The roofs were covered with precast concrete roof tile, except over the discharge area enclosure and the inner horizontal rod room. Over these areas, the roofs were composed of 6-foot-thick reinforced concrete. The massive reinforced concrete walls around the reactor core at the lower levels provided additional radiation shielding.

Figure 2-1.8. *Example of Reactor Building, 105-DR Building in 1995*

The reactors rested on 23-foot thick concrete foundations topped with cast iron blocks that served as a thermal shield. The walls consisted of reinforced concrete in the lower portions and concrete block in the upper portions, varying from 3 to 5 feet thick. The graphite cores were surrounded by a cast iron thermal shield layer. The entire thermal shield was surrounded on all sides (except the bottom) by a 52-inch thick biological shield that consisted of alternate layers of masonite and steel. For an example of the concrete buildings containing the reactors, see Figure 2-1.8.

Figure 2-1.9. *Example of Chemical Separations Building, 221-T Plant*

200 Areas

Construction features and materials of the 200 Areas are similar to those used in the 100 Areas and other production areas. Due to the 200 Areas chemical separations function, however, the Manhattan Project/early Cold War period facilities used considerably more concrete, fewer and smaller windows, symmetrical plans, and foundations constructed of poured concrete or reinforced concrete piers with spread footings. Floors, walls, and ceilings were made of reinforced concrete or concrete block. Roofs were likewise constructed of reinforced concrete and covered with built-up felt, tar, and gravel material.

Five general types of building construction were initially used in the 200 Areas. In addition to the three types found in the 100 Areas, the 200 Areas also had reinforced concrete framing with concrete block and structural steel framing with wood siding (Du Pont 1945a, Vol. 3). The common materials and stylistic features found in the 200 Areas were similar to those in the 100 Areas.

The 200 Areas separations buildings used by far the most concrete of any other area. The amount is unknown but can be inferred by the dimensions of the buildings. The Manhattan Project separations buildings (U, T, B Plants) were massive reinforced concrete structures, measuring over 800 feet long, 65 feet wide, and 80 feet high. Because of the radioactivity present during the separation process, the concrete walls surrounding the separation cells were constructed 7 feet thick to provide necessary protective shielding. Cell covers were constructed of removable 7-foot-thick concrete blocks. For an example of a separations building built during the Manhattan Project, see Figure 2-1.9.

REDOX, like the other separations buildings, is a monolithic, rectangular (467 feet by 161 feet by 82 feet high), flat roof structure constructed almost entirely of reinforced concrete. Built in 1952, the Reduction-Oxidation Plant with its chemical separation process was chosen to replace the bismuth phosphate process employed at B and T Plants.

"As with other canyon [separations] buildings, its process equipment is contained in small rooms, called cells, which are arranged in rows in an area spanned by a traveling crane. The cells are topped with 4-foot concrete blocks that are removable by crane to provide access to the cell beneath.... Heavy concrete shielding walls... are up to the level of the crane rails, giving the appearance of a canyon.... From a process view point, the REDOX facility is divided into a canyon area and a silo area. The canyon consists of nine process cells arranged in two parallel rows running east to west and separated by a pipe tunnel. The five-foot thick concrete shielding walls protected REDOX workers from the intense radiation found in the process cells. The silo, located on the west end of the building is 84 feet by 41 feet by 132 feet high and contains a process area and an operating area. Its extraction column shaft is 12 feet by 69 feet by 86 feet high and has eight floor levels." - DeFord and Carpenter 1995, p. 2-1

> ***"The frame is of structural steel with an outer sheathing of aluminum panels over rock wool insulation and 16-gauge sheet steel. The first floor is concrete slab; the duct level is sheet metal roof decking, and the second level is a concrete slab. The roof is insulated metal decking. Interior walls are reinforced concrete steel, metal studs, metal lath, and plaster. The vault and process area doors are constructed of steel...*** **- see "Plutonium Finishing Plant" in Appendix B on the Internet**

Because the REDOX process used gravity flow columns, the columns were housed in a silo. By the time PUREX was constructed, pulsed columns were used instead of gravity flow columns so the silo required for the REDOX process was not needed for the PUREX process. PUREX is a reinforced concrete, rectangular-shaped facility measuring 1,005 feet long, 104 feet high (with approximately 40 feet below grade), and 61.5 feet wide. The Plant's main canyon portion was originally 860 feet long. The reinforced concrete cell blocks were fabricated outside PUREX prior to installation (Gerber 1993f).

The Plutonium Finishing Plant (Z Plant) is 180 feet wide by 500 feet long, extending 9.5 feet below grade to 46.8 feet above grade.

Except for wood framing in the Manhattan Project administrative/non-production facilities, most of the smaller, non-chemical separations buildings were constructed of structural steel framing with outer sheaths of aluminum panels, corrugated metal, or transite shingles. Some roofs were constructed of insulated metal decking. Interior walls were made of reinforced concrete, sheetrock/plasterboard, or general plaster covering. Early wood frame facilities had symmetrical features with repetitive, multipane, industrial-style windows, and gable roofs. Cladding consisted of corrugated metal or transite/asbestos shingles over the original horizontal wood siding.

Figure 2-1.10. *Example of Concrete Block Structure, 3745A Building in 1994*

300 Area

The three general types of building construction in the 300 Area were the same as those in the 100 Areas (Du Pont 1945a, Vol. 3). Most of the materials and styles used in the 300 Area facilities constructed during the Manhattan Project and Cold War era were steel frames with reinforced concrete or concrete block, asymmetrical features, flat roofs, and corrugated metal cladding. The hazardous materials storage facilities, like the 303 Fresh Materials Storage buildings for instance, were constructed of reinforced concrete and concrete block. For an example of a building constructed only with concrete block, see Figure 2-1.10.

Figure 2-1.11. *Example of a Temporary Wood Frame Structure, 3762 Building in 1995*

Figure 2-1.12. *Example of Structure with Metal Paneling, 309-E Building in 1995*

Figure 2-1.13. *Example of Transite Shingle Cladding, 3707-B Building in 1994*

Figure 2-1.14. *Construction of the Fast Flux Test Facility in the Early 1970s*

Some of the Manhattan Project and early Cold War period buildings were of wood frame construction with symmetrical rectangular plans (see Figure 2-1.11), transite (asbestos) shingle or corrugated metal cladding (over original wood siding), gable roofs, and repetitive, industrial style double-hung sash windows, and swamp coolers. Such buildings were non-manufacturing facilities, like the Manhattan Project administration facilities (for example, the 3702 and 3703 buildings). The 303-J Building, however, has a wood frame construction even though it contained hazardous materials.

In the larger buildings composed of several wings or which are of considerable length (for example, administration and service buildings), fire walls of common red brick with salt glazed tile caps were used to reduce fire hazards involved in this type of construction. The 3706 Radiochemistry Laboratory has such a fire wall. Access openings through these walls were equipped with metal clad fire doors set in steel framing. Vaults attached to these buildings were also of brick and tile (Du Pont 1945a).

By the early 1950s, 300 Area building construction had adopted additional asymmetrical design features. Framing for the most part was constructed of concrete and steel, and cladding consisted of metal paneling (see Figure 2-1.12), asbestos transite shingles (see Figure 2.1-13), corrugated transite or corrugated metal. By the 1970s and 1980s, new buildings, especially in the central and southern part of the 300 Area, were rectangular shaped, steel and wood framed, with a single story and aggregate pebble/stucco siding over plywood sheathing. Many of these facilities were prefabricated units, installed for office and laboratory use with premanufactured sidings and roofs constructed of fluted metal panels, sheet metal, or corrugated metal.

400 Area

Because the 400 Area buildings were built in the 1970s, their construction is similar to the post-1970 facilities in the 300 Area. The primary materials used in the construction of the 400 Area Fast Flux Test Facility facilities included extensive applications of concrete, textured concrete facades, steel framing, fluted metal panels, sheet metal, and corrugated metal. Premanufactured construction materials were used in the bolted steel (Butler-style) buildings in the 400 Area. Several of the wood or steel frame, flat roof buildings have fluted concrete block walls or exterior finishes of stucco fascia material/aggregate stucco over plywood sheathing with tar and gravel built-up roof coverings. See Figure 2-1.14 of the Fast Flux Test Facility being constructed.

The Fast Flux Test Facility complex includes a containment building, reactor, heat removal equipment and structures, and core component, handling/examination, instrumentation, and control facilities (Mayancsik 1988). The containment building, which houses the reactor and plant operating equipment, is "a cylindrical carbon steel shell. Steel-lined reinforced concrete cells occupy the lower portion of the containment building, from grade level to approximately 78 feet below grade.... A structural steel mezzanine above the operating perimeter provides additional work area.... A 200-ton polar gantry crane and a jib crane are located above the mezzanine for handling large equipment and materials. The central portion of the operating floor is occupied by a steel operating deck" (Mayancsik 1988, p. 1-8).

The adjacent 403 Fuel Storage Facility has a below-ground cell that contains a carbon steel storage vessel 21 feet in diameter that provides storage in liquid sodium of the spent fuel assemblies from the Fast Flux Text Facility.

The 400 Area Fuels and Materials Examination Facility (427 Building) was designed in the mid-1970s and constructed during the early 1980s as a major addition to the breeder technology development program at the Hanford Site. The 427 Building, a poured concrete building measuring 175 feet by 270 feet by 98 feet high, is typical of industrial, asymmetrical, flat roof facilities in the Fast Flux Test Facility complex. The building extends 35 feet below grade and has a total of 188,000 square feet of operation space. The facility is divided into six operating floors with an attached mechanical equipment wing on the west side.

600 Area

The Nike missile sites were some of the facilities designated as 600 Area facilities, which were not all located geographically together. The thirteen Nike-period structures at the H-52 launch site, including the underground missile storage facility, are constructed mainly of concrete. Common stylistic features and construction materials include one story, flat roof, concrete block structures erected on concrete footings and on-grade concrete floor slabs. Wood or metal, multipane, double-hung sash windows, or fixed sash and glass block windows, are symmetrically placed. Roofs are constructed of wooden joists covered with wood sheathing and built-up cover of tar and gravel. Smaller support facilities (water pumping, sewage, storage) are constructed of concrete block or wood frame, with flat or shed roofs, minimal windows and metal doors. The fueling and missile maintenance and assembly areas were surrounded by dirt and grass berms for safety protection purposes. The only visible structures at the surface of the concrete underground missile storage facility are two entrance doors set in the mounds of earth that cover the concrete superstructure. Concrete pads cover the former missile firing area.

The structural remains of the five most intact anti-aircraft artillery sites, situated in the Hanford Site's central plateau south and southeast of the 200 Areas, include artillery emplacement revetments and other sandbagged/cobblestone structures and the remains of concrete pathways/sidewalks, building slabs, entry pads, and flooring.

Other 600 Area facilities include fire stations, atmospheric and environmental monitoring facilities, weather station structures, and magazine storage facilities. The wood and steel frame buildings mainly have flat roofs and are constructed of concrete block walls with horizontal wood siding, while the weather and environmental monitoring (meteorology) towers and radar facilities are constructed of steel and set in poured concrete slabs. The 213 J and K magazine storage facilities, which stored purified plutonium-nitrate paste, contaminated sodium, and soil samples for fallout studies, are constructed of massive reinforced concrete. The remaining 600 Area facilities are Butler-type, storage/support buildings constructed of bolted steel and corrugated sheet metal.

700 Area

Four DOE buildings/structures remain in the 700 Area, all constructed during the Cold War period. These one-to-two story, steel and reinforced concrete frame buildings have flat roofs, concrete block/reinforced concrete walls erected on concrete footings or on-grade concrete floor slabs. The premanufactured structures have aggregate stucco and corrugated metal cladding, flat roofs, and wood framing.

1100 Area

Common construction characteristics and materials in the 1100 Areas include steel framing, concrete block, corrugated metal cladding, and gable truss roofs with corrugated metal. Foundations and floors are mainly reinforced concrete.

Some facilities have flat roofs, precast concrete walls, steel columns and roof beams. Pre-manufactured facilities include Butler-style buildings with corrugated metal siding and roofing, steel frames, situated on concrete foundations or ground level concrete slabs.

TEMPORARY CONSTRUCTION FACILITIES

Because of the emphasis on speed and cost-efficiency during the Manhattan Project and Cold War Era, Hanford Site contractors constructed numerous temporary construction facilities. Many of the temporary construction facilities still exist. To accomplish speedy and low cost construction, standard design and assembly line construction were adopted and used throughout the site, especially in the construction camp at the former Hanford townsite, White Bluffs Shops, Central Shops, 3000 Area Camp, Camp Hanford, and in the 100-B, D, F, 200 East and West, 300, 700, and 1100 (Richland Village) areas. While the structures appear to be standard issue shells on the outside, the interiors were typically redesigned to suit a particular operation. In assembly line construction, all materials for a certain building or group of similar buildings (for example, barracks, bathhouses) were prefabricated at the various craft shops and then sent to the building site for installation. Modified versions of the Navy's 800 series B-2 barracks were used extensively to house workers at the Hanford construction camp and the Richland Village. After the war, a considerable number of these barracks were moved to the 100, 200 and 300 areas for use as offices and other administrative purposes.

Four common types of wood framing were used in the construction of temporary buildings (Du Pont 1945a):

- Post-and-girder construction for large flat-roofed sheds, wide gable-roofed buildings, and multi-story structures having large deadweight loads. This type of framing was seen in many administration and service buildings, garages, warehouse, and shops.
- Column-and-truss construction for large roof spans and high ceiling clearances.
- Studded-and-gable roof construction for smaller width structures having dead-weight roof loads. This type was mainly used for barracks, small area construction offices, and for all buildings brought in on skids.
- Open-shed construction with the roof supported by post and girders. This type was used in the construction of storage buildings and field craft shops.

Common types of roofs used included gable, hipped, shed, and flat roofs. Corrugated, galvanized sheet iron roofs and plywood roof and wall sections were used for prefabricated sheds. Reinforced concrete roofs were used for vault construction.

The siding of temporary facilities in the Hanford construction camp, 3000 Area Camp, and administration buildings in the process areas consisted either of 1/2-inch Gypsum board applied horizontally in 2-feet by 8-feet sheets with V-matched top and bottom edges, 1-inch by 6-inch tongue and groove, or 1-inch by 8-inch ship-lap siding (Du Pont 1945a).

Figure 2-1.15. *Example of a Butler Building, 352-D Building in 1994*

In the process areas, interior building lining was used in administration/office construction. Walls, partitions, and ceilings were covered on only one side of the studs with 1/2-inch Celotex, Firtex, or Sheetrock. The doors were wood with the exception of the metal-clad fire doors and fireproof vault doors. Window sash and frames were wooden throughout. Two standard types of windows were used: double-hung or single-pivot/slide windows, all with either six or eight panes. Ventilation outlets consisted of gable louvers, roof and wall evaporation coolers, sheet metal roof ventilators, wooden roof ventilators, and wall and ceiling exhaust fans (Du Pont 1945a).

PREFABRICATED UNITS

To accommodate the larger than expected work load during the Manhattan Project and Cold War Era and for speed and cost-efficiency, prefabricated units were shipped to the Hanford Site for a wide variety of construction uses. All the process areas had prefabricated sheds that included Hobb's Pacific huts and Butler type huts. Pacific huts were used mainly for barracks, commercial facilities, and offices at the Hanford Site construction camp. The larger Butler huts were used as special storage warehouses, especially in the 200, 300, and 1100 areas. The smaller Butler huts were used for offices, warehouses, and small shops in most areas. The Butler-type buildings were constructed of corrugated metal or sheet metal and bolted steel, commonly resting on a poured concrete slab (see Figure 2-1.15).

Figure 2-1.16. *Example of a Quonset Hut, 3746-D Building in 1994*

Several types of prefabricated units, which included Quonset huts (see Figure 2-1.16), were transferred from other government projects and used principally for field offices and warehouses. Quonset huts (still in use in the 200, 300, 400, 600, and 1100 areas) are prefabricated units constructed of semi-cylindrical roofs of galvanized, corrugated sheet metal attached to metal purlins supported by steel ribs. Hook bolts with nuts and washers connect the purlins to the ribs. The semicircular ends of the building are of board and batten construction. The windows and flooring are constructed of wood.

HIGH STYLE ARCHITECTURAL FORMS

While the vast majority of Hanford Site buildings are industrial vernacular structures, several of the facilities exhibit high style architectural features, such as the following.

Brutalism: The 337/337-B buildings and certain features of the 331 Building exhibit the architectural characteristics of Brutalism. The term comes from the French "Breton Brut," meaning rough or untreated concrete. The goal of this style is an honesty in structural, spatial, organizational, and material concepts that result in buildings characterized as rude and ruthless, where the scale or relation of mass and detail to human beings is often referred to as brutal. The distinguishing characteristics of the style (Kirk 1997), which experienced its peak popularity from the years 1955 to 1970, include:

- Concrete exposed at its roughest
- Exaggerated structural members
- Raw and unfinished simple materials
- Formalist (building forms clearly express their function)
- Grand scale
- Exposed mechanical systems, ducts, pipes
- Functional elements doubling as decorative elements

Art Deco/Art Moderne: The architecture of the 3760 Building, the former Hanford Technical Library, is a vernacular adaptation of Art Deco/Art Moderne features and styles. Distinctive elements include the building's streamlined but angular, hard-edged form and a projecting concrete entrance/frontispiece placed in a modified front facade setback. The symmetrical design of the front facade is highlighted by parallel concrete piers/pilasters. A strong, decorative horizontal banding is expressed in the repetitive ribbon windows.

LANDSCAPE MODIFICATIONS

The Hanford Site's flat, arid environment was perfectly suited to the needs of the federal government for the Manhattan Project and wartime mobilization effort. Like other desert environments in the American West used by the federal government, the Hanford Site was viewed as an isolated wasteland, remote from population centers, to be used indiscriminately for national defense or natural resource extraction purposes. The resources of the desert landscape were seen as inexhaustible, like the Hanford Site's abundant water supply from the Columbia River and the area's glacial sediments that provided sand and gravel aggregate for constructing large concrete structures.

The establishment of the Hanford Site with restricted public access because of the production of plutonium and nuclear technologies research had a two-fold effect on the area landscape. While the Hanford Site landscape was radically altered by construction activities associated with the Manhattan Project/Cold War period, the construction of the Hanford Site also resulted in removal of most pre-1943 buildings from the Hanford and White Bluffs townsites as well as other rural structures. Most of the early settlements and farmsteads were situated along the Columbia River as were later the plutonium production reactors and ancillary facilities.

Soon after the end of World War II, many of the wartime mobilization structures or temporary construction facilities were removed. For example, the Hanford construction camp located at the former Hanford townsite was totally removed within a year after the end of the war. At its peak occupancy in 1944, the camp housed approximately 45,000 construction workers, a city in itself that consisted of hundreds of temporary construction facilities, trailers, stores, schools, churches, barracks, medical, and tent units. During the life of the camp, a total of 1200 buildings and 9 service facilities were constructed. While no structures remain at the Hanford construction camp, aerial photographs detect the grid system of the former townsite.

Other similar modifications to the landscape from the Manhattan Project and Cold War period are scattered over the Hanford Site landscape. The former Central Shops facility (a major maintenance and servicing area for Manhattan Project construction activities located north of 200 West and removed shortly after the end of World War II) left a landscape dotted with concrete slabs, foundations, and pits. Similarly, the concrete slabs, foundations, walkways, roads, and grids of former anti-aircraft artillery and Nike facilities are still evident in stark contrast to the surrounding desert landscape.

The Hanford Site's distinctive Manhattan Project and Cold War landmarks still standing include the reactors, separations buildings, water towers and stacks, yellow radiation markers to identify surface and sub-surface contamination, nuclear waste burial grounds for contaminated material, waste storage ponds, railroads and roads, and the production/process areas themselves.

Since 1987 when the last of the nine reactors was shut down, operational activities have shifted toward cleanup of areas contaminated by radioactive and/or chemical wastes. The mission change to environmental restoration/remediation has had a significant impact on the industrial landscape. Decommissioning and decontamination activities have led to extensive alteration and demolition of many of the remaining Manhattan Project/Cold War buildings and structures. For example, the site's military landscape (located in the 600 Area) has also been dramatically altered with demolition of most of the Nike and anti-aircraft artillery sites. Of all the production areas, the 100-B, C, D, DR, F, and H areas have been the focus of the most extensive demolition activities.

While the industrial landscape at the Hanford Site has been significantly affected by cleanup activities through the demolition of numerous buildings and structures, the basic design and configuration of the site production areas has remained relatively intact.

PRIMARY CONTRACTORS/SUBCONTRACTORS

Throughout its existence the Hanford Site has been a government-owned, contractor-operated facility. Ownership of the Hanford Site has been passed between four different government agencies, and a total of seventeen different contractors have been responsible for a wide array of site operations, including construction. The Hanford Site's management and contractors' history can be organized into three phases: single contractor, diverse contractors, and consolidation of contractors. Because the construction history of the Hanford Site was affected by the various contractors, it is important to know who the contractors were.

SINGLE CONTRACTOR

U.S. Army Corps of Engineers and E. I. Du Pont de Nemours and Company

The Corps, Manhattan Engineer District, entered into a cost-plus-fixed-fee contract with Du Pont for the design, engineering, construction, equipping, training personnel, and operation of the Manhattan Project at the Hanford Site. Du Pont, the prime contractor at the Hanford Site from December 21, 1942 to September 1, 1946, had general supervision of all construction activities. Du Pont designed the Hanford Site structures on the basis of technical and process information developed and furnished to Du Pont by the University of Chicago and other scientists under government contracts.

The Hanford contract during the Manhattan Project allowed the work to be subcontracted. The two principal attractions of subcontracting work were economy and speed. "Many subcontractors specialize in certain types of work and maintain well-trained and highly-skilled organizations which are available for transferring to the project site and starting work shortly after acceptance of the subcontracts.... Hence the time and cost are saved which would be spent in recruiting skilled workmen...or in training unskilled employees...." (Du Pont 1945e, Vol. 2, pp. 289-290).

Whenever practical, Du Pont subcontracted work to use the administrative expertise and personnel of as many firms as possible. Du Pont subcontracted portions of the work considered to be of a specialized nature, such as masonry work, roofing, construction of roads and railroads, and erection of tanks, structural steel, boilers, stacks, and coal handling equipment. For example, a subcontractor mixed the concrete at the five concrete plants, but Du Pont poured the concrete in the buildings. On a larger scale, Du Pont and the subcontractors jointly constructed the central administration area known as the 700 Area.

Other subcontracted work included the construction of infrastructure facilities. For example during the Manhattan Project, a subcontract was awarded to the Guy Atkinson Company "for the construction of the Permanent Railroad System and the Temporary Railroad Facilities... Approximately 30.8 miles of temporary railroads were constructed..." (Du Pont 1945a, Vol. 4, p.1092).

In a few cases, critical labor shortages were important factors considered in reaching decisions to subcontract parts of work instead of having it performed by Du Pont employees. "Some of the well-known contractors, because of their past experience and many contacts in the contracting field, had large labor followings from which available workmen could be drawn" (Du Pont 1945e, Vol. 2, p. 290).

The types of work subcontracted for the Hanford Site were practically the same as those subcontracted in the construction of other war projects handled by Du Pont. "However, the remote geographical location and mammoth size of the Hanford plant necessitated the provision of housing and eating facilities for practically all of the construction and operation employees, and resulted in the awarding of some contracts that were not common to the other projects... Because of these abnormal conditions, subcontracts were awarded for the operation of commissaries and eating facilities and for the design of Richland Village" (Du Pont 1945e, Vol. 2, p. 290). It was practical for Du Pont to subcontract out commissary responsibilities due to their lack of previous experience in this field, as it was desirable to secure the services of a local architect familiar with local conditions and lessen the war-imposed load on the Architectural Section of the Du Pont Design Division.

Atomic Energy Commission and General Electric Company

On September 1, 1946, the General Electric Company replaced Du Pont as the prime site contractor. General Electric remained the prime contractor until 1965. Soon after the arrival of General Electric at the Hanford Site, ownership of the Hanford Site was transferred from the Corps, Manhattan Engineer District, to the newly created Atomic Energy Commission on January 1, 1947.

General Electric was directed to build two new production areas, 100-DR and 100-H. The Design and Construction Divisions of General Electric were responsible for the design, specifications, material and equipment lists, and direction of the work for the process units. For 100-H Area alone, approximately 2,500 drawings were made by General Electric and their prime architect-engineer and construction subcontractor.

As prime site contractor, General Electric relied upon numerous subcontractors to conduct their site-wide operations and management responsibilities. The construction procurement division of General Electric and the purchasing section of Atkinson-Jones Company were jointly responsible for the procurement of building materials, construction equipment, and miscellaneous materials and supplies. General Electric often procured cement from the Spokane Cement Company and the Lehigh Company, both of Washington State. Curtis Sand and Gravel Company, a Hanford Site subcontractor, supplied the concrete aggregate (GE 1952a).

J. A. Jones and Vitro Engineering were two contractors who arrived at the Hanford Site in 1953 to provide engineering and construction services. The J. A. Jones Company remained at the Hanford Site until 1987 and was responsible initially for construction of three plutonium production reactors, complete with support facilities that included fuel preparation stations, water treatment and storage plants, retention basins, powerhouses, and associated roads, rail lines, and utilities. J. A. Jones provided a variety of services at the Hanford Site that included construction, maintenance, fabrication, and property management services. Their administrative offices in north Richland included fabrication shops, warehousing, a weld test laboratory, and heavy equipment and automotive shops. Vitro Engineering was initially awarded the Hanford Engineering Services contract and provided design-engineering functions at the Hanford Site until 1981.

The architect-engineering firm Giffels & Vallet, Inc., of Detroit, Michigan was a major subcontractor during the 1940s and 1950s. The firm had a large specialty group of architects and engineers performing a wide range of design and engineering services. For example, Giffels & Vallet was responsible for the design and field supervision of all buildings and service facilities in the 100-H Area, including the 105-H Reactor facility. Because Giffels & Vallet already had staff with security clearances, they were selected during the late 1940s due to the emergency nature of the contract that had considerable national security implications.

DIVERSE CONTRACTORS

By the end of 1963, all nine Hanford Site reactors and their supporting fuels preparation, reprocessing, and waste management facilities were operating at full capacity. Employment was nearly 9600 people: 8300 with operating and service contractors, 1000 on construction and miscellaneous contracts, and 290 with the Atomic Energy Commission's Richland Operations Office. General Electric held the primary plant operating contract. Vitro Engineering provided engineering services, and J. A. Jones Construction Company provided minor construction services. In 1964, the Atomic Energy Commission decided the operation of the Hanford Site should be transferred from General Electric to several other contractors. By the end of 1965, the transition from essentially a single contractor to multiple contractor operation had been completed, and a new era had started.

The period from 1965-1987 was a time of contractual diversification at the Hanford Site. Between 1965 and 1967, seven contractors would take over from General Electric, who left the Hanford Site in 1967. For example in 1965, Battelle Memorial Institute assumed responsibility for managing the Hanford Site laboratories that had been operated by General Electric. These were renamed the Pacific Northwest Laboratory.

The selected contractors included (UW 1977):

- Pacific Northwest Laboratory: Battelle Memorial Institute-Pacific Northwest Division (BMW)
- Reactors and Fuels: Douglas-United Nuclear, Inc. (DUN), joint venture between McDonnell Douglas Aircraft Company and United Nuclear Corporation
- Chemical Processing: ISOCHEM, Inc. which transferred to Atlantic Richfield Hanford Company (ARCHO) in 1967
- Support Services:
  - General Services: ITT Federal Support Services (transferred to ARCHO in 1971)
  - Computer Services: Computer Services Corporation (replaced by Boeing Computer Services in 1975)
  - Bioassay Dosimeter Analysis: U.S. Testing Company
  - Industrial Medical: Hanford Environmental Health Foundation
  - Plant Engineering: Vitro Engineering, a division of Automation Industries, Inc. (replaced by Kaiser Engineers in 1982)
  - Minor Construction: J. A. Jones Construction Company

The Atlantic Richfield Hanford Company succeeded General Electrical as prime operator at the Hanford Site in 1967.

Rockwell Hanford Operations, a component of the Atomics International Division of Rockwell International, was responsible for operating one of the world's largest radiochemical processing complexes at the Hanford Site. Rockwell succeeded Atlantic Richfield Hanford on July 1, 1977, assuming control of the reprocessing operations until 1987 (UW 1977).

The Westinghouse Hanford Company arrived at the Hanford Site in 1970 to build and operate the Fast Flux Test Facility, a state-of-the-art research reactor, and operate the Hanford Engineering Development Laboratory, a center for breeder technology.

In 1975, ownership of the Hanford Site was transferred from the Atomic Energy Commission to the newly established Energy Research and Development Agency. Two years later, DOE was established and assumed control of the Hanford Site.

CONSOLIDATION OF CONTRACTORS

In 1985, the DOE announced it would consolidate the operations/management and engineering/construction contract at the Hanford Site, a process completed in 1987 with the Westinghouse Hanford Company awarded the consolidated management and operations contract that included control over reactor operations, fuel manufacturing, chemical engineering, and waste management activities. Kaiser Engineers took over the consolidated engineering/construction contract, and Boeing Computer Services was awarded the computer subcontract providing data processing and software development services.

AREAS FOR FURTHER RESEARCH

Constructing facilities for processes, which were in the midst of being developed, was a major challenge. Add to that the schedule pressures to win the war and the need for secrecy and it is a wonder that the Hanford Site went from desert to the startup of the B Reactor in 1-1/2 years. That initial period of construction and the subsequent years of increasing production capacity and keeping the facilities and equipment in step with the latest research findings still has many facets to be explored. Because this narrative is a history of facilities and not intended to be the definitive history of construction at the Hanford Site, the details of the construction challenges, successes, and failures and the construction workers' perspectives could not be examined. Hopefully, this narrative will serve as a guide for future researchers to pursue more extensive study of important issues noted in this section, such as

- Relating the architecture or at least the construction materials to the function for the main facilities, especially the nuclear reactor and chemical separations facilities. For example, was reinforced concrete required for buildings with potentially explosive structures in them, such as the reactor buildings, or was wood used only for buildings with no hazardous operations in them?
- Studying how the Hanford Site represents scaled-up and industrialized versions of reactors at the Oak Ridge Reservation
- Exploring how Lieutenant General Leslie Groves persuaded E.I. Du Pont de Nemours to accept a prominent role at the Hanford Site

SECTION 2

FUEL MANUFACTURING

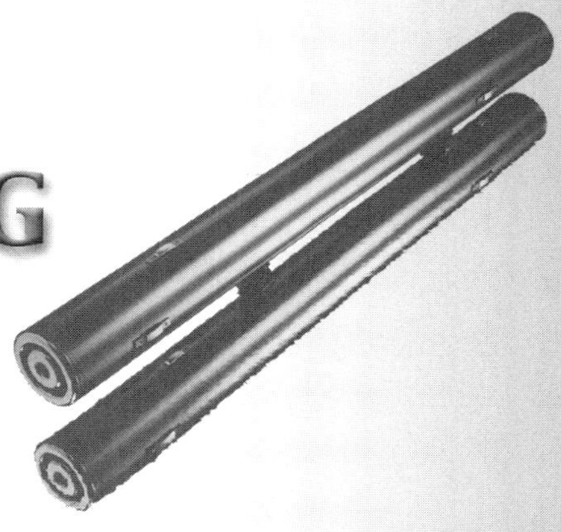

D.W. HARVEY
PACIFIC NORTHWEST NATIONAL LABORATORY

SECTION 2 - FUEL MANUFACTURING

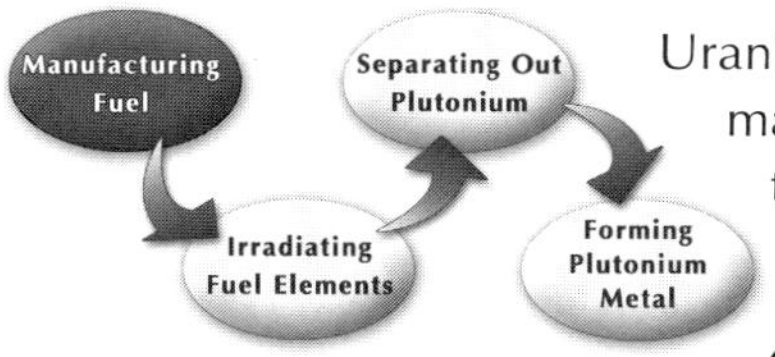

Uranium fuel elements were the heart of the plutonium production process. The manufacturing of the elements was the first step in the process at the Hanford Site and one that illustrates the magnitude of the uncertainties that accompanied every step.

"The manufacturing of fuel elements represented the expenditures of vast amounts of research and development efforts to extrapolate on a commercial scale what... had been purely research data. The employment of uranium, prior to its inception in the... program, had been confined generally to its application in small amounts, measured in grams, in the laboratory. Therefore, the immediate fabrication of amounts ultimately aggregating hundreds of tons imposed an extremely difficult assignment on the Du Pont company, and necessitated the inauguration of...developmental programs to determine...the influence on these of microscopic quantities of impurities present in uranium. Because the factor of speed dominated all phases of the project...it was necessary that all developmental and experimental work be performed simultaneously with research, design, fabrication and actual construction in the field." - Du Pont 1945a, pp. 356-357

Fuel fabrication workers manufactured nuclear fuel in the Hanford Site's 300 Area in the form of solid uranium pipe-like cylinders or fuel elements encased in aluminum cans. The fuel elements contained fissionable uranium-235, the source of neutrons needed to produce plutonium-239. "Slugs" was another term used for fuel elements at the Hanford Site, at least until the early 1950s (Weakley 1999).

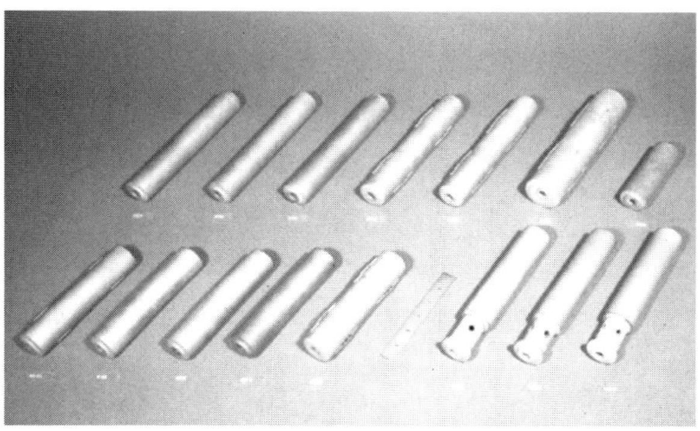

Figure 2-2.1. *Finished Fuel Elements, 1964*

The process of manufacturing fuel consisted of foundry and machine shop operations that converted uranium billets into fuel elements. Figure 2-2.1 shows the finished fuel elements. When completed, the fuel elements then went on to the second step in the plutonium production process to be irradiated in the reactors located in the 100 Area.

The fuel manufacturing process was a step-by-step operation beginning with the delivery of uranium billets to the Hanford Site. The uranium ore had been mined from Africa, Canada, and the United States. After the uranium ore was processed and delivered to the Hanford Site, workers stored the billets in one of the nine 303 buildings (303-A, B, C, D, E, F, G, J, and K) within the 300 Area. Before the billets were sent to the 313 and 314 buildings to begin the fuel manufacturing process, inspectors tested samples of these billets at random in the test reactor in the 305 Building to ensure purity and quality. The 313 Building was the Metal Fuels Fabrication Building, and the 314 Building was the Metal Extrusion Building. For information on the buildings, see the Expanded Historic Property Inventory Forms for the 305 and 314 buildings and the Historic American Engineering Record for the 313 Building in Appendix B on the Internet.

CONTENTS

Divider Page Photo: Fuel Elements (Neg. 8201457-3cn)

The fuel manufacturing process involved selecting billets of suitable quality to be extruded into rods, machined into cores, cleaned and canned in aluminum or zirconium cladding, machined and welded into finished fuel elements, and, finally, subjected to quality assurance testing. Fuel fabrication workers produced aluminum-clad fuel for the single-pass reactors in the 313 and 314 Buildings from 1944-1971, while zirconium-clad fuel production (co-extrusion process) for N Reactor took place in the Fuel Cladding Building (333 Building) from 1961-1988. Researchers developed the co-extrusion process in the 1960s in the 306-E Development, Fabrication and Test Laboratory, and fuel fabrication workers applied the co-extrusion process in the 333 Fuel Cladding Facility. For more information on the 306-E and 333 buildings, see the Expanded Historic Property Inventory Form in Appendix B on the Internet.

The mission of the 333 Building was to manufacture fuel elements for the N Reactor using the newly developed co-extrusion process. Co-extrusion was an unprecedented process in which the exterior of the zirconium alloy (Zircoloy-2) was pressed together (co-extruded) with the uranium interior using extrusion machinery. This process fundamentally altered the way fuel elements had been made since World War II at the Hanford Site and throughout the rest of the world. Co-extrusion provided a more complete, uniform bond that was superior to the bond used in previous canned elements, which were subject to bubbles and other flaws between the layers. Zircoloy-2 was superior to aluminum as an outer layer because it could tolerate greater temperatures inside the pressurized system of the new reactor with a lower rate of corrosion.

Problems in obtaining a successful bonding and, therefore, canning of fuel elements had plagued the project from the beginning. There was real concern that not enough canned elements would be ready for the startup of B Reactor in 1944. Even though researchers found solutions for successful bonding, distortions of the cores and blistering of the fuel elements continued during the post World War II era. Nevertheless, research and development efforts progressed to improve the fuel manufacturing process and metallurgical properties of the uranium metal (Weakley 1999).

Knowing that worldwide supplies of uranium were finite, the United States embarked upon a large research effort to find alternate supplies of reactor fuels. Researchers sought methods to stretch or diversify the uranium fuel supply for commercial nuclear reactors by creating oxide fuel blends. This was the result of President Eisenhower's Atoms for Peace program, unveiled in 1953, which envisioned non-defense utilization of nuclear technology to fuel civilian and industrial power needs. The Atomic Energy Commission pursued a program of developing alternative nuclear fuels for industrial and commercial uses. Military defense needs still generated numerous research and development programs at the Hanford Site, including the initial production of tritium, which was used in hydrogen bombs.

From the beginning, the Hanford Site attempted to recover expensive and non-renewable resources such as uranium. As the demand for weapons-grade plutonium grew in the late 1940s and early 1950s, it was important that uranium materials be recovered, recycled, and reused whenever possible. Development of the nuclear fuel production process as well as recovery and recycling of uranium occurred concurrently, first at the Manhattan Project sites, later at the Atomic Energy Commission sites and at private companies under contract with the federal government.

Public scrutiny and environmental concerns in the 1970s and early 1980s were primarily focused on reactors and their operation as opposed to the fuel production process. Every step in the plutonium production process at the Hanford Site, however, generated a considerable amount of waste and other by-products. Forty years of plutonium production produced a legacy of waste that led to safety and environmental upgrades and an accelerated program to reduce radioactive discharges in every operation to the lowest practicable levels. This was later expanded to cover all hazardous wastes. Fuel manufacturing activities, especially in the 313, 314, and 333 Buildings, generated acidic liquid (chemical) waste from cleaning, testing, and uranium scrap processing. The environmental issues woven together with the end of the Cold War shutdown the N Reactor in 1989 and with it the era of fuel manufacturing at the Hanford Site came to a close.

Significant Events: Fuel and Target Fabrication
(adapted from DOE 1997b, p. 22)

- During the Manhattan Project, fuel for the Clinton X-10 reactor (later Oak Ridge National Laboratory) and the Hanford B, D, and F production reactors was manufactured by companies in Detroit, Michigan; Columbus, Cleveland, Toledo, Warren, and Hamilton, Ohio; Fort Wayne, Indiana; Reading, New Kensington, and Springdale, Pennsylvania; Bridgeport, Connecticut; and Chicago, Illinois.
- By the spring of 1945, Hanford's 300 Area had assumed all of the fuel fabrication responsibilities for the site's reactors except extrusion. Hanford extruded uranium rods onsite from 1946 to 1948, then shifted to rolled rods supplied by offsite private contractors. Hanford rolled uranium rods from 1950 to 1952.
- Hanford manufactured lithium targets for tritium production from 1949 to 1952 and again from 1965 to 1967. The site also made bismuth targets for polonium-210 production and lead-cadmium rods used as a neutron-absorbing "poison" to control reactors.
- The M Area at the Savannah River Site was built in 1952 to clad and assemble fuel elements for the five production reactors located there.
- Facilities at the Savannah River Site M Area manufactured lithium-aluminum targets for tritium production and targets for manufacturing americium, curium, plutonium-238, and other isotopes.
- Uranium slug machining for Hanford and the new Savannah River Site reactors was taken over by FMPC at Fernald, Ohio, which opened in 1952, and the Weldon Spring plant in Missouri which opened in 1956. Extrusion was performed by private contractors in Adrian, Michigan, and moved to Ashtabula, Ohio in 1961. Fernald produced rolled uranium rods onsite.
- To meet the demands of supplying fuel for 13 operating production reactors, private contractors continued to support Fernald and Weldon Spring by machining uranium slugs in the 1950s.
- In the 1950s, production reactor fuel changed in several respects: natural uranium was replaced by LEU (low enriched uranium), solid cylinders were replaced by tubes, and, with the opening of the N Reactor at Hanford in 1963, aluminum-clad fuel was supplemented by fuel clad with zirconium.
- By the time N Reactor started up at Hanford in late 1963, there were sufficient stocks in LEU at Fernald to supply the reactor without requiring additional LEU from the gaseous diffusion plants.
- Weldon Spring shut down in 1966, and Fernald subsequently assumed all of the fuel fabrication mission.
- In 1968, the Savannah River Site converted to HEU (highly enriched uranium) fuel and DU (depleted uranium) targets. The HEU was supplied by recycling research, Naval and production reactor spent fuel and recovering the HEU at the Savannah River (Site H Canyon) and Idaho National Engineering Laboratory (Idaho Chemical Processing Plant). Weapons-grade HEU stored at Y-12 was also used to supply some fuel for the Savannah River Site reactors. Fernald continued to supply LEU slugs for the N Reactor and DU targets for the Savannah River Site reactors.
- Hanford's 300 Area made only N Reactor fuel after 1971. The facilities shut down in 1972 and later resumed production on N Reactor fuel in 1981.
- Fuel and target fabrication at Hanford's 300 Area ceased permanently in 1987 with the closure of N Reactor. Production at the Savannah River Site M Area and Fernald ended in 1989 with the shutdown of the last Savannah River Site reactor.

URANIUM PREPARATION

The manufacture of minute amounts of plutonium required significant quantities of uranium ore. "For each kilogram of plutonium made for the U.S. arsenal, miners took roughly 1,000 tons of uranium ore from the ground" (DOE 1995e, p. 13). The uranium had to be converted into metal before it could be used in the production of nuclear fuel. Metallurgical properties of uranium were unknown before the Manhattan Project. Most of the early uranium metallurgical research was accomplished from 1942-1943 at various research facilities in the United States, including the University of Chicago Metallurgical Laboratory. Iowa State University researchers developed the process that became the standard for converting uranium ore into metal (DOE 1997e).

> ***"Workers at the Fernald uranium foundry... converted hundreds of tons of uranium hexafluoride gas (the tails from the enrichment process) into uranium 'green-salt' crystals. These crystals were blended with magnesium granules and cooked in a furnace. The mixture ignites, converting the green-salt crystals into uranium metal. Some of this metal was made into reactor fuel...for plutonium production reactors at Hanford and Savannah River."*** **- DOE 1995e, p. 15**

In addition to investigating the metallurgical properties of uranium, the scientists faced the question of enrichment. The early research efforts concluded that "because enriched uranium was not available in large quantities until 1945, and enrichment focused on producing highly enriched uranium for weapons, Manhattan Project reactors used fuel made of unenriched natural uranium metal" (DOE 1997e, p. 150).

> *"Three basic types of production reactor fuel and targets were manufactured. Some of the production reactors used natural or low-enriched uranium as fuel. The uranium-235 in the fuel sustained the chain reaction while the uranium-238 in the fuel captured neutrons to produce plutonium. Other reactors used 'driver fuel' (made with highly enriched uranium) and separate targets (made of depleted uranium) for the same purposes."* - DOE 1997e, p. 150

To have a substance that could be used by the reactors to create plutonium involved mining the uranium, refining it, and shaping it into a form for insertion in the reactors.

MINING

During the Manhattan Project, the United States purchased high-assay uranium ore from the Belgian Congo (Zaire), supplemented with ore concentrate from sites in Canada and the Colorado Plateau in the United States. Various contractors in several states refined the wartime uranium. The procurement of uranium ore in quantities sufficient for large-scale production of fissionable materials began when the U.S. Army Corps of Engineers obtained the option to purchase some 1200 tons of Belgian Congo ore already in the United States and another 3000 tons in the Belgian Congo (Hewlett and Anderson 1972a).

By the end of 1944, the Corps had received Congo ore containing approximately 370 tons of uranium oxide. At the same time, the Eldorado Mining and Refining Company in Canada had produced about 400 tons of uranium oxide. "The first imported uranium, high-grade pitchblende ore containing up to 65 percent uranium oxide by weight, was milled both in Canada and by domestic contractors" (DOE 1997e, p. 19).

Uranium mining expanded dramatically after World War II, from 38,000 tons of ore in 1948 to 5.2 million tons in 1958 – nearly all of it for nuclear weapons production. The domestic uranium industry began to supply significant quantities of uranium ore to plutonium production facilities, like the Hanford Site, shortly after the end of World War II. All total, the United States mined about 60 million tons of ore to produce uranium during the Cold War era. By the early 1950s, more than 400 uranium mines had opened in the United States, primarily in Arizona, Colorado, New Mexico, Utah, and Wyoming. Imported uranium was purchased in the form of already mixed concentrates and high-grade ores. Domestic uranium was purchased as either ore or concentrate (DOE 1997e).

REFINING

Mining and refining the uranium involved "extracting uranium ore from the earth's crust and chemically processing it to prepare uranium concentrate, sometimes called uranium octaoxide or yellow" (DOE 1997e, p. 19). Uranium ores and concentrates were sampled and assayed to determine uranium content, as well as impurities and the existence of other constituents. Uranium mills shipped the yellowcake to plants that refined the concentrate into forms suitable for several different roles in weapons production, roles different from the Hanford Site mission. One facility was the Fernald Plant in Ohio, where thousands of tons of uranium were refined and sent to the enrichment plants at Oak Ridge, Tennessee; Paducah, Kentucky; and Portsmouth, Ohio. (DOE 1997e).

Improvements in the efficiency of uranium ore production progressed during the Cold War period. Uranium accountability was improved in order to upgrade environmental and worker safety, economical usage of the substance, and national security. For instance, by the early 1970s, approximately 40-60 pounds of uranium a year were recovered from used pre-braze acid solutions and recycled back to the vendor (Vinther 1996, Tokarz 1984).

During the 1950s, the Atomic Energy Commission built government-owned uranium refineries in Fernald, Ohio and Weldon Spring, Missouri. The two sites produced ingots of natural, low-enriched and depleted uranium. The ingots were extruded into billets for shipment to the Hanford and Savannah River sites for cladding and assembly. The extrusion was performed by Bridgeport Brass Company in Adrian, Michigan from 1954-1961 and then by its corporate successor, Reactive Metals, Inc., in Ashtabula, Ohio (DOE 1997e).

Uranium concentrates were refined or chemically converted to purified forms suitable as feed materials for the next step in the process. "Examples of these feed material [sic] were uranium hexafluoride for enrichment at gaseous diffusion plants, and uranium tetrafluoride, or metal, for fuel and target fabrication. Refining...also involves the recycling of various production scraps, production residues, and uranium recovered from fuel processing" (DOE 1997e, p. 20).

DELIVERY OF BILLETS

The form of the uranium arriving at the Hanford Site changed in the early years. Initially in October 1943, the uranium arrived in the form of extruded rods from offsite suppliers. Beginning November 1944 with the expectation of installing a press to extrude rods from billets, the uranium was shipped to the Hanford Site as billets (Gerber 1992c). The press was installed early 1945.

To transport the uranium billets from the Fernald Plant to the Hanford Site's rail yard (the Riverland Classification Yard) near the present site of Vernita Bridge, E.I. Du Pont de Nemours and Company contracted with the Chicago, Milwaukee, St. Paul and Pacific Railroad to move them under military guard. (See the section on Site Security for information on security procedures to protect shipments). The billets, which measured 4.5 inches in diameter, 12-20 inches in length, and weighed at a minimum 125 pounds, were then transported by government railroad from the Riverland Yard to the 300 Area. Until the railroad line from the Riverland Yard to the 300 Area was completed, the uranium billets were transferred to the 300 Area by semi-trailer trucks. In the 300 Area, the billets were unloaded onto flatbed trucks and moved to the 303 Fresh Metal Storage buildings.

During storage, the rods and after November 1944 the billets were inspected for correct measurements and components. Sample billets were taken from the material in storage and tested at the 305 Test Reactor/Hot Cell Verification Building for purity. The test reactor functioned as a quality assurance tool to test samples of graphite, uranium, aluminum tubes and canning materials (for fuel rod jackets), and other materials to be used in the Hanford Site production reactors. "Measurements of material purity using instrumentation and calculation were too time consuming for wartime schedules. Therefore, actual performance tests of material purity using samples in a reactor were deemed necessary" (Gerber 1992c, p. 63).

The reactivity of uranium samples tested in the 305 Building against samples of known purity could be used to predict a sample's efficiency if loaded in a full-scale production reactor. Fuel impurity tests consisted of artificially poisoning the uranium fuel elements with one to four layers of copper foil to determine the effect on radioactivity. As needed, the billets that passed the test would be transferred to the 313 and 314 buildings to begin the fuel manufacturing process. Billets found not fit for irradiation in the 100 Area reactors were brought back to the 303 Buildings to be cleaned, re-measured and inspected for possible reuse. For information on the 303-A Building, see the Historic Property Inventory Form in Appendix B on the Internet.

MATERIAL ACCOUNTABILITY

"As might be expected of any program involving the fabrication of more than 1500 tons of a rare and expensive metal, strict and adequate controls had to be instituted and maintained to account for all uranium on hand at the various locations, whether it be in the form of cast billets, fabricated pieces, finished elements, or scrap" (Du Pont 1945a, p. 364).

"In late 1953, General Electric instituted a statistical analysis system that allowed a review of the vendor's uranium fuel core [uranium ingot] weights to be compared with those of the receiver's uranium fuel core weights. That year a comparison of over one million uranium fuel elements revealed a weight difference of .014% between the fuel elements shipped from Fernald, Ohio to the same fuel elements received at Hanford." - **GE 1953l, p. 2**

When the boxes of uranium rods/billets were delivered to the 303 buildings, they were weighed and checked against the shipping invoices. "Each billet was assigned a number that would be a part of the subsequent fuel rod as it moved through the entire fabrication process" (Kubik 1997, p. 21). The uranium was weighed to make sure it matched the weight given by the vendor.

During machining, small chips and fines of uranium (tiny, dust-like particles from the milling and other processes) were produced as scrap. The weight of such scrap, rejected cores, and processed cores was carefully tracked against the billet's original weight. At the end of each day, reports were submitted to the various Hanford Site departments showing yield summaries for each uranium billet and accounting for all of the uranium (Kubik 1997).

FUEL MANUFACTURING PROCESS

Because fuel manufacturing was the least likely of the production processes to experience a serious accident, it was considered safe enough to be located near populated areas like Richland. Therefore, the fuel manufacturing facilities were built much closer to Richland than the areas where the reactors and chemical separations facilities were located.

Figure 2-2.2. *Aerial View of the 300 Area in 1944 Looking Northeast*

The manufacturing of nuclear fuel took place in the 300 Area (Figure 2-2.2). It was known initially as the Metal Fabrication and Testing Area and consisted primarily of facilities for manufacturing uranium fuel elements and testing samples of graphite (DOE 1997e). The fuel manufacturing process included the milling of metallic uranium billets obtained from offsite production facilities into fuel elements in the form of pipe-like cylinders. The uranium billets were extruded into rods. The rods were extruded, outgassed, and straightened in the 314 Building (Figure 2-2.3). The rods were transferred to the 313 Building (Figure 2-2.4) for machining into cores and canning the cores into fuel elements. The fuel elements were trimmed to the specified diameter and cut to the required length in a series of turret lathes. Then they were subjected to a final testing in the 305 Test Reactor prior to being sent to the 100 Areas for irradiation in the reactors. Figure 2-2.5 shows cross sections of two fuel elements.

EARLY FUEL MANUFACTURING EFFORTS

Early on, scientists came to the conclusion that "...uranium could be extruded, and though little else was known of its physical characteristics, it was agreed that production of the slugs [fuel elements] should start from extruded material. By June 1943, extrusion had been demonstrated to a point where it was no longer experimental, but nevertheless was still capable of considerable improvement" (Du Pont 19451945a, pp. 366-367).

Figure 2-2.3. *Metallurgical Engineering Laboratory (314 Building) in 1944/1945 Looking Northeast*

Along with the extrusion process, researchers needed to address how to package the uranium for it to be irradiated. Uranium is very reactive with water. Water both corrodes the uranium metal and forms rapidly expanding uranium oxides that distort its shape. To prevent corrosion and distortion and, more importantly, to "prevent the release of highly radioactive fission products into the coolant" and subsequently into the Columbia River, the uranium fuel needed to be "canned" (DOE 1997e, p. 150).

Figure 2-2.4. *Metal Fuels Fabrication Facility (313 Building) in 1943/1944 Looking Northeast*

Early research efforts focusing on different methods for canning the fuel elements took place at the Metallurgical Laboratory in Chicago. The laboratory was "actively engaged in developing...aluminum-silicon bonding, zinc bonding and the unbonded canning processes. The process most favored was the aluminum-silicon bond. In addition, a request was made...to install equipment...for the canning of unbonded slugs [fuel elements]. This line...was to be...installed for developmental purposes and as insurance against failure to produce a satisfactorily bonded slug" (Du Pont 1945a, pp. 218-219). As time went on, the aluminum-silicon process improved to the point where the zinc bonding process was abandoned by October 1944 (Du Pont 1945a).

> *"Crawford Greenewalt, Du Pont's technical liaison with the Metallurgical Laboratory and later president of Du Pont, "agreed to build a small slug-production unit at Wilmington in an effort to translate the more promising canning techniques into a process which would be practical on a large scale. In the meantime...several bonding schemes showed promise...[including] the possibility of developing an unbonded, canned slug or even a coated slug which would withstand the extreme operating conditions in the Hanford piles [reactors]. During 1943, the Metallurgical Laboratory investigated a variety of bonding methods. The two principal methods involved using an aluminum-silicon alloy as the bonding material between slug [core] and can and incorporating a zinc bond in a special canning technique. In each process, there were hundreds of variables in applying the bond and inserting the slug [core] in the can. By the beginning of 1944, some line of development had to be established if the company was to have enough slugs to start the B pile [reactor] in August, [but] it was extremely difficult*

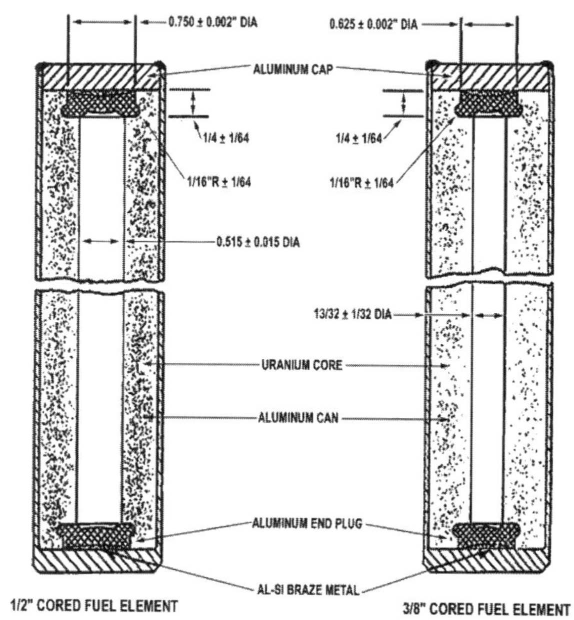

Figure 2-2.5. *Cross Sections of Two Fuel Elements*

to achieve an uniform distribution of the bond, which was necessary to avoid hot spots on the slug surface. In fact, some of the best minds at Chicago were convinced that the unbonded slug would be the most easily fabricated and the most reliable. Greenewalt readily admitted the complications of bonding, but he saw its distinct advantages too. Not only would the bond facilitate heat removal from the slug, but the bond itself would protect the uranium from corrosion and swelling even if the can did leak." - Hewlett and Anderson 1972a, pp. 223-224

Despite intensive experimentation at laboratories across the country to develop a process to manufacture fuel elements, progress at coming up with a reliable production system was discouragingly slow during the winter of 1943-1944. Du Pont concluded that the widely dispersed laboratories were amassing a lot of data but not developing a process. Accordingly in March 1944, they transferred to the Hanford Site all Wilmington experimental fuel element operations as well as most of the Du Pont engineers at the Metallurgical Laboratory in Chicago. Du Pont's plan was to eventually locate all fuel manufacturing steps, from uranium billets to finished fuel elements, at the Hanford Site.

"The first experimental canning operations started in the 313 building at Hanford on March 20, 1944. The development line consisted of little more that a series of open tanks in which scores of operators dipped clusters of machined slugs [fuel elements]...in molten bronze, tin, and Al-Si. Since the temperature, composition, and duration of each dip were extremely critical, the operators had great difficulty in achieving uniform results or detecting faulty slugs. After the final dip, the slugs [cores] were forced into the aluminum cans with hydraulic presses, a tricky process that produced a large number of rejects. The next step involved end-trimming and the complicated task of arc-welding the aluminum cap [to the can] in an argon atmosphere. Completely a manual operation, the welding step required weeks of training to achieve reasonable results. When the end had been faced and machined, the slugs were subjected to a series of tests to detect weld failures, pinholes, or lack of bond uniformity. In the first two weeks they succeeded in canning a total of thirty-six slugs, and none of these looked acceptable." **- Hewlett and Anderson 1972a, pp. 224-225**

Initially, though, work was limited to the experimental canning processes. The first fuel jacketing equipment to go into operation was known as the experimental line. This equipment included an electric heater press to heat and bond the uranium fuel cores to their aluminum jackets. However, the heaters burned out frequently, did not heat the cores and cans to consistent temperatures, and did not produce a uniform bonding. This problem was serious because non-uniform bonding caused thin places in the jacketing that, under irradiation, heated up more than other places. These hot spots could cause fuel element ruptures in the reactors (Gerber 1992c).

Du Pont and the Corps were deeply concerned over the lack of progress in solving the problems of fuel manufacturing in the 300 Area. In retrospect, it is hard to imagine how tens of thousands of fuel elements needed for the first loading of the reactor were ever completed on time. Walter Simon, Hanford Site chemist, and the Site's first operations manager, described the situation: "One of the most difficult problems...was the making of uranium fuel slugs [elements]. The uranium was held in an aluminum can, a slug [fuel element], about eight inches long and [an] inch and five-eighths in diameter. The can had to fit very tightly with no air space or bubbles. They couldn't leak because if water got into the uranium it destroyed the ability to react" (Sanger 1995, pp. 153-154).

Frustration increased at the Hanford Site over the lack of progress in solving the fuel element manufacturing problems. The crisis over fuel element manufacturing was not unexpected. It had been long assumed that the development and production of fuel elements was the most critical job facing the project, but the future of the Hanford Site was uncertain as long as the success of the canning process was not determined. "As a safety measure, the Army [Corps of Engineers] had contracted with a Chicago manufacturer to fabricate one charge of unbonded slugs for B pile [reactor]." However, Du Pont "finally succeeded in concentrating at Hanford the last remnants of the canning operations previously located in Chicago, Cleveland, and Wilmington" (Hewlett and Anderson 1972a, pp. 303-304) and with that came a solution to the uniform bonding problem of the fuel elements.

Dr. Raymond Grills, a chemist from Illinois, and his associate Ed Smith, were assigned to the Hanford Site by Du Pont to study uranium canning and come up with a solution.

Grills *"noted that the unsuccessful fast oxidation of the uranium slugs [cores] prevented a perfectly even coating, no matter how quick the canning. He decided to try the whole canning operation in a solder bath, working beneath the surface of the molten solder. This way no air could get in, he reasoned, and all sides of the slug [core] would thus be uniformly exposed to the coating liquid. Grills submerged the slug [core] in one of the four-foot-deep, round tanks and tried to can it beneath the surface...using long tongs. ...they obtained perfectly good, uniform heat transfer. But then the two chemists discovered they had burned a hole through the aluminum can. Apparently the temperature of the liquid in the bath was so high that it melted the aluminum. ...the canning operation would have to be done more swiftly. They tried again, working faster this time; the second slug looked good. After several more tests and experiments, a canning technique had been found that assured [sic] 100 percent perfection!"* - Groueff 1967, p. 300

Leona Marshall Libby, one of the few female scientists on the Manhattan Project, toured the fuel element canning operations in the 300 Area with Enrico Fermi, who had achieved the world's first self-sustained nuclear reaction, and observed:

> *"We passed great vats of melted tin, eutectic, and cookbook flux in which the slugs [cores] were being heated, after which they were plucked out with forceps, fitted to the open end of empty thin-walled aluminum cans, inserted in presses, and crump, the can folded up like an accordion. The problem was solved over the next 24 hours – by one of the production line crew. The metal solder into which the uranium slug [core] was dipped melted a couple degrees below the temperature at which the aluminum can melted. The temperature of the solder was held constant at these few degrees difference, the slug [core] and the can were soaked in tin, and then the slug [core] was slid into the can while both were immersed in liquid solder. It was called underwater canning."* - Libby 1979, p. 176

Once the problem of canning the cores was solved, the production line for canned fuel elements made a round-the-clock effort and produced thousands of fuel elements needed for the start-up of B Reactor. Fortunately for the fuel manufacturing team, minor adjustments and changes to the completion of B Reactor delayed the entire project for several weeks, which gave the team additional time to catch up. The yields of acceptable fuel elements went up, and the number of failures in autoclave tests declined even more sharply. Thus, an adequate number of bonded fuel elements were ready for the first reactor charge. The canning crisis was over, and Du Pont felt confident they had enough fuel elements to supply all three reactors.

313 AND 314 BUILDINGS

The first steps in the fuel manufacturing process occurred in the 313 Metal Fuels Fabrication Building and the 314 Metal Extrusion Building where the fuel cores were manufactured and jacketed. For information about these buildings, see the Historic American Engineering Record for the 313 Building and the Expanded Historic Property Inventory Form for the 314 Building in Appendix B on the Internet. The two buildings had equipment for the extrusion, outgassing, and straightening of the uranium rods, machining them into short cylinders (cores), canning the cores, and testing the resulting fuel elements in preparation for use in the transmutation process.

Completed in the autumn of 1943, the 313 Building was the Hanford Site's original fuel manufacturing facility and produced fuel for the Hanford Site's eight single-pass reactors from 1944-1971. Its primary mission originally was to extrude and machine uranium rods to specific dimensions required for the cores, jacket or can the cores, and test the jackets for proper bonding and sealing. Until the 314 Building was completed and the press for extruding billets into rods installed in it, the first uranium billets were extruded into rods offsite and the rods sent to the Hanford Site for completion of the process. With eight additions in 1944, the 313 Building contained three fuel-jacketing areas, a welding area, a fuel jacket (can) cleaning area, and a control room (Gerber 1993b).

The 314 Building was completed in 1944. After the 1000-ton extrusion press was installed in the 314 Building in early 1945, the 313 and 314 buildings assumed all of the fuel manufacturing responsibilities. The press allowed the Hanford Site to process raw uranium billets into extruded rods that were suitable for manufacturing into fuel cores. The uranium billets were taken to the 314 Building to be extruded into rods, then outgassed and straightened. The rods were then transferred to the 313 Building for machining the rods into cores. Bare uranium rods were lathed into cores that were then canned (called the jacketing or canning process) in a triple dip bath procedure that involved first coating them with bronze, then tin, and finally an aluminum-silicon braze (Gerber 1992c). They were then marked with a fluoroscope, end-machined, stamped with identification numbers, and had end caps welded on. A final etching in a compound of hydrofluoric acid and nitric acid, followed by a methanol rinse, completed the fabrication process. Three final quality assurance tests were conducted. The first one (the frost test) took place in the 313 Building, and the final two, autoclave and radiograph testing, took place in the 314 Building (Gerber 1992c). If a fuel element passed the autoclave test, it then underwent a final radiograph (x-ray) test in the 314 Building to detect porosity in the end weld bead. Any porosity could have become a pathway for water to contact the uranium core and cause the fuel element to rupture (Gerber 1992c).

The extrusion of uranium rods took place in the 314 Building until 1948 when the process shifted to rolled uranium rods supplied by offsite private contractors. Although its mission and equipment changed, the 314 Building continued to straighten the uranium rods and took over the autoclave and radiograph equipment from the 313 Building to test the canned fuel elements. In addition, the 314 Building staff conducted uranium scrap processing operations until the closure of the single-pass reactors and end of fuel element preparation activities in the 300 Area (Gerber 1992c).

By 1954, the canning method had changed to a lead-dip process (Weakley 1999). By the late 1960s, a Hot Die Size (nickel-plating) canning process had just been developed when fuel manufacturing operations ceased in the 313 and 314 buildings in 1971 when the Atomic Energy Commission ordered the shut down of all reactors.

FUEL MANUFACTURING PROCESS - STEP-BY-STEP

The fuel manufacturing process for the single-pass reactors at the Hanford Site consisted of extrusion, outgassing, straightening, machining, triple dip baths, cleaning, canning, final machining, quality assurance, and autoclaving. Figure 2-2.6 shows the fuel manufacturing process for reactors at the Hanford Site from 1945-1954.

Extrusion, Outgassing, and Straightening

The uranium billets were taken from storage in the 303 buildings and transferred to the 314 Building for the first steps of the fuel manufacturing process: extrusion, outgassing, and straightening. Metallic uranium, in the form of billets about 4-1/2 inches in diameter and 12 to 20 inches in length, were heated in a muffle-like furnace in an inert gas (argon) atmosphere to about 1700 degrees Fahrenheit. The uranium billets were extruded into rods measuring 1.5 inches in diameter by 12 to 14 feet in length by the high powered hydraulic press, and then through a die (Du Pont 1945c). Figure 2-2.7 shows a billet being loaded onto an extrusion press.

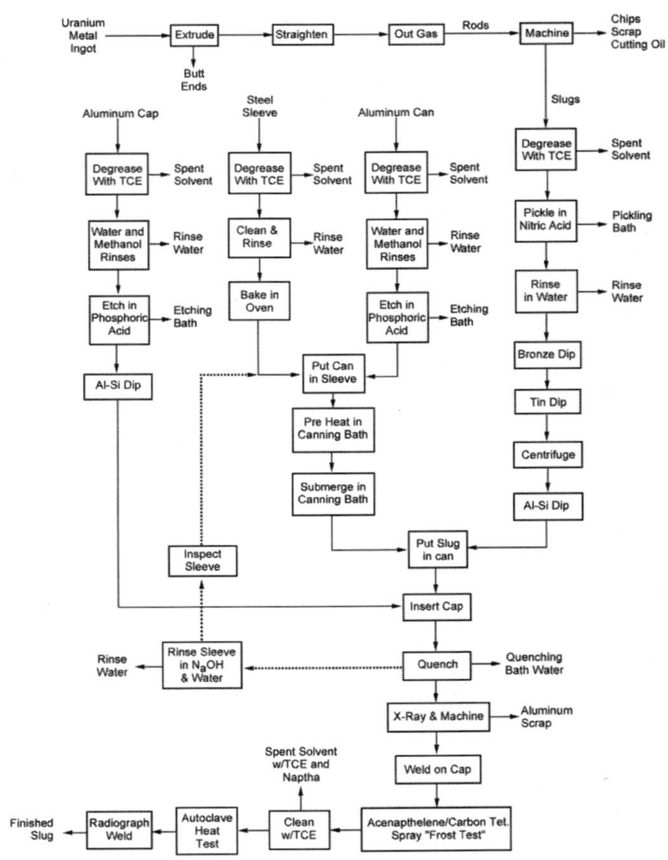

Figure 2-2.6. Fuel Manufacturing Process for Reactors at the Hanford Site from 1945-1954

"At the start of the [extrusion] process, workers preheated uranium billets in a rotary electric resistance furnace. They quickly placed the hot billet into the extrusion container after brushing or flattening it as necessary to remove rough or swollen spots. A hydraulic ram pushing against a block at the back of the billet forced the hot metal through a die at high pressure. In about ten seconds, the extrusion press formed a 20 inch long, 200 pound billet into a 14 foot long rod. Workers quickly straightened and quenched the finished rod in water. They removed the unextruded butt end of the billet from the press and recycled it as scrap." - DOE 1997e, p. 151

After the extrusion, the rod was outgassed and straightened. The outgassing and straightening procedures, which began in the 314 Building in September 1944, were essential to prevent the cores from distorting in the reactors. The rods received a preliminary straightening manually, were quenched in water, and then outgassed to remove hydrogen from the metal to prevent formation of gas pockets or bulky chemical compounds of uranium and hydrogen, either of which might result in serious difficulties in the reactor operation. After outgassing, the rods received a final straightening in a mechanical device and were then ready for machining (Du Pont 1945c).

Figure 2-2.7. *Loading a Billet onto an Extrusion Press, 1984*

Machining

The Hanford Site began machining uranium rods on turret lathes in December 1943 (Du Pont 1944) in the 313 Building. The rods came from outside sources until the 314 Building was completed and could begin supplying the rods. The uranium fuel rods fabricated from billets in the 314 Building were machined in the 313 Building. If the fuel rods were too rough, they were scrapped or returned to the 303 buildings for storage and possible future use (Weakley 1999). Extruded or rolled uranium fuel rods, 12-14 feet long and averaging 1.5 inches in diameter, had to be cut and finished into 8-inch long, 1.36-inch diameter cores with tight tolerances. To do this, a machinist finished one rough end of an outgassed, straightened uranium core on a lathe. Then he used the lathe to reduce the core to its proper diameter, cut off the core to its proper length, finished the cut end of the core, and rounded off the corners. During this process, a large flow of coolant prevented the uranium chips from igniting.

After the cores were machined and visually inspected for surface flaws, "they were placed in baskets and moved by an electric hoist and a monorail to the degreasing tanks, where the dirt and oils from shipping, handling and machining were removed. After inspection to assure [sic] that there were no serious defects which would adversely affect pile [reactor] operations and to make certain that the slugs [cores] were within the dimensional limits of tolerance, they were given an acid bath for removal of any grease which may be on the surfaces" (Du Pont 1945a).

The cores were then dipped in a boiling vapor degreasing solvent, rinsed in cold solvent and dried in hot solvent vapors. As each core was degreased, it became coated with a thin oxide film that had to be removed by immersion in a hot pickling solution of 80% nitric acid. The cores were then water rinsed, dried, and given a visual quality assurance and quickly moved to the triple dip bath (UPOA 1977a).

Triple Dip Baths

Beginning in August 1944, the uranium cores in the 313 Building were jacketed in a triple dip method that consisted of bathing them in molten bronze, tin, and then a molten aluminum-silicon mixture. Each of the three baths randomized the shape and the size of the uranium grains. This process helped prevent core distortion and swelling in the reactor (Gerber 1992c).

The first of the three metallic baths in the triple dip process was a bronze bath. An operator used tongs to dip and agitate each core in the bronze bath for 45 seconds. The bath was used to bring the core up to a uniform temperature, called the uranium beta phase. At this temperature, the core would alloy with the bronze, and a thin coat of the bronze would adhere to the bare and cleaned uranium core (UPOA 1977a, Gerber 1992c).

Within 5 seconds of the bronze bath, the core was taken to the second bath where it was washed free of molten bronze from the bronze bath by dipping it into a tin bath for 45 seconds. The tin bath was made of 99.9 percent tin with 0.1 percent aluminum added to prevent the tin from oxidizing (Kubik 1997, UPOA 1977a). The tin bath cooled the core to the uranium alpha phase and removed the excess bronze (Gerber 1992c, Kubik 1997). Following the tin bath, the core was taken to the centrifuge, set vertically in a rack and spun to remove the excess tin (UPOA 1977a).

The third and last bath was a blend of 88 percent aluminum and 12 percent silicon. This blend was a standard eutectic alloy that Du Pont purchased in ingots. The aluminum-silicon bath acted as the key bonding agent between the uranium core and the aluminum can and cap (UPOA 1977a, Kubik 1997).

Triple-Dip Canning Process

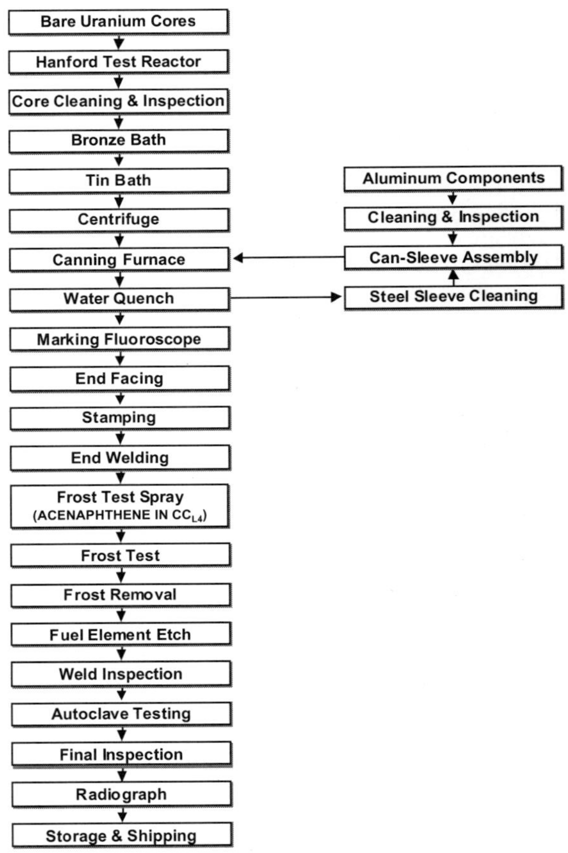

The lead-dip process, a simpler canning scheme, replaced the triple dip process in 1954. "The lead dip process had two principal advantages over the triple dip process: (1) canning costs were lower because fewer operations and less equipment were required, and (2) the structure and properties of the uranium metal were essentially unchanged in the canning process, thus the uranium metal retained its desired properties obtained during fabrication and heat treatment of the uranium" (Ballinger and Hall 1991, pp. 2.2-2.3).

Canning

The next step in the process was the vital one of coating and canning each core. The resulting fuel elements had to meet the "most rigid requirements with respect to materials and tightness. The materials of which the can is made must have a very low absorptive capacity for neutrons; they must possess a high resistance to corrosion by the action of water; and they must be able to transmit the tremendous amount of heat generated in the uranium slug [core] to the surrounding cooling water" (Du Pont 1945c, p. 7).

In 1943, the only materials known that would meet the canning requirements were aluminum and an aluminum-silicon alloy. The can had to be perfectly tight and bonded because uranium is very reactive with water. The failure of a single fuel element would stop the flow of cooling water, necessitating the complete abandonment of an entire reactor area. A tight fit or bonding was needed to prevent the fission by-product elements formed in the core from being washed away by the cooling water resulting in serious radioactive contamination of the reactor area and the Columbia River.

Development of sealed cans that would allow sufficient cooling of the uranium cores was a difficult task. The steps in the canning process were performed under precisely controlled conditions of temperature, time, and materials purity. "Each slug [core] is coated with bronze and then with tin by dipping in baths; this is followed by centrifuging for removal of excess tin. After a dip in the aluminum-silicon bath, the canning operation is performed by inserting the fuel element

[core] into an aluminum can below the surface of another aluminum-silicon bath followed by the placing of an aluminum top cap" (Du Pont 1945g, p. 8). The aluminum can was contained in a steel sleeve that held the shape and dimensions of the core during the high temperatures of the canning phase and protected the outer surface from the molten aluminum-silicon. The assembly was quenched in cool water and the steel sleeve and excess aluminum-silicon were removed (UPOA 1977a). Figure 2-2.8 shows the cleaning area with the uranium fuel elements and aluminum cans inside sleeves on a conveyor belt.

Figure 2-2.8. *Cleaning Area with the Uranium Fuel Elements (left) and Aluminum Cans in Sleeves (right) on a Conveyor Belt, 1964*

The aluminum cans, caps, and steel sleeves were cleaned and inspected before use in the canning process. Aluminum caps and cans were vapor degreased, cleaned in a sodium dichromate solution; followed by a methanol dip and air drying (Weakley 1999). Steel sleeves surrounding each can were cleaned in sodium hydroxide (Kubik 1997). Each can, cap and sleeve then underwent a final visual quality check before being sent to the canning lines (UPOA 1977a).

To improve the bond between the core and the can, the inside of the aluminum can was degreased, cleaned, and etched (Weakley 1999). A solution of 20 percent hydrogen phosphoric acid was poured in the can and allowed to stand for five minutes before it was rinsed. Next it was cleaned, first with a degreasing solvent and then with an aqueous wash solution. After that, the aluminum end cap was machined and brazed on, and the finished fuel element was etched in nitric acid.

Final Machining

Following canning, the ends were machined to marks made under a fluoroscope, which ensured an equal amount of aluminum at each end for heat transfer and a minimum of aluminum to prevent excess neutron absorption (Du Pont 1945c). Fluoroscopic examination of the fuel elements also measured the precise thickness of the aluminum cans and caps and the aluminum-silicon bonding inside the can, and compared the total thickness to the placement of the core in the can (UPOA 1977a). "After degreasing, the top cap is then welded to the can by means of an argon shielded arc to assure [sic] maximum freedom from leaks" (Du Pont 1945g, p. 8).

Quality Assurance

At each step of the fuel manufacturing process, the cores, canning components, materials, tools, equipment, and baths were all carefully subjected to a rigorous, visual check. After final machining and welding, inspections, testing, and a quality assurance of each canned fuel element were performed to "assure [sic] acceptance only of those which meet the specifications for surface defects, dimensions and warpage, penetration of aluminum-silicon through the aluminum can, and bonding of the uranium slug [core] to the can" (Du Pont 1945g, p. 8). The fully automatic, bond and penetration testers separated out the fuel elements not passing the test. These testers examined the fuel elements "internally and externally for completeness of bonding and for residual aluminum jacket wall thickness" (Weakley 1958, p. A-44).

The finished fuel elements then underwent three tests. The first was the Frost Test, used to detect defects in the aluminum-silicon bonding. The test consisted first of cleaning and spraying canned fuel elements with acenaphthene mixed with carbon tetrachloride and heated to test the bond between the fuel element and the can. The acenaphthene left a white film or frost on the fuel element that gave the test its name. The fuel elements were then passed through a heated induction coil. Defects in the fuel element or in the aluminum-silicon bonding caused the fuel element's surface temperature to rise, melting the acenaphthene's frosty film and visually revealing any defects as a shiny area (Du Pont 1945d, Gerber 1992c, Weakley 1999). Once the fuel elements were defrosted, the acenaphthene was removed with trichloroethylene. The fuel elements then underwent a second test, an Etch Test, where the fuel elements were etched for 10 minutes in a solution of nitric acid. If this procedure revealed gas bubbles or nonbonded spots on the aluminum can where the aluminum-silicon bonding material might have penetrated the can and weakened it, the fuel element then would be rejected and sent back through a recycling process.

The canned fuel elements were then given a visual inspection for dimensional specifications, weld quality, and surface flaws, and if the bonding had been completed successfully, they were transferred to the steam autoclaves (Du Pont 1945d, Kubik 1997).

Autoclaving

The third and final test was insertion into a pressurized autoclave. The fuel elements were heated in a steam autoclave in the 314 Building to test for leaks. The fuel elements were loaded into wire baskets and lowered into the autoclaves where they were subjected to steam pressure of 100 pounds per square inch for 40 hours. The steam pressure in the autoclave, in combination with undetected defects in any part of the fuel manufacturing process, caused the defective fuel elements to burst (Du Pont 1945d). The fuel elements passing the autoclave test were then radiographed to check the porosity of the weld and then subjected to a final visual assurance test before the canned fuel elements were taken to the 100 Area reactors (UPOA 1977a). By the end of World War II, canning quality was such that fewer than one failure occurred per 2000 fuel elements surviving the inspections and tests (Du Pont 1945g).

Quantity Requirements

Along with strict quality assurance requirements, there were several quantity requirements. The following indicates the magnitude of the fuel manufacturing and canning processes: "more than 1250 tons of billets were required for the initial charging of the three piles [reactors]; more then 130 tons of billets were required for normal placement each month; more than 210,000 canned slugs [fuel elements] were required for the three initial charges; and, finally, more than 23,000 canned slugs were required each month after routine operations had been established" (Du Pont 1945g, p. 8).

POST WORLD WAR II FUEL MANUFACTURING FOR SINGLE-PASS REACTORS

FUEL MANUFACTURING CHANGES

In the 5 years following the successful production of the first plutonium at the Hanford Site, Du Pont and its successor, General Electric, continued to improve the fuel manufacturing process. Some of the changes were meant to improve the process, some were meant to improve plutonium production, and others were meant to use the fuel manufacturing materials more economically. One means was by attempting to vary the fuel element dimensions to minimize the distortions (see Figure 2-2.1).

The fuel manufacturing process went through numerous changes after World War II, most notably in the 313 and 314 buildings. The Hanford Site stopped extruding uranium rods in 1948, switching to rolled rods from offsite suppliers as the rolling of uranium rods was found to be a less expensive process and possessed metallurgical advantages over earlier fabrication efforts (DOE 1997b).

By 1950, the Hanford Site began making rolled uranium rods when a rolling mill was installed in the 314 Building. But the process in the 314 Building was short-lived. The following year, the Atomic Energy Commission shifted the rolling work to Fernald's Feed Materials Production Center and its supporting contractors (DOE 1997e).

During the late 1940s and early 1950s, the number of plutonium production reactors greatly expanded. By 1955, there were a total of thirteen production reactors, eight at the Hanford Site and five at Savannah River. To supply fuel for all these reactors, the Atomic Energy Commission established the Feed Materials Production Center at Fernald in 1951 and the Weldon Spring Plant in 1953 (Weakley 1999). In addition, commercial contractors continued to support the Atomic Energy Commission's fuel manufacturing needs throughout the 1950s (DOE 1997e)

In 1954, the Hanford Site switched to a new, lead-dip process for canning the fuel (Weakley 1999).

> *"The process consisted of immersing the uranium fuel cores [elements] in a bath of molten lead covered with molten aluminum, followed by a molten aluminum-silicon bath. At about the same time, the bonding test was changed, eliminating the use of acenapthelene [sic] and carbon tetrachloride. Between 1955 and 1964, about 30,000 single-pass reactor fuel elements were canned each week. A 'hot die size' process involving nickel plating which incorporated nickel sulfate, nickel chloride, and boric acid was developed in the early 1960, but never implemented on a large scale."* - DOE 1997e, p. 157

> ***"To increase plutonium production capacity, Hanford began adding low-enriched uranium fuel slugs [fuel elements] to its reactors as early as 1950. Enriched uranium fuel also allowed Hanford engineers to even the reactor's temperature and power distribution, reducing problems caused by uneven thermal expansion and radiation-induced swelling of the graphite core. Neutron absorbing...slugs...also helped to even out the reactor's power distribution. Most of these enriched uranium slugs were manufactured using the same techniques as the natural uranium slugs. However, some of the fuel elements were made of highly enriched uranium alloyed with aluminum, which required special fabrication techniques to prevent accidental criticalities. These "driver" elements were often used in combination with special targets such as the lithium targets used to make tritium. Improvements in fuel slug design gradually reduced the tendency of the fuel slugs to become misaligned inside the reactor."*** **- DOE 1997e, p. 154**

During the 1950s, the 300 Area was the locale for numerous research and development efforts to improve the fuel manufacturing process. The research and testing was a daily process, meant to take advantage of operator and technicians' suggestions, safety inspections, and the latest technologies and creative scientific thought. Some of the research and testing explored new methods of converting the uranium billets to fuel rods. Others altered the times, temperatures, and contents of the baths or the temperature and length of the welding process. As problems with core distortion in the reactors continued, scientists explored ways to alter the dimensions of the fuel elements in an attempt to alleviate the swelling, blistering, and warping. Other research and testing projects sought ways to reuse fuel manufacturing materials through various recycling programs (Kubik 1997). A 1957 report, *Production Tests: 300 Series – A Bibliography*, documents over 1,000 research tests that occurred in the 300 Area during a 13-year period from 1944-1957 (Schmidt 1957). See the Research and Development Section for more information on research tests.

FUEL ELEMENT DISTORTIONS AND MODIFICATIONS

In October 1945, the first fuel element distortions occurred during the irradiation process in a Hanford Site reactor. Investigations and research efforts were undertaken to comprehend why some fuel elements warped, others swelled and burst the aluminum cap or can, and others acquired surface blisters or bubbles (Dahlen 1946).

The General Electric Company was convinced that part of the problem with fuel element distortion lay with human error. In their annual report for 1952, General Electric noted: "The incentive to develop mechanical methods stems primarily from the need for canned fuel elements of constant, uniform quality, devoid of the infrequent but costly deviations due to the human element. Mechanized canning is expected to assist in the development of an even better slug" (GE 1953b, p. 35).

To combat human error, General Electric sought to automate as much of the fuel manufacturing process in the 313 Building as possible. "The building was to contain the most modern semi-automated and automated fabrication process lines, non-destructive quality assurance testing, material recovery programs and conveyance systems available.

Plans were made to install radiography equipment to inspect the cap-can welds, ultrasonic testing equipment to replace the frost test and ultrasonic transmission testing equipment to check the uranium fuel elements' [cores'] grain size and orientation" (Kubik 1997, p. 35).

Automation affected the canning line, where an automatic spray unit replaced the rinse tank. A mechanical quench unit was installed, and the welding equipment was converted to a semi-automatic system (GE 1953b).

A system was devised, although not entirely successful, to assist in tracking down the causes of core distortions. At the start of the fuel manufacturing process, each fuel element was assigned a number. As the fuel elements were fabricated, careful and detailed records were kept of each process step. It was easy to trace a distorted fuel element back through the manufacturing process. It was more difficult to ascertain the cause of the distortion (UPOA 1977a).

Operations and technical personnel in the 100 Area and the 300 Area worked together to resolve the problems of fuel element distortion. Valuable production time was lost each time a uranium fuel element failed in a reactor because the reactor had to be completely shut down to remove the damaged fuel element (Burley 1956, Kubik 1997).

For example in 1951, a total of 102 fuel elements failed in the five reactors, resulting in a loss of 2,282 hours of reactor time. That year, the average removal time for fuel elements was nearly 24 hours (Warren 1952). The following year, the aluminum cans on seventeen fuel elements failed, resulting in a loss of 268 hours of reactor time to remove nine of those fuel elements (Kubik 1997). It was not unusual for each reactor to be shut down at least once a month due to fuel element failures.

Careful examinations of the distorted fuel elements led researchers to believe fuel element length was a factor in the ongoing problem of fuel element distortion in the reactors (Kubik 1997). In an attempt to better control the fuel element's temperature in the reactor and to improve the flow of cooling water around the fuel elements, the fuel element's configuration and dimensions were changed, and the length varied (Kubik 1997). Figure 2-2.1 shows fuel elements with varying configurations and dimensions.

INCREASED POWER LEVELS - IN PURSUIT OF THE PERFECT FUEL ELEMENT

Ideally, the perfect fuel element would produce record amounts of power in the reactor without warping, swelling, or blistering. Plutonium production was costing the United States a great deal of money, so support for increases of reactor power levels was growing. Increased power levels allowed more plutonium to be produced per hour, which achieved greater economic efficiency as well as increased production.

During the late 1940s, the power levels of the Hanford Site reactors were increased to generate additional production. While the subsequent increase in temperatures kept the graphite from swelling further, and in some cases, even shrank it to its original size, the hotter conditions limited the already short time in which fuel elements could remain in the reactors. The higher heat also caused blistering or distortion of the fuel elements and sometimes rupturing of the cans. "By 1947, operators at Hanford developed a series of methods for detecting fuel element swelling and failures. One of these was a simple optical list. The decrease in light intensity through the process tube indicated the presence of blisters on the cans" (Carlisle and Zenzen 1996, p. 62).

Additional research conducted during the late 1940s and early 1950s investigated fuel element failures, "looking specifically at how uranium acted in fission reactions. The results of experiments provided important information for improving the designs of the fuel elements...several relatively simple mechanical expedients adapted the reactors to higher operating temperatures and higher power levels" (Carlisle and Zenzen 1996, pp. 62-63).

Pushing the operating levels of the Hanford site reactors entailed a number of risks, including the effect on the source of power and the fuel elements that *"needed constant redevelopment in order to withstand the intense heat to which they were subjected. By 1950,...General Electric substantially increased exposure time for slugs, up to three times the level feasible in 1946, without encountering blistering and warping. One method, involving the use of 2% zirconium alloyed to the slug [core], helped stabilize the slug, even under very high exposures. Improved inspection techniques, new watering treatment processing, and enhanced instrumentation also extended fuel slug lifetime and increased overall production levels."* - Carlisle and Zenzen 1996, p. 64

The Atomic Energy Commission "supported increased power levels in part because these tended to lower plutonium production costs per gram. ...improved slug designs had allowed higher power levels and cut raw material requirements by 50%. Not only did the procedures allow for better use of scarce uranium, but the steps led to dollar savings..." (Carlisle and Zenzen 1996, p. 64).

It was not until 1954 that the fuel manufacturing processes and reactor operations were perfected to the point where a reactor could run longer than 1 full month. During that year, DR Reactor completed a record run of almost 50 straight days. In February 1959, all eight single-pass reactors ran continuously for almost 9 days (DeNeal 1965).

During the mid 1950s, new facilities were constructed in the 313 Building that increased Hanford's capability to produce weapons-grade plutonium by 230 percent. The expanded and automated fuel manufacturing process could produce 615 tons of uranium fuel elements each month (GE 1956a). The following year fuel manufacturing and production capabilities had increased to 750 tons per month, producing eight new fuel elements with a total of 21 variations (Kubik 1997).

To produce the perfect fuel element, researchers in the mid-1960s *"began to experiment with a new canning procedure called the Hot Die Size process...to replace the Al-Si [aluminum-silicon] fuel fabrication process. ...the new process was more economical and more productive, and viewed as the best fuel fabrication process replacement for the two single-pass reactors still in operation, KE and KW. The Hot Die process would provide a solid state diffusion bond between the fuel core [element] and the aluminum can, using an electro-nickel diffusion barrier, and plated with nickel; the nickel plating would create a better barrier between the aluminum can and the uranium fuel core; a barrier that would prevent the rapid diffusion between the aluminum and the uranium that was believed to be one of the causes of slug swelling and distortion. The fuel assemblies were preheated to the proper bonding temperature in an inert atmosphere furnace to bond the nickel plating to the uranium core. Overall, the bonding between the uranium core and the aluminum can would be better and the cladding more uniform. The process was more flexible; uranium fuel cores could be made in ten- and twelve-inch lengths, instead of four- and eight-inch lengths. In October 1968, one Hot Die Size fabricating line was installed in Building 313; 7.5 tons of uranium fuel elements were produced using the new method."* - Kubik 1997, pp. 43-44

Efforts to increase the power levels of the Hanford Site reactors were concurrent with core distortion research that continued into the 1960s. It was agreed that part of the problem laid with "the demand for more weapons-grade plutonium [which] was accompanied by a demand for increased power in the reactor. This in turn led to performance demands for the uranium fuel elements and stress from increasing levels of power. ...scientists, engineers and technicians ascribed the problems associated with slug distortion to the changing crystalline structure of the uranium fuel rods [cores], to the fuel fabrication process in Building 313, to the dimensional specifications of the slugs and to human error on the fabrication line" (Kubik 1997, p. 33).

Internally and externally cooled fuel elements were tried in 1957 to minimize core distortion (Weakley 1999). "They had a tubular hole down the middle, allowing cooling water to run both around and through them in the reactors'" (Gerber 1992c, p. 5).

N REACTOR FUEL ELEMENT PROCESS

The co-extrusion process, or the N Reactor fuel manufacturing process, was developed in the 1960s to accommodate the fuel needs of the N Reactor. The co-extrusion process was developed in the 306-E Development, Fabrication and Test Building and implemented in the Fuel Cladding Facility (333 Building) (Figure 2-2.9).

Figure 2-2.9. *Fuel Cladding Facility (333 Building), 1996*

ENRICHED URANIUM

N Reactor used slightly enriched uranium fuel. The fuel was a tube-in-tube design consisting of a metallic uranium fuel element surrounded by a thin Zircoloy-2 cladding. The inner fuel tube was enriched to 0.95-percent uranium-235. The outer fuel tube was enriched to either 0.950- or 1.25 percent (DOE 1997e).

Fernald and Weldon Spring facilities produced the enriched uranium ingots and sent them to the Ashtabula Plant to be extruded into tubular billets. The Fernald Plant then shipped the billets to the Hanford Site where in the 333 Building they were clad with Zircoloy-2 into finished fuel assemblies using the co-extrusion process. By the time of the startup of the N Reactor in late 1963, the Fernald Plant had sufficient stocks of enriched uranium to supply the N Reactor without additional uranium from gaseous diffusion plants (DOE 1997e). "The Hanford Site also chemically recycled enriched uranium from its own fuel, and enriched 'mined' uranium for reactor fuel from the high level waste tanks on site, using the U Plant. Enriched uranium from the gaseous diffusion plants was not needed for the Hanford Site until 1985, shortly before N Reactor was shut down" (DOE 1997e, p. 157).

306 FUEL ELEMENT PILOT PLANT

The 306 Building, the Fuel Element Pilot Plant, was completed in 1956 to assist the 313 Building operations and to pilot process improvements in fuel manufacturing methods for single-pass reactors. In 1960, the facility was expanded to approximately double its original size to contain the co-extrusion process for N Reactor fuel elements. The addition became known as the 306-E Building, and the older section became the 306-W Building. For more information on the 306-E and 306-W buildings, see the Expanded Historic Property Inventory Forms in Appendix B on the Internet. The 306-E Building housed a complete fuel element manufacturing line, including component preparation, canning, finishing, and all inspections except autoclaving.

333 FUEL CLADDING BUILDING

The 333 Fuel Cladding Building was completed in 1960 as the new fuel cladding facility for the N Reactor. For more information on the 333 Building, see the Expanded Historic Property Inventory Form in Appendix B on the Internet. The original mission of the 333 Building was to manufacture fuel elements for N Reactor using the newly developed co-extrusion process. This process fundamentally altered the way that fuel elements had been made at the Hanford Site since World War II. Essentially, all of the fuel element components, including the uranium fuel element and all of the cladding materials, were cleaned, assembled, and extruded together. This method

Figure 2-2.10. *Billet Ovens (lower left) and 2700-Ton Extrusion Press in the Fuel Cladding Facility (333 Building)*

provided a more uniform bond than the earlier process of jacketing or cladding single-pass reactor fuel elements. Figure 2-2.10 shows the 2700-ton extrusion press in the 333 Building.

Fuel element manufacturing in the 333 Building was an unprecedented type in which the Zircoloy-2 exterior and the internal cladding were pressed together with the uranium core by the extrusion machinery. Because co-extrusion provided a more complete bond, it was superior to previous canned fuel elements that were subject to bubbles and other flaws between the layers. Zircoloy-2 provided a superior outer layer than the previously used aluminum because it could tolerate greater temperatures inside the pressurized system of the new reactor with a lower rate of corrosion. N Reactor's light-water recirculation design configuration required a fuel element with a tougher exterior resistant to the higher pressure necessary to operate its cooling system.

In the older single-pass reactors, corrosion had been a principal cause of fuel element failure. Zircoloy-2 lasted longer than aluminum and allowed the reactor to function more efficiently. Problems associated with the canning method for the single-pass reactors were a lack of uniform temperatures and thickness of layers in the metal dipping process, extensive training of personnel to achieve the high level of weld necessary, the lack of bonding between the layers, and pinholes (Hewlett and Anderson 1972a).

SEQUENCE OF OPERATIONS

The sequence of operations for the co-extrusion process for N Reactor fuel elements included component cleaning, billet assembly and preheating, extrusion, machining to length, cleaning, and welding of the end caps. Figure 2-2.11 shows the manufacturing process for fuel elements for the N Reactor. Specifically, the

"Zircoloy-2 cladding materials (an alloy of zirconium with nickel, tin, chromium and iron) were degreased in an organic solvent, rinsed with nitric and hydrofluoric acid, and air-dried. The uranium billets were degreased with perchloroethylene, etched with nitric acid, rinsed with water, dried and inspected. The uranium, copper, and zirconium parts were assembled and welded, tested, heated and extruded together. The extruded elements were cooled, cut, and machined. Nitric acid rinses removed copper and silicon residues and nitric sulfuric acid chemically milled away excess uranium on the ends of the slugs. A final nitric and hydrofluoric acid etching preceded the brazing on the ends of the caps. The end caps were degreased and etched as well. After additional finishing, the parts were given a final etching in nitric hydrofluoric acid, tested and assembled. The process reached a peak volume of 250 fuel elements per week in the mid-1980s." - DOE 1997e, p. 157

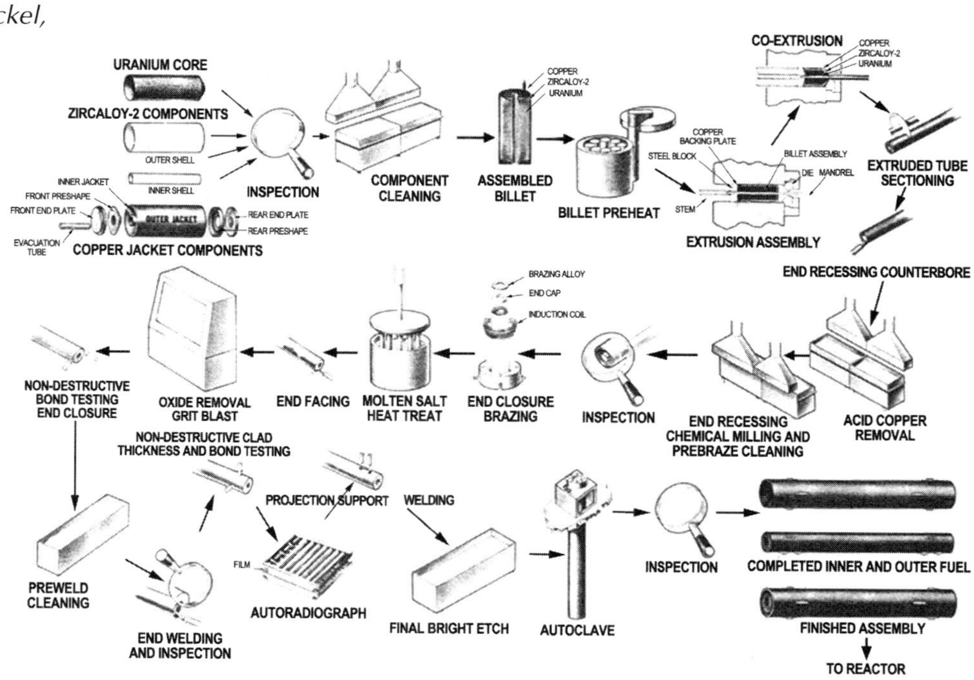

Figure 2-2.11. *Manufacturing Process of the Fuel Elements for the N Reactor*

Figure 2-2.12. *Zircoloy Element Being Cleaned, 1985*

N REACTOR FUEL MANUFACTURING PROCESS - STEP-BY-STEP

Most of the fuel element components were manufactured offsite and then assembled in the 333 Building. Exceptions were the front and rear preshapes, which were produced in the 313 Building. The components manufactured offsite were the pre-shaped uranium billets, Zircoloy-2 inner and outer shells and end caps, inner and outer jackets, front and rear end plates, evacuation tubes, and braze rings. Jack Miller, former 300 Area employee, noted that "...the uranium billets [for N Reactor] ranged 250 to 350 pounds. They were round, and not very big since uranium is heavier than lead..." (Sanger 1995, p. 174). Following delivery to the 333 Building, each component was visually inspected to make sure that no obvious flaws existed. Then each component was degreased with a vapor agent to remove any oils remaining from handling and cleaned before assembly (Clemens 1988). Figure 2-2.12 shows a Zircoloy-2 element being cleaned.

Billet assembly occurred next. This process consisted of welding the rear end plate to the inner and outer jackets, placing the uranium billet, Zircoloy-2 shells, and pre-shapes inside the jackets, and welding on the front end plate with an evacuation tube attached. The tube was connected to an evacuation system, which removed air from the assembly to help ensure a bond between the Zircoloy-2 shells and the uranium billet. The billet assembly was then lubricated and placed in the preheated ovens (Figure 2-2.10). Over a period of approximately 12 hours, the billet assembly was gradually brought up to a temperature of about 1400 degrees Fahrenheit through a stepped process that allowed for the varying expansion rates of the metals involved. The billet assemblies were removed from the preheated equipment one at a time by a crane and loaded onto the extrusion press feeder (Clemens 1988).

The billet assembly then was fed into the 2,700-ton extrusion press and pressed through a die (mold) to form it into the proper shape and size. Prior to co-extrusion, the uranium billet was about 20 inches long and 8 inches in diameter. After co-extrusion, the resulting tube was about 32 feet long, 2-3/4 inches in diameter. The tubes, which had a temperature of about 750 degrees Fahrenheit, were rotated on the roll-out table for about 5 minutes each to promote even cooling and reduce warp (Clemens 1988, Weakley 1999). The heat and pressure of the co-extrusion process not only changed the size but also, more importantly, facilitated the formation of a metallic bond between Zircoloy-2 and uranium. This decreased the likelihood of a bond failure (one of the most common fuel element problems) and increased the ease of heat transfer once the fuel was within the reactor.

Once cooled, the approximately 32-foot long outer tubes were sawed two at a time into about 12 sections. The inner tubes were 40 feet long, cut into about 18 sections. The length of the sections depended on the needs of the reactor at the time and the uranium enrichment involved. The reactor process tubes, which held the fuel elements and carried the cooling water, tended to warp within the reactor – the more warp, the greater the need for the shorter elements, which could be pushed through more easily (Vinther 1996).

During the sectioning process, the ends that extended beyond the parts suitable for use as fuel elements were cut off and returned to the 313 Building for recycling into preshapes – one of the many instances of recycling integrated into the process.

Copper-silicon was used for many of the parts (preshapes, jackets, end plates, and braze rings) and copper for the tubes. Characteristics of copper-silicon that made it an important and useful part of the process were its softness that helped it slip through the die with relative ease and superior welding properties. End cropping also removed sections that did not meet the standards for wall thickness of the various layers and hence may have contained some uranium as well.

Various cut-up patterns were used over the years to try to achieve the maximum use of uranium. The uranium was so expensive that saving even a small amount of it meant a significant reduction in production costs (Remaize 1996).

The cutting process included one of the many quality-control checks made throughout the manufacturing cycle. A 1/4-inch slice was removed from the front, middle, and end of the coextruded tube. The small pieces were then cut on a band saw to check on the bond between the Zircoloy-2 and the uranium. If the bond was found to be incomplete, the tube was scrapped and sent to the Fernald Plant for recycling. All tests for unbonded fuel elements were especially important as a lack of proper bond was the single largest cause of rejects (Remaize 1996).

After the desired length was achieved, uranium was milled out from both ends of the pieces to create a recess. The recess provided a space into which the end cap would later be inserted. The uranium fines (tiny, dust-like particles from the milling and other processes) and chips were originally collected, encased in concrete in the 304 Concentration Building, and shipped to the Fernald Plant for recycling. Beginning in 1983, however, the fines and chips first went to the new 303-M Oxide Burning facility and were then shipped to the Fernald Plant (Remaize 1996, Weakley 1999).

The fuel elements were transported to the 333 Building's chemical bay, where, in four separate processes, the now unnecessary copper-silicon was chemically stripped and the ends etched with acid to clean the uranium and Zircoloy-2 surfaces. The acid used to etch the uranium ends was kept separate so that it could be pumped to the 313 Building for separation of the uranium, which was then shipped to the Fernald Plant for recycling.

A quality control inspection was the next step. Each fuel element was visually inspected as well as measured to determine whether proper wall thickness, outer diameter, and inner diameter had been achieved (see Figure 2-2.13).

Figure 2-2.13. *Inspection of the Fuel Rods in the 333 Building, 1985*

The next step, the end closure brazing process, closed off each end of the fuel element with a Zircoloy-2 end cap. The end cap was placed in the recess with the brazing alloy, a mixture of Zircoloy-2 and beryllium, placed on top of it. The fuel element was then inserted into the brazing chamber. A glass jar of sufficient height and width was placed over the vertical fuel element and sealed to the brazing chamber. The air in the chamber was removed by a vacuum pump. An induction coil came down over the glass enclosure and was heated so that the brazing alloy would melt and begin to seal the end. Final end closure was achieved by pressurizing the brazing chamber with argon gas to push the now liquid alloy into any remaining voids. The induction coil was turned off and the fuel element allowed to cool for a few minutes before removing the jar. The fuel element was turned over and the process was repeated on the other end (Weakley 1999).

A molten salt bath followed to heat treat the uranium to randomize the grain structure that had formed during the co-extrusion process (Weakley 1999). Were the grains left aligned, they could push the end cap off in the reactor, resulting in contaminated cooling water within the process tubes and a reactor shutdown. For the salt bath, the fuel elements were hooked vertically onto a carousel mechanism and lowered into the salt bath. Another quality control check was done here to ensure that the fuel pieces were within the acceptable tolerances for straightness. After being checked in the warp gauge, a fuel section was either deemed straight enough and passed on to the rest of the process or rejected for further straightening. A fuel element could be re-straightened and re-heat-treated twice before being completely rejected and sent to the Fernald Plant for recycling.

A machining process followed, whereby the excess brazed metal on both ends was removed and a small cut made into both the inner diameter and the outer diameter. The cuts were used to help align the welding of the end cap. Vacu-blasting came next, which used a grit blast to remove the zirconium oxide scale that built up during the salt bath.

The end brazing was then tested using a non-destructive, ultrasonic method to make sure the brazing alloy had completely filled all the voids in the ends. Improper brazing was the second largest cause of fuel element rejects. However, if the brazing was incomplete on one of the longer elements, the cylinder could be sent back to be cut to the next shorter length and proceed through the other steps again. If the fuel element was already as short as it could be, it was rejected and recycled.

The next step was an acid cleaning that removed grit left on the Zircoloy-2 surface during the vacu-blasting. This etching prepared the fuel element for end welding, during which a weld bead was put over the step cuts for the second and final closure of the end. The fuel enrichment and unique identification number was stamped into the end of the fuel. The completed fuel elements then moved into another series of non-destructive tests to measure clad thickness and find areas that had not bonded (see Figure 2-2.14). The fuel elements were submerged into a tank of water for testing. After the fuel element was lowered into the water, mechanical devices, called transducers, checked for cracked bonds and thickness of the cladding (Figure 2-2.15). These tests were a major source of error detection.

Figure 2-2.14. *Non-Destructive Testing of Fuel Elements, 1984*

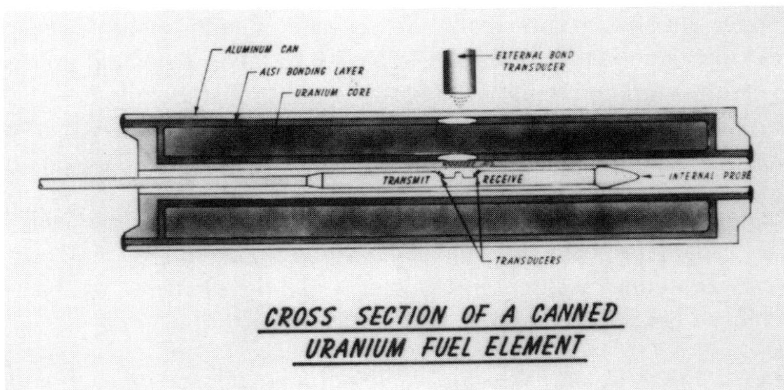

Figure 2-2.15. *Cross Section of a Fuel Element Showing the Transducers, Which Check for Cracked Bonds and Wall Thickness*

The spot welding of the Zircoloy-2 projection supports and springs comprised the next step. The support hardware was welded onto the outer diameter of each fuel element. Those designated as outer fuel elements also had a locking clip welded into its inner diameter. Those designated as inner fuel elements also had a locking clip welded to their outer diameters. The locking clips allowed for the inner and outer fuel elements to be locked together into a completed fuel assembly. The springs helped position the inner fuel element within the outer fuel element. The supports on the outer diameter of the outer fuel element helped position the completed fuel assembly within the reactor process tubes. The spot welding was followed by a final acid bath.

Figure 2-2.16. *Autoclave Area in the 333 Building, 1984*

To test the fuel elements, autoclave operators placed the completed fuel elements in baskets and inserted the baskets into autoclaves.

The autoclaves subjected the fuel elements to reactor-like conditions in terms of pressure and temperature (Figure 2-2.16), especially testing the fuel elements for potential ruptures. Also, the high pressure and temperature of the autoclaves caused the Zircoloy-2 to form a black oxide on the surface. This precluded further oxidation and, therefore, provided a protective coating over the Zircoloy-2.

One of the causes of ruptures was water. Autoclave operators heated the fuel elements in the autoclave for 72 hours to allow enough time for any water to evaporate. Ruptured fuel elements were detected by checking for hydrogen within the autoclave. The presence of hydrogen indicated that water had penetrated the Zircoloy-2 cladding and had reacted with the uranium core. Ruptured fuel elements were rejected and recycled (Remaize 1996, Weakley 1999). As the quality of the fuel elements increased, the testing time was reduced to 48 and sometimes as low as 32 hours (Peterson 2000). Figure 2-2.17 shows the inspection of the fuel elements in the 333 Building.

Figure 2-2.17. *Inspection of the Fuel Elements in the 333 Building*

A final quality assurance inspection was made following the autoclave test. The completed outer and inner fuel elements subsequently were placed in a hydraulic press, pushed together, turned, and locked to form a completed fuel assembly. A thin steel shoe was applied to the outer supports on the outer fuel element at this time to aid the assembly's passage through the reactor process tubes, which were also made out of Zircoloy-2. Without the steel in between, the Zircoloy-2 fuel slugs and Zircoloy-2 process tubes would tend to cohere, metallurgically speaking, therefore impeding the progress of the fuel assembly during the charging/discharging operation and possibly creating unwanted grooves or scratches in the process tubes (Remaize 1996). The finished assembly was steam cleaned and packaged for shipment to N Reactor.

N Reactor Fuel Manufacturing Process Modifications

There were no radical changes in design of the coextruded fuel assemblies from their first use in the N Reactor in 1963 to final shutdown of the reactor in 1986. However, numerous small but significant modifications were made to continually refine the design and production process in an ongoing quest to improve reactor performance, achieve worker safety and environmental quality goals, promote cost-effective production, and accommodate experiments. For example, a system of grading fuels according to their uranium enrichment level was instituted in 1963 so that the higher grade fuels, which could withstand more severe reactor conditions, could be arranged selectively and strategically within the reactor to achieve the most efficient level of operation. Several enrichments of uranium were used in the reactor at any given time – the lower the number of high-grade assemblies, the lower the cost.

Recycling of billet components and testing and etching substances were other common methods of controlling costs and addressing environmental concerns as well. Solutions to environmental questions were sought constantly, especially during the 1970s-1980s. Many of the cleaning and stripping chemicals used in the production process were highly toxic and/or carcinogenic. A study of one such substance used in stripping copper, hexavalent chromium, was successfully concluded with the finding that it could be rendered harmless by mixing it with sodium sulfite, sulfuric acid, and water. That finding eliminated a potential groundwater pollutant.

Often, experiments were conducted to test hypotheses regarding improvements to the production process. For example, tests were carried out in 1984 to investigate possible factors affecting the success, or lack thereof, of the co-extrusion bond. Four tests were run on 36 billets using slight variations on the thickness of the front end plate, a faster than standard speed for pushing the billets through the press, using a front preshape with a different silicon content, and using a different drying process on the front and rear preshapes (UNI 1984).

One particularly difficult problem to solve was that of the design of and materials for the inner supports. The supports, which had shown promise and were actually used in the early coextruded fuel slugs, were determined to be inadequate due to their tendency to crack (Lewis 1996).

SAFETY AND QUALITY ASSURANCE

Because successful reactor operation was dependent on properly functioning fuel slugs, a major emphasis was placed on quality assurance throughout the manufacturing process. Nuclear and worker safety also were overriding concerns. It was infinitely more economical and safer to uncover flaws in the fuel slugs before they were placed into the reactor. The sooner a flaw could be detected, the less expensive it was to either fix the fuel element or reject it for recycling. If a fuel element ruptured in a process tube, thereby releasing uranium into the cooling water, the reactor had to be shut down and the contamination removed. Shutting down and re-starting the reactor was quite an expensive proposition with ramifications for meeting deadlines on plutonium and power production (Krieg 1996).

Another means of ensuring safety was the practice of coding every fuel element with an identification number that indicated its level of enrichment, from which extrusion it originated, and from which part of the extrusion. Logs were maintained that indicated at which point in the production process any given fuel element was and exactly how it had been made. These life histories not only allowed staff to keep track of where the uranium was at any given time within the 333 Building and thereby avoid a criticality but also to document the progress of a fuel element through the production process. When a fuel element ruptured within a process tube and was pushed out of the reactor, the number could be read by remote means and the cause of the rupture possibly traced back to the production process (Vinther 1996). If a number of fuel elements that shared a common characteristic ruptured, such as all having been heated in the same autoclave, the autoclave operators would examine the chart paper of the suspect autoclave for irregularities, such as abrupt temperature increases (Peterson 2000).

Similarly, problems that occurred prior to a fuel element being inserted into the reactor could be traced to their source. For example, a high rate of unbonds between the interior uranium and exterior Zircoloy-2 cladding on the outer diameter of the outer fuel extrusions was detected in early 1976 following their co-extrusion. The problem was the direction in which the ingots were fed into the dies. A slight change in the fabrication process in 1972 had resulted in the reversal of the direction in which the uranium ingots were fed into their extrusion dies. The vendor changed the production process to have the ingots fed through the other direction so that problem was traced and solved. Through a series of tests and experiments, the fuel fabrication workers determined that the most cost-effective way of dealing with the subject billets already on hand was to send them off-site for reworking into a different size billet (Bowles 1976).

TRITIUM PRODUCTION

A major campaign to diversify the production possibilities of the N Reactor, and therefore the 333 Building, was a mid-to-late 1960s trial co-production of tritium. Tritium was the key component of hydrogen bombs. Tritium "was used to boost the plutonium detonation in both nuclear and thermonuclear weapons" (Carlisle 2000). While tritium was not the only co-product within the capabilities of the N Reactor, it represented one of the most attractive ongoing financial opportunities. Tritium isotopes decay at a much faster rate than plutonium isotopes. It cannot be stockpiled or recycled in the same manner as plutonium and therefore must be produced in an essentially continual fashion to maintain weapons capability. The N Reactor could not be operated economically on a power-production only basis. The reactor had to produce plutonium, tritium, or some other product to perform on a profit-making basis.

Tritium was created in the N Reactor during a major campaign from 1965-1967 using fuel assemblies that consisted of an outer, or driver, fuel element manufactured in the 333 Building, and an inner target element made in the 3722 Area Shop (Gerber 1993b, Lewis 1996, Williams 1966). For further information on the 3722 Building, see the Historic Property Inventory Form in Appendix B on the Internet. Instead of uranium, the inner target elements contained lithium aluminate, a ceramic material. Lithium aluminate powder was formed into a solid through a sintering process, which used heat and pressure to fuse the material into a pellet. The pellets were then placed into an aluminum can clad in Zircoloy-2, which replaced the usual inner tube of the standard fuel assembly (Dickeman 1962).

Once completed, the targets were transferred from the 3722 Building to the 333 Building for autoclaving, cleaning, inspection, and insertion into the driver fuel element. The driver fuel element was made in the 333 Building by following the same procedures for fabricating the outer tube of a fuel element for producing plutonium (Gerber 1993b). The only difference was the outer driver tube was constructed with a higher than normal uranium enrichment.

The N Reactor was occasionally used to produce tritium in the 1970s and again during the late 1980s. Associated alterations and upgrades were made to the 333 Building between 1987 and 1989 in preparation for once again producing the target assemblies but in a different configuration than in the 1960s. N Reactor was placed in cold standby status in 1988, and when the final shutdown occurred in 1991 the fuel production life of the 333 Building was discontinued (Remaize 1996, Gerber 1993b, Stapp and Marceau 1995).

Commercial/Non-Defense Fuel Manufacturing

Knowing that worldwide supplies of uranium were finite, the United States embarked upon a large research effort to find alternate supplies of reactor fuels. Research was undertaken to find methods to stretch or diversify the uranium fuel supply for commercial nuclear reactors by creating oxide fuel blends. This was the result of President Eisenhower's Atoms for Peace program unveiled in 1953 that envisioned non-defense utilization of nuclear technology to fuel civilian and industrial power needs. In response, the Atomic Energy Commission pursued a program of developing alternative nuclear fuels for industrial and commercial uses. The Hanford Site was the locale of an extensive research effort to demonstrate the effectiveness of fuels blended from the combinations of plutonium oxide, uranium oxide and other mixed oxide materials.

The 300 Area's Plutonium Fabrication Pilot Plant was completed in 1960 to provide research laboratories and fuel manufacturing facilities for the development of reactor fuels containing plutonium. The earliest fuels manufactured in the Plutonium Fabrication Pilot Plant (308 Building) were irradiated in the adjacent Plutonium Recycle Test Reactor (309 Building). Both buildings were constructed at the same time to demonstrate the effectiveness of the various plutonium oxides containing blends of plutonium, uranium, and other metals. For more information on the 308 and 309 buildings, see the Expanded Historic Property Inventory Forms in Appendix B on the Internet.

Later in the decade, the 308 Building " produced neptunium-aluminum alloy fuel target slugs [cores] clad in Zircoloy-2 for a Plutonium-238 production test that was conducted in N Reactor. ...lithium aluminate fuel targets with a rod-in-tube configuration and Zircoloy-2 cladding also were manufactured in the 308 Building for coproduct (tritium) production testing in N Reactor. Over 2000 fuel configurations were produced in the 308 Building before 1976, many for a variety of offsite reactors" (Gerber 1993b, p. 36).

Additionally, "in the late 1970s, a Training Research Isotopes, General Atomics (TRIGA) reactor was installed in the building to perform neutron radiography QA testing of fuel elements and fuel jackets" (Gerber 1993b, p. 36). At the same time a high bay addition was completed to the 308 Building that contained laboratory facilities for the storage, handling, quality assurance testing, assembly and instrumentation of Fast Flux Test Facility fuel.

Recovery, Reuse, and Recycling Methods

The materials used in the fuel manufacturing process in the 313 Building, including the uranium fuel elements and the copper, tin, and aluminum-silicon used in the triple dip process, were expensive, non-renewable resources. As the demand for weapons-grade plutonium grew in the late 1940s and early 1950s, the fuel manufacturing process consumed even greater quantities of these costly materials. It was important that these materials be recovered and reused or recycled whenever possible.

Difficulties with early fuel canning techniques produced thousands of rejected fuel elements by mid-1944. "Scrap slugs were utilized whenever possible in the experimental work...in order to release equal quantities of acceptable slugs for production purposes" (Du Pont 1945a, p. 365).

By the mid 1950s, "between 10 and 30 percent of the fuel elements do not meet specifications and are rejected. These rejected fuel element cores are recovered by a chemical process and prepared for recanning" (Weakley 1958, p.126).

Along with rejected fuel elements, there were also uranium scraps to be recovered. During the earliest fuel manufacturing operations at the Hanford Site, uranium scraps consisted of lathe turnings, rod ends, and rejected fuel elements from the machining and canning operations in the 313 Building.

Beginning with the startup of the extrusion press tests in January 1945, extrusion butt ends, oxides, and container residues were collected. By early 1946, however, the volume of uranium scraps accumulated from operations in the 313 and 314 buildings and the resulting expense and fire and security hazards brought a change in policy. A chip recovery operation began in the 314 Building. It involved collecting all chips and turnings from machining operations, "sorting them, breaking them into small pieces, washing, drying, and then pressing them into briquettes" (Gerber 1992c, p. 7).

In the spring of 1946, an additional scrap recovery operation known as the oxide burner began on the north side of the 314 Building. All uranium-bearing dust and particulate matter that could be collected, as well as the tailings or settlings from washes and quenches, was burned to convert it to oxide (powder) form (Kubik 1997).

The General Electric Company refined the uranium core recovery process, making it safer and less labor-intensive. In the mid-1950s, the 313 Building experienced extensive remodeling and expansion to accommodate recovery processes along with new fuel canning technologies (Gerber 1992b). "A process for the recovery of rejected slugs [cores] was developed in the 300 Area. These rejected slugs [cores] represented a form of recoverable uranium that was jacketed in an aluminum can bonded with aluminum-silicon, and it was desired to remove the coating so that the slug [core] could be recanned or reclaimed as scrap metal" (Du Pont 1945a, p. 218).

Canned uranium fuel elements that failed any one of the non-destructive quality assurance tests in the 313 Building were immediately sent to Recovery Operations in the building's south end. Here, the aluminum cap and can base were removed. The aluminum can and the compound layers were dissolved with a series of three baths, first in sodium hydroxide and sodium nitrate solution, second in hydrofluoric acid, and finally in nitric acid. The cores were then cleaned, rinsed, dried, and returned to the canning line to be recanned (Du Pont 1945a, Weakley 1999). In 1953, the hydrofluoric acid and nitric acid baths were replaced with a boiling 50-percent sodium hydroxide solution that does not release explosive hydrogen gas (Weakley 1999).

A canned, rejected uranium core could be sent through the recovery process a maximum of three times before it no longer met the dimensional specifications (weight, length, and diameter) for Hanford's single-pass reactors. Cores that lost their dimensional integrity during the recovery process were shipped back to the original vendor for reprocessing into new uranium billets. The recanned uranium cores were designated Z fuel elements, and as with all steps of the fuel manufacturing process were carefully checked for flaws prior to and during the recanning process (Burley 1956, Kubik 1997).

By mid-1947, there was a large stockpile of copper-contaminated tin from the triple dip baths. General Electric began to cool and strain this contaminated molten tin to collect and remove the copper-tin crystals that formed in the cooling process. The recovered tin could be recycled back to the canning line several times. Each year, the tin recovery process became more economical and efficient, with fewer impurities (Kattner 1949).

According to Kattner (1949, p. 5), "the principal value of the process lies in the fact that it eliminates the necessity for using virgin tin. During this time (November 1, 1947 to March 3, 1948) about 24,000 lb. of tin from the scrap stockpile have been recovered, equivalent to an estimated 40,000 lbs. of virgin tin."

There were continuous discussions over the years dealing with the problems associated with the large accumulation of scrap, which included canned, rejected uranium cores, aluminum-silicon, and tin. The focus was on reclaiming all possible scrap for reuse in the fuel manufacturing process, and in doing so, reduce the quantities of accumulated scrap. Cores and tin were given a top priority (Kattner 1949, Kubik 1997).

Over the decades, the ability to recover, recycle and reuse the rejected canned uranium cores was a key factor in each proposed new step of the fuel manufacturing process. Any recovery method had to be cost effective, simple, and relatively waste-free (Kattner 1949). The materials used to recover the canned and rejected uranium cores could not be so strong as to corrode the core or cause it to lose its dimensional integrity (Kubik 1997).

MANAGEMENT OF FUEL MANUFACTURING WASTE

Every step in the plutonium production process at the Hanford Site generated a considerable amount of waste and other byproducts. Until 1973, all low level, radioactive solid waste from fuel manufacturing activities in the 300 Area was disposed of in the 300 and 600 Area burial grounds. After 1973, 300 Area low level, radioactive solid waste was transported to the 200 Area for disposal. From the time of their construction in 1944, the numerous 300 Area laboratories and fuel manufacturing facilities were connected to a common process sewer that collected low level radioactive liquid waste and disposed of it in a single pond located east of the 300 Area near the Columbia River. The original process pond dike broke in 1948, spilling most of the pond's contents into the Columbia River. Following this accident, a new process pond was built to the north of the original pond, and the two ponds were used in tandem. Periodically, the sludge from the ponds was dredged to recover uranium (GE 1953l). The process ponds were phased out in 1974 and replaced by the 300 Area Process Trenches (Kubik 1997). For additional information see the Waste Management Section.

For four decades, fuel manufacturing activities, especially in the 313 and 314 buildings, generated acidic liquid (chemical) waste from canning, capping, sleeve cleaning and testing, and uranium scrap processing. The liquid chemical waste from the 313 and 314 buildings that did not contain recoverable uranium was discharged to the process sewer until the shut down of the single-pass reactors. From the early 1960s to 1973, the 333 Building discharged waste acids (nitric, sulfuric, hydrofluoric, and chromic nitric sulfuric) containing uranium, zirconium, copper, beryllium into an underground tank containing limestone. The waste acids trickled through the limestone and entered percolation ponds through a process sewer (Weakley 1999).

Public scrutiny and concerns in the 1970s-1980s were primarily focused on reactors and their operation as opposed to the fuels production process. In 1970, the Atomic Energy Commission instituted an accelerated program to reduce radioactive discharges in every operation to the lowest practicable levels. The program was later expanded to cover all hazardous wastes (Vinther 1996, Weakley 1976). One of the results of this program was the installation, beginning in 1973, of the Waste Acid Treatment System, which routed toxic and carcinogenic materials left over from the N Reactor fuels fabrication process into tanks for treatment and storage rather than into settling ponds from which the materials could leach into the Columbia River (DOE 1996c).

The Waste Acid Treatment System process was developed in the 333 Building and placed in service in 1973 to process waste acids and chemical wastes disposed of by various 300 Area laboratories and fuel manufacturing facilities (Weakley 1999). Specifically, the treatment system caught and neutralized waste acids from the 333 Building's fuel manufacturing activities. Waste acids were collected in the Acid Pump House (334-A Building) and then pumped to tanks in the south end of the 313 Building for neutralization. "The uranium sludge collected from the automatic rinse tank is also added to the neutralizer tank and the uranium recovered" (Weakley 1958, p. 129).

AREAS FOR FURTHER RESEARCH

One of the major achievements in the area of fuel manufacturing was the development of the co-extrusion process, which fundamentally altered the way fuel elements had been made since World War II at the Hanford Site and throughout the rest of the world. Because this narrative is a history of facilities and not intended to be the definitive history of fuel manufacturing at the Hanford Site, the details of such achievements as co-extrusion, issues concerning waste, and other areas of interest could not be explored. Hopefully, this narrative will serve as a guide for future researchers to pursue more extensive study of important issues noted in this section, such as

- According to Kubik (1997, p. 33) "...scientists, engineers and technicians ascribed the problems associated with slug [fuel element] distortion to the changing crystalline structure of the uranium fuel rods [cores], to the fuel fabrication process in the 313 Building, to the dimensional specifications of the slugs and to human error on the fabrication line." Of interest to future researchers would be to uncover how they arrived at these conclusions and to uncover if possible the cause of the dramatic increase in fuel element failures from 3 in 1950 to 115 in 1951.
- Future researchers may also want to look into the invention of the co-extrusion process: its origin, inventor, etc.
- One theme that deserves more coverage is the test reactors and fuel manufacturing pilot facilities in the 300 Area. During the Cold War era several test reactors and pilot plants were constructed and used for non-defense purposes. With the worldwide uranium supplies limited, efforts were undertaken to develop and test alternate fuels in these facilities, focusing on the effectiveness of plutonium oxide and mixed oxide fuel blends. What research and development programs were initiated to improve the fuel manufacturing process in these test facilities?
- More in depth research on the effect of the Hanford Site's fuel manufacturing technology on the development of the nondefense/commercial nuclear industry could be undertaken. What has been the connection between commercial nuclear research at the Hanford Site with offsite nuclear power companies like Siemens?
- Another theme that could be the focus of future research efforts is the relationship of the Hanford Site's fuel manufacturing technology with other U.S. Department of Energy plutonium production sites, such as the Savannah River and Oak Ridge sites.
- Also to be explored is what protective measures the U.S. Department of Energy undertook for workers while they manufactured fuel? What industrial health and safety issues were associated with fuel manufacturing activities? What numbers of workers were employed in fuel manufacturing at the Hanford Site? What were the names and types of worker positions? What was the organizational structure in regard to the fuel manufacturing workforce?
- More extensive research is needed into the wastes generated by fuel manufacturing and what technologies were developed to recover or at least neutralize these wastes.

SECTION 3

REACTOR OPERATIONS

D.C. STAPP
PACIFIC NORTHWEST NATIONAL LABORATORY

SECTION 3 - REACTOR OPERATIONS

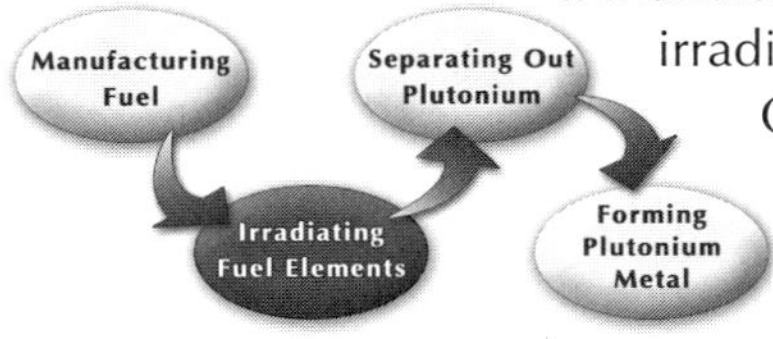

If manufacturing the fuel elements was the heart of the plutonium production process, then irradiating those fuel elements was the soul. Irradiating the fuel elements as part of Reactor Operations was the second step in the process at the Hanford Site and one that continues to illustrate the magnitude of the uncertainties that accompanied the plutonium production process.

> ***"Beginning in the Manhattan Project, and during the twenty years following it, Hanford's work and role were defined by its production reactors." -*** **Findlay and Hevly 1995, p. 141**

To make plutonium for the atomic bomb (the Hanford Site's primary purpose), the U.S. Army Corps of Engineers constructed the world's first full-scale nuclear reactors to irradiate uranium. Over a 25-year period, the government constructed and operated nine different production reactors at the Hanford Site to produce plutonium. With this capability, the Hanford Site served the nation as the government's major source of plutonium during World War II and the Cold War era.

The expertise Hanford staff gained in constructing and then operating the early reactors (B, D, and F) as well as the later reactors made the site one of the primary centers of nuclear reactor technology. From 1945 through the mid-1950s, the Hanford Site was the only full-scale reactor center in America (Goldschmidt 1982). Using knowledge and experience gained during these years, Hanford staff contributed significantly to the field of nuclear technology by helping to develop reactor technology for energy production, develop strategies for managing nuclear waste, and understand health effects of reactor-related radiation. During the 1950s, General Electric, Combustion Engineering, and Westinghouse were the three primary companies involved in building nuclear energy plants. Because General Electric was the prime contractor at the Hanford Site (1946-1964), they were in an optimal position to apply the knowledge and experience gained at the Hanford Site to their commercial nuclear division in San Jose, California.

The achievements, however, did not come without a price. Because of the pressure to produce large quantities of plutonium, the power levels of the reactors were increased beyond design expectations, which led to such problems as leaks and ruptures in the fuel elements. Because the first eight of the nine reactors did not have closed-loop systems for the cooling water, environmental problems arose when irradiated cooling water was returned to the Columbia River.

Beginning in 1944, the U.S. Government started construction of the first full-scale nuclear reactors. The Hanford Site provided the plutonium for the implosion bomb, while its sister site, the Oak Ridge Reservation in Tennessee, provided the enriched uranium for the gun-type bomb. The Corps shipped the finished product to Los Alamos, where staff constructed the bombs. The Hanford Site also produced polonium, an important material used in the final assembly of the bomb.

As more fully described in Chapter 1, the roots of plutonium trace to work conducted in the early 1940s at Ernest O. Lawrence's Radiation Laboratory at the University of California at Berkeley. Of the two main isotopes of uranium (element 92), namely uranium-235 and uranium-238, only uranium-235 fissions in a way that

CONTENTS

Divider Page Photo: B Reactor, 1945 (Neg. P8015)

makes it suitable for a bomb. In 1940, however, scientists at the Lawrence Laboratory confirmed that the other isotope, uranium-238, could absorb a neutron and decay into the transuranium element, neptunium (element 93). Glenn Seaborg then discovered that neptunium decayed into element 94, which he named plutonium. When Seaborg's team showed that plutonium was more likely under some circumstances to fission than uranium-235 atoms, the seeds of the Hanford Site were sown (Gosling 1994).

Seaborg's discovery was important because it confirmed another technological path leading to the atomic bomb. Scientists already believed they could make a bomb with uranium-235 atoms, the isotope comprising 0.7 percent of the atoms in natural uranium, but separating uranium-235 from its natural uranium state required immense effort (hence the construction of the massive gaseous diffusion plants at the Oak Ridge Reservation). The discovery of plutonium, in particular the work of Enrico Fermi and his graphite reactor at the University of Chicago, spurred government support. Table 2-3.1 shows the evolution of the reactor technology that resulted in the work at the Hanford Site.

Table 2-3.1. *Evolution of Reactor Technology in the 1940s*

Date	Event	Location
1940	Enrico Fermi shows a chain reaction possible using large amounts of uranium in a reactor	Columbia University
1941	Glenn Seaborg discovers plutonium-239 more likely to fission under some circumstances than uranium-235	University of California at Berkeley
1942	Fermi achieves first self-sustaining chain reaction with CP-1Reactor at Stagg Field	University of Chicago
1943	Air cooled pilot-scale graphite reactor completed	Oak Ridge Reservation, Tennesse
1944	Water cooled graphite production reactors completed	Hanford Site, Washington
1945	First atomic explosion – Trinity test	White Sands Proving Ground,New Mexico

In September 1942, Lieutenant General Leslie R. Groves took command of the Manhattan Project. At the time, the federal government was sponsoring research on four major processes for producing or separating isotopes (Gosling 1994, p. 14):

- Electromagnetic methods for separating uranium-235 isotopes being researched at the University of California at Berkeley under the direction of Ernest O. Lawrence
- Centrifuge methods being researched at the University of Virginia under the direction of Jesse R. Beams
- Gaseous diffusion methods being researched at Columbia University under the direction of John R. Dunning
- Graphite reactor technology at the Metallurgical Laboratory in Chicago under the direction of Enrico Fermi.

On December 2, 1942, Fermi confirmed that he could create a self-sustaining nuclear reaction with graphite. Fermi used the graphite as a moderator to slow down the neutrons from the high speeds caused by the fission process. The moderator made it possible to keep the chain reaction going. Fermi's graphite moderator was one among the three moderating methods being explored, which paved the way for producing plutonium. By late December 1942 after much debate by various committees, the Manhattan Project was developing a full-scale gaseous diffusion plant, a pilot-scale graphite reactor as well as full-scale graphite reactors, and a small electro-magnetic plant, which later was enlarged (Gosling 1994, pp. 14-16).

Designing and Constructing the Hanford Reactors

The design for the Hanford reactors came out of the Metallurgical Laboratory in Chicago under the general direction of Arthur Compton. Building upon Fermi's basic graphite moderator concept, designers were faced with a major challenge: how to design a reactor that would 1) be safe, 2) be controllable, and 3) have enough neutron multiplication to sustain a reaction. Three primary designs were the major contenders in 1942 (Gosling 1994, pp. 26-27). The contenders differed according to the method used to carry away the heat produced by fission in the reactors:

- Helium-cooled design pursued by Thomas V. Moore
- Water-cooled design developed by Eugene Wigner and Gale Young
- Liquid metal cooled design pursued by Leo Szilard.

In October 1942, Groves pushed Arthur Compton to make some decisions regarding the reactor design to be pursued. Out of this came the decision to build a pilot-scale graphite reactor constructed at the Oak Ridge Reservation in Tennessee even before Fermi had demonstrated that he could achieve a self-sustaining chain reaction. The pilot scale reactor was not really a prototype for the Hanford reactors. It was air-cooled rather than water-cooled, but it was useful for various engineering design purposes. According to Alvin Weinberg, one of the principles in the Metallurgical Laboratory, the main significance of the graphite reactor was that it produced significant quantities of plutonium for the first time ever (1 gram per day), which made it possible to test separations processes on a larger scale than had been done heretofore (Sanger 1995, p. 35).

In early 1943, the Hanford Site was selected as the site for the full-scale plants authorized by President Roosevelt. By this time, the Metallurgical Laboratory staff were pursuing two major reactor designs: water cooled and helium-cooled. A tri-partite system was in place to make design decisions. The conflicts and synergies resulting from this tri-partite system were a major reason for the success of the Manhattan Project. The tri-partite system involved

- Scientists, mostly at the Metallurgical Laboratory in Chicago, who provided the theoretical expertise
- Du Pont engineers, who provided the practical construction, operations, and engineering expertise
- U.S. Army Corps of Engineers Manhattan Engineering District, who made the high level decisions and any necessary to resolve disagreements between the Metallurgical Laboratory scientists and Du Pont engineers

In February 1943, Crawford H. Greenewalt of Du Pont was debating which reactor cooling design to pursue:

> *"Greenewalt's initial response to the water-cooled design was guarded. He worried about pressure problems that might lead to boiling water in individual tubes, corrosion of slugs [fuel elements] and tubes, and the one-percent margin of safety for k [neutron multiplication]. But he was even more worried about the proposed helium-cooled model. He feared that the compressors would not be ready in time for Hanford, that the shell could not be made vacuum-tight, and that the pile [reactor] would be extremely difficult to operate. Du Pont engineers conceded that Greenewalt's fears were well-grounded. Late in February [1943], Greenewalt reluctantly concluded that the Met Lab's model, while it had its problems, was superior to Du Pont's own helium-cooled design and decided to adopt the water-cooled approach."* - Gosling 1994, p. 30

While the basic water-cooled graphite moderated design was agreed upon, numerous other design decisions awaited resolution. Du Pont followed an approach in which they would freeze the design of a key component, thereby freeing them to design associated components and subcomponents (Carlisle and Zenzen 1996, p. 80).

The basic components of the Hanford Site design were the following:

- Fuel – The aluminum-clad fuel elements contained metallic uranium. They measured 1.5 inches in diameter and 8 inches in length (see Chapter 2, Section 2, Fuel Manufacturing).
- Moderator – A graphite moderator was used to moderate or slow the neutrons so that a chain reaction could occur.
- Control System – The control system initiated and controlled the nuclear reaction. Boron rods were moved in or out of the reactor as necessary. The rods could also be used to shut down the reaction.
- Coolant – River water was the liquid coolant, which circulated through the reactor to carry away the heat produced in fission. It was necessary to treat the cooling water to control corrosion.
- Shielding – To protect workers, a combination of heavy iron and concrete was placed around the reactor core to absorb the radiation created by fissioning.

Construction of the first three Hanford reactors (B, D, and F) began in March 1943. Over time, decisions related to the shielding, water flow control, loading and unloading devices, coating of the uranium fuel, and water purification emerged. With these decisions, the reactor plants began taking shape. The three reactors were completed one after another: B Reactor in September 1944, D Reactor in December 1944, and F Reactor in February 1945. Plutonium production commenced almost immediately upon completion of each reactor. Eventually, six more reactors were constructed at the Hanford Site. The DR, H, C, KE, and KW reactors were quite similar in design with the two K Reactors differing from the older reactors mainly in the number, size, and type of process tubes, the size of the moderator stack, and the type of shielding employed. The ninth reactor, N, while based on the original graphite design, differed considerably. Information on reactor block size and construction materials used for all nine reactors are provided in Table 2-3.2. The events surrounding the operation of these and the later reactors are discussed in the sub-section, Reactor Operating History. First, an overview of the function and operation of reactor system components is provided.

Table 2-3.2. *Reactor Core Characteristics for Hanford Reactors*

Characteristics	**B (1944-1968)**	**D (1944-1967)**	**F (1945-1965)**	**H (1949-1965)**	**DR (1950-1964)**	**C (1952-1969)**	**KW (1954-1970)**	**KE (1955-1971)**	**N (1963-1987)**
Graphite Stack (Core + Reflector)									
Axial (m)	8.53	8.53	8.53	8.53	8.53	8.53	10.21	10.21	11.88
Vertical (m)	10.97	10.97	10.97	10.97	10.97	10.97	12.50	12.50	10.06
Width (m)	10.97	10.97	10.97	10.97	10.97	10.97	12.50	12.50	10.06
Mass of Graphite									
Moderator (tons)	1089	1089	1089	1089	1089	1089	1542	1542	726
Reflector (tons)	544	544	544	544	544	544	907	907	907
Lattice Pitch (cm)	21.27	21.27	21.27	21.27	21.27	21.27	19.05	19.05	20.32x 22.86
Carbon/Uranium Atomic Ratio (Typical Loads)	101	99	100	109	106	100	77	78	34
Number of Process Tubes	2004	2004	2004	2004	2004	2004	3220	3220	1003
Material of Process Tubes	Al	Al	Al	Al	Al	Al	Al/Zr	Al/Zr	Zr-2
Horizontal Control Rods	9	9	9	15	9	15	20	20	84
Vertical Control Rods/Ball Channels	29	29	29	45	29	44	41	41	107
Boron Ball Channels	29	29	29	45	29	45	51	51	0

REACTOR OPERATING HISTORY

Between 1943 and 1987, the government constructed and operated nine graphite-moderated, water-cooled, plutonium production reactors at the Hanford Site (Table 2-3.3). Over these 44 years, international, national, and regional events impacted the production reactors at the Hanford Site in various ways. Many of these were political in nature, some were technical, and others social (see sidebar box). The purpose of this sub-section is to illustrate the effect of these events on the reactor facilities and operating systems. To facilitate the discussion, the Reactor Operating History sub-section is divided into three phases: 1) the Manhattan Project, 1943-1945, 2) an initial Cold War period, 1946-1957, and 3) a final Cold War period, 1958-1989.

PHASE 1 - MANHATTAN PROJECT (1943-1945)

From the time the reactor program was approved until the plutonium for the first bombs was produced, those working on the Manhattan Project were pressured to get the job done. The initial stimulus for getting the plants built as fast as possible was to beat the Germans at doing the same. Later, after the Army learned that Germany was far from producing a bomb, the stimulus became ending the war as soon as possible. Once German military leaders surrendered on May 7,1945, the focus of the atomic bomb effort shifted from Germany to Japan. With emotions against Japan running rampant across the country, pressure to deliver the bomb stayed high.

Table 2-3.3. Hanford Reactor Dates of Operation

Reactor	Construction Start	Operations Start	Operations Stop
B	Aug 1943	Sep 1944	Feb 1968
D	Nov 1943	Dec 1944	Jun 1967
F	Dec 1943	Feb 1945	Jun 1965
DR	Dec 1947	Oct 1950	Dec 1964
H	Mar 1948	Oct 1949	Apr 1965
C	Jun 1951	Nov 1952	Apr 1969
KW	Nov 1952	Dec 1954	Feb 1970
KE	Jan 1953	Feb 1955	Jan 1971
N	May 1959	Mar 1964	Jan 1987

By February 1945, the B, D, and F reactors were up and running. The Corps placed well-defined production goals on the Hanford Site reactor department, which the staff at the three reactors worked hard to meet. After 6 months enough plutonium had been produced for two bombs. The first was used for the Trinity Test at what is now the White Sands Missile Range in New Mexico on July 16, 1945. The Trinity Test confirmed that the plutonium-type bomb would work, and planning commenced for the bombing of Japan.

The Little Boy bomb, fashioned from uranium-235, was produced from the gaseous diffusion plant at the Oak Ridge Reservation in Tennessee and was dropped on Hiroshima on August 6, 1945. Three days later, the Fat Man bomb, made with plutonium from the Hanford graphite reactors, was dropped on Nagasaki, Japan. Japan surrendered on August 14, 1945, thereby eliminating the planned drop of a second Fat Man bomb, which was the third and final bomb left in the arsenal.

The events of World War II established a role for the Hanford Site that would define its future over the next three decades. Had more bombs been needed for the war effort, plutonium bombs were the only option available to the United States. The uranium-235 supply was exhausted and additional supplies of uranium-235 would not be available for some time. While both plutonium-239 and uranium-235 bombs worked, Hanford reactors could produce material for the plutonium bombs faster than the gaseous diffusion plants at the Oak Ridge Reservation could produce material for the uranium bombs. This fact influenced the post-World War II atomic weapons strategy.

PHASE 2 - INITIAL COLD WAR PERIOD (1946-1957)

Following the end of World War II, America had an atomic weapons monopoly and no enemy. In 1946, the weapons program tended to flounder. The Atomic Energy Commission took control of the atomic weapons complex from the U.S. Army Corps of Engineers late that year following passage of the Atomic Energy Act on August 1, 1946. The Atomic Energy Commission conducted an assessment of the weapons program in early 1947, only to learn that the nuclear arsenal everyone assumed existed, in fact, did not. The Commission found that Belgian supplies of uranium to feed the Hanford Site reactors were not adequate, that problems with the reactors such as graphite expansion (discussed in the Technological Challenges and Developments sub-section) were emerging, and that no ready supply of bombs existed. As Los Alamos engineer Jacob Wechsler explained:

> *"We had lots of capsules – nuclear cores – I guarantee you. But we didn't have weapons, we had lots of pieces. The idea was, if there was a threat, you would start putting them together. The fusing systems weren't there, the initiators needed to be changed, the detonators needed to be stored in desiccated boxes and you put them in when you needed them and then put them on again. And it went on and on this way. We didn't have any weapons, we had piles of pieces."* - Rhodes 1995, p. 284

The Chairman of the Atomic Energy Commission, David Lilienthal, recalled how President Truman responded when informed about the sorry state of affairs: "He turned to me, a grim gray look on his face, the lines from his nose to his mouth visibly deepened. What [did] we propose to do about it? He realized the difficulties" (Rhodes 1995, p. 284). Churchill's Iron Curtain speech March 1946 and the Truman Doctrine, announced in March 1947, gave the Atomic Energy Commission further justification to move aggressively in developing a nuclear production system to maintain American supremacy.

Over the next decade the Atomic Energy Commission developed the weapons program as it saw necessary to stay ahead of the Soviet Union. The Atomic Energy Commission assigned three objectives for the reactor program: a) prolong useful

Key Events Affecting the Reactor Program at the Hanford Site from the United States' Perspective

- July 16, 1945 – Trinity explosion confirmed success of plutonium bomb.
- August 9, 1945 – Plutonium bomb was used on Nagasaki, Japan.
- March 5, 1946 – Premier Minister Winston Churchill gave Iron Curtain speech recognizing the state of a cold war.
- July 1946 – Weapons testing program began on Bikini Atoll, which created a requirement for continued quantities of plutonium.
- March 12, 1947 – President Truman asked Congress to support the Truman Doctrine to help "free peoples who are resisting attempted subjugation by armed minorities or outside pressures."
- April 1, 1948 – Berlin Blockade began and lasted 321 days.
- September 23, 1949 – President Truman announced that the Soviet Union exploded an atomic bomb.
- February 1950 – McCarthyism movement began.
- April 1950, – National Security Council 68 called for nuclear weapons build-up in response to Soviet design for world domination.
- June 25, 1950 – North Korea invaded South Korea.
- November 1, 1952 – Atomic Energy Commission exploded hydrogen bomb on Enewetok, using plutonium for a key component.
- August 14, 1953 – Soviet Union exploded first hydrogen bomb.
- July 1955 – U.S. feared Soviet Union possessed significantly more bombers than the United States.
- November 1956 – Soviet Union crushed Hungarian revolt.
- November 17, 1956 – Khrushchev made "we will bury you" statement.
- October 4, 1957 – Soviet Union launched Sputnik.
- December 1957 – Gaither Report to the National Security Council identified missile gap with Soviet Union (determined in the 1960s to not be accurate but not reported to the public).
- October 23, 1962 – Cuban Missile Crisis began.
- October 7, 1963 – Kennedy signed Limited Test Ban Treaty outlawing nuclear tests in the atmosphere, under water, and in outer space.
- August 7, 1964 – Congress approved Gulf of Tonkin Resolution, which led to an escalation of U.S. Involvement in Vietnam.
- July 1, 1968 – Nuclear Arms Nonproliferation Treaty signed by the United States, the Soviet Union, and 58 other nations.
- October 17, 1973 – Arab oil producers began embargo against the United States.
- March 28, 1979 – Accident at Three Mile Island nuclear plant occurred.
- July 1980 – President Carter signed Presidential Directive 59 retracting capacity to wage limited and protracted nuclear war.
- April 26, 1986 – Chernobyl nuclear power plant exploded in Soviet Union.
- October 1989 – Cold War ended.

life of existing equipment through rehabilitation and efficient operation, b) build replacement reactors and additional facilities for increasing production, and c) develop new and more efficient techniques for operating the reactors and processing their products.

Tensions between the Soviet Union and the United States continued to escalate and reached a threshold in the spring of 1948 when the Berlin blockade began. One response from the Atomic Energy Commission was to authorize the restart of B Reactor, which had been shut down since 1946 due to concerns about graphite expansion. Also during this time, the Atomic Energy Commission debated whether to build reactors at new areas of the Hanford Site or replace reactors at existing areas. When it appeared that a solution to the graphite expansion problem might be at hand, the Atomic Energy Commission scaled back its plans, authorizing one replacement reactor at D area, called DR for D Replacement, and one new reactor at a new area, H (Figure 2-3.1), located between F and D areas.

In 1949, D, F, and B reactors were operating, DR Reactor was unloaded and standing by because it had no water system, and H Reactor was still under construction. The Hanford Site began running the reactors at higher neutron flux and thus higher temperatures, which led to more production and less graphite swelling. When the announcement came in September 1949 that the Soviets had exploded their first nuclear weapon years earlier than most had predicted, the order came to complete the waterworks for the DR reactor.

Figure 2-3.1. *H Reactor at the Hanford Site Around 1950*

The Cold War intensified in the 1950s as the Korean war began. With concerns over another world conflict mounting, the Atomic Energy Commission began constructing its sixth Hanford reactor, C Reactor, June 1951. January 1952, President Truman decided to increase the ratio of plutonium over enriched uranium supplies, which in turn resulted in authorization and construction of the jumbo KE and KW reactors at the Hanford Site.

The K reactors represented a second generation reactor. Although still using the same basic graphite-moderated, water-cooled design, they differed from the earlier reactors in several significant ways (Carlisle and Zenzen 1996, pp. 96-99). First was the enormous increase in power levels. K reactors were designed for 1,800 megawatt-thermal, while the first Hanford reactors were designed for 250 megawatt-thermal. (These numbers represent heat generation levels used to define the capacity of the production reactors and should not be confused with numbers used to describe electric-power reactors, which used megawatt-electrical, a figure representing electrical output.) The increase in power levels were made possible by advancements in technology and reactor design and many improvements in reactor fuel technology, hydraulics of the reactors (not to mention greatly increased cooling flow). K reactors also marked a first by capturing heat from the cooling water to heat the facilities. Improved control systems enabled the K reactors to operate with fewer staff than the other reactors, thereby increasing cost efficiency. And finally, they used new materials as well as new types of equipment.

While it was the relations between the Soviet Union and the United States that conceptually drove decision making regarding Hanford reactors, it was the concept of the nuclear arsenal that drove the specific requirements. Weapons testing at the Nevada Test Site (Ranger, Buster-Jangle, and the Tumbler-Snapper series) demonstrated the viability of the plutonium-weapons program. The thinking within the congressional Joint Committee for Atomic Energy was to continue to increase the stockpile of plutonium weapons, which increased the demand for plutonium. Even the hydrogen bomb developed in 1952 required plutonium. Decisions, such as the Atomic Energy Commission's in 1947 to build 150 Fat Man bombs (Rhodes 1995, p. 298), were what drove the plutonium production requirements. The pressure to meet these requirements had three important effects on the Hanford reactor program:

- A constant tug of war between production staff, who needed to keep the reactors running at the highest power levels possible, and safety staff in Washington D.C. and at the Hanford Site, who were responsible for ensuring the safety of workers, surrounding populations, and the environment (Findlay and Hevly 1995, pp. 147-150). Thus in the drive to deliver plutonium and polonium for the arsenal, additional risk to human and environmental health were taken.
- A consistent return to the basic graphite-moderated, water-cooled reactor design each time a new Hanford reactor was authorized. Although it was understood that other designs such as a beryllium or heavy water moderator might be more efficient, decision makers also understood the risk that developing new technology

entailed. The Hanford reactor design might not be the most efficient, but it could guarantee plutonium. This topic is explored further in the sub-section on Technological Challenges and Developments.

- By using the same basic design, reactor staff were able to dramatically increase the efficiency of the existing reactors. Moreover, staff were able to make a dramatic increase in scale with the design of the jumbo K reactors as well as incorporate several technological advancements to improve efficiency and safety. These advancements were part of a $500 million expansion program at the Hanford Site, which included reactor improvements, new reactors, and new separations facilities.

PHASE 3 - FINAL COLD WAR PERIOD (1958-1989)

As the Korean War Armistice was signed in 1953, the demand for plutonium began to diminish. Because plutonium has a half-life of 24,000 years, the only threat to plutonium inventories was actual use in weapons. By projecting the amount of plutonium being produced and the amount that was needed for the weapons testing program, planners in the mid-1950s could see that plutonium supplies would be adequate before long and, therefore, the Hanford production capacity in 1958 was sufficient for the foreseeable future.

Several other factors, however, led to a proposal in 1957 for a new production reactor at the Hanford Site. These included increasing interest in President Eisenhower's Atoms for Peace initiative of 1953, increasing concern for human and environmental safety, and the drive to reduce plutonium production costs. In response to these factors, General Electric proposed a dual purpose reactor for the Hanford Site in 1957. The fundamental difference of the N Reactor was, in addition to producing plutonium, the heat from the cooling water could be used to produce electricity for commercial purposes, thus promoting the goal of Eisenhower's Atoms for Peace initiative. Key features of the new reactor were the following:

- Closed-loop, primary cooling system to eliminate cooling water discharges to the Columbia River, a cause of extensive contamination, and with it some of the concern for human and environmental safety
- Heat exchangers to generate steam that could power electricity-generating turbines
- Confinement system to limit the spread of radioactivity in the event of an accident
- Advanced materials and instrumentation to improve safety and efficiency as practicable

Congressional approval and funding for the new reactor was not automatic as the funding for the first eight reactors at the Hanford Site had been. President Eisenhower, for example, did not support the new reactor because he did not see the need for more plutonium production capability. Others questioned the need since the required quantities for the plutonium stockpile were easily achievable with existing capacity. Additional opposition related to the potential for electricity generation. Coal and hydropower companies were threatened by potential competition from nuclear power and used their political connections to oppose the new reactor.

"After several attempts to obtain funding for the steam plant and associated infrastructure, a measure was finally approved in 1962, thanks largely to support from President John F. Kennedy. When President Kennedy came to the Hanford Site in September 1963 for the ground breaking of the steam plant (an event that opened the Hanford Site to the public for the first time ever and was attended by about 30,000 people), he stressed the important role that the dual purpose reactor was playing in maintaining worldwide nuclear technological superiority." **- Carlisle and Zenzen 1996, pp. 115-123**

One of the major advocates for the N Reactor, as the new reactor came to be called, was Henry "Scoop" Jackson, Congressman and later Senator from Washington State. He argued that the new production reactor at the Hanford Site was an important step in keeping the United States at the forefront of nuclear technology. The support from the Atomic Energy Commission and the Joint Committee on Atomic Energy was sufficient to obtain congressional funding for the new reactor with one important change: the electricity producing portion of the reactor would not be built, although the reactor portion would be designed so that the steam plant could be added in the future. Design of the N Reactor began in 1958 and construction in 1959. The reactor officially started up December 31, 1963 to meet the deadline.

Mark Jensen, a member of the operating crew who shut the N Reactor down, talked with some of the early operators of N Reactor. His understanding from those discussions is "the reactor was fueled enough to go critical, then immediately shut down to complete construction, and started up again for testing in March 1964" (Jensen 1999).

The N Reactor began producing plutonium in 1964. The steam plant began producing electricity in 1966, and for many years it was the largest nuclear power plant in the world (Figure 2-3.2). Producing 800 megawatts of electricity, it roughly equaled the total output of all nuclear power producing reactors in America at the time (Carlisle and Zenzen, note 40, p.248). While the government operated the reactor that produced the steam, a consortium of public and private utilities operated the turbines that produced the electricity. "The fence was physical, but it nicely symbolized the intellectual and ideational division between the weapons and the civilian sides and between the government preserve and the commercial world" (Carlisle and Zenzen 1996, p. 113).

Figure 2-3.2. *N Reactor and Electricity Generating Plant Around 1985*

N Reactor and the steam plant construction occurred during the same time that the commercial nuclear power industry was starting to take off. As a result, the international and national industry showed considerable interest in N Reactor. Many industry and governmental groups came to the Hanford Site to tour the facility. The project helped put the Hanford Site on the nuclear industry map, which would have important implications in the future for non-defense related work.

Because by the 1960s the plutonium stockpile was much greater than deemed necessary, the requirements for plutonium decreased. Most significant was the Cuban Missile Crisis in October 1962, a showdown between the two superpowers that nearly led to deployment of nuclear weapons. The absurd size of the nuclear arsenal combined with the realization that nuclear war could happen helped change the direction of the Cold War. In 1963, Britain, the United States, and the Soviet Union signed the Limited Test Ban Treaty, which outlawed nuclear tests in the atmosphere, under water, and in outer space.

On January 8, 1964, President Johnson, announced that some Hanford reactors would be shut down. The Atomic Energy Commission selected H, DR, and F reactors for closure based upon their age and condition (Carlisle and Zenzen 1996, p. 148). In 1967, another announcement came for a closure at the Hanford Site (D Reactor), and in 1968, the Atomic Energy Commission announced that B Reactor was to close.

The economic impact of these closures was severe and resulted in efforts to find new missions for the Hanford Site. The Tri-City Nuclear Industrial Council, later called the Tri-City Industrial Development Council, formed and began looking for ways to leverage Hanford assets. In 1965, the General Electric contract, which had been in place since 1946, was spread among six different companies, each of which was to diversify. Douglas-United Nuclear, Inc. (a joint venture between McDonnell Douglas Aircraft Company and United Nuclear Corporation) won the contract to operate the fuel manufacturing and reactor complexes. A brochure prepared to illustrate potential uses of Hanford assets (Holsted and Albaugh 1964b) indicates that reactors and their associated facilities were thought to have various uses to the burgeoning nuclear industry. Thus, in addition to diversifying the Hanford Site away from a single contractor, the Hanford Site was also attempting to diversify away from the Atomic Energy Commission (Carlisle and Zenzen 1996, p. 156-160).

In 1969, another closure announcement reverberated across the Hanford Site. All remaining production reactors (C, KE, KW, and N) were to be shut down. Fuel manufacturing and the separations plants would close as the need for their services ended. This decision would have effectively ended the mission of the Hanford Site. Recognizing the economic consequences to the local community, local economic leaders undertook a major effort to keep the N Reactor operating to provide steam for the Hanford Generating Plant. An eleventh hour agreement kept N Reactor running so at least

Contractors Responsible for Hanford Reactors

1943-1945	E. I. Du Pont de Nemours
1946-1964	General Electric Company
1965-1987	Douglas United Nuclear, Inc.
1987-1994	Westinghouse Hanford Company
1995-present	Bechtel Hanford, Inc.

electricity could be produced. This required several modifications to operations. The conversion was completed in 1971, and N Reactor generated steam for use in producing electricity throughout the 1970s.

As the Hanford Site entered the 1980s, several events suggested that the diversification efforts might pay off. The concept of using the Hanford Site as a dedicated nuclear facility for experimental and commercial reactors never fully materialized, but it did arrive in a fashion. See Holsted and Albaugh (1964a) for their proposal for the Hanford Site to be an experimental reactor site. The Washington Public Power Supply System, which had provided the funding for the Hanford Generating Plant at N Reactor, chose Hanford as a site for three new power reactors. However, because of rising interest rates, cost overruns, and decreased use of energy by the public, only one power reactor was completed, and the investors suffered major losses. Other events though made diversification look promising. The Atomic Energy Commission chose the Hanford Site for a new prototype breeder reactor, the Fast Fuel Test Reactor, later renamed the Fast Flux Test Facility. Also, Exxon Nuclear Fuels constructed a commercial fuel manufacturing facility, later sold to Siemens.

Several events, external and internal to the Hanford Site, that affected the Hanford reactors began in the early 1970s. The 1970s for the nuclear industry was not a good decade. Increasing concerns over nuclear technology, public media events such as "The China Syndrome," and actual accidents such as Brown's Ferry and Three Mile Island led to increasing scrutiny, more regulation, more procedures, and more training. The N Reactor, the sole operating Hanford production reactor at the time, was only operating to produce electricity. Because its long-term operation was tenuous, maintenance budgets tended to shrink during this decade, and, as a result, the condition of N Reactor declined somewhat. Procedures and training, however, increased dramatically as the federal government promulgated new rules and regulations. Security was upgraded during this period due to the worldwide increase in terrorism.

In the late 1970s, relations between the United States and the Soviet Union began to deteriorate. President Carter started a military buildup, which accelerated when President Reagan took office. The federal government decided to reinstate the production of plutonium at N Reactor. As plutonium production commenced, numerous modifications and adjustments were needed to comply with safety and other regulatory requirements. A major upgrading was required and new programs commenced. By 1982, weapons-grade plutonium was once again being produced at N Reactor.

The N Reactor continued producing weapons-grade plutonium throughout the 1980s. Irradiated fuel elements continued to be deposited in the fuel basins where most of the fuel processed in the 1970s also was located. The Chernobyl explosion in 1986 led to a shutdown of N Reactor and national assessments of its ability to operate safely. Because of concerns over safety, electrical production ceased as well. While it may have continued to operate had plutonium been needed, when the Cold War ended in 1989 there was simply no need to continue production. With N Reactor's transition to cold standby, the Hanford Site's production mission ended, and the cleanup mission began.

REACTOR OPERATING SYSTEMS

The purpose of the reactor operating systems at the Hanford Site was to produce plutonium and other radioisotopes. As described in the Construction History Section, each reactor area at the Hanford Site was essentially self-sufficient, containing all the buildings and processes needed to operate a reactor (Figure 2-3.3). The reactor operating systems and examples of the buildings and structures that represent them are presented in Table 2-3.4 and discussed in greater detail below.

IRRADIATION SYSTEM

The irradiation system was the primary system. Virtually all other reactor systems supported the irradiation system in one way or another. To effect irradiation, reactor control operators placed natural uranium into a graphite reactor stack where the uranium-235 atoms fissioned and uranium-238 atoms were transformed into plutonium-239.

The majority of irradiation related activities took place in the reactor building, designated as a 105 building. Each of the 105 buildings contained a reactor block, a reactor control room, a spent-fuel discharge area, fuel storage basin and associated fuel handling equipment, fans and ducts for the ventilation and recirculating gas systems, water cooling

Table 2-3.4. *Production Reactor Systems and Representative Buildings Selected for Mitigation*

System	Function	Example Buildings
Irradiation	Irradiate fuel to make plutonium	105B, 105KW, 105N
Water Treatment	Provide treated water to cool reactor	163N, 181B,181KW, 181N, 182B, 182N,183KW, 183N, 183NA, 183NB, 190KW
Waste Management	Manage reactor-related wastes	105KW-RTC, 107KW, 107N,116B, 116KW, 116N, 117B, 117KW, 117N, 118B-1, 119KW, 119N, 183H, 218-E-16, 1301N, 1310N, 1313N, 1314N, 1325N
Operations Support	Provide support to reactor operations	212N, 436, 1116N, 2101M, 4703
Facilities Support	Provide support to reactor facilities	108N, 1705N, 1705NA, 1712N, 1714N, 1717K, 1722N, 1908KE, 4621W
Administrative Support	Provide administrative services	1720K, 4710
Security	Provide security to reactor area	181NA, 1112N, 1701K
Power Generation	Provide power to or from reactor	105N, 109N, 155N, 184N, 185N
Research and Development	Provide research and development	190D Complex, 305, 305B, 309/309E, 318 Complex, 321 Complex, 326, 337B 405 1706KE, 1706KER

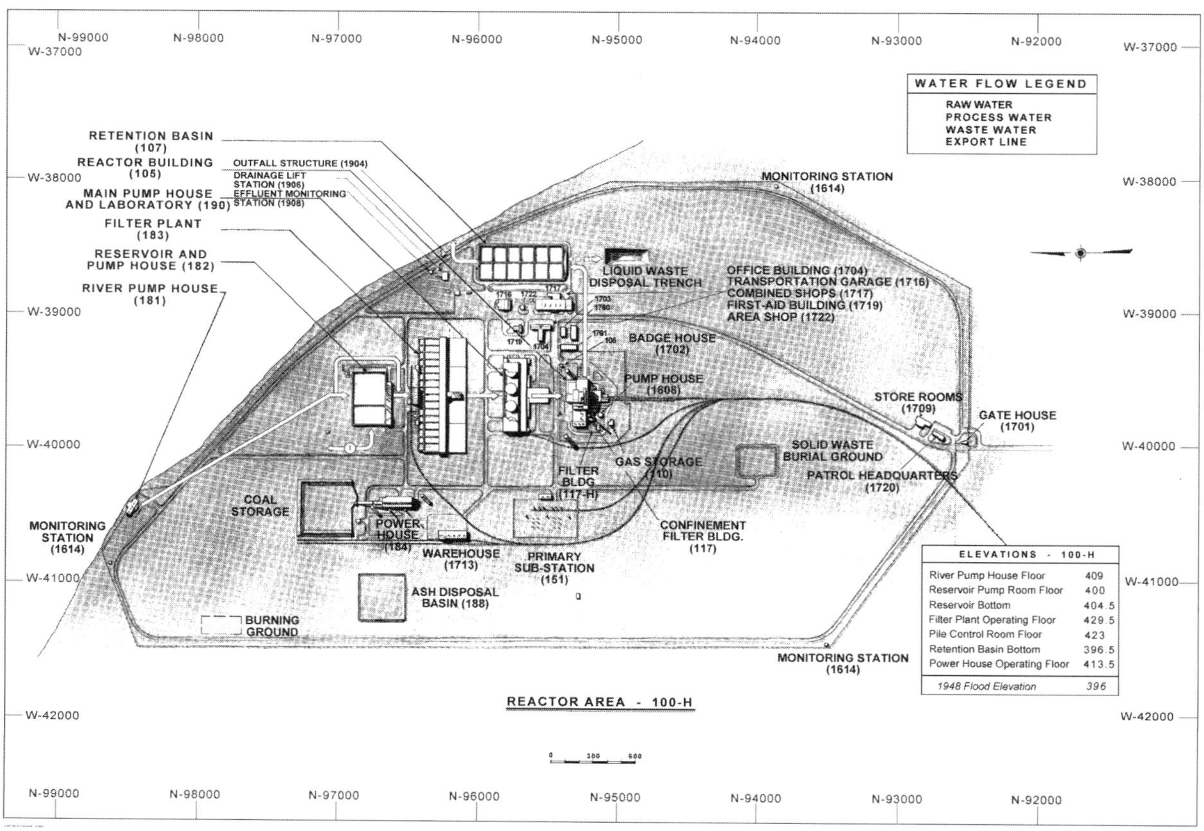

Figure 2-3.3. *Map of a Typical Hanford Reactor Area (100-H) Showing Key Buildings and Structures*

systems, and supporting offices, shops, and laboratories. A typical reactor building (Figure 2-3.4) was a reinforced concrete and concrete block structure roughly 250 feet long by 230 feet wide by 95 feet high (75 by 70 by 30 meters). The building had massive reinforced concrete walls 3-5 feet thick (1-1.5 meters) around the reactor block to provide radiation shielding with lighter construction above (either concrete block or corrugated asbestos-cement). Roof construction was primarily precast concrete slab or poured insulating concrete.

Figure 2-3.4 illustrates several key areas of a reactor building. Fuel operators would load the uranium fuel elements into the process channels located in the front face of the reactor block, a process referred to as charging. As operators loaded the reactor with new fuel elements, the irradiated fuel elements already in the reactor were pushed out the back, a process referred to as discharging. Fuel discharge and storage areas were located adjacent to the rear face of the reactor.

To start the chain reaction, reactor control operators would first withdraw vertical safety rods and then horizontal, neutron-absorbing control rods on the left side of the reactor block (when facing the front of the reactor). The neutron-absorbing control rods absorbed neutrons and thereby suppressed the chain reaction when in the reactor. To slow the chain reaction or stop it if necessary, control operators would move the control rods back into the reactor block. To stop a reactor quickly, that is to SCRAM the reactor, control operators would trip the vertical safety rods, dropping them into the reactor block from the top of the reactor. An additional shutdown system was also located on top of the reactor, that being the boron ball safety system in which neutron-absorbing steel-and-boron balls were dropped into vertical safety channels for emergency reactor shutdown.

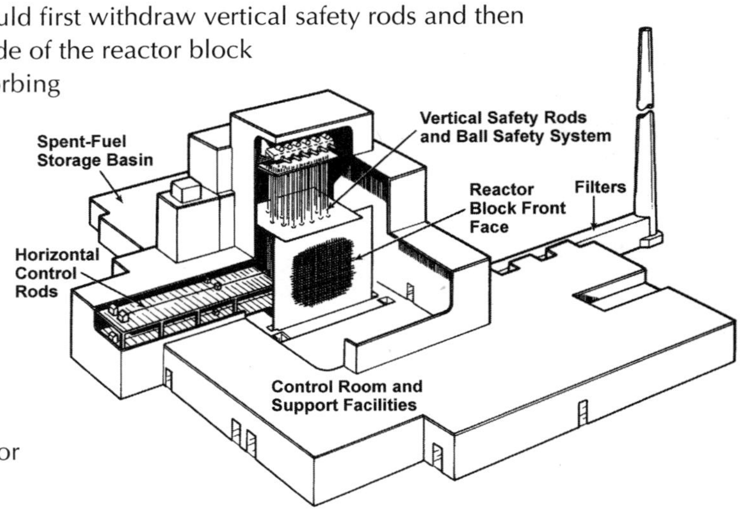

Figure 2-3.4. *Layout of a Typical Reactor Facility*

To determine when the proper amount of plutonium was ready and stop the fission process, physicists made calculations daily based on the heat generated by the reactor and the length of time the fuel elements had been in the reactor. These calculations were important because if the fuel elements were irradiated too long, too much plutonium-239 would absorb further neutrons and turn into plutonium-240 and plutonium-241. Plutonium-240 was undesirable. It had a short half-life of 12.4 years, after which it decayed into americium. Americium, because of its high toxicity, created logistical problems related to handling. Plutonium-241 spontaneously fissions, which made it undesirable for weapons use (Albright et al. 1993).

The physicists made calculations for each process channel because the reaction varied depending upon the location. Areas central to the reactor block tended

The Meaning of SCRAM

"By December 1942, the world's first nuclear reactor had been secretly constructed at the University of Chicago. The device, intended to produce a controlled chain reaction, consisted of a 'pile' of graphite blocks in which uranium was embedded. It was assembled in a squash court under the west stands of Stagg Field, the University's football stadium. To limit the reaction, control rods were inserted into the pile. Since neutrons brought about fission, the control rods were made of material that would absorb neutrons, like a blotter, and thus regulate the rate at which fissioning took place. As a safety device, a special emergency control rod was suspended over the pile. It was held by a rope so that safety officers – 'the suicide squad,' they were called – could cut it in an emergency and drop the rod into the reactor. An ax was provided for this purpose. (The emergency shutdown of reactor, which on today's models is accomplished by elaborate automated equipment, is still referred to as a SCRAM – short for safety control rod ax man.)" **- Ford 1982, p. 28**

to have a higher neutron flux and thus required less time. Generally, about 25 percent of the process channels would be discharged at a time. When the physicists informed reactor staff that enough fuel elements were ready for discharge, an outage was scheduled, the reactor shut down, and the fuel elements discharged. When operating the reactor to only produce electricity, the irradiation time could be longer, resulting in fewer discharges and, therefore, less down time.

As the fuel operators loaded the reactor with new fuel elements, the irradiated fuel elements were pushed out of the rear face, falling into a pool of water. When the B Reactor started up, boxes of fuel elements and dummy elements (used for spacing fuel within the reactor) were brought to the reactor front face from the 103 Building. Using the C-elevator (charging), workers positioned themselves in front of the tubes to be charged and uncoupled the endcaps. At the same time, workers on the D-elevator (discharging) would uncouple the endcaps of the same process tubes at the rear face. These workers would then raise the D-elevator to a level higher than the highest row of process tubes to be charged/discharged and exit the rear face. Operators on the front face would then use the charging machine to push new fuel and dummy elements into position while pushing the irradiated and dummy elements out the rear face.

This was the most problematical and dangerous step in the operation as exposure to an irradiated element was lethal. The B Reactor workers used periscopes to monitor the discharging elements as they exited the rear face. In theory, the elements would exit the rear face by dropping into a discharge funnel through an aiming tube, enter an underwater discharge chute, pass through an escapement gate, and come to rest in the discharge area of the fuel storage basin. In practice, the process was too complicated with too many snag points that had to be cleared. Within 2 months of commencing operations at B Reactor, the removal of the escapement gates simplified the process somewhat, and 3 months later the aiming tubes and discharge funnel were eliminated. With these improvements and added experience, charge/discharge efficiency rose from 10 tubes per 8-hour shift to 50.

Shielded by 20 feet of water, the B Reactor workers used long tongs first to remotely separate the irradiated fuel elements from the dummy elements resting together on the bottom of the discharge basin, then to place the elements into their designated steel buckets, each with a holding capacity of approximately 1000 pounds. Working from the wooden floor suspended over the basin, workers then used long rods to attach each bucket to an overhead monorail system that allowed them to position the bucket according to its content within the fuel storage area of the basin. Dummy elements were retained for reuse in the reactor. For further details, see the B Reactor Historic American Engineering Record in Appendix B on the Internet.

Once the fuel elements had cooled for a sufficient time, about 30 days, the buckets were pulled out of the basin and the elements placed in railroad cars for transport to the cooling basin where they stayed until ready for shipment to the 200 Area. In addition to producing plutonium and other isotopes for weapons, some of the Hanford reactors had experimental test penetrations located on the right side of the reactor block where irradiation experiments could be performed.

Three aspects of the 105 buildings warrant further description:

- Reactor block
- Control room
- Fuel storage basin.

Reactor Block. A typical reactor block (Figure 2-3.5) consisted of a graphite moderator stack encased in cast iron thermal shielding 8-10 inches thick (20-25 centimeters) and a biological-shield (alternating layers of steel plate and Masonite or heavy aggregate concrete 40-80 inches thick – 100-200 centimeters).

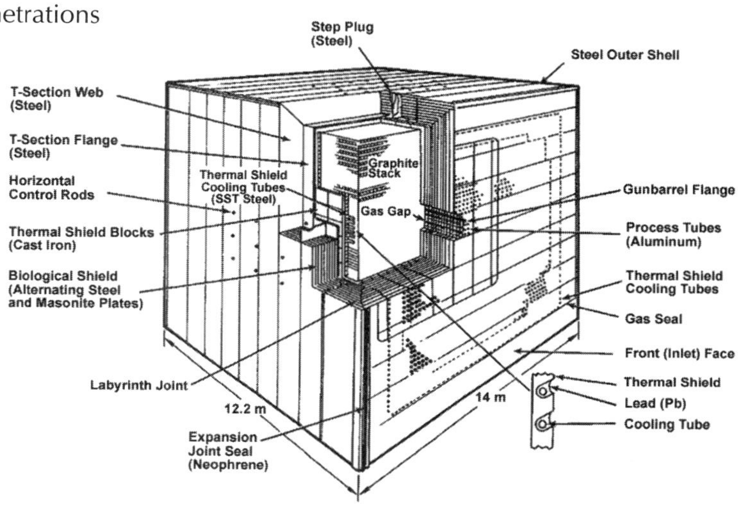

Figure 2-3.5. *Cutaway of a Reactor Block*

The entire block rested on a massive concrete base and foundation. A typical reactor block assembly weighed approximately 9000 tons (8000 metric tons including the weight of the base) and had overall dimensions of 46 feet high by 46 feet wide by 40 feet deep (14 x 14 x 12 meters). Additional details concerning the reactor block are provided in Tables 2-3.5 and 2-3.6. The principal components of a production reactor block were:

- Reactor moderator stack, an assembly of graphite blocks cored to provide channels for process tubes, control rods, and other equipment
- Thermal and biological shields, surrounded by a heavy, vault-like steel outer shell equipped with gas-tight seals for the reactor block penetrations
- Process tubes, which held the uranium fuel elements and carried the cooling water
- Foundation
- Horizontal control rods
- Vertical safety rods and boron ball safety system
- Monitoring and experimental/test equipment

Table 2-3.5. *Components of a Typical Hanford Nuclear Reactor Block (100-F)*

Graphite Moderator Stack	The encapsulated graphite moderator stack measures 36 feet high, 36 feet wide, and 28 feet long (11 x 11 x 8.5 meters). Its total volume is 36,288 cubic feet (1,028 cubic meters). Individual graphite blocks, stacked tightly in a crisscross pattern, are 4.2 inches square by 44.5 inches in length (10.5x122 centimeters).
Thermal Shield	The graphite stack is bordered underneath by a layer of 10-inch thick (26-centimeter) cast-iron blocks (the bottom wall of the thermal shield). A massive steel-reinforced, high density concrete base completely underlies the thermal shield, serving as a support pad. The full six-sided thermal shield is composed of a single layer of approximately 3,300 cast-iron blocks, varying greatly in size and weight. The top, bottom, and side thermal shields contain slots that hold cooling tubes. The cooling tubes are held in place by lead poured into the slots around the tubes.
Biological Shield	The next layer of shielding is the biological shield. It is 52 inches thick (132 centimeters) and forms an integral encasement on the top and four sides. On the top and two sides, steel plate T-section flanges, each approximately 46 feet long, 4 feet wide, and 4 feet high (14 x 1 x 1 meters), form the inner wall and ribbing of the biological shield. An outer shell of 0.25-inch (0.635-centimeter) steel plate was welded in place to completely seal the top and side walls of the biological shield. The top biological shield is not welded to the side walls of the reactor block. The shield was laid in place with edges of all four walls meeting in diagonal, stair-step labyrinth joints. All seams are enclosed by gas-tight expansion seals. The reactor's front and rear face biological shielding was fabricated in the same alternately layered pattern of Masonite and steel but in sectioned blocks rather than solid walls. There are a total of 264 such blocks (called B-blocks), 132 on each face.
Process Tubes	Extending through the reactor block from the front to the rear face are 2,004 symmetrically located 1.7-inch diameter (4.4-centimeter) aluminum process tubes. These tubes are connected to the cooling water piping on both the front and rear face.
Foundation	The reactor block rests on a massive reinforced concrete foundation. The foundation is 64 feet long, 56.5 feet wide, and 22.5 feet deep (20 x 17 x 7 meters). The concrete base of the reactor block is separated from the foundation by a 0.25-inch thick (0.635-centimeter)steel membrane plate.

Table 2-3.6. *Estimated Weight Comparison of the Six Oldest (B, D, F, DR, H, and C) and the K Reactor Blocks*

Component	Estimated Weight (pounds)	Estimated Weight (kilograms)
Graphite moderator stack	3,607,565	1,636,364
Thermal shield blocks (top and bottom)	680,728	308,773
Thermal shield blocks (sides)	562,178	255,000
Thermal shield blocks (front and rear)	1,002,602	454,773
Biological shield (top and sides)	5,411,345	2,454,545
Biological shield (front and rear)	4,415,056	2,002,636
Process tubes and gun barrel	433,909	196,818
Reactor base	1,780,733	807,727
Estimated total weight, 6 oldest reactors	17,895,000	8,117,000
Estimated total weight, K Reactor	24,251,000	11,000,000

Control Room. The reactor irradiation and water cooling systems contained numerous instruments to monitor performance. The reactor control room was the information center for most of these monitoring systems (Figure 2-3.6). Unless otherwise indicated, the information for the control room came from Wahlen (1989). Each reactor building contained the following six types of instrumentation:

Figure 2-3.6. *B Reactor Control Room Showing Instrumentation Gages*

- Reactor Safety Circuits – Visual readout devices provided information on the status of the irradiation process. The instruments were connected directly to reactor safety circuits so that the reactor would shut down if limits were exceeded.
- Reactor Process Controls – The reactor process instruments provided information so the operators could control the efficiency of the reactor.
- Temperature Monitoring Instruments – These instruments monitored the temperature of the effluent water, the graphite moderator, and the thermal shield.
- Pressure Gauges – Many miscellaneous pressure gauges in the control room indicated the pressures at various points in the primary and secondary cooling systems.
- Rupture Monitor – The rupture monitor system detected fuel element failures in the reactor. It continually sampled effluent water from both ends of the rear cross header before the effluent entered the rear riser. Visual and audible alarms annunciated when trip points were exceeded.
- Reactor Power Calculator – The total power level of the reactor was computed by a calculator using total flow to the reactor and the bulk coolant temperature increase across the reactor.
- On the initial startup, three types of circuits could initiate a SCRAM (automatic insertion of control and/or safety rods) if conditions warranted: manual trip circuits, power failure circuit, and the Beckman-Neutron Flux Monitor.
- To increase production, operators had to run the reactors closer to the safety limits. This was made possible by adding more devices, such as the pressure monitor, seismoscope, vertical safety rods/horizontal control rods interlock, zone temperature monitor, bypass switches, and mercoid switches.

The nuclear instrumentation range requirement was from normal background or shutdown flux levels to approximately one hundred billion times normal background. A single instrument was not available with a range this large. This required instruments divided into different process control systems to monitor reactor operation at various flux levels. Other systems monitored the temperatures in the reactor of the moderator, thermal shield, and effluent water.

Fuel Storage Basins. Each reactor building contained a spent-fuel storage basin, which served as a collection, storage, and transfer facility for the fuel elements discharged from the reactor. The storage, located behind the reactor, was typically divided into three zones: the spent-fuel discharge area, the storage area, and the transfer area. Each basin had a wash pad, and some had an underwater fuel-inspection facility. The above-grade structures in the storage basin were constructed with concrete block walls and precast concrete (panel) tile roofs from 10-39 feet (3-12 meters) in height. The below-grade portion was 20 feet (6 meters) deep with reinforced concrete columns and walls. The average thickness of the outside walls of the basin was 1.5 feet (50 centimeters). The bottom of the basin was about 6 inches (15 centimeters) thick. The total volume of concrete in each basin was about 750 cubic yards (573 cubic meters). The total spent-fuel storage basin area ranges from 650 square meters to 929 square meters, depending on the specific reactor.

Under 20 feet (6 meters) of water in the spent-fuel discharge area, irradiated fuel elements were sorted into storage buckets by long, hand-operated tongs. The buckets were then transferred by an overhead monorail system to the storage area where they were held for a time to allow the decay of short-lived radionuclides. Following the storage period, the buckets of fuel elements were moved on an overhead monorail system to the transfer area.

The transfer pits, which measure 6 by 9 feet (2 by 2.5 meters), were located at one corner of the fuel storage basins and at the inner end of the fuel transfer area. The transfer pits were 5 feet (1.5 meters) deeper than the basins and were connected to the basins by a canal over which the monorail system ran. Here the irradiated fuel elements were loaded into casks, then raised by a crane and placed in special railroad cars for shipment to the chemical reprocessing facilities in the 200 Area.

WATER COOLING SYSTEM

Nuclear reactors produce an enormous amount of energy in the form of heat during fissioning. Nearly 94 percent of the heat generated in the reactor is generated in the uranium fuels with the remaining 6 percent generated in the graphite moderator. The temperatures must be kept below certain thresholds or damage will occur, creating the potential for a severe accident. Coolants used to control reactor temperatures can be gases (for example, carbon dioxide, helium), liquids (for example, ordinary water, heavy water, and fluid organic compounds), and liquid metals (for example, sodium) (Gaines 1970, p. 20). The Hanford approach was to use ordinary water as the coolant, passing it through the fuel-bearing process tubes in the reactor. A system was devised to deliver a continuous supply of cooling water to avoid overheating. Severe overheating could lead to the melting of the fuel core and release of fission products to the environment.

"A characteristic thing about Hanford is that the site is enormous, but the reactor is a very small thing and all the rest is water treatment. Most of the money went into water treatment." **- Herbert Anderson as quoted in Sanger 1995, p. 46**

Three major factors drove the design of the water cooling system. The first was the volume of water required. By the late 1950s, when the Hanford reactors were operating at maximum power levels, the average flow of cooling water through eight reactors combined was over 500,000 gallons per minute (Foster 1957, p. 6). C Reactor, for example, was designed to handle 62,000 gallons per minute and could handle up to 80,000 gallons per minute if necessary. Providing this amount of water required an enormous pumping, piping, and storage infrastructure.

The second was the need to purify and treat the water for use in the reactor. A large-scale system of settling basins, mixers, pumps, sampling stations, and storage tanks was constructed to ensure the massive volumes of water met the stringent requirements for purity and chemical composition. For example at B Reactor, a coagulant was added to the raw river water to aid in filtering out suspended solids. Solids in the cooling water could become activated (made radioactive) as they passed through the reactor, creating an additional radioactive waste for disposal. Sulfuric acid was added before the coagulation and filtration process to reduce the pH below 7.5. (The Columbia River's pH ranges from 7.6 to 8.4). After filtration, lime was added to raise the pH to 7.5 to 7.8. To inhibit the corrosion on process tubes and fuel elements, 1.8 to 2.2 parts per million of sodium dichromate was added (Gerber 1993a, p. 16).

The third was the need for multiple backup and emergency systems. The integrity of the water supply had to be maintained, which was no small task given the number of pumps, pipes, storage vessels, and valves involved. Each of the older type of reactor facilities at the Hanford Site incorporated three systems to maintain adequate flow of cooling water. (N Reactor, although water cooled, was a different design from the earlier Hanford reactors and, therefore, incorporated a modified water treatment system that nonetheless shared many of the same components as the older reactor facilities). The

Hanford Site Specifications for Cooling Water
(Gerber 1993, p. 16)

Turbidity	0 to 2 parts per million
Iron	not to exceed 0.1 part per million
pH	7.5 to 8.2 (7.5 to 7.8 preferred)
Total Chlorine	not to exceed 2 parts per million
Free Chlorine	not to exceed 0.2 part per million
Manganese	not to exceed 0.01 part per million
Aluminum	not to exceed 0.05 part per million

Primary Cooling System consisted of an electric motor driven pump system. The Secondary Cooling System consisted of a steam power pumping system. The third, the Last Ditch Cooling System, consisted of a dual system of elevated storage tanks and an interconnecting pipeline that connected all of the reactor plants. A flow chart of the water cooling system can be seen in Figure 2-3.7.

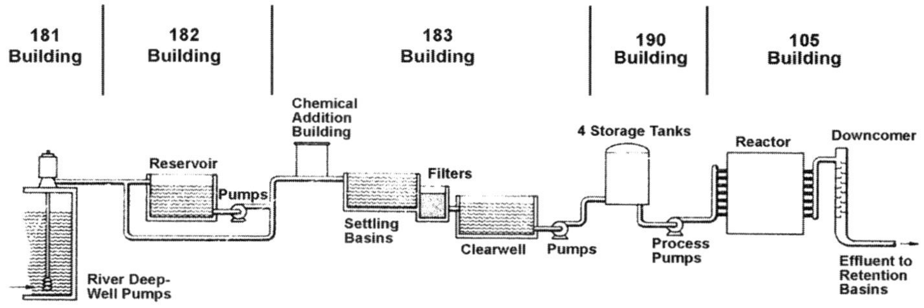

Figure 2-3.7. *Water Flow for a Typical Cooling System in a Production Reactor*

The Primary Cooling System consisted of five steps (Wahlen 1989, pp. 5-13):

Step 1 – Pump Water from Columbia River. Each reactor area had a river pump house (referred to as the 181 buildings) with the exception of the 100-K Area, which had two pump houses. At 100-B and 100-D areas, the pump houses served as the water supply for two reactors each: B and C reactors, D and DR reactors, respectively. Water was drawn from the Columbia River by vertical deep well pumps. The impellers and submerged pump bowls extended to about 11 feet (3.5 meters) below the normal low water point of the river. The intake channel at each of the pump houses was dredged and lined with boulders and concrete to form a forebay. The water entered the deep well of the pump house through traveling screens that precluded the entrance of large river debris and fish. The water was pumped to a reservoir, referred to as the 182 buildings.

Each 181 building had a transformer station that reduced the 13.8 kilovolts to 2,300 volts, which then supplied a minimum of two 2,300-volt buses. A bus is a set of two or more electric conductors that serve as common connections between load circuits and each of the polarities (in direct-current systems) or phases (in alternating-current systems) of the source of electric power. Each bus carried about one-half the electrical load of the facility. The electric motors driving the submerged pump impellers were in a squirrel cage and had three phases, 2,300 volts, induction motors, and manual control panels. The 181-B and 181-D buildings had remote control systems. The pumps in the 181-B Building could be controlled from the 183-C Head house, and the pumps in the 181-D Building could be controlled from the 183-D Building.

Step 2 – Store Water and Pump to Treatment Facilities. The 182 buildings typically provided primary and reserve water to cool the reactors, water for the steam condensers, and water for the facility. The 182 buildings for the B and C reactors also provided raw water for the 200 Area separations plants. Pumps transferred water from the 182 buildings to the 183 buildings for treatment. The 182 buildings also provided water for emergency cooling of the reactor by the Last Ditch Cooling System.

Step 3 – Add Chemicals, Settle, and Filter. Before use in a nuclear reactor, Columbia River water had to be purified to prevent film build-up in the reactor process tubes. These treatments took place in a series of buildings and structures referred to as the 183 buildings. The 183 building facilities consisted of the Chemical Addition Building, flocculators and settling basins, filters, clear wells, pump room, and filtered water storage tanks (Wahlen 1989)

Step 4 – Store Treated Water and Pump to Reactor. Four large tanks were provided at each of the reactors, excluding 100-N, which was a different design. The filtered water storage tanks supplied water to the coolant pumps for each reactor. The tanks for the B, D, F, and H reactors were housed within the 190 buildings. For the DR and C reactors, the tanks were external to the buildings. The tank's discharge lines tied to a common header that supplied the pumps in the 190 annex facilities. At the DR Reactor, the filtered water flowed directly to the storage tanks by gravity from the effluent flume in the 183-DR Building.

Step 5 – Pass Water through Reactor, Cool in Retention Basins, and Return to River. From the pumps in the 190 buildings at each reactor, cooling water passed through the reactor once in a single pass. The water was collected in a large discharge piping system that discharged to retention basins and ultimately to the Columbia River or to the soil column near the river. A notable exception was the N Reactor, which incorporated recirculation of the primary coolant and used demineralized water on a feed and bleed system. The bleed portion of this system was discharged exclusively to the soil column near the reactor buildings. Additional discussions on the flow of effluents from the reactor are included in the section, Waste Management.

Because the Primary Cooling System was run on electricity, the Secondary Cooling System was a steam powered system, designed in case the electric power at the Hanford Site failed. In the Secondary Cooling System, steam-driven turbines and pumps kept the water flowing through the Primary Cooling System facilities. The Last Ditch Cooling System was available should both the electricity and steam powered systems fail. The Last Ditch Cooling System was comprised of two parts: the reactor high tanks and the export water system. The reactor high tanks were two 300,000-gallon high tanks that automatically supplied water when conditions warranted.

Safety Control Systems

When a Hanford reactor was fully charged (loaded with fuel) and experiencing a self-sustaining chain reaction, it presented a serious potential hazard. If an accident occurred that led to an explosion and meltdown, the effect on the local environment and population would have been severe. Because of this great threat, safety control systems were put in place to reduce the likelihood that such an accident would occur as well as the potential effect of an accident. Also in place were various monitoring systems, dress codes, and procedures to protect staff working in the reactor building.

Reducing Effects of a Severe Accident. As discussed in the Construction History Section, placing the Hanford reactors in remote areas, away from population centers, was one mechanism for reducing the impact of any severe accident on the local populace. Still, if there were an explosion or meltdown, the release of radionuclides would have been extensive. The first eight Hanford reactors were constructed with no way to contain the spread of radionuclides should an accident occur. Gerber (1992a, pp. 101-103) comments on the attention given this matter by the Advisory Committee on Reactor Safeguards (called the Reactor Safeguards Committee from 1947-1953), which reported to the Atomic Energy Commission. Their concerns escalated throughout the 1950s as the Hanford Site continued to increase the power levels for the reactors. Finally in 1958, the pressure from the Advisory Committee resulted in the addition of filters and fog sprays to the reactor safety systems at a cost of about $4 million. Fog sprays condensed the steam around the reactor core, cooled the atmosphere, and rained out fission products.

When it came time to design and construct the ninth and final production reactor, N, the General Electric Company opted for a confinement system rather than a containment system, primarily for cost considerations. The Advisory Committee on Reactor Safeguards again took issue with this decision. By this time, all commercial reactors being constructed in the United States had containment systems, essentially steel and concrete shells constructed around the reactor core, which would prevent the release of any materials during an accident. With the N Reactor confinement system, the initial burst of steam resulting from a postulated reactor coolant-pipe break would be released to the outside. Once the confinement pressure subsided and before any fission products were released, the steam vents would be closed and ventilation valves opened. The ventilated steam would be filtered through charcoal and high-efficiency filters to prevent any release of fission products from fuel failure (WHC 1989). Eventually, the Advisory Committee approved the confinement system and allowed N Reactor to begin operations.

Reducing the Risk of System Failures. The Hanford reactor design incorporated the notion of triple redundancy, meaning that there were two backup systems for any system whose interruption could lead to a severe accident. The systems and their backup are described below:

- Shutdown System – If it were necessary to shut down the reactor because the cooling water supply was interrupted or the core temperature was too high, control operators could automatically drive the control rods

into the reactor and stop the reaction. If the automatic shutdown did not perform, operators could inject the rods manually. If that did not work, they could drop the boron balls into the graphite core channels of the vertical safety rods. The boron ball technology was developed in 1952-1953 and used thereafter. Existing reactors were retrofitted.
- Cooling System – The water system, described in greater detail above, consisted of the main water system, a secondary system, and the emergency cooling (last ditch) system (water towers).
- Electrical System – The two main electrical supplies into the Hanford Site were commonly referred to as A and B buses. Any important system had to be connected to both buses in case one bus failed. An additional backup was provided by diesel generators at key functions.

Emergency Procedures. The Atomic Energy Commission established procedures for reactor staff to follow in the event of an emergency. Types of emergencies prepared for in the mid-1950s included the following (GE 1955c, 01.1):

- Evacuation – If a severe radiation hazard occurred at a reactor area, operations would be discontinued and the area would be evacuated. The evacuation signal was a 5-minute steady siren or whistle blast. The events were classified as Class I if the threat to other areas was considered remote and Class II if other areas could be affected.
- Radiation Warning – When an area was evacuated because of radiation hazard, the Control Center in the 7211 Building in the 700 Area sent out a radiation warning. Upon receipt of the warning, area security personnel would relate wind conditions to the Meteorology Station (Building 622) every 15 minutes until the all clear message was given.
- Civil Defense Alert – This evacuation would arise from the possibility of enemy activity at the Hanford Site. Upon issuance of the alert, all lights/perimeter lights would automatically be dimmed, which was the signal that staff had 1 hour to evacuate to a staging area 10 miles distant from all operating facilities. Before 1955, facilities were ordered to turn off all lights during an alert, a condition referred to as a blackout.
- Civil Defense Take Cover – This action would result from enemy activity occurring in the region. Personnel would seek immediate cover in the designated facilities near their workplace. All lights/perimenter lights would automatically be dimmed. The signal was a 3-minute wailing siren or short whistle.

Reducing Risk to Workers. The health and safety of workers was a major theme at the Hanford Site. Much effort focused on general worker safety such as wearing gloves to avoid cuts, wearing goggles to avoid eye accidents, wearing hard hats, etc. Lost work time due to such routine industrial accidents was a problem with an easy solution. Thus, one can find an article in nearly every weekly Hanford General Electric News reviewing these common industrial safety practices. Reducing routine industrial accidents was one strategy for meeting production quotas.

The major focus of the health and safety programs was radiological. Radiological issues were not, however, commonly discussed in the weekly Hanford paper. This was presumably both for security reasons and to avoid alarming people. General work procedures involving radiation zones and radiation areas are discussed more fully in the Health and Safety Section. Safety measures specific to Hanford reactors during the mid-1950s (GE 1955c, 06.2.1) are presented below.

Many aspects of work in a reactor area were driven by the radiological classification of the area. The 105 buildings and area within their perimeter fences were defined as Radiation Areas. All individuals entering a Radiation Area faced the following requirements:

- Record no more than 2 millirem on Exposure Estimate Card for each shift during which 2 hours or more were spent inside
- Take a hand and foot count before leaving
- Eat only in lunchrooms and offices
- Smoke only where permitted, not in Radiation Zones and other posted areas
- Wear personnel meters at all times

Within Radiation Areas, certain areas were defined as Radiation Zones. These were defined in the mid-1950s as 1) areas where the radiation level exceeded 1 millirem/hour, 2) where surface contamination was significant, or 3), where airborne contamination exceeded 1/10 of the applicable permissible concentration. Within the Radiation Area, however, the 1 millirem/hour requirement was raised to 5 millirem/hour "...because it is necessary for personnel to routinely pass through these areas it would be decidedly inconvenient to either permanently or intermittently establish them as Radiation Zones" (GE 1955c, 06.2.1).

Normal work clothes were sufficient for work in Radiation Areas. In Radiation Zones, clothing requirements depended on specific conditions, which were posted for each reactor area on a Radiation Zone Status Board located in the Monitor Room. Areas with low-level contamination required workers to wear a cap, coveralls, gloves, and shoe coverings. Areas with gross contamination allowed no personal outer clothing and required two pair coveralls, a hood, shoe covers, British Leggings, gloves, and coverall openings taped. Under certain conditions, respiratory equipment was also necessary.

The Atomic Energy Commission required supervisors to be notified whenever the following radiological incidents occurred:

- Contaminated injury in a Controlled Injury Zone (certain zones in the 105 reactor buildings)
- Skin contamination of 10 millirem/hour or greater
- Skin contamination that cannot be reduced to less than 100 counts/minute
- Personnel exposure above 100 millirem
- Any unexpected or unplanned contamination spread or radiation exposure
- Any discharge of radioactive material to the air, river or ground beyond the customary applicable limits
- Unsafe working conditions, practices, or incidents arising from work with radioactive material that might be the subject of an investigation
- Ingestion or inhalation of contamination in amounts greater or suspected to be greater than the permissible limit
- Any other incident involving radiation or radioactive materials or of obvious immediate interest to management, such as fires, explosions, etc.
- Emergency reactor shutdowns, including SCRAMs

To determine radiation levels and identify unexpected contamination in the workplace, a series of radiation monitors and alarm systems were put in place. Fixed type instruments were commonly used in reactor facilities to protect individual workers, determine the radioactivity of various samples, and to control radiation in working areas. Ion chambers were widely used to provide continuous monitoring of the radiation fields. Such continuous monitoring would indicate abnormalities and long range trends.

Fixed instruments were in all radiation monitoring laboratories to radioanalyze air samples, water samples, and other samples required in the hazards evaluation program. The major fixed instruments commonly used in the Hanford reactors were the following:

- HM chambers were used in areas where radiation levels were so high, personnel entry could not be allowed. HM chambers also were used to confirm continuous low radiation levels at locations where portable instrument monitoring was not usually required.
- Five Fold Hand Counters were used to take hand and shoe readings of workers after leaving a radiation zone, before eating, and before leaving a radiation area. These instruments were placed at convenient locations within the 105 buildings.
- The Kanne Gas Ionization Chamber was used in connection with a Beckman Amplifier and recorder to measure and record the amount of radioactivity in the stack exhaust air as well as certain exhaust air from certain specific locations in the 105 buildings. For example, in the 105-C, H, KE, and KW buildings, Kanne Chambers were used to monitor the air from the reactor front face, the rear face, and the top of the reactor. Locations varied for each reactor.

Hanford reactor staff also made use of a large number of portable detection and dose rate instruments in the mid-1950s, although as the following statement indicates, such was not always the case:

"The inception of the atomic energy program found us sadly lacking in portable instruments to properly evaluate the hazards associated with the program. Ten years of never ceasing development have produced a very large number of portable instruments, which, when properly used and interpreted, give a concise and clear picture of the radiation hazards in and around the reactors ." - GE 1955c, 05.4, p. 1

By the mid-1950s, the following portable-type instruments were available and commonly used:

- Juno (beta, gamma, and alpha) and C.P. (Cutie Pie) (beta and gamma) meters
- Zueto and Alpha meters used primarily for low energy beta radiation
- Several types of Geiger-Mueller (GM) counters: EGM, TGM, VGM, and NGM (beta and gamma)
- BF_3 Neutron Counter used to measure slow and intermediate neutron fluxes
- Neut meter, a portable, fast, neutron monitoring instrument
- Samson meter used to measure gamma, beta and alpha radiations but primarily for determining beta activity of reactor effluent water

Two types of air samplers were used widely in the reactor areas. These air samplers were portable or semi-portable types that sample only one location. The Moto-aire samplers were used for long or continuous samples and were semi-permanently installed in some locations for such sampling. The Staplex was a high volume portable air sampler also commonly used but only for sampling in less than 30 minutes.

Key work areas in the 105 buildings that required radiological controls included the following:

- Work Area – The portion of the reactor building in front of the reactor proper, used for storage of process material and equipment used for charging and other front face shutdown work.
- Discharge Area – The area in back of the reactor proper, used for handling of irradiated materials discharged from the reactor. It was the most hazardous location, radiation wise, for workers.
- Reactor Charge Face – The front portion of the reactor proper, where fuel elements are loaded, process tubes replaced, and a wide range of operational and maintenance activities occurred.
- Experimental Levels – Experimental level installations were used for special research and experimental projects and for special irradiations.
- Top of the Reactor – Located on top of the reactor were the vertical safety rods and the channels in which they operated, the boron ball safety system hoppers containing the balls, parts of the ball recovery equipment and ball transfer equipment. Because of the high neutron flux that was present when operating, this area was generally only accessed during shutdowns.
- Inner Rod Room – This room contained racks that supported the horizontal rods and the openings in the reactor in which the rods operate. During reactor operation, high gamma and neutron dose rates prevented entry by personnel.
- Outer Rod Room – The outer rod room contained the racks that supported the horizontal rod assemblies, the rod drive mechanism, and the cooling water system for the horizontal rods. Few entries were made during reactor operations.
- Transfer Area – This area contained facilities for loading irradiated process material in well (rail) cars for transfer to the separations facilities. Empty shipping cases were removed from the well cars, lowered into the loading pit, loaded with material, and returned to the well cars.
- Storage Area – Located directly behind the reactor's discharge area, this area consisted of a large water-filled basin, pick-up chutes, metal handling facilities and equipment to remove material from the basin.

Reducing Risk to the Environment from Ongoing Operations. The major environmental risk of reactor operations concerned the discharge of reactor cooling water to the Columbia River. As discussed above, the water was diverted to retention basins for several hours to reduce both the temperature and radionuclides with short half-lives. Nevertheless, substantial volumes of radionuclides entered the river every day. Little effort was made over the years to reduce the number of radionuclides.

As the Hanford Site began increasing the power levels of reactors, fuel element failures increased in frequency. Due to the high amounts of environmental contamination created by these fuel failures, various improvements were made over the years to reduce failures, improve detection of failures, and limit the impact of failures. These are reviewed in greater detail in the sub-section on Technological Challenges and Developments.

WASTE MANAGEMENT SYSTEM

Operating reactors and supporting functions generated various gaseous, liquid, and solid wastes during operations. The systems designed to manage these wastes are described in the section, Waste Management.

SECURITY SYSTEM

The government considered any information concerning reactor technology or production of radioisotopes as extremely sensitive and instituted a variety of measures to protect the information. One method involved limiting access to different areas of the site. All staff had to have either "P" approval, permitting entrance to the Plant Controlled Area (the area bounded by the perimeter fence surrounding the plant); a Formal P ("FP") clearance allowing unescorted access to production limited areas (such as the individual reactor areas); an "L" clearance, which allowed access to certain classified information; and a "Q" clearance, allowing access to all classified information, except "Top Secret." Only individuals with FP clearance were allowed into the reactor area. To ensure that no one went in who did not have the proper clearance, guards were posted at the entrances to the reactor areas to inspect the badges of those entering.

A second form of security pertained to documents containing restricted information that were considered classified. The security classifications for such documents included "Top Secret," "Secret," "Confidential," "Confidential – Undocumented," and "Restricted Data." Strict guidelines governed the handling of such documents, for example, what security level one needed to handle the documents, where they needed to be stored, and how they were to be destroyed. Even waste materials containing classified information had to be protected. Staff were directed to tear classified scrap into small pieces, deposit the scrap in an envelope, seal the envelope, mark it "Classified Scrap," and deliver it to the Security Patrol for destruction (GE 1955c, 03.4).

TECHNOLOGICAL CHALLENGES AND DEVELOPMENTS

Hanford reactor staff faced major technological challenges during the four decades of operation. Some challenges were critical to ensure plutonium production. Other challenges were necessary to enhance safety. And finally other developments were related to the drive to improve cost efficiency. Each of these areas is reviewed below.

CRITICAL PROBLEM AREAS

Three critical problem areas encountered immediately following startup of the B Reactor were:

- Understanding and reducing corrosion
- Xenon poisoning
- Graphite expansion

Understanding and Reducing Corrosion. With the large amount of water used in association with metals to operate the reactor systems, corrosion was a constant battle. For example, as cooling water passed through the process tubes, film composed of metallic oxides and other corrosion products in the water would build up. This film affected cooling water performance and increased corrosion buildup on the process tubes. The need to avoid corrosion was a major driver in the design of the cooling water treatment system. As a result, much work was done to understand the relationship of water, water treatment processes, and the way reactor systems were corroding (see B Reactor Historic American Engineering Record in Appendix B on the Internet). Trying to reduce corrosion is a good example of the dilemma the Hanford Site regularly faced when trying to solve a problem. A modification might fix one problem while worsening another. Hanford systems were so complex that a beneficial change in one system inevitably created a negative impact on another part of the system. Hanford staff routinely had to balance the costs and benefits of technological and procedural changes and make a decision. The following water corrosion example illustrates this point:

> In the 1950s *"the Hanford Works [former name for the Hanford Site] began an intense review of intake water treatments. Sodium dichromate, a key corrosion inhibitor that had been added to reactor water since World War II, was evaluated closely. Because sodium dichromate was known to have detrimental effects on the fish of the Columbia River, much experimentation with other corrosion blockers was undertaken. However, due to the dramatic rises in tube and fuel element corrosion when sodium dichromate was withdrawn, site scientists decided to continue using it."* - Gerber 1995, p. 34

Higher power levels increased corrosion and failure of the reactor process tubes. The extraction process technology improved for aluminum process tubes over the years, resulting in tubes with more uniform thickness, helping improve performance (GE 1963d), p. 117). By the late 1950s, new applications for materials such as zirconium were developed and incorporated into Hanford reactors. Process tubes, for example, started being fabricated from Zircoloy-2. Zircoloy-2 had its own problems, however, and efforts to inhibit the hydriding commenced (DeNeal 1970, pp. 32-39). For descriptions of corrosion facilities, problems, studies, and solutions, see the 190-D Building Historic American Engineering Record in Appendix B on the Internet.

Xenon Poisoning. The Hanford reactor chapter would be incomplete without retelling the story of the initial operations failures, commonly referred to as the xenon poisoning problem (Rhodes 1986, pp. 557-560). Walter O. Simon, the first Hanford operations managers recalled the event:

> *"The night B Reactor went critical [September 26, 1944] we had a lot of high-ranking technical people watching this startup, and when it went critical and then shut itself down, the silence was deafening. It was complete consternation. As background to this, the scientific people, the Chicago people, the Nobel Prize winners, Wigner and Fermi and Szilard, were all much more on the risk taking side than Du Pont was. Du Pont was a conservative organization. For instance, if someone asked for a two-story building, Du Pont design would put enough steel in it for four stories, being convinced that sooner or later someone would add a floor or two. Du Pont conservatism paid off on the reactors. Enough extra tubes had been added to overcome the fission product poisoning."* - Walter O. Simon in Sanger 1995, p. 152

The Princeton theoretician, John A. Wheeler, is credited with solving the xenon problem. Wheeler had been concerned for some time about fission product poisoning. As the chain-reaction occurred, a non-neutron absorbing mother fission product would decay within hours to a neutron-absorbing daughter product. As the volume of the daughter product increased, the chain reaction would slow and eventually stop. As the daughter product decayed into a non-neutron absorbing third element, the chain reaction would start again. The solution to overriding this poisoning effect was to increase the reactivity by feeding more uranium into the reactor.

Fortunately, Wheeler had counseled Du Pont about this problem and convinced them to add about 500 channels in the corners of the reactor. At considerable cost and schedule delay, this had been done over the objection of other scientists. Once Wheeler and Fermi were convinced that more uranium was the solution, Du Pont only had to connect the channels to the water supply, add process tubes, and load with uranium. Had these channels not been constructed into the graphite core during the original construction, it would have taken many more dollars and many more months to modify the reactor. This event illustrated the value of having experienced engineers working with theoretical physicists to design and construct something as new and complex as a nuclear reactor.

Graphite Expansion. Reactor staff identified the graphite expansion problem, also referred to as graphite creep or the Wigner effect, soon after the reactors started operating (Carlisle and Zenzen 1996, pp. 58-59). Physicist Eugene Wigner had hypothesized in 1942 that fast neutrons in the reactor could induce changes in the graphite. The intense radiation from the reactor's neutron flux caused the carbon atoms in the graphite's crystal lattice to realign themselves.

Reactor staff recognized by late 1945 that the exterior dimensions of the reactor were changing. In January 1946, pre-measured samples were removed from the graphite stack of one of the reactors. Post-irradiation measurements indicated that the graphite had expanded significantly. Staff also observed that the process tubes were bowing as a result of graphite growth, particularly at the upper portions of the stack (DeNeal 1970, p. 52).

The Atomic Energy Commission first responded by shutting down B Reactor in 1946 to preserve its capability, especially for producing polonium, an important material used in the final assembly of the bomb. The assumption was that the first three reactors would be shut down and new ones would be constructed with a design that avoided graphite expansion. DR was the first such reactor to be constructed.

General Electric researchers in 1947 found that annealing graphite helped. They accomplished annealing by operating the reactors at higher temperatures and allowing them to cool (Nightingale 1962). This finding enabled the Atomic Energy Commission to scale back their original expansion plans (Gerber 1996, pp. 2-4). In 1947, reactor staff began experimenting with carbon dioxide environments. By 1949, researchers were convinced that higher graphite temperatures created by the carbon dioxide environment limited the graphite growth, thus enabling the reactors to keep operating (DeNeal 1970, p. 52).

> ***"Because it had a lower heat removal capacity than helium, CO^2 allowed the carbon atoms in the graphite crystal, displaced by irradiation, to heat up, become active, and thus realign themselves. By 1954 the CO^2 additions were working so well that the oldest reactors operated with a gas atmosphere composed of 40 percent helium and 60 percent CO^2."*** **- Gerber 1995, p. 33**

The graphite expansion issue affected aspects of reactor technology beyond justifying new reactors. For example, as graphite growth distorted the process tube channels making charging and discharging difficult, especially in the upper portions of the reactor, the length of fuel elements was reduced from 20 to 10 centimeters (DeNeal 1970, p. 6). This distortion also affected the performance of the vertical safety rods when placed in the reactor, which lead to the development of the boron ball safety system being installed in 1952 (DeNeal 1970, p. 9).

IMPROVING HEALTH AND SAFETY CONCERNS

Human and environmental health and safety were significant issues at the Hanford Site from the beginning (see the Health and Safety Section). For reactors, these concerns fell into the following areas:

- Reducing fuel element failures
- Understanding and reducing the effects of reactor effluents on the Columbia River
- Reducing the likelihood and effect of a serious reactor accident
- Improving worker safety

Reducing Fuel Element Failures. As shown in Table 2-3.7, fuel elements began to fail beginning in 1947. By 1951, the Atomic Energy Commission understood the severity of the problem. The higher operating temperatures caused the fuel elements to blister and rupture. Fuel element failures were a serious problem because harmful fission products entered the water cooling system following a failure. If undetected, the radionuclides passed to the Columbia River, thereby creating an environmental problem. When radionuclides were detected in the water cooling system, the reactor would be shut down and the cooling water would be diverted to a trench or crib, thereby avoiding environmental damage. Frequent shutdowns impacted production schedules because it could take 24-36 hours to restart the reactor.

Improvements in three major areas helped reduce the frequency of fuel element problems: manufacturing of fuel elements, detection of fuel element failures, and reactor performance. The design and materials were improved in the manufacture of the fuel elements. For example, uranium powder began to be forged into cylindrical elements in such a way that the crystal structure throughout was anisotropically oriented (aligned in different directions). Also using 2-percent zirconium alloyed to the fuel element reduced the number of failures. (See the Fuel Manufacturing Section for details.)

New methods to detect fuel swelling and failure improved detection of fuel element failures.

> *"One of these was a simple optical test: the decrease in light intensity through the process tube indicated the presence of blisters on the cans. Another method involved measuring the level of radioactivity discharged from the process tube water. An alarm sounded when the reader encountered a sudden increase in neutrons, indicating that a slug [fuel element] had ruptured and released some of its radiation to the cooling water. The Panellit gauge allowed immediate tracing to the particular offending tube and slug."* - Carlisle and Zenzen 1996, p. 62

Another problem related to fuel failures, known as boiling disease, also required a technological solution to improve reactor performance and thereby reduce fuel element failures.

As shown in Table 2-3.7, improvements helped reduce fuel element failures but did not completely eliminate them. The problem continued at N Reactor even with a new type of fuel element (see the Fuel Manufacturing Section) and higher operating temperatures. Studies and modifications to procedures and equipment enabled reactor staff to minimize the number of fuel failures (DOE 1997a).

Table 2-3.7. *Frequency of Hanford Fuel Element Failures*

Year	Number of Failures
1948	2
1949	0
1950	3
1951	115 (includes 13 lithium targets elements)
1952	142
1953	93
1954	211
1955	242
1956	191
1957	201
1958	174
1959	71
1960	130
1961	86
1962	95
1963	68
1964	97
1965	89 (includes 23 thoria target elements)
1966	17 (includes 3 thoria target elements)
1967	29
1968	20 (includes 1 thoria target element)
1969	16 (includes 7 neptunium and 1 thoria target elements)

Understanding and Reducing the Effects of Reactor Effluents. The initial concern was the effect of reactor effluents on fish in the Columbia River. The Manhattan Project commissioned studies before the first reactors were built. These studies, conducted by Richard Foster, began at the University of Washington laboratories. In 1945, Foster came to the Hanford Site and set up a biology program in facilities at the 100-F Reactor. As the biology program grew, additional short- and long-term studies on fish, animals, and the environment commenced (Becker 1990, Stannard 1988). When the Hanford Site began to increase power levels in the early 1950s, the Advisory Committee on Reactor Safeguards increased its focus on the impact to the river and called for more river studies.

The initial strategy for reducing the effects of reactor effluents focused on two areas:

- First, delay the disposal of effluents to the river by holding the effluents in tanks for several hours
- Second, prevent disposal of additional contamination from fuel element failures by detecting such failures early and diverting the water from the holding tanks to cribs and trenches

This strategy stayed in place until the N Reactor. By using a closed primary cooling loop in the N Reactor, the Atomic Energy Commission was able to effectively eliminate all effluent flow to the

Boiling disease *"referred to accidental raising of the water temperature in particular fuel-loaded process channels to 212 degrees F. A resulting pocket of steam presented problems when an alternating phase steam-water system developed in the tubes and caused increased resistance, decreasing cooling, and increased slug [fuel element] ruptures. The thin aluminum-zirconium coating on the slugs would expand, blister, crack, and spill the contents into the tubes. Ruptured slugs required shutting down the reactor and then recovering the damaged slug, which might be stuck in a tube. Dealing with the problem led GE to design new control instruments. At the onset of boiling disease, these devices scrammed the piles [reactors], reducing the chance of greater potential damage."* - Carlisle and Zenzen 1996, pp. 64-65

Columbia River. When fuel elements failed in the N Reactor, the water was diverted to the 1301 trench. Some of the contaminated water in the 1301 trench did eventually start to reach the Columbia River through N Springs, requiring construction of another trench, 1325. The contamination problem created at N Springs notwithstanding, the closed primary cooling loop eliminated an enormous amount of contamination going to the Columbia River. The contaminated effluent going to the river from the single-pass reactors is shown in Tables 2-3.8 and 2-3.9.

Reducing the Likelihood and Effect of a Serious Accident. Because of the great amount of energy contained in a reactor and the radioactivity that could be released if a fire or explosion occurred, much attention was given to reducing risk as much as possible. Isolation was the initial strategy used at the Hanford Site. Technology-based strategies were also incorporated in the initial designs and were enhanced over the years. Technologies focused on the following areas:

- Detecting a problem as early as possible so that the reactor could be shut down or, in the case of a fuel element failure, the water diverted
- Shutting down the reactor quickly and completely
- Providing backup systems to ensure all critical functions continued to perform, thereby reducing the risk that problems would escalate
- Confining the spread of radioactivity in the event of an explosion or fire
- Determining the risk of an accident

Concern for environmental safety and health intensified beginning in the 1960s, largely in response to the commercialization of reactor technology to generate electricity. As commercial reactors began to appear across the nation, the public began to express concerns, such as the impact on local water temperatures and the harm to neighboring populations from radioactive releases. But the overriding concern was the threat of an accident, such as a melt-down, which would potentially lead to a large release of radioactivity.

The large increase in size of the K reactors (1800 megawatt-thermal) led to increasing discussions on safety measures. One item pursued by the Advisory Committee on Reactor Safeguards was development of a comprehensive startup program to be conducted before initial operations. Such a startup program would help alleviate the concern about auxiliary process tube cooling and graphite wetting for handling loss of cooling water accident. The Advisory Committee also wanted safety devices installed to warn of approaching criticality and to shut down the reactor whenever cooling water was lost. General Electric disagreed because such instruments did not exist nor were they likely to (Carlisle and Zenzen 1996, p. 100).

Differences in philosophy on risk started emerging in the 1950s. The Advisory Committee on Reactor Safeguards focused on the worst case scenario (a catastrophe), avoiding that scenario, and the possible consequences. Their focus was on safety devices. General Electric and those in the emerging field of nuclear engineering defined risk as a combination of probability and magnitude of an event. General Electric would modify procedures to minimize probability even though operations were increasing in scale. The following incident in 1955 at KW Reactor showed General Electric that procedures could not always eliminate the chance of an accident (Carlisle and Zenzen 1996, p. 101).

On January 4, 1955, after 17 hours of operation at low levels, the reactor cooling water was blocked by a plug. That problem should have been avoided if startup procedures had been followed. At the same time, the pressure gauge for measuring flow through the process tube turned out to have been improperly set and calibrated. The combination of events showed that measures of reactor safety had to be based on more than just the chances of a single catastrophic incident. The KW Reactor event had shown that two unlikely events could occur in series, thereby compounding each other and causing a serious problem.

In 1960, confinement facilities were installed in the first eight reactors to control the release of radioactive matter from the reactor buildings in the event of a nuclear incident (GE 1963d). By keeping the confinement zone at slightly negative pressure, using a fog spray system, and ventilating through filters, radionuclide release was reduced. Although the N Reactor was the first Hanford reactor to incorporate a confinement system in the original design, all commercial

Table 2-3.8. *Relative Abundance of Radionuclides in Reactor Effluent*

MAJOR (Accounted for 90% of radioactivity in effluent)	MINOR (Accounted for 8% of radioactivity in effluent)	TRACE (Accounted for 2% of radioactivity in effluent)		
Sodium-24	Phosphorus-32	Carbon-14	Cadmium-115	Neodymium-149
Silicon-31	Zinc-65	Calcium-45	Iodine-131	Promethium-149
Chromium-51	Zinc-69	Scandium-46	Iodine-132	Promethium-151
Manganese-56	Gallium-72	Scandium-47	Scandium-137	Europium-152
Copper-64	Yttrium-90	Manganese-54	Barium-140	Samarium-153
Arsenic-76	Yttrium-92	Iron-59	Lanthanum-140	Europium-156
Neptunium-239	Strontium-91	Cobalt-60	Lanthanum-141	Samarium-156
	Strontium-92	Strontium-85	Cerium-141	Europium-157
	Yttrium-92	Strontium-89	Cerium-143	Terbium-160
	Yttrium-93	Strontium-90	Praseodymium-142	Tungsten-187
	Niobium-97	Yttrium-91	Praseodymium-143	Polonium-210
	Iodine-133	Zirconium-95	Cerium-Praseodymium-144	Actinium-227
	Iodine-135	Molybdenum-99	Praseodymium-145	Uranium-238
	Uranium-239	Ruthenium-103	Neodymium-147	Plutonium-239
		Silver-111	Promethium-147	

Table 2-3.9. *Total Releases to the Columbia River from Eight Single-Pass Hanford Production Reactors, Curies per Year (Median Values of 100 Realizations in Curies per Year)*

Yr	Na24	P32	Sc46	Cr51	Mn56	Zn65	Ga72	As76	Y90	I131	Np239	Beta
44	905	216	57	4211	18530	708	4820	1178	300	2	17060	34300
45	34900	2890	2533	44920	95930	10518	75150	20340	3279	54	192100	114230
46	28360	2190	2058	29090	142281	8688	64780	14212	2452	41	153400	68140
47	25450	1860	1839	17785	57049	7457	57360	12319	1436	35	127840	39990
48	33970	2219	2161	30180	98137	8362	69960	15772	2351	47	151100	51630
49	46870	3175	2961	49520	204630	11713	96070	24660	2617	75	214600	105130
50	72590	4027	3762	66090	256090	14546	131730	30510	3636	116	279900	282200
51	99020	3333	2495	86430	232300	11153	113690	23660	6044	694	261700	549300
52	132690	5050	1799	127380	317000	9037	98400	34740	9909	1091	259000	989600
53	202500	8688	2253	173700	962300	8691	120410	98940	13102	821	316200	1481100
54	243400	7261	6545	153860	815000	21910	206900	91380	9938	1562	391600	1702400
55	318000	7184	5269	276600	1471000	26670	267400	139530	15008	3070	419400	2158000
56	407800	7722	6522	252400	1611800	31940	317500	134280	16569	2645	450300	2264000
57	644700	12325	6334	381300	4602000	27560	268600	212130	24230	3500	500100	5065000
58	751400	18484	5931	405800	4081000	27180	258420	293300	32180	5029	422300	4613000
59	1019000	17993	4753	418400	5068000	32030	271978	218400	31010	1986	275100	5321000
60	1382900	19490	9069	592800	7160000	42720	299650	236900	53270	4944	354800	6411000
61	1096300	21526	3313	634400	6370000	47110	236170	166890	49670	4825	243910	6551000
62	1094200	13845	3988	1085600	7965000	56010	92580	86660	37760	4302	257100	5407000
63	887900	11738	6432	592400	5586000	14920	101660	100630	29810	3115	211800	4733000
64	960000	12311	7469	536600	5873000	15710	144080	114480	24980	2827	247500	4425000
65	764500	12126	7968	390200	6364000	13379	94400	124600	21200	2076	168400	3554000
66	613000	7365	6485	256220	4000700	9656	80190	74600	13421	778	78950	2712700
67	671900	10118	8616	298200	5324000	15360	77260	94010	17889	1407	114970	2853000
68	499500	8632	3589	148940	4207000	7734	59090	71670	9798	1150	99790	2397000
69	359200	5478	4571	75730	5618500	6451	41602	61250	5095	473	59820	1523300
70	178041	1759	1135	55297	1026430	3394	32720	20253	7198	1194	36879	822860
71	13200	235	31	7040	85400	386	2660	2440	566	20	3540	72100
Sum	12582196	229239	119937	7191093	79613077	490993	3685230	2519734	444718	47878	6309159	66300980

reactors being constructed at this time had containment systems. Thus, the Advisory Committee on Reactor Safeguards struggled with approving General Electric's reactor design but eventually did (DOE 1997a).

All of these efforts related to the differing methods used to determine the risk of an accident. See Carlisle and Zenzen (1996, pp. 138-142) for an overview of the difference between the deterministic approach, which characterized reactor design in the 1940s and 1950s, and the probabilistic approach, which became vogue in the 1960s.

Improving Worker Safety. Worker safety was also a major concern as workers regularly needed to perform tasks in radiation zones. The government and its contractors generally felt that the risk was known and manageable and could be minimized by good working conditions and procedures. In the early reactors, shielding was placed around the reactor core to afford a minimum level of protection. Protective clothing improved over the years (see the History of Workers Section), as did methods of detecting radiation in the workplace (see the Health and Safety Section). Also over time, the Hanford Site started incorporating remote systems into the day-to-day operations. For example at N Reactor, the charging and discharging of fuel was semi-automatic.

IMPROVING COST EFFICIENCY

The cost of making plutonium became an issue in the early 1950s as costs associated with the nuclear weapons program started to escalate. This led to numerous efforts to increase efficiency of the Hanford reactors, which accounted for a significant portion of the cost associated with producing plutonium. As shown in Figure 2-3.8, these efforts to improve efficiency were successful. Specific efforts in the areas of increasing power levels, enlarging area of neutron flux, and using waste heat are described below.

Increase Power Levels. The increasing cost of plutonium production led the Atomic Energy Commission to support higher power levels because it lowered the cost per gram. General Electric showed that improvements in the fuel elements allowed higher power levels, which in turn cut raw material requirements 50 percent. By reducing the amount of uranium needed, the Hanford Site was able to produce 40 percent more plutonium per dollar of operating cost in 1949 when compared to 1947 (Carlisle and Zenzen 1996, p. 65; Gerber 1996, pp. 2-4).

When DR and H reactors were planned, it was recognized that power levels would be increased. To increase safety, the numbers of horizontal control rods and vertical safety rods were increased. Graphite blocks were machined with slightly concave faces (Carlisle and Zenzen 1996, pp. 60-61). General Electric experimented with different concentrations of carbon-dioxide/helium gas. Helium was used to prevent graphite from catching fire as oxygen levels associated with operating at high temperatures (needed for higher power levels) increased. Experimenting showed that carbon dioxide was also inert, but it had poorer heat transfer properties and thus allowed greater control of graphite temperature. The solution was to use a mixture. Annealing with other modifications allowed reactors to operate at temperatures higher than design levels (Carlisle and Zenzen 1996, p. 62).

By 1951, when the C Reactor was authorized, General Electric was able to increase power levels still farther by enlarging the

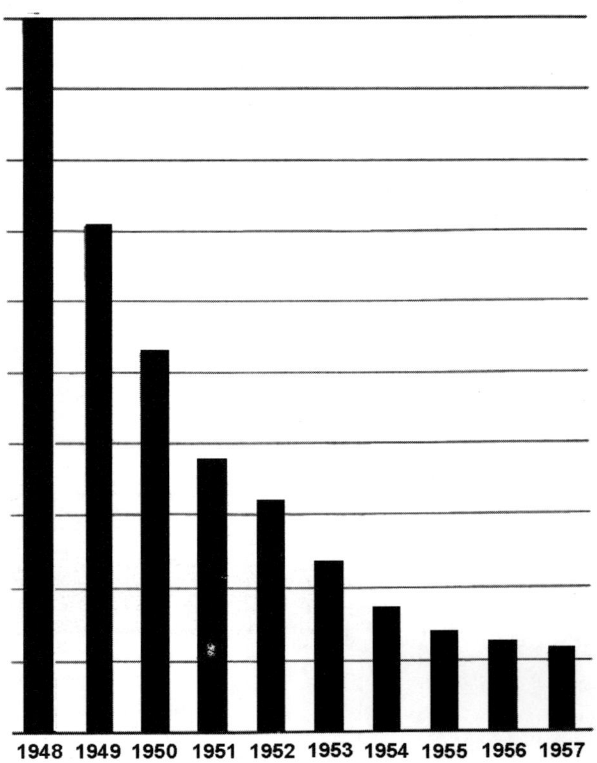

Figure 2-3.8. *Comparative Costs Demonstrating Improved Efficiency of Producing a Gram of Plutonium Nitrate*

plumbing facilities to increase the flow of cooling water and by reducing the ratio of graphite to uranium, thereby increasing the probability of neutron absorption. The C Reactor was rated at 750 megawatt-thermal when it started November 1952 compared to 250 megawatt-thermal for the earlier reactors.

By the time of the K Reactors, designers increased the power levels to 1,800 megawatt-thermal. Changes in the water systems allowed higher power levels. Improved pump designs reduced the number of pumps while increasing the amount of water fourfold: 125,000 gallons/minute in the K reactors and capable of increasing to 140,000. Table 2-3.10 identifies the original design levels and maximum levels eventually reached for each reactor.

Table 2-3.10. *Increases in Hanford Reactor Power Levels*

Reactor	Year Built	Design Level (MWt)	1963-64 Level (MWt)
B	1944	250	1,940
D	1945	250	2,005
F	1945	250	1,935
DR	1949	250	1,925
H	1950	250	1,955
C	1952	750	2,310
KW	1955	1,800	4,400
KE	1955	1,800	4,400
N	1963	4,000	3,950

Running the reactors at higher power levels had several impacts. On the beneficial side, plutonium production rates improved. In addition, higher power levels helped reduce graphite expansion, a good example of an inadvertent beneficial effect. On the negative side, increased power levels led to increased corrosion, fuel element failures, and required many modifications throughout the reactor systems. The increased risk to the health and safety of the workers, surrounding populations, and the environment posed by increased power levels resulted in considerable attention to this issue by the Advisory Committee on Reactor Safeguards.

Enlarge Area of Maximum Flux and Stabilize Temperature. The variation in neutron flux within the reactor core during irradiation was another target of inefficiency and adverse effect. The neutron flux within the reactor graphite core was greater in the center portions of the reactor than it was toward the sides, front, and back. On a graph, the plot resembled a bell curve in both directions (front to back and side to side). This meant that the fuel elements in the central portions of the middle process tubes would be irradiated quicker than those along the perimeters. The impact could be partially reduced by leaving the fuel elements in the perimeter process tubes longer and discharging the central process tubes more frequently. Within the central process tubes, however, little could be done to eliminate the variation between central and perimeter fuel elements, because they all had to be discharged at the same time. In addition, the neutron flux pattern caused heat temperature variations, which led to increased graphite expansion in certain portions.

Reactor staff recognized these problems early on and explored various mechanisms for leveling out the neutron flux distribution. Much progress was made by shifting the positions of the neutron absorbing control rods during operation to enlarge the area of maximum flux (Gerber 1996, p. 2-1). These measures helped flatten the curve and create a more even neutron flux distribution, though the problem could never be totally eliminated. Even with the later N Reactor and its advanced instrumentation, staff continued to search for ways to reduce this phenomenon (Burnside et al. 1986).

Use Waste Heat. An innovation in the K Reactors was the use of waste heat as energy to heat and cool the building's work areas. According to Carlisle and Zenzen (1996, p. 97), this was the first time this concept had been demonstrated at a nuclear facility. To accomplish this, once the water left the graphite block, it was pumped to a heat exchanger, which transferred heat from the cooling water to an ethylene glycol water solution. This solution was then pumped through the air ducts to different buildings. General Electric was proud of this innovation, which cost $614,000 but saved about 1.5 million gallons of fuel oil per year. The Atomic Energy Commission thought it could have been done earlier, but the higher power levels of the K reactors finally made it economical for General Electric to do (Carlisle and Zenzen 1996, pp. 97-99).

The ultimate use of waste heat was at the N Reactor, as discussed previously. In this case, heat generated during the irradiation process was converted into steam to power turbines, which generated electricity for use off the Hanford Site. The revenue generated by selling this power to utilities through the Bonneville Power Administration helped offset the cost of plutonium production. The N Reactor was built just as the nuclear power industry was beginning to build commercial-size plants and thus it attracted much national and international industry exposure (DOE 1997a).

CONTRIBUTIONS OF HANFORD REACTORS

Hanford reactors made significant contributions across two major fronts. First and foremost, Hanford reactors produced the plutonium and other isotopes needed for the nuclear arsenal. While the military, moral, and economic debates can question the need for nuclear materials, the point is that Hanford staff managed to deliver the product. The second significant contribution, admittedly less definitive, relates to the contribution of Hanford reactors to the nuclear industry in general. Viewed in the context of technology development spectrum, Hanford reactors played an important, if largely indirect, role in the development of the industry.

HANFORD SITE'S CONTRIBUTION TO THE NUCLEAR STOCKPILE

The Hanford Site managed to construct the world's first full-scale plutonium production reactors and make enough plutonium in time to significantly impact the ending of World War II. Following World War II, when military leaders chose the plutonium bomb to be the hallmark of the nuclear weapons program, the Hanford Site constructed more reactors and delivered the ordered amounts of plutonium (see Figure 1.7 in Chapter 1 and Table 2-3.11). In all, over 67,000 kilograms of plutonium were made at the Hanford Site of which 13,000 kilograms were fuel-grade plutonium. The Hanford Site produced 100 percent of the plutonium between 1945 and 1953 and over 65 percent of the plutonium during the history of U.S. plutonium production.

Table 2-3.11. *Total Production of Plutonium at the Hanford Site*

Reactor	Total Plutonium kg	Fuel-Grade Plutonium kg	Number of Production Years
B	5,532	606	22
C	6,471	571	17
D	5,489	654	12
DR	3,836	51	14
F	4,499	55	20
H	4,136	99	15
KE	13,498	1,042	16
KW	12,836	1,789	15
N	10,965	8,173	22
Total	67,262	13,040	153

In addition to plutonium, Hanford reactors also produced other isotopes for the nuclear weapons program. For example, polonium, with a half-life of 138 days, was needed for bomb initiators in the early bombs. The need to ensure polonium supplies was the major reason for shutting down B Reactor following the World War II to maintain one reactor in good operating condition (Carlisle and Zenzen 1996, pp. 55-56).

The Hanford Site also produced 10.6 kilograms of tritium (Roblyer 1994, pp. 7-8), a key ingredient of the hydrogen bomb. In general, while tritium production at the Hanford Site was more in the experimental than production vain, it demonstrated the capability. Other defense-related isotopes that were experimented with included the following (DeNeal 1970):

- Thorium was irradiated to produce uranium-233 from 1954 into the 1960s
- Depleted uranium was irradiated to produce plutonium-240 from 1958 into the 1960s
- Neptunium target elements were irradiated during the 1960s
- Lithium-aluminum splines were tested to demonstrate a method for making tritium
- Studies involving uranium-oxide fuels were conducted as part of the NASA Lewis Research Center program for the development of the space propulsion system
- Samples of americium were charged to produce medical grade plutonium-238

Had the Hanford Site not been so consumed with requirements to stockpile plutonium, alternative isotope production may have played a bigger role. The significant fact, perhaps, is that the Hanford Site had the capability to produce a wide range of nuclear materials.

HANFORD SITE'S CONTRIBUTION TO THE DEVELOPMENT OF THE NUCLEAR INDUSTRY

> ***"I never thought of Hanford in terms of being a factory. There was a sense of adventure about it. I associate it with pioneering. I would think it was like the first steamship, that must have been exciting. The first airplane was exciting, the first locomotive was exciting. I feel these comparisons are the right comparisons."*** **- Wheeler in Sanger 1995, p. 157**

Findlay and Hevly (1995, pp. 143-146) make the point that the Hanford Site was a production facility, and that the story has more to do with making technology work than anything else. When viewed within the context of technology development, the Hanford Site does belong towards the production or applied end of the spectrum. But the Hanford Site was far from simply a production facility insofar as that term implies a factory.

The Hanford Site was the first full-scale reactor complex for a decade and that gives it a special place in the history of the nuclear industry. Technology development continues to occur even after production begins because frequently critical issues surface that must be resolved. The experience gained by those involved in the early stages of operation can be instrumental in the development of an industry, and that was the case for the Hanford Site's effect on the nuclear industry.

By the mid-1950s, a limited amount of non-production-related reactor research and development was ongoing across the country. National Laboratories involved in developing breeder reactors and other types of power reactors included Argonne National Laboratory, Brookhaven National Laboratory, and Oak Ridge National Laboratory. The major industrial concerns included Battelle Memorial Institute; the Carbide and Carbon Chemical Company; E. I. Du Pont de Nemours; the General Electric Company; North American Aviation, Inc.; Pratt & Whitney Aircraft Division of the United Aircraft Corporation; Sylvania Electric Products, Inc.; and the Westinghouse Electric Corporation (ASME 1952, pp. 18-23).

By the late 1950s, with the advent of the Atoms for Peace Program and the Atomic Energy Commission's prototype program, reactor research and development were beginning to move out of the laboratory. But even as late as 1960, only two centers in America had reactors in full-scale operation with proven records of operability: the Hanford and Savannah River sites. Thus it is not surprising that the staff at the Hanford Site served as a resource for non-production related research and development efforts— which at least had an indirect if not a direct effect on the nuclear industry. William J. Dowis was an electrical engineer, who worked at the Hanford Site from 1943-1983, supervising the maintenance and design of electric systems for the reactors.

> ***"In the 1950s and early 1960s before the General Electric Company, Westinghouse, and Combustion Engineering had fully developed their choices of reactor types for electric power generation, the Atomic Energy Commission Division of Reactor Development sponsored studies of reactor types suitable for power-only operations. The types of reactors in use at Hanford and Savannah River were considered, of coarse [sic], since they were the only types that had been in use very long, with proven records of operability."*** **- Dowis 1986, p. 28**

Dowis (1986, pp. 28-33) goes on to identify several studies that the Division of Reactor Development requested Hanford staff to pursue:

- Use converted Hanford reactors to serve as electric central stations
- Use converted Hanford reactors to provide power for pumping water from the mouth of the Columbia River to California
- Use converted Hanford reactors to provide heat in Los Angeles to cause a rising column of warm air that would cut through the inversion air layer and relieve smog conditions
- Use converted Hanford reactors to provide heat for desalination of seawater
- Prepare a critique of a Russian nuclear-electric plant design
- Provide interpretations to the U.S. Central Intelligence Agency about the nature of buildings, pipelines, roads, and substations shown in an aerial spy photograph of a Russian installation
- Evaluate the potential for nuclear thermal plants in the Pacific Northwest, an area dominated by hydroelectric plants

Hanford staff also conducted studies for outside commercial and other governmental entities. Dowis (1986, pp. 28-40) recalled the following projects:

- In 1954, Hanford staff helped General Electric evaluate reactor technologies in search of the best one to commercialize. The graphite-moderated types were considered, eventually losing out to the light-water-moderated boiling-water reactor.
- In 1955, Hanford staff prepared the McGurdy Report, which resulted from their investigation of the legal, financial, engineering, and administrative aspects of commercializing the Hanford type of reactor. This work was done for the General Electric Atomic Power Study Group in Schenectady, New York and for a number of privately owned utilities in the Pacific Northwest.
- In 1965 and 1966, staff with Hanford experience (W. J. Dowis and Art Carson) served on General Electric's Division Safeguards Evaluation Function, which evaluated General Electric's procedures in producing atomic equipment. Dowis noted, "The work was extremely interesting; it covered all safety aspects of work in progress, and included discussions with the forward-thinking men in Division Safeguards Evaluation Function with respect to how atomic products might affect the health or well-being of customers and the general public. The committee made recommendations, but much of its value probably resulted from the GE Section managers knowing that their procedures would be scrutinized, so they kept a tight rein on their people" (Dowis 1986, p. 36).
- In the 1960s, many Hanford staff were involved in a safety review of the reactor design for KRB, a General Electric nuclear plant to be sold in West Germany. This was one of the first boiling water reactors.
- In 1970, reactor staff did a general Hanford Site study for the City of Richland to evaluate the best locations for new power reactors.
- In 1970, reactor staff did a study to evaluate the use of warm water discharged from nuclear power plants for irrigation. Such a use could make the nuclear plant more attractive economically for certain locations and would remove the objectionable environmental effect of discharging warm water to cold water streams and lakes.

As development of the nuclear power industry accelerated in the 1960s, the Hanford Site continued to play an important role. The N Reactor immediately became the largest power producing reactor in the world. Newspaper accounts and photographs from the mid-1960s show a steady stream of national and international industry people touring N Reactor, and contemporary stories in professional journals confirm the station that N reactor held within the emerging power industry at the time. Many Hanford reactor staff then moved on to construct the Fast Flux Test Facility, a major development in the field of breeder reactor technology. Finally, the concept of the Hanford Site as a nuclear park captured the imagination of many in the nuclear industry although it did not come to fruition.

AREAS FOR FURTHER RESEARCH

The story of nuclear development is one of the major technology stories of the twentieth century. The nuclear weapons part of the story is of interest for its technological achievement. The Hanford experience contributes to this topic. As a result, scholars and the public will likely continue to express interest in the history of nuclear weapons development in general and the Hanford Site in particular. The non-defense aspect of nuclear technology is equally interesting from a scientific, technology, and cultural standpoint.

As shown in this section, the Hanford Site also played a significant, if small, role in the development of the nuclear industry. As historians, social scientists, and others seek to understand why the non-defense nuclear industries failed to reach their promise in the late twentieth century, experiences from the Hanford Site will be sought.

With the basic facts now well documented through the scholarship of researchers such as Carlisle and Zenzen (1996), Findlay and Hevly (1995), and Gerber (1992a), scholars can now place the Hanford reactors into a theoretical framework that provides important insights into topics such as technology and culture, war and bureaucracy. Inevitably, as researchers pose new questions, additional research will be needed. While it is difficult to predict when the next wave of research will come or the research paradigms that will be in vogue, the following Hanford-specific areas could benefit from further research:

- When budget information is released, what did it cost to operate the reactors? What factors and events increased and decreased the budget?
- What was the decision making process for determining how the reactors operated and the quantity of plutonium produced?
- It was argued in this section that the operating experience of Hanford staff and the technology were important to the nuclear industry, since the Hanford Site was the only full-scale operating plant into the mid-1950s. Are there any specific examples of contributions from the Hanford Site? Did any managers, engineers, or scientists involved in reactor-related work rise to prominence within the industry? Are there cases where industry standards may have been different if Hanford staff had had more influence?
- What is known about Soviet espionage directed toward Hanford reactors? Are records from the Federal Bureau of Investigation available? What is known about the 305 reactor designs identified in Rhodes (1995) as having been stolen?
- How did Hanford reactor design, safety, and performance compare with other plutonium production reactors, and how did they compare to Soviet and other graphite moderated reactors used around the world?
- What is the history of improving worker safety for reactor employees? For instance, which remote systems were incorporated in which reactors? What was the relationship between production and risk analysis?
- The decision to shut down the PUREX facility and not process the fuel produced in the N Reactor resulted in the processed fuel being stored at the 100-K fuel basins for periods much longer than originally planned. The fuel canisters stored at 100-K emerged as Hanford's number one safety hazard in the early 1990s. Did the decision makers have all the facts when that decision was made? Was it clear that the fuel would be stored in a condition that would become a major site safety problem within a few years?

SECTION 4

CHEMICAL SEPARATIONS

B.J. FREER
CH2M HILL HANFORD, INC.

AND

C.A. CONWAY
RADIO-CHEMICAL PROCESS CONSULTANT

SECTION 4 - CHEMICAL SEPARATIONS

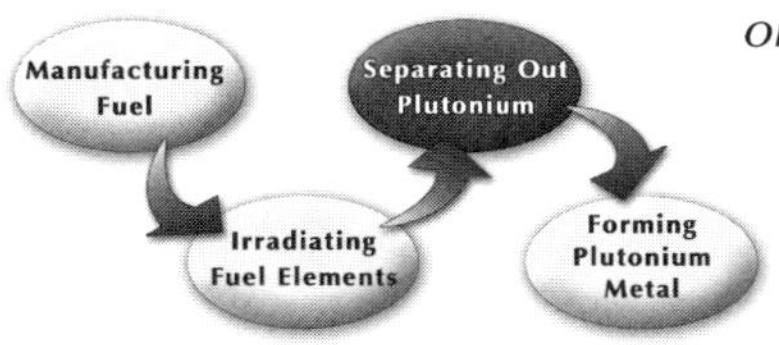

"Plutonium production at Hanford depended as much on chemical separations as it did on chain-reacting piles." - Rhodes 1986, p. 603

Once the fuel elements were irradiated, the plutonium had to be separated out. Separating plutonium for use in the atomic bomb was known as chemical separations because of the chemicals used in the process. Chemical separations was the third of four steps in the plutonium production process used at the Hanford Site—a step that added its own degree of uncertainty to the plutonium production process.

Figure 2-4.1. *First Chemical Separations Plant (T Plant) Built at the Hanford Site*

The day after Christmas 1944, the operators for E. I. Du Pont de Nemours and Company started the first production run at the T Plant (see Figure 2-4.1) to separate plutonium out for use in an atomic bomb. Forty-six years later in 1990, the U.S. Department of Energy (DOE) ordered the last chemical separations plant at the Hanford Site, the PUREX Plant, shut down.

During the Manhattan Project and throughout much of the Cold War era, the central purpose of chemical separations was to extract plutonium from the uranium and fission products contained in irradiated fuel elements. During the Cold War, the Atomic Energy Commission and Hanford Site contractors expanded the scope of chemical separations. The expanded scope included saving the uranium as a by-product during production as well as recovery and reprocessing of uranium and fission products already in underground waste storage tanks.

From 1945 to 1990, nuclear operators and engineers at the Hanford Site witnessed dramatic increases in plutonium produced at the chemical separations plants. Cold War pressures to increase plutonium output led to the use of new processes designed to maximize plutonium recovery. The results were more plutonium for America's nuclear weapons arsenal and more wastes. "Contaminated environmental media and facilities from chemical separations of irradiated reactor materials pose unusual and severe restoration problems" (DOE 1997e, p. 171). Of the four steps in the plutonium production process, chemical separations created the most and the worst waste. "The plants [T and B plants] sent approximately 10,000 gallons of waste to the tank farms for every metric ton of uranium fuel they processed" (DOE 1997e, p. 174). Nuclear waste from the chemical separations processes at the Hanford Site comprise 58 percent of the total volume of the DOE nuclear weapons waste from all sites (DOE 1997e, p. 36). Therefore, nuclear waste management at the Hanford Site was inextricably tied to chemical separations.

The chemical separations process began when irradiated fuel elements from reactors in the 100 Areas were loaded into shielded cask cars and began their rail journey toward processing plants in the 200 Areas. The chemical separations process ended when plutonium nitrate paste was shipped to

CONTENTS

Divider Page Photo: T Plant (Neg. 8503592-80)

the Plutonium Finishing Plant for finishing, storage, and subsequent transfer to other sites in the DOE complex such as Los Alamos and Rocky Flats for the next phase in the process of building an atomic bomb.

RESEARCH IN CHEMICAL SEPARATIONS AND THE DISCOVERY OF PLUTONIUM

> ***"I designed the Oak Ridge separations plant but it was smaller and much simpler than Hanford because of the vastly smaller amounts involved. In fact, we had the basic design of the Hanford project done before we started at Oak Ridge. This is the reverse of what generally happens. Usually, you do a semi-works in order to get some dope for a major plant."*** **- Raymond P. Genereaux, Du Pont's designer of chemical separations plants at the Hanford Site, quoted in Sanger 1989, p. 40**

By 1941, Glenn Seaborg had used chemical separations methods to discover plutonium. When the discovery was verified, scientists continued experimenting with chemical separations processes with an eye toward applications. Chemical separations methods used from 1940-1943 refer to processes that chemically separate, extract, and purify plutonium from materials such as uranium. Manhattan Project officials used information from chemical separations experiments to determine industrial requirements for chemical separations at the Hanford Site.

During 1941, a rush was already on to develop military weapons that used atomic energy. At this time, only a select group of scientists were privy to knowledge of the destructive potential of a bomb made with plutonium. Although the verdict was mixed when it came to the ethics of making weapons capable of such broad destruction, many scientists were worried if the United States did not develop this technology, Germany would. The chemical separations plants at the Hanford Site are testimony to the urgency. They became an integral part of the quest to make a bomb with plutonium. Specifically, the first full-scale architectural expressions of a total commitment to extract plutonium exist in the form of the chemical separations plants at the Hanford Site during World War II: T Plant (1944) and B Plant (1945). U Plant was built but never used during World War II.

GLENN SEABORG AND ELEMENT 94: APPLIED RADIO-CHEMISTRY

> ***"Seaborg would name Element 94 for Pluto, the ninth planet outward from the sun, discovered in 1930, and named for the Greek god of the underworld—the god of earth's fertility, but also the god of the dead: plutonium."*** **- Rhodes 1986, p. 355**

In an entry from his journal dated Monday, June 3, 1940, a young American named Glenn Seaborg noted an important challenge facing chemists. The challenge, as Seaborg saw it, was to master a new form of chemistry and apply this knowledge to the development of military weapons for America. In this entry, Seaborg writes that he was, "eager to work in this exciting field" (Seaborg 1994, p. 7). The exciting field Seaborg spoke of is known as applied radio-chemistry. Applied chemistry concerns itself with problems faced by industry. Applied chemists often work towards the development and application of technology. The chemical separations processes used at the Hanford Site were forms of applied radio-chemistry, which concerns itself with radioactive chemicals.

On February 25, 1941, while conducting experiments with colleagues in his Berkeley laboratory, Seaborg is credited with the discovery of a new element, plutonium. In his journal entry for that day, Seaborg summarizes an on-going series of chemical separations experiments by writing, "alpha activity is due to the new element with the atomic number 94" (Seaborg 1994, p. 29).

Seaborg used carrier chemistry, similar to Marie Curie's and Otto Hahn's fractional crystallization (Rhodes 1986, p. 603), to make the discovery. But Seaborg took carrier chemistry from the laboratory into the realm of operations. He writes of putting chemistry into industrial practice in a related journal entry. "One of the major earlier developments in plutonium chemistry was the discovery of the 'oxidation-reduction principle' whereby plutonium can be reversibly placed in different oxidation states whose chemical properties differ markedly. It is this principle that forms the basis for several of the extraction and decontamination processes that have been evolved" (Seaborg 1994, p. 260). Carrier chemistry and the "oxidation-reduction principle" formed the basis for specific radio-chemical processes and facilities at the Hanford Site.

THE METALLURGICAL LABORATORY: A SECOND ROUTE VIA THE HANFORD SITE

Physical separation plants, located at the Oak Ridge Reservation, divided isotopes of uranium one by one to produce material for a bomb (see the Reactor Operations Section). The goal of this route to an atomic bomb was to separate the fissionable isotope of uranium, uranium-235, from the 99.3 percent of uranium-238 that occurs naturally. The physical separation methods developed for this route were still posing problems in 1942. When Manhattan Project officials sized up the situation, they decided a second route using chemical separations should proceed full speed ahead.

> ***"While Met Lab physicists chafed under Du Pont domination, a smoother and quieter relationship existed between the chemists and Du Pont." -*** **Gosling 1994, p. 30**

Manhattan Project decision makers thought plutonium promised a viable second route because physicists calculated it was 1.7 times more likely to fission than the uranium-235 frontrunner (Gosling 1994, p. 7). Reflecting on this, John A. Wheeler, the physicist who co-developed the general theory of the mechanism of fission, said, "Plutonium is like, what is the best grade of gasoline, super? Plutonium is the super-premium of fissile material" (quoted in Sanger 1989, p. 127). Additionally, under the proper conditions ordinary uranium would produce the new element plutonium. Seaborg's laboratory studies spurred the mission on by proving that plutonium could be extracted and purified through chemical separations. However, it still was not known if Seaborg's laboratory chemical separations could guide the complex transitions to industrial radio-chemical processing of plutonium.

To speed up and concentrate research in one place, the U.S. Army Corps of Engineers set up the Metallurgical Laboratory at the University of Chicago. Metallurgical Laboratory officials pinned their hopes on the idea that an element more likely to fission was also more likely to provide the massive burst of energy the Corps was hoping to deliver in a bomb.

Based on Seaborg's experiments, bismuth phosphate and lanthanum fluoride emerged as the two best chemical methods for separating plutonium from uranium. However, chemical separations were not as far along as the physical separations methods. "The lack of actual irradiated uranium for experimentation, ignorance of the chemical properties of plutonium, and engineering inexperience with high radiation levels, complicated the development effort tremendously" (DOE 1997e, p. 171). As a result, one of the central missions of the Metallurgical Laboratory in Chicago was to provide a baseline of plutonium chemistry and validate the Corps' plan to operate massive industrial plants that would extract and purify plutonium after it was created from ordinary uranium. Some of the best minds in chemistry were exploring a method for isolating plutonium via chemical separations, and military officers noted for their singularity of purpose targeted successful results as a goal.

On a date much noted by historians, December 2, 1942, Enrico Fermi "proved the chain reaction" (Rhodes 1986, p. 496) and added momentum to the Manhattan Project. Glenn Seaborg witnessed this event and wrote about it from the standpoint of chemical separations. "One thing is certain; although Fermi has demonstrated that we now have a means of manufacturing 23994 [plutonium] in copious amounts, it is the responsibility of the chemists to show that the 94 can be extracted and purified to the degree required for a working bomb" (Seaborg 1994, p. 217). Seaborg and his group of applied radio-chemists continued the tedious work of extracting infinitesimal samples of plutonium and attempting to purify them "to the degree required."

On October 5, 1942, almost two months before Fermi proved the chain reaction, Du Pont asked a chemical engineer named Raymond P. Genereaux to design the chemical separations plants for plutonium. "I went out to the Met Lab that month. It turned out they didn't know what chemical process they would use, wet, dry, you name it. We realized immediately we had to design flexibility into it if we were going to move ahead and get any designing done before we knew for sure what we were doing, before we knew what the process would be" (Genereaux, quoted in Sanger 1989, p. 40). Genereaux completed design work for the original Hanford Site chemical separations plants in March 1943. That same month people who lived within the newly annexed area were told to leave for good by the War Department.

In May 1943, Du Pont pressed their head representative on the Manhattan Project, Crawford Greenewalt, to decide whether bismuth phosphate or lanthanum fluoride would be the chemical separations method used at the Hanford Site. "Greenewalt chose bismuth phosphate, though even Seaborg admitted he could find little to distinguish between the two" (Gosling 1994, p. 30). However, it should be noted that Greenewalt and Seaborg did wind up using lanthanum fluoride for the final concentration steps. Meanwhile, the Manhattan Project was pressing ahead at the Hanford Site no matter what method was chosen. "Construction of the separation plants began even before the process had been selected" (DOE 1997e, p. 173).

A Billionfold Scale Up: From Test Tube to Hundred Gallon Centrifuge

> ***"Seaborg's challenge seemed even more daunting, for while piles had to be scaled up ten to twenty times, a separation plant for plutonium would involve a scale-up of the laboratory experiment on the order of a billion-fold." -* Gosling 1994, p. 27**

May 1944, Seaborg visited the chemical separations plants being constructed in the 200 Areas at the Hanford Site. "To reach these areas, we drove over some of the flattest, most lonesome territory I have ever seen" (Seaborg 1994, p. 463). On that same day, he noted that it was an "awe-inspiring experience to see the thousands of workmen busily engaged in the building of these complicated edifices" (Seaborg 1994, p. 463). Seaborg's response was partly based on the land and space occupied by the chemical separations plants, but his awe must have also stemmed from a corporeal sensation that gargantuan plants were being built before his eyes to work out test tube chemistry he had formerly held in his hands. Figure 2-4.2 shows a chemical separations plant (B Plant) during construction.

Seaborg was not alone in being awestruck by the magnitude. Du Pont's engineers were confounded by the scale up from Seaborg's earlier experiments. So were the hired hands who dug the dirt and poured concrete for the windowless rectangular forms. These guys held shovels, wore hard hats, and quickly learned to quit asking "why?" They shared a secret with Du Pont's radio-chemical design engineers. The immense gray rectangular blocks actually did rise up out of the desert. They shared knowledge that a substantial portion of the buildings was below ground level. Workers soon started personalizing the strange buildings and deflecting the desire to guess their purpose – and risk being fired – by calling them Queen Mary's because of their shape and size. The nickname stuck even though they were much larger than the ocean liner of that name.

***Figure 2-4.2.** Construction of the B Plant in the 221 Building for Chemical Separations*

Chemical separations began in the laboratory with specks of plutonium in a petri dish. The billion-fold scale-up of chemical separations at the Hanford Site resulted in 8-gallon batches of plutonium nitrate, which then went on to the plutonium finishing process.

OPERATING HISTORIES OF CHEMICAL SEPARATIONS PLANTS

During the Manhattan Project, three chemical separations plants were built along with their related bulk reduction facilities: T Plant (221-T and 224-T buildings), B Plant (221-B and 224-B buildings), and U Plant (221-U and 224-U buildings). U Plant never operated as a facility for separating out plutonium. Instead the 221-U Building was used for training, decontamination, and repair, and was later modified to recover uranium. Its bulk reduction facility, the 224-U Building, was not used until 1956 when it was modified to convert uranium to uranium trioxide. The main architectural forms of these original complexes were massive rectangular reinforced concrete canyons. Raymond P. Genereaux explained the term "canyons." "Incidentally, do you know why we called the main separations buildings a canyon? Originally, we were going to build them with walls, no ceiling. It became apparent there would be radiation to the sky, then reflected. We added a roof, three feet of concrete" (Raymond P. Genereaux quoted in Sanger 1989, p. 41).

Table 2-4.1. *Chemical Separations Facilities at the Hanford Site*

Years of Operation	Official Name	Vernacular Name	Building Number
1944-1956	T Plant	T, T Canyon	221-T
1945-1956	B Plant	B, B Canyon	221-B
1952-1958	U Plant	U, U Canyon	221-U
1952-1967	Reduction Oxidation Plant	REDOX, S Plant	202-S
1956-1988	Plutonium-Uranium Extraction Plant	PUREX, A Plant, A Canyon	202-A
1956-72/1983-91	Uranium Trioxide Plant	UO_3 Plant	224-U

Table 2-4.2. *Chemical Separations Facilities Modified for Other Functions*

Modified Facilities and New Mission	Operations Start	Operations Stop
U Plant (221-U) Training, Decontamination, and Repair	1945	1951
U Plant (221-U) Uranium Recovery, also known as tri-butyl phosphate	November 1952	March 1957
Uranium Trioxide Plant/UO_3 Plant (224-U5)Conversion of Uranium to Uranium- Trioxide	1956 1983	1972 1991
T Plant (221-T) Decontamination	1958	Operational through the time period covered in this document, 1990
B Plant (221-B) Fission Products Processing	1968	Cesium recovery completed September 1983. Strontium recovery completed February 1985.

Two additional chemical separations complexes were built during the Cold War. They were known as REDOX (S Plant) and PUREX (A Plant). The PUREX Plant resembled the architectural shape of the T and B plants, while the REDOX Plant was more compact because of a high silo at one end. Table 2-4.1 shows the chemical separations facilities and their years of operations. Table 2-4.2 shows which facilities were modified to take on new roles.

T, B, AND U PLANTS

During the Manhattan Project, the number 221 designated the main processing structure within each chemical separations complex. Individual plants were distinguished by a single letter following 221, such as 221 B. The letters T, B, or U referred to the location of groups of buildings that, together, were basically self-sufficient, shared resources, typically worked toward a similar general mission, and were located in the same area.

The T, B and U plants shared an identical architectural form. The U and B Plants were approximately 810 feet long, 85 feet wide, 102 feet high (20 feet below grade). The T Plant had an additional head end area that made it 65 feet longer than the B Plant. The head end contained scaled-down versions of all equipment. The purpose of the head end was to conduct applied research for modifications to the chemical separations processes. The plants were composed of two separate areas: gallery and process. Galleries consisted of electrical, pipe, and operational floors. The galleries were separated from the process areas by thick reinforced concrete shielding that was approximately 7 feet thick. The process side consisted of an open canyon area with a floor called the canyon deck. Standing on the canyon deck, a worker saw big yellow hairpin hooks rising from the floor. B Plant had 40 of these hooks, and T Plant had 42. These hooks allowed the canyon bridge crane to lift the huge cell block covers that shielded the equipment. Typical cells were approximately 20 feet deep and about 30 percent below grade. Figure 2-4.3 shows workers installing equipment in a cell in the PUREX Plant in 1955.

Figure 2-4.3. *Workers Installing Equipment in a Cell at PUREX, 1955*

REDOX PLANT

The REDOX Plant was 467.5 feet long, 161 feet wide, 82 feet 9 inches tall in the canyon area, and 131 feet 10 inches high in the silo area. Two parallel galleries consisted of operating, pipe, and sample areas. These galleries were separated by the canyon area. The process area of the canyon contained cells A through J. These cells fed the 60-foot tall solvent extraction columns in the silo. Operational control galleries were also located in the silo area. The silo area had eight floors, two elevators, and a narrow internal staircase.

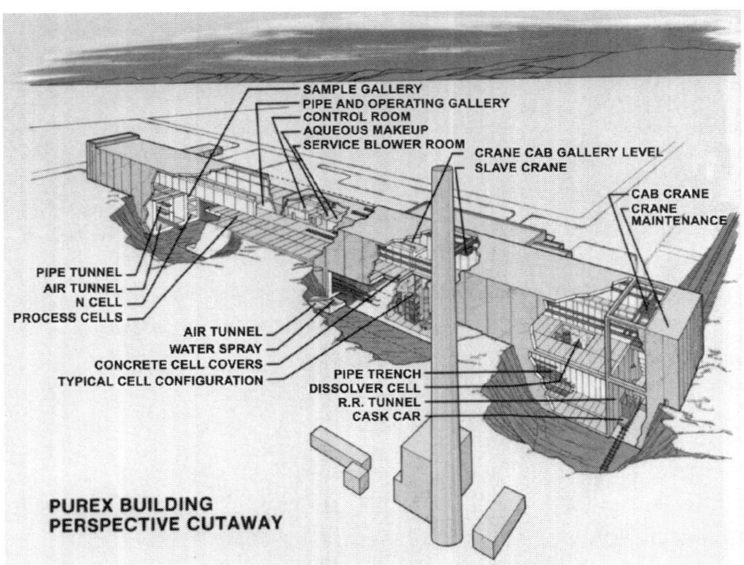

Figure 2-4.4. *Cutaway of the PUREX Plant*

PUREX PLANT

The PUREX Plant was 1,005 feet in length, 104 feet high (40 feet below grade), and 61.5 feet wide (Gerber 1996, p. 4-12). Like the other chemical separations plants, the PUREX Plant was divided into process and gallery areas. Twelve process cells were arranged in a single file in the canyon (A to H and J to M). "Each cell was 14 feet wide, 42.5 feet deep" (Gerber 1996, p. 4-12). Cells contained process vessels. A 5-foot thick concrete wall shielded the process area from the galleries. Galleries at the PUREX Plant included storage, sample, pipe, and operations area, and crane cab. Figure 2-4.4 shows a cutaway of the PUREX Plant. Because the REDOX process used gravity-flow columns, the columns were housed in a silo. By the time PUREX was constructed, pulsed columns were used instead of gravity flow columns so the silo required for the REDOX process was not needed for the PUREX process.

CHEMICAL SEPARATIONS PROCESS

The purpose of chemical separations at the Hanford Site was to extract and purify plutonium from fuel elements irradiated in the 100 Area reactors. As described in the Construction History Section, each chemical separations plant at the Hanford Site was a self-sufficient entity. This means each complex contained all the necessary buildings, equipment, and labor to conduct chemical separations. Among the primary reasons for the self-sufficiency of the various operational plants was compartmentalization of information and processes, physical security, and redundancy to ensure that a shutdown at one facility did not compromise operations at another. The exception to the redundancy principle at the Hanford Site was the Plutonium Finishing Plant of which there was only one. Table 2-4.3 shows the main processes in chemical separations, the support operations, and the facilities in which they were housed.

Table 2-4.3. *Chemical Separations Processes, Support Operations, and Related Facilities*

System	Role/Function	Facilities
Transportation (Operations Support)	Receive irradiated fuel elements from reactors	202-A, 202-S, 221-T, 221-B, 221-U
Dissolving (Main Production)	Dissolve irradiated fuel elements	202-A, 202-S, 221-T, 221-B, 221-U
Aqueous Makeup (Main Production)	Prepare chemicals for processing	202-A, 202-S, 271-T, 271-B, 271-U, 276-B
Laboratory Support (Operations Support)	Manage process	222-B, 222-T, 222-U, 2703-E
Precipitation/Centrifugation (Main Production)	Extract and purify product from feed	221-T, 221-B, 221-U
Solvent Extraction (Main Production)	Extract and purify product from feed	202-A, 202-S
Ion Exchange (Main Production)	Extract and purify product from feed	202-A, 233-S
Ventilation (Operations Support)	Provide fresh air flow to process cells and out stack	221T-A, 222S-B, 291-A, 291-B, 291-T, 292-T, 293-A, 294-A,
Wastes (Waste Treatment)	Release and/or store wastes	202-A, 216-U10, 218-E14, 218-E15, 218-WR, 241-AW, 241-T, 241-TX, 241-TY, 242-T, 244-UR, 291-A, 291-U, 292-T, 293-A, 2706-T, 2706-TA, BC Cribs, B5 Reverse Well
Steam Heat	Generate steam for processing and heating	284-E, 284W
Research and Development	Provide research and development	209-E, 222-S, 321, 323, 325, 325-A, 3706-A
Water Treatment	Treat and supply process cooling and sanitary water	284-E, 284W

OVERVIEW OF THE PROCESSES

At its most basic level, chemical separations existed because as plutonium was created in the reactors other new elements were also created. This happens through a phenomenon called transmutation. To envision the task of chemical separations, imagine a cylindrical fuel element, about 8 inches by 1.5 inches, that was made of heavy uranium metal. While sitting at just the right distance from other fuel elements in a reactor, atoms were transmuted. In other words, they were instantaneously changed from one element into another through the fission process. Although the nuclear composition of the fuel element changed, it was still a solid cylinder of metal. New elements, including plutonium, were created but they were stuck in the fuel element. It was the job of chemical separations to get the plutonium out of the fuel element.

Government contractors used three types of chemical separations processes at the Hanford Site: 1) bismuth phosphate batch, 2) solvent extraction through reduction oxidation, and 3) solvent extraction through plutonium-uranium extraction. Processes that were common to the three forms of chemical separations included transportation, storage, dissolving, preparing chemicals, ventilation and filtration systems and are described in this sub-section. Processes that were unique to these three technologies, such as precipitation for bismuth phosphate batch processing at the T and B Plants, or the use of pulse columns at the Plutonium-Uranium Extraction Facility, are discussed in following sub-sections.

The discussion that follows is not intended to address the range of operational conditions encountered through nearly 50 years of chemical separations at the Hanford Site. For source material, the writer of this document relied upon technical manuals, operator training manuals, interviews with retired operators, and secondary historical accounts. This narrative reflects the perspectives of these sources and draws on them to relate technological function to built structure. Potential sources of interest to the reader are Foote (1997), Barthel (1996, pp. 67-77), Linenthal (1995), Bodnar (1992), Johnson and Schene (1987), Lowenthal (1985), King et al. (1977, pp. 190-195). Readers interested in pursuing actual operational conditions in radio-chemical processing history may wish to consult 200 East and West Area plant log books, and the "off normal" and "unusual occurrence" reports for these areas. Most of these documents are available for review at the U.S. Department of Energy's Public Reading Room at the Consolidated Information Center, Washington State University-Tri-Cities, Richland, Washington.

Loading, Transportation, and Storage of Irradiated Fuel Elements

The chemical separations process began with irradiated fuel elements. Reactor workers used high pressure water to eject irradiated fuel elements out of the reactor core through the rear face of the reactor. During this step, reactor workers remotely guided the irradiated fuel elements into fuel storage basins filled with water. At this point, the fuel elements emitted extremely high levels of electromagnetic radiation beams, and the water in the storage basins functioned to shield workers and equipment from this hazard. Fuel elements were strewn about on the bottom of the storage basin after they were expunged from the reactor core, and the first step to ready them for the next leg of their journey was to group them. This process, called loading, began when reactor operators used long tongs to remotely move fuel elements into metal transport buckets. These buckets were about the size of a 2-foot cube. The loading process for the irradiated fuel elements continued when reactor crane operators placed full buckets inside heavy shielded casks. Next, reactor crane operators placed the casks in well sections of special railroad cars called cask cars. The cask cars were constructed of concrete and steel and were filled with water. Figure 2-4.5 shows a rail cask car.

Figure 2-4.5. *Rail Cask Car Used to Transport Irradiated Fuel Elements*

From 1945 through the mid 1950s, the transportation of the irradiated fuel elements was a two step process. Step one was temporary storage. Step two was transportation to a chemical separations plant. In step one, railroad crews delivered irradiated fuel elements, housed inside the cask cars, to temporary storage facilities called lag storage buildings. For information on the 212-N Lag Storage Building, see the Historic Property Inventory Form in Appendix B on the Internet. The lag storage buildings were located in the 200 North Area about midway between the reactors and chemical separations plants. Operators unloaded the cask cars at a lag storage building by lifting the casks with an overhead crane, removing the buckets from the casks, and placing the buckets at the bottom of a pool of water called a cooling basin. Attached to each cart was a long metal post with a cork at one end. Information on this piece of cork was secret and pertained to the history of that particular load of irradiated fuel elements.

When Hanford Site managers decided that it was time for a load of irradiated fuel elements to continue toward a chemical separations plant, operators loaded the railroad cask cars by placing the buckets inside the casks. The casks went inside the water filled wells within rail cask cars. Step two commenced when the cask cars began to move along the rails toward a chemical separations plant. Although the exact distances vary for each reactor and chemical separations plant, irradiated fuel elements traveled an average of 10 miles by rail to reach a chemical separations plant on the Hanford Site.

In the mid 1950s, the lag storage buildings in the 200 North Area began to be used for other miscellaneous purposes such as storage of equipment (DOE 1998b). They eventually became surplus because Hanford management decided the temporary fuel storage basins in the reactor areas were sufficient and because of increased capacity at the chemical separations plants. They based their determination of sufficiency on guidelines and later regulations as to how much iodine contamination was tolerated on vegetation. The on-going consideration was between the pressure for more plutonium coupled with concern for how much iodine was going out of the chemical separations facility stacks. This juxtaposition was re-evaluated as new pressures arose or new filtering technology was made available.

Figure 2-4.6. *Railroad Cut Where the Rail Line Enters the Plant*

The decision to skip temporary storage in the lag storage buildings meant the transportation of irradiated fuel elements was one step instead of two from the mid 1950s through the end of the Cold War. The fuel elements were shipped directly from reactor areas to chemical separations plants. Railroad operators used diesel powered engines to push cask cars along rails toward one of the chemical separations plants. At all Hanford Site chemical separations plants, the entry point for the rail cask cars was known as the railroad cut (see Figure 2-4.6). The railroad cut was characterized by high earth berms. These berms were roughly 100 feet high and extended out about a quarter mile on either side of the rail tracks. The tracks ran into one end of the chemical separations buildings through huge doors on a rail entry to the canyon. The area behind the door was called the tunnel, and the entire plant was called a canyon.

What Was the Purpose of Storing Irradiated Fuel Elements?

Regardless of whether irradiated fuel elements were stored at the 200 North Area or in fuel storage basins near the reactors, the purpose of temporary storage was to reduce high radiation levels by allowing some of the short-lived radioisotopes of biological concern, like iodine131, to partially decay. Radioactive decay is also known as half-life. The radioisotopes pertained to the chemical separations process because the radioisotopes vaporized during an initial step called coating removal and then exited 200-foot plant stacks as gaseous airborne radio-chemical emissions. The general principle behind storage and cooling of the irradiated fuel elements was simple: longer storage meant less radioactivity. This principle was in constant conflict with the pressure to turn out more plutonium in shorter periods of time. Shortening the storage time of the irradiated fuel elements was a method for increasing daily production output. This was the case because chemical separations facilities could process fuel elements faster than it took reactors to irradiate them. Especially during the Manhattan Project but also at times during the Cold War, chemical separations facilities were ready to process but were without feed because of the need to "cool off" fuel. Thus, the historical parameters for cooling times varied depending on a number of factors, including the military's perception of their need for plutonium, operating norms during a particular production time period, and experiments conducted during operations at the plants (Goldberg 1998, pp. 61-65). Stack filtration systems designed to capture radioactive off-gas emissions were also part of this equation during the Cold War. According to one historian's reading of the declassified historical record, the number of days irradiated fuel elements were allowed to cool before August 9, 1945 ranged from 15 to fewer than 30 days. However, this historian does note that "the exact length of cooling time during this period is unknown or is classified" (Gerber 1992a, p. 44). In the days following the use of an atomic weapon made with plutonium from the Hanford Site on Nagasaki, Japan, irradiated fuel element cooling times were fewer than 50 days. Later, during the Cold War, storage periods averaged 90 to 120 days. For additional information, see HEDR (1994a).

Dissolving Process: Coordination between Dispatcher, Crane Operator, and Others

Each chemical separations plant had a dispatcher, who controlled all access to the plant. Among other control duties, the dispatcher maintained a log of all canyon entries and exits. The train crew would contact the dispatcher of the chemical separations plant and inform them the train was waiting at the railroad cut with a load of cask cars. Once the dispatcher verified that the plant was ready to receive the cars, they unlocked the tunnel door from the dispatch office with a remote switch. The dispatcher picked up an internal phone and placed a call to an operator posted near the tunnel door. The dispatcher told the operator when the door could be opened to receive cask cars and informed the rail crew when to proceed into the tunnel. The rail crew used the train to push the cask cars into the tunnel to a specific spot, unhooked the cask cars being delivered, and departed. An operator then closed the huge tunnel door.

Figure 2-4.7. *Bridge Crane in the B Plant Canyon during Modification*

The dispatcher informed an operations supervisor that the cask cars were inside the tunnel. The supervisor would request the crane operator to unload the cask cars into dissolver tanks. Before this process began, the dispatcher ensured no personnel were inside the canyon. The dispatcher unlocked the crane cab entry and gave the crane operator permission to enter the crane cab and load the dissolver. A crane operator began by positioning the bridge crane over the dissolver cell. The operator removed the dissolver cell cover block and placed it on the canyon deck. Next, the lid was removed from the dissolver tank and buckets of fuel elements were emptied into this water-filled tank. Each tank held one "charge," which consisted of three buckets of fuel elements, weighing a total of 3 tons. After the crane operator charged the dissolver, the lid was replaced on the dissolver tank and the cover block on the dissolver cell. The crane operator informed the dispatcher that the cell was closed and charged.

Figure 2-4.8. *Crane Operator Lou Walker*

What Was a Canyon Bridge Crane?

Because of the need for remote operation due to high radiation fields, each chemical separations plant relied upon a canyon bridge crane for some operational steps, removal and replacement of equipment (see Figure 2-4.7). The bridge portion of a canyon bridge crane straddled the width of the canyon. The bridge crane moved along overhead rails attached to both sides of the canyon. The bridge crane was able to move the full length of the canyon (approximately 800 feet). A winch motor was attached to the bridge. The winch either reeled up or sent out a metal cable with a hook or other device, such as an impact wrench, at the end. The winch motor was able to move side to side along the crane bridge. Canyon areas were approximately 75 feet wide. The crane was controlled by an operator in the crane cab. The crane cab was a heavily shielded operating area that moved along with the bridge crane. The operating room in the crane cab was separated from the canyon area by a thick parapet wall. Submarine like periscopes allowed crane operators to see while controlling the crane (see Figure 2-4.8).

What Was a Dissolver Tank?

Dissolver tanks were encased in protective cells near the tunnel. They were large cylindrical vessels with steam coils on the side, called tube bundles, and a lid. When a crane operator filled a dissolver tank with a load of fuel elements it was said to be charged. Two separate processes occurred after a dissolver tank had been charged. The first process was coating removal, the second dissolution. During coating removal, the aluminum cans surrounding the fuel elements were chemically dissolved. During dissolution, the fuel elements were transformed from a solid to a slurry. Figure 2-4.9 shows a dissolver tank.

When the dissolver was charged the dispatcher contacted the weather group at the Meteorological Tower and informed them that operations intended to start the dissolver. The weather group provided the current conditions to the dispatcher, such as dilution factor of air, wind velocity, wind direction. A forecast of weather conditions was also provided. The dispatcher then forwarded the weather data to the supervisor, and the supervisor determined if the weather data fell within operating parameters. According to one historian's reading of the record, dilution ratios of 1000:1 were favorable, 500:1 to 1000:1 moderately satisfactory, and 500:1 or lower unfavorable (Gerber 1992a, pp. 80-81). These ratios referred to the difference between the concentration of a substance when it exited the stack and when it deposited on the surface of the earth. If the weather was not within operating parameters, the dissolving process may have been placed on hold until favorable conditions emerged. If the weather data fit dilution specifications, then dissolving operations continued.

Figure 2-4.9. *Dissolver Tank at B Plant*

One of the primary reasons the chemical separations plants obtained weather data before dissolving was to control the dilution of emissions to specified operating levels. The emissions posed health and safety concerns because of the radioactive and chemical fumes produced during the dissolving process. For additional information see DOE (1998c, p. 2.5), Hales (1997, pp. 144-150) and Gerber (1992a, pp. 81-82).

To begin dissolving operations, the chief operator at the process control board unlocked each piece of equipment required for the dissolving steps. The dissolver operator would then follow the procedures in the run book and fill out the data sheet as each procedure was accomplished (see Table 2-4.4). The process engineering group prepared the run books, which defined the sequential steps necessary for the chemical separations process. Instructions contained in the run books were scaled-up from data provided by laboratory experiments and pilot plant testing. In conjunction with using a run book, the operator filled out a data sheet. While the run books were hardbound documents that remained at the process control board, data sheets recorded

Table 2-4.4*. Examples of Entries in a Run Book and a Data Sheet for Decladding Operations*

Run Book for Decladding	Data Sheet for Decladding	
1. Turn on steam to A-1 tube bundle to heat wash water and fuel elements.	1. Run No.	______
	Approval	______ (signature)
	Date	______
2. Jet wash water from A-1 dissolver to A-3 tank.	2. Time jetted	______
	A-3 W.F.	______
	Sp. Gr.	______
3. Add 4200 lbs 50% caustic to A-1	3. Time added	______
	A-1 W.F.	______
	Sp. Gr.	______
4. Turn on A-1 air sparger	4. Time on	______
5. Turn on steam to A-1 tube bundle	5. Time on	______
6. Heat A-1 to 185 degrees F	6. Time	______
7. Hold at 185 degrees F until A-1 Sp. Gr. Reaches 1.15	7. Time	______

Note: This table only pertains to decladding (removal of the metal coating around the fuel element) and is meant as an example of the steps an operator might take in this process. Actual operational practices differed for each facility and time period. Please refer to the text for general information on decladding.

Figure 2-4.10. *Operators at Process Control Boards in the B Plant*

What Is Specific Gravity?

"Specific gravity is defined as the weight of a given substance compared to an equal volume of pure water at 4 degrees C. An instrument called a manometer was one way operators measured specific gravity of a solution in a vessel. A manometer read out in increments of specific gravity divisions. It functioned with two dip tubes that ran vertically into a process vessel. One dip tube was 10 inches shorter than the other. Pressure from the short tube pushed down on manometer oil while pressure from the long tube pushed the oil up into the manometer tube. The manometer reading would indicate the difference in the hydrostatic pressure of 10 inches of solution." **RHO 1979a,p. 11-30**

information about the process as it was happening. Figure 2-4.10 shows a process control board. Often all steps were not completed before the end of a shift. When this happened, the operators would initial and draw a line under the step they completed last.

To remove the aluminum can surrounding the fuel element, a caustic agent such as sodium hydroxide was added to the water in the dissolver tank. The addition of a caustic caused a chemical reaction inside the sealed dissolver tank. The caustic agent reacted with the aluminum but did not react with the uranium. As the aluminum changed from a solid to a liquid, it increased the weight per volume (the specific gravity) of the caustic solution in the tank. Instruments, such as manometers, measured the specific gravity, which indicated when the aluminum had been liquefied. When the aluminum was liquefied, the dissolver tank was cooled and the liquid aluminum solution jet transferred to a waste tank.

After the aluminum can was removed, the irradiated fuel elements were dissolved. The objective was to turn solid irradiated uranium into a slurry. The operators rinsed the fuel elements with water and then added an acid, such as nitric acid. As the acid reacted with the fuel elements, the acid dissolved a portion of the fuel element into slurry. As the reaction continued, the acid became saturated and lost its effectiveness. Typically, it took three separate additions or "cuts" of fresh acid to completely dissolve the fuel elements. Each charge (three buckets of fuel elements weighing 3 tons) slurried into approximately 3,000 gallons of liquid feed. Although fuel elements were completely dissolved, a heel was often left on the bottom of dissolving tanks. This happened when a portion of the dissolved fuel solution was not transferred out of the dissolver tank. Upon cooling, the remaining feed solution turned into a solid heel.

Preparing Chemicals: Aqueous Makeup

All chemical separations plants had units called Aqueous Makeup Units. These units prepared chemicals in liquid forms needed for the various steps in the chemical separations process. The term Aqueous Makeup Unit "indicates that chemicals are dissolved in an aqueous (water) solvent. Although some plants use organic solvents that are not water, the term AMU is still widely used and accepted for operations involving any type of chemical makeup or chemical handling" (RHO 1979a, p. 3-27). Examples of the tasks performed by an Aqueous Makeup Unit operator included (RHO 1979a, pp. 3-37, 3-38):

- Received and stored chemicals, both dry and in liquid form
- Made up chemical solutions and had them available for use in the plant processes
- Inventoried chemicals in storage
- Conducted some of the chemical analytical tests required for makeups
- Prepared equipment for maintenance
- Took necessary samples
- Housekeeping

Operations personnel requested chemicals from the Aqueous Makeup Unit by calling up and asking them to "send some K-O-H" (potassium hydroxide), for example. An Aqueous Makeup Unit operator would open a valve and send potassium hydroxide to a header. A header was a device that held liquids for future transfer. The Aqueous Makeup Unit operator would let the process operator know when the potassium hydroxide was in the header and ready to be drawn into a process vessel (RHO 1979a, pp. 3-37 to 3-41). Figure 2-4.11 shows the Aqueous Makeup Unit for the Plutonium-Uranium Extraction Facility.

Figure 2-4.11. *Aqueous Makeup Unit for the Plutonium-Uranium Extraction Facility*

Ventilation and Filtration Systems

All the chemical separations plants relied on ventilation systems to operate. Ventilation systems brought washed, filtered, and dust-free outside air in the direction of the most contaminated portions of the plant and exhausted this air from the building. Ventilation systems operated to protect electronic equipment and to control the spread of radioactive contamination. The intake system brought air into the facility and the exhaust system sent it up and out the stack.

What Was the Purpose of Dilution?

The purpose of diluting the exhaust was to bring the concentration of various contaminates in line with the release limits. Dilution specifications were based on calculations made by the Meteorological Department and the Health Instruments Section. Dilution functioned by dispersing exhaust streams into a large atmospheric spatial area. Release limits for airborne exhaust streams changed over time as information on health effects and biological activity was studied and estimated. Factors which affected the selection of the release limits included: 1) production quantity directives from the military and later Atomic Energy Commission, 2) status of scientific knowledge on the emission problem, 3) knowledge of filtration technology, 4) scientific knowledge of the relationship between contaminants and biological response, 5) design constraints of the plants. This complex calculus of dilution became more complicated when the pressure to produce plutonium came face to face with the limited ability of the plants to contain contamination during the Manhattan Project. As World War II ended, information affecting release limits continued to be collected and interpreted by Hanford Site scientists. Hanford Site management, in turn, was able to draw upon this information during the Cold War to make decisions about the implementation of additional emission control technologies and formulate airborne release programs (see Gerber 1992z, pp. 77-112; Hales 1997). Often, decisions regarding dilution rates were not based on new understandings of the dangers but rather from concerns regarding the increasing number of people being exposed to those dangers. A report issued by the Technical Steering Panel for the Hanford Environmental Dose Reconstruction Project contains the following statement on this topic: "During the first few years of operations, large amounts of radioactive materials – primarily iodine-131 – were released to the atmosphere from the separations plants. Better filter systems, new knowledge about radiation hazards, and stricter operating procedures all but eliminated these releases by the mid-1950s." - HEDR 1994a, p. 3

The air flow in a building was controlled by varying the static pressure, causing the air to flow in the direction of the lowest pressure. In chemical separations plants, the highest pressures were maintained in the areas occupied by workers. Potentially contaminated areas were maintained at levels lower than the occupied areas. Finally, the lowest pressures were maintained in the cells and inside process vessels, which were the most contaminated areas.

Four ventilation systems operated simultaneously in a typical chemical separations plant. The service ventilation system handled the areas housing administrative staff and Aqueous Makeup Units where airborne radioactive contamination was not a hazard. A second system ventilated the occupied work spaces in electrical, pipe, and operating galleries. A third system ventilated air in the process areas. The final system, known as a vessel vent system, ventilated air from the vessels themselves.

Between 1944 and 1947, the sole function of the ventilation systems was to dilute contaminates to tolerance levels. Tolerance referred to an allowable amount of contamination as designated by Hanford Site officials at a particular point in time. See the Worker Health and Safety Section for more information on tolerance levels. Dilution in these ventilation systems worked by adding fresh air to the exhaust stream. For more information on these ventilation systems, see the Historic Property Inventory Forms in Appendix B on the Internet for the 291-T Exhaust Stack, 291-Z Air Filter and Exhaust Stack, and 292-T Fission Products Release Laboratory.

After World War II, three exhaust filtration devices were used in conjunction with dilution. The purpose of these filtration systems was to remove the majority of target radioactive and chemical contaminates from the process and vessel exhaust systems. The three filter systems were:

- Caustic scrubbers
- Sand filters and fiberglass filters
- Silver reactors

Caustic scrubbers were the earliest filtration systems used in vessel vents to remove radio-chemical vapors from the process vessels. They were first installed in 1947 and 1948.

Sand filters were also installed in 1947 and 1948. Sand filters used at the chemical separations plants were columns of graded earth. Earth was arranged in these columns from coarse (pebbles) to fine grades of sand. Exhaust streams entered at the coarse end and exited out through the fine end. Figure 2-4.12 shows a model of a sand filter system. Fiberglass filters operated on the same principle as sand filters but were constructed of fiberglass.

Figure 2-4.12. *Model of a Sand Filter System*

Off gases from dissolver operations were passed through a silver reactor. The function of a silver reactor was to remove iodine-131 from exhaust. "The silver nitrate bed in the filters would chemically react with radioiodine to form silver iodide thus binding the I-131 more tightly than could physical capturing devices" (Gerber 1992a, p. 92). The silver iodide was disposed of along with the other wastes from the chemical separations plants. Silver reactors were first installed in 1950 and were used through the remainder of Cold War.

The effectiveness of these ventilation and filter systems was investigated by the Hanford Environmental Dose Reconstruction Project. The Hanford Environmental Dose Reconstruction Project estimated the radiation doses people may have received from the release of radioactive material from the Hanford Site from 1944-1992. The results showed "about 92 percent of Hanford radioactive iodine-131 releases to the air occurred from December 1944 through December 1947. In all, an estimated 740,000 curies of radioactive iodine-131 were released to the air" (HEDR 1994b, p. 4). See Table 2-6.4 in the Waste Management Section for the annual amounts of six radionuclides released to the air from the chemical separations plants at the Hanford Site from 1944-1972. Since this table was published, corrections for the releases starting August 1951 and after have been made. The corrections increase the releases from August 1951-1972 by about 25%. For details concerning the corrections, see http://www.pnl.gov/eshs/cap/hra/pub/hedr.html on the Internet.

BISMUTH PHOSPHATE BATCH PROCESS: T AND B PLANTS

The bismuth phosphate process produced plutonium nitrate. The chemical separations plants built by Du Pont workers during the Manhattan Project were designed to produce plutonium in batch runs. A batch was based upon one fully charged dissolver (three buckets of fuel elements). A basic rule of thumb in chemical separations at the Hanford Site was that as the concentration of plutonium increased in a batch, the concentration of uranium and other fission products decreased.

At a general level, the bismuth phosphate batch process was used to separate plutonium, as plutonium nitrate, from dissolved irradiated fuel. Before a batch process could start, operators transferred the dissolved fuel feed solution from the dissolver tank into another tank for precipitation. The first cycle extracted most of the plutonium from the uranium and fission products in the dissolved fuel feed through two steps of precipitation and centrifuging. The uranium and fission products were considered waste and were sent to waste tanks. The plutonium was the product. As the plutonium was transferred down the process line, it became progressively more concentrated. This product solution continued down the process line into a second cycle that was similar to the first cycle. The goal was to concentrate the plutonium and remove the uranium, fission products, and impurities. The plutonium product then entered a decontamination stage that used a process similar to the first two cycles. The result was a 330 gallon batch of plutonium nitrate. This 330 gallon batch was, in turn, concentrated down to 8 gallons of plutonium nitrate paste. Figure 2-4.13 shows the bismuth phosphate process at the Hanford Site.

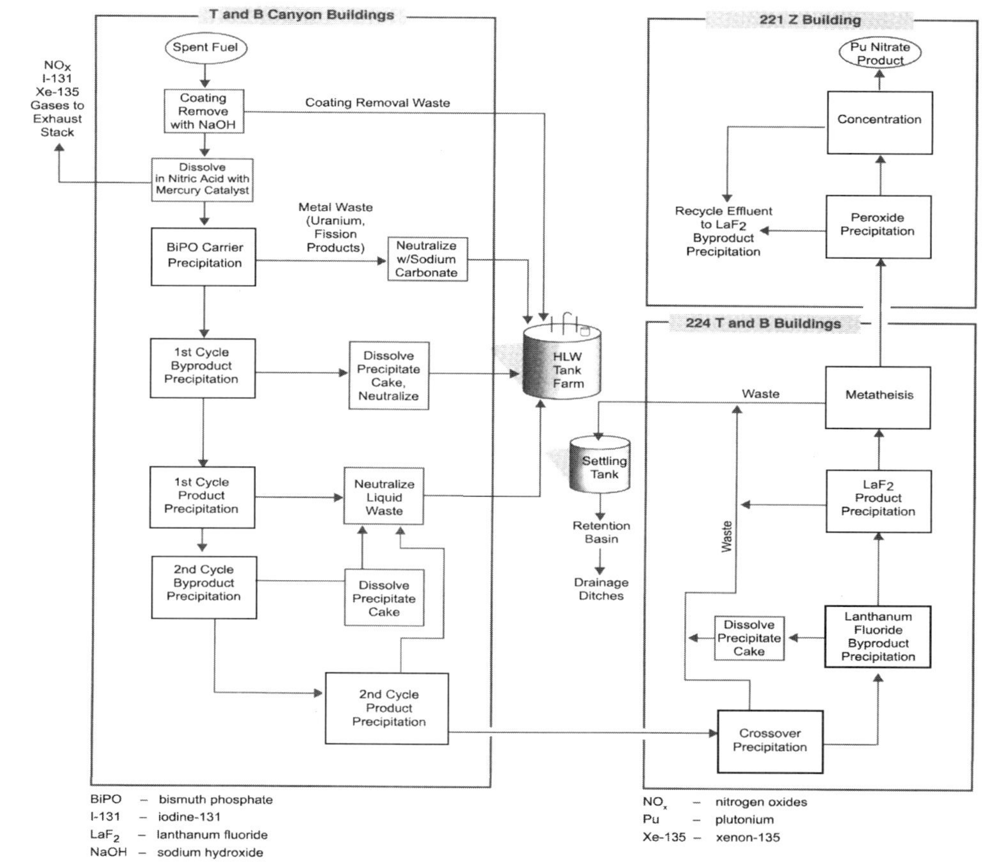

Figure 2-4.13. *Bismuth Phosphate Process at the Hanford Site*

The bismuth phosphate chemical separations process at Hanford's B and T Plants worked by alternating steps of precipitation and centrifuge. Altering the chemistry of bismuth phosphate created a precipitation of either plutonium or waste. During a plutonium precipitation, plutonium was carried out of the solution as irradiated fuel solution along with the bismuth phosphate as semi-solids. Centrifuging caused the plutonium to stick to the bowl while uranium and fission products went out the top of the bowl as supernate, and into waste tanks. The plutonium cake left on the bowl was then dissolved and transferred into a product holding tank for further purification and concentration. During byproduct precipitation, uranium and fission products turned into semi-solids by reaction with another chemical form of bismuth phosphate. Centrifuging caused the plutonium to exit out the top of the centrifuge bowl as supernate and into a product holding tank, while the uranium, fission products, and impurities remained on the bowl.

The wastes on the centrifuge bowl were then dissolved and transferred to a waste tank. In the final concentration step, lanthanum fluoride was used in the place of bismuth phosphate, but the process steps remained similar.

The bismuth phosphate chemistry worked by repeatedly alternating the valence state of plutonium-239 from plus four to plus six and back. Valence refers to the combining capacity of an element in a compound. In the plus six valence state, plutonium-239 remained in solution, while the uranium and fission products precipitated with the bismuth phosphate. In the plus four valence state, plutonium-239 precipitated. This means it turned into semi-solids along with the bismuth phosphate. Supernate was the clear liquid above a precipitate. Oxalic acid or ferrous ions were added to the bismuth phosphate solutions to achieve the plus four state. Sodium bismuthate was used for the plus six state (Gerber 1996, p. 4-6). Table 2-4.5 shows how the valence state of plutonium-239 affects its reaction with bismuth phosphate.

Table 2-4.5. *Bismuth Phosphate and Plutonium-239*

Plutonium-239 Valence	Plutonium-239 Reaction with Bismuth Phosphate	Plutonium-239 Status in Centrifuge
+IV (plus four)	Precipitates, turns into semi-solid	Cake, stuck on bowl
+VI (plus six)	No reaction, stays dissolved in solution	Supernate, liquid, out top of bowl

Precipitation

Precipitation refers to a process in which, under controlled conditions, a chemical may be added to liquid that causes the targeted material to turn into a semi-solid and become suspended in liquid. A household experiment illustrates precipitation: add a couple teaspoons of vinegar to a glass of milk. The mixture will separate into solids and liquid.

Precipitation was used for two phases: extraction and decontamination. The extraction phase separated plutonium from uranium and "about 90% of the fission products" (Gerber 1996, p. 4-6). Decontamination separated plutonium from the majority (about 99 percent) of the remaining fission products. Extraction purified plutonium to the specifications required for the plutonium finishing process. Among other reasons, decontamination was required to reduce the gamma and beta radiation levels in the plutonium-239. For information on gamma and beta radiation, see the Worker Health and Safety Section.

For extraction, the dissolved fuel solution from the dissolver tank was jet transferred to an agitator tank. The operator turned on the agitator (a propeller like device at the bottom of a tank) and chemicals were added. In the byproduct precipitation cycle, bismuth phosphate and oxalic acid were added to change the valence of plutonium-239 to a plus four state. To do this, an operator added the chemicals called for in the run book at a specified rate. For the product precipitation cycle, bismuth phosphate and sodium bismuthate was added to obtain the plus six valence for plutonium-239. For decontamination, the dissolved feed solution was reduced in volume, and the byproduct and product precipitation cycles were repeated.

What Is a Centrifuge?

A centrifuge was a vessel with a case and a bowl. The bowl held approximately 100 gallons of liquid and was mounted on a vertical shaft inside the case. An electric motor sat atop the case and spun the bowl. As the bowl spun with a precipitate inside, centrifugal force caused semi-solids to stick to the side of the bowl. These semi-solids were called "cake." The remainder of the solution, called supernate, overflowed the bowl and was either retained for further use or sent to waste tanks. Figure 2-4.14 shows a centrifuge on a canyon deck, not in its operational position inside a cell.

Centrifugation

Next the centrifuge was started and, depending on the exact specifications, held at 870 or

1740 revolutions per minute (rpm). After verifying the empty centrifuge was spinning at the appropriate rpm, the operator began to jet the precipitation mixture at a precise rate into the centrifuge.

Figure 2-4.14. Centrifuge on a Canyon Deck (Not in Operational Position Inside a Cell)

The precipitate slurry entered the centrifuge bowl from the bottom. In the first byproduct precipitation with plutonium in a plus six valence, the plutonium remained in solution and overflowed the spinning centrifuge bowl into a catch tank. The operator then washed the uranium and fission product cake and removed it from the side of the bowl. This step was called cake removal in which acid was transferred to the centrifuge and rapidly moved up and down by manipulating the speed of the bowl. The centrifuge was then stopped and the uranium and fission product solution jetted to a waste tank. Following this, a skimmer was used to remove excess water. A skimmer was a curved piece of pipe mounted on the inside of the centrifuge bowl. Using a remote lever on a panel control board, an operator activated the skimmer to move it from the center until it was near the wall of the bowl. The remaining uranium and fission product solution created by using the skimmer was then jet transferred into a waste tank.

When the plutonium was in its plus four valence state, the cake consisted of plutonium. Following a cake wash and skimming to remove the plutonium from the bowl, the plutonium was sent down the process line as plutonium nitrate. As the plutonium was sent down the line, the extraction and decontamination steps were repeated. The original design called for two extraction steps followed by three decontamination cycles (DOE 1994a, pp. 2-10 – 2-11).

Concentration

After the final decontamination steps, the plutonium nitrate was sent in 330 gallon batches, via underground pipe, to a receiver tank in one of the bulk reduction buildings (224-T and 224-B). There the 330 gallons of plutonium nitrate were concentrated to 8 gallons. Concentration also purified the plutonium nitrate by removing "the lanthanides (cerium, strontium, lanthanum, etc.) that the BiPO4 [bismuth phosphate] could not carry out of the stream" (Gerber 1996, p. 4-7).

Plutonium-239 is an alpha emitter. Alpha emitters require different industrial protection than when mixed with gamma and beta. The main difference for industrial protection from alpha sources is that measures are designed to avoid ingestion, inhalation, and injection. The plutonium nitrate solution sent to the bulk reduction buildings was relatively free of gamma and beta sources. This meant electromagnetic beam radiation was relatively absent and, therefore, the massive shielding used in the chemical separations buildings was not needed in the bulk reduction buildings. Additionally, process vessels (centrifuges, agitator tanks, etc.) were available for direct contact maintenance. Therefore, while the processes in the bulk reduction buildings were roughly similar to processes in the chemical separations buildings, the architectural structure of each was dramatically different (DOE 1992d).

The concentration steps in the bulk reduction buildings were similar in principle to the bismuth phosphate process in the chemical separations buildings. The key difference was that hydrogen fluoride and lanthanum salts were used instead of bismuth phosphate. For this process, the basic steps of precipitation and centrifugation remained the same. However, instead of extraction and decontamination being the primary goals, purification and volume reduction were the objectives.

Loadout

When the plutonium nitrate had been purified and the volume reduced, it was ready for "loadout." Loadout referred to the transfer of the concentrated plutonium nitrate to the plutonium finishing plant. Loadout of the plutonium in the bulk reduction buildings took place at the F-10 Tank inside the PR Cage. PR Cage stood for product receiver cage.

The F-10 Tank was the last tank in the chemical separations process. It was approximately 4 inches wide, 24 inches long, and 24 inches high. The shape of the tank was determined by criticality prevention requirements. One of these requirements was the absence of any moderating material that would slow down neutrons. Therefore, the F-10 Tank was very thin and designed to allow neutrons to escape.

The loadout process began by preparing the F-10 Tank, containing 8 gallons of concentrated plutonium nitrate, for transfer to a product receiver can. To do this, the loadout operator verified the amount of plutonium nitrate in the tank raised the product receiver can to the proper height near the drain spigot on the F-10 Tank. A special absorbing material called diaper paper was placed around the Product receiver can to absorb any excess plutonium nitrate. Following this, the operator pushed the product receiver can underneath the F-10 spigot. The spigot on this tank was equipped with a quick opening ball valve. The spigot came straight out from the F-10 Tank an inch or so and then went straight down for about 6 inches. Next, the product receiver can was slowly raised up until the spigot was inside the product receiver can. At this point, the operator opened the ball valve and drained the contents of the F-10 Tank into the product receiver can. The full product receiver can was placed on a truck, which transported the product receiver can to the plutonium finishing plant for further processing of the plutonium nitrate.

Continuous Solvent Extraction Processes: REDOX and PUREX

Continuous solvent extraction replaced the bismuth phosphate batch process for plutonium production at the Hanford Site in the 1950s (DOE 1997e, p. 175). A former Hanford chemical separations engineer, Charles A. Rohrmann, explained the switch from batch to continuous processing in a June 1956 *Nucleonics* article: "Economic considerations favor the continuous solvent extraction processes over batch processing. Batch processing, generally more costly in industry, is unusually high for the processes employed in Pu [plutonium] handling because of the costly raw materials used and the large volumes of waste for which costly storage facilities must be provided" (Rohrmann 1956, p. 66). Two forms of solvent extraction were used at the Hanford Site: reduction oxidation (known as REDOX) and plutonium-uranium extraction (known as PUREX). According to a DOE historical overview, the REDOX process was developed at the Hanford Site in the 3706 Building and tested in the 321 Building (DOE 1997e, p. 175). For information on the 3706 and 321 buildings, see the Expanded Historic Property Inventory Forms in Appendix B on the Internet.

The REDOX Plant (S Plant/202-S Building) was built in 1952 to house the REDOX process. The REDOX process was the only process used at that facility, and no other Hanford Site chemical separations plants used the REDOX process. The PUREX process was developed by Knolls Atomic Power Laboratory (DOE 1997e, p. 175). The PUREX Plant (A Plant/202-A Building) was built in 1955 to house the PUREX process. The PUREX process was the only process used at the A Plant, and no other Hanford Site chemical separations plants used the PUREX process.

> **What Is a Product Receiver Can?**
>
> **The product receiver cans (known as "PR cans") were used for transporting concentrated plutonium solutions from the Bulk Reduction buildings to the Plutonium Finishing Plant. They were made of stainless steel and resembled common 5 gallon propane tanks. These criticality safe product receiver cans were transported inside a large cylindrical metal shipping container, 3-foot diameter and 4 feet tall, as a backup in case of an accident and to ensure adequate geometric spacing for criticality safety.**

One main difference between the solvent extraction and bismuth phosphate batch processes was that the former used a constant process stream flow. Instead of constituting a batch, irradiated fuel elements were dissolved and added to an on-going stream. Another significant difference between batch and continuous processing was that the latter was based

"The solvent extraction processes are based on the fact that both uranium and plutonium can be put in valence states in which they are soluble in appropriate organic solutions, while in other valence states they are not. Since plutonium is more easily reduced than uranium, the solvent extraction processes have the ability to recover plutonium and uranium separately. Because the organic and aqueous phases can be readily mixed and separated, the processes lend themselves to a continuous, multi-stage operation in the extraction columns. By successive oxidations, reductions, extractions, and back extractions, highly purified uranium and plutonium products are obtained." - DOE 1991a, pp. 4-9 to 4-10

on a different kind of chemistry. This chemistry, solvent extraction, used countercurrent flow to take advantage of the transfer capacity of two immiscible solutions. Immiscible refers to a condition in which two liquids do not mix but form separate layers. In the process, a solution with greater density flowed to the bottom of a vessel. To illustrate, consider that water has a greater density than oil and when combined, oil rests on the top of water. Countercurrent flow was a process of sending the solution with the greatest density (called the aqueous) through the top of a column, flowing downward, while a solution with less density (called the organic) was fed into the column from the bottom, flowing up. As these immiscible solutions passed by one another, selected isotopes were transferred from the aqueous to the organic. This was called extraction. If the selected isotopes were transferred from the organic to the aqueous, the process was called stripping (DOE 1994a, pp. 2-24–2-25). The logic of the chemistry formed the basis for continuous solvent extraction at the Hanford Site.

What Is a Solvent Extraction Column?

A solvent extraction column was a long pipe with enlarged bowls, called disengaging sections, at each end. The disengaging sections had inlets for aqueous and organic solutions to enter. Aqueous entered at upper points and organic entered at lower points. Aqueous exited at the bottom disengaging section, and the organic exited through the top of the disengaging section. A cartridge, consisting of a pole with flat perforated plates spaced along the pole, fit inside the pipe. The perforated plates had a hole in the center where they were placed onto the pole. The purpose of a solvent extraction column was to create an emulsion whereby two immiscible solutions were forced into close contact with one another. The goal of such a column was to maximize the surface area of contact between the two solutions. At the Hanford Site, solvent extraction columns were of two types: gravity flow and pulsed. Gravity flow columns were used at the REDOX Plant. Countercurrent flow in gravity columns operated exclusively on the tendency for solutions with more density to flow down and solutions that are less dense to flow up. Pulse columns, on the other hand, were initially designed for the U Plant though never used to produce plutonium there. They were used for plutonium production at the PUREX Plant. They operated on the same principle as gravity flow columns except they had a piston driven pump. This pump, called a pulser, caused the contents of the column (aqueous and organic streams) to surge up and down in strokes. Pulse columns provided more surface area for contact between streams than gravity flow columns. Because pulse columns created more surface contact per volume of liquid, they were substantially shorter than gravity flow columns. This difference in function explains why the architectural structure of the REDOX Plant does not resemble that of the PUREX Plant. The REDOX Plant operated with gravity flow columns that were housed in an 8-story high bay silo located at one end of the plant. Figure 2-4.15 shows the high bay silo at the REDOX Plant. Because of the silo, the REDOX Plant represents a brief stylistic detour in Cold War chemical separations structural design at the Hanford Site. The PUREX Plant did not have a high bay silo. Its pulsed columns were roughly the same height as the original Manhattan Project process vessels and so mimicked the architectural form of the original chemical separations plants (T and B). From an aerial perspective, it could not easily be distinguished from the T and B plants, which enhanced its security. Figure 2-4.16 shows an aerial view of the PUREX Plant.

Figure 2-4.15. *High Bay Silo at the REDOX Plant, 1981*

Figure 2-4.16. *Aerial View of the PUREX Plant*

REDOX Process

The REDOX process produced plutonium nitrate and uranium. One way of contrasting the REDOX process with the bismuth phosphate batch process is through the notion of efficiency. Efficiency was defined from the standpoint of maximizing plutonium production. Among a host of other engineering criteria, efficiency related to the capacity for processing fuel elements, percentage of plutonium extracted from a given tonnage of irradiated uranium, and overall output of plutonium. With these criteria in mind, the REDOX process was more efficient than the bismuth phosphate batch process. The REDOX process also differed from the bismuth phosphate batch process in that it saved uranium as a byproduct (AEC 1951a). From a production operation standpoint, the REDOX process had a distinct disadvantage because it used an undiluted solvent, hexone (methyl isobutylketone), that had a low flash point. The low flash point required substantial precautions to be taken against explosion (DOE 1998b, DOE 1993, DOE 1992c). Hexone also posed difficulties in attempts to reuse it in the REDOX process because of its solubility in water and instability with nitric acid (DOE 1991a, p. 4-10). Wastes from the REDOX process also greatly complicated the management of underground tanks because of their intense radioactivity, noxious chemicals, and thermal heat (DOE 1997e, p. 180, DOE 1992c).

The three main steps of the REDOX chemical separations process took place in these columns. These columns were the crux of the REDOX process. Columns at REDOX range from 27 to 57 feet tall and in diameter from 3 to 16 inches. They are located at the silo in the west end of the facility.

Radio-chemistry at REDOX was based on the transfer of a salt from one immiscible solvent to another solvent. Uranium and plutonium dissolve readily in nitric acid. The resulting nitrate salts are able to be manipulated chemically. Therefore, uranium and plutonium nitrate were chosen for the REDOX process. For example, if hexone (an organic solvent) is added to a beaker of uranyl nitrate, some of the uranyl nitrate will transfer to the hexone. After adequate mixing, the amount of uranyl nitrate will reach equilibrium in each phase. The addition of a third chemical, such as aluminum nitrate, sodium nitrate, or calcium nitrate will break the equilibrium and cause more uranyl nitrate to shift into the hexone. Reversing the process by adding an aqueous to the hexone will cause the equilibrium to shift again, resulting in more uranyl nitrate exiting the hexone and entering the aqueous phase. The REDOX process is based on successive alternations of these processes. The specific chemistry that allows this to happen is based on a process called "salting out." In the REDOX process, hexone is salted with aluminum nitrate nonahydrate to allow extraction of plutonium into the hexone. After the plutonium is extracted into the hexone, it is then sent to another column and stripped into the aqueous stream because no salt (aluminum nitrate nonahydrate) is present in the aqueous. In the REDOX process, columns 1C, 2E, 2B, etc. are stripping columns and columns 1A, 1S, 2D, 2A, etc. are known as extraction columns.

After irradiated fuel elements had been turned into liquid feed, the REDOX process had four main steps: 1) precycle, 2) partition, 3) uranium cycles, and 4) plutonium cycles.

The precycle cycle partitioned the feed into three separate streams. These three streams were the plutonium, uranium, and fission products streams. The extraction cycle began when the feed solution, as an aqueous stream, was introduced into the 1A and 1S columns. Aluminum nitrate was added to this aqueous stream to alter the valence of the uranium and plutonium to a plus six. As the hexone flowed upward through the aqueous stream, it extracted the uranium and plutonium. The plutonium and uranium continued through the process. The aqueous stream, traveling down through the column, retained the remaining fission products. The fission products in this aqueous stream were then transferred to the D-13 Waste Receiver waste tank and thus eliminated from the process. The plutonium and uranium product from columns 1A and 1S formed the feed line for column 2E. In column 2E, the plutonium and uranium were stripped into an aqueous phase and transferred to the F-2 Concentration Tank for further concentration. The hexone, traveling up through column 2E and carrying fission products, was sent to column 10. Hexone was recovered in column 10, sent to the G-3 Hexone Still for further processing, and fission products were sent to the D-13 Waste Receiver Tank. This completed the precycle cycle.

The partition cycle began when plutonium and uranium were transferred from the F-2 Concentration Tank to the F-1 Feed Tank, entered the 2D column, and were extracted into an up flowing hexone stream. The plutonium/uranium in the organic (hexone) solution was then sent to the IB column. In this column, the plutonium and uranium were separated from one another. To accomplish this, the plutonium was stripped downward into an oncoming aqueous solution, and the uranium continued up the column with the hexone. The 1B column was unique because fission products remained during this phase, split equally between the uranium and plutonium streams. Plutonium was stripped into the aqueous phase by altering its chemistry from plus-six valence down to plus-three valence. In the plus-three valence, plutonium was no longer dissolved in hexone. Changing plutonium's valence to plus three was accomplished by adding ferrous sulfamate. The uranium was unaffected by the addition of ferrous sulfamate.

The uranium, still in the organic (hexone) solution, was then extracted into an oncoming aqueous stream in the IC column. "The aqueous stream was then stripped of the dissolved hexone, concentrated, and simultaneously adjusted to the required pH in the effluent concentrator" (DOE 1994a, p. 3-13). The uranium, then in an aqueous phase, was transferred to the F-5 Concentration Tank and finally transferred to the F-4 Feed Tank. The plutonium, then in an aqueous phase, was transferred to the E-6 Rework Receiver and finally transferred to the E-5 Oxidizer Feed Tank.

The uranium cycles began with the transfer of the uranium aqueous phase into column 3D. The uranium was then extracted into hexone and sent to column 3E where it was stripped and sent to the E-10 Concentration Tank. From the E-10 Concentration Tank, uranium was prepared for transfer to the 224-U Building for concentration. Fission products and plutonium in the salt waste flowing from the bottom of the column were not typically sent to waste tanks at the Hanford Site. Rather, they were sent to the F-2 Concentration Tank in preparation for the 2D column. This completed the uranium cycle.

The plutonium cycles decontaminated the plutonium by removing the fission products still remaining in the plutonium. To do this, sodium dichromate and nitric acid were added to change the valence of the plutonium into a plus four or six. This aqueous solution was then sent from the E-5 Tank to the 2A column where the plutonium was extracted into the upward moving organic flow. Plutonium was stripped back into the aqueous stream in the 2B column. The plutonium was then loaded out at the Plutonium Concentration Facility or at the 3B column and transferred to the plutonium finishing plant. Fission products were sent to waste tanks, and hexone was sent for recovery and reuse.

The function of the Plutonium Concentration Facility was to concentrate plutonium nitrate. Although the operational life of this facility was shortened due to an accident that occurred after the installation of an anion exchange column, the original equipment concentrated plutonium nitrate from the REDOX Plant by boiling off unwanted solvents and water.

A process called solvent recovery functioned to reuse hexone at the REDOX Plant. The main steps involved decontaminating the hexone by removing plutonium, uranium, and fission products. The hexone was then tested for purity, combined with fresh hexone, and added to on-going processes at the REDOX Plant.

PUREX Process

The PUREX process was similar to the REDOX process. To avoid repetition, the PUREX process is presented in a simplified form. The primary process steps both used solvent extraction and operated with continuous countercurrent flow to isolate radioisotopes. The differences were the PUREX process used pulsed solvent extraction columns and a different solvent, tri-butyl phosphate.

The solvent extraction process at the PUREX Plant produced weapons-grade plutonium, fuel-grade plutonium, depleted uranium, slightly enriched uranium, neptunium, and thorium (DOE 1994a, p. 3-1). Six steps produced these products: 1) co-decontamination cycle, 2) partition cycle, 3) uranium cycle, 4) plutonium cycle, 5) neptunium cycle, 6) N cell cycle. Figure 2-4.17 is a flow chart of the PUREX process.

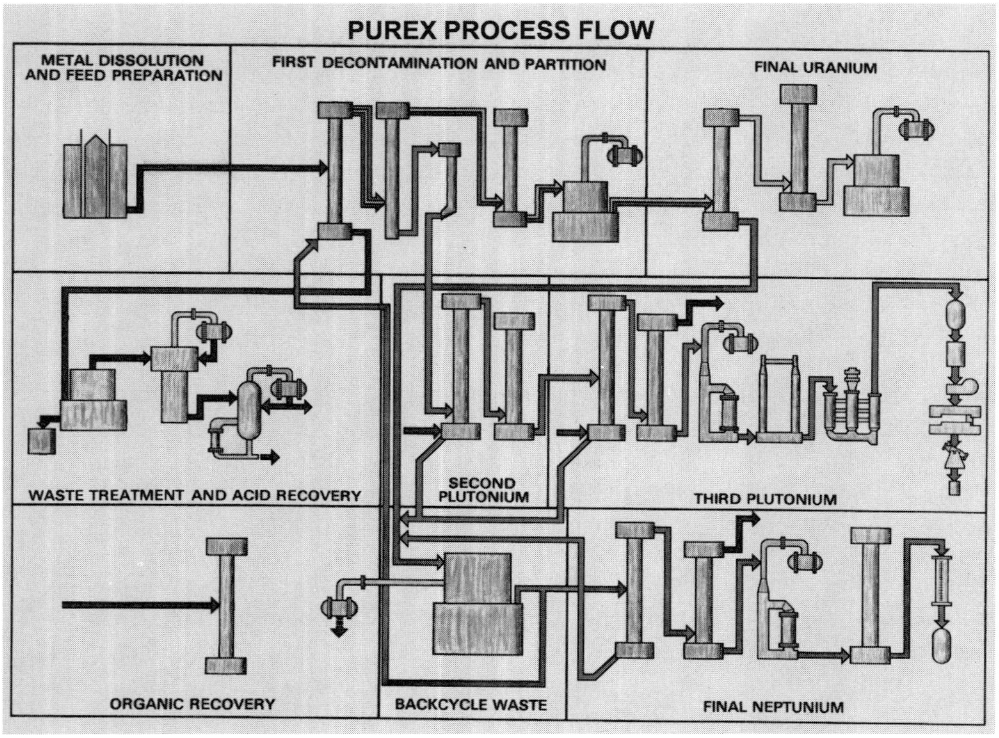

Figure 2-4.17. *Flow Chart of the PUREX Process*

The co-decontamination cycle was the beginning of the solvent extraction process at the PUREX Plant. In this cycle, 99 percent of the fission products remained in the aqueous stream while plutonium, uranium, and neptunium were extracted into an oncoming organic stream. The feed tank for this first step was TK-H1. The extraction took place in the HA Column. At this point, the organic stream was laden with plutonium, uranium, and neptunium as it flowed upward and out of the column as feed for the partition cycle.

Plutonium was removed from the uranium and neptunium in the partition cycle. This cycle took place in the 1BX Column. Plutonium was stripped from the organic stream in this step. Plutonium then entered the aqueous stream transferred to a feed tank (TK-J5). Uranium and neptunium overflowed the 1BX Column with the organic stream and entered the bottom of the 1C Column where they were stripped from the organic into the aqueous stream.

In the uranium cycle, the uranium product was decontaminated by removing fission products through solvent extraction. Uranium entered as an aqueous stream into Column 2D, was extracted into an organic stream, and entered Column 2E where it was stripped into the aqueous stream again. The uranium was then concentrated in tank E-K4-1 and stored in tank TK-K5 for shipment by tanker truck to the Uranium Trioxide Plant.

The second and third plutonium cycles further separated plutonium from the uranium and fission products. Plutonium entered the 2A Column as an aqueous stream, was extracted into an oncoming organic stream, and overflowed into the 2B Column. Plutonium was then stripped from the organic phase back into an aqueous stream that flowed into tank TK-L3. The third plutonium cycle further decontaminated the plutonium through solvent extraction. The plutonium product ended up in tank TK-L8.

The neptunium cycle used a back-cycle waste stream as feed. The repeated extraction/stripping steps were used to remove neptunium. When the neptunium levels reached a desired concentration, an ion exchange column was used to isolate this target isotope.

N Cell was the plutonium product loadout area for the PUREX Plant.

CHEMICAL SEPARATIONS PROCESSES MODIFIED FOR RECOVERY MISSIONS

Several challenges for chemical separations emerged at the Hanford Site as the United States entered the nuclear arms race with the Soviet Union. One response to a perception of increasing tension in the nuclear arms race resulted in on-going discussions between the Atomic Energy Commission and Hanford Site officials concerning the efficient use of chemical separations resources. Of the many topics discussed, the idea of using existing facilities to tackle new missions was underscored. As a result, officials at the Atomic Energy Commission and the Hanford Site made three major modifications to the chemical separations plants during the Cold War. Their first modification altered the U Plant to enable it to reprocess uranium that was previously stored in underground waste tanks. A second modification allowed the Uranium Trioxide Plant to convert uranyl nitrate hexahydrate to a dry form of uranium trioxide for shipment off-site. In the third case, the B Plant was modified to reprocess and encapsulate fission products from underground waste tanks and the waste streams of the PUREX Plant.

SOLVENT EXTRACTION FOR URANIUM RECOVERY: A GOLD MINE IN THE TANKS

> ***"The guys with the high foreheads looked at those waste tanks and realized that they had a gold mine with uranium."*** **- Knight 1999b**

During the Manhattan Project, uranium ore from the Belgian Congo was used to manufacture fuel elements at the Hanford Site. After the plutonium had been extracted from the irradiated fuel in the original bismuth phosphate chemical separations plants, the uranium was sent to underground waste tanks. This original process did not anticipate or envision a need to use uranium at a later date. With the Cold War, the demand for uranium increased in the 1950s. Uranium supplies became viewed as an important strategic security issue by military leaders and high-ranking politicians in the United States. As a result, an original bismuth phosphate chemical separations plant that had been held in reserve as a backup, U Plant, was modified in 1952 to recover uranium from huge underground waste tanks (AEC 1951b).

To modify the U Plant, however, precluded using the REDOX solvent extraction process that made its debut the same year in 1952. Because the REDOX process used gravity flow, columns were housed in a silo eight stories high. According to Roger Hultgren, a former Hanford Site manager of operations in chemical separations, it was out of the question to use gravity feed columns based on the REDOX process for the uranium recovery mission at U Plant because there was no room for nearly 60 foot columns in the existing structure. "They [REDOX Plant columns] were so tall that none of them would have fit into the old canyon facilities" (Hultgren 1999a).

Instead of using gravity flow columns that would have required construction of a new structure, the solvent extraction process for uranium recovery employed pulsed columns at U Plant. In their Industrial and Engineering Chemistry article, "Scale-Up Problems in the Plutonium Separations Program," O. F. Hill and V. R. Cooper consider the application of the pulsed column a noteworthy event at the U Plant. "A significant development of this program was the pulse column as a solvent extraction contractor" (Hill and Cooper 1958, p. 602). The significance Hill and Cooper assigned to the implementation of the pulse column was primarily defined in terms of increased efficiency.

Pulsed columns were designed to fit into existing process cells at the U Plant. The process cells were 20 feet deep. The reason operators and engineers were able to fit vertical pulsed columns into existing cells designed for tank like vessels is explained by their function. The function of a pulsed column is defined in the Hanford Site General Radio-Chemical

Training Manual: "Some columns are equipped with reciprocating pumps that move the solutions up-and-down as they migrate through the column. This up-and-down motion is called 'pulsing' and the pumps are called 'pulsers'" (RHO 1979a, p. 12-17). The pulsing action allowed for shorter columns and eliminated the need for the high silo that characterized the architectural form of the REDOX Plant. "If you use pulse columns for solvent extraction process you no longer need 60 foot high columns, they can be 15 feet high" (Hultgren 1999a).

Feed for the uranium recovery process at the U Plant came from underground waste tanks. This feed consisted of "uranium bearing effluents" (DOE 1994a, p. 3-8) that were sluice mined from the bottom of the waste tanks. Sluicing refers to spraying water to change a solid into a slurry. For further information on the composition of materials inside the waste tanks, see the Waste Management Section.

Before the actual sluicing operation could commence, a pre-set amount of water was sent to an accumulating tank. Operators used a strain gauge to determine when the accumulating tank had the right amount of water. This step prepared the tank to hold the sluiced uranium. Next, water from the accumulator tank was pumped to the sluice nozzle in the waste tank. As the uranium was sluiced and became a slurry, it was continuously sent to the accumulator tank. In the accumulator tank, the uranium settled out and the water recirculated back through the sluicing nozzle.

The preparation of feed for the uranium recovery process at the U Plant was described by two longtime veterans of chemical separations at the Hanford Site, Roger Hultgren (1999a) and Russ Knight (1999a).

Russ Knight: *"The feed for U Plant was prepared by sluicing the uranium that was left in the tank farms. Most of the uranium was collected in the first tank in a cascade of three. I do recall that there was from 7 to 9 feet of uranium in [tank] 101 T. I think that was sort of a typical situation. The T farm tanks consisted of three tank cascades, this is similar to other Hanford tank farms. Twelve tanks were in the farm and each one is about 75 feet in diameter."*

Roger Hultgren: *"The liner of the tank was 16 feet high. The cascade was a gravity feed. You would fill the first tank up to that 16 foot level and it would overflow into the next tank and so forth. In most cases there was minimal amount of carry over of uranium. The thing that was very significant here concerned how to actually extract it for uranium recovery. A number of methods were proposed, but sluicing eventually won out. The sluicing operation consisted of first pumping the liquid on top of the uranium to another tank. The sluicing itself was done with a high pressure gun."*

Russ Knight: *"The sluicing operations consisted of three pits being put on the tanks. The first was called a sluice pit, the other was called a pump pit, and the third was a pump out. Sluicing operation consisted of a sluice nozzle, which was a high pressure nozzle. It was an actuating eccentric cam at the top and we could set a given position that we wanted to sluice in the tank. The nozzle was adjustable and we could direct it at one quarter of the tank, the eccentric cam at the top would oscillate the nozzle back and forth."*

What Is a Strain Gauge?

A strain gauge was an electrical coil mounted beneath an accumulator tank. As the coil was compressed, a signal was transmitted to a recorder that read out in total pounds. For example, 1,000 gallons of water at 8.33 lbs/gal = 8,330 lbs.

The water-uranium slurry flowed to a slurry pump and was continuously pumped back to the accumulator tank. Because sluicing is a continuous process, the slurry pump functioned to keep the surface of the uranium free from liquid. This made the high-pressure sluicing operation much more efficient at removing uranium.

The sluicing process continued to a specified strain gauge reading. For example, the accumulator tank may begin with a reading of 8,330 pounds water and be finished when the strain gauge read 10,330 pounds. This example reading would indicate 2,000 pounds uranium in the accumulator tank.

After sluicing was completed, acid was added to change the slurry into a feed solution. Nitric acid was added while the contents were circulated through a photo-electric chamber. As the slurry dissolved, the solution clarified.

Roger Hultgren: *"Once the solution was collected in the tank farms it was pumped into U Plant into a large receiver tank. These tanks actually received the sluiced uranium, it was the feed for this uranium recovery process through solvent extraction."*

The uranium recovery process at the U Plant began by centrifuging the uranium slurry to remove solids. The next step was to send this uranium slurry to a set of columns. Roger Hultgren described this process. "Solvent extraction with a pulse column takes place through countercurrent flow. The uranium is brought in from the top of the column. This is called the aqueous stream. The pulser would pulse the organic as [tri-butyl phosphate] from the bottom of the column and that was used as the extractant" (Hultgren 1999a). A general overview of the technical specifics of the uranium recovery process highlights these steps. "The decontamination column used TBP [tri-butyl phosphate] to selectively extract uranium from the plutonium, fission products, and other ions in the aqueous phase. An aqueous scrub stream of nitric acid was introduced at the top of the column to scrub residual traces of the fission products and plutonium from the uranium containing TBP stream... The plutonium and fission products were retained by the aqueous effluent while the uranium was left in the TBP stream" (DOE 1994a, p. 3-8).

Roger Hultgren was present at the startup of U Plant's recovery mission for uranium. "Startup at U Plant was a headache for everybody because we were the first ones to get deeply involved with solvent extraction... The U Plant startup had a shakedown period, we took our lumps" (Hultgren 1999b).

One of these "lumps" concerned a problem that caused uneven mixing of the aqueous and organic streams. Hultgren (1999a) recalls,

> *"We had the columns on the deck of U Plant before we had the process going. We walked out on the deck because it was a cold run. The columns they had built did not work properly. We set up a dummy run and they put thermometers on the columns with duct tape. The thermometers would tell you if you were getting hot spots. They wanted it evenly spread out and constant pulsing. You could change the stroke over here on the pulser from ten strokes to thirty strokes a minute. It looked just like a horses tail until we got good extraction. That was for uranium."*

Hultgren sums up his work experiences with the uranium recovery process by noting, "basically it recovered the material and we ended up with a process that met all expectations" (Hultgren 1999b).

URANYL NITRATE HEXAHYDRATE CONVERSION AT THE URANIUM TRIOXIDE PLANT

During the Cold War, chemical separations both salvaged uranium from waste tanks and produced it as a byproduct from the plutonium production operations. Uranium typically arrived at the Uranium Trioxide Plant as uranyl nitrate hexahydrate. Concentration and calcination processes were used at the Uranium Trioxide Plant to convert uranyl nitrate hexahydrate to uranium trioxide powder because uranium trioxide powder could be used for armor for tanks, base material for military projectiles, and fuel for nuclear submarine and other reactors in the nation's plutonium production complex. The uranium trioxide powder was stored in shipping casks at the Hanford Site until it was shipped off-site for various defense purposes.

Three steps in the uranium trioxide process were:

- Concentrate the uranyl nitrate hexahydrate by boiling off excess liquid from the solution. The goal was to increase the uranium concentration from 4 lbs./gal to 10 lbs./gal.
- Calcine the concentrated uranyl nitrate hexahydrate. This was accomplished by submitting the uranyl nitrate hexahydrate to 800 F degree heat. At this point, uranyl nitrate hexahydrate spontaneously oxidizes into uranium trioxide powder.
- Load milled uranium trioxide powder through hoppers into shipping containers.

FISSION PRODUCTS PROCESSING: CESIUM AND STRONTIUM ENCAPSULATION AT B PLANT

According to the B Plant Operations Training Manual the purpose of fission products processing was "to isolate the longer-lived fission products by fractionalization for storage in a safer form, and to return the short-lived fractions to underground storage and ultimate solidification" (RHO 1979b, p. 1-1). To accomplish these goals, facilities at the B Plant were modified to remove two isotopes (strontium-90 and cesium-137). Figure 2-4.7 shows the B Plant during modification.

Three main steps for isolating strontium-90 and cesium-137 from other fission products were:
1) receive feed and partition,
2) purify strontium-90, and 3) purify cesium-137.

What Is an Anion Exchange Column?

"An anion exchange column was a pipe packed with negatively charged absorbing resins. An ion exchange column works on the same principles but is packed with positively charged absorbing resins. Anion exchange columns were similar in function to common home water softeners. Typical home water softeners remove minerals like calcium and iron from water. In the case of anion exchange columns, they both absorbed and released selected radioisotopes. This was accomplished with the resin by passing chemical solutions through the column. Anion exchange columns operated in three stages. In the first stage, loading, the feed stream was passed through the resin bed and selected radioisotopes were transferred into the resin. Second, during a phase called scrubbing, a chemical solution was passed through the column to remove unwanted radioisotopes from the resin. In the elution phase, a chemical solution was passed through the resin removing the target radioisotope and carrying it along in the solution. Anion exchange columns typically required regular chemical treatments to restore the absorption capacity of resins." **- Cartmell 1959**

Nuclear waste was feed for the fission products processing. The fission products processing operations at B Plant received feed from underground storage tanks and from the PUREX Plant. The feed arrived as liquid, with some suspended semi-solids of strontium-90. A centrifuge was used to remove strontium-90 from the cesium-137 and the remainder of the liquid. Suspended strontium-90 stuck to the side of the bowl as cake product. The supernate, which still contained strontium-90 and cesium-137, became feed for a precipitation process. In this precipitation process, phosphotungstic acid was added to the supernate. Approximately 98 percent of the cesium-137 precipitates or turns to a semi-solid. In turn, this solution was jet transferred to a spinning centrifuge. The supernate from this step contained the remaining strontium-90. The cake was cesium-137.

Purification of strontium-90 took place through solvent extraction. Strontium-90 feed for this process was the supernate from the above partitioning step. Four pulsed columns, 1A, 1S, 1B, 1C, alternately extracted and stripped strontium-90 from the remainder of fission products in the feed solution. The feed entered the 1A Column as an aqueous stream and the strontium-90 was extracted into an organic stream consisting of di (2-ethylhexl) phosphoric acid with a tri-butyl phosphate modifier and normal paraffin hydrocarbon diluent (RHO 1979b, p. 6-4). The aqueous stream continued down the pulsed column with some cesium-137 and other fission products and was sent to underground waste tanks. Strontium-90 was stripped into an aqueous stream in the 1B Column. The strontium-90 was then removed from the aqueous and made ready for isolation and eventual encapsulation.

After cesium-137 cake was removed from the centrifuge, purification of cesium-137 began in a 2,000 gallon ion exchange column (RHO 1979b, 6-4). Cesium-137 and some sodium were loaded into the resin bed in the column. Next, a dilute ammonium carbonate solution was passed through the resin bed to scrub out the sodium. In the eluting step, concentrated ammonium carbonate was passed through the resin bed. Eluting carries the cesium-137 with the solution. The cesium-137 and strontium-90 were then transferred to the 225-B Waste Encapsulation Storage Facility. In the 225-B facility, operators converted cesium-137 and strontium-90 to a solid, encapsulated the radioisotopes, and placed them into storage. For information on the 225-B Facility, see the Expanded Historic Property Inventory Form in Appendix B on the Internet.

TECHNOLOGICAL CHALLENGES

Chemical separations posed a number of challenges for Hanford Site management, engineers, and operators. One of these challenges, the pressure to continually maximize the efficiency of production, is reflected quantitatively in the output figures. Figure 1.7 in Chapter 1 shows the plutonium produced from 1947-1987. However, the qualitative aspects of the pressure to produce are less tangible. Qualitative aspects refer to social experiences in the past that were not often recorded. The three areas of technological challenge discussed below were chosen because they reflect both quantitative and qualitative aspects of responses to the pressure to produce plutonium during the Manhattan Project and the Cold War.

RADIATION EXPOSURE TO WORKERS ON CANYON DECKS

The canyon deck was where workers received the most exposure in the original bismuth phosphate plants (T and B plants). Figure 2-4.7 shows the canyon deck. Samplers were required to be on the canyon deck to take samples of process solutions from risers on the canyon deck. Over-exposure while taking these samples was a job hazard for the samplers.

This exposure was eliminated in the REDOX and PUREX plants because they had remote sampling stations. This meant that samplers no longer had to go onto the canyon deck to take samples. Because workers were not anticipated on the decks of the REDOX and PUREX plants during normal operations, the canyon decks were not shielded from and so generated high electromagnetic and alpha radiation levels. As a result, the canyon decks of both facilities were much more radioactive than those of T and B plants.

Figure 2-4.18. *Worker Taking a Sample of Radioactive Solutions*

Russ Knight (a chemical separations technician, 1945-1989) (Knight 1999), Joe Soldat (a health physicist, 1948-1994) (Soldat 1999), and Roger Hultgren (chemical separations technician 1944-1989) (Hultgren 1999b) discussed the differences between taking process samples from the canyon deck and from remote stations.

Russ Knight: *"The process that was going on in the canyon was a remote process. It was controlled in the facility as to what the intent was to be in that facility. But in order to understand [what] they really had going on they had to take samples, and those samples were brought into the laboratory. We did radioanalysis on the materials to assure them that they were getting the product that they were looking for at that time, which was plutonium, in the specific phase that they wanted it, in the process.*

The 222 B and T Laboratories were direct support facilities for the B Plant and T Plant chemical processing, which was bismuth phosphate batch processing... The reason I mention the importance of batch processing is that in order to know what position the materials were in the plant, they would sample [see Figure 2-4.18] and bring those samples to the laboratory. The laboratory would provide the analysis back to operations so that they knew exactly where they were and if they needed to do anything additional [which] was required at a given point. That is basically what the process control laboratory emphasizes. As the process continues, it needs to be monitored, and the laboratory provided the monitoring services based on the samples that were brought to us.

A true evaluation with what was going on in the process would indicate that 1) we knew what product we had and where it was in the process, and 2) what was going off in the waste streams to ensure that we were not putting any product out that we were in so dire need of.

The T, B, and U buildings all required sampling to take place on deck. Samplers had to enter the canyon building and that was why these controls were placed on the doors. The dispatcher's job was to ensure that safety procedures were followed for entry into the canyon."

Joe Soldat: In *"1949 at T Plant , I took a reading and turned around and left. They were taking samples out and putting them on diaper paper on the deck. They said, "we want to go in there and do contact maintenance." They had this labyrinth you went into and you call the operator, he unlocked the door."*

Brian Freer: *"This is on the canyon deck at T Plant?"*

Joe Soldat: *"Yes. And I stuck the meter around the corner, it went off scale, I turned around and left. I said, 'forget it.'"*

Russ Knight: *"To enter the canyon, samplers had to put on a breathing mask, it was called an assault mask then. It had a canister on it to absorb vapors and particulates. Usually the samplers brought their own masks over from 221, because they would suit up in the 271 and then go to the 222 building.*

They would go in, go to the specific cell, and engage the samplers into the sample port. In the sampling team there was a health physics person who monitored for dose rate of the samplers, both for entry and for unusual conditions. If they encountered an unusual condition they would exit the canyon and alert the dispatcher that there was something unusual and this would be corrected before they would proceed. Under normal circumstances they would proceed to the position of requested sample.

The sample ports are installed between cells on the canyon deck. In between the cell blocks there is a sampler for each cell. It is recessed into the concrete and has a cover shield over it. It has a small pipe coming up with a cap. They had tools for removing the cover plate, they had tools for removing the cap from the cover plate, also had tools for operating the valves. Before they would actually take the sample they would turn the sampler on, and circulate out of the tank up into this small sampler cup and that was a free flow for 10 minutes to ensure they got an ambient mixture for the sample.

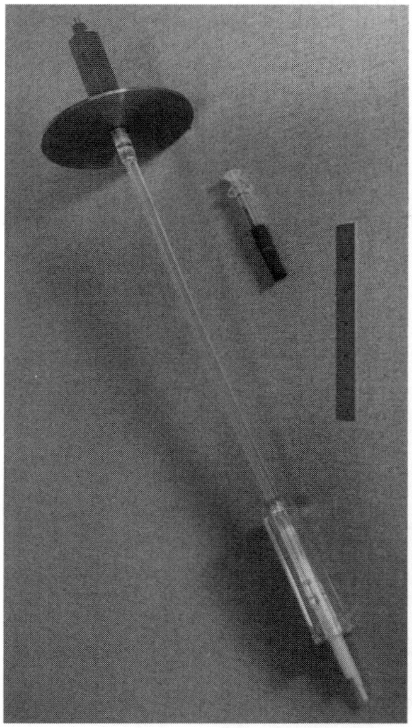

Figure 2-4.19. Trombone Sampler Used with the Doorstop Sample Carrier in T and B Plants

Exposures to personnel were a concern in the canyons and as we improved our ability for processing, getting away from bismuth phosphate batch into continuous extraction processing. Two new plants came on line, first REDOX. The design of the plant was unique because we reduced the exposure to personnel by not having them go directly onto the canyon deck to take samples. Instead, sampling took place in sample galleries. The sample gallery was a much better approach to sampling because of adequate shielding and because of the type of sampling that we did. Instead of the

Figure 2-4.20. Doorstop Sample Carrier Used in T and B Plants

trombone sampler [see Figure 2-4.19] and doorstop sample carrier [see Figure 2-4.20], we used a Gilmont sampler. A Gilmont sampler gave us a direct aliquot of sample. This eliminated double handling. We had the dilution made up, we would send the dilution flask with the sampler and they would put the aliquot directly into the sample vile and it was diluted at a safer level to work at.

REDOX was unique because it had a north and a south sample gallery. The north sample gallery was the one that was primarily used on the plutonium extraction side and the south sample gallery was used for the uranium operation. The sampling gallery was designed similar to the operating gallery except that it was designed for sampling only. You had sample boxes made up and shielded at stations. Sample ports and sample identifications indicated the process.

This was safer for two reasons. One, the volume of samples was small and dilute. The radiation levels were much reduced. The sample carriers were smaller and not as heavy to carry. This was a major improvement over the bismuth phosphate process. Many times today they talk ALARA, as low as reasonably achievable. We were ALARA conscious in the early days as well."

Roger Hultgren: *"Thinking back to what we learned over the years. One of these things relates to the exposures that people were receiving doing the sampling at T and B Plants. About 1954 I left the B and T Plants and went into scope and design for PUREX. At that time, we discussed the merits of, from a design standpoint, how do you get away from exposure?*

What actually occurred was this sample gallery. It was parallel to the canyon itself. The samplers could actually walk, they would have to wear clothing for going into the canyon but all the samplers were in this particular sample gallery. The samplers could actually take a sample in a hood, remotely. We did not use doorstops there, except on rare occasions. It was all pipettes through small aliquot samples. So the exposure to samplers was reduced in PUREX."

REDOX Ruthenium Problem

The REDOX Plant commenced plutonium production operations in 1952 (AEC 1951b). Right after operations began, process problems developed with the solvent, hexone, and its inability to remove an isotope of the element ruthenium.

George Backman (personnel protection supervisor for chemical separations at the Hanford Site, 1948-1988) (Backman 1999), John Jech (health physicist at the Hanford Site, 1950-1988) (Jech 1999), and Joe Soldat (a health physicist, 1948-1994) (Soldat 1999) discussed the REDOX Plant during the time when this problem occurred. They described their experiences.

George Backman: *"One of the major problems they had occurred right after operations began. The solvent extraction material was hexone. It did a pretty good job except it didn't take out the ruthenium as well as they wanted. So they had a special cell that was used after they dissolved the fuel, they sent the solution there and oxidized with potassium permanganate in there and it volatized the ruthenium into RUO4 [ruthenium- tetraoxide], which is a gas. Then they ran it through a scrubber and then they routed that line directly out to the stack. They ended up with failures in this caustic scrubber system."*

John Jech: *"That was a J-1 system."*

Joe Soldat: *"Yes, the caustic scrubber failed."*

George Backman: *"Consequently, once that failed the ruthenium tetraoxide was free to go out the stack, which it did."*

Joe Soldat: *"It was one of those things that seemed to always happen on a holiday. It was New Years 1954."*

George Backman: *"That was one of the bad times, 400 curies."*

Joe Soldat: *"We had guys trying to pick up specks all over the project. There ended up being one big release and two smaller ones, about what you said George. The first release was either 300 or 360 but there was another 400. When the ruthenium was going up an aluminum nitrate would go out when they dissolved the cans off the fuel elements. Ruthenium was occluded, which means inside of aluminum nitrate flakes."*

Brian Freer: *"For those who weren't there, what did these flakes look like?"*

Joe Soldat: *"They looked like snow."*

George Backman: *"There were some big ones and some tiny ones. A lot of them you couldn't even see. But the size had nothing to do with the amount of radiation. Sometimes, the ones you couldn't see would be hotter than the ones you could see, one couldn't tell by just looking."*

Joe Soldat: *"What they did to help solve the problem of this continuing slough off from inside the stack, they got a crane and installed a wash ring at the top of the stack – a big round pipe with holes in the bottom and nozzles. They hoisted it over the top of the stack and sprayed it down inside so that it would wash the lining out and into the pit sump and then it was pumped into a tank somewhere."*

George Backman: *"They had to use an extra crane with an extension to prop up the first one because the stack was too high to install the washer. The second one had to lift up the extension until the angle didn't have so much strain. They pretty much solved that problem in that manner. They ran that stream through the sand filter. We could find, from taking measurements from this filter, ruthenium from this stream.*

The stuff that hit the top of the stack as RUO4 [ruthenium-tetraoxide] changed to RUO2 [ruthenium-dioxide] and it was so small that it almost went like the wind. It made these big long streaks."

Joe Soldat: *"We found it on the roof of the 100 D building."*

John Jech: *"All the way to Spokane."*

George Backman: *"I remember one big streak went right across to 100 B."*

Joe Soldat: *"I have got a map that shows the path."*

George Backman: *"That is why they had this big cleanup area around B Plant and T Plant, mainly B Plant, because of this."*

Joe Soldat: *"July 4th weekend all the monitors were out there picking up particles at the PUREX construction site because the contractor had a no radiation zone contract and here is all this screaming hot stuff."*

Brian Freer: *"What were the occupational health and safety concerns with this problem."*

Joe Soldat: *"They posted all of the 200 Areas and some of the roads. [The signs said] "stay on the road, don't go in the field." The roads were washed off or blown off or the wind took them off, but the field."*

John Jech: *"The Fire Department also helped here."*

George Backman: *"And it washed off pretty good."*

Joe Soldat: *"I monitored people from transportation that put up the signs, they would go into the fields to put up these signs."*

Brian Freer: *"Were these called speck hunts?"*

John Jech: *"Even before this there were many speck hunts."*

George Backman: *"This was a special one."*

Joe Soldat: *"We learned from the old ones."*

CRANE MAINTENANCE

> ***"REDOX could not run without the crane. The crane loads from the railroad. There is a railroad and they have to lift out the fuel and put it into dissolvers – without the crane you are dead. You just can't do anything."*** **- George Backman, personnel protection supervisor for chemical separations at the Hanford Site, 1948-1988**

Many operations and radiation protection personnel who worked at the REDOX Plant during the early to mid 1950s recall it as an intense time. The REDOX Plant was in its heyday as a plutonium production workhorse. The pressure to maintain plutonium production at the REDOX Plant is illustrated by challenges that arose when it came to servicing and repairing the bridge crane. George Backman (personnel protection supervisor for chemical separations at the Hanford Site, 1948-1988) (Backman 1999) described the problem.

George Backman: The crane maintenance problem was *"related to the ruthenium problem. ...when they opened up H Cell, ruthenium gas would come out, particles and all. [This was] very hot and it would contaminate the crane badly. The inside [crane cab control area] was still all right. [Management] decided one time that they were going to clean up that crane. We used to have electricians lined up from all over the plant to go in and work for a few minutes fixing any crane problems that we had.*

It was the contamination on the outside of the crane. One day they decided that they wanted to steam clean the crane. They decided, "By golly, we are going to clean this crane up." They got up there and burned all the motors out. So they had to take all the motors off. Then they took them all down into the pipe gallery to re-wind them. And there was all kinds of pressure, at that time, from the military to get the plutonium out. This was 1954. That was a real mean and wild time."

After this incident, the crane was routinely decontaminated by spraying it with a non-electrolyte organic solution.

AREAS FOR FURTHER RESEARCH

Chemical separations posed a number of challenges for Hanford Site management, engineers, and operators. Despite that chemical separations was a newly discovered field, all workers were continually under pressure to chemically separate out more and more plutonium. To address both the production demands and subsequent waste disposal problems, all involved thought outside of the box by developing processes to mine plutonium from the waste. Of the four steps in the plutonium production process, chemical separations created the most and the worst waste. Because this narrative is a history of facilities and not intended to be the definitive history of chemical separations at the Hanford Site, neither could we fully explore the successes and failures of mining plutonium from the waste nor the correlation between the waste produced, how it was stored, and the environmental and health consequences, nor numerous other aspects of interest. Researchers investigating chemical separations at the Hanford Site in the future have numerous avenues to travel, a few of which are:

- Correlating the chemical separations production process to the waste produced and that to the disposition of the waste and environmental and health consequences. The Hanford Environmental Dose Reconstruction study of emissions from the chemical separations plants and radionuclide pathways and the subsequent health studies contributed to our knowledge of environmental and health consequences, but uncovering information, such as whether any attempts were made to reduce waste by changing the chemical separations process, has yet to be done.
- Investigating what was done with the desorbed materials from filters and silver reactor
- Seeking more information on the technical problems PUREX experienced in the 1980s that resulted in an 11-year shutdown because of it inability to comply with environmental and safety measures
- Gathering a better understanding of the effect the chemical separations process developed at the Hanford Site had on chemical separations processes in the industry and other countries

SECTION 5

PLUTONIUM FINISHING

M.S. GERBER
FLUOR HANFORD, INC.

SECTION 5 - PLUTONIUM FINISHING

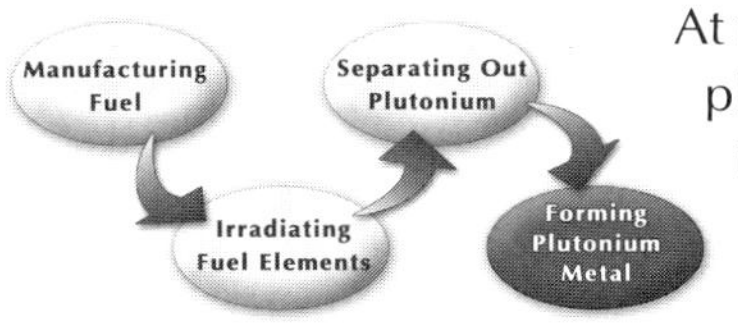

At the end of the chemical separations process, the plutonium was in the form of plutonium nitrate. This nitrate solution had to be converted into plutonium metal for use in weapons. This process of converting plutonium nitrate into plutonium metal is called plutonium finishing and was the fourth step in the process.

Plutonium finishing began at the Hanford Site on January 16, 1945 (Du Pont 1946a, Section 12) in the 231-Z Plutonium Isolation Building located in the 200 Area (GE [sic Du Pont] 1944, Section C). The 231-Z Plutonium Isolation Building was originally named the 231-W Building. When the plutonium arrived at the 231-Z Building for finishing, it was in a nitrate solution. Nuclear operators further purified the plutonium by adding hydrogen peroxide to 8-gallon batches, which separated the plutonium from the nitrate solution. The code name for the resulting plutonium was the "AT solution." Through World War II and until July 1949, the rest of the plutonium finishing process at the Hanford Site consisted of drying the plutonium and shipping it in containers to the Los Alamos Site in New Mexico. At the Los Alamos Site, it was made into metallic plutonium and then formed for use in nuclear weapons (Du Pont 1946a, Section 12).

When World War II ended and the rush for immediate production slowed, the U.S. Army Corps of Engineers, who managed both the Hanford and Los Alamos sites, began to investigate process improvements and production efficiencies. In 1946, they asked the new operating contractor at the Hanford Site, the General Electric Company, to design a new facility that could handle the entire plutonium finishing process. The process consisted of converting plutonium nitrate first into oxalate, then oxide, next into a metallic form for use in atomic weapons. With this conversion, the plutonium would then be in a safer and more stable form to ship the long distances to the weapons assembly facilities (GE 1953j, Streid 1949).

Figure 2-5.1. *Plutonium Finishing Plant Complex, 1949*

On January 1, 1947, the Hanford Site, along with the other federal atomic defense production sites, passed to the control of the newly created Atomic Energy Commission. The Atomic Energy Commission embarked on a large expansion program in response to projected military needs and new tensions in the Cold War. As a key part of the expansion program at the Hanford Site, the Atomic Energy Commission accelerated design and construction of the Plutonium Finishing Plant (Hewlett and Duncan 1969, Streid 1949). The code name for the plant was MJ-2. A description of the Plutonium Finishing Plant can be found in the Expanded Historic Property Inventory Form for the 234-5Z Building in Appendix B on the Internet.

CONTENTS

Divider Page Photo: Plutonium Button
(Neg. 88080280-6cn)

The Plutonium Finishing Plant commenced operations with radioactive materials on July 5, 1949. Figure 2-5.1 shows the Plutonium Finishing Plant the year it began production. For many years after that time, however, the Hanford Site continued to ship small amounts of plutonium nitrate to the Los Alamos Site for use in special research projects. A remote, mechanized system was incorporated in 1952. The Plutonium Finishing Plant made weapons-grade plutonium until May 1989, with some interruptions in service for upgrades and modifications. The facility also processed fuel-grade plutonium for use in power reactors from the early 1960s through the 1970s. Significant events in the history of plutonium finishing at the Hanford Site may be seen in Table 2-5.1. For a definitive description of the history of the Plutonium Finishing Plant, see Gerber (1997).

Table 2-5.1. *Significant Events: Plutonium Finishing*

Date	Event	Project No.	Reference
Aug 1942	**U.S. Corps of Engineers Manhattan Engineer District**		
Dec 21, 1942	E.I. Du Pont de Nemours - Site Contractor for plutonium manufacturing operations including plutonium finishing		
Jan 16, 1945	Receipt of plutonium nitrate in 231-Z Building to begin finishing operations		Du Pont 1946a, Section 12; GE 1944, Section C
Sep 1, 1946	General Electric Company - Site Contractor for plutonium manufacturing operations including plutonium finishing		
Jan 1, 1947	**Atomic Energy Commission**		
Jul 5, 1949	Start up of Rubber Glove (RG) Line in the newly completed Plutonium Finishing Plant (Z Plant)	Project 234-5	Work 1948, Bell 1952, GE 1953j
Mar 18, 1952	Startup of the Remote Mechanical A (RMA) Line in the Plutonium Finishing Plant (Z Plant)	Project C-198 combined with C-413	GE 1952e, Stark 1952
1954-1955	Installation of Task I in the RMA Line and Expansion of the 234-5Z Building	Projects CG-549 and CG-551	GE 1955d, 1955f
1956	Moving chemical processing work from the 231-Z Building to the Plutonium Finishing Plant		GE 1953g, Thomas 1956
1957	Combination of Tasks I and II in the RMA Line	Project CG-691	Ketzlach 1955b, Fritz 1955, Baker 1957
Oct 1960	Startup of the Remote Mechanical C (RMC) Line	Projects CG-734 and CG-745	Fritz 1958; Swain 1959; GE 1960i, 1960j
May 1964	Startup of the Plutonium Reclamation Facility (236-Z Building)	Project CAC-880	Low 1967
Jan 1, 1966	Isochem, Inc. - Contractor for chemical processing facilities including plutonium finishing		GE 1965b, 1965c, 1966
Sep 5, 1967	Atlantic Richfield Hanford Company - Contractor for chemical processing facilities including plutonium finishing		Isochem 1967, ARHCO 1967a
1975	**Energy Research and Development Agency**		
1977	**U.S. Department of Energy**		
Jul 1, 1977	Rockwell Hanford Operations - Contractor for chemical processing facilities, including plutoniumfinishing		
Jul 1987	Westinghouse Hanford Company - Site Contractor, including plutonium finishing		
May 1989	Shutdown of Plutonium Finishing Plant		

Design, Layout, and Construction

The design and layout for the new plutonium finishing facility at the Hanford Site were based on the facilities at the Los Alamos Site (GE 1954j) with notable improvements. During World War II, the plutonium finishing operators at the Hanford and Los Alamos sites used rubber gloves inside a protective box to manually manipulate the plutonium. In conjunction with designing a new plutonium finishing facility, General Electric soon decided that to meet the health and safety standards as well as to process the increased amounts of plutonium that were forecast a remote, mechanized operation would be necessary. The mechanized system would also have one containment barrier more than that used at the Los Alamos Site (GE 1947b, Bell 1952). Other technical challenges that affected the design, layout, and construction of the new plutonium finishing facility involved the use of slightly enriched irradiated uranium rather than the strictly natural uranium used during World War II and working with a plutonium nitrate resulting from evolving chemical separations processes.

During the years the Plutonium Finishing Plant was being designed and built, the United States responded to early Cold War events by increasing plutonium production. Two methods of increasing production included placing slightly enriched uranium in Hanford's production reactors and developing the REDOX (reduction oxidation) chemical separations process to replace the bismuth phosphate process that had been used in Hanford's T Plant and B Plant during World War II. Like most building projects, the original design for the Plutonium Finishing Plant continually changed as planners strove to make the process more productive and safe. Such changes began with the selection of the location.

Throughout the initial design period, planners of the Plutonium Finishing Plant assumed that the facility would be located in an area of the Hanford Site known as the 400 Area (originally planned near the 200 West Area and not the same location as the present 400 Area). Hence, they often referred to the feasibility examination effort as the "400 Area Study." However in August 1947, to be efficient and economical with the use of existing utility and support facilities, General Electric recommended that the new structure be located within the fence of the 231-Z Plutonium Isolation Building. They realized that existing boiler, water filtration, electric, telephone, warehouse, road and guard facilities would have to be expanded to support the new plant, but they decided that this course would be cheaper and faster than building an entirely new support infrastructure. Because the new facility would house the very last steps in plutonium metal production, Hanford personnel began calling it the Z Plant, the plant at the end of the line (Streid 1949, GE 1947c).

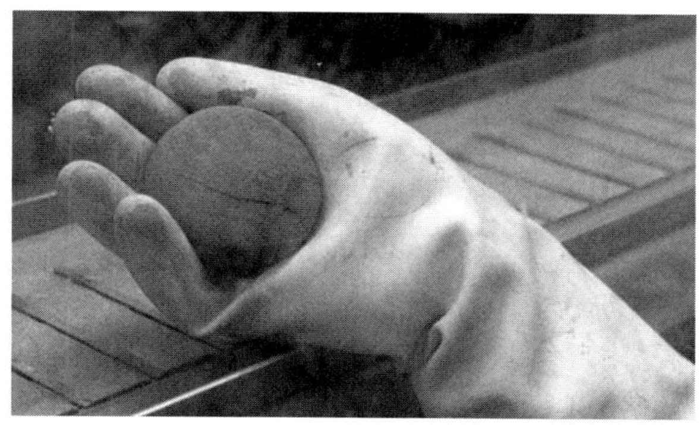

Figure 2-5.2. *Plutonium Button Produced on the Remote Mechanical C Line, 1988*

Along with the selection of a location, General Electric also established the layout of the process, parameters, and procedures for the analytical and developmental laboratories, waste disposal practices, ventilation needs, temperature cycles for the plutonium finishing processes, and various other operating boundaries. Parts I and II were terms used to denote the two key phases in the plutonium finishing process. Part I was the production of plutonium buttons—metal chunks of plutonium (see Figure 2-5.2). Part II was fabrication of weapons components by casting, grinding and lathing the plutonium into pits. General Electric initially planned to place Parts I and II of the plutonium finishing operations in separate portions (sections 234 and 235) of the new building. However in November 1947, General Electric planners decided to combine Parts I and II into one line to keep the plutonium confined within the glove boxes (then known as dry boxes or hoods) at all times. When they combined the lines, they began to designate the structure as the 234-5Z Building. Hanford workers soon coined the vernacular name "Dash 5" for the facility (Work 1947a, GE 1953j, Prudich 1947, GE 1947d).

Another aspect of the layout that deviated from the original design was the retreat to a manual instead of mechanized process. In February 1948, the Atomic Energy Commission communicated its need to further accelerate the Plutonium Finishing Plant schedule. As a result, the planners decided to install a manually operated process similar to the one used at the Los Alamos Site during World War II on a rush basis, while continuing to design the mechanized, remote process. Because in the manual process operators handled the plutonium through ports in the glove boxes outfitted with rubber gloves, Hanford planners named it the Rubber Glove (RG) Line. Designers then designated the RG Line, along with the basic service structures and facilities essential to operate it, as Phase I of the 234-5 Project (Work 1948, Bell 1952, GE 1953j).

In June 1948, the construction crew poured the foundations and footings for the 234-5Z Building (Streid 1949) and had the bulk of the building ready to turn over to the operations crew by the end of April 1949. By mid-May, twenty-seven operators were on staff. Operations personnel conducted simulated runs to test process and laboratory equipment and practiced moving material in and out of gloveboxes using electronically sealed plastic bags (GE 1949a, p.4).

According to a technical contractor report, the 234-5Z Building was designed to be a "self-contained and independent plutonium fabrication plant with all necessary supporting services and facilities" (Streid 1949, p. 12). The Atomic Energy Commission had called out a key specification stating that the chemical and metallurgical processes in use at Los Alamos be used. However, a main difference from Los Alamos was that at the Hanford Site the entire manufacturing process was carried out within enclosed glove boxes. At Los Alamos, some steps of the process at that time were not executed in completely enclosed glove boxes, and not all of the glove boxes in the process line were interconnected. The initial RG Line in the 234-5Z Building had three separate air zones for contamination control. Zone 2 was the air in the operating areas. Zone 3 was air in the equipment areas. Zone 4 was inside the glove boxes. Zone 1 was the name reserved for air in the office area. The facility operated such that air from all of the contaminated zones was drawn to a common air exhaust manifold, passed through Chemical Warfare Service filters, and discharged out the 200-foot high 291-Z Stack at a velocity up to 30 miles per hour (Streid 1949, Work 1947a, Christensen and Maraman 1969).

In addition to the 234-5Z Building itself, Phase I of the construction of the Plutonium Finishing Plant built the 2704-Z Office Building, the 2705-Z Temporary Technical Office Building, the 2719-Z First Aid Building, the 291-Z Stack, the 2901-Z Elevated Water Storage Tank, 216-Z-1 and 216-Z-2 waste cribs, the 241-Z Waste Storage Facility, an exclusion area badge house, a gas bottle storage building, and the 272-Z shop and warehouse. The project also extended the 284-W Boiler House to the east with the addition of a boiler, combustion control equipment, coal conveyor extension, and stack interface. Likewise, it extended the 283-W Filtered Water Plant to the east with the addition of another settling basin, pumps, and piping. (Plock 1949, Work 1947b, GE 1953j).

Approximately 180 staff began processing plutonium nitrate in the 234-5Z Building on July 5, 1949 at 11:00 p.m. using the RG Line. At the same time, the General Electric Laboratory in Schenectady, New York was working on the design of Phases II and III of the 234-5 Project. Phases II and III together were known as the 432 Project, a reversal of numbers for security purposes. Phase II consisted of completing some rooms for the analytical laboratory, a crucible shop, a chemical development laboratory, maintenance shop "D" (a shop for repair of contaminated equipment and tools), instrument repair rooms, and two maintenance glove boxes. Additionally, Phase II included preparation work for the installation of the remote mechanical processing line (GE 1953j).

The Remote Mechanical (RM) Line itself was the major component of Phase III. In September 1949, the Atomic Energy Commission specified that the RM Line would need to handle the output of REDOX (the reduction oxidation facility - a new chemical separations plant being constructed at the Hanford Site). Phase III of the 234-5 Project also included the completion of a mechanical development shop, men's locker room, a special products shop, a recovery area for the re-work of scrap and out-of-specification metals and oxides (see the Recovery and Recycling Sub-Section), and miscellaneous rooms (Gross and Greninger 1949).

On October 24, 1949 in response to events in Korea, the Atomic Energy Commission issued dramatic calls for increased production in the RG Line and authorized construction of two more RM lines within the Plutonium Finishing Plant, along with a building addition to house the third line, even though the first remote mechanical line had not started production. The second RM Line would be located on the first floor of the 234-5Z Building at the west end of the south-center corridor (GE 1950d, GE 1951d). At that time, plutonium finishing workers began calling the first RM Line the RMA Line, and they termed the second RM Line the RMB Line. The Atomic Energy Commission soon dropped the third RM Line from the work scope along with the building addition to house it. Years later in 1957, a third, different RM Line was authorized.

On April 30, 1951, the original Plutonium Finishing Plant construction project was closed out administratively, although none of the RM Lines were operational nor were the crucible shop, some of the laboratory rooms, and some other features of Phase III completed. The Atomic Energy Commission and General Electric combined the remaining work on the original project with the more recently authorized RMB Line project and incorporated the design changes for increased capacity, which had been developed at the General Electric Laboratory (Bell 1952, GE 1954j).

Finally on March 18, 1952, the RMA Line began to process the plutonium(IV) oxalate prepared in the 231-Z Building. The initial operations of the RMA Line performed all of the process steps in plutonium metal reduction and fabrication except for the first step. The first step, precipitating the plutonium-nitrate feed solution with oxalic acid and other agents (GE 1952e, Stark 1952), was still being done in the 231-Z Building. Additionally on May 17, 1952, cold shakedown tests began on the RMB Line (GE 1952h). Builders finished other Phase II and III facilities internal to the 234-5Z Building later in 1952. Figure 2-5.3 shows the layout of the Plutonium Finishing Plant as it was in 1995.

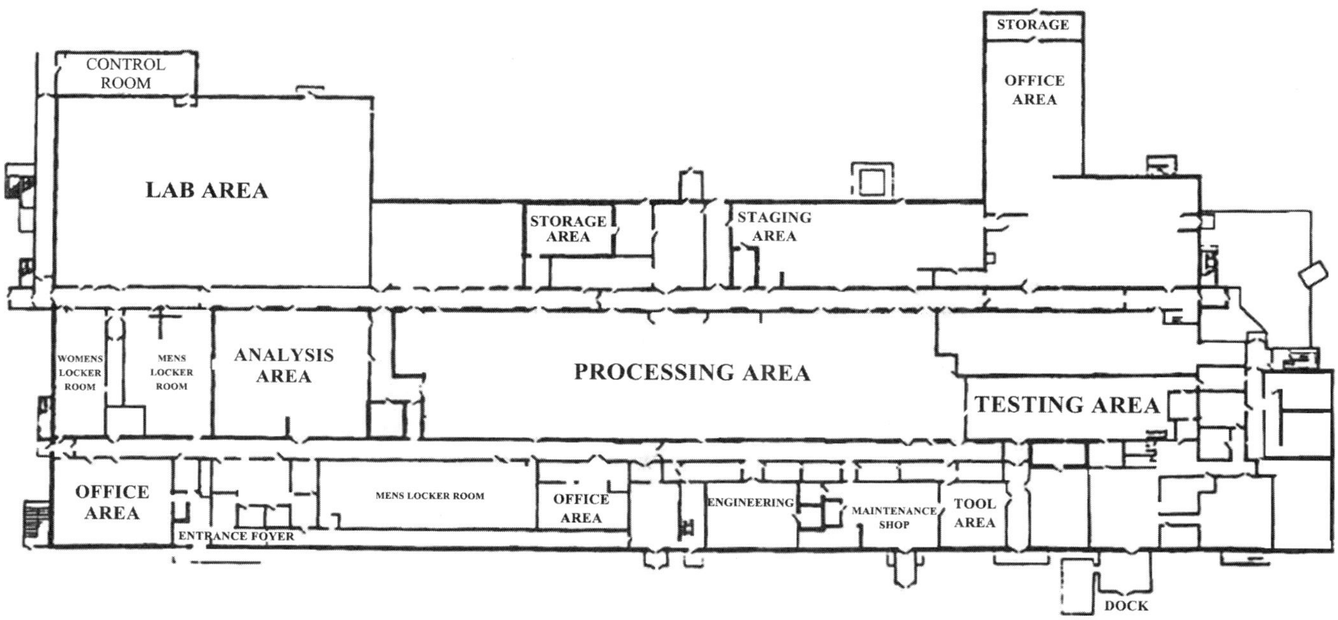

Figure 2-5.3. *Simplified View of the Plutonium Finishing Plant, 1995*

OPERATIONS

The history of plutonium finishing at the Hanford Site is one of continually being pressed to expand production during the Cold War era. The production capacity of the Plutonium Finishing Plant is reflected in the overall production of plutonium at the Hanford Site. As shown in Figure 1.7 ("Total Annual Production of Plutonium at the Hanford Site" in the Reactor Operations Section), post World War II production went from fewer than 500 kilograms per year in the late 1940s and early 1950s to around 4500 kilograms in the mid 1960s. To produce those amounts, operators and equipment in the Plutonium Finishing Plant had to keep pace with those in the reactors and chemical separations plants.

Key to keeping pace was the move from the manual to mechanical production lines. The year the RG Line began operations in 1949, the Hanford Site produced approximately 250 kilograms of plutonium. The year the first remote mechanical line (the RMA Line) began in 1952, Hanford Site production increased to approximately 750 kilograms and climbed steadily to over 3500 kilograms in 1959. The year the second and greatly improved remote mechanical line (the RMC Line) began in 1960, Hanford Site production increased to over 4000 kilograms, topping out at over 4500 kilograms in 1965.

As the Cold War subsided in the late 1960s, plutonium production switched from that used for weapons to that used for commercial reactors. For 10 years from 1972-1982, mostly fuel-grade plutonium was produced. Consistent with the defense build up of the 1980s, only weapon-grade plutonium was produced from 1983-1987. The final campaign took place from July 1988 through May 1989, making weapon-grade plutonium. Approximately 475 staff were employed at the Plutonium Finishing Plant during its last year.

With the demise of the Cold War and increased concern over the effect of contamination from the Hanford Site on human health and the environment, production of plutonium stopped at the Hanford Site. Beginning in 1989, the Hanford Site's mission turned from production to waste cleanup with the signing of the Hanford Federal Facility Agreement and Consent Order (Tri-Party Agreement - TPA), the cleanup pact between the U.S. Department of Energy (DOE), the U.S. Environmental Protection Agency, and the Washington Department of Ecology (Ecology 1994). Also at that time, the Cold War ended and DOE began to develop final plans to achieve cleanout and shutdown of most of the Plutonium Finishing Plant. Ironically, the Plutonium Finishing Plant's highest staffing periods came during the cleanout and shutdown years of the early 1990s when about 600 people were employed in the complex.

Staff at the Plutonium Finishing Plant were basically three types. The supervisors were typically chemical engineers. The workers were mostly health physics technicians (sometimes called radiation control technicians) and nuclear operators with nuclear training and experience. Other workers were carpenters, pipe fitters, and those with skills needed to maintain any operating plant.

Rubber Glove (RG) Line

Operators began to process plutonium on the RG Line July 5, 1949 on a three-shift schedule, Mondays through Fridays (Bell 1950). The initial RG Line was composed of 28 stainless steel glove boxes, measuring 180 feet in total length. Plutonium handling activities took place in Zones 2, 3, and 4 with Zone 4 (inside the glove boxes) having the most negative pressure for contamination control. Metal plates extended from the floor to the glove boxes and then from the glove boxes to the ceiling, separating Zones 2 and 3 so as to minimize cross-contamination from high-hazard to low-hazard activities (Streid 1949, Work 1947a).

The basic plutonium finishing operations carried out in the RG Line consisted of several standard steps known as tasks. Tasks I through VIII are shown in Table 2-5.2. Tasks IX through XIV were identified as topics not actual process steps. Instrumentation was Task IX, Control was Task X, Ventilation was Task XI, The Conveyor System was Task XII, Maintenance of equipment was Task XIII, and Sampling was Task XIV (Gerber 1996). A flow chart of Tasks I through III is shown in Figure 2-5.4.

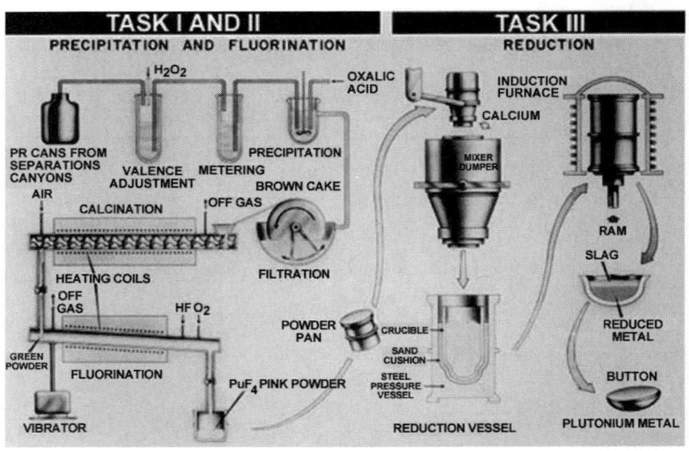

Figure 2-5.4. *First Three Tasks of the Plutonium Finishing Process*

Soon after processing began, General Electric reported that "operational and mechanical difficulties...did not permit the forecast July production schedules to be met...Innumerable leaks were found in the [vacuum] system" (GE 1949b, pp. 39, 52). By autumn, the contractor had found "a number of design deficiencies...that contributed to a production capacity bottleneck" (Gross and Greninger 1949, p. 2). General Electric concluded that the "air locks...were wholly inadequate" (Bell 1950, p.15). Indeed, the RG Line did not meet the Atomic Energy Commission production schedules for the remainder of 1949 (GE 1949c, 1949d, 1949e, 1949f). In December, the Atomic Energy Commission relieved the facility of production commitments to make necessary repairs (GE 1950c).

Table 2-5.2. *Tasks in Plutonium Finishing Operations*

Task	Name	Function
I	Purification, Oxalate Precipitation, Wet Chemistry, or Feed Preparation	Precipitate the plutonium-nitrate feed solution with oxalic acid and other agents
II	Hydrofluorination or Dry Chemistry	Diffuse hydrogen fluoride gas through the precipitate
III	Reduction	Combine the plutonium tetrafluoride with calcium; fire at high temperature until fused into a plutonium metal chunk called a button
IV	Casting	Render plutonium buttons into pits
V	Machining	Machine the pit to specified dimensions
VI	Cleaning	Chemically cleanse the pit
VII	Coating	Coat the pit for radiation safety
VIII	Final Inspection	Inspect the pit for correct specifications

Although the technology had been tested on the rubber glove line at the Los Alamos Site, their volume was much smaller than that at the Hanford Site. The operation of the RG Line at the Hanford Site was continually plagued with problems. Its history is one of solving problems by making improvements either to the process or the equipment.

Airborne contamination was a problem in the new building. Throughout 1949, the monitoring organization, the Health Instruments Division, obtained positive air samples in some of the processing rooms and in the analytical laboratory at high enough levels that the site standards required personnel to wear assault masks of the same type as those worn by the Army's Chemical Warfare Service. During the first year, engineers, chemists, and operators began work on the daunting problems in the new line. General Electric reported that "it was necessary on several occasions to suspend [operations] completely...and devote all efforts to decontamination work."

Although first hand accounts by process operators present in the 234-5Z Building for the startup of operations in July 1949 are not available, the building superintendent did commit to writing the opinion that the work force was very discouraged. The superintendent blamed the very rapid construction and startup period, driven by production quotas, for the "continual mechanical and process difficulties" encountered. His report, he stated, was "intended to serve not only as a review of 234-5Z operations during this [initial operations] period but also as a documentary evidence of the difficulties experienced and the manner in which they impeded startup." When operations began, he wrote, the "training program was not fully complete…[the] equipment….could hardly have been considered as functioning satisfactorily…[and the] operating procedure [was] lacking in detail in several respects" (Bell 1950, p. 2). Thus, he concluded, the continual equipment failures and contamination events "were to the dismay of many, but to the surprise of none" (Bell 1950, p. 7).

The many contamination incidents lowered production and dampened the morale of facility workers as the contractor acknowledged at year's end: *"The lost processing time, the hours of hard, unpleasant work under uncomfortable conditions, and the ever-present potential health hazard have had serious effect on the progress of this operation and on the morale of employees concerned."* - Bell 1950, p. 15

Throughout 1950, the vacuum systems on various glove boxes within the RG Line, especially the coating Glove Boxes 25 and 26, experienced recurring difficulties. Corrosion failures of parts within glove boxes likewise took place repeatedly (Hildreth 1950). Table 2-5.3 outlines the major modifications made to the RG Line.

Table 2-5.3. *Modifications to the Rubber Glove Line*

Date	Need/Problem	Modification	Reference
1951	Difficulties with vacuum systems	Operators installed a continuous gas analyzer detection system for the Zone 3 area as a safety measure in mid-year. As a result of safety problems, Plutonium Finishing Plant (PFP) management discontinued the practice of preparing buttons directly from lathe turnings late in the year.	
1951-1952	Meet production goals and augment safety	Workers installed a 100-line intercommunication phone system within the 234-5 Building, duct level floor work and safety showers, altered laboratory facilities to provide for Plutonium metallurgy studies in the building, and constructed new waste cribs. They replaced air lock doors, constructed the 711 Waste Storage Hut, designed fire protection equipment for the 234-5 Building filter room, emplaced a soundproof operator's booth for the 234-5 Building control panel, and installed a vacuum leak detection equipment for improved contamination control.	GE 1951e, 1952c, 1952d, 1952g, 1952j, 1952k, 1952m, 1952n; Schalliol and Wick 1951
July 1952		Rubber Glove (RG) Line process engineers decided to retire the RG Line's purification glove boxes (Hoods 5, 6 and 7) and to prepare all PFP feed material in the 231-Z Building. That same summer, they emplaced a new casting furnace in the RG Line and designed and fabricated a pit machining unit for the line.	GE 1952j
October 1952		In October, RG Line process engineers investigated improved drying procedures for Task II, including the use of heat lamps while pulling air through the oxalate before hydrofluorination.	Kerr 1952a
1952		Facility engineers undertook several process changes to improve the performance and expand the production. In December, they adopted a pressurized argon purge (displacing air from the crucible assembly in Task III by admitting argon under pressure), instead of a vacuum purge, as a standard operating procedure. Fabricators completed an experimental coating unit that had been started in 1951, and operators began testing a variety of new procedures to lower the high coating rejection rate being experienced at PFP.	Quinn 1952a, 1952b; Chapman and Mobley 1952; GE 1953d, 1952d, 1952i

In early 1951, the Atomic Energy Commission authorized multiple RG Line improvement projects, including installation of an additional coating unit for Glove Box 26, installation of auxiliary stainless steel ventilation ducts for Glove Boxes 4-8, fabrication of an experimental coating glove box, and installation of an additional casting unit for Glove Box 13 (Bell 1951a, 1951b). Plutonium Finishing Plant engineers also instituted several process changes in the RG Line to expand and improve production. During the summer and autumn, they continued efforts to standardize the chemical addition step in Task III so as to simplify operations and save money (Cooper 1951, Endow and Kendall 1951). However throughout 1951, operators of the RG Line continued to experience difficulties with the vacuum systems of several glove boxes.

During 1951 and 1952, the Atomic Energy Commission continued to call for increased production of weapon-grade plutonium. In November 1952, C Reactor, the fifth production reactor at the Hanford Site started up, and the Atomic Energy Commission notified the 200 Area facilities of which the Plutonium Finishing Plant was a part that they should plan to handle a major increase in reactor output (Carr 1958). As a result, plutonium finishing planners studied ways to increase 234-5Z Building capacity to meet the needs of the expansion (GE 1952m, 1952n).

REMOTE MECHANICAL A (RMA) LINE

The remote mechanical line was scheduled to start up in 1951. However, the Atomic Energy Commission had General Electric modify the line to increase production and to produce a second model pit in addition to the one for which the RMA Line was already designed (GE 1953l, 1953m). To accomplish this, plutonium finishing engineers made 46 pre-startup modifications to the RMA Line. Additionally, the Atomic Energy Commission directed that both the RG Line and the RMA Line acquire the capacity for "reconversion of the plutonium parts in stockpile to the required design for new weapons models" (GE 1954j, p. 11).

March 18, 1952, the RMA Line began to process plutonium from the 231-Z Building. At the end of that month, General Electric described operations as "successful...No serious equipment problems have been evident, although minor maintenance and revision were necessary" (GE 1952e, p. F1).

Tasks II through VIII of the RMA Line identified the same process steps and equipment as those in the RG Line. Tasks IX through XIV were added to the RMA Line to identify components that affected the entire line as opposed to a step in the process: instrumentation, control, ventilation, conveyor system, maintenance, and sampling (GE 1954j, Stark 1952).

The RMA Line at the Hanford Works consisted of a row of 30 interconnected stainless steel glove boxes, 30 control desks, 10 control cubicles, 24 instrument panels, 9 resistance furnaces, 5 induction furnaces, a sample can handling assembly, a 110-foot long general conveyor and manipulator, other smaller conveyors and furnace loaders, and miscellaneous support equipment. It was located in rooms 232, 233, 235-A1, 235-A2, and 235-A3 of the Plutonium Finishing Plant. A "mechanically unpierced solid barrier...made of steel...[with windows for]...observing operations in Zone 4 (the space in the hoods where actual handling of product occurs)" was a key feature of the line (GE 1954j, pp. 13, 18). The barrier separated Zone 2 (the operating side) from Zone 3 (the equipment side). Office space and corridors in the 234-5Z Building occupied Zone 1, the designation for the area with highest positive air pressure for purposes of contamination control. For a detailed description of the RMA Line, see the Expanded Historic Property Inventory Form for the 234-5Z Building in Appendix B on the Internet.

Production

As would be expected of a mechanical versus manual operation, the RMA Line out produced the RG Line. For example during the closing months of 1952, the RG Line's share of production hovered between 30-50 percent, and the RMA Line's share varied between 50-70 percent (GE 1952j, 1952l).

Throughout 1953, the Atomic Energy Commission issued major, overall expansion directives to begin construction of the PUREX (plutonium uranium extraction) Plant, the KE and KW jumbo reactors, five new laboratories, other significant facilities in the 300 Area, and Phase I of a multi-year REDOX Capacity Increase Project. Hanford workers were asked to substantially increase the plutonium production over that of 1952 (GE 1954g). The plutonium finishing operators fulfilled and often exceeded their production commitment to the Atomic Energy Commission with the RMA Line doing the vast majority of the processing, despite the continual problems (GE 1953e, 1953f, 1953g, 1953h, 1953i, 1953k, 1953m).

In early 1953, plutonium finishing planners decided not to modernize the RG Line as planned to achieve increased production but to modify the RMA Line instead. In the opening months of 1953, the RG Line processed only 5-6 percent of the total plutonium-nitrate sent to the 234-5Z Building and did no fabrication work at all (GE 1953e, 1953f, 1953g). At the end of March, the Atomic Energy Commission placed the RG Line in standby condition with the exception of the coating units (Glove Boxes 25 and 26). The coating glove boxes continued to operate during April and the Atomic Energy Commission ordered them into standby status in May. The RG Line conducted no further processing with the exception of 2 weeks of Task III operations (plutonium tetrafluoride reduction) in September when a serious contamination event closed parts of the RMA Line for decontamination activities (GE 1953i, 1953l). In 1955, plutonium finishing operators and crafts personnel removed and buried the equipment for Tasks I and II of the RG Line (GE 1955j). In an immediate move to increase production in the RMA Line, plutonium finishing engineers doubled the batch sizes in

Task III in March 1953 (Brown 1953, GE 1953g). By year's end, they were conducting 40 percent of the reductions as double-batches.

To allow for simultaneous fabrication of two different models of weapons components plutonium finishing engineers moved the primary machining glove box of the RMB Line (Glove Box 200-B) into the Zone 3 area of the RMA Line (GE 1953l). Throughout 1954-1956 as they tried more experiments and continued modifications in the RMA Line, plutonium finishing personnel realized that the RMB Line was becoming obsolete. In fact, it never processed any plutonium. Construction forces removed the RMB Line still unused when they installed the RMC Line in 1957-1959 (Travis 1958a).

September 1953, the Atomic Energy Commission raised the production quotas sharply as the Hanford Site began to ship plutonium buttons as well as plutonium nitrate to the new Atomic Energy Commission site (Rocky Flats) in Colorado. Defense production facilities at the Rocky Flats plant functioned solely to fabricate pits for weapons from metallic plutonium, and so they depended on a steady supply of buttons as feed material from the Hanford Site. A year later, however, the Hanford Site's new sister plant in South Carolina, the Savannah River Plant, began reactor operations, plutonium separations, and plutonium finishing. Finishing operations at the Savannah River Plant were confined to making plutonium buttons not pits for weapons. The Savannah River Plant began shipping buttons to the Rocky Flats Site autumn 1954, thus relieving some pressure on the Hanford Site's Plutonium Finishing Plant. However, the Atomic Energy Commission issued calls for further production increases, thus rendering this reprieve temporary (Bebbington 1990, p. 66).

Plutonium finishing workers fabricated weapons components for two different designs throughout 1954. Commitments to the Atomic Energy Commission for the production of buttons and finished weapons components were either met or exceeded for most of the year. One exception to the high production performance occurred in the weeks after July 27 when a serious fire of plutonium turnings in the RMA Line curtailed the fabrication of pits for 3 weeks. The fire occurred when plutonium metal turnings ignited while being removed from a machining hood. General Electric reported that the "entire rear area" of the RMA Line (air zone 3) was "grossly contaminated and all production operations were suspended" (GE 1954f, 1954h, 1954k, 1954l, 1954n, 1954o, 1954p). The same held true in 1955 when nuclear operators and engineers set new records for high plutonium output with the exception of the early months when RMA Line operations were closed for the installation of new equipment.

By 1956, nuclear operators and engineers produced pits for a variety of weapons models (GE 1956f). During 1957-1961, operators and engineers manufactured pits for 9 different weapons designs in the 234-5Z Building. In addition to producing new models, plutonium finishing operators also started processing plutonium nitrate from the new PUREX separations plant in February. In April, they then processed the last plutonium nitrate from T Plant (GE 1956h) and started accepting the first shipment of plutonium nitrate from the Savannah River Plant in July (GE 1956k).

Over the next several years, the pattern continued. As the Atomic Energy Commission expanded plutonium production capacities at the Hanford Site and across the nation, the plutonium finishing workers increased their production by continually improving the process and equipment. For example, a significant process improvement in 1957 was to combine Tasks I and II on the RMA Line (Baker 1957). The result was almost immediate. Plutonium Finishing Plant workers achieved the largest single leap in production in February 1958 when they manufactured by a large margin more plutonium metal than had been produced during any previous month in the history of the 234-5Z Building (GE 1958c). Management deemed the continuous Task I-Task II process so proficient they decided not to maintain the older, standard glove boxes for back-up (Harmon and Reas 1957, GE 1957h). By 1959, General Electric found the overall process and equipment improvements in the RMA Line's Tasks I and II resulted in a vast rise in plutonium metal production over 1958, and 1960 ended with an annual record for the Plutonium Finishing Plant. In January 1960, they again exceeded all previous monthly records for the production of buttons, and, in April 1960, they exceeded the January record itself (GE 1960d, 1960f).

Problems

The record setting production levels are most impressive when set against the background of equipment and process failures. The RMA Line, like the RG Line, was continually plagued with problems. Whereas the RG Line was tested technology, though, the RMA Line had ground for problems. It was the first large scale, remote mechanical system used in the world to finish plutonium. The history of RMA Line operations is basically one of learning from the problems, changing the process and equipment to address the problems, and applying what was learned to the next remote mechanical line design. For details of the modifications made to solve the problems, see Table 2-5.4.

Table 2-5.4. *Modifications to the Remote Mechanical A Line*

Date	Need/ Problem	Modification	Project No.	Reference
1952		The 234-5 Building personnel fabricated and began testing new mixing equipment for Task III of the Remote Mechanical A (RMA) Line. Additionally, to help improve performance, designers developed a new general conveyor for the line, as well as a new hood conveyor for Task III. Plutonium Finishing Plant (PFP) personnel also fabricated a new pit machining unit, emplaced additional waste gas disposal lines for Task II, and built a new chip cutter for the RMA Line. Late in the year, facility engineers authorized a new method of argon purging for Task III to allow equipment simplification and thus save wear and tear on some of the pumps and valves.		Benoliel 1952
1952		Facility engineers undertook several process changes to improve the performance and expand the production of both lines. In December, they adopted a pressurized argon purge (displacing air from the crucible assembly in Task III by admitting argon under pressure), instead of a vacuum purge, as a standard operating procedure. Fabricators completed an experimental coating unit that had been started in 1951, and operators began testing a variety of new procedures to lower the high coating rejection rate being experienced at PFP.		Quinn 1952a, 1952b; Chapman and Mobley 1952; GE 1952c, 1952h, 1953d
May 1953		General Electric Company engineers combined various design strategies for RMA improvements into two key projects. Project CG-549 focused on building a Task I process into the 234-5 Building, capable of bypassing the precipitation and purification activities then being carried out in the 231-Z Isolation Building. Project CG-551, entitled "Expansion of Building 234-5," focused on a sweeping series of revisions to Tasks II and III, including the addition of a seventh furnace to Task II. This project also provided for the relocation and expansion of final product inspection facilities within the PFP Building.	CG-549 CG-551	GE 1953i, Ingalls 1953, Helgeson 1953
May 1953	Furnace failures	Facility engineers instituted a cylinder sampling and testing program with the requirement that cylinders be purged to remove hydrogen before they were placed in service.		GE 1953h, 1953i
Dec 1953		After a period of testing, facility engineers adopted a faster, hotter fluorination cycle as the regular cycle at PFP.		GE 1954e
1954		The new Task III equipment consisted of a welded, stainless steel reduction vessel, with redesigned heating coils, packing glands, valve stems, and seals. Designers realized that a key need in the new Task III equipment was to ensure a completely tight seal to prevent water from entering the reduction crucible. Additionally, PFP personnel worked on site preparation (ventilation duct work and electrical power connections) to install a new Final Inspection Area in Room 192 of the PFP, and on expansions to the process air requirements for the RMA Line. In attempts to provide for better contamination control, and for more efficient re-use of Plutonium-bearing scraps, PFP engineers began design on a new casting storage hood, to replace the many individual storage hoods then being used in the RMA Line. They also designed and emplaced another briquetting hood in the RMA Line to package plutonium-bearing scraps for recycle.	CG-549 CG-551	Wirta 1954; GE 1954k; Shaw 1953a, 1953b, 1953c, 1954

Date	Need/ Problem	Modification	Project No.	Reference
1954		Additionally in 1954, safety experts decided to emplace Cambridge™ "fire-proof" filters in the filter boxes of the RMA Line and analytical and technical laboratories of the 234-5 Building. However, the bonding material around these filters soon failed, causing increased emissions from the 291-Z Stack. Safety experts then proposed a new fire detection system for the building, but managers deferred a decision on installation of such a new system. Engineers and construction forces also undertook a complete re-balancing of the 234-5 Building ventilation system. Workers in the 234-5 Building's Crucible Shop tested several new crucible alloy and fabrication methods, for use in the Plutonium processing lines.		Mobley 1957; GE 1954k, 1954n, 1954r, 1954s
Apr 1955		PFP personnel emplaced new Task II and Task III equipment in the RMA Line in Project CG-551. In April, engineers began hot shakedown runs with Plutonium and reported that "generally the [new] equipment is operating satisfactorily." In Task IV of the RMA Line during 1955, engineers directed efforts at developing a permanent mold casting process and at increasing production and reducing the amount of waste metal turnings through double-casting. In Task V, they focused on designing a remote, automatic lathe. In Task VII, they conducted trials early in the year to compare various new thermocouple designs. At the same time, they took an intense look at the overall RMA Line conveyor system. In the autumn, they issued a recommendation that a slat-type conveyor be substituted in the area serviced by two sections of the folding arm conveyor.		GE 1955f, 1955k, 1955l; Shaw 1955
Apr 1955		Nuclear operators and chemists occupied the new Final Inspection area "with no difficulties encountered." Later in the year, chemists and physicists undertook studies to improve and mechanize this area with emphasis on faster and more accurate gauging methods and improved cleaning and polishing equipment. At the same time, Health Instruments Division personnel used improved radiation monitoring instruments and techniques to reveal that the surface gamma dose rates of the Plutonium metal was 40 percent greater than previously had been thought. Also in 1955, construction workers completed installation of the RECUPLEX (recovery of uranium and plutonium by extraction) Facility in Room 221 of the 234-5 Building.		Reas 1956, Helgeson 1956a, Ketzlach 1955b
1956		PFP designers began work to develop equipment to combine Task I-Task II in the RMA Line.		Fritz et al. 1956
1956		PFP personnel began studies on the conversion of RECUPLEX to a manufacturing facility and designed an office annex to the 234-5 Developmental Laboratory. They completed design for a new mechanized cleaner and polisher for the final finishing operations and developed equipment for mechanizing radiography operations in the inspection facility. PFP personnel began several other projects to improve operations, including a project entitled Permanent Mold Casting that modified Task IV equipment, designs for equipment to fabricate test and experimental weapons components, and designs for yet more changes to the conveyor system. They also drafted a prototype, manually operated vacuum pick-up device for handling uncoated plutonium metal pieces, a remotely operated final machining lathe in order to reduce worker exposure, a remotely operated Task V briquetting press, improved process ventilation machinery, a prototype transfer can for transporting material from one hood to another through Zone 3 of the line without the use of a plastic bag, and changes to the duct systems and drive mechanisms of Task VII units to provide better distribution of nickel plating around Plutonium pits made in the RMA Line.		Harmon 1957; GE 1956e, 1956f, 1956g, 1956i, 1956j; Travis 1956; Helgeson 1956b
1956		Task III operations continued in a fairly routine manner throughout 1956, with some process changes and repairs. In mid-year, engineers allowed an increase in Task III capacity through the use of larger crucibles and larger charges. In Task IV, maintenance crews replaced the heating elements in all four units in January along with two thermocouples and a furnace lift mechanism. In May, June, July, August and December, they had to perform still more maintenance in the casting furnaces. Also in 1956, crews executed additional repairs on equipment in Tasks V, VII, and the RMA conveyor system.		GE 1956e, 1956f, 1956g, 1956i, 1956j, 1956k, 1956l, 1957e

Date	Need/ Problem	Modification	Project No.	Reference
1957		Completion and startup of the continuous Task I-Task II process	CG-691	Baker 1957
1957-1958		Other projects undertaken by PFP personnel to support the production requirements of the late 1950s and early 1960s included the installation of Cambridge Type G fire resistant exhaust filters during 1957-1958. At the same time, building managers instituted a program of close housekeeping control of flammable solvents in the PFP complex buildings.		Palmer 1957; Finch 1957; Mobley 1957
1958-1960		Building managers installed still more fire protection improvements, including fire detectors throughout the 234-5 Building, carbon dioxide fire suppression systems in the hydraulic lines, and a series of automatic wet-pipe sprinkler systems in some of the corridors and vestibules.		Ager and Knights 1963
1958-1959		In late 1958, designers and engineers began working toward a vacuum system improvement project in the 234-5 Building with completion in autumn 1959. This project consisted of modifications to the 26-inch vacuum source located in the 291-Z Building and to the connection and service points in the analytical laboratory, development laboratory, the RECUPLEX Facility, and the RMA Line. Many feet of the old vacuum lines were replaced.		Ciccarelli 1958, Tabasinske 1959, LaRiviere 1959
1958	Excessive exposure	Facility engineers installed additional barrier-shielding between operators and Zone 3 in the Task I-IV space, extending the new vacuum cleaning system being installed in the Remote Mechanical C (RMC) Line to RMA Line Tasks I-IV, and providing a new storage hood for Plutonium powder produced in Task II.	CGC-800	Braden 1958
1958	Excessive exposure	Another Atomic Energy Commission (AEC) project also reduced exposure in the RMA Line by increasing the vacuum system's power to sweep powders and other small debris from RMA Line hood floors. In the Vacuum System Improvements project, construction forces installed new vacuum system connections from the 291-Z Stack to service points in Tasks I, II and III of the A-Line, and to other points within the 234-5 Building.	CGC-826	Tabasinske 1959, LaRiviere 1959, Fritz 1957
1959-1962		Beginning in 1959, PFP engineers and weapons experts began a series of modifications to both the RMA and RMC Lines, and designed a specialized, vault-type cell inside a 3,600 square foot addition on the northwest corner of the 234-5 Building. Even though the RMC Line was still being constructed, AEC needs were changing so rapidly that pre-startup modifications were necessary. PFP personnel completed this project, known as Additional Fabrication Facilities, in 1962. The project provided equipment flexibility capable of processing fuels that had been exposed to much higher levels of irradiation and of fabricating many new weapons. Essentially, this equipment gave the RMA Line particularly the capability to participate in special, experimental weapons fabrication programs. This project also completely overhauled casting operations in both lines. The reinforced, poured concrete and metal addition to the 234-5 Building contained a heavily shielded cell with 4-foot thick walls that measured 20 feet wide by 30 feet long by 30 feet high on the inside. It housed a new, consolidated final inspection facility with a million-volt (MEV) x-ray unit. The facility also ontained an 8-foot by 5-foot by 5-foot deep covered pit for servicing the x-ray unit, a lead-shielded dark room, and a personnel air-lock entrance.	CGC-811	Ciccarelli 1959a; McKee 1959; GE 1960f, 1960g, 1960h, 1960i
1960		The platinum liner of the fluorinator tube of the continuous Task I-Task II equipment in the RMA Line failed and was replaced.		GE 1960f
1962		In October, the AEC funded two additional projects to fit the line for a role in the fabrication of Plutonium components for specialized weapons tests. In the Plutonium Ingoting and Auxiliaries Facility project, construction forces installed facilities in the RMA-Line for handling ingoting feed (buttons), briquettes, and miscellaneous metal scrap, as well as casting mold preparation, make-up and breakout, and weight density determination of castings and ingots produced. Workers completed this project in August 1963.	CGC-983	Travis 1962

Date	Need/ Problem	Modification	Project No.	Reference
1962		In the Electrical and Hydraulic Equipment Mezzanine - RMA Line project, construction forces installed mezzanine space along "D" wall of the RMA Line area to house motor control centers to provide for systematic distribution of electrical power, electrical cubicles, as well as hydraulic equipment for prototype equipment used in research and development in special weapons fabrication. The new, individual prototype units were not connected to regular RMA production systems.	CGC-984	Fillmore et al. 1964
1963		In the Metal Stabilization Facility project, workers emplaced a unique cell for research and development work for experimental weapons fabrication.	CAC-987	Travis 1963
1962-1963	Equipment difficulties	PFP personnel completely renovated Task I-Task II equipment in Hood 9-A during the winter. They placed new filters in the lines leading to the filtrate pumps and in the off-gas lines, and they installed a new stainless steel reactor vessel and new valves in mid-1963. However, they had to perform more renovation work, particularly to the vacuum system, at year's end.		
1972		Facility planners began designing another fire protection upgrade that eventually was completed in the PFP during 1976-1977. In that project, construction forces emplaced stainless steel filter screens, smoke and heat detectors, and automatic shut-off dampers in the E-3 (operating area) and E-4 (glove box) HEPA filters of the 234-5 Building. In 1973, they also installed a new gamma indicating radiation monitoring system in the PFP, as well as criticality overflow drains in some glove boxes.		Cochrane 1977; Lee 1972a, 1972b; Jones 1973; Laws 1973
1978		PFP engineers replaced lucite panels remaining on some of these hoods with Lexan.		

In mid-1952, the Atomic Energy Commission authorized research to recommend improvements to the RMA Line. By late summer, internal contractor recommendations focused on both the technical and mental aspects of operating the new system (Collins 1952, p. 2). The research committee concluded that "remote operation of all plutonium purification and fabrication equipment is ultimately desirable. However, until reliable, tested, remote-control equipment is available, the remote control requirement may have to be compromised on some steps of the process to allow non-remote or rubber glove operation in order to obtain equipment with sufficient operating reliability" (Ingalls 1952, p. 3). By then, the plutonium finishing engineers had directed a majority of the facility's Task III work to the RG Line because of the RMA Line operating problems. These problems were eventually resolved to varying degrees over time.

REMOTE MECHANICAL C (RMC) LINE

From 1957-1961, Plutonium Finishing Plant designers undertook many projects in an effort to accommodate the vast production increases. The most significant of these began in 1957 when the Atomic Energy Commission authorized construction of a new remote mechanical line in the 234-5Z Building to incorporate all of the improvements and changes that had been made over the previous years of RMA Line operations. The new line, designated the "C" Line, was installed where the former RG and RMB lines had stood. General Electric completed removal of the majority of RG Line equipment in early 1957 (GE 1957f, 1957g; Ciccarelli 1959b).

The original Atomic Energy Commission authorization for the RMC Line called for it to operate only Tasks I-III of the plutonium finishing process and not to fabricate any weapons pits. The RMC Line was "to provide for forecasted increases in Pu [plutonium] production, to assure [sic] production continuity, and to provide for short-term peak loads from the primary solvent extraction plants" (GE 1955i, p. Ff-5). Soon after the PUREX Plant, which began operations in 1956, proved its vast production capacity, the Atomic Energy Commission decided to add Tasks IV-XII to the RMC line. In April 1957, the agency officially authorized two projects: CG-734, RMC Button Line, and CG-745, RMC Fabrication Line. The RMC Line was essentially similar to the RMA Line, but the RMC Line had an automatic vacuum cleaning system that served Tasks I-IV, greater radiation shielding, and improved radiography facilities (Fritz 1958, Swain 1959). General Electric completed the entire installation effort in 1959 at a cost of nearly $1.9 million (Moorman 1960, Shaw 1960).

The RMC Line consisted of a completely self-contained, remotely operated series of glove boxes fabricated from high-grade Austenitic 304L stainless steel with all welded construction. (Austenitic 304L is stainless steel composed of austenite made stable by alloying it with nickel.) It was installed with seamless Austenitic 304L stainless steel piping and castings for corrosion resistance. All glove boxes except the conveyor glove boxes were fabricated with 1/4-inch thick steel floors and ceilings, 1/4-inch bent angles for the framework, plexiglas^tm panels, and molded neoprene^tm gaskets as air-tight seals around all windows and ports. Lucite or homalite^tm ports equipped with plastic bags were provided for maintenance and equipment removals. The conveyor glove boxes were fabricated of 3/16-inch thick stainless steel plate, with 3/16-inch bent angle framework. The button line portion of the RMC conveyor was known as HC-1, and served 17 glove boxes and various tanks. It was 60 feet long and was constructed of stainless steel slats mounted on a stainless steel frame. Figure 2-5.5 shows a nuclear operator in the control room for the RMC Line in 1988.

Figure 2-5.5. *Nuclear Operator in the Control Room of the Remote Mechanical C Line, 1988*

Leaded glass, carbon steel, and plexiglas shielding were designed for the RMC Line in 1957 and strengthened in 1959 because new information showed some line operators could be over exposed. Causal factors included increased production, materials with higher activity levels, and more precise instrumentation in measuring neutron dose rates (Swain 1959). Designers added carbon steel plates to the internal floors of RMC Line glove boxes as well as supplemental shielding external to the glove boxes at many points (Fritz 1960, Ciccarelli 1959b). For a detailed description of the RMC Line, see the Expanded Historic Property Inventory Form for the 234-5Z Building in Appendix B on the Internet.

Figure 2-5.6. *Nuclear Operators Check Idle Equipment on Remote Mechanical C Line, 1982*

After many months of cold startup tests and resultant adjustments, plutonium finishing engineers and operators began processing plutonium nitrate feed in the RMC Line in October 1960. By November, the new line was operating so well that it was performing all of the Task I-II work and 80 percent of the firing of buttons (GE 1960i, 1960j). In December, the contractor described the new line as "generally satisfactory" (GE 1961c, p. D-2). However, during the early months of 1961, the RMC Line produced buttons of poor quality with high reject rates. After engineers adjusted equipment, production rose (GE 1961d, 1961e, 1961f, 1961g). For details of the modifications made to solve the problems on the RMC Line, see Table 2-5.5. The RMC Line operated with increasing success and by 1964 had taken over the bulk of plutonium metal processing and was operating on a 7-day work week schedule due to the high demand for production. An elite crew of four to six operators with one line manager per shift operated the line (Rodgers 1986). Figure 2-5.6 shows operators checking the RMC Line equipment.

Table 2-5.5. *Modifications to the Remote Mechanical C Line*

Date	Modifications	Project No.	Reference
1962	The Atomic Energy Commission (AEC) authorized the project, Remote Mechanical C (RMC) Button Line Filtrate Handling Facility, in mid-1962. This project emplaced equipment to concentrate filtrate resulting from Task I operations.	CGC-978	Braden et al. 1962
1962	The AEC authorized the Personnel Protection, Storage Hood Shielding - 234-5 Building project. This project provided for fabrication and installation of ten-inch thick, water-filled shielding tanks around the three operating sides of storage glove boxes HC-22-SR and HC-45-S in the RMC-Line. General Electric Company justified the project in February 1962 by stating that these two storage glove boxes were being used at four times their design capacity to satisfy weapons specifications. This situation was increasing neutron radiation exposure to personnel working nearby, to the point where the site limit of five rem per year would be exceeded if corrective actions were not taken.	CGC-953	GE 1962c
1963	Plutonium Finishing Plant (PFP) engineers and safety experts instituted a major change in the warning systems in the 234-5 Building, when they changed the siren alarm signaling a nuclear excursion to the "ah-oo-gah" sound. That December, they placed a new air sample counter in service in the facility, making possible faster detection of airborne contamination.		GE 1963c; GE 1964c
1964	PFP personnel equipped the RMC Line with a new paging system.		Patton 1964
1966	PFP personnel overhauled receiving glove box HC-4. Facility personnel discovered and removed Plutonium that had built up in duct work and filter boxes downstream of Hood 9B. At the same time, PFP engineers and safety experts engaged in a number of critical mass control studies and safe conditions studies in various aspects of Hanford's Plutonium finishing and transport.		Knights 1966; Sloat 1963a, 1963b, 1963c
1972	Also, they began designing another fire protection upgrade that eventually was completed in the PFP during 1976-1977. In that project, construction forces emplaced stainless steel filter screens, smoke and heat detectors, and automatic shut-off dampers in the E-3 (operating area) and E-4 (glove box) HEPA filters of the 234-5 Building.		Cochrane 1977; Lee 1972a, 1972b
1972	AEC modified the fluorinator off-gas water jet system in the RMC Line, by installing a recycling jet scrubber system in glove box HA-46.		McKnight 1972a, 1972b
1973	Construction forces also installed a new gamma indicating radiation monitoring system in the PFP, as well as criticality overflow drains in some glove boxes.		Jones 1973; Laws 1973
1973	PFP engineers retrofitted the fluorinator off-gas water scrubber system in the C-Line with a recirculating water system for effluent reduction.		Fecht et al. 1973
1984	PFP personnel inspected instrumentation and electrical systems on RMC glove boxes, repaired or replaced equipment as needed, and dismantled furnace assemblies for inspection and upgrading.		Yoo 1985
1986	Shut down once again for a series of mechanical and safety upgrades.		RHO 1986a, 1986b, 1986c
1987	Westinghouse Hanford personnel braced the HF cylinder banks and Hoods HC-15-A, B, and C of the RMC Line to withstand seismic disturbances; upgraded hood lights; modified the Hood HC-9B transfer mechanism and graphic display panel; replaced the hydrofluorination gas line penetration into Hood HC-9B; obtained new pressure gauges; temperature indicators, liquid level controllers, tubing, hydrofluorination gas sensors, HEPA filters, the Hood HC-9B calciner filters, and circuit breakers for the breathing air system; redesigned the Hood HC-15 furnaces and of the hydrofluorination gas interlock system; installed conveyor upgrades; changed out some lead glass glove box panels; and modified the argon gas supply and the Hood HC-9B fire damper. They also upgraded the RMC Line continuous oxalate precipitator, changing it to a mixed-suspension mixed-product removal draft tube precipitator. They made this change to improve the particle size of the plutonium oxalate product. They also modified the feed blending systems in the 234-5 Building, and upgraded other support equipment for the RMC Line and for building systems in general.		

DUAL OPERATION OF THE REMOTE MECHANICAL A AND C LINES

With two lines running, the production and problems of one line affected the other. Therefore, jointly describing their production and problems from the time the RMC Line started gives the reader a broader perspective of the plutonium finishing process.

Production

On January 8, 1964, President Lyndon Johnson announced a decreased national need for special nuclear materials due to the sufficiency of the existing stockpile and to the recent treaty banning atmospheric nuclear weapons testing. Soon afterward, the Atomic Energy Commission announced plans to shut down the Hanford production reactors in a phased fashion. The agency began the reactor closures in December 1964 with the DR Reactor, followed by the F and H Reactors in 1965. Ironically, the Plutonium Finishing Plant produced its all time record number of pits in December 1964 (DeNeal 1970). As a result of decreased production needs and equipment problems, the Atomic Energy Commission ordered the RMA Line out of service completely near the end of 1964.

During the second half of 1965, the Atomic Energy Commission also announced a decision to remove the weapons fabrication work from the 234-5Z Building and to concentrate that work at the Rocky Flats Site. The agency continued small-scale, experimental fabrication projects at the Los Alamos Site. The weapons fabrication program at Hanford concluded in mid-December of 1965. Immediately, facility personnel undertook layaway (but not yet removal) of the fabrication portion of the RMA and RMC Lines, and they essentially completed these tasks in March 1966. Building staff levels fell from a Cold War high that had approached 300 people to 240 people in 1966. The plutonium metal conversion portions of the RMC Line continued to operate, and engineers added a button bag-out and sealing operation.

As a result of the Atomic Energy Commission decision to move plutonium fabrication work out of the 234-5Z Building, General Electric undertook a diversification study late in 1965 to plan alternative future uses for the facility. The resulting report envisioned an expanded role for the Plutonium Finishing Plant in non-defense work (GE 1965b). Throughout the period 1966-1972, plutonium finishing management did pursue non-defense work, and operations in the facility were split between defense and non-defense work.

At midnight on December 31, 1965, General Electric turned over the operation of all its Hanford Site chemical processing facilities, including the 234-5Z Building, to a new operating contractor, Isochem, Inc. (GE 1965a, 1965c, 1966). In mid-1967, the Atomic Energy Commission awarded a new operations contract for Hanford Site chemical processing operations to the Atlantic Richfield Hanford Company (ARHCO), and operations in the Plutonium Finishing Plant were interrupted by the contractor changeover between August 18 and September 5 of that year (Isochem 1967i, ARHCO 1967a).

Figure 2-5.7. *Oxide Produced on the Remote Mechanical A Line, December 1979*

In September 1967, the Atomic Energy Commission authorized plutonium finishing engineers to reactivate glove box HA-39 (part of the closed RMA Line's fabrication equipment) in order to grind plutonium-aluminum scraps into turnings suitable for plutonium recovery (ARHCO, 1967a). A year later, the Atomic Energy Commission ordered the RMA Line's Tasks I-III cleaned out and re-activated to participate in expanded programs for commercial (non-defense) nuclear experiments and development (ARHCO 1968a). For the remainder of 1968, the RMA Line produced oxides for early experiments in the development of Hanford's Fast Flux Test Facility (the nation's prototype breeder reactor that would be built at the Hanford Site). ARHCO reported mechanical efficiency as 91.6 percent and stated that product purity was within specifications (ARHCO 1968b). Figure 2-5.7 shows the oxide produced on the RMA Line.

Meanwhile, the RMC Line operated on a mixed schedule, producing some weapons-grade plutonium and an increasing amount of fuel-grade plutonium and oxides. For example in 1966, the RMC Line produced weapons-grade plutonium buttons throughout the early months of the year, but operators cleaned and inventoried the line on April 25. Three days later, they began processing fuel-grade plutonium from N Reactor and continued this work throughout the rest of the year (ARHCO 1967a). In general, the RMC Line performance was excellent. Three campaigns in March 1969 produced buttons of average yields of 98.4 percent, 97.8 percent, and 98.2 percent, respectively (ARHCO 1969a).

Because the RMA Line produced only oxides from 1968 on and the RMC Line mostly metals, Hanford Site personnel referred to the RMA Line as "the oxide line" and to the RMC Line as "the metal line" (ARHCO 1968a).

During 1968, the Plutonium Finishing Plant began participating in an International Safeguards Experiment in which some of the plutonium nitrate in the 234-5Z Building and some stored in the 224-T Building were accessed and inspected by the International Atomic Energy Agency, a multi-national, non-proliferation group headquartered in Vienna, Austria (Elgert 1968). Throughout 1969, cooperation with the International Atomic Energy Agency expanded as did attention to waste effluents and scrap receiving, identification, sorting, and re-packaging.

In 1970, Plutonium Finishing Plant engineers implemented the Plutonium Retrieval System, an inventory system to account for nuclear materials. The computerized system fed into the National Nuclear Materials Management and Safeguards System maintained for the Atomic Energy Commission at the Oak Ridge Reservation in Tennessee. Unfortunately, the Plutonium Retrieval System was plagued with inaccuracies and discrepancies for many years (Murphy 1976, 1978; Brown et al. 1980; Glasscock et al. 1982).

The Energy Research and Development Agency, which had taken over for the Atomic Energy Commission in 1975, decided to replace ARHCO with a new chemical processing contractor at the Hanford Site. Rockwell Hanford Operations took over the management of the Plutonium Finishing Plant and the other ARHCO facilities July 1, 1977. Among its first acts, Rockwell initiated reviews of "all procedures" for the Plutonium Finishing Plant and measurements of plutonium build-up in all glove boxes throughout Z Plant operating areas (Cockeram 1977, p. 11).

DOE, which replaced the Energy Research and Development Agency late in 1977, directed Rockwell in January 1978 to study the possibility of decontaminating and decommissioning most of the Plutonium Finishing Plant (Hopkins and Jones 1978, Graves and Kusler 1978). Rockwell formulated a scenario in which the plutonium storage vaults (2736-Z and 2736-ZA facilities) would remain in use for decades, the plutonium finishing laboratories would remain operational to support the PUREX Plant, and the Plutonium Reclamation Facility (see Waste Management Section) would continue to process scraps from the Plutonium Finishing Plant cleanout and other ongoing Hanford Site activities (RHO 1979a).

In response to a build-up of nuclear arms ordered by President Ronald Reagan in the early 1980s, DOE restarted the Hanford Site's PUREX Plant in November 1983. That same year, DOE considered tentative plans to reactivate the RMA Line to process the plutonium nitrate coming from PUREX. The PUREX facility had a new internal facility (N Cell) that was supposed to process plutonium nitrate into plutonium oxide before shipment to the Plutonium Finishing Plant. However, no one knew for certain if N Cell could perform adequately. Plutonium finishing personnel prepared a new flow sheet for RMA Line operations along with a safety analysis (Crocker 1984, Wilson and McClusky 1983). As it happened, N Cell operations performed well, and DOE decided in 1984 to keep the RMA Line in standby status. It never again operated. Today the RMA Line in the 234-5Z Building consists of 25 glove boxes and 3 conveyers in a complex of 4 rooms (235-A3, 235-A2, 235-A1, 235-B) and the control room in Room 233.

In 1984, DOE ordered the RMC Line to prepare to restart production of weapons-grade components. Plutonium finishing personnel mounted an intensive project to upgrade the RMC Line, bringing in a team from the Los Alamos Laboratory to review the startup safety analysis.

Plutonium Finishing Plant personnel achieved restart of the RMC Line on July 1, 1985 and operated it until October of that year. According to a contractor report, "the initial startup period...extended until the last week of September due to repeated operational problems and key equipment failures...The production run started in earnest on September 30, 1985 with the last [plutonium nitrate] ...load-in occurring on October 18" (Rodgers 1986, p. 14). In total, the RMC Line produced about 63 percent of DOE's target goal. Plutonium finishing personnel then conducted cleanout and shutdown activities in the RMC Line through October 31.

The RMC Line restarted after multiple repairs in April 1986 and operated for 6 months making weapons-grade plutonium. Plutonium finishing personnel then shut it down once again for a series of mechanical and safety upgrades (RHO, 1986a, 1986b, 1986c). In July 1987, the Hanford Site's operating contract was consolidated, and the chemical processing work, along with other work, was awarded to a new contractor, Westinghouse Hanford Company. The RMC Line conducted one more campaign after that and was shut down permanently in May 1989.

Problems

Although the frequency of problems in the Plutonium Finishing Plant continually dwindled, they did continue to appear. For example, from 1962-1963 nuclear operators and engineers at the Plutonium Finishing Plant struggled with off-standard feed, high in plutonium-240 isotopic content from the REDOX and PUREX facilities (Bruns et al. 1963). They also had difficulties in the inspection facilities with new, million-volt x-ray equipment. Additionally, personnel were hampered by the effects of a criticality accident at RECUPLEX (RECovery of Uranium and PLutonium by EXtraction) and the consequent loss of much of the 234-5Z Building's plutonium scrap recovery capabilities. The April 7, 1962 accident occurred when batches of plutonium solution came together in an amount large enough to cause the solution to go critical. The resulting flash of radiation contaminated the entire room housing the RECUPLEX equipment and brought about the permanent closure of RECUPLEX operations.

ARHCO conducted evaluations at the Plutonium Finishing Plant from 1976-1977, looking at worker safety and health issues. These evaluations revealed multiple conduct of operations violations in the areas of chemical storage, material storage, and other hazards (Draney and Hill 1976).

ARHCO summarized its overall assessment of the Plutonium Finishing Plant conditions and capabilities in mid 1977 by stating: "During the early 1970s, it became increasingly obvious that corrosive process solutions were exacting a toll in terms of deterioration of piping and equipment... Considering the obsolescence and general deterioration of equipment, as well as the evolutionary upgrading of regulatory requirements...extensive repairs and revisions are necessary..." (Szempruch and Szulinski 1977, p. 1).

In January 1978, DOE completed an engineering study that found the 234-5Z Building and the 291-Z Stack had "insufficient capacity to withstand tornado or SSE [safe shutdown earthquake] loadings ... tornado forces could overstress [these facilities].... Lateral restraints of the ventilation ductwork and piping...are lacking. Critical equipment items and components (glove boxes, primary filters, electrical panels, criticality alarms, master fire alarm boxes, fire protection systems controls, continuous air monitors, etc.) do not have good connections and anchorages to the building ... the emergency utility systems for the 234-5Z Building are not separate, self-contained systems.... Hence the loss of normal power may also prohibit the use of emergency power" (Cockeram 1978, pp. 13-14).

In response to the need to control contamination, plutonium finishing engineers de-pressurized the 234-5Z Building in 1978 for the first time in its history. Since its construction, the building had been operating at a positive pressure relative to the outside atmosphere to prevent the inflow of "unconditioned dust and air." To operate potentially contaminated areas at negative pressure relative to adjacent areas as now required, Rockwell reconfigured the 234-5Z Building's air flow systems "in such a manner to cause air flow from areas of low contamination potential to areas of high contamination potential." In the changeover, engineers rearranged ventilation ducts within the building, provided air locks between clean zones and potentially contaminated areas, sealed the exterior of the building adjacent to contaminated areas to prevent leakage into the building, placed new dampers and fans in the 291-Z Building, installed new air monitoring and warning systems and new zone pressure control instruments (Glover 1978b, p.3).

Production problems also continued in the later years. In 1985, Rockwell found that only 25 percent of the plutonium buttons produced by the RMC Line were within specification limits. Rockwell then recommended several modifications to the RMC Line equipment as well as entire replacement of the vacuum drum filter, the fluorinator off-gas scrubber, and several lines within the RMC system. It prepared a list of 125 maintenance items and identified the most important 40 items. By December 1985, it had received DOE approval to proceed and had initiated engineering design to rebuild or replace the 40 items (Rodgers 1986). For details of the modifications made to solve the problems on both the RMA and RMC Lines, see Table 2-5.6.

Table 2-5.6. *Modifications to Both the Remote Mechanical A and C Lines*

Date	Need/ Problem	Modification	Project No.	Reference
1969	Replace existing plastic panels	The Atomic Energy Commission (AEC) commissioned a study of the safety and fire resistance of glove box windows at the Hanford Site. The study indicated a need to replace existing plastic panels, usually a methyl methacrylate known as plexiglas. Eventually, during 1974-1975 and in the 1980s, the Energy Research and Development Administration (ERDA - successor agency to the AEC), and the Department of Energy (DOE - successor agency to ERDA) ordered several such replacements with Lexan polycarbonate plastic on A-Line, C-Line and the Plutonium Finishing Plant (PFP) laboratories glove boxes.		FMRC 1969, Olson 1974, Corbell 1975
1970		Both contractor and federal management agency personnel spent a great deal of time planning effluents control initiatives during this period. Also in 1970, they initiated planning for a major fire protection, fire alarm and electrical improvements project in many areas of the PFP complex.		Teeple and Tabasinske 1970
1975	Upgrading of instrumentation	Meanwhile in 1975, the new ERDA ordered reviews of many practices, policies and conditions in the PFP complex, particularly safeguards and security policies, plutonium storage practices, and the integrity of plutonium storage and shipping containers. As a result, the PFP personnel wrote a complete, new plutonium processing manual. They identified the need for an overall upgrading of instrumentation throughout PFP facilities. ERDA also produced a new manual, revising mandates for the safety of non-reactor nuclear facilities. PFP engineers then undertook a series of comparisons between ERDA's plutonium storage criteria and the physical conditions and accoutrements of rooms storing plutonium in the 234-5 Building. When they learned that several of the storage rooms in the old structure did not comply with the new ERDA criteria, they recommended a series of SAS upgrades and special nuclear materials accountability improvements to make the facility as secure as possible. On August 30, 1976, a serious accident in the 242-Z facility shut down all PFP complex operations.		Uebelacker 1975; ERDA 1976b; Gimpel et al. 1975a, 1975b, 1975c, 1975d, 1975e, 1975f, 1975g, 1975h; Hinkson 1977
1976-1980		During 1976-1980, designers, engineers and construction forces implemented the new recommendations in several projects. In Project B-157, Safeguards and Security Upgrading, they built the 2701-ZA Central Alarm Station (CAS), located just inside of the PFP protected area fence. (A description of the CAS can be found in the Expanded Historic Property Inventory Form for the 234-5 Building in Appendix B on the Internet.) They also installed intrusion detectors at various locations in the complex, additional alarming systems in the existing door surveillance system, and a secondary perimeter fence. In Project B-158, Z Plant Calorimeters, they installed a multi-calorimeter system, along with associated hardware and software. In Project B-195, Materials Management Information System, they provided an on-line nuclear materials tracking system that served as the first phase toward the goal of a real-time inventory. In Project B-169, Criticality Alarm Systems, they installed new criticality alarms in the 234-5, 236-Z and 224-T Buildings.	B-157, B-158, B-169, B-195	Gallant 1977a, 1977b, 1977c; Williams 1976; Garfield and Clausen 1978; Corbell 1978, 1979

Date	Need/ Problem	Modification	Project No.	Reference
1976		During the same years, in Project B-184, they added improved fire protection equipment and systems to PFP facilities and the 224-T Building. During 1976, PFP personnel replaced all of the E-4 filters in the 234-5 Building. In 1977, in response to ERDA audits, they instituted preventive maintenance program upgrades in the A-Line, the PRF, the PFP laboratories, and the 234-5 process systems in general.	B-184	Cochrane 1977; Pinto 1979; ARHCO 1976a, 1976b, 1977; Berube 1977a, 1977b, 1977c, 1977d
1978		Due to the directive to plan for D&D of most PFP complex facilities, DOE in 1978 modified the PFP ventilation upgrade project that had been designed in 1977, to include only replacement of the galvanized metal filter boxes for the E-4 filter system with stainless steel boxes, replacement of the exhaust collection headers associated with filter boxes, replacement of some duct work, and upgrading of selected circuit breakers. Construction personnel implemented this downsized version of the project during 1980-1981. They also completed an upgrade to the 291-Z Building's 17" vacuum system. During 1979-1980, in a separate effluents upgrade project, construction workers relocated the Plutonium Finishing Plant's 26" mercury vacuum pumps and made other ventilation modifications.		Cowley 1978; Heubach 1979a, 1979b, 1979c; VEC 1980
1979	Contingency for criticality prevention	They also installed additional lead glass shielding, and installed glove box drain screens "to provide an additional level of contingency for criticality prevention."		Walser and VanDerCook 1979a, 1979b; Cahow et al. 1979a, 1979b
		In the 234-5 Building itself, personnel installed a new criticality alarm, upgraded the gamma alarm system, installed new hand and foot counters, leak-tested fire suppression systems, verified and corrected as-built drawings, upgraded waste storage and staging facilities, wrote new procedures, safety analyses, process and operating specifications, and performed many other modifications. They re-roofed the 234-5 Building in 1983. Additionally, DOE authorized construction of the 270-Z Office Building just north of the 234-5 Building.		Veneziano 1983, RHO 1984

NON-DEFENSE WORK

Beginning in the 1960s, the Atomic Energy Commission undertook new initiatives to interface with the burgeoning commercial nuclear industry. During subsequent years, the Hanford Site produced approximately 13 metric tons of fuel-grade plutonium (plutonium-239 in various blends with plutonium-240 and other plutonium isotopes). The RMA Line in the Plutonium Finishing Plant processed much of this amount as well as much fuel-grade plutonium shipped to the Hanford Site from off-site. Among the customers for fuel-grade RMA Line oxides during these years were some of the prime developers of commercial nuclear fuels and experimental projects in the nation.

In 1962, the Atomic Energy Commission asked General Electric to investigate the feasibility of receiving, storing, and disbursing fuel-grade plutonium at the Plutonium Finishing Plant. The Atomic Energy Commission wished to buy back plutonium produced by power reactors that operated on uranium fuel supplied by the Atomic Energy Commission. The Atomic Energy Commission then planned to sell the economically valuable plutonium at a later date for use in research programs. Nuclear Fuels Services facilities in West Valley, New York would separate Plutonium from spent fuels and purify it for shipment to Hanford as plutonium nitrate (Warren 1963, Kendall 1964, Braden 1965, Gifford 1965). The Atomic Energy Commission authorized Project CGC-180, Plutonium Buy-Back Facility, and construction forces installed it in the 234-5Z Building (Shaw 1966).

The Plutonium Finishing Plant began production of specially blended, fuel-grade plutonium oxide for EURATOM (European Atomic Energy Community - a consortium of the governments of France, West Germany, Italy, the Netherlands, Belgium and Luxembourg) for commercial nuclear power development in September 1964. That same year, the 234-5Z Building began supplying fuel-grade plutonium (also known as unclassified plutonium) to Hanford's 209-E Critical Mass Laboratory for experimental purposes and to a commercial firm for fuel development tests for the Southeast Experimental Fast Oxide Reactor (GE 1966, Walser and Lyon 1965, Hewlett and Holl 1989).

When the Atomic Energy Commission announced its decision to withdraw weapons-grade plutonium fabrication work from the Plutonium Finishing Plant in late 1965, facility personnel undertook a diversification study for the building. The study concluded that the Atomic Energy Commission should encourage "the growth of a competitive commercial capability in the field of plutonium processing and fabrication ... comprehensive plutonium fabrication service [should be made available] to industry and government." (GE 1965b, p.2).

Beginning in 1966, the Plutonium Finishing Plant's interaction with the commercial nuclear industry, nuclear research customers, and foreign nuclear customers expanded rapidly. The RMC Line began processing fuel-grade plutonium from N Reactor. For the remainder of the year, plutonium finishing personnel shipped both weapons-grade and fuel-grade shipments of plutonium. Many national laboratories, private laboratories, and companies in the commercial nuclear industry became customers.

As the commercial nuclear industry continued to expand, plutonium finishing workers spent about 30 percent of their time and effort in non-defense work in 1968. ARHCO anticipated further non-defense work in the facility: "Plutonium oxide or mixtures of plutonium-uranium oxides for reactor programs will probably be supplied in large quantities" from the 234-5Z Building (Knights 1968, p. 1).

In 1976, a serious accident occurred at the 242-Z facility when a chemical explosion took place in a column that contained plutonium and americium. (This accident is discussed further later in the text.) Shortly after that time, the Energy Research and Development Agency began a long evaluation of the continued viability of the Plutonium Finishing Plant facilities. In late 1976, the agency gave the contract to produce oxides for Fast Flux Test Facility Cores 3 and 4 to the Savannah River Plant. These oxides then were shipped to the Hanford Site for sampling, blending, and re-packaging in part of the inactive fabrication equipment in the RMC Line. However, this work was a small fraction of the work that plutonium finishing workers had been expecting. They completed this work January 31, 1979 (Reberger 1977, Cockeram 1978, RHO 1979a).

At the same time, national policy regarding nuclear energy and nuclear fuel re-processing was changing in significant ways. The federal breeder reactor program, for which the Fast Flux Test Facility was a prototype, was cancelled in 1979. Policy preferences of the Carter Administration, as well as contamination events at the West Valley facility, led to a decision to stop re-processing and blending defense materials with commercial nuclear materials. In 1978, DOE ordered the deactivation of the Plutonium Finishing Plant facilities. Non-defense work in the facility was halted, although some fuel-grade plutonium was still produced as a by-product of shutdown.

QUALITY CONTROL: ANALYTICAL LABORATORIES

As originally constructed, the 234-5Z Building contained only one analytical laboratory, also known as the sample laboratory or the process control laboratory. All of the building's safety bases and criticality assumptions depended upon knowing the isotopic content and amounts of plutonium being handled. This laboratory functioned to analyze samples of the plutonium product from various stages of the finishing process to ensure the materials were within safety and isotopic specifications. The laboratory also supported chemical analysis of materials used in the finishing process.

The analytical laboratory was located at the east end of the Plutonium Finishing Plant toward the south side. Phase II of the plutonium finishing construction project included the completion of two more rooms in the analytical laboratory. Chemical laboratory assistants known as "laboratorians" (later called laboratory technicians) initially worked in open-faced hoods, but for safety purposes, federal and contractor management converted many hoods to glove boxes over the years (Roemer 1970).

The analytical laboratory began operation when the RG Line started up in July 1949. By 1951, the analytical laboratory went from a two-shift to a four-shift operation 5 days a week to accommodate plutonium production increases (GE 1951g). By April 1952, the analytical laboratory operated on a two-shift, 7-day per week basis to accommodate even further production increases (GE 1952f). In March 1953, the 231-Z analytical laboratory closed and sent its work to the Plutonium Finishing Plant analytical laboratory, thus increasing the workload on 234-5Z chemists performing sample analyses (GE 1953g).

Throughout the late 1980s and into the early 1990s, the plutonium finishing laboratories worked to support the Plutonium Reclamation Facility, the UO_3 Plant and RMC Line campaigns as well as the modifications and upgrades associated with them. Chemists revised laboratory procedures and sample and analysis plans for these campaigns (RHO 1985a, 1985b, 1986a, 1986b, 1986c). During 1994-1996 as plutonium finishing and other personnel wrote the Plutonium Finishing Plant Environmental Impact Statement, laboratory chemists studied ways to analyze, verify, and stabilize the nearly 8000 containers of multi-faceted plutonium-bearing scrap stored in the Plutonium Finishing Plant complex. DOE plans to deactivate the laboratories once the stabilization campaigns are completed. For more information, see *Plutonium Finishing Plant Stabilization: Final Environmental Impact Statement* (document no. DOE/EIS-0244-F), dated May 1996.

RESEARCH AND DEVELOPMENT: DEVELOPMENTAL AND METALLURGICAL LABORATORIES

The developmental laboratory also was known as the Plutonium Chemistry Laboratory, process improvement laboratory, research and development laboratory, and in later years the Chemical Technology Laboratory. It was located at the southwest corner of the Plutonium Finishing Plant. Previous to the availability of a developmental laboratory in the 234-5Z Building, chemists carried out early research and development for plutonium finishing in the nearby 231-Z Building (GE 1953j, Bell 1950).

In addition to the developmental laboratory, a plutonium metallurgy laboratory operated in the Plutonium Finishing Plant from late 1952 to late 1956 when it moved to the 231-Z Building. The metallurgy laboratory examined the material properties of plutonium in its metallic state, studying alloys, tensile properties, stability, and fabrication and coating methods (GE 1953d).

As soon as processing began in the Plutonium Finishing Plant in 1949, personnel in both the developmental and analytical laboratories directed their efforts to improving the accuracy of measurements, reducing product losses, and trouble-shooting process difficulties that arose. They were intensely interested in the development of methods to recover and recycle plutonium from scrap and waste materials (Kendall 1949). The other field they mainly pursued was how to couple the plutonium products of the separations plants directly to the RG Line process, thus bypassing 231-Z Building operations (Groswith 1949, Barton 1950a, 1950b). Throughout the operation of the Plutonium Finishing Plant these laboratories contributed to process improvements that enhanced both the quality of the product and the productivity of the plant, improvements in product measurement techniques, plutonium recovery and recycling techniques and support to PUREX and the UO_3 Plant (Gerber 1997).

RECOVERY AND RECYCLING OF PLUTONIUM

Nearly as soon as they began processing plutonium nitrate in the RG Line in 1949, building personnel began to look for ways to recycle and recover plutonium from wastes generated in the process. Atomic Energy Commission and contractor personnel considered these studies important because of the high economic value of plutonium and the desire to implement better waste management practices. In fact, over the years of plutonium finishing operations, they gave more attention to recovery activities than to anything else except production.

Early recycling efforts included skull recovery, recycling analytical laboratory wastes, and the recovery of Task I supernatant. Skull recovery was the retrieval of plutonium from metallic plutonium-bearing scraps (GE 1952n). These scraps, known as skulls, were the hardened residues left on the molds in which the plutonium metal was cast. Operators commenced operations in the facility November 1952 and performed so successfully that they were able to recapture 1000 grams of plutonium for use in the process of preparing plutonium (IV) oxalate in the 231-Z Building during the first month (Facer and Weidenbaum 1951; GE 1952n, 1953d).

Major recycling efforts included the operation of the RECUPLEX Facility, the 232-Z Incinerator , the 242-Z Waste Treatment Facility and the Plutonium Reclamation Facility

RECUPLEX FACILITY

In 1951, plutonium finishing chemists piloted the RECUPLEX (RECovery of Uranium and PLutonium by EXtraction) process in the developmental laboratory (GE 1951h, Cooper 1951b). The process produced a plutonium solution in a concentration range from 50 to 60 grams per liter as suitable feed for the 234 Building metal line. By the end of 1951, plutonium finishing engineers were designing a RECUPLEX process for placement in Room 221 and duct level Room 263 of the 234-5Z Building. Demonstrations showed plutonium nitrate produced in the RECUPLEX process had a high purity (99.9+ percent) and a good yield (approximately 99.9 percent) (Hill and Leitz 1952, p. 5). On May 1, 1953, construction forces began work, and they completed the facility in March 1955. Engineers began processing with plutonium feed material (supernates from Task I) on July 11 (GE 1953i, 1955h, 1955h; Lane 1965).

The RECUPLEX Facility consisted of three large, stainless steel glove boxes with lucite panels. Glove box operations were largely manual. Operators manipulated a control panel in the RECUPLEX room as well as gloves attached to the boxes themselves. The majority of RECUPLEX feed was solid waste in the form of slag and crucible fragments, scrap powders from RMA Line operations, oxidized plutonium turnings from Task V, and remnants from metal samples. Operators placed this material in the slag and crucible glove box, where equipment dissolved and filtered it. Pumps then sent the liquid product from this glove box to the receiving and blending glove box to be blended with other contaminated liquid feed streams from the 234-5Z Building. The receiving and blending glove box functioned as a feed preparation facility for the solvent extraction glove box where the plutonium-bearing liquid feed was mixed with 15-20 percent tributyl phosphate in carbon tetrachloride solution and processed through a solvent extraction system. The end product was a plutonium-nitrate stream concentrated to between 80-100 grams per liter plutonium (Judson 1956).

During 1958-1960, the plutonium finishing operators and engineers frequently set new processing records at RECUPLEX. In March 1959, they recovered 94 kilograms of plutonium. That July, they set another record and then surpassed that in December. Due to increasing production needs, plutonium finishing managers changed RECUPLEX operations from a 5-day to a 7-day work week in late 1960 (Smith 1958; GE 1959c, 1959d, 1960c, 1960j). However, high production at RECUPLEX was accompanied by a series of serious radiation events and a mounting number of equipment failures. Throughout 1958-1960, plutonium production levels at the Hanford Site continued to increase. As a consequence, throughput in all facilities rose, and the RECUPLEX equipment was subject to increasing stress from the high volume of very corrosive solutions. Also, the buildup of nuclear materials in equipment joints, crevices, and other inaccessible areas brought about constantly rising radiation readings in the RECUPLEX room itself. General Electric stated that because of these high levels, RECUPLEX shut down for "an extensive cleanup of hoods and equipment" in February 1960. When a criticality accident (nuclear excursion) resulted in widespread contamination throughout the RECUPLEX room on April 7, 1962, the Atomic Energy Commission decided in May 1962 to deactivate the RECUPLEX Facility on a permanent basis.

232-Z INCINERATOR

Spring 1958, plutonium finishing engineers proposed construction of the 232-Z Incinerator as a way to recover plutonium from miscellaneous solid wastes generated in the Plutonium Finishing Plant. Candidate wastes would include cartons, filters, rags, paper, gloves, clothing, tools, and small equipment components. In the vision of the engineers, plutonium-bearing materials first would be hand-sorted into burnable or leachable fractions. The combustible materials would be incinerated, and then the plutonium would be recovered by a wet leaching of the ash. Non-combustible materials would be leached in nitric acid and then transferred to RECUPLEX (later the Plutonium Reclamation Facility) for solvent extraction recovery of plutonium.

Plutonium finishing personnel carried out design revisions and construction of the 232-Z Facility during 1959-1961. They executed pre-start acceptance tests and additional modifications during late 1961. On January 8, 1962, they began full-scale operations with waste materials. (As soon as it is published, a description of the 232-Z Facility can be found in Appendix B on the Internet.) Early leaching operations proceeded smoothly. At the incinerator itself, however, the conveyor belt began to fail within just 2 months of startup. Furthermore, it continually jammed or slipped and needed tightening, and the chopper blades also needed frequent replacement. Because of these problems, plutonium finishing operators were able to log only small amounts of time on the 232-Z Incinerator (from zero to a maximum of 12 hours per week) during the first 3 months (Bruns et al. 1963).

Operators continued with sporadic incinerator performance throughout the rest of the year, never achieving more than 17 hours per week of operating time (Bruns et al. 1963). Yet, despite all of the problems, the 232-Z Facility did recover enough plutonium to achieve economic pay-back of its costs within the first operating year (Reinker 1964).

In 1973, ARHCO shut the incinerator down as part of an overall review of the Plutonium Finishing Plant's interfaces with the environment. Although plutonium finishing workers made many suggestions for the improvement of incinerator operations throughout 1974-1976, the Atomic Energy Commission never restarted the 232-Z Facility (Felt 1974).

242-Z WASTE TREATMENT FACILITY

As a part of the overall effort to reclaim plutonium from wastes at the Plutonium Finishing Plant, the Atomic Energy Commission authorized Project CGC-912, 242-Z Waste Treatment Facility, in 1960. While the RECUPLEX Facility recycled solid wastes and organic liquid wastes and the 232-Z Facility recycled combustible and leachable solid wastes, the 242-Z Facility recovered plutonium from aqueous waste streams produced in the Plutonium Finishing Plant.

The 242-Z Facility is described in Appendix B on the Internet. In the spring of 1962 after the criticality event occurred at RECUPLEX, the Atomic Energy Commission undertook intensive criticality reviews of the 242-Z Facility. The agency decided to add neutron monitors, samplers, and filters as well as interlocks and flow rate regulators. The facility began operations in late 1963 (Battey 1962, Graf et al. 1962, Braden 1963, Braden et al. 1963).

In 1964, construction forces installed an americium-241 recovery system in a glove box in the 242-Z Facility. It became operational in May 1965. The product was americium nitrate. In 1969, plutonium finishing workers installed equipment in the 242-Z Facility to convert the americium-241 recovery operation from a batch to a continuous process and to increase the purity and concentration of the americium product (Braden and Ciccarelli 1963, Lane 1965, Krigens and Laws 1968, ARHCO 1969b).

On April 30, 1976, a strike by the Hanford Atomic Metal Trades Council stopped virtually all processing at the Hanford Site. While operators were restarting the americium recovery equipment in the 242-Z Facility on August 30, 1976, an explosion occurred in a column of one of the glove boxes (the cation exchange column W14A within the WT2 glove box). This column contained about 100 grams of americium-241 that had accumulated on the resin before shutdown. The chemical operator who was standing in front of the column was grossly contaminated and suffered facial cuts, burns, eye damage, and some americium-241 deposition.

An Environmental Research and Development Agency investigation determined that the probable cause of the explosion was an exothermic reaction between the 7 molar (a measure of acid concentration) of nitric acid and the radiolytic degradation products from the resin in the column. The column's pressure relief system was not adequate. Following the accident, the Environmental Research and Development Agency closed the 242-Z Facility permanently

(ERDA 1976a, Heid et al. 1979). The accident also had a far-reaching effect on operations throughout the Plutonium Finishing Plant because it prompted a series of evaluations that led DOE to decide in 1978 to decontaminate and decommission most of the Plutonium Finishing Plant facilities.

PLUTONIUM RECLAMATION FACILITY

During 1960, General Electric proposed to the Atomic Energy Commission the construction of a multi-purpose plutonium reclamation facility at the Hanford Site. The reclamation facility would be an enlarged, safer, and more flexible version of RECUPLEX. The Atomic Energy Commission agreed and approved construction the same year (GE 1960h, Travis 1960). After the April 1962 criticality closed the RECUPLEX Facility permanently, the rapid completion of the Plutonium Reclamation Facility became a virtual necessity for plutonium finishing operations (Zangar 1962, GE 1962d).

Construction of the Plutonium Reclamation Facility began June 1962 and concluded December 1963. After initial test runs, operators began processing in May 1964. The Plutonium Reclamation Facility (236-Z Building) differed from RECUPLEX in that it used a continuous organic treatment and recycling process, operated remotely, and contained geometrically favorable equipment rather than relying on administrative controls to ensure nuclear safety (Low 1967). Geometrically favorable equipment was equipment built in sizes and configurations that would avoid an accumulation of plutonium large enough to cause a criticality. Figure 2-5.8 shows a schematic of the Plutonium Reclamation Facility.

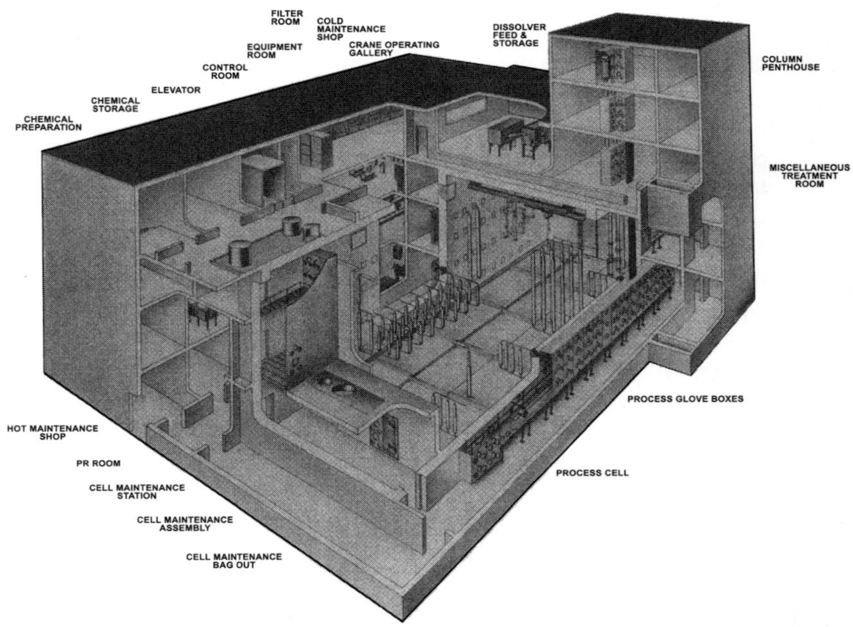

Figure 2-5.8. *Schematic of the Plutonium Reclamation Facility, 1984*

The Plutonium Reclamation Facility (236-Z Facility) is described in Appendix B on the Internet. The Plutonium Reclamation Facility process began with the Miscellaneous Treatment system, a collection of five (later six) glove boxes located on the fourth floor of the facility. Operators dissolved or leached plutonium-bearing scrap in a Miscellaneous Treatment glove box. Plutonium-bearing scrap that already existed in aqueous solution bypassed the Miscellaneous Treatment system and entered the next phase of processing (solvent extraction) from a transfer station located in the Plutonium Reclamation Facility. In upgrades completed in 1983, construction forces installed equipment for the transfer of plutonium-bearing scrap already in aqueous solution in a glove box located in the Plutonium Finishing Plant (Upington 1968, Ludowise 1980).

Operators accomplished solvent extraction in the Plutonium Reclamation Facility through remote operation of equipment in a large central cell known as the canyon. They used a solution of tributyl phosphate diluted with carbon tetrachloride to strip metal nitrates and other wastes out of the plutonium-bearing feed. Then they used a nitric acid-hydroxylamine solution to separate the product. They concentrated the product stream as plutonium nitrate to be ready as feed for the RMA and RMC Lines (Ludowise 1980).

By 1969 according to ARHCO, the Plutonium Reclamation Facility was processing approximately 4000 scrap items per month. The facility routinely processed ash from the 232-Z Incinerator, dissolved solids (plutonium metal and plutonium oxides), performed solvent extraction, and operated several unique processes for materials "not adaptable for direct input to other processing" (Kuklinski and Olson 1969, p. 7). One glove box processed chloride scraps, and another began burning polystyrene cubes embedded with plutonium for recovery of product material. During June and July 1969, operators executed another successful process test that paved the way for routine dissolving of the magnesium oxide crucibles used in the reduction step of the RMC Line. Again, they were burdened with ongoing equipment breakdowns and corrosion caused by the high corrosivity of the Plutonium Reclamation Facility process solutions (ARHCO 1969b).

Maintenance outages and repairs continued to impact the operation of the facility. In mid-1975, ARHCO reported that the Plutonium Reclamation Facility operated only about 25 percent of the time because of the need to replace corroded pumps, valves, and lines. The Energy Research and Development Agency shut down the Plutonium Reclamation Facility in late December 1975 for maintenance upgrades and reviews of criticality prevention specifications and procedures. After restart, the Plutonium Reclamation Facility operated only a short time until the explosion in the 242-Z Facility resulted in the shutdown of the entire Plutonium Finishing complex. The facility did not again resume full operations.

EFFLUENTS AND EFFLUENT CONTROL

CRIBS, DITCHES, AND TANKS

Over the years, the various federal agencies and contractors who managed the Plutonium Finishing Plant built 22 soil disposal facilities, including cribs, ditches, tile fields, reverse wells and French drains to receive liquid wastes from the various Plutonium Finishing Plant facilities. The facilities were designated as 216-Z structures. Suffix numbers were applied in sequence according to when the structures were constructed. During the early years of plutonium isolation and finishing activities at the Hanford Site, plutonium finishing workers disposed all wastes not designated for plutonium recovery or recycling to the soil column. They fed these wastes to the soil through the 241-Z Facility, a system of holding and sampling tanks and associated piping and auxiliary facilities (Smetana 1976, Marshall 1977, Pinto 1978).

Throughout the operating years at the Plutonium Finishing Plant, personnel replaced cribs when plutonium breakthrough occurred beyond limits defined as acceptable at the time. For example in 1958, the Atomic Energy Commission authorized replacement of crib 216-Z-3 when soil samples indicated that plutonium had reached a depth of 50 feet above the water table. The various federal management agencies authorized many other crib replacements over the years. As the result of crib disposal of organic solvents from plutonium finishing processes for many years, a large amount of carbon tetrachloride entered the ground just south and east of the 236 Z Building.

Z-9 CRIB MINING

One unique aspect to the management of effluents from the Plutonium Finishing Plant was the mining of the Z-9 Crib. The 216-Z-9 Crib was a 21-foot deep underground soil disposal facility with an enclosed cavern (a wooden void-space) at the head end, located just southeast of the 234-5Z Building. It was covered by a 9-inch thick concrete slab measuring 90 feet by 120 feet, but the crib sloped inward toward its bottom and was only 30 feet by 60 feet at the floor. The crib was built during 1954 to serve the RECUPLEX Facility, and it received the current acid waste stream of both aqueous and organic wastes from RECUPLEX operations as well as unwashed fabrication oil from the RMA Line from 1955-1962. Figure 2-5.9 shows the above ground equipment used to mine plutonium from the 216-Z-9 Crib.

Figure 2-5.9. *Equipment Used to Mine Plutonium from the 216-Z-9 Crib, 1976*

General Electric took samples of crib soil in 1964 and 1965 and concluded that the crib posed a "potential criticality hazard" (Fillmore 1965, p. 1). The company developed a soil mining and leaching plan as early as 1966. Essentially, the plan called for the use of modified mining equipment to remove the top 1-foot layer of crib soil and to leach out the plutonium by dissolving batches of the soil in mixtures of nitric-hydrofluoric acids. The plan further specified that the leached soil would be packaged into drums and buried in specified Hanford Site solid waste burial grounds (Bierman 1964, Borgeson 1966).

In 1967, sampling personnel revised estimates of the plutonium content of the Z-9 Crib from the 27.5 kilogram figure given in 1965 to 196 kilograms. Wide variations existed in soil sample results based on where in the crib (position and soil depth) samples were taken. Hanford scientists advanced a number of theories to explain the differing results, including postulating that gelatinous precipitates had formed when RECUPLEX waste contacted local soils, and that ion exchange reactions between solvents and the soil were "holding up" plutonium in the soil's upper layers. At the time, neither cation, anion, nor organic residue analyses had been run, nor had pulsed neutron reactivity measurements been taken (Anderson 1972). Because of the widely varying soil sample results, Hanford scientists modified project proposals for the mining activity as well as the chemical flow sheets for the leaching process many times. These changes, as well as the changeover in the Hanford Site chemical processing contract from Isochem, Inc. to ARHCO in 1967, delayed the soil mining project (Ingalls 1967, Borgeson 1967, Knights 1969).

After a number of other delays, on August 17, 1976, plutonium finishing personnel began the first removal of contaminated soil from the Z-9 Crib. Nine days later, the Energy Research and Development Agency shut down the operation to resolve differences in two methods of soil analysis being used. On January 11, 1977, they authorized the resumption of mining operations. Plutonium finishing personnel removed the top 1 foot of soil from the north half of the crib and produced a total of 28.5 kilograms of plutonium. When they mined the south half, they produced 29.5 kilograms of plutonium and completed the operation July 14, 1978 (Malody 1976; Hanson 1977, 1978).

EFFLUENT CLEANUP PROJECT

By the late 1960s, Hanford Site records indicated that about 80 percent of the plutonium discharges to Hanford Site soils came from Plutonium Finishing Plant facilities. The Atomic Energy Commission directed ARHCO to recommend program upgrades to correct this situation (Sloat 1967). In response to ARHCO proposals, the Atomic Energy Commission authorized Project HCP-669, Plutonium Effluent Cleanup, in March 1972. This project was designed to "provide final disposal of the collected organic wastes [produced by the plutonium finishing operations and] to isolate and immobilize the associated radioactivity" (Hall 1972, p. 2). The project entailed three components: conversion of high-salt aqueous waste (mostly from the Plutonium Reclamation Facility) into a salt cake, using the 242-T Evaporator; recovery of carbon tetrachloride from spent organic extractant at the Plutonium Reclamation Facility through a steam distillation process; and storage of the degraded organic phosphate residue from the extractant in existing tanks (Hanson 1971, 1972; Chiaramonte 1972; Nemzek 1972).

STABILIZATION AND DEACTIVATION

After the Plutonium Reclamation Facility shut down in 1987 and the RMC Line shut down in 1989, the Plutonium Finishing Plant entered a waiting period while the world witnessed some of the most historic events of the twentieth century. In late 1991, the former Soviet Union began to dissolve into fifteen separate and independent republics. January 1993, President George Bush signed the second Strategic Arms Reduction Treaty (START II) with Russian President Boris Yeltsin. The result was that U.S. policy makers decided not to manufacture more weapons-grade plutonium.

DOE signed a Record of Decision on the final Environmental Impact Statement for the Plutonium Finishing Plant. This Record of Decision outlined the manner in which the agency would complete stabilization. Unlike traditional approaches, the Plutonium Finishing Plant Record of Decision identified a toolbox of multiple stabilization and cleanup alternatives that could be used as needs arise during the various processes. Among the alternatives were the cementation of some plutonium bearing solids, thermal stabilization of some plutonium-bearing liquids, and various treatment methods for plutonium oxide polycubes and other unusual items. Selection of the alternatives would be determined more precisely as laboratory characterization and knowledge of treatment alternatives went forward. Monitoring and control systems could be changed as various nuclear inventories were moved, removed, and stabilized. The various degrees of cleanout over time at the 232-Z Incinerator, the Plutonium Reclamation Facility, and the 231-Z Plutonium Isolation Building nearby were not prescribed completely, but options were bounded that would allow stabilization methods and time frames to change. There are no agreements to remove the plutonium vaults that are located in buildings separate from the main plants in the Plutonium Finishing Plant (Vogt 1996; Nuclear Waste News 1996a, 1996b).

AREAS FOR FURTHER RESEARCH

As with the other operations in the plutonium production process, the history of plutonium finishing is one of continually being pressed to expand production during the Cold War era. Key to keeping pace was the move from the manual to mechanical production lines. That move, however, was fraught with problems, which researchers and workers raced to solve. This narrative addresses many of the remote mechanical line problems and solutions, but because this narrative is a history of facilities and not intended to be the definitive history of plutonium finishing, there are areas not addressed, which future researchers may want to explore, such as:

- How did plutonium finishing work at the Hanford Site compare with plutonium finishing operations at the Los Alamos and Rocky Flats sites?
- Did the needs and requirements of the Plutonium Finishing Plant at the Hanford Site have an effect on the production rates, separations chemistry, or other operations? If so, what changes were made at other facilities to accommodate the plutonium finishing needs? Or was the Plutonium Finishing Plant able to be responsive to the other Hanford production facilities, handling whatever materials were sent to it?
- How can the decision to construct and begin operating the Plutonium Finishing Plant without connecting it to existing underground, high-level waste storage tanks or to construct new tanks specifically for the Plutonium Finishing Plant, be understood? At what level and by whom was this decision made, and what evidence can be found for the rationale? Did anyone argue in favor of high-level waste tanks for the early Plutonium Finishing Plant? If so, who and what arguments did they advance?
- Storage of finished plutonium and plutonium scraps, leftovers from experiments, and other concentrated, non-waste, alpha emitting substances at the Hanford Site took place largely at the Plutonium Finishing Plant. The entire topic of the storage of plutonium and plutonium bearing materials should be explored. What facilities and standards were used? How were these facilities upgraded over time? Were the facilities and upgrades adequate? What inventory of plutonium bearing materials is left in the Plutonium Finishing complex today, and what are the future plans for stabilization and storage of these materials?

SECTION 6

WASTE MANAGEMENT

D.H. DEFORD
HANFORD SEARCH

SECTION 6 - WASTE MANAGEMENT

Manufacturing fuel, irradiating the elements, separating out the plutonium, and finishing the end product for use in weapons required numerous industrial, chemical, and radiological processes. Each process generated huge quantities of hazardous and/or radioactive waste at the Hanford Site. The character of this waste and the manner in which it was handled, transported, stored, and disposed of varied depending on the source of waste, era, location, and evolving knowledge, practices, and regulations.

Given the near lack of knowledge at the start of Hanford's operations, the early managers had to develop waste management practices based on a limited understanding of nuclear physics, engineering, radiochemistry, and the effects of ionizing radiation on living organisms. E.I. Du Pont de Nemours and Company, the contractor who designed and constructed the industrial complex necessary to manufacture plutonium, acted in concert with the U.S. Army Corps of Engineers to establish the initial waste management practices necessary to bring the early facilities on-line and to operate them through the final years of the World War II. Management by an experienced explosives manufacturer, exigencies of the war, pressure for plutonium production, practical expertise, and trial-and-error contributed to the initial family of waste management methods employed at the Hanford Site.

Some of the initial practices proved undesirable or impractical, and adjustments were made. Open ponds and trenches gave way to reverse wells, which yielded, in turn, to cribs. Burial containers that collapsed under the weight of cover materials were replaced with metal drums and concrete boxes. Segregation and retrievable storage was implemented as the special hazards of transuranic waste became apparent. Double-shell storage tanks replaced single-shell tanks for storage of high level waste. Record keeping improved.

Early waste management planners appear to have relied heavily on future technology to resolve problems that seemed to have no immediate solution. Herbert M. Parker, Hanford's chief health physicist during the Manhattan Project, stated in 1948 that "Projection of the problem to [the] future...appears to be irrelevant in terms of the technological progress in corrective measures that can be anticipated" (Parker 1948a). With this reliance on future technology to resolve problems, waste management planners stored high level waste, stabilized contaminated soil by covering it with fresh soil, disposed of low level liquid waste in the soil, and buried solid waste in shallow trenches. The waste has not gone away, although radioactive decay has reduced the magnitude of the radionuclide content. Only in the latter 1990s had Parker's expected "technological progress" begun to be applied in large scale to clean up the Hanford Site.

On the regulatory side, the Atomic Energy Act of 1946 gave the Atomic Energy Commission control over radioactive waste and standards for managing the waste. The *Leaf versus Hodel* decision in 1984 opened the legal door for increased environmental oversight by the U.S. Environmental Protection Agency and the Washington State Department of Ecology. In 1989, the U.S. Department of Energy (DOE), U.S. Environmental Protection Agency, and the State of Washington signed an agreement (the Tri-Party Agreement), which dictated a schedule for the many tasks needed to cleanup the Hanford Site (Ecology 1994).

Although managed, much of the waste at the Hanford Site remains. Some forms of waste have at times crossed the boundaries of the Hanford Site

CONTENTS

Divider Page Photo: Double-Shell Tank Farm (Neg. 7813231)

and impacted neighboring communities. With the full disclosure policy established by DOE in 1986 and the release of 19,000 pages of previously classified waste management and other data, much public criticism has been directed toward DOE and its predecessor agencies for allowing the waste to accumulate and, in some cases, to escape its artificially imposed boundaries. This criticism has served to improve cleanup efforts, openness, and accountability for managing the waste at the Hanford Site.

To appreciate the scope of the waste management task, Table 2-6.1 shows what Hanford's current inventory of waste and nuclear materials includes.

Table 2-6.1. *Hanford Site Waste Types and Nuclear Materials*

	Volume	Depth if Spread on a Football Field	Curies*	Chemicals
Tank Waste	54 million gal	150 feet	200 million	240,000 tons
Solid Waste	25 million ft³	500 feet	6 million	70,000 tons
Soil and Groundwater	40 billion ft³	200 miles	1 million	100,000 to 300,000 tons
Facilities	175 million ft³	4000 feet	10 million	trace
Nuclear Material	25,000 ft³	0.5 inch	200 million	none

* A measure of the amount of radioactivity

E0001111_10

DOE recognizes several distinct classes of waste that have been generated by the nation's nuclear industry and that must be managed. These include hazardous waste, low level waste, high level waste, transuranic waste, uranium mill tailings, and various combinations of classes such as mixed low level waste. Because uranium mill tailings are not present at the Hanford Site, they are not addressed further (DOE 1992a). Most countries define high level and low level waste according to their level of radioactivity. In the United States, however, DOE defines the classes of waste according to the waste source. This leads at times to "low level waste" having more radioactivity than waste classified as "high level" (Gephart and Lundgren 1997). Table 2-6.2 provides definitions used in the United States for each of the classes of waste.

Table 2-6.2. *Classes of Waste Generated at the Hanford Site*

Waste Type	Definition
Hazardous Waste	Non-radioactive solid waste, such as lead and other heavy metals, solvents, paints, oils, other hazardous organic materials, or materials that pose a risk to the environment and human health because they exhibit certain characteristics of corrosivity, ignitability, reactivity, or toxicity (see 40 CFR Parts 260-279 for details)
Low Level Waste	Any waste containing radioactive materials, either solid or liquid, that are not classified as high level waste, transuranic waste, spent nuclear fuel, or by-product material
Transuranic Waste	Waste contaminated with alpha emitting radionuclides with half-lives greater than 20 years in concentrations more than 1 ten-millionth of a curie per gram of waste (DOE 1997e)
High Level Waste	Solid and liquid waste from the reprocessing of spent reactor fuel to extract plutonium and containing some elements that decay slowly and remain radioactive for thousands of years, requiring permanent isolation. Due to intense levels of radioactivity, high level waste must be handled by remote control from behind protective shielding and is typically stored at the Hanford Site in underground tanks (see 42 USC 10101 et seq. for details).
Mixed Waste	Combination of hazardous waste and low level or transuranic waste
Sanitary Waste	Non-regulated waste

The above waste categories represent the end-product of many years of waste management regulation and are useful in managing today's waste inventory and waste currently being generated. However, these distinctions among waste types have not always existed. In the early years of Hanford Site operations, radioactive waste was more simply treated, and only two categories were generally recognized: high level and low level. High level waste was stored in underground tanks, and low level waste was disposed of in the soil. For practical purposes, low level waste was divided into liquid and solid waste categories with different disposal methods for each. Over time, additional waste categories were recognized, and associated waste treatment, storage and/or disposal methods were mandated.

Waste management practices at the Hanford Site have been characterized by evolving concepts of the physical behavior and environmental safety of radioactive and hazardous waste. This section is organized around the historical waste disposal practices employed at the Hanford Site rather than by modern waste categories.

WASTE MANAGEMENT LEGISLATION

Legislation that has impacted the management of Hanford's waste has evolved over time and reflects the needs and desires of an interested and vocal populace.

No legislation relating specifically to waste management can be identified for the early years of operations at the Hanford Site. The U.S. Army Corps of Engineers' Manhattan Project conducted its wartime affairs under a policy of strict secrecy. In the absence of knowledge, the public could only be minimally interested, demanding little or no regulatory legislation. Not until the Atomic Energy Act of 1946 did the federal government begin to develop legislation dealing with atomic energy. The following chronology provides a partial characterization of legislation that has impacted the management of nuclear processes and the resultant waste.

1942 The Manhattan Project was the U.S. Government's top-secret program to produce an atomic bomb for use during World War II. Internal Army and contractor standards and procedures were developed to dictate the management of waste.

1946 The Atomic Energy Act (42 USC 2011 et seq.) placed production and control of nuclear materials within a civilian agency, originally the Atomic Energy Commission created in 1947. The Atomic Energy Commission was succeeded by the Energy Research and Development Administration (1975), which, in turn, was replaced by the U.S. Department of Energy (1977).

1948 The Federal Water Pollution Control Act (33 USC 1251 et seq. 1948) provided the original framework for water pollution control. It was later amended and renamed the "Clean Water Act."

1955 The Clean Air Act (42 USC 7401 et seq.) was enacted to protect and enhance the quality of the Nation's air resources. Its primary application is through permits to regulate new and existing facilities.

1969 The National Environmental Policy Act (42 USC 4321 et seq.) established the requirements for conducting environmental reviews of federal actions that have the potential for significant impact on the human environment.

1970 The Atomic Energy Commission Immediate Action Directive 0511-21 defined transuranic waste and directed that transuranic waste could no longer be buried and should instead be segregated and placed in retrievable storage.

1972 The Maritime Protection, Research, and Sanctuaries Act (33 USC 1401 et seq.) was enacted to regulate ocean dumping of waste and, thereby, prohibited the dumping of any waste from the Hanford Site in the ocean.

1973 Atomic Energy Commission Order 0511 amplified and enlarged on the Immediate Action Directive (IAD) 0511-21 of 1970.

1974 The Safe Drinking Water Act (42 USC 300f et seq.) was enacted to protect drinking water supplies. Standards developed under these amendments are used in groundwater monitoring programs.

1976 The Resource Conservation and Recovery Act (RCRA) (42 USC 6901) established a national policy to reduce or eliminate hazardous waste and treat, store, and dispose to minimize the threat of pollution. The Hazardous and Solid Waste Amendments amended RCRA in 1984 to expand RCRA's scope and add detailed requirements.

1976 The Toxic Substances Control Act (15 USC 2601 et seq.) was enacted to protect human health and the environment from exposure to and manufacture, distribution, use, or disposal of substances containing toxic chemicals.

1977 The Clean Water Act (33 USC 1251 et seq. 1977) amended the Federal Water Pollution Control Act. Its objective is to "restore and maintain the chemical, physical, and biological integrity of the nation's waters." The act addresses surface water only.

The National Pollutant Discharge Elimination System is a provision of the Clean Water Act, which prohibits discharge of pollutants into U.S. waters unless a special permit is issued by the U.S. Environmental Protection Agency, a State, or a Tribal Government of an Indian reservation.

1980 The Comprehensive Environmental Response, Compensation, and Liability Act (CERCLA) (42 USC 9601 et seq.) was enacted as a federal statute (also known as the Superfund) that provides the statutory authority for cleanup of hazardous substances that could endanger public health, welfare, or the environment. It was reauthorized and amended by the Superfund Amendments and Reauthorization Act in 1986.

The National Priorities List is a list of the most serious uncontrolled or abandoned hazardous waste sites identified for possible long-term remedial action under CERCLA.

The Formerly Utilized Sites Remedial Action Program is a DOE program that addresses the cleanup of sites and adjacent properties contaminated by activities of the Manhattan Project.

1982 The Nuclear Waste Policy Act (NWPA) (42 USC 10101 et seq.) was enacted. It directs DOE to design, site, and construct a geologic repository for the disposal of defense, high level radioactive waste and for spent fuel from civilian (commercial) nuclear reactors.

NWPA established a Nuclear Waste Fund to fund the geologic repository by assessing fees on the public utilities that will use the repository. The fund resulted in the construction of the Waste Isolation Pilot Plant near Carlsbad, New Mexico.

1984 The *Leaf versus Hodel* decision required DOE to comply more fully with RCRA and CERCLA. Neither RCRA nor CERCLA had a major impact on DOE production systems before the *Leaf versus Hodel* decision. With that decision, DOE applied the ruling department wide, and the U.S. Environmental Protection Agency and Washington State Department of Ecology became more involved as regulators of RCRA and CERCLA.

1989 The Hanford Federal Facility Agreement and Consent Order (Tri-Party Agreement) (Ecology 1994) was signed. An agreement among DOE, the U.S. Environmental Protection Agency, and the Washington State Department of Ecology, it legally requires DOE to safely manage and dispose of liquid and solid waste at the Hanford Site. It contains milestone deadlines by which specific waste management and cleanup activities are to be accomplished.

1991 National Defense Authorization Act for Fiscal Year 1991 (Public Law 101-510) was enacted, requiring DOE to identify and monitor certain categories of underground high level waste tanks.

In addition to the legislation listed above, the following current DOE Orders, listed by order number, regulate today's waste management at the Hanford Site.

1540.2 Hazardous Material Packaging for Transport outlines administrative procedures for the certification and use of radioactive and hazardous material packaging.

5400 Series	General Environmental Protection Program ensures that all DOE environmental restoration program activities will comply with applicable local, state, and federal environmental protection laws and regulations.
5440.1C	National Environmental Policy Act establishes DOE procedures to implement NEPA.
5600 Series	Management of DOE Weapons Program and Weapons Complex assigns responsibilities and authority for the packaging and transportation of nuclear materials.
5820.2A	Radioactive Waste Management establishes policies to manage high level, low level, and transuranic waste.

PROCESS FACILITIES AND RESULTING WASTE STREAMS

Several industrial-scale process facilities operated in concert to produce Hanford plutonium. The production process began with the receipt of purified uranium billets at the 300 Area fuels manufacturing facility. Through a series of extrusion and milling processes, operators transformed the raw uranium into 6-inch long, rod-shaped reactor fuel elements, referred to as slugs. Because the fuel elements were to be in direct contact with cooling water in the reactor core and since uranium oxidizes rapidly in water, the fuel elements had to be jacketed in a material that did not react with water. The operators used a dipping process to apply an aluminum alloy that fully coated each element. Operators followed strict quality control procedures to weigh and measure the fuel elements and check the integrity of the jacket. These fuel manufacturing processes produced uranium-contaminated liquid and solid waste that were disposed of in 300 Area waste sites.

Trucks transported the fuel elements to one of nine production reactors where the elements were loaded into a reactor core and irradiated for a specified period of time and power level. In this process, neutrons generated in the reactor core interacted with uranium atoms to form plutonium. The fuel elements were then discharged from the reactor core into fuel storage basins where they were allowed to cool under 20 feet of water. The cooling process reduced the temperature of the elements and allowed for decay of short-lived radioisotopes created in the irradiation process.

Reactor operations generated several waste streams including solid waste that was disposed of in burial grounds, low level liquid waste that was disposed of in the soil, and reactor cooling water, another form of low level liquid waste. After passing through the reactor, the cooling water (reactor effluent) from the reactors without closed loop systems was allowed to cool in holding basins before release to the Columbia River.

After the now intensively radioactive fuel elements cooled in the reactor fuel storage basin, operators transferred them by remote handling methods into heavily shielded shipping casks and transported them by rail to one of the chemical separations plants located in the 200 Areas. There, the fuel elements were dissolved in acid, and through a wet chemical process the plutonium was separated from the uranium matrix of the fuel elements. The processes involved in chemical separations resulted in high level liquid waste that was transferred through underground lines to underground storage tanks. Low level liquid waste was also generated and disposed of in ponds, trenches, French drains, and cribs (see Glossary for definitions). Solid waste was shipped by truck or rail to 200 Area burial grounds for disposal. Airborne contaminants were vented up stacks and released to the atmosphere.

The separated plutonium, a gelatinous semi-solid, was transferred to the Plutonium Finishing Plant in the 200 West Area where it was converted into plutonium metal and then formed for use in weapons. The Plutonium Finishing Plant generated small quantities of high level waste as well as larger quantities of low level liquid and solid waste, all of which were stored or disposed of in 200 West Area disposal or storage facilities.

Scientists conducted extensive research and development in support of the plutonium production technologies, primarily in the 300 Area laboratories. The low level liquid and solid waste generated from this research was disposed of in the

300 Area ponds, trenches, burial grounds, and at waste disposal facilities in other areas. Some 300 Area laboratory solid waste was high level waste that required special handling and burial facilities. Low level solid waste was also received from offsite locations for disposal at the Hanford Site.

All program areas generated conventional solid waste that was disposed of in landfills and sanitary liquid waste that was disposed of in conventional septic tank and other sewage treatment facilities.

The agencies in charge of the Hanford Site have monitored groundwater on the Hanford Site since the 1940s to help determine what chemical and radiological contaminants have made their way into the groundwater and the vadose zone—the soil between the ground surface and the water table. Regulatory requirements for monitoring increased in the 1980s.

HIGH LEVEL RADIOACTIVE LIQUID WASTE

High Level Radioactive Waste

***"Highly radioactive waste material resulting from the reprocessing of spent nuclear fuel, including liquid waste produced directly in reprocessing and any solid material derived from such liquid waste that contains fission products in sufficient concentrations; and other highly radioactive material that is determined, consistent with existing law, to require permanent isolation."* - DOE 2001**

Five chemical separations plants primarily produced Hanford's high level radioactive waste, although the Plutonium Finishing Plant and other operations contributed small quantities. T and B Plants were the pioneer plants that operated from the mid 1940s through 1956. Improved technologies led to the Reduction-Oxidation Plant (REDOX) that operated from 1952 to 1967, and to the Plutonium-Uranium Extraction Plant (PUREX) that operated from 1956 to 1988. U Plant (221-U) operated from 1952 until 1958 to recover uranium from T and B Plant high level radioactive waste. All plants used complex, toxic and corrosive chemicals in their separation processes.

At each separations plant, operators dissolved the irradiated uranium fuel rods in nitric acid, which removed both the protective outer jacketing and reduced the uranium-plutonium metal to a liquid state. They then extracted plutonium from the uranium nitrate solution and sent it to the Plutonium Finishing Plant to be purified into plutonium metal.

Each step of the plutonium extraction process produced high level radioactive waste that was transferred through underground lines to large underground storage tanks located in tank farms (see Figure 2-6.1 and the Historic Property Inventory Form for the Chemical Handling Facility/Tank Farm, 334 Building, in Appendix B on the Internet) near each separations plant. Much of this high level radioactive waste continues to be in storage today.

Figure 2-6.1. *Double-Shell, High Level Radioactive Waste Tank Farm under Construction, 1978*

CHARACTERIZATION

High level radioactive waste is difficult to characterize. The differing chemical separation processes employed over time produced twenty-six different waste streams that were chemically dissimilar but were combined in the underground tanks to form even more complex waste combinations. Exposure to air, heat, cascading, settling, radioactive decay, kinetics, and subsequent chemical treatments further complicated the nature of the high level radioactive waste with radical differences developing from tank to tank. The addition of sodium hydroxide, an acid neutralizer, caused portions of the waste to form solid particles and to separate into layers. Evaporation of liquids led to salt cake (see Figure 2-6.2) and slurry formation, causing further uneven distribution of chemical compounds and radioisotopes. Waste was transferred from tank to tank as it went through subsequent treatment and isotope recovery campaigns. All of these factors contributed to waste complexity and to the difficulty of sampling and characterizing the high level radioactive waste to plan for its remediation.

Figure 2-6.2. *High Level Radioactive Waste in Underground Storage Tank with Salt Cake Visible on Surface, 1971*

High level radioactive waste characterization is further complicated by the activities of U Plant, which was used from 1952 to 1958 to recover valuable uranium from T and B Plant high level radioactive waste. U Plant was originally constructed as a bismuth-phosphate chemical separations plant, identical to its sister plants, B and T. But its capacity proved unnecessary and it served only as a training facility until 1952 when it was redesigned for its new mission.

Since B and T Plants extracted only plutonium from the uranium fuel elements, the resulting high level radioactive waste remained rich in uranium. The high value of uranium drove a decision to convert U Plant to a uranium recovery mission. Using tributyl phosphate and saturated kerosene chemistry, U Plant reprocessed T and B Plant high level radioactive waste to recover uranium. Its waste stream included tributyl phosphate, saturated kerosene, nitric acid, oxalic acid, and other chemicals. Unfortunately, the U Plant uranium recovery process generated about twice as much waste as it processed, placing additional strain on the storage tank farms that were nearing capacity.

As each chemical separations plant came on line, the designers improved the technologies to reduce the quantity of high level radioactive waste per unit of irradiated fuel processed. For instance, PUREX generated only about 250 gallons of high level radioactive waste per ton of irradiated fuel, a major improvement over the 10,000 gallons per ton generated by T and B Plants. However, production capacity also increased with PUREX processing 33 tons of fuel per day, compared with T or B Plant's mere 30 tons per month in their best and final years. Ultimately, nearly 245 million gallons of high level radioactive waste were transferred to Hanford's 177 underground storage tanks from the five separations plants (Anderson 1990, Gephart and Lundgren 1997).

B and T Plants used bismuth-phosphate chemistry in a batch process employing repeated dissolution, precipitation, and centrifugation, which produced about 10,000 gallons of high level radioactive waste per day. Process chemicals included nitric acid, phosphoric acid, bismuth phosphate, sodium dichromate, bismuth nitrate, potassium permanganate, sodium nitrate, and others. All of these, plus aluminum, uranium, plutonium, and numerous fission products were included in the high level radioactive waste. The highly acidic waste was acid neutralized with large quantities of sodium hydroxide before transfer to the storage tanks (Anderson 1990).

REDOX came on line in 1952 with an improved separation process and different chemistry. Its solvent extraction process used methyl isobutyl ketone (hexone), aluminum nitrate, nitric acid, sodium dichromate, and other such chemicals. REDOX high level radioactive waste included all of these chemicals plus solutions containing ferrous ions.

PUREX became operational in 1956 as Hanford's final and most advanced separations plant by using a continuous flow extraction process, involving tributyl phosphate, saturated kerosene, nitric acid, oxalic acid, ammonium fluoride, ammonium nitrate, and numerous other chemicals. All became part of the PUREX high level radioactive waste stream. Concentrators reduced the waste volume to 250 gallons per ton of processed uranium. The concentrators evaporated liquids to reduce the volume. This did not reduce the radioactivity. Production, however, increased to 33 tons per day, largely offsetting the net waste volume decrease that might otherwise have been realized (Anderson 1990).

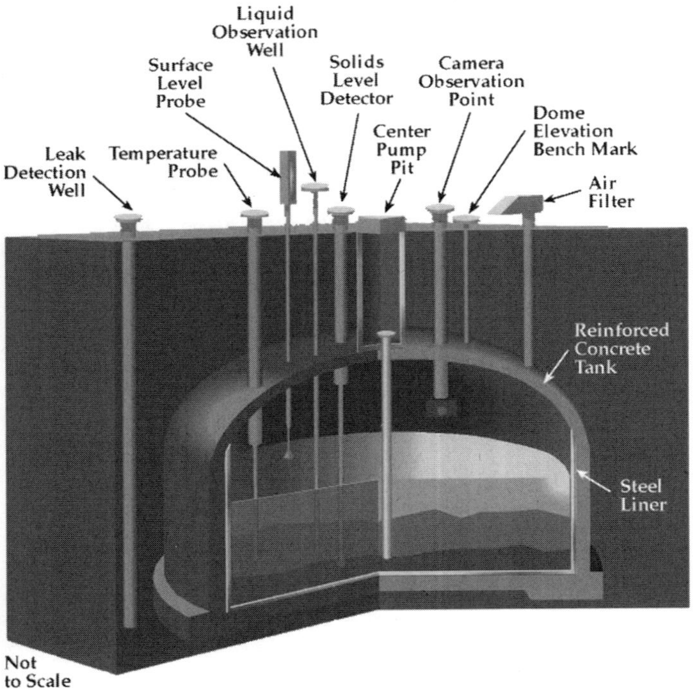

Figure 2-6.3. *Typical Single-Shell Tank at the Hanford Site with Numerous Access Ports and Risers for Monitoring and Waste Sampling*

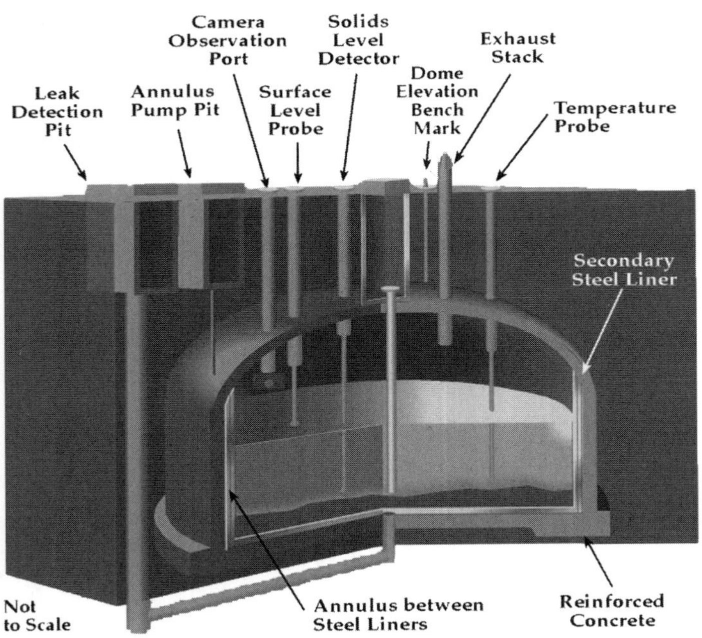

Figure 2-6.4. *Typical Double-Shell Tank at the Hanford Site with Numerous Access Ports and Risers for Monitoring and Waste Sampling*

WASTE HANDLING FACILITIES

Facilities associated with high level radioactive waste include the five chemical separations plants, the lines and encasements used to transfer high level radioactive waste about the 200 Areas, diversion stations, storage tanks, and evaporator buildings. For a description of the chemical separations facilities, see Section 4, Chemical Separations.

The primary method used to transport high level radioactive waste across the Hanford Site was through underground process lines that ran, for the most part, through concrete encasements. Most process lines are 3-inch diameter, stainless steel pipe with welded joints. Encasements are underground concrete fixtures designed to contain and protect from one to seven process lines. The encasements vary in width, depending on the number of lines within them. Diversion stations located at the process facilities and tank farms permit rerouting of high level radioactive waste through different lines to alternate locations.

For high level radioactive waste, the lines carried the liquid to underground storage tanks, clustered into tank farms. Storage tanks are of two types. Both types, single-shell and double-shell tanks, are cylindrical shaped concrete structures with carbon steel inner liners (see Figures 2-6.3 and 2-6.4 and the Historic Property Inventory Forms for the 241 Buildings in Appendix B on the Internet). Single-shell tanks have a single inner liner while double-shell tanks have two liners with a space between them. This space provides a pathway by which leakage from the inner liner may be trapped, collected, detected, and recovered. Leakage has occurred only from single-shell tanks.

Single-shell tanks are generally smaller but more numerous than double-shell tanks. Starting in 1945, 149 single-shell tanks were constructed, providing 94 million gallons of storage capacity. They range in capacity from 55,000 to 1 million gallons and are buried with their upper surfaces 6 to 11 feet below grade. The earth cover provides shielding from the intense radiation of the high level radioactive waste, permitting tank farm workers to move about the areas above the tanks. Numerous and various openings exist in the tank dome to accommodate vents, filters, detection tubes, probe access wells, pumps, cameras, and the many kinds of instruments necessary to monitor content of the tanks. For information on the instruments used to monitor the tanks, see Gephart and Lundgren (1997).

Between 1968-1986, 28 newer double-shell tanks were built and now provide a storage capacity of 31 million gallons. They range in size from 1 to 1.1 million gallons and are buried 7 feet below grade. Similar openings exist in the tank dome to provide instrumentation, process, and sampling access to tank contents.

To reduce the volume of waste, operators sent the waste from the tanks to the waste evaporator buildings (see Figure 2-6.5), one of which is still operating today. Waste evaporator buildings are large, rectangular concrete structures. Evaporators are steam-heated containers, which boil the liquid out of the particulates, leaving a condensate from the evaporated liquid and a concentrated slurry. The condensate can be treated as low level liquid waste and so be disposed of in nearby trenches or cribs. Evaporation generated a thick, high level radioactive waste slurry that was returned to the high level radioactive waste underground storage tanks where it cooled and formed salt cake. Well over 200 million gallons of liquids were evaporated from high level radioactive waste storage tanks in this manner (Agnew 1997). In the course of evaporation, a small amount of radioactivity was vented to the atmosphere and as liquid to the ground (Gephart and Lundgren 1997).

Figure 2-6.5. *Building Containing Evaporator Used to Reduce Volume of High Level Waste by Boiling Off Water*

KEEPING TRAC

The extreme complexity of Hanford high level radioactive waste presents an unprecedented record keeping and modeling challenge. A high level radioactive waste inventory program called TRAC (Tracks Radioactive Components) was developed in the 1970s to provide managers with inventory tracking and modeling capability. All existing high level radioactive waste transfer records were incorporated into the TRAC database. TRAC serves to track and predict waste inventory by tank, estimate radionuclide content, estimate solids precipitation, and correct for radionuclide decay. It can assist remediation planners by modeling and simulating waste recovery processes. Before TRAC, managers used non-automated transfer records to track the waste and manage the tank farms.

REDUCING WASTE

Reducing Hanford's high level radioactive waste gained in importance over the years as the problems caused by the waste became more apparent. Reducing the volume of high level radioactive waste took several forms, the most successful was the more efficient extraction technology, which generated less waste in the REDOX and PUREX plants. The following chronological description identifies the waste reducing and other waste management activities undertaken to reduce as much high level radioactive waste to be stored in the limited underground tanks as possible.

1944 – 1945 First alkaline slurries of high level radioactive waste were transferred from T and B Plants to the 64 original, underground tanks. Waste was cascaded through a series of tanks to allow solids to settle out in the earlier tanks and only the supernatant liquid (liquid left on the surface) to reach tanks at the end of the cascade. T and B Plants originally produced over 10,000 gallons of high level radioactive waste for each ton of uranium processed.

1948	As underground tank storage capacity became strained, authority was granted to dispose of a portion the high level radioactive liquid waste from the T and B plants in the soil after cascading through three tanks.
1947 – 1948	30 new single-shell tanks constructed with capacities ranging from 530,000 to 758,000 gallons.
1949	Hot Semiworks (C Plant) began operations to pilot technologies for REDOX and PUREX processes and generated high level radioactive waste.
1950 – 1952	30 new single-shell tanks constructed, each with a capacity of 758,000 gallons.
1951	242-T and 242-B Evaporators began processing T and B Plant high level radioactive waste stored in tank farms to reduce volume by boiling off the liquid. Together they processed over 16 million gallons of waste, achieving over 80 percent reduction in volume. The concentrated waste was returned to the storage tanks and the evaporated condensate was disposed of in the soil. For more information on the 242-T Waste Evaporator, see the Historic Property Inventory Form in Appendix B on the Internet.
1952	U Plant began a 2-year uranium recovery campaign, generating between 5,500 and 9,200 gallons of waste per ton of uranium in 1952. Waste evaporators reduced this rate to 3,600 gallons but could process only about 700,000 gallons per month, while U Plant generated over 1.7 million gallons per month of high level radioactive waste. Scavenging of U Plant high level radioactive waste with potassium ferrocyanide and nickel sulfate to settle out cesium-137 permitted disposal in the soil of 11.6 million gallons of relatively inactive supernatant liquid. Sodium ferrocyanide was used as well because it was less expensive to manufacture than potassium ferrocyanide. Another 29 million gallons of waste were disposed of in the soil after scavenging at U Plant removed long half-life fission products (Anderson 1990). Cesium-137 scavenging continued for several years with about 41 million gallons of supernatant liquid discharged in the soil. About 150 tons of ferrocyanide were added to 18 underground tanks (Gephart and Lundgren 1997). Studies show that over time ferrocyanide breaks down into less reactive chemicals (Meacham et al. 1996). The concentration of ferrocyanide in the Hanford tanks is 10-40 times lower than that needed to sustain a fire or explosion. For this reason, the ferrocyanide safety issue was closed out in 1996 and the remaining tanks were removed from the Watch List established by Public Law 101-510, Section 3137 (Public Law 1990).
1952	REDOX began operation, generating high level radioactive waste at a rate of 4,378 gallons per ton of uranium in 1952.
1953	Self concentration was approved for REDOX high level radioactive waste at S Tank Farm. Self concentration involved disposing of vapor condensates in the soil instead of returning it to the underground tank. Over the life of the REDOX Plant, self concentration and REDOX process improvements reduced REDOX high level radioactive waste generation rates from 4378 gallons per ton of uranium in 1952 to 594 gallons per ton of uranium by 1964 (Anderson 1990).
1953 – 1955	21 new single-shell tanks constructed.
1954 – 1956	In addition to disposing of the low level liquid waste in the soil, the sediment on the bottom of the evaporators was also disposed of as low level waste. From 1954-1956, 4.8 million gallons (10,200 curies) of high level radioactive supernatant liquid waste and 1.9 million gallons (11,600 curies) of evaporator sediment from the B and T plants were disposed of in the soil. Disposal was limited to areas of soil capable of retaining liquid without its penetrating through to the water table (Anderson 1990).

Chemical scavenging of high level radioactive waste from the B and T plants was also accomplished during this period. Ferrocyanide compounds and nickel sulfate were added to high level radioactive waste to cause cesium-137 to settle out, thus allowing the resultant cesium-free supernatant liquid to be disposed of in the soil. From 1954-1956, 3.1 million gallons were disposed of in this manner.

In total, these waste volume reduction efforts reduced the high level radioactive waste volumes in underground storage from the B and T plants by a factor of 3.25, reducing net waste generation from 17,000 gallons per ton of uranium to 5,240 gallons per ton of uranium (Anderson 1990).

1956 PUREX began operating, eventually generating high level radioactive waste at a rate of 250 gallons per ton of uranium and processing 33 tons of uranium per day. PUREX high level radioactive waste was sufficiently concentrated that its heat producing fission products caused the waste liquid to boil. This self boiling accelerated to a point in 1957 that boil off reached 10 gallons per minute, and it became necessary to add water to the waste liquids to maintain a reasonable level (Anderson 1990). Self boiling ended the practice of cascading.

All PUREX high level radioactive waste was acid neutralized with large quantities of sodium hydroxide, greatly increasing the waste volume.

1963 – 1964 4 new single-shell tanks constructed.

1966 Thorium recovery campaign generated 1.4 million gallons of high level radioactive waste.

1968 – 1988 28 new double-shell tanks constructed.

1968 – 1978 B Plant operated to recover cesium and strontium. This process recovered strontium-90 from high level radioactive waste sludge and cesium-137 from high level radioactive supernatant liquid waste stored in underground tanks. Removal of these isotopes from the high level radioactive waste had the effect of reducing the heat generating capacity of the waste. The strontium and cesium isotopes were encapsulated for the Oak Ridge National Laboratory isotope program. Remaining waste was evaporated with condensates disposed of in the soil and the concentrates returned to underground storage. A large inventory of encapsulated strontium and cesium remains stored at the 200 East Area Waste Encapsulation and Storage Facility (see the Expanded Historic Property Inventory Form for the Waste Encapsulated and Storage Facility, 225 Building, in Appendix B on the Internet).

1973 – 1976 Two new evaporator plants, 242-S and 242-A came online. 242-A remains operational today.

Over the history of the Hanford Site, the volume of high level radioactive waste that was stored was reduced by almost 80 percent. Of the approximately 245 million gallons of high level radioactive waste generated and stored from all processes, approximately 55 million gallons remain in the underground storage tanks today, awaiting treatment and permanent disposition. Approximately 190 million gallons were removed through the methods such as evaporation and scavenging or through tank leakage (Gephart and Lundgren 1997, Anderson 1990).

INCIDENTS

Numerous incidents relating to high level radioactive waste have occurred over the operating history of the Hanford Site. Two examples of incidents are given here to demonstrate the challenge and complexity of managing high level radioactive waste: a chemical reaction at a tank farm and leaking transfer lines.

The chemical reaction that occurred in 1953 in an underground vault at the U Tank Farm is an example of incidents resulting from undesired chemical reactions. The chemical reaction caused surface contamination in the area adjacent to the vault. Cleanup crews placed lead sheeting over the contaminated area and covered the sheeting with clean soil to prevent spreading the contamination (GE 1953a).

Another type of incident was leaks from transfer lines. Transfer line leaks occurred when line sections separated or when temporary connecting jumpers loosened, causing contamination of the soil adjacent to the leak sites. Because the resulting liquid contaminate was quickly absorbed by the gravel soil, cleanup crews typically remediated the problem by removing the contaminated soil for burial elsewhere or covering the contaminated area with an overburden of clean soil (Maxfield 1979).

An example of such a line leak occurred near T Plant in 1955 when leaking contaminants formed a pool on the surface that produced radiation readings of 100 rads/hour near the surface. The spilled liquids were allowed to soak into the soil, which was then covered with fresh soil and then with asphalt (Maxfield 1979).

All spill sites, no matter the cause, were marked with warning devices and are included in waste site remediation planning.

STORAGE TANK LEAKAGE

Due presumably to the corrosive nature of high level radioactive waste, the temperature ranges generated by decaying fission products, and the aging of the storage vessels, some of the single-shell underground storage tanks have begun to leak. While the actual mechanism of leakage is uncertain, it seems probable that the causes include stress, corrosion, cracking, and mechanical tearing of the liner (Anderson 1990). Sixty-seven single-shell tanks (but no double-shell tanks) are known or suspected to have leaked.

The single-shell tanks are steel-lined concrete vessels with a design life of 20 years. That life has been exceeded by more than 30 years and will have been exceeded by 50 years before workers can remove all waste by 2018 in accordance with the Tri-Party Agreement (Ecology 1994). Despite the 20-year life expectancy, leakage of the single-shell tanks began with a 55,000-gallon leak in 1956, only 10 years after the first tanks were built. The most severe tank leak occurred at the T Tank Farm in 1973 when 115,000 gallons of high level radioactive waste were released in the soil.

As tanks continue to age, additional instances of leaking are likely to occur. As of 1995, as much as 1 million gallons of waste may have entered the soil beneath the tanks. This volume is estimated to contain as much as 1 million curies of radioactivity, mostly from cesium-137 (Gephart and Lundgren 1997).

Detecting leaks is difficult. The depth at which the underground tanks are buried, the hazardous nature of high level radioactive waste, and the impossibility of fully emptying tanks makes direct tank inspection nearly impossible. Workers must rely on indirect means to detect and gauge tank leakage by comparing measurements over a period of time. Approximately 800 vertical monitoring wells (dry wells) have been drilled near the single-shell tanks. Radiation readings taken at different depths in the wells give indications of leakage that may have occurred. Wells drilled laterally to extend beneath the tanks are also used (Gephart and Lundgren 1997). Figure 2-6.6 shows a worker monitoring a high level radioactive waste tank.

In addition, groundwater monitoring wells are placed at hundreds of locations to detect and measure contaminants that may reach the underlying aquifer. Environmental monitoring determined that the groundwater under the B-BX-BY tank farms, located in the 200 East Area, has been contaminated with chemical and radiological contaminants either from leaking tanks or spills (DOE 1998e).

Figure 2-6.6. *Worker Monitoring the High Level Radioactive Waste Tank at the Hanford Site, 1988*

When tanks are discovered to have leaked, cleanup crews applied several remedies. Waste was transferred to sound, usually double-shell tanks. Diatomite was added to some tanks to solidify the remaining liquids. Finally, tank evaporators were used to further reduce the fluid levels and minimize further leakage to soil.

LOW LEVEL RADIOACTIVE LIQUID WASTE

Low Level Radioactive Waste is typically contaminated with small amounts of radioactivity dispersed in large amounts of material, but some high activity and/or high concentrations can exist. Low level radioactive waste may exist in liquid or solid form.

> **Low Level Radioactive Waste**
>
> ***"All radioactive waste not classified as high level waste, transuranic waste, spent nuclear fuel, or natural uranium and thorium byproduct material defined under section 11e(2) of the Atomic Energy Act as amended."*** **- 42 USC 2011 et seq.**

All Hanford program areas generated low level radioactive liquid waste that was disposed of into the soil near the facility generating the waste or into the Columbia River. Several methods were devised to dispose of low level radioactive liquid waste in the soil, including open ponds and trenches, reverse wells, French drains, and cribs. Some methods were discarded due to adverse environmental impacts. The method used was dependent on the degree of waste concentration, the presence of radionuclides that emitted high-energy radiation, and the prevailing disposal standards at the time.

100 AREAS

Hanford production reactors generated low level radioactive liquid waste that was disposed of into the soil through cribs and trenches or directly into the Columbia River through effluent lines. The eight original reactors used vast quantities of Columbia River water to cool the reactor cores to safe operating temperatures. This cooling water created the greatest volume of low level radioactive liquid waste. Hanford's ninth reactor, N Reactor, used a different cooling system design that did not discharge cooling water directly into the river.

The purified water entered the reactor cores at flow rates up to 200,000 gallons per minute. Passing in direct contact with the reactor fuel elements, the cooling water carried away heat created by the fission processes occurring in the fuel, and maintained the core temperature at desired levels. At exit, the water temperature had been raised to near boiling (90-95 degrees C).

The high neutron flux of the reactor core activated elements in the cooling water, creating a number of radionuclides, including calcium-41, chromium-51, and zinc-65. The water was also contaminated with activation products from the reactor core and with fuel element fission products such as cesium-137, strontium-90, tritium, and transuranics, such as uranium and plutonium when fuel cladding failures occurred (DeFord and Einan 1995). The radionuclides of greatest environmental impact from this process were sodium-24, neptunium-239, arsenic-76, zinc-65, and phosphorus-32 (HEDR 1994a).

After passing through the reactor, the cooling water was transferred through effluent lines to a retention basin located near the river where it was held for 2.5 to 4 hours, allowing it to cool and short-lived radionuclides to decay. From the retention basin, the water passed through an outfall structure and large underground line followed by underwater pipes to the river bottom at mid channel. The outfall structure was designed to divert effluent to an open concrete spillway if the underground line were to become plugged.

The quantity of radionuclides released to the Columbia River in reactor effluent began high and increased over time. From five radionuclides alone (see Table 2-6.3), more than 200,000 curies were released to the river in each of the first two full years of reactor operations, 1945 and 1946. This rate doubled three times between 1945 and 1960 to more than 2 million curies during 1960, the peak year. The volume of reactor effluent had increased ten fold as cold war demands for plutonium production led to reactor power level increases and the construction of five additional reactors. After 1960, discharges of low level radioactive liquid waste to the river began to diminish as the reactors without closed-loop systems began to shut down at an approximate rate of one per year (Clukey 1957, HEDR 1994a). Table 2-6.3 is a summary of curies released to the Columbia River from five radionuclides between 1944-1971. For a more complete listing of radionuclides released to the Columbia River, see Table 2.3-9 in Section 3, Reactor Operations.

Table 2-6.3. *Annual Summary of Curies of Five Radioactive Materials Released to the Columbia River from the Hanford Site, 1944-1971*

Year	Sodium-24	Phosphorus-32	Zinc-65	Arsenic-76	Neptunium-239
1944	905	216	708	1178	17060
1945	34900	2890	10518	20340	192100
1946	28360	2190	8688	14212	153400
1947	25450	1860	7457	12319	127840
1948	33970	2219	8362	15772	151100
1949	46870	3175	11713	24660	214600
1950	72590	4027	14546	30510	279900
1951	99020	3333	11153	23660	261700
1952	132690	5050	9037	34740	259000
1953	202500	8688	8691	98940	316200
1954	243400	7261	21910	91380	391600
1955	318000	7184	26670	139530	419400
1956	407800	7722	31940	134280	450300
1957	644700	12325	27560	212130	500100
1958	751400	18484	27180	293300	422300
1959	1019000	17993	32030	218400	275100
1960	1382900	19490	42720	236900	354800
1961	1096300	21526	47110	166890	243910
1962	1094200	13845	56010	86660	257100
1963	887900	11738	14920	100630	211800
1964	960000	12311	15710	114480	247500
1965	764500	12126	13379	124600	168400
1966	613000	7365	9656	74600	78950
1967	671900	10118	15360	94010	114970
1968	499500	8632	7734	71670	99790
1969	359200	5478	6451	61250	59820
1970	178041	1759	3394	20253	36879
1971	13200	235	386	2440	3540
Sum	12582196	229239	490993	2519734	6309159

N Reactor used a closed-loop cooling system that did not discharge cooling water directly to the river. Instead, water heated by the reactor core passed through a heat exchanger, yielding much of its heat, and then returned to the core in a continuous, closed, loop. As cooling water became contaminated, it was bled off and routed to a covered trench for disposal in the soil (see Figure 2-6.7) (DOE 1997a).

Workers disposed of low level radioactive liquid waste in numerous cribs from sources such as overflow of fuel storage basins, decontamination activities, drainage from building filters, and floor drains. Some reactor sites also used a crib to dispose of highly contaminated cooling water from a single process tube that had been contaminated by a ruptured fuel element. In some cases, this crib was merely an open pit into which effluent was directed through a hose attached to the rear face of the reactor (WIDS 1998).

Fuel element failures caused by a rupture in the cladding were an all-too-frequent occurrence at Hanford production reactors with nearly 2000 occurring between 1951 and 1965 (DeNeal 1965). Some were no more than a pin-hole-sized breech in the outer, protective cladding of a fuel element. Others were more severe and exposed much of the fuel element's irradiated uranium fuel to the cooling water, which became grossly contaminated with uranium, fission products, and trace quantities of plutonium.

In the early years of fuel rupture experience, water from only the affected process tube was diverted from the rear face of the reactor to a crib where it was disposed of in the soil. In later years, the cooling water from the affected process tube was allowed to mix with that from all other tubes and, after cooling in the retention basin, was either released to the river or to a large trench that had been excavated for that purpose. Water diverted to these riverside trenches was allowed to percolate into the soil rather than being released directly to the river. This provided for ion exchange with the soil and for additional radioactive decay before the effluents eventually reached the river through the soil.

EXPERIMENTAL ANIMAL FARM

An additional source of low level radioactive liquid waste was the Experimental Animal Farm located at the 100-F Area. The Experimental Animal Farm operated from 1945-1976 to measure the effects of ionizing radiation on living organisms. Early researchers focused on the effects of reactor effluents on Columbia River fish. Later, researchers included the use of swine, sheep, dogs, rats, and plants.

Operations varied widely depending on the nature and scope of the particular experiment but generally involved the introduction of radioisotopes into living tissue with follow-up study to identify and measure its effect over time. Researchers tested the effect of iodine-131, strontium-90, cesium-137, isotopes of plutonium and uranium and other isotopes by having the animals inhale, ingest, and absorb the isotopes or by inoculating the animals with them (DeFord 1993).

Low level radioactive liquid waste streams included contaminated cleaning water resulting from animal pen cleaning, waste from the radio-botany and pharmaceutical labs, and various liquid waste from the biology lab. Most low level radioactive liquid waste went to 100-F cribs or trenches. Some was injected into reactor effluent and flushed to the Columbia River through the reactor outfall or passed directly to the Columbia River through a separate waste line and outfall (DeFord 1993).

Figure 2-6.7. *One of Two Covered Trenches Used at N Reactor to Dispose of Reactor Effluent in the Soil, 1981*

200 AREAS

Numerous sources of low level radioactive liquid waste have existed in the 200 Areas. Five chemical separations plants generated many forms of low level radioactive liquid waste. The Plutonium Finishing Plant added its share, including hazardous chemical waste such as carbon tetrachloride. Waste evaporators, used to reduce the volume of high level radioactive waste, produced large quantities of low level radioactive waste in the form of condensate that was released in the soil. Scavenged waste from some high level waste tanks was also disposed of in the soil as low level radioactive liquid waste.

Figure 2-6.8. *Low Level Radioactive Liquid Waste Pond with Overflow Lobes and REDOX (Center) in the Background, 1973*

In 1945-1946, all 200 Area low level radioactive liquid waste that did not meet the prevailing criteria for high level waste was disposed of in the soil in open trenches and ponds (see Figure 2-6.8). Narrow, radioactively contaminated beach zones developed, which were intruded upon by insect, bird, and animal life. These along with high winds spread contaminants to nearby downwind locations. After 1946 to protect personnel and the environment, only low level radioactive liquid waste with low activity continued to flow to open ponds and trenches. The deactivated ponds and trenches were flushed with fresh water and eventually backfilled (WIDS 1998).

Rather than dispose of low level radioactive liquid waste with higher activity in open ponds and trenches, in 1946 workers increasingly disposed of it below grade where contamination would be confined to sub-surface soil. To do this, numerous reverse wells were placed into service. These were deep, well-like shafts with perforated casings drilled to a depth above groundwater. The wells had a tendency to become plugged with solids suspended in the low level radioactive liquid waste and had the disadvantage of introducing the waste liquids near groundwater. Reverse well use was discontinued after about a year and no new ones were constructed.

French drains were constructed about the same time as reverse wells. French drains are shallow shafts used to dispose of low volumes of low level radioactive liquid waste with low activity. The shaft is typically a 3-6 foot concrete pipe about 36 inches in diameter buried vertically and filled with rock or gravel. The liquid usually entered the shaft from a waste line at the top of the pipe, which was at ground level and flowed through the pipe to the soil below. French drains continued to be used through the operational history of the Hanford Site for disposal of low level radioactive liquid waste with low activity. Numerous French drains are scattered throughout the Hanford production areas.

Low level radioactive liquid waste was also disposed of in cribs. Cribs were constructed with their upper surfaces only a few feet below grade to maximize the distance to groundwater, yet avoid the surface contamination problems associated with open ponds and trenches. Cribs are underground chambers used to dispose of large volumes of low level radioactive liquid waste. They were usually constructed of loosely spaced timbers, creating a chamber of more than 1000 cubic feet. The liquid would percolate through the chamber to the underlying soil. Numerous cribs exist in the 100, 200, and 300 Areas. Some cribs were used in conjunction with settling tanks in which waste solids would settle out before the supernatant liquid waste passed on to the cribs.

Over time, some cribs began to fail due to silt or chemical deposits, and overflow drain fields were attached to improve dispersal in the soil. These were not unlike the drain fields used in sanitary sewage systems, and they proved equally effective for low level radioactive liquid waste. Since 1960, many cribs have been constructed with drain fields, and a few drain fields have been constructed without cribs (WIDS 1998).

Open trenches and ponds as large as 50 acres continued in use for low level radioactive liquid waste with low activity or that was highly diluted, such as steam condensate and cooling water from certain contaminated facilities (WIDS 1998).

300 Area

The management and disposal of low level radioactive liquid waste at the 100 and 200 Areas was largely decentralized, with each facility manager responsible for waste generated at that facility. Most facilities were supported by dedicated cribs, French drains, ponds, and trenches.

The 300 Area differed from its sister areas in that the management and disposal of low level radioactive liquid waste was centralized. From their construction in 1944, the many 300 Area laboratories and fuel manufacturing facilities were connected to a common process waste sewer that collected low level radioactive liquid waste and discharged it in a single process pond located east of the area near the river.

The original process pond, a 353,000 square foot by 5-foot deep facility, overflowed its banks in 1945, and a retaining dike was added. The cause was determined to be poor percolation to the soil resulting from buildup of impermeable precipitants on the bottom of the pond. Waste level rose over time and the dike failed in 1948, releasing more than 14 million gallons of low level radioactive liquid waste into the Columbia River. The dike was repaired, and the pond returned to service (Young and Fruchter 1991).

A second pond was constructed immediately north of the original, and the bottoms of both were periodically scraped with the scrapings piled onto the dikes to further reinforce them. The ponds continued in operation until 1974 when they were replaced by a pair of 1500-foot process trenches.

Additional low level radioactive liquid waste processing improvements occurred in the mid-1950s when the Radioactive Liquid Waste System was constructed to replace the aging and by then leaking process waste sewer. Of modern design and stainless steel materials, the Radioactive Liquid Waste System fed a low level radioactive liquid waste retention and neutralization facility capable of sampling and diverting high activity waste to tanker trucks for disposal at 200 Area cribs. Low level radioactive liquid waste with low activity continued to flow to the 300 Area ponds and trenches. A railroad load-out facility with shielded railcars was added in 1965 to replace tanker trucks (DeFord 1994).

The Waste Acid Treatment Facility became operational in 1975 to process waste acids disposed of by various 300 Area laboratories. This waste was sampled, neutralized, centrifuged, and otherwise processed before being released for disposal to various locations including 200 Area waste disposal sites. Before 1975, 300 Area acids were disposed of in 300 Area ponds and trenches along with all other low level radioactive liquid waste (DeFord 1994).

A few cribs and French drains were used in the 300 Area as needed to meet specific and usually temporary needs of laboratories and other facilities (DeFord 1994).

400 Area and Fast Flux Test Facility

The Fast Flux Test Facility is a sodium-cooled breeder reactor, with associated facilities located in the 400 Area. It began operations in 1980.Although several French drains exist for disposal of cooling water and floor drain waste, no low level radioactive liquid waste is known to have been disposed of in the soil at this facility.

Incidents

Several incidents involving low level radioactive liquid waste have already been mentioned in the preceding paragraphs. A few additional descriptions will help to demonstrate the scope and complexity of managing low level radioactive liquid waste.

- The groundwater has been contaminated by waste from U Plant. The 216-U-1 and 2 Cribs are side by side facilities that received U Plant cell drainage and other low level radioactive liquid waste during the operation of U Plant for its uranium recovery mission. About 4000 kilograms of soluble uranium were drained to the cribs and became insoluble as it reacted with the soil beneath the cribs. Acid waste, inadvertently discharged to the cribs, combined with the uranium, making it both soluble and mobile in the soil column but in insufficient volume to move the uranium far enough in the soil to reach groundwater.

  A high volume, low activity waste crib, 216-U-16, was installed nearby. Low level radioactive liquid waste discharges to the 216-U-16 crib were sufficient by 1985 to collect atop an impermeable caliche layer located about 165 feet below the crib. The low level radioactive liquid waste moved laterally along the caliche layer, passing beneath the 216-U-1 and 2 Cribs and transporting the now soluble uranium through openings in the caliche layer to groundwater beneath. Groundwater contamination was detected soon thereafter and pump-and-treat techniques were used in 1985 to remove uranium contaminants from the groundwater. Uranium contamination levels were thereby reduced from 72 nanocuries/liter to 17 nanocuries/liter (Baker et al. 1988).

- The Process Waste Sewer has leaked in the 300 Area. Portions of the stainless steel lines in the 300 Area Process Waste Sewer have corroded and leaked low level radioactive liquid waste into the underlying soil.

- The reactor retention basin has leaked and overflowed. Each Hanford production reactor, except the N Reactor, used large metal or concrete retention basins to hold contaminated cooling water for a few hours before releasing it to the Columbia River. Each of these basins has developed leaks, allowing low level radioactive liquid waste to escape to the underlying soil. Also, basin outflow gates have become blocked, causing the basin to overflow and form, in one case, a deep erosion ditch flowing to the nearby riverbank.

- Cribs have overflowed and vegetation has become contaminated. Occasional overflows would occur at cribs when low level radioactive liquid waste was introduced at rates greater than the underlying soil could absorb. In these cases, low level radioactive liquid waste would sometimes overflow through the crib vent pipes, contaminating the surface area at and near the crib. Deep-rooted vegetation, such as Russian thistle, would sometimes intrude into contaminated soil, resulting in the plant becoming contaminated. Burrowing animals such as rabbits and gophers would also intrude into underground contaminated soil, becoming contaminated in the process and spreading that contamination about in the wake of their movements.

LOW LEVEL RADIOACTIVE AND TRANSURANIC SOLID WASTE

Solid waste includes non-liquid, non-soluble material ranging from municipal garbage to industrial waste that contains complex and sometimes hazardous substances. Radioactive solid waste contains materials that are contaminated with radionuclides. Low level radioactive solid waste is frequently found mixed with non-radioactive hazardous waste and is called mixed low level waste. When mixed with transuranic waste, it becomes transuranic mixed waste.

> **Transuranic Solid Waste**
>
> ***"Waste contaminated with uranium-233 or transuranic elements having half-lives of over 20 years in concentrations more than 1 ten-millionth of a curie per gram of waste"*** **- DOE 1997e, p. 228**

Radioactively contaminated solid waste has been disposed of by shallow earth burial since the beginning of Hanford operations in 1944. Although incinerators disposed of low level radioactive solid waste at other sites, they were not a part of a waste management program at the Hanford Site. Burial grounds were established in support of each of the Hanford production programs, including fuels manufacturing, reactor operations, chemical separation, plutonium purification, and research and development. Each burial ground has received vast quantities of low level radioactive solid waste. At least 28 sites in the 100 Areas, 28 in the 200 Areas, and 11 in the 300 and 600 Areas have been used for this purpose (WIDS 1998, Miller and Wahlen 1987). (By 1990, the "600 Area" defined all land within the Hanford Site not occupied by the 100, 200, 300, 400, 1100, or 3000 Areas.) Management practices for Hanford burial grounds may be generally described as follows (Dorian and Richards 1978):

1944 – 1954
- No intensive waste segregation program. Hazardous waste, low level radioactive solid waste, and transuranic waste were commingled for disposal
- Combustibles and non combustibles buried in the same trench
- Burial records contain minimal information
- Decentralized disposal with virtually all waste buried near point of origin

1955 – 1965
- Alternate disposal methods and sites studied, documented, and, in some cases, implemented
- Intentional burning of combustible low level radioactive solid waste in burial trenches began and ended in 1955
- Records improved

1966 – 1973
- Burial grounds centralized. Central landfill constructed for sanitary solid waste.
- Measurement of burial materials improved
- Burial records much more complete
- Some segregation of waste by category
- Beginning in 1968, increasing amounts of low level radioactive solid waste transported to 200 Areas for disposal
- After May 1970, all transuranic waste stored or disposed of in the 200 Areas

Post 1973
- Sanitary solid waste disposed of at the Hanford Site Solid Waste Landfill
- All low level radioactive solid waste disposed of in the 200 Area burial grounds
- Environmental Restoration Disposal Facility constructed in 1995

LOW LEVEL RADIOACTIVE SOLID WASTE

Some types of low level radioactive solid waste, such as contaminated protective clothing, are common to all production programs and, therefore, to all burial grounds, while other waste types are unique to the facility that generated them. Although general descriptions of the contents of low level radioactive solid waste burial grounds are available, detailed content descriptions are typically available only for the latter years of Hanford operations.

100 Areas

Reactor operations generated significant quantities of low level radioactive solid waste, which workers disposed of in the 100 Areas burial grounds and, after 1973 along with the Fast Flux Test Facility waste in the 200 Areas. Twenty-eight major reactor area burial sites exist in the 100 Area (Miller and Wahlen 1987). The greatest portion of low level radioactive solid waste from the reactor area consists of large quantities of obsolete or failed reactor hardware. To this was added contaminated protective clothing, tools, and miscellaneous process related materials.

200 Areas

Low level radioactive solid waste generated by the chemical separations and related processes was disposed of in 200 Area burial grounds. As with reactor operations, the majority of solid waste was contaminated process hardware made useless by failure or obsolescence. To these were added construction and demolition waste resulting from upgrades to the chemical system tools and protective clothing, and miscellaneous process related materials. Also included were soil and other materials contaminated by various chemical plant spills and leaks.

Some large, highly contaminated items are stored on flatbed railcars in the PUREX railroad tunnels that will require future decontamination or disposal in the 200 Area burial grounds (Anderson 1996).

The product of the chemical separations plants was an impure plutonium nitrate solution that was delivered to the Plutonium Finishing Plant in the 200 West Area for purification. Its processes generated low level radioactive solid waste that was disposed of in 200 West Area burial grounds. Again, obsolescent process hardware and materials such as tanks, pumps, piping, valves, and instrumentation provided the greatest quantity of waste. To these were added contaminated tools, protective clothing, rags, filters, and other materials.

300 Area

Fuel elements for the Hanford reactors were manufactured at Hanford's 300 Area from uranium stock. Using extrusion and cladding processes housed in the 306, 313, 314, and 333 Buildings, workers produced vast quantities of rod-shaped, reactor fuel elements and shipped them to Hanford's nine production reactors for irradiation. Numerous types of contaminated solid waste resulted from the fuel manufacturing processes, including industrial hardware and tools, protective clothing, equipment, glassware, swipes, and miscellaneous process related materials. The waste stream also included quantities of uranium millings, shavings, and dust resulting from fuel manufacturing processes. Large quantities of contaminated demolition waste were also disposed of during renovation activities at the fuel manufacturing buildings. Until 1973, all low level radioactive solid waste from fuel manufacturing was disposed of in the 300 and 600 Area burial grounds. After that year, 300 Area low level radioactive solid waste was disposed of in the 200 Area burial grounds.

Research and Development

Research and development played an essential role in improving the processes at all stages. Most research and development activities were housed in the 300 Area in laboratory facilities such as the Pile Technology, Radiochemistry, and Radiometallurgy laboratories, Plutonium Recycle Test Reactor, High Temperature Lattice Test Reactor, High Temperature Sodium Test Facility, and others.

All of these labs generated low level radioactive solid waste, most of which was buried in 300 and 600 area burial grounds. Some research and development labs, especially Radiochemistry, produced low level radioactive solid waste with remarkably high dose rates that required special handling and disposal procedures, discussed later in this section.

Two additional major research and development laboratories, Hot Semiworks and the Critical Mass Facility, were located in the 200 East Area. Low level radioactive solid waste generated there was disposed of in the 200 Area burial grounds. Hot Semiworks created an additional quantity of low level radioactive solid waste when in 1987 the laboratory was decommissioned and its contaminated demolition rubble was disposed of in the 200 East Area burial grounds (DOE 1998f).

Off Site Waste

Not all solid waste in the burial grounds was generated at the Hanford Site. While early records are incomplete, at least 58 off site waste generators are known to have shipped low level radioactive solid waste to the Hanford Site for disposal, essentially all going into the 200 Area burial grounds. These waste generators include universities, other DOE sites and laboratories, the military, and commercial companies that have been involved in government programs resulting in low level radioactive solid waste.

Types of programs that have generated low level radioactive solid waste disposed of at the Hanford Site include (Anderson 1996):

- Accelerator studies
- Animal studies in DOE facilities and universities
- Basic research
- Cleanup and restoration projects
- Department of Defense waste (U.S. Army and Navy)
- Reactor studies
- Irradiators and sources
- Fuel fabrication facilities
- Laboratory waste

An unusual form of off site waste received at the Hanford Site is more than 50 reactor compartments from nuclear submarines, which are disposed of in a 200 East Area burial ground. When U.S. Navy nuclear submarines are decommissioned, the spent fuel is removed and the entire reactor section of the hull is cut from the rest of the vessel, removed, sealed, and shipped to the Hanford Site for disposal. The intact section is transported by barge to a Hanford receiving dock where the section is loaded onto a massive wheeled trailer for transport to the burial ground (see Figure 2-6.9).

Figure 2-6.9. *U.S. Navy Nuclear Submarine Reactor Section Arriving by Barge at Hanford Receiving Dock, 1989*

TRANSURANIC WASTE

Transuranic waste contains radioactive materials contaminated with alpha emitting radionuclides with half-lives greater than 20 years in concentrations of more than one ten-millionth of a curie per gram of waste (DOE 1995e). Before 1970, transuranic waste was not recognized as a separate waste category and was disposed of in burial grounds as routine waste. Nearly all of it was buried in the 200 Area burial grounds with only small amounts entering a 100-F Area site and two 600 Area sites (Anderson 1996).

Atomic Energy Commission Immediate Action Directive 0511-21 defined transuranic waste and directed that after May 1970 waste with known or detectable contamination of transuranic waste radionuclides must be segregated from other waste categories. No further burial of transuranic waste was permitted, and it must instead be placed in retrievable storage. This was accomplished through use of a new style burial trench designed to permit transuranic waste recovery at a later date (see Figure 2-6.10). AEC Order 1511, issued in 1973, established a segregation limit of 10 nanocuries/gram. Transuranic waste with concentrations of radioactivity greater than 10 nanocuries/gram must be segregated and retrievably stored. Transuranic waste with lower concentrations need not be segregated. Since 1988, transuranic waste has been stored above ground in the Central Waste Facility.

Before 1980, liquid organic waste was sometimes disposed of in low level radioactive solid waste burial grounds. This type of waste was banned from burial ground disposal in 1980 because of its potential chelating effect on other waste in the burial trench and due to its potential damage to ion exchange properties of the soil. Liquid organics were subsequently stored in burial grounds in the same retrievable manner as transuranic waste.

To conform with Washington State Administrative Codes on waste storage, an additional low level radioactive solid waste category was defined in 1987. Hanford's Extremely Hazardous Waste has been segregated and stored as retrievable waste after this date.

Figure 2-6.10. *Transuranic Waste Drums on Floor of Hanford Burial Trench, a Burial Method Permitting Later Retrieval, 1980*

To meet the need for storage of extremely hazardous waste in buildings, the Central Waste Facility buildings were constructed. The first was completed in 1988 with another dozen completed shortly thereafter. These are 4000 square foot buildings that meet Washington State Code, RCRA (42 USC 6901), and other requirements for extremely hazardous waste and low level radioactive solid waste storage. Another seven, 35,000-50,000 square foot buildings have subsequently been constructed or are under construction (Anderson 1996).

RECORD KEEPING

Burial ground record keeping was minimal in the early days of the Hanford Site with little information on the amounts and types of low level radioactive solid waste buried. During the 1950s and 1960s, some documents were issued on waste disposal activities, but these records are not complete. Studies have estimated the volume and radioactivity of previously unrecorded buried waste based upon the ratio of the radionuclides present in the fuel elements and on other known and deduced waste generation and disposal information.

Beginning in the late 1960s, routine reports of low level radioactive solid waste in the 100 and 200 areas became more complete and included the amount of land area used, volume of waste, curie content of the various waste radionuclides, and the coordinates of the burial location.

Since 1971, the contents of burial grounds have been tracked on automated databases. A succession of automated information systems have been used, culminating in the current system, the Solid Waste Information and Tracking System. The Solid Waste Information and Tracking System is a state-of-the-art waste inventory and projection system that can calculate the activity of all radionuclides in each waste container. Its query capabilities include waste generator, date, location where waste was generated, burial location, activity at time of burial, and current activity (Anderson 1996).

REDUCING WASTE

An essential element of waste management is reducing waste. This includes those practices calculated to minimize the quantity or volume of waste materials requiring disposal. Chief among these are manufacturing practices that are improved to produce less waste and waste volume reduction by compaction, incineration, or other methods. Hanford operations have used all of these and other waste minimization practices.

Low level radioactive solid waste was burned at some Hanford burial grounds to reduce the volume before the Clean Air Act Amendments of 1970 and other environmental legislation precluded the practice.

From 1962 to 1973 to salvage plutonium, workers incinerated plutonium contaminated waste that would otherwise have been packaged and buried. Such waste included contaminated filters, rags, paper, protective clothing, gloves, and other combustible items used at the Plutonium Finishing Plant, REDOX, and PUREX. Once incinerated, the plutonium was recovered from the residual ash. The incineration and removal of the plutonium resulted in a 98 percent reduction of this particular waste stream (Anderson 1996). See the sub-section on the 232-Z Waste Incinerator in the Plutonium Finishing Section for more information on recovering plutonium via incineration and the Historic American Engineering Record in Appendix B on the Internet.

A novel waste minimization activity at Hanford involves the regulated fleet of trucks, tractors, locomotives, railcars, and other high value vehicles that have become radioactively contaminated but continue in use as regulated equipment. Such contaminated equipment would otherwise be disposed of by burial and replaced with new equipment when it was no longer repairable. Instead, the Hanford Site has established a program that keeps these regulated vehicles available for work in radiation zones where they are subject to additional contamination. Such vehicles continue in use as long as they may be safely used by operating personnel in protective clothing.

BURIAL GROUND DESCRIPTIONS

Burial grounds for low level radioactive solid waste vary widely in their size and configuration from small pits at reactor sites to mammoth trench excavations in the 200 Areas. Burial ground designs vary from simple, unlined excavations to lined excavations that permit recovery of pit contents to burial grounds that use underground caissons and burial pipes or vaults.

Burial grounds in the 100 and 300 areas are close to the Columbia River and are underlain with permeable materials. The depth to groundwater of some burial grounds is as little as 5 meters. The 200 Areas burial grounds lie on a geologic plateau and are underlain by a considerable thickness of materials with low permeability. Waste buried there is 55 meters or more above groundwater. Groundwater movement beneath these 200 Areas burial grounds has extremely low flow rates and the soil has a large capacity for ion exchange. The slow movement allows time for ion exchange, which acts to remove and retain radionuclides (Anderson 1996).

Initially, almost all of the radioactive solid waste was buried near the area in which it was generated. Beginning in 1968, however, increasing amounts of waste were transported to the 200 Areas for disposal. Because of increased attention to all radionuclides discharged to the environment, by 1970 most low level radioactive solid waste was disposed of in 200 Area burial grounds. The last 300 Area burial ground closed in mid-1972 and only minor use was made of 100 Areas burial grounds until they fully closed in 1973. In all, 75 solid waste burial grounds exist at the Hanford Site (DOE 1997c).

100 Areas

With the exception of the 100-N Area, all reactor sites used burial grounds to dispose of low level radioactive solid waste. Each site had at least one large burial ground and several smaller ones that varied from small burial pits to large burial trenches. For example, the main burial ground at the 100-F Reactor site measured 600 x 500 x 20 feet deep, while the 118-H-5 Thimble Pit, used to bury a single waste item, was only 30 x 2 x 5 feet deep (WIDS 1998).

The vast majority of 100 Area low level radioactive solid waste is reactor components and hardware that became irradiated and required disposal due to wear, failure, or obsolescence. These include items such as aluminum spacers, boron splines, graphite, process tubes, lead shielding, control rods, nozzles, and cadmium sheets. Also included are contaminated tools and protective clothing, glassware, swipes, and miscellaneous process materials (Miller and Wahlen 1987).

The quantities of low level radioactive solid waste from all reactors except N Reactor from 1944 through the end of reactor operations in 1971 are (Miller and Wahlen 1987):

• 1,700,000 aluminum fuel spacers	425 tons
• Lead-cadmium poison elements	1,059 tons lead 44 tons cadmium
• 142,000 boron-aluminum splines	652.5 tons aluminum 8.4 tons boron
• Carbon-14 (graphite and desiccant)	15,500 pounds
• 28,000 process tubes	266 tons aluminum
• Lead (brick, sheet, wool, casks)	258 tons
• Miscellaneous hardware	146 tons
• Cadmium sheeting	800 pounds
• Thermocouple Wire	372 pounds
• Soft waste (paper, plastic, clothing) that contains a very small percentage of total radionuclide inventory but makes up more than 75 percent of the total low level radioactive solid waste volume from all reactors except N Reactor	129,000 cartons

A few small burial sites in the reactor area were established for disposal of single, highly contaminated or irradiated waste items, such as a reactor test loop or a collection of contaminated fuel spacers. An unusual form of low level radioactive solid waste from the Experimental Animal Farm was buried at 100-F burial grounds. Extensive research was conducted on living animals and many were sacrificed as part of the research process. Contaminated animal carcasses were disposed of in a large, vented, steel tank that was periodically topped off with diesel oil and set afire to reduce waste volume. When the tank was no longer needed, it was dragged to the 118-F-6 Burial Ground and buried (DeFord 1993).

After 1973, most low level radioactive solid waste from the reactors, including all waste from the 100-N Reactor, was disposed of in the 200 Area burial grounds.

All 100 Area burial grounds are simple, unlined, trench or pit excavations. None remain in service and all have been backfilled to grade with native soil. Each is identified with signs and radiation warning devices. Most are unfenced with access only controlled through Hanford Site access security.

200 Areas

Because it was expedient to dispose of low level radioactive solid waste near the generating source, burial grounds were opened at nearly all Hanford areas. However, environmental considerations such as water table and proximity to the Columbia River have made 100 and 300 area sites less desirable. The 200 Areas burial grounds were recognized as superior sites in terms of environmental protection. In 1973, 200 Areas sites became the centralized burial grounds for all low level radioactive solid waste at the Hanford Site (Anderson 1996).

Operating since 1944, these burial grounds are massive excavations that have received low level radioactive solid waste items as large as contaminated trucks and railroad locomotives. Some burial grounds are as long as 1500 feet and over 90 feet wide at the surface. Most are at least 18 feet deep. Bottom treatment varies with era and type of waste to be disposed. Most have bottoms of native sediment with no special treatment. Those designated for retrievable storage may have asphalt paving. For a time, transuranic waste drums were placed in V shaped burial grounds rather than the traditional flat bottom excavations to expedite retrieval of waste (DeFord 1991).

All burial grounds are identified by permanent concrete markers that denote the burial ground by number and flag it as a disposal site for radioactive waste. Most but not all 200 Areas burial grounds are protected by security fencing. Hanford Site drawings further delineate the location of burial grounds with each individual trench identified.

Before 1970, 200 Area burial grounds could be divided into six waste type categories.

- Dry waste burial grounds received low level radioactive solid waste packaged primarily in cardboard boxes containing various forms of dry waste. These included contaminated protective clothing, rags, paper, wood, and soil. Hardware contaminated with plutonium and various forms of off-site waste were also included.
- Industrial burial grounds received low level radioactive solid waste packaged in large wooden or concrete containers (see Figure 2-6.11) containing obsolete equipment from chemical plants and 300 Area labs that were heavily contaminated with fission products.
- Construction burial grounds received low level radioactive solid waste with low activity from construction and demolition work.
- Railroad tunnels at PUREX are used to store numerous large pieces of contaminated equipment on railroad flatcars.
- Underground vaults located at 222-B, T, and S Analytical Laboratories received small quantities of low level radioactive solid waste with high activity and dose rate from contamination by mixed fission products and plutonium.
- Caissons located at two burial grounds were used to dispose of low level radioactive solid waste with high activity from the 300 Area hot cells and low level radioactive solid waste with high plutonium content. Caissons are either tubular metal pipes buried vertically in the trench or larger circular tank-like containers made from corrugated metal or concrete, buried in the trench with offset feed chutes reaching above grade (DeFord 1994).

Figure 2-6.11. *Bulldozer Pushes Soil onto Concrete Burial Box on Floor of Hanford Burial Trench, 1980*

Environmental Restoration Disposal Facility

The Environmental Restoration Disposal Facility began operations July 1996. Located between the 200 East and 200 West areas on Hanford's central plateau, the Environmental Restoration Disposal Facility is a disposal facility for waste generated during the environmental restoration, deactivation and decommissioning efforts at the Hanford Site and is notable as Hanford's most advanced and most recent disposal site for low level radioactive solid waste. The specific site on the central plateau was selected because of its distance from the Columbia River, its elevation above groundwater, and its centralized access to transportation and utility systems.

The facility is a deep trench that will be the central disposal facility for large quantities of low level radioactive solid waste generated during environmental restoration work. Upwards of 12 million cubic yards of contaminated materials may be disposed of at the facility. Hazardous and mixed waste will also be disposed of there. The facility is not authorized to receive transuranic waste.

The Environmental Restoration Disposal Facility has been designed as a series of cells, each of which is 500 feet by 500 feet at the bottom with a ratio of 3:1 sloped walls, 70 feet deep, and will hold about 800,000 compacted cubic yards of material. Each cell is double lined and built with a leachate system to collect and remove moisture. Each will be filled to the brim with low level radioactive solid waste, covered with a 15-foot-thick engineered barrier (consisting mainly of native soil), and will be planted with native habitat when waste capacity has been reached. Four such cells have been constructed to date. Two are full to their operating capacity and capped with an interim cover, and the other two are receiving waste. Additional cells will be constructed to meet the disposal capacity needed for environmental restoration, deactivation, and decommissioning (Crigler 2001).

300 Area and 600 Area

Fuel manufacturing processes and research and development activities in the 300 Area generated large quantities of low level radioactive solid waste that before 1973 were buried in 300 and 600 area burial grounds. After 1973, 300 Area low level radioactive solid waste was transported to the 200 Areas for disposal.

The distinction between 300 and 600 area burial grounds is sometimes confused. Essentially all waste disposed of in both area burial grounds was generated in the 300 Area. However, where other areas located their burial grounds within the area security fences, 300 Area did not. The 600 Area designation was devised to encompass all locations outside of area security fences and so includes those 300 Area burial grounds located adjacent to, but outside of, the 300 Area security fence. For that reason, 300 and 600 area burial grounds are included in this section.

The first burial ground to support 300 Area activities, 618-8, operated for only about a year in 1944. It was located north of the area under what is now an employee parking lot. A second burial ground, 618-1, was also north of the 300 Area and operated for 6 years beginning in 1945. The 618-2 Burial Ground followed in 1951 and operated until it caught fire in 1954. Another, 618-3, was excavated in 1954 but was filled to capacity the following year with demolition rubble from remodeling activities at the fuel manufacturing buildings (DeFord 1994).

Contamination repeatedly being spread, high dose rates, and waste fires in burial grounds near the 300 Area led to planning for burial grounds farther removed from facilities and employees. The 300 North Burial Ground, 618-10, was excavated in 1954 several miles north of the 300 Area and was followed by the 618-11 or Wye Burial Ground even farther north. For more information on the 618-10 Solid Waste Burial Trench, see the Historic Property Inventory Form in Appendix B on the Internet.

In addition to routine low level radioactive solid waste, these burial grounds received waste with a high dose rate from the Radiochemistry and Radiometallurgy laboratories. Exposure rates as high as 5 roentgen/hour were measured during waste disposal activities, and revised disposal methods were developed. To reduce dose rate to burial ground workers who handled the waste during disposal, waste was transported in various kinds of truck-mounted, shielded casks. These casks were designed so the waste containers could be dumped directly into metal caissons. The metal caissons had previously been buried in the disposal trench with only their openings visible at the surface. Once the waste with the high dose rate was dropped into the caisson, sand or gravel and sometimes concrete was placed in the caisson to reduce the dose rate at the surface.

Numerous instances of burial ground contamination were recorded at 300 and 600 area burial grounds. Most occurred during disposal when waste containers would rupture and contaminants would settle over the site. High dose rates also continued to be a problem until the 300 Area low level radioactive solid waste was transported to the 200 Areas for disposal, mostly in shallow burial grounds in the 200 West Area.

The 300 Area burial grounds ranged in size from 4000 to 40,000 square feet. The largest, 618-11, consisted of three burial trenches (each 900 x 50 x 25 feet deep), 50 pipe caissons, and at least 4 large diameter caissons. The pipe caissons were constructed by welding together five bottomless 55-gallon drums. The resulting 15-foot-long cylinders were buried vertically and spaced 10 feet apart in rows. Large diameter caissons were 10 feet high, 8-foot diameter corrugated steel tanks, each buried 15 feet below grade. The caisson was connected to the surface by an offset, 36-inch-diameter pipe through which waste could be dropped into the tank (DeFord 1994).

INCIDENTS

Three categories of incidents involving low level radioactive solid waste occurred with some frequency, mostly in the 1950s and 1960s, and all of which resulted in contamination being spread to burial grounds and adjacent areas. These included fires in the burial grounds, burial containers that collapsed, and contamination spread when waste was in transit to burial grounds.

Fires occurred in 100, 200, and 300 area burial grounds. Most appear to have been caused by spontaneous combustion in waste materials with the fire spreading to adjacent refuse. The spread of contamination was sometimes severe, covered large areas adjacent to the burial grounds, and required major decontamination efforts by Hanford personnel. Fires in 100 and 300 Area burial grounds and the associated spread of contamination contributed to the decision to centralize the disposal of low level radioactive solid waste in the 200 Areas.

One of several burial ground fires in the 300 Area occurred in 1954 when a fire at the 618-4 Burial Ground spread contaminated particles as far as 1500 feet to the northeast with individual particles reading as high as 4.5 roentgen/hour (Anderson 1996).

The collapse of burial containers was an occasional problem at the 200 Area burial grounds where large boxes were frequently used to transport and bury contaminated hardware and apparatus from process facilities. The containers were placed on the floor of the burial trench and the trench backfilled. The weight of the backfill soil would sometimes collapse the burial box, causing a rush of contaminated air from the box interior to transport contaminants to the area of the trench and sometimes beyond. Areas as large as 4 square miles were contaminated in this manner (Anderson 1996).

Similar problems occurred at the burial ground in the 300 Area, especially the 618-10 and 618-11 burial grounds, when paper cartons or metal cans containing waste with some activity would rupture when dropped into pipe caissons, resulting in a column of contaminated air that rose upward in the pipe and onto the surface of the burial ground (Gerber 1992c).

Occasionally, radioactive releases would result during transport and contaminate the roadway, railroad, and/or transport vehicles, requiring significant decontamination over large areas. One such incident occurred in 1957 when a burial box containing REDOX hardware was pulled from its railcar when one of the box swing cables came loose and caught on a railroad tie. The ground where it landed was contaminated to a level of 2 roentgen/hour (Anderson 1996).

Incidents were not the only source of hazardous conditions at the burial grounds. Even the routine handling of the waste could result in exposure rates as high as 200 roentgen/hour (Anderson 1996). To dispose of the waste, employees often had to use heavy equipment with all the hazards incident to bulldozer and crane operations. Backfilling required skillful use of tractors working on steep slopes in soft soil conditions.

AIRBORNE RADIOACTIVE EFFLUENT

Especially in its early years of operation, the production of plutonium at the Hanford Site generated large quantities of airborne radioactive effluent. Being airborne, the contaminants affected both employees working on the Hanford Site as well as neighboring populations. Table 2-6.4 is a summary of six radioactive materials released to the air from 1944-1972. Since this table was published, updated data for the releases starting August 1951 have been published.

Table 2-6.4. *Annual Summary of Curies of Six Radioactive Materials Released to the Air from the Hanford Site, 1944-1972*

Year	Iodine-131	Ruthenium-103	Ruthenium-106	Strontium-90	Plutonium-239	Cerium-144
1944	2,139	0.49	0.035	0.021	0.0012	1.57
1945	555,089	87	12	6.94	0.39	460.15
1946	96,298	87	18	10.53	0.58	650.03
1947	31,909	51	12	7.38	0.41	451.30
1948	1,842	12	4.6	2.94	0.16	166.82
1949	8,694	0.42	0.19	0.12	0.006	6.15
1950	5,379	0.81	0.35	0.20	0.010	10.40
1951	27,397	2.4	0.58	0.28	0.014	16.92
1952	5,108	32	11	0.40	0.013	23.33
1953	1,747	266	89	0.52	0.015	31.05
1954	827	485	168	0.67	0.021	40.66
1955	671	6.87	2.00	0.78	0.025	43.30
1956	118	5.76	2.63	1.24	0.014	72.88
1957	274	12.6	4.58	1.90	0.015	118.73
1958	822	16.6	4.38	1.98	0.011	130.08
1959	227	15.2	5.27	2.30	0.008	144.41
1960	232	16.6	5.84	2.58	0.007	161.88
1961	92	15.0	6.16	2.81	0.011	171.68
1962	29	9.13	5.39	2.59	0.009	150.80
1963	78	7.53	5.10	2.47	0.011	139.79
1964	11	9.26	5.77	2.85	0.015	160.78
1965	6.08	7.84	5.40	2.71	0.014	154.25
1966	9.10	5.41	4.75	2.45	0.018	127.84
1967	1.28	3.82	4.35	2.12	0.001	107.41
1968	0.021	1.97	4.78	2.18	0.001	103.48
1969	0.0013	0.73	3.61	1.74	0.0004	72.81
1970	0.0011	0.19	0.89	0.42	0.0001	17.19
1971	0.000063	0.11	1.14	1.09	0.0003	30.34
1972	0.000000	0.00	0.12	0.11	0.00003	3.55
SUM	738,999	1,160	388	64.32	1.78	3,770

The updated data increase the releases from August 1951-1972 by about 25%. For details concerning the updated data, see http://www.pnl.gov/eshs/cap/hra/pub/hedr.html on the Internet.

CHARACTERIZATION

Air contaminated by the chemical separation processes passed from the separations buildings through 200-foot high stacks into the atmosphere (see Figure 2-4.1 in the Chemical Separations Section). During the first 3 years of operation (1945-1947), the predominant radionuclide was iodine-131 released from the fuel during the nitric acid dissolution steps. Over time, improved technology and revised procedures reduced iodine-131 stack emissions to a small fraction of the original levels.

Other contaminants were ruthenium-103 and 106, which tended to become concentrated on the inner surfaces of the separations plant ventilation ducts. After a time, paint and metal flakes contaminated with ruthenium from the fans and vent ducts began to break free and in the absence of filtration before 1947 were carried out the stacks into the atmosphere. Being of greater mass than the iodine-131 gasses, these contaminated flakes fell to earth near the stacks, contaminating the area around the separations plants. Additional contaminants, including iodine-131, cesium-144, strontium-90, and others, were also found to have contaminated these small flakes. Flake contamination became as concentrated as 50 flakes per square foot near the stacks. During that period of high concentration, plant personnel were required to wear protective masks to avoid inhaling the contaminants (HEDR 1994a, Roberts 1958).

TECHNOLOGICAL IMPROVEMENTS

The primary technological improvements that reduced air contaminants were air filtration systems. Because no adequate filtration systems were available before 1947, none were used. Sand filters were devised and installed at the T Plant and B Plant in 1947. The sand filters were effective in reducing the levels of contaminants reaching the atmosphere. Some were replaced with improved, high efficiency fiberglass filters in 1950. Silver reactor filters were also added to further reduce airborne contaminants.

PROCEDURAL CHANGES

Equally important in reducing airborne contaminants were procedural changes that increased the length of time the irradiated reactor fuel was allowed to cool before being processed at the separations plants. Due to its short half-life of 8.04 days, much of the primary contaminant, iodine-131, would decay away if given sufficient cooling time. If the irradiated fuel elements were processed too soon, increased quantities of the gaseous isotope were liberated into the air as the fuel elements were dissolved in acid (HEDR 1994a, Roberts 1958).

The rush to produce plutonium during World War II tended to minimize the time that reactor fuel was permitted to cool in the reactor fuel basins before being processed at the separations plants. Initial schedules allowed only 32 days and occasionally less cooling time. As the war ended, cooling times were extended to 35-80 days and by 1949 to 90-125 days. Iodine-131 emissions decreased accordingly (HEDR 1994a, Roberts 1958).

Other procedural remedies in the mid-1940s were schedule related. Wind dispersal forecasts were consulted before dissolving operations were begun at separations plants. Dissolving operations were sometimes delayed due to unfavorable forecasts. Meteorologists determined in 1945 that night time wind conditions were generally more favorable than daytime, and dissolving operations were moved to the night shift.

Technological and procedural improvements did much to reduce the iodine-131 atmospheric release problem by 1948. However, Table 2-6.4 shows the result of an isolated release occurring in 1949 from the "Green Run." To measure how radioactive contaminants were spread, an experimental batch of green reactor fuel (fuel allowed to cool for a reduced period of time) was dissolved at T Plant. Normal air filtering was bypassed to ensure that measurable levels of contaminants were released into the atmosphere. More than 8000 curies of iodine-131 were released during the Green Run. For more information on the Green Run, see Jenne and Healy (1950).

Table 2-6.4 also shows the result of another set of iodine-131 releases in 1951 from additional experiments in reduced fuel cooling time and from failure of silver reactor filters at B and T Plants. After 1951, iodine-131 releases generally decreased and have not exceeded 100 curies per year after 1960.

SANITARY LIQUID WASTE (SEWAGE)

With its thousands of World War II era construction workers, the Hanford Site quickly grew into one of Washington State's largest cities with all of the infrastructural needs of a community of such size.

The predecessor communities of Hanford, White Bluffs, and Richland did not have municipal sewage treatment plants, so provisions had to be made to meet the needs of the Hanford Site. State of the art sewage treatment plants were planned and constructed at a time when primitive out-house technology was more the rule than the exception. Each Hanford program area was supported. The Hanford construction camp alone had three modern, if temporary, sewage treatment plants, and numerous septic tanks—the envy of any 1940's era city of comparable size. Reactor sites and chemical separations plants were similarly provided with indoor plumbing connected to septic tanks and drain fields (Du Pont 1945b).

As most of the Hanford Site was blessed with excellent drainage due to deep surface layers of alluvial sands and gravels, septic tank and drain field technology proved adequate for all Hanford production areas. Each reactor site had three to five septic tanks, and the 200 and 300 Areas had more due to larger employee concentrations.

Central Shops and temporary housing areas had much larger concentrations of personnel, including housing for thousands of construction workers and many of their families. Sewage treatment plants were required to supplement septic tanks at those areas. As construction at the Hanford Site was completed, the construction camps were abandoned and their sewage treatment plants shut down. Many of their remnant foot-prints are still visible today (DeFord 1996).

New septic tanks and drain fields were constructed as sanitation needs increased with plant growth and expansion. A single, recent, exception to septic tank/drain field technology is the 124-N-10 Central Sewer System, constructed near the 100-N area. It is a lagoon and aeration facility that remains operational to support current 100-N Area sanitary waste requirements.

Today, 85 septic tank/drain field systems are in operation (WIDS 1998). More than 100 others have been abandoned, some because of the failure of the system to meet sanitary needs. However, most were abandoned because the facilities they supported were shut down. Septic tank deactivation is typically accomplished by crushing the top of the tank and/ or filling the void with sand and gravel.

FACILITIES

Hanford sewage systems may be divided into two general classifications: temporary, construction era systems and permanent, production era systems.

The 1940's construction era facilities at the Hanford and 3000 Area construction camps and at Central Shops were considered to be temporary facilities established to support the construction of the permanent production plants at the 100, 200 and 300 Areas. As temporary facilities, they were provided with sewage treatment systems of temporary design.

The Hanford construction camp was equipped with 76 temporary septic tanks and drain fields. Septic tanks were standard design, three-pass wooden baffle boxes of varying sizes and were located throughout the camp area. Additionally, three gravity-fed sewage treatment plants were constructed adjacent to the Columbia River. Each treatment plant included sewage lift sumps, chlorinators, flocculators, and settling basins. Sewage reaching these plants was chlorinated to control bacteria and flocculated to remove grease. Solid matter was settled out before the supernatant liquid was released to the river. During the peak month of July 1944, these facilities processed more than 4 million gallons of sewage daily, using 1400 pounds of bottled chlorine gas per day (Du Pont 1945b).

The 3000 Area construction camp and the Central Shops were also equipped with temporary sewage treatment plants. These were gravity-fed systems consisting of main trunk lines flowing into large septic tanks, thence by means of an open ditch to open settling basins. Two such basins were provided for the Central Shops plant and three for the 3000 Area plant (Du Pont 1945b).

Richland Village, now the City of Richland, was also provided with temporary sewage facilities while a municipal sewage treatment plant was constructed. These temporary facilities consisted of two Imhoff type septic tanks and 1600 feet of sewer line. One tank was located east of Goethals Drive between Williams and Jadwin and the other east of George Washington Way between Swift and Williams. These operated until April 1944 and were replaced by a municipal sewage treatment plant located south of town and east of George Washington Way (Du Pont 1945b).

Permanent, septic tank/drain field sewage treatment systems were provided for all of Hanford's production areas. A typical permanent septic tank was unremarkable in its construction, meeting industry standards in all respects. All were of concrete construction and varied in size from 500 to 15,000 gallons with overflow capability into perforated concrete, vitrified clay or PVC drain fields. Drain fields were typically of herring-bone design and were buried to a depth of 3 to 4 feet below grade. Actual system size and configuration was determined by capacity demand and local soil conditions.

Septic tanks were sometimes pumped out and their contents disposed of elsewhere. Designated trenches at the Sanitary Waste Landfill received this waste until 1987 and the 100-N Sewage Lagoon has been used since. Commercial off-site sewage disposal facilities are now being used (DOE 1996e).

INCIDENTS

No incidents have occurred in the sanitary waste septic tanks when used for the purpose of their design. However, process area septic tanks have become contaminated with improperly disposed hazardous or radioactive waste. Such septic tanks are provided an appropriate waste site status that flags them as having additional regulatory and remediation requirements. Examples are the 1607-F-2 Septic Tank in the 100-F Area and the 3607 Septic Tank in the 300 Area. Radioactive contaminants in the sludge from the 3607 Septic Tank were measured at 25,000 counts per minute (DeFord 1994, Gerber 1992c).

All sanitary waste treatment facilities are documented in the Waste Information Data System (WIDS 1998). The Waste Information Data System is the single, official repository of information on the existence and character of all the Hanford waste sites and may be accessed on the Internet at http://www.bhi-erc.com/projects/p_m/eis/wids/wids.htm.

SANITARY SOLID WASTE (TRASH)

When established in 1943, the Hanford Site inherited several municipal sanitary solid waste landfills (garbage dumps) from the predecessor communities of Hanford, White Bluffs, and Richland. These were primitive, shallow dump sites where household and community trash was dumped and periodically burned to reduce volume. Some back filling is thought to have occurred, but this was a costly and uncommon community practice in that era. Numerous additional, small, private dumping grounds existed at farm sites and elsewhere within the Hanford Site boundaries (DeFord 1996).

The municipal landfills continued in use for a short time to support Hanford construction activities but were eventually backfilled with adjacent native soil. Larger and deeper sanitary landfill sites were excavated at the Hanford construction camp, the City of Richland, the 1100 Area, 300 Area, at each reactor site, and at the 200 East and 200 West Areas. At least one existing gravel pit, Pit 11, was also used for sanitary solid waste disposal. Waste volume was reduced by burning at most sites until the practice was discontinued in accordance with modern air quality standards. Before hazardous waste was defined by the regulatory agencies, a few small burial pits were designated for disposal of paints, solvents, and other volatile materials, most of which were burned in the pit before being covered over with backfill. Most of the above remained in use until the Central Landfill was constructed in 1973.

Construction debris was sometimes buried in excavations adjacent to the major Hanford construction sites. Construction debris included scrap lumber, concrete, piping, roofing materials, siding, metal, and other scrap building materials. No hazardous or radioactive materials are known to have been included.

Landfill management was decentralized to program area managers until 1973 when site-wide responsibility for sanitary solid waste management was assigned to a site contractor. A centralized, trench-type landfill was then constructed near the geographic center of the Hanford Site. This solid waste landfill operated from 1973 until 1996 as the primary sanitary solid waste disposal site. Since 1996, all Hanford sanitary solid waste has been transported to the City of Richland Landfill located south of the Hanford Site (Rohay 1997).

SOLID WASTE LANDFILL

The Solid Waste Landfill, often referred to as the Central Landfill, has served as the Hanford Site's primary disposal site for non-hazardous and non-radioactive solid waste from 1973 until its closure in 1996. Hazardous and radioactive waste was disposed of at other Hanford areas designated for that purpose and are described in previous sub-sections of this section.

The Solid Waste Landfill received common office waste, construction debris, asbestos waste, and other types of waste common to a large industrial community. It is a conventional, excavated, trench-type land disposal facility that covers approximately 66 acres. It is 2977 feet in length, 965 feet in width and was divided into five units, each consisting of a series of parallel trenches (DOE 1996e).

Trenches were of two sizes, single and double. Single trenches measured 16 feet wide at the base, 46 feet wide at the top, and 15 feet deep and were filled with waste to a depth of 11 feet. Double trenches were 52 feet wide at the base, 92 feet wide at the top, and 20 feet deep and were filled with waste to a depth of 16 feet. Trenches varied in length from 463 feet to 657 feet. Single and doublewide trenches were backfilled with a 1-foot layer of cobble or gravel covered with 3 feet of native soil (Rohay 1997).

WASTE VOLUME AND CHARACTERIZATION

Hanford sanitary landfills were reserved for non-hazardous and non-radioactive plant waste. Although landfill logs were not maintained before 1982, estimates have been developed for the Solid Waste Landfill based upon trench geometry and the thickness of the waste layer. Over 524,000 cubic yards of solid waste, 19,000 cubic yards of asbestos, and 1,226,000 gallons of sewage and other liquids were disposed of at the Solid Waste Landfill. Volumes for other sanitary landfill sites are not available (DOE 1996e).

Based upon volume estimates for Solid Waste Landfill, approximately 40 percent of landfill content is office waste, consisting largely of waste paper products. Asbestos waste accounts for 10 percent and bulky items such as appliances and furniture account for another 10 percent. Construction and demolition debris, consisting of waste wood products, concrete scrap and soil, account for another 30 percent. The final 10 percent is made up of all other materials, including food materials, industrial waste, medical waste, and various inert materials. Other than for asbestos, these percentages probably approximate the contents of other Hanford solid waste landfills (DOE 1996e).

An estimated 1-1.5 million gallons of sanitary sewage was disposed of at the Solid Waste Landfill from 1975 to 1987. The sewage originated from portable toilets and septic tanks and was dumped into designated trenches along the east and west perimeter of the Solid Waste Landfill.

Approximately 10,000 gallons of liquid organic waste from the 1100 Area motor pool catch tanks was also disposed of in the sewage trenches from 1985-1987. These volatile organics were used for cleaning and degreasing motor vehicles and for other purposes. Trace amounts of each have been found in the groundwater below the Solid Waste Landfill. No sewage or other liquid waste has been disposed of at the Solid Waste Landfill since 1987. The volatile organics found in the groundwater include the following (Rohay 1997):

- Chloroform
- Carbon Tetrachloride
- 1,1-Dichloroethane
- 1,1,1-Trichloroethane
- Trichloroethane
- Tetrachloroethane

NON-RADIOACTIVE, DANGEROUS WASTE LANDFILL

The Solid Waste Landfill was originally divided into three units for operational purposes. In 1975, the northern most unit, about 10 acres, was isolated for the disposal of asbestos waste materials and non-radioactive chemical waste. Because of the presence of regulated dangerous waste, this northern most unit was designated formally as the Non-Radioactive Dangerous Waste Landfill and is regulated under RCRA (42 USC 6901) as a land disposal facility.

Asbestos waste is defined as asbestos or material containing asbestos that is non-radioactive and non-hazardous. It accounts for approximately 10 percent of all waste disposed of at the Solid Waste Landfill. Disposal has been in accordance with 40 CFR 61, Subpart M.

The Non-Radioactive Dangerous Waste Landfill continued to receive asbestos until 1988 at which time it was operationally closed. Since then, all Hanford Site asbestos waste has been disposed of in other dedicated trenches in the Solid Waste Landfill until its closure in 1996.

AREAS FOR FURTHER RESEARCH

The process of producing plutonium generated huge quantities of hazardous and radioactive waste at the Hanford Site. Early waste management planners hoped future technology would resolve disposal problems that had no immediate solution. The waste has not gone away, although radioactive decay has reduced the magnitude of the radionuclide content. Public criticism has been directed toward DOE and its predecessor agencies for allowing the waste to accumulate and, in some cases, to escape its artificially imposed boundaries. This criticism, however, has served to improve cleanup efforts, openness, and accountability for managing the waste at the Hanford Site. While much is known about waste management at the Hanford Site, much remains to be discovered and documented. Scholars may find any of the following subjects suitable for additional research:

- Develop a guide to all waste at the Hanford Site: its source, present location, planned final disposal, and regulators, indicating the level of assurance that all Hanford waste sites are known and documented
- Research how the initial waste management standards and practices were arrived at for the Manhattan Project and how they have evolved over time. Who were the key players in their development? What was the quality of initial waste management practices in light of 1940's science and engineering?
- Research the history of risk assessment in relation to determining how the waste would be managed. What were the procedural and managerial issues and how was risk evaluated whenever waste was deposited into the soil or released into the air or the Columbia River?
- Research how the historic funding levels affected the waste management legacy problems. To what extent were requested waste management improvements rejected by legislatures due to funding constraints
- Develop a calendar of waste management regulations, decisions, results of delaying decisions (providing temporary storage), and implementation dates, including the numbers of waste management structures built each year
- Develop a history of all means of projecting, processing, and inventorying waste
- Develop a comprehensive list of incidents
- Research the effectiveness of waste management at the Hanford Site. For instance, while low level radioactive liquid waste was centrally controlled in the 300 Area, it was decentralized in the 100 and 200 areas. What effect did this difference have, if any, on contamination and waste characterization problems in the respective areas?
- Research the best industry practices for managing waste for each time period. How have Hanford waste management practices compared with those of other nuclear defense production sites and those of other nations?
- Research how the Atomic Energy Commission arrived at the decisions defined in their Immediate Action Directive 0511-21
- Research how the Washington State Administrative Codes on waste storage came about
- Research to what extent and in what manner public disclosure of waste data has impacted subsequent waste management practices at the Hanford Site. How has disclosure affected those who live in neighboring communities and may have been impacted by waste releases into the environment? What has been their response?
- Research to what extent waste management decisions were politicized (location of the Waste Isolation Pilot Plant, for example)

SECTION 7

RESEARCH AND DEVELOPMENT

C.F. NOONAN AND D.C. STAPP
PACIFIC NORTHWEST NATIONAL LABORATORY

SECTION 7 - RESEARCH AND DEVELOPMENT

The majority of the research and development (R&D) conducted at the Hanford Site, especially during the early decades, was connected directly or indirectly to the production of plutonium. From the outset, the Hanford Site's main contribution was not innovative research but the production of fissionable material for bombs as quickly and cost efficiently as possible (Findlay and Hevly 1995). R&D efforts focused on increasing the amount of plutonium being produced, increasing the life of equipment, improving health and safety for the workers and public, and reducing costs. As the years progressed, R&D staff steadily increased their work away from plutonium production and began working for other clients outside the Hanford Site.

Types of Research
(Adapted from AEC 1960)

Basic Research: Primary aim is fuller understanding of the subject under study rather than any practical application or immediate useful goal.

Applied Research and General Development: General research undertaken with the clear probability that it will have a useful end result.

Project Development: All efforts devoted to the solution of specific problems and producing an end result.

Process and Product Improvement: Concerned with lowering costs and increasing the efficiency of products or systems that have been developed and put into production or operation.

R&D efforts at the Hanford Site covered a wide variety of topical areas including the physical sciences (metallurgy, chemistry, atmospheric sciences, etc.); medicine, biology, and related life sciences; improved nuclear reactors and reactor systems; and energy applications. Within each of these areas, R&D could fall anywhere from basic research to process and product improvement. The type of research directly impacted the staff and facility requirements. Basic research would typically require theoretical scientists and laboratory settings, while process and product improvements might involve engineers, technicians, and test facilities such as test reactors and high bays.

The federal government has historically been a major funding source for R&D of national importance. Major areas include national defense, energy, and basic science. The U.S. Department of Energy (DOE) and its predecessor organizations have contributed significantly to these efforts across the nation. The Hanford Site has played a small but significant role in this national research agenda. Understanding the research problems undertaken and the accomplishments achieved at the Hanford Site contributes to the historical record on government funding of R&D.

This narrative relates the history and accomplishments of R&D activities at the Hanford Site between 1944 and 1990. The focus is the work the R&D staff assigned to the Pacific Northwest National Laboratory (as it is presently known) and its predecessor organization, the Hanford Laboratories, conducted. The narrative also includes the R&D work Hanford staff conducted before the Hanford Laboratories was formed.

Our goal in this narrative is to give the reader a basic understanding of the types of R&D performed

CONTENTS

Divider Page Photo: Early Researchers at the Applied Fisheries Laboratory, circa 1944 (Becker 1990, p. 63)

at Hanford (see sidebar box), especially those not related to the production of plutonium. Research directly related to plutonium production is also discussed here and more fully described in the sections on production: Fuel Manufacturing, Reactor Operations, Chemical Separations, and Plutonium Finishing. Our secondary goal is to link R&D activities to buildings and properties contributing to the Hanford Site Historic District. The physical structures that housed research teams play a vital role today in comprehending R&D at the Hanford Site.

HISTORICAL CONTEXT

The nature of R&D at the Hanford Site changed over the years in response to the national and international context in which events occurred. This context helped define eras different from those described in the other sections of this book, which are defined by the production of plutonium. The R&D eras were affected by such events as the rise of civilian nuclear power, the environmental movement, the energy crisis, and the information revolution. The dates of the eras are not precise but are instead indicators of generally when events had the most effect on R&D at the Hanford Site. Some of the events overlapped as indicated by the dates. The eras that affected R&D at the Hanford Site were:

- Pre-World War II (Pre-1942)
- Manhattan Project Era (1942-1946)
- Weapons Complex Era (1946-1988)
- Energy Era I (1955-1970)
- Energy Era II (1971-1988)
- Cleanup Era (1989-present)

PRE-WORLD WAR II (PRE-1942)

It was not until the late 1800s that a significant scientific community emerged in the United States. This community was primarily supported by industry, universities, and independent research centers with industry in the forefront. By the beginning of the 20th century, industrial research was decentralized and competitive, major institutions performed research and technical services, and government work was contracted out, usually to universities. Basic research and applied research, especially military, were in dire straits on the eve of World War II (Kleinman 1995, Smith 1990).

The story of the development of nuclear technology is one of the major R&D stories in the history of the world and is central to the story at the Hanford Site. Chapter 1 sketches this historical tale and the introduction of fission to the United States. Although the U.S. Government was interested in the subject of fission and wanted to be kept informed about any progress in the field, it would not devote any funding for research until more information could be obtained. It was not until 1940, after several research facilities in universities across the country had made significant progress on the fission problem, that federal funding for research was anticipated (Hewlett and Anderson 1990).

MANHATTAN PROJECT ERA (1942-1946)

All efforts of developing atomic energy as a power source were diverted toward developing an atomic weapon when the United States entered World War II (Groves 1983). In 1942, President Roosevelt agreed the U.S. Army Corps of Engineers should be in charge of designing plans to develop atomic weapons. Many physicists in America were soon involved in solving the complex problems surrounding production of radioactive isotopes, isotope refinement and separation, and weapon design.

By 1945, this massive R&D effort had proven successful. The R&D teams at the University of Chicago and Los Alamos developed the technology. The engineers and others at the production sites of the Oak Ridge Reservation and the Hanford Site produced the isotopes. The scientists and engineers at Los Alamos designed and assembled the test bomb, Trinity, and those used at Hiroshima and Nagasaki.

WEAPONS COMPLEX ERA (1946-1988)

Following the end of World War II, the Cold War began and efforts commenced to increase and improve the nuclear arsenal. Weapons designers first increased atomic bomb yields, then developed the new more powerful hydrogen bomb (Rhodes 1995). As nuclear weapons technology advanced, the government embarked on an aggressive weapons testing program that released large amounts of radionuclides into the environment (Hacker 1994). These events, in concert with a generally increasing awareness of the dangers of radiation, stimulated considerable R&D on human and environmental health effects of radiation.

Developing the atomic bomb had started out as a theoretical academic question and ended in a tangible result. This result of the Manhattan Project led the federal government to place an increasing priority on science, not only for defense but in the other federal agencies as well.

ENERGY ERA I (1955-1970)

In early 1946, just a few months after the bombings of Hiroshima and Nagasaki, the U.S. Congress passed the Atomic Energy Act. This legislation called for the development of federally funded research programs to secure scientific and technical accomplishments in the field of nuclear energy. This was a new and exciting field, and private industry expressed interest in using nuclear technology for power production. Private industry, however, made little advancement in the nuclear energy field until the 1950s. Two reasons commonly cited are the control the government maintained on nuclear materials and the great technological risk that industry was unprepared to assume (Hewlett and Holl 1989).

President Eisenhower gave a boost to nuclear energy with his Atoms for Peace Initiative of 1954. The U.S. Congress amended the Atomic Energy Act in 1955 to provide funds for a series of demonstration projects. Demonstration plants were constructed, the Navy began using nuclear technology to power its submarines and ships, and the nuclear energy industry began to develop. By the 1960s, utilities across the nation were constructing nuclear power plants, and expectations were high throughout the world for this new source of power. Before 1966, electric utilities had ordered fewer than ten reactors total. In 1966-1967, that number quadrupled. After declining slightly in 1969, nuclear power reactor orders peaked in 1972-1973. Government and industry funded substantial R&D efforts to increase the efficiency and safety of nuclear energy.

Also during this era, alternative uses for nuclear technology were being sought. Project Plowshare represented one of the more commercial efforts related to alternative uses. R&D efforts within this program were associated with using nuclear explosive devices for large-scale construction and other non-military purposes, such as using nuclear devices to construct canals, to improve oil exploration efforts, and for mining. The U.S. Congress committed $3M to the program in 1958 and 1959, but the program fizzled shortly thereafter as a result of President Eisenhower's moratorium on nuclear testing (Hewlett and Holl 1989; O'Neill 1995, 1998).

ENERGY ERA II (1971-1988)

This era was marked by the energy crisis of the early 1970s, which was related to the rapid increase in oil prices from Middle East suppliers. The crisis set off an unprecedented growth in R&D work related to non-carbon energy sources, such as nuclear, wind, and solar. The breeder reactor program dominated R&D efforts related to nuclear energy.

Also characteristic of this era was the increase in environmental awareness and demands for environmental protection (see Balogh 1991, D'Antonio 1993, Pope 1998, Wellock 1998). One significant impact of the environmental movement on the Hanford Site was a rapid rise in the number of laws and regulations governing nuclear technology and radioactive waste management. R&D efforts were increasingly needed to assist in developing regulatory guidelines as well as developing waste management technologies. The environmental protection movement placed pressure on the Atomic Energy Commission, which had taken over from the U.S. Corps of Engineers in 1947 to manage the Manhattan Project sites. This pressure led to the splitting of the Atomic Energy Commission in 1975 into the Energy Research Development Agency and the Nuclear Regulatory Commission.

Nuclear power reactor construction peaked in 1974 as the industry faced continued pressure from new regulations, high interest rates, increased public scrutiny, and large cost overruns. This marked the start of a downward spiral for nuclear energy that was furthered by the Three Mile Island incident in 1979 and the Chernobyl meltdown in 1986. Budgets for nuclear-related R&D suffered accordingly.

Recognizing that earlier projections for recycling waste were grossly exaggerated and because of the lack of public confidence in industry's ability to deal with waste, the U.S. Congress passed the Nuclear Waste Policy Act in 1982 (42 USC 10101). It relieved industry of the responsibility for waste disposal and identified the need for long-term storage of high level radioactive waste. This Act initiated a variety of efforts to develop a national repository, first in Kansas, and later at a variety of locations, including the Hanford Site. R&D efforts were needed to determine locations acceptable for long-term storage of waste as well as to develop systems for managing and containing the wastes.

Finally, the environmental movement had another important effect on R&D: the public's growing interest in and concern about nuclear technology. This increased the number of studies relating to the potential effects and risk of radionuclides and nuclear power generation.

CLEANUP ERA (1989-PRESENT)

During the 1970s and 1980s, environmental concerns heightened regarding contaminant releases to waterways and lands and their impact on people. Contamination discoveries such as those at Love Canal in New York State, where residents were exposed to harmful chemicals that had been dumped in the nearby Love Canal, and Times Beach, Missouri, where dioxins had been mixed with tar and sprayed on roads throughout the community, alarmed many. These events contributed to the passage of the *Comprehensive Environmental Response, Compensation, and Liability Act,* better known as the Superfund, and a rampant increase in cleanup of contaminated lands across the nation. However, because the DOE complex was still focused on nuclear weapons production and its high security prevented outside scrutiny and pressure for cleanup, the DOE sites experienced little impact from the environmental cleanup movement until the late 1980s.

By the end of the 1980s, events converged to move DOE into its own era of cleanup. The 1986 Chernoybyl nuclear power plant explosion in the Soviet Union led to a general opening of DOE's nuclear production facilities to outside inspection, especially at the Hanford Site where design similarities with the N Reactor gave great cause for concern. By the late 1980s, the cold war was waning and the need for nuclear weapons production, especially plutonium from the Hanford Site, disappeared. Pressure from anti-nuclear advocates led to the eventual shutdown of the Hanford Site production facilities and demand for cleanup.

During this period, the Environmental Protection Agency began pressuring federal agencies to sign federal facility compliance agreements, which were legally binding agreements between regulators and agencies that set standards and schedules for compliance with environmental statutes. The Hanford Site, under the leadership of Michael J. Lawrence, was the first DOE site to sign a compliance document in 1989, which is now in the third revision (Ecology 1994). As cleanup activities commenced, recognition of the enormity and complexity of the cleanup problem associated with radioactive wastes identified a need for new technologies to characterize, monitor, and remediate contamination (Congress 1991).

ORGANIZATIONAL AND OPERATIONAL SETTING

Over the years, the Hanford Site has had several contractors manage the day to day operations as well as conduct R&D. E.I. Du Pont de Nemours and Company and the General Electric Company were responsible for all activities at the Hanford Site. In 1965, however, the Atomic Energy Commission decided to disperse the activities among several contractors. The Battelle Memorial Institute assumed responsibility for the R&D activities at the Hanford Site. The sidebar box shows the directors responsible for R&D at the Hanford Site.

E.I. DU PONT DE NEMOURS AND COMPANY

Directors of the Research Laboratories

Name	Company	Years
Herbert M. Parker	General Electric Company	1953-1965
Sherwood L. Fawcett	Battelle Memorial Institute	1965-1967
Fred W. Albaugh	Battelle Memorial Institute	1967-1971
Ronald S. Paul	Battelle Memorial Institute	1971-1973
Edward L. Alpen	Battelle Memorial Institute	1973-1975
Tommy W. Ambrose	Battelle Memorial Institute	1975-1979
Douglas E. Olesen	Battelle Memorial Institute	1979-1984
William B. Wiley	Battelle Memorial Institute	1984-1994

As has been thoroughly documented elsewhere (Thayer 1996), the success of the construction and startup of the Hanford Site by E.I. Du Pont de Nemours and Company can be accredited to Du Pont's organizational and managerial characteristics. The most important, for purposes of this section, was Du Pont's emphasis on industrial research and development. In the 40 years before its wartime assignment at the Hanford Site, Du Pont had made research the principal basis for its products and resulting income (Du Pont 1946a, Thayer 1996). This organizational emphasis on R&D encompassed all aspects of Du Pont's construction and management of the Hanford Site.

The Du Pont departments listed below were the key organizations responsible for R&D at the Hanford Site during the Manhattan Project.

Instrument Department was responsible for the maintenance of all project instruments as well as new installations and alterations to existing equipment and instrument development (Du Pont 1946b).

P Department was in charge of process improvement, provided the consultation and checking services for the Construction Division and Pile Operations, and developed and expanded the Fish Lab. Operations of the P Department were located at a variety of buildings including 107, 146-F, 303, and 305 (Du Pont 1946d, 1946e).

S Department was concentrated in the 200 Areas. This department was concerned with all aspects of separation facilities and isolation buildings. Another very important function of this group was the emphasis on meteorological and climatological research. The meteorology group worked closely with the chemical separations personnel to schedule dissolver operations when weather conditions were the safest for releases from those operations. They also studied wind direction and velocity, precipitation, conducted forecasts, and determined stack-to-ground dilution factors. This departmental role expanded when airborne waste began to be linked to vegetation contamination. The S Department was concentrated at the 622 Meteorology Complex (Du Pont 1946e). For information on the 622 Meteorological Complex, see the Historic Property Inventory Form in Appendix B on the Internet.

T Department was responsible for all plant process technology, operated the plant control and development labs, maintained the Technical Manual and the Operating Standards, issued reports of technical progress, provided isolation technology and plant assistance, and produced canning, reactor, and separation technology. The T Department, along with the Fish Labs, also carried out Columbia River studies to ensure adequate control of effluent materials for the protection of fish and other forms of aquatic life. These studies were later taken over by the Health Instruments Section (Du Pont 1946f, 1946g).

Medical Department/Health Instruments Section was a specialized safety and technical group primarily concerned with the problems of occupational and public health hazards. This group was responsible for reporting hazard conditions, assisted the Production and Technical Departments in determining safe working conditions from radiation, recorded personnel exposure, and maintained medical records. A variety of radiation monitoring devices (Betty Snoop, Cutie Pie, pencils, etc.) and medical examinations (urine samples, thyroid checks, etc.) were used to ensure Hanford worker radiation protection. This group was concentrated in a variety of buildings such as 231-Z, 292-T, 321-Complex, 3706, 3745-A/B, and 3746 (Du Pont 1946c). For information on these buildings, see their Historic Property Inventory Forms in Appendix B on the Internet. For more information on the Medical Department and its Health Instruments Section, see the Worker Health and Safety Section.

GENERAL ELECTRIC COMPANY

After the General Electric Company assumed the primary contract for operation of the Hanford Site in 1946, R&D activities were reorganized and expanded (see Table 2-7.1). During the first postwar expansion from 1947-1949, plutonium production was of primary importance at the Hanford Site. However, because of the Atomic Energy Act regulations, which were issued in 1946, the peaceful uses of atomic energy also had to be explored.

Table 2-7.1. *Hanford Site Research and Development Personnel, 1946-1957*

Year	Percent of R&D Personnel
1946	<1
1947	1
1948	5
1949	7
1950	7
1951	9
1952	9
1953	10
1954	9
1955	8
1956	9
1957	9

R&D activities carried out by General Electric included radiobiology; radioactive isotope production; design and development of reactors, water treatment plants, separation facilities and equipment; radiological and meteorological monitoring; aquatic life studies; and biological studies of animals using radioactive elements.

In 1953, all departments conducting R&D were consolidated under one new organizational name–the Hanford Laboratories – to reflect their primary mission, research. Research facilities were not necessarily identified as "laboratories" before this time but rather were distributed throughout the production organization facilities. Construction on the Hanford Laboratories complex in the 300 Area was completed in 1953 (GE 1954a). This concentration of research groups in one primary area created better communication and eliminated duplicate research. The Hanford Laboratories continued research and development programs in support of irradiation processing, chemical processing, waste management, and reactor fuels preparation. Laboratory personnel also provided advanced research in the physical and life sciences as well as radiation monitoring. These studies took place onsite at a variety of buildings, laboratories, and farms because the 300 Area initially did not have the capacity to handle all research activities.

BATTELLE MEMORIAL INSTITUTE

In 1963, General Electric, which had operated the Hanford Laboratories since 1946, announced that it wished to end its contract, citing potential conflict with its desire to pursue work in the commercial nuclear field. In 1964, the Atomic Energy Commission decided to split the Hanford Site operating contract among several different contractors. Battelle Memorial Institute, a research and development organization from Ohio submitted a proposal to operate the Hanford Laboratories.

Battelle had a longstanding relationship with the Atomic Energy Commission and had played a significant role in the Manhattan Project with 400 staff working round the clock on fuel rod fabrication, alloy development, and other projects. These wartime projects were followed by numerous Atomic Energy Commission projects, several of which were in collaboration with the Hanford Laboratories (Boehm and Groner 1986).

In its proposal, Battelle committed to invest the net fee resulting from the contract plus an initial appropriation of $5 million of its own funds into laboratory facilities to promote research and development in the Tri-City and Pacific Northwest areas (Boehm and Groner 1986, p. 72). In return, Battelle asked for the ability to use the laboratories for private research. It would eventually take 2 years to negotiate the finer points of this private use contract with the Atomic Energy Commission, but it would prove to be well worth the effort in terms of benefitting both the government and private industry by sharing facility costs and expertise.

On May 28, 1964, Senators Jackson and Magnuson announced that Battelle had been awarded the contract to operate the Hanford Laboratories. When Battelle took over in 1965, the research facility was separated from Hanford Site operations and renamed the Pacific Northwest Laboratory. The government also issued Battelle a Use Permit, which enabled Battelle to conduct research for government agencies as well as private companies (PNNL 1999). Figure 2-7.1 shows the announcement of Battelle's selection to operate the laboratories at the Hanford Site.

The initial mission of the Pacific Northwest National Laboratory, as it is currently called, was to focus on nuclear technology and the environmental and health effects of radiation. The Use Permit enabled Battelle to do research for others, obtain patents, license technology, and develop partnerships with industry. By the 1970s, the Pacific Northwest National Laboratory's mission had evolved to include a wide range of services and research in the areas of energy, environment, and the economy (PNNL 1999) (see sidebar box for listing).

IT'S BATTELLE!
- To Operate Laboratories At Hanford
Offer Described As Outstanding; Objectives Met
Tri-City Growth Seen
Chance Seen As Challenge
Tri-City Herald

Figure 2-7.1. *Article Appearing in the Tri-City Herald Announcing Battelle as the New Operator of the Laboratories at the Hanford Site, May 1964*

OTHER RESEARCH CONTRACTORS

After 1965 when numerous contractors took over various operations at the Hanford Site, each of these contractors had their own R&D organizations to experiment with new materials, process improvements, and methods. Because these advancements were typically restricted to production, we have not covered them in this section.

In one case, however, an organization other than Battelle did take on a major R&D effort not related to production. The Hanford Site was selected as the location for the Fast Flux Test Facility, a prototype breeder reactor. Westinghouse Corporation won the contract in 1970 to construct and operate the Fast Flux Test Facility. Approximately 800 Battelle employees who had been working on the Fast Flux Test Facility Project up to the time the contract was awarded transferred to Westinghouse.

IMPORTANT AREAS OF RESEARCH

When the Hanford Site was chosen to produce plutonium, questions immediately arose about the possible environmental impacts. Radiation tolerance limits were not completely understood in World War II America. No tolerance limits were established for the hazards that would be encountered during production or the number of radionuclides that would be produced. Scientists did not comprehend how these isotopes would behave in soluble or insoluble form. It was uncertain if they entered the body through inhalation, ingestion, or absorption through the skin. They did not know how the isotopes would concentrate in the food chain in plants, insects, fish, birds, mammals, and eventually humans. Furthermore, very few studies of the sandy earth and extreme winds of the Columbia Basin had been undertaken. This compounded the uncertainties of the entire project and forced almost immediate inquiry into broad biomedical and ecological research (Gerber 1992a, Stannard 1988). Thus, research was performed concurrently with construction and production at the Hanford Site. Indeed, wastes were present in the area's air, ground, and water. Emissions occurred repeatedly above levels that were then defined as permissible or tolerable. Radioactive substances entered the food chain through both the air and river and were exacerbated by dust storms and construction activities. To ameliorate some of these problems, Hanford researchers first focused on worker health and safety followed by environmental studies.

An immense array of research and development has been conducted at the Hanford Site over the years. Hanford's depth and breadth of nuclear expertise created a demand for Hanford staff as the nuclear industry evolved. Studies in the physical and biological sciences have been carried out at the Hanford Site since the early 1940s. Human health and the health of the surrounding environment provided the primary focus for these early studies. Potential hazards to these groups provided the foundations for extensive biological and radiological programs at the Hanford Site.

Research and Development Capabilities in the 1970s of the Pacific Northwest National Laboratory
(PNNL 1999)

Air & Water Pollution
Alloy Development
Analytical Chemistry
Applied Mathematics
Aquatic Biology
Atmospheric Sciences
Biomedical Science & Technology
Biophysics
Catalysis-Surface Chemistry
Cellular & Molecular Biology
Ceramics & Cements
Chemical Development & Processes
Comparative Pathology
& Physiology
Composite Materials
Computer Simulation & Systems Analysis
Computer Technology
Corrosion Technology
Ecology-Radioecology
Electrochemistry
Electrochemical Devices, & Systems
Electronic Devices, Circuits
& Systems
Energy Conversion Processes
Environmental Sciences
Fluid Mechanics
Food Technology
Geosciences
Graphite Technology
Heat Transfer
Hydraulics
Hydrology
Industrial & Applied Chemistry
Industrial Economics
Inhalation Toxicology
Inorganic Chemistry
Instrumentation
Laser Technology
Marine Science & Technology
Materials Fabrication
Materials Separation & Purification
Mechanical Components & Devices
Microscopy: X-Ray, Optical & Electron
Mineralogy
Nondestructive Testing
Nuclear Economics
Nuclear Fuels Development,
Processing & Separations
Nuclear Instrumentation
Nuclear Physics
Nuclear Reactor Technology
Nuclear Reactor Materials
Nuclear Safeguards Evaluation
Operations Research
Optics
Organic Chemistry
Process & Physical Metallurgy
Physical Chemistry
Plant Nutrition
Process Development
Radiation Biology
Radiation/Irradiation Effects on Materials
Radiation Shielding & Protection
Radioactive Wastes Processing & Disposal
Radio & Radiation Chemistry
Radiobiology
Radioisotope Technology & Applications
Radiological Science & Technology
Regional Economics
Seismology
Semiconductors-Solid State Devices
Solid State Physics
Spectroscopy
Statistics
Systems Development & Analysis
Theoretical & Applied Mechanics
Theoretical & Mathematical Physics
Thermodynamics
Toxicology
Ultraprecise Measurement Methods
Ultrasonics
Water Treatment, Use & Recovery
Welding-Joining Technology

Major areas of R&D were conducted at the Hanford Site outside of producing plutonium for weapons. Charts at the end of this section list both the biological and ecological research projects as well as the physical research projects at the Hanford Site from 1951-1990.

Eight main areas of R&D were important at the Hanford Site: detecting radiation, human radiation experiments, meteorology and atmospherics, applied fish studies, radioactivity field and reactor effluent studies, environmental and biological monitoring, radiobiology, and inhalation (pulmonary radiotoxicology) studies.

DETECTING RADIATION

Detecting radiation was one of the first areas of research initiated at the Hanford Site and was essential to secure the health and safety of Hanford workers and area residents. Although good plant construction and design was necessary to protect workers and the environment from any possible dangers, Hanford managers were aware that an onsite team of health physics experts was needed for their radiation detection program (Gerber 1992b).

The first health protection and instrumentation team arrived at Hanford in early 1944. They researched the biomedical effects of ionizing radiation and devised a variety of methods of shielding and monitoring radiation workers (see the Health and Safety Section). They quickly established themselves in an organization known as the Health Instruments Section. The Health Instruments Section was instrumental in defining and measuring radiological hazards, safety procedures, and developing and calibrating instruments (Gerber 1992a, 1992b; Stannard 1988). The Health Instruments Section was incorporated into Du Pont's Medical Department. The Medical Department/Health Instruments Section specialized in safety and the problems of occupational and public health hazards.

The Health Instruments staff placed a number of new and unique radiation detection instruments into use at the Hanford Site (Gerber 1992a, 1992b; Howell et al. 1989). These included integrons (air monitors), gas and liquid sampling vessels, Geiger-Mueller counters, and other devices for personnel monitoring. A variety of radiation monitoring devices and medical examinations (urine samples, thyroid checks, etc.) were used to ensure worker radiation protection.

To assist in improving the quality and decreasing the expense of radiation protection, many portable health physics instruments were designed, modified, or fabricated onsite (see Table 2-7.2). Portable health physics instruments have always been at the forefront of maintaining radiation control programs at the Hanford Site. The two functional types are those designed to monitor contamination and those designed to measure exposure rates. Most R&D of portable health physics instruments occurred in the 318, 3745, 3745-A, 3745-B buildings (Howell et al. 1989, Roesch 1964).

HUMAN RADIATION EXPERIMENTS

In early 1942, officials of the Manhattan Project realized they would need a special medical program to monitor the exposure of scientists and workers to materials produced during plutonium production. Determining adverse health effects were a priority for safety as well as the risk that any excessive radioactivity spread outside the facilities might be detected, which could compromise the security and with it the schedule of weapons production.

Human radiation experiments were one of the most controversial R&D activities undertaken. Government sponsored human radiation experiments occurred within the broader context of biomedical science, which encompassed experimental research as well as the diagnosis and treatment of disease. The Corps and its successors also had related interests in radioecology, radiological warfare, civil defense, and nuclear fallout. The chief objectives were to determine the biological effects of radioactive material, measure doses from injected, ingested, or inhaled radioactive substances, or measure the time it took such substances to pass through the body (also called uptake studies). At Hanford, these were manifest in a variety of formats, and research subjects included Hanford Site employees, medical patients, prisoners, and routine monitoring of area children. The experiments at the Hanford Site were divided into two main subject areas, environmental releases and human plutonium injections. A few of the Hanford studies will be discussed below. For more information on complex-wide experiments, see DOE 1995a, 1995b; CEC 1986; and Welsome 1999.

Table 2-7.2. *Portable Instruments Used to Monitor Contamination and Measure Exposure Rates at the Hanford Site, 1944-1988*

Name of Instrument	Type of Detection	Date Intro.	Designed at Hanford	Modified at Hanford	Fabricated On/Off Site	Date Retired
Contamination Monitoring Equipment 1940s						
Victoreen GM	Beta-gamma	1944-49		Yes	Off	1950-59
Eltronics GM	Beta-gamma	1944-49				1950-59
Victoreen 389C (Thyac)	Beta-gamma	1948		Yes	Off	
Zeus	Beta-gamma	1947				
Sandy	Alpha	1944				1950-59
Portable Poppy	Alpha	1949				1950-59
Exposure (Dose) Rate Monitoring Equipment 1940s						
Beckman	Beta-gamma	1944				1950-59
Hanford Cutie Pie (CP)	Beta-gamma	1948	Yes (1944)		Off	1950-59
Juno	Alpha-beta	1945				1980-89
Epsey Junos	Alpha-beta	1949				1950-59
BFP Instrument	Neutron	1948				
The Neut	Neutron	1948				1950-59
Contamination Monitoring Instruments 1950s						
Nuclear Chicago GM	Beta-gamma	1950				1950-59
Victoreen 389 CD GM	Beta-gamma	1953				1960-69
HAPO GM	Beta-gamma	1956	Yes		Off	1970-79
Scintillation Portable Poppy	Alpha	1955				1970-79
Victoreen Alpha Meter	Alpha-beta	1950				
Zeuto	Alpha-beta	1950				
Chicago Samson	Alpha-beta gamma	1955				
Exposure (Dose) Rate Monitoring Instruments 1950s						
Red Face Juno	Alpha-beta	1950	Yes		On	1980-89
Yellow Face Juno	Alpha-beta	1950	Yes		On	1980-89
Improved Cutie Pie (CP)	Beta-gamma	1954	Yes		Off	
Improved Juno	Alpha-beta	1954	Yes		Off	
Totem Pole		1950s	Yes		On	
TPC			Yes			
LPC			Yes			
BFP		1958				1960-69
Contamination Monitoring Instruments 1960s						
Eberline E-120 CRM	Alpha	1968	Partial		Partial	
Eberline E-140 CRM	Alpha	1969	Partial		Partial	
Radeco 225 CRM	Alpha	1969	Partial		Partial	1970-79
Exposure (Dose) Rate Monitoring Equipment 1960s						
HPC	Gamma	1961	Yes		On	
BFQ	Neutron	1963	Yes		On	1970-79
Snoopy	Neutron	1969				
Contamination Monitoring Instruments 1970s						
Eberline PAC-6	Alpha	1977				
GM detector	Beta	1972				
GM detector (pancake probe)	Beta	1972				
Exposure (Dose) Rate Monitoring Instruments 1970s						
Eberline Teletector	Gamma	1972				
Eberline R0-3B	Beta	1975				
Bumblebee (modified Eberline R0-3B)	Alpha, Beta Gamma	1976				
Contamination Monitoring Equipment 1980s						
Bicron "Surveyor"	Alpha	1984				
Exposure (Dose) Rate Monitoring Instruments 1980s						
Johnson "Extender"	Gamma	1984				
Victoreen Model 325		1985	Yes		Off?	
"Black Widow"	Alpha-beta gamma	1980				

In 1949, a voluntary tissue sampling program was initiated to analyze the deposition of plutonium in humans. Bone, liver, and lung tissue were obtained from former employees during autopsies. This program grew into the current U.S. Transuranium Registry administered by Washington State University. In 1951, fourteen subjects participated in a study that examined the percutaneous absorption of tritium oxide. As a result of this study, researchers recommended a 50 percent reduction in the maximum level for atmospheric tritium oxide (DeLong et al. 1954). To determine uptake by the human thyroid, eight Hanford workers consumed milk during 1963 from dairy cows fed iodine 131 (Watson et al. 1964). From 1963-1971, 64 inmates at the Washington State Penitentiary received x-rays of their testes to examine the effects of ionizing radiation on human fertility and testicular function (CEC 1986, Welsome 1999). And children were subjected to routine monitoring in whole-body counters to determine what kinds of radionuclides may be present in their bodies from living in the vicinity of the Hanford Site and eating locally grown produce (Honstead 1967). Many of the human radiation experiments performed with research monies by the Atomic Energy Commission and its predecessor agencies would not currently be allowed under more rigorous research guidelines. Many of these studies were developed in conjunction with larger research programs such as meteorology and atmospherics, inhalation studies, health physics, and radiobiology.

METEOROLOGY AND ATMOSPHERICS

Because process gas would be released from the chemical separations plants, temperature and other atmospheric conditions were environmental safety concerns for the Hanford Site from the beginning. A meteorologist was sent to inspect the Hanford Site and its topographical features when construction first began in 1943. Soon after the Hanford Site designed a comprehensive meteorological and atmospheric research program.

Preventing harmful effects of plutonium production by-products required the continued study and monitoring of meteorological conditions. The S Department, located in the 622 Complex, was responsible for Hanford's meteorological and atmospheric research program. The Meteorology Group within the S Department worked closely with the chemical separations staff to schedule when process gas could be released during dissolver operations. This depended on the most favorable weather conditions. For information on the 622 Meteorological Complex, see the Historic Property Inventory Form in Appendix B on the Internet.

Du Pont's S Department built the first meteorology tower in 1944. The tower was used to study wind direction and velocity, precipitation, forecasts, and determine the stack-to-ground dilution factors. This departmental role expanded due to concerns with airborne releases and resulting contamination of vegetation.

Du Pont's meteorological and atmospheric research program became more comprehensive over time. Between the years 1959 and 1974, more than 300 atmospheric studies were completed involving surface and elevated releases of tracers (Nickola 1977, Nickola et al. 1983, Ramsdell et al. 1985). Tracers were used to track the movement of stack effluents across the landscape. These planned experiments were necessary to analyze atmospheric conditions and develop tracer techniques. Charles E. Elderkin measured the dispersion characteristics of the atmosphere with respect to particles in one of the largest atmospheric diffusion experiments ever conducted. In addition to Elderkin, Jake Hales was another eminent researcher in the field of meteorology and atmospherics (see sidebar box).

Before the 1950s, most radioactive material releases into the air were the result of operating processes or were accidental. However, one release called the "Green Run" was a planned experiment. It occurred in December 1949, 3 months after the Soviet Union detonated its first nuclear device. The purpose of the experiment was to measure how an airborne release of radioactive materials spread through the environment in hopes of monitoring the Soviet nuclear weapons program. During this experiment, "green" fuel still containing large amounts of radioactive iodine was discharged through the stacks of T Plant. Weather conditions at the time of the test were unfavorable and radioactive gases were dispersed over populated areas. The radioactive iodine came to rest on surrounding ground, vegetation, and water. A 1993 review of classified materials revealed that this test was not "intended to be a radiation warfare experiment or a field test of radiobiological effects on humans" (GAO 1991).

CHARLES E. ELDERKIN - HANFORD SITE 1959-PRESENT

Charles Elderkin has contributed to science at the Hanford both as a research scientist and as the Manager of the Atmospheric Sciences Program. His research has focused the physics and chemistry of clouds and precipitation, evaluations of wind characteristics to plan sites and designs for energy-producing wind turbines, climatic effects from energy developments, modeling and field observations of wind dispersion characteristics for pollution control, and meteorological forecasting. He has made significant research contributions in the field of boundary layer dispersion. In Operation Green Glow, an experimental release of radionuclides in the atmosphere, he measured the dispersion characteristics of the atmosphere with respect to the particles. This was the largest atmospheric diffusion experiment ever conducted, and information from this test helped Hanford scientists assess the impact of placing nuclear power plants in climates of differing type and terrain. Elderkin was also the project manager for the Wind Characteristics Program Element of the Federal Wind Energy Program, for which he evaluated wind characteristics to plan sites and help design energy-producing wind turbines (Elderkin and Fuquay 1965).

JAKE HALES - HANFORD SITE 1962-1991

Jake Hales is an atmospheric chemist who came to the Hanford Site in 1962. He quickly rose to senior scientist and Manager and Director of Atmospheric and Earth Science Programs. His research interest has been in the application of chemical engineering to environmental sciences, atmospheric sciences in particular. He has focused on particle transport and reaction-rate phenomena in his research. Hales research at PNNL includes the Multistate Atmospheric Power Production Pollutant Studies, for which he performed fundamental research on the removal of material by clouds, which is later released as rain. This acid rain is the wet deposition of sulfate particulate matter. He performed theoretical modeling, along with field observations, to determine how removal of particulate matter, such as sulfates and nitrogen oxides, affect rain. Hales created the Pluvious Model, created to model storm, cloud and atmospheric chemistry at small scales. From this model, the GChM-Global Chemistry Model grew, which modeled the same parameters but on a global scale. This model was designed for multi-day simulations over a wide range of spatial scales to solve air pollution problems.

As a result of public concerns about these releases causing adverse health effects, the exposures from the Green Run as well as other releases were estimated by the Hanford Environmental Dose Reconstruction Project, which began in 1988 with the work performed by Battelle under the supervision of the U.S. Centers for Disease Control and Prevention. The U.S. Centers for Disease Control and Prevention also funded the follow-on study, the Hanford Thyroid Disease Study, in 1989. The mission of the Hanford Thyroid Disease Study was to determine whether an increased incidence of thyroid cancers existed in the populations living at or around Hanford in the 1940s and 1950s. The results of these studies are not covered within this document, but information can be obtained from the U.S. Department of Energy Public Reading Room located in Richland, Washington (http://reading-room.pnl.gov) and the Hanford Site web page (http://www.hanford.gov).

Public concerns about the hazards of atmospheric nuclear fallout testing and the siting of proposed nuclear facilities were mounting between the 1950s and mid 1970s. This prompted researchers at Hanford to conduct a large number of diffusion grid experiments involving surface and elevated releases of radioactive tracers. The first and best-known group of experiments was conducted in the summer of 1959 and called Operation Green Glow. It consisted of 28 releases of fluorescent pigment during relatively stable night-time atmospheric conditions (Nickola et al. 1983; see also sidebar box on Elderkin). In 1962 and 1963, Hanford intentionally released small amounts of iodine 131 from the REDOX Plant to study the dispersion of radioactive iodine into the air and soil. During the emission, plume trajectories were plotted and samples collected at several altitudes and up to distances 50 miles away. In addition, vegetation and milk samples were collected. This study was designed to evaluate the spread and behavior of iodine released into the atmosphere and provide data to estimate the hazards potentially associated with weapons testing or nuclear fallout. To obtain data on inhalation uptake, two Hanford employees and several beagles stood directly in the path of the release (Gamertsfelder 1963, DOE 1995c).

During the 1970s experiments were conducted to determine whether radioactive particles from various study sites on the Hanford Site were resuspended and transported by wind. Interrelationships between wind speed, direction, airborne soil, and radioactivity levels on airborne particles indicated that airborne concentrations increased with wind speed, the weathering half-life was greater than previously reported, and concentration rates were more concentrated than fallout levels. These studies were useful for determining the extent of cleanup operations, i.e., soil cover-up, areas to clean up first, and modeling concentrations for down wind and offsite. The meteorology and atmospherics group continues to provide information on weather forecasts, preparing climatological summaries for the Hanford region, and observe and record weather conditions.

Just as researchers were concerned with atmospheric releases of radioactive materials, Hanford scientists were also concerned about radiation and heat effects on the Columbia River and plants, algae, insects, fish, and introduction to the food chain. These concerns led to the development of comprehensive research programs focusing on applied fish studies, radioactivity field studies, studies on reactor effluents, and environmental and biological monitoring.

APPLIED FISH STUDIES

Initially very little was known about the interactions of ionizing radiation and the immediate environment, therefore, it was essential to monitor the safety of reactor operations through a number of scientific studies. The secretive nature of activities at the Hanford Site necessitated that initial investigations occur offsite. In August 1943, the Fisheries Department at the University of Washington was contracted to undertake special studies on the effects of radioactivity on aquatic organisms, especially fish. The Fisheries Department soon thereafter organized as the Applied Fisheries Laboratory and became an active component of Hanford research. Figure 2-7.2 shows early researchers at the Applied Fisheries Laboratory, circa 1944 (Becker 1990, p. 63).

Initially the Applied Fisheries Laboratory used x-rays to study the effects of radiation on fish. However, those involved in the project believed the specific isotopes involved in reactor operation had to be studied because the isotopes were flushed into the Columbia River, thereby, affecting the fish (Becker 1990). As a result, Du Pont opened a small onsite facility in 1945 in the 100-F Area, the Aquatic Biological Laboratory. For more information on the Aquatic Biological Laboratory, see the 108F Building in Appendix B on the Internet.

Figure 2-7.2. *Early Researchers at the Applied Fisheries Laboratory Around 1944*

The results of the few classified studies the Applied Fisheries Laboratory conducted between 1943-1945 were not published until 1947-1948 (Becker 1990, Bonham et al. 1948, Welander et al. 1948). The Applied Fisheries Laboratory connection with the Manhattan Project was acknowledged after World War II when the group was renamed the Laboratory of Radiation Biology to better reflect their expanding line of investigation (Becker 1990). Later studies involving the Columbia River and its organisms were carried out within the reactor effluent program, biological monitoring, and radioactivity field studies.

RADIOACTIVITY FIELD STUDIES AND REACTOR EFFLUENT STUDIES

Radioactive material releases to the Columbia River came primarily from the Hanford Site's first eight production reactors. These reactors, called single-pass or once-through reactors, used river water for cooling. Once circulated through the reactors, hot effluent water was discharged into cooling ponds and eventually returned to the river. The result was a release of fission products and other radioactive materials into the river. In concurrence with the applied fish studies, collection of Columbia River water began in 1944. However, routine analysis of radioactivity in river organisms began in 1946. From these studies, researchers learned that the amounts and distribution of radioactivity in the Columbia River were controlled by complex ecological factors including seasonal variations in aquatic communities, food chain relationships, and river temperature and flow. The studies on artificial radioactivity and river ecology were interrelated and led to the development of the scientific field known as radioecology (Becker 1990).

Controlled reactor effluent studies were related to both the applied fish and radioactivity field studies. All three were concerned with the impact radioactive isotopes would have on the area's main source of water used for drinking and agricultural purposes in addition to cooling the reactors. The most effective way to examine the biological effects of reactor cooling water effluent was in laboratory experiments. Fish were reared in tanks filled with varying dilutions of reactor effluent. The objective of these studies was to determine what conditions (radioactivity, chemical toxicity, temperature, or any combination of factors) caused adverse effects on the fish. Effects were evaluated by comparing mortality and growth rates between control and exposed groups of fish (Becker 1990).

Other field studies involving use of Columbia River water included use as irrigation on controlled vegetable plots. The data were used to predict the relative hazard of different isotopes associated with Hanford operations. Migratory fish were also monitored to study how radionuclides would disperse from the Columbia to the Yakima and Snake rivers (Kornberg 1964).

ENVIRONMENTAL AND BIOLOGICAL MONITORING

First implemented in 1943, environmental monitoring was essential to secure the health and safety of Hanford workers and area residents. Detection instruments monitored workers, residents, and the environment for radiation (see previous sub-section on Detecting Radiation). In view of what we know today, unvoiced concerns for the health and welfare of Richland residents and Hanford Site employees were masked with consumption and behavioral recommendations by the Medical Department (Gerber 1992a, pp. 72-73). For instance, the use of iodized salt was encouraged because animal experiments in 1945 suggested inert iodine reduced thyroid uptake of carrier-free radioiodine. In mid-1947, atmospheric tests determined airborne radioactive contamination deposited readily on sagebrush. The Richland Public Health Section then advised residents to remove them on the premise that they aggravated allergies and harbored ticks. And in 1949, traditional hunting areas near the Hanford Site were closed due to the discovery of high radioactivity levels of iodine-131 in game animals.

The rush to successfully complete construction and start plutonium production at the Hanford Site necessitated minimizing potential health and safety problems. Consequently, public suggestions were the only way to alert residents and employees to potential health concerns at the Hanford Site without causing too much alarm and panic. In the early phases of environmental monitoring and radiation protection, it was often unclear what acceptable tolerance doses were and what the limits should be. Initial studies at the Hanford Site helped establish these guidelines and studies have continued to the present day. For more information on tolerance doses and limits, see the Health and Safety Section.

Monitoring of various species has been of primary importance at the Hanford Site. Transmission of radioactive materials through biological chains revealed information about radionuclide transfer. Controlled laboratory experiments on animals fed or injected with radioactive compounds aided the development of exposure levels for humans and provided a baseline for comparing anomalies of wildlife on the Hanford Site. Field studies on various biological species (plant and animal) have contributed to basic knowledge of the processes of radionuclide uptake and transport (GE 1951c, Kornberg 1964, Pearce 1967). Area livestock were also examined and though positive but low thyroid readings were found in almost all animals, researchers felt comfortable in stating that it was very "unlikely that animals or persons consuming milk and meat products grown in the vicinity of Hanford were in any case damaged" (Herde quoted in Stannard 1988, p. 762).

RADIOBIOLOGY

When the production plants at the Hanford Site and other areas started operations in the 1940s, scientists did not understand well the effects that radiation could have on workers, the public, or the environment. The programs that began initially with the applied fish studies expanded over the years to include aquatic life studies, plant life studies, ranch and wildlife studies, the development of an experimental animal farm, human radiation studies, therapy with radioactive materials, and general biology (Parker 1948b, GE 1951c). Stannard (1988) summarizes the range of radiobiological research across the weapons complex and at the Hanford Site, identifying three major areas at the Hanford Site:

- Fish studies (described in a previous sub-section) to understand the biological effects of reactor effluents released to the Columbia River on indigenous fish.
- Large and small animal studies to introduce radioisotopes into living tissue of animals such as sheep, pigs, dogs, and rats, and study them over time to identify and measure effects. Inhalation, inoculation, ingestion, and adsorption methods were used to introduce radioisotopes such as iodine-131, strontium-90, cesium-137, and plutonium and uranium isotopes into the animals.
- Radioecology studies to understand the effects of soil containing radionuclides. Greenhouses and garden plots were maintained to grow cereal grains, alfalfa, and other crops in soil containing controlled amounts of strontium-90 and cesium-137.

> ***"Researchers at the Hanford Site conducted pioneering radiobiological studies during these years [1945-1965]. They examined the effects of I-131 [iodine-131] on sheep, the biomedical effects of tritium and plutonium, and the reactions of plant life to radiation, and they participated in the quest for radiation therapies. Hanford's chief health physicist, Dr. Herbert M. Parker, also helped to establish tolerance doses for various forms of exposure to radiation. Additionally, the Hanford Site played a seminal role in tracking and studying the effects of fallout, from distant atomic bomb tests, on the Pacific Northwest and on other regions, including the South Pacific and Alaska. As information developed about the biomedical effects of ionizing radiation in the two decades after the founding of the Hanford Plant, researchers in chemistry, physics, medicine, and agriculture, as well as members of the public who read popular scientific journals, came to recognize the magnitude and intransigence of the effects of radioactivity on living tissue,"*** **- Gerber 1993b, p. 172**

The Radiobiology Program at the Hanford Site along with the parallel programs at the Oak Ridge Reservation in Tennessee, and Brookhaven, New York, played an important role in the nation's efforts to understand the impact of ionizing radiation on living tissues (Gerber 1993b, pp. 172-199).

Three of the central researchers of radiobiology at the Hanford Site were Herbert M. Parker, William Bair, and Ethel Gilbert (see sidebar box).

INHALATION (PULMONARY RADIOTOXICOLOGY) STUDIES

Inhalation experiments at the Hanford Site were a major area of research within the general Radiobiology Program. They began by placing mice in the atmosphere of a chemical processing plant expected to contain radioactive particles. In the early 1950s, the inhalation program was more formally organized under the title Pharmacology Operation. The program then concentrated on introducing radioactive elements to animals via intratracheal instillation (Stannard 1988). Hanford researchers reported the first clear-cut demonstration of the development of lung cancer from plutonium in 1955 (GE 1956b). To further address health and safety concerns regarding worker inhalation of radioactive particles and

HERBERT M. PARKER - HANFORD SITE 1944-1984

Herbert M. Parker was a medical physicist who started his career in Enrico Fermi's Metallurgical Lab. Before coming to Hanford, he researched the new and unstudied field of the effects of beta-particle radiation in fission products. One of Parker's earliest concerns was the effect of inhaled radiation. In the early days of the Hanford Site, he was the first chief health physicist and Head of the Health Instruments Section. Parker was responsible for expanding what had been a primarily monitoring and control operation into a world-renowned radiobiology and health physics research program. In the days of the General Electric Company, Parker was made Director of the Hanford Laboratories. He was one of the leading researchers into dilemmas of radon lung dosimetry and modeling for dosimetry (measuring the number of roentgens absorbed due to exposure to radiation). He wrote the earliest summary on the probable mechanisms of the biological action of beta particles, assisted in the development of the first standards for maximum permissible plutonium concentration levels in biological tissue, and subsequently devised standard units of ionizing radiation dose, first known as the rep or roentgen equivalent physical. Parker was a member of the National Academy of Engineering, and a U.S. representative for multiple national and international committees on radiation standards, including the first Atoms for Peace Conference in 1955 (Parker 1977 and Parker et al. 1986).

WILLIAM J. BAIR - HANFORD SITE 1954-PRESENT

Bill Bair was the first in the world to earn a Ph.D. in the brand new field of Radiation Biology from the University of Rochester in 1954. He was a health physicist, who spent his entire career at the Hanford Site, first as a staff scientist and ultimately as the manager of the Life Sciences Program, studying the respiratory and non-respiratory effects of toxins (radioactive and non-radioactive) on biological organisms. Bair developed the inhalation toxicology program at the Hanford Site into a major center for all aspects of inhalation toxicology. Bair designed and developed the means for aerosolizing radioactive materials safely by devising exposure chambers, aerosol generators, measuring systems, and administration devices to administer toxic aerosols to test animals. He studied the effects of the toxins on the animals, including the development of tumors and safe means of removing cancerous lumps, and was the first person to demonstrate the production of a lung tumor by plutonium. He also studied the relative toxicity of plutonium isotopes, including their role as cancer agents. The results of this study helped deduce the relative long-term risk associated with each isotope. His research led to devising new admissible radiation standards for the Natural Resources Defense Council. In addition to radionuclide research, his studies focused on the role of toxic agents other than radon and its associated daughter products in animal research, including cigarette smoke, diesel exhaust, and uranium ore dust (Bair 1965, 1988; Thompson and Bair 1972).

ETHEL S. GILBERT - HANFORD SITE 1973-1994

Ethel Gilbert was a senior scientist in radiobiology and epidemiology. Her career has focused on the health effects of high level and low level radiation on humans, including modeling cancer risk from radiation. While at the Hanford Site, she conducted an analysis of health data and mortality data from 44,500 Hanford Workers from 1944-1978 and studied the effects on survivors of the 1945 atomic bomb dropped on Hiroshima. During Gilbert's career, she also developed models for estimating cancer risks, including a model on risk in the event of a nuclear disaster and the risk of lung tumors from toxic substances (Gilbert and Marks 1979, Gilbert and Sever, 1987).

determine safety guidelines, controlled research experiments were performed on mice and later involved beagles, sheep, and swine bred and raised in laboratory facilities. These studies aided investigations of levels of toxicity in various organs and the blood, concentrated deposition throughout the body, and other defects on the organism (Pearce 1967). Studies on the inhalation and ingestion of iodine in laboratory animals concluded that iodine was usually rapidly absorbed and transported, quickly excreted, and primarily deposited in the thyroid (Stannard 1988).

During the 1960s, the Pharmacology Group concentrated most of its research on the inhalation toxicology of plutonium. The ultimate concern was its long-term toxicity on workers, which eventually led to other studies extending into the 70s and 80s on the inhalation of uranium ore dust, diesel exhaust fumes, and cigarette smoke. The inhalation toxicology program also grew in size to accommodate studies on transplutonics, plutonium-sodium aerosols, and radioisotope toxicology.

ENVIRONMENTAL RESEARCH

During the creation of the Hanford Site, the formation of a large buffer zone around the central facilities was imperative for security and public safety. As a result, several portions of the Hanford Site have remained relatively undisturbed by human activities. Approximately 4 percent of the site's 560 square miles has been used to produce plutonium and contains waste treatment and research facilities.

A reserve was established to *"Preserve portions of vegetation types that once covered a great expanse of the West... ALE [Arid Lands Ecology Reserve] is intended to maintain significantly large remnants of the native plant and animal communities... Provide an undisturbed system where ecologists can examine environmental questions... Provide an area large enough that ecologists can conduct controlled manipulations without destroying the integrity of the natural system... Provide field study areas for the scientific and educational community, particularly graduate students in the environmental sciences."* - O'Farrell 1973

In 1967 the Atomic Energy Commission set aside 120 square miles of shrub-steppe land at the Hanford Site to preserve a natural ecosystem for research and educational use (Hinds and Rogers 1991). This outdoor laboratory, originally called the Arid Lands Ecology Reserve, is managed by the Pacific Northwest National Laboratory . The Arid Lands Ecology Reserve was later named after Richard E. Fitzner, and Lester E. Eberhardt, who had spent many years observing animals at the reserve before they died in a plane crash while studying wildlife near Yakima, Washington in 1992. The Fitzner-Eberhardt Arid Lands Ecology Reserve has remained primarily undisturbed and fits the classification used by federal land management agencies to designate lands on which various natural features are preserved for research and educational purposes (PNL 1993).

In 1977, the Hanford Site was designated a National Environmental Research Park by the U.S. Department of Energy. William H. Rickard was the driving force behind the establishment of the park (see sidebar box). The Hanford National Environmental Research Park, one of seven national environmental research parks (see Figure 2-7.3), includes (see Figure 2-7.4):

- Fitzner-Eberhardt Arid Lands Ecology Reserve (formerly the Arid Lands Ecology Reserve)
- Industrial zone of the Hanford Site
- Buffer zones on the opposite shore of the Columbia River – the U.S. Department of the Interior's Saddle Mountain National Wildlife Refuge and the Washington State Wahluke Wildlife Area.

The national environmental research park concept is an outgrowth of the National Environmental Policy Act of 1969 and the public desire for a clean environment. The Hanford National Environmental Research Park provides the foundation for present and future studies in the areas of environmental restoration, waste and land management, and the maintenance of plant and animal species. A vast variety of studies have been completed at Fitzner-Eberhardt Arid Lands

WILLIAM H. RICKARD - HANFORD 1960-PRESENT

William Rickard is a terrestrial ecologist whose research has centered around field sampling to measure primary productivity, soil-plant mineral relations and man-imposed perturbations, especially cattle grazing, severe soil disturbances, and airborne chemical contaminants in the shrub-steppe desert of the Columbia Basin. As a senior staff scientist, he has dedicated his life to the long-term monitoring of vegetation and wildlife. He joined Hanford's young terrestrial ecology program to study the effects of radioactive fallout on the environment, and to do a pre- and post-shot assessment of the proposed Project Chariot, which was a plan to excavate a harbor in Alaska with a nuclear device. His in-depth research has included monitoring the effects of fallout and soil contamination on the terrestrial systems, including root uptake and the effects on the transfer of radionuclides through the food chain. Rickard's greatest contribution to the field of ecology, in his estimation, is his role in establishing the Hanford Site as a National Environmental Research Park (Rickard 1972, 1979, 1988; Vaughan and Rickard 1977; Sackschewsky et al. 1992).

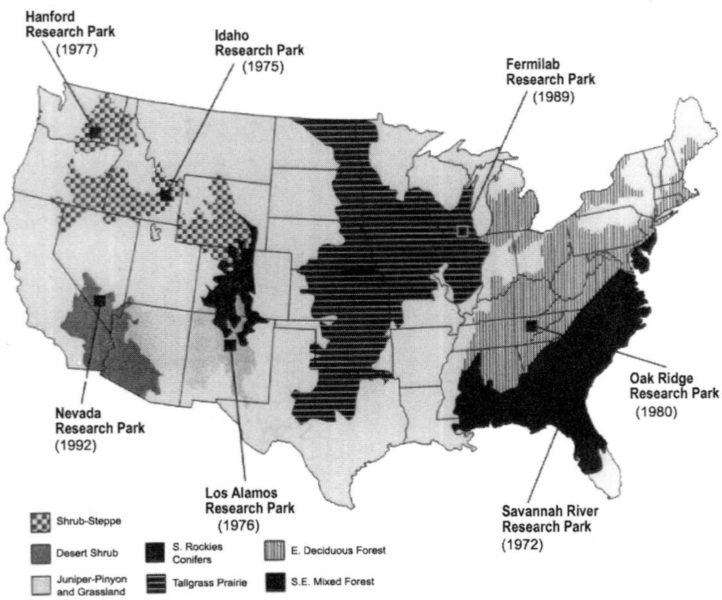

Figure 2-7.3. *Seven National Environmental Research Parks Designated by the U.S. Department of Energy*

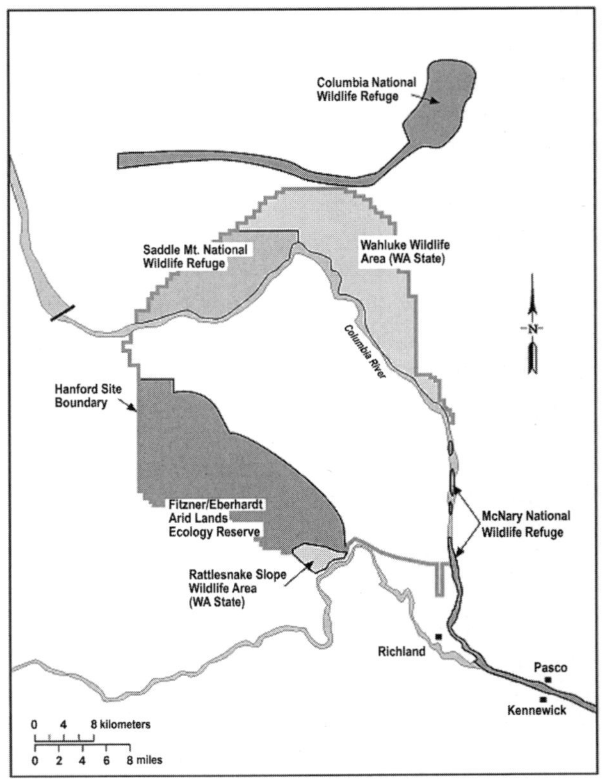

Figure 2-7.4. *Hanford National Environmental Research Park*

Ecology Reserve, including many in the disciplines of geology, microclimatology, ecology, botany, soil science, and animal ecology. As of 1990, over 40 species of mammals, 187 species of birds, 3 species of amphibians, and 9 species of reptiles had been documented, some of which spend part or all of their life-cycle on the Reserve (Fitzner and Gray 1991).

NUCLEAR PRODUCTION

Following World War II, the Hanford Site played an increasing role in R&D activities in response to demands from the government for greater weapons production and demands from private industry to commercialize nuclear energy. As nuclear technology expanded and became better understood, the government increased its research on means of controlling the atom, from the mining of uranium to the safe disposal of spent fuel.

Much of the R&D initially related to plutonium production are found in the respective sections on Fuel Manufacturing (Section 2), Reactor Operations (Section 3), Chemical Separations (Section 4), and Plutonium Finishing (Section 5). This section focuses on R&D activities that extended beyond the boundaries of Hanford plutonium production.

By the early 1950s, nuclear technology was in a growth mode. Although no technology beyond weapons production had yet been developed, scientists were eager to explore the atomic energy and its place in every day living. The utility industry was preparing for its arrival, universities were starting programs to train students, and politicians were pressuring the Atomic Energy Commission to begin transferring the technology (Hewlett and Holl 1989b, American Nuclear Society 1992).

By the mid-1950s, the nation, especially the government, faced a great need for nuclear expertise to fulfill its growing list of responsibilities. The Hanford Site with its staff and facilities represented one of the major centers of nuclear technology in the world, and the only national resource with seasoned reactor operating experience. As a result, much R&D work came to the Hanford Site. One of the main nuclear researchers was Spencer H. Bush (see sidebar box). An indication of the value of this expertise is demonstrated by the number of Hanford staff who GE took with it when it left the Hanford Site to develop its commercial business.

SPENCER H. BUSH - HANFORD SITE 1953-PRESENT

Spencer Bush began his career at the Hanford Site. Coming to the Hanford Site was natural for him. He had been involved in the Manhattan Project at the Los Alamos Laboratory while an enlisted man. He was the driver of the truck that delivered the bomb from Los Alamos to the Alamogordo Bombing Range, New Mexico for the world's first atomic explosion, the Trinity test in 1945. His interest in nuclear technology led him to the Hanford Site after finishing his Ph.D. in 1953. He first contributed to Hanford research as a senior staff scientist, Physical Metallurgy and Fuel Fabrication Program and later as a Consultant to PNNL. His main research contributions have been in the fields of physical and mechanical metallurgy of nuclear materials and reactor safety and more efficient reactor and fuel design. He has studied and solved problems concerning the effects of irradiation on metal components, and stress corrosion on pipes, and subsequent cracking. His research has also considered the effect of fabrication variables on properties of zirconium alloys, irradiation effects in uranium alloys and reactor structural materials and stress corrosion, all aimed at improving safety and efficiency of reactors, specifically N Reactor (Bush 1965).

Nuclear research at the Hanford Site that went beyond the immediate production needs focused on:

- Fuel Performance
- Reactor Performance
- Nuclear Waste Management
  - Waste forms
  - Interim storage
  - Long-term storage
  - Transportation
  - Regulations
  - Environmental Restoration

- Policy and Regulatory Analysis
- Environmental Restoration

Fuel Performance

Hanford staff gained considerable expertise in manufacturing fuel, assessing fuel failures, and improving fuel performance. Hanford reactors used metallic fuel, recognized early on as an inefficient form of uranium compared to oxide fuel blends when used to produce energy. Much of this work was in response to fuel failures, which increased as the Hanford Site increased the performance of the reactors.

When it came time to design the advanced K Reactors at Hanford, much work went into exploring alternatives for the standard Hanford fuel assembly. Again, when the N Reactor was being designed, a substantial amount of R&D went into the manufacturing of fuel elements using a coextrusion process (see the Fuel Manufacturing Section). All of this work established Hanford staff as a resource for nuclear fuel technology.

Because of Hanford's fuel expertise, the Atomic Energy Commission chose the Hanford Site to develop methods to increase the efficiency of uranium fuel by creating oxide fuel blends. Hanford's expertise relative to uranium fuel was limited to metallic fuel, conducive for producing weapons grade plutonium but not efficient for producing energy. Nevertheless, in the mid-1950s, few had the operating experience of the Hanford Site. Thus, the Hanford Site was chosen for a large fuels diversification program known as the Plutonium Fuels Utilization Program begun under President Eisenhower's Atoms for Peace initiative. The hallmark of this program was the Plutonium Recycle Test Reactor, which was located in the 309 Building (Purcell 1966). Tests on a variety of powdered and pelletized fuels continued throughout the 1960s using plutonium oxide blended with uranium oxide and other metallic oxides. For information on the 309 Building, see the Expanded Historic Property Inventory Form in Appendix B on the Internet.

As a result of this and other Hanford research on the performance of fuel, nuclear power plants began using solid, ceramic-like pellets of enriched uranium, slightly larger than pencil erasers, stacked atop each other and sealed in strong metal tubes, called fuel rods. By the early 1970s, Hanford fuels researchers were called upon to work on breeder fuels. This included work at the Chalk River Reactor in Canada. And in 1986, Hanford fuels researchers were chosen to play a new role in a new space technology development program known as SP-100. The SP-100 nuclear reactor, designed by General Electric, would produce continuous power for at least 7 years. The 309 building was modified to prepare for this work, although the program fizzled after a few years.

Reactor Performance

By the mid 1960s, public interest in containing a reactor accident increased in large part due to suspicions about regulatory problems (such as the use of mill tailings for construction purposes), thermal pollution, radiation standards, environmental protection, and waste disposal. When the Atomic Energy Commission's reactor development studies were going on in the mid-1950s, Hanford staff were busy working on the K Reactors to address the Korean Expansion and did not play a substantive role in the various studies. The K Reactors had, however, despite the fact that they were still using the basic Hanford graphite moderator technology, introduced many new technological aspects, especially in the area of new materials, such as zirconium.

Following the construction of the K Reactors, considerable R&D went into the design of N Reactor, which was still a graphite reactor but incorporated many new elements of technology, such as enhanced core design, new fuel making technology, new materials, and a closed-loop system (DOE 1997a). See the Reactor Operations Section for more information on the design of the N Reactor. These developments attracted much attention from the rapidly emerging nuclear energy industry. N Reactor quickly became one of the world's best known, power-producing reactors. While everyone recognized the limitations of the core technology, it was the various elements of the N Reactor system that attracted interest – the piping, safety systems, fuel rupture detection system, and controls.

Hanford staff also did a variety of lattice testing for the design of different light-water-cooled reactors, including those at the Hanford Site. Tests were conducted in the Plutonium Recycle Critical Facility, also in the 309 Building, to determine which geometrical arrangement of fissionable materials would work best in a reactor. Spacing, fuel size, number of fuel elements, choice of material, and other factors all had benefits and drawbacks. The challenge to researchers was to determine which combination worked best. Much of this work was done under the auspices of the Nuclear Regulatory Commission for commercial companies such as Westinghouse Corporation, Babcock and Wilcox Corporation, and Combustion Engineering.

By the mid-1960s, interest in containing a reactor accident increased. The Atomic Energy Commission focused its attention on the emergency core cooling system. The thought was that if the emergency core cooling system was properly designed and constructed, the system would prevent the core from melting when overheated and contain any radioactive releases. Hanford staff were called upon to assist in this effort by conducting containment integrity experiments (Walker 1992).

An indication of Hanford's reactor development expertise is the contribution of the Hanford Site to the Nation's Breeder Reactor Program. The breeder reactor was a fast reactor, not requiring a moderator. Liquid metal, such as sodium, was used to control the heat generated during the reaction. The breeder technology was desirable because the reaction actually produced more plutonium than it used. Research into breeder reactors began in the 1950s and was pursued by the United States, United Soviet Socialist Republic, France, the United Kingdom, and Japan. The Clinch River Breeder Reactor Project in the United States built a prototype breeder. The Hanford Site supported the Clinch River Project in a variety of ways.

As a result of the Clinch River Breeder Reactor Project, the U.S. Government decided to build a full scale breeder reactor. The Hanford Site was chosen, largely though the efforts of its powerful Senators, Senators Magnuson and Jackson. The Atomic Energy Commission commissioned Battelle to start designing the new breeder reactor known as the Fast Flux Test Facility in the late 1960s. Westinghouse Corporation won the contract to both construct and operate the Fast Flux Test Facility and with that approximately 800 R&D staff who had been working on the design for Battelle transferred to Westinghouse. Construction began in 1970, and the Fast Flux Test Facility began operations in 1980.

"Fast flux" means the neutrons move faster in a sodium-cooled reactor than they would if the reactor were cooled with water (WHC 1994). Heat was removed from the reactor by circulation of liquid sodium through three primary loops including primary pumps, piping, and intermediate heat exchangers, all located within the containment building housing the reactor.

The Fast Flux Test Facility was designed specifically to test the irradiation of fuels and materials. Reactor activities were later expanded to include long-term testing and evaluation of reactor components and systems, fusion power materials, passive safety systems as well as the production medical isotopes and research on space power systems (Mayancsik 1988). In 1984, the future of the facility became uncertain. The U.S. Government abandoned the program because of cost over runs, design failures, a vote by the U.S. Congress, and the belief it could undermine non-proliferation treaties.

In the late 1960s, PNNL researchers also worked on how to derive nuclear power. Much of this R&D was funded through the electric power industry consortium known as the Electric Power Research Institute (EPRI). EPRI was created in 1965 to address an increasing concern over electric power capability. The great northeastern blackout of 1965 exposed serious vulnerabilities in the nation's electricity supply system. The Middle East war in 1967 increased awareness of the dangers of relying on foreign sources for energy sources, such as oil. The National Environmental Policy Act of 1969 kicked off decades of environmental legislation that would challenge utilities' scientific and technical capabilities. These and related issues led many to question the ability of the existing energy infrastructure.

In response, the industry formed a collaborative science and technology organization to serve the entire power industry from energy conversion to end use. EPRI began offering scientific expertise and innovative technology to help its industry partners increase efficiency and prepare for the future. U.S. electric utilities established EPRI as a nonprofit membership corporation to manage a national research program on behalf of its funders, the industry, and society. In forming one of the first industry-wide research consortia, electric utilities pioneered the concept of pooling their resources for maximum benefit.

EPRI funded an enormous amount of R&D over the years, utilizing R&D organizations such as the Pacific Northwest National Laboratory. Researchers developed computer models and designed changes to increase the efficiency of cooling towers used by the nuclear power industry.

Nuclear Waste Management

Managing the waste created by nuclear facilities was a consideration from the beginning. Although low-level, liquid radioactive wastes were originally disposed of directly in the ground or in open ditches, they were believed to be temporary holding areas until safer and more effective means of disposal could be developed. High level radioactive waste was stored in underground tanks, and low level radioactive waste was disposed of in the soil, much of which would not be permitted by today's regulations. Little in the way of R&D efforts were expended at the Hanford Site during the first few decades. The means of managing waste were considered effective, and little pressure was applied to improve methods. Nuclear regulation began to intensify in the 1960s, though these new measures tended to focus on reactor siting and performance issues (Walker 1992).

As utilities started building and operating nuclear power plants in the 1970s and spent fuel began piling up, the government recognized the need to address the long-term management of spent nuclear fuel. The spent nuclear fuel assemblies, which were highly radioactive from the fission process, had to be isolated from the environment for long periods of time. Radionuclides in spent fuel assemblies have half-lives ranging from hours to millions of years. Ten half-lives must lapse for the radiation to be deemed safe by reaching acceptable levels.

The Nuclear Waste Policy Act of 1982 established the Office of Civilian Radioactive Waste Management within the U.S. Department of Energy to develop, construct, and operate a system for spent nuclear fuel and high level radioactive waste disposal. This included a permanent geologic repository, an interim storage capability, and a transportation system.

Scientists and engineers, some of them from the Hanford Site, had studied many options over the years for disposing of spent fuel. These options included leaving the waste at the reactor sites, burying it in the ocean floor or a deep geologic repository, putting it in polar ice sheets, and rocketing it into outer space. Based on a final Environmental Impact Statement prepared in 1980 that evaluated all of these options, deep underground geologic disposal was determined to be the safest solution (DOE 1980). The Nuclear Waste Policy Act established a specific funding mechanism for developing the waste disposal system, the Nuclear Waste Fund. This mechanism included collecting funds from those who used the electricity produced by nuclear reactors rather than the general taxpayer.

In the search for a site for the permanent repository, the Office of Civilian Radioactive Waste Management had initially conducted a national search for a potential site. Eventually the agency chose nine sites to be studied in six different states. President Ronald Reagan approved three sites from a candidate list of five for more detailed study. These potential locations for permanent repositories for spent nuclear fuel and high level radioactive waste were the Hanford Site, Deaf Smith County in Texas, and Yucca Mountain in Nevada.

To evaluate the Hanford Site for a long-term waste storage repository, the Basalt Waste Isolation Project was initiated. The Hanford Site was chosen for its 15,000 feet of basalt, believed to be viable for a geological repository. The Basalt Waste Isolation Project generated an enormous amount of R&D at the Hanford Site in the areas of site characterization and deep surface repository engineering. The project ended December 1987 when Congress amended the Nuclear Waste Policy Act and directed the U.S. Department of Energy to study only one site, Yucca Mountain, Nevada, to determine whether it was a suitable site for a repository.

The 1987 Amendments Act also authorized the Secretary of Energy to site a Monitored Retrievable Storage facility (DOE 1981b, 1983). This facility differed from the geologic repository in that it was an above-ground facility that would store a limited amount of spent nuclear fuel temporarily before sending it to a permanent repository. Hanford researchers, in this case primarily those at the Pacific Northwest National Laboratory, were very involved in the design work, workforce analysis, and siting requirements for the Monitored Retrievable Storage facility. The temporary storage method used by utilities was to place fuel in large pools of water contained in steel-lined concrete basins. Rather than using water for the Monitored Retrievable Storage facility, spent nuclear fuel assemblies would be stored in a dry environment, above-ground, using heavy containers or casks made of steel and/or concrete. Casks would be either placed upright on concrete pads or stored horizontally in metal canisters in concrete bunkers. The advantage of dry storage was a more stable environment with lower maintenance costs. Like pool storage, dry storage was not intended to be a permanent solution. However, funding to build the Monitored Retrievable Storage facility never materialized.

The Nuclear Waste Policy Act, as amended, also included requirements for spent nuclear fuel and high level radioactive waste to be transported in containers certified by the Nuclear Regulatory Commission. To safely and efficiently move radioactive waste around the country, PNNL researchers helped develop special containers for highway and rail transport. Each shipping container was designed to maintain its integrity under normal transportation conditions and during hypothetical accident conditions. The designs had to demonstrate protection against radiological release to the environment under the following hypothetical accident conditions:

- A 9 meter (30-foot) free fall on to an unyielding surface
- A puncture test allowing the container to free-fall 1 meter (40 inches) onto a steel rod 15 centimeters (6 inches) in diameter
- A 30-minute, all-engulfing fire at 800 degrees Celsius (1475 degrees Fahrenheit)
- An 8-hour immersion under 0.9 meter (3 feet) of water

Compliance with this sequential series of tests was demonstrated by computer modeling, scale-model or full-scale tests, parts of which were developed or evaluated by PNNL researchers.

In addition to the spent fuel from commercial reactors, the Nuclear Waste Policy Act also covered disposal of high level radioactive waste from the government's plutonium production facilities. In addition to storage and transportation issues, research had to address how to convert radioactive waste into a stable waste form, one capable of encapsulating the radionuclides for thousands of years. While Hanford researchers pursued a variety of stabilization technologies for high level and low level radioactive waste, vitrification technology was a major focus. Vitrification was an adaptation of melter technology developed in the glass making industry (Larson 1996).

Other waste-related R&D focused on the decommissioning of the Shippingport Reactor near Pittsburgh, Pennsylvania. Shippingport was one of the early demonstration reactors funded by the Atomic Energy Commission and private industry in the 1950s. The Shippingport plant was a pressurized water reactor design, based on Argonne National Laboratory's submarine reactor technology. This first commercial light water reactor plant operated from 1958-1982 (Beaver 1990).

Once operations at Shippingport ceased, the plant had to be decommissioned. Decommissioning this reactor was seen as an opportunity to learn much about reactor decommissioning, a process every reactor would eventually have to undergo. PNNL researchers assisted the U.S. Department of Energy in a wide range of studies surrounding materials performance, extent of contamination, cost models, and decontamination approaches.

Policy and Regulatory Analysis

As a result of concerns about the dangers of nuclear fallout and exposure to radiation in general, scientists, health professionals, and the public demanded standards for the protection of public health and safety. In 1954, the Atomic Energy Commission set up a regulatory program to do just that (Walker 1992). PNNL researchers assisted the Atomic Energy Commission for many years both in the development of regulations and in the implementation of policies and regulations. Such activities occurred for both the commercial and defense industries. This work continued under the Nuclear Regulatory Commission when the Atomic Energy Commission was split up in 1977.

Environmental Impact Statements were required under the National Environmental Policy Act. PNNL researchers were often called upon to supply scientific evidence for Environmental Impact Statements, such as those for the Hanford Defense Waste and the Commercial Wastes Environmental Impact Statements.

Hanford R&D staff also conducted similar work for the International Atomic Energy Agency. The International Atomic Energy Agency, created in 1957, is a specialized agency within the United Nations. Headquartered in Austria, the International Atomic Energy Agency serves as the world's central intergovernmental forum for scientific and technical co-operation in the nuclear field, and as the international inspector for the application of nuclear safeguards and verification measures covering civilian nuclear programs.

Environmental Restoration

As the Cold War started to wane, the U.S. Department of Energy came under increasing pressure to clean up its legacy wastes created during four decades of plutonium production. In 1986, the U.S. Department of Energy released 19,000 pages of documents on the history of the Hanford Site. From these, the public learned that radioactive and chemical wastes had been released in greater quantities than had been previously assumed. The issue of waste disposal became more prominent and once again, R&D staff were called upon to develop and evaluate cleanup methods and technologies that could assist with the cleanup efforts.

In situ vitrification is a good example of a technology developed by Hanford R&D staff, which may have widespread industrial applications. In 1988, Battelle Memorial Institute created Geosafe Corporation for the purposes of developing and commercializing advanced vitrification technologies for site remediation and waste treatment worldwide. The Corporation acquired rights to the U.S. Department of Energy's patent for the In Situ Vitrification technology, which was developed to treat TRU-contaminated soil in place.

OTHER MAJOR STUDIES

Space nuclear system studies were designed to test the potential biological problems connected with the utilization of reactors or radionuclide power sources in space. The spread of radioactive contamination of operations in space could constitute a human hazard. These studies employed swine, rats, and dogs to explore possible dangers of inhaled and ingested particles. This research was an offshoot from the basic inhalation research program discussed above (PNL 1967a, Stannard 1988). In 1957, the Atomic Energy Commission launched proposals for the development of industrial uses of nuclear explosives, such as power generation by explosions, oil and mining operations, and large-scale earth excavations. Dubbed Project Plowshare, these proposals stemmed directly from the Atoms for Peace Program of 1953. Project Chariot, a program within Plowshare, was a planned excavation of a harbor near Cape Thompson, Alaska. Because the project site had not been thoroughly investigated before the planned work, bioenvironmental and public safety programs were initiated. The Hanford Biology Program contributed to three Project Chariot related environmental studies in Alaska: limnological investigations, ecology of terrestrial invertebrates, and radioecology of terrestrial and fresh water communities. The controversial Project Chariot program was suspended in the early 1960s due to public concerns about health hazards (O'Neill 1995, 1998). Other work in Alaska included studying the repercussions of radioactive nuclear fallout. A fairly large ecological program was started and food chain studies were undertaken for a number of years (GE 1960a, Stannard 1988).

RESEARCH AND DEVELOPMENT FACILITIES

The Hanford area most associated with R&D is the 300 Area. Although the 300 Area was originally known for fuel manufacturing, as R&D increased at the Hanford Site, research facilities were located in the 300 Area for three main reasons. The infrastructure was in place. The risk of accident was less than the areas where the reactors and chemical separations plants were in the 100 and 200 Areas, and the 300 Area was closer to residential areas for easier accessibility.

During the Cold War period, several test reactors and fuel fabrication pilot facilities in the 300 Area were used for non-military purposes. With the worldwide uranium supplies limited, research efforts were undertaken to develop and test alternate fuels. The most ambitious efforts were focused on the effectiveness of plutonium oxide and mixed oxide fuel blends. The Physical Constants Test Reactor and Thermal Test Reactor in the 305-B Building, the Plutonium Fabrication Pilot Plant, the Plutonium Recycle Test Reactor in the 309 Building, and the High Temperature Lattice Test Reactor in the 318 Building were constructed to develop and test alternate fuels as well as conduct other experimental testing programs.

During the period of considerable growth of the 300 Area in the early 1950s, many buildings and facilities were constructed under the Hanford Laboratories Operation Program for R&D activities. The most prominent developmental laboratories and shops included:

- 108-F Biology Laboratory – In 1949, the Biology Laboratory was completely remodeled to provide office and laboratory space for the Hanford Site Biology Program (DOE 1992b, DeFord 1993). The building was expanded in 1953 and again in 1962 to provide additional space for biological experiments. For over a quarter of a century, the building served as the main biology lab at the Hanford Site to study the effects of animals and plants. The Biology Program moved to the 337 Building in 1977, and the 108-F building was abandoned. In addition to the 108-F Laboratory, numerous other facilities, such as laboratories, barns, pens, pastures, gardens, and kennels, supported the Hanford Biology Program. For more information on the 108-F Biology Laboratory, see the Expanded Historic Property Inventory Form in Appendix B on the Internet.
- 305 Test Pilot/Hot Cell Verification Building – Built to house a small reactor that tested samples of the graphite, uranium and other materials used in essential Hanford operations. The 305 Reactor functioned as a quality assurance tool to house testing of samples of each lot of graphite, uranium, aluminum jacketing material, and other materials used in the large production reactors. The reactor was removed and buried in 1977-1978. For more information on the 305 Test Pilot/Hot Cell Verification Building, see the Expanded Historic Property Inventory Form in Appendix B on the Internet.
- 320 Low Level Radiochemistry Building – Completed in 1966, was virtually the last Hanford facility dedicated solely to national defense missions. The building's original mission was to house analytical chemistry services and provide plant support for work involving low level and non-radioactive samples. Twenty-two laboratories were located in the structure, including several devoted to wet chemistry, a large radiochemistry laboratory, an analytical R&D laboratory (Gerber 1993b, p. 52). Several pioneering techniques were researched in the building. These included radiometric techniques, new mass spectrometric techniques, combined atomic absorption analysis, and laser-based spectrometric techniques. The 320 Building currently provides a low level radiochemistry facility in which very sensitive radiochemical analysis, sample preparation, and methods development can be performed. For information on the 320 Low Level Radiochemistry Building, see the Historic Property Inventory Form in Appendix B on the Internet.
- 324 Waste Technology Engineering Laboratory – Completed in the mid-1960s as the Fuel Recycle Pilot Plant, designed partially to support PRTR operations by housing chemical reprocessing and metallurgical examination on the PRTR's fuel elements. As such, it was built as a dual facility with both radiochemical and radiometallurgical hot cells and laboratories. It also was designed to house the Waste Solidification Engineering Project, one of the first high level waste vitrification demonstration programs in the world (Gerber 1993b, pp. 48-50).
- 325 Radiochemistry Laboratory/Cerium Recovery Building – Completed in 1953, built to safely house and handle multi-curie level chemical development work with high activity substances. Initial building missions listed as high priorities included production support and process improvement (Gerber 1993b). The High Level Radiochemistry wing addition in 1959-1960 made the 325 Building the largest among the laboratories at the Hanford Site (Gerber 1992a). The addition, with its three larger hot cells, housed isotope research activities. For more information on the 325 Radiochemistry Laboratory/Cerium Recovery Building, see the Expanded Historic Property Inventory Form in Appendix B on the Internet.

- 326 Physics and Metallurgy Laboratory – Completed in 1953, designed to ensure the maintenance and improvement of reactor operations while developing new technologies for future reactors. The primary physics mission was to conduct the exponential reactor physics development work. The primary metallurgical mission was to examine reactor components and fuel elements to evaluate characteristics and performance (Gerber 1993b). For information on the 326 Physics and Metallurgy Laboratory, see the Expanded Historic Property Inventory Form in Appendix B on the Internet.
- 327 Post Irradiation Test Laboratory – Completed in 1953, built to house the examining and testing of irradiated materials, particularly fuel elements and fuel cladding materials from and for the production reactors, and of the effects of higher power levels of irradiation on new and different reactor structural materials. For information on the 327 Post Irradiation Test Laboratory, see the Expanded Historic Property Inventory Form in Appendix B on the Internet.
- 328 Mechanical Development Building (later called the Engineering Services and Safety Shop) – Built as a central shop to support the above laboratories. For information on the 328 Engineering Services and Safety Shop, see the Expanded Historic Property Inventory Form in Appendix B on the Internet.
- 329 Biophysics Laboratory – Built to support the pioneering environmental monitoring and bioassay programs that were developed at the Hanford Site during the 1940s and 1950s. The initial mission of the facility was to house the preparation and counting of radioactivity levels in air, vegetation, soil, wildlife, river and well water samples (Gerber 1993b, p. 24). Missions ranged from monitoring nuclear atmospheric fallout from nuclear bomb tests to fallout deposition studies on terrestrial organisms in Alaska to examinations of Mount St. Helens volcanic ash. For more information on the 329 Biophysics Laboratory, see the Expanded Historic Property Inventory Form in Appendix B on the Internet.
- 335 and 336 Sodium Test Facilities – Completed in 1968 and 1969, respectively, built to conduct sodium-related tests for the development of the Fast Flux Text Facility and cold sodium purification and characterization systems used in Fast Flux Text Facility studies. The sodium test loops were deactivated in 1977. Building 335 was known initially as the Fast Reactor Thermal Engineering Facility and Building 336 as the Core Segment Development Facility.
- 337 Technical Management Facility – Completed in five segments from 1970-1972, was the preeminent 300 Area structure built to support the development of the Fast Flux Text Facility. Originally, it housed engineering studies, including sodium loops and large mechanical mock-ups, and was a technical support facility for the mock-ups performed in the adjacent 337-B (high bay) Building. The 337-B Building initially housed R&D activities that included a core mechanical mock-up, a test bed for selected Fast Flux Text Facility components. For more information on the Technical Management Facility, see the Historic Property Inventory Form in Appendix B on the Internet.
- 427 Fuels and Materials Examination Facility – Completed in 1983 in the 400 Area, was a major addition to the breeder reactor technology program at the Hanford Site. Its function was to inspect irradiated fuels and materials from the Fast Flux Text Facility.
- Battelle Complex – When the Battelle Memorial Institute took over the Hanford Laboratories in 1965, it committed to build its own facilities on a square mile of land that it purchased at the southern end of the Hanford Site. The research complex was originally known in local lore as "Fort Courage," because "the building style and color reminded one a bit of something out of Beau Geste, and because the whole idea seemed wildly optimistic at that time" (Moore 1992, p. 122). Since this is a private facility, it is not being documented as the others are.

National and International Interactions of Hanford Research and Development Staff

Disseminating research results and sponsoring symposia and seminars are indicators of the interactions of the Hanford R&D staff with their national and international counterparts.

PUBLISHING RESEARCH RESULTS AND PARTICIPATING IN SEMINARS AND CONFERENCES

The Hanford Site had a strong tradition of participating in regional, national, and international conferences. One of the most important conferences was the First Atoms for Peace Conference held in Geneva, Switzerland in 1955 (Fermi 1957). According to the list of participants, only Herbert Parker attended the conference from the Hanford Site (IC 1956). When the second conference occurred 2 years later, however, many Hanford staff attended and gave papers, especially in the area of potential health effects (IC 1958).

Participation in regional, national, and international conferences increased as national security concerns lessened over the years.

SPONSORING SEMINARS AND SYMPOSIA

Beginning in 1962, the Hanford Laboratories and its successor, the Pacific Northwest National Laboratory, sponsored annual seminars and symposia on subjects directly related to research activities at the Hanford Site (Stannard 1988, p. 447). Before this time, seminars and symposia were not held on a regular basis. Table 2-7.3 shows between 1962-1990 symposia and seminars were held every year except in 1969 and 1982. Seminars may have been held in 1969 and 1982, but no references have been found to indicate as much. Both national and international scientists and researchers attended these meetings, which provided many opportunities for peer review. A significant number of symposia were held on radioisotope toxicology, a subject that contributed to Hanford's reputation as one of the prime centers for biological research in the subject (Stannard 1988, p. 450).

Table 2-7.3. *A Sampling of Symposia and Seminars Held at the Hanford Site, 1948-1990*

Year	Symposia and Seminars
1948	Fish and Fish Problems of the Hanford Reservation (Olson 1948)
1949	The Present Status of Polonium Tolerance Estimation (Hackett 1949)
1955	Symposium on the Iodine Problem (Hill 1955)
1962	Biology of the Transuranics (Thompson 1962)
1963	Foliar Sorption of I-131 by Plants: Biology of Radioiodine (Bustad 1964)
1964	Inhaled Radioactive Particles and Gases (Bair 1965)
1964	Biology of Radioiodine (Watson et al. 1964)
1965	Radiation and Terrestrial Ecosystems (Hungate 1965)
	Swine in Biomedical Research (Bustad and McClellan 1966)
1966	Symposium on Gastrointestinal Radiation Injury (Sullivan 1968)
1967	Diagnosis and Treatment of Deposited Radionuclides (Kornberg and Norwood 1968)
1968	Myeloproliferative Disorders of Animals and Man (Clarke et al. 1968)
1970	Pollution and Lung Biochemistry (Menzel 1970)
1971	Biological Implications of the Transuranium Elements (Thompson and Bair 1972)
1972	Radionuclide Carcinogenesis (PNL 1972a)
1973	Cell Cycle in Malignancy and Immunity (Hampton 1973)
1974	Radiation and the Lymphatic System (Ballou 1974)
1975	Biological Implications of Metals in the Environment (Drucker and Wildung 1975)
1976	Pulmonary Macrophage and Epithelial Cells (Sanders 1976)
1977	Developmental Toxicology of Energy-Related Pollutants (Mahlum 1977)
1978	Biological Effects of Extremely Low Frequency Electromagnetic Fields (Phillips 1978)
1979	Pulmonary Toxicology of Respirable Particles (Sanders 1979)
1980	Coal Conversion and the Environment: Chemical, Biomedical, and Ecological Considerations (Mahlum et al. 1980)
1981	Biological Availability of Trace Metals (Wildung and Jenne 1981)
1983	Life Span Radiation Effects Studies in Animals: What Can They Tell Us? (Thompson and Mahaffey 1983)
1984	Interaction of Biological Systems with Static and ELF Electric and Magnetic Fields (Anderson et al. 1984)
1985	Health and Environmental Research on Complex Organic Mixtures (Gray 1985)
1986	Radiation Protection: A Look to the Future (Bair 1988)
1987	Modeling for Scaling to Man: Biology, Dosimetry, and Response (Mahaffey 1989)
1988	Multi-Level Health Effects Research: From Molecules to Man (Park and Pelroy 1989)
1989	Environmental Monitoring, Restoration, and Assessment: What Have We Learned? (Gray 1990)
1990	Indoor Radon and Lung Cancer: Reality or Myth? (Cross 1992)

AREAS FOR FURTHER RESEARCH

In the early years, the Hanford Site was the dominant client for the research staff. As the need for nuclear weapons subsided and other research needs arose, sites such as Hanford pursued diversification. Researchers found their knowledge of materials, physics, mathematics, the environment, etc. could be applied to new areas, particularly those in industry. In this regard, Hanford R&D underwent a fundamental shift from being a specialized research institution supporting a clear-cut national mission – nuclear weapons production – to a generalized research institution supporting a combination of national missions such as energy efficiency, economic competitiveness, environmental quality, and education. Government needs also changed as the government increasingly called on research staff not just to develop technology but to evaluate technologies, develop protocols, and provide independent oversight. Therefore, virtually limitless areas exist for future research of R&D activities at the Hanford Site and in relation to academia, other nuclear installations, and the history of science and technology in the United States.

- R&D activities at the Hanford Site have undergone a variety of changes during the past 50 years. In many fields, the history of basic research has gone undocumented and passed over in favor of the more lucrative commercial applications of applied research, project development, and process and product improvements. A thorough analysis and chronology of Hanford's basic research programs since the late 1940s should be undertaken in the future.
- With this in mind, oral interviews with pioneering researchers at the Hanford Site may also prove of interest to the general public and historians of science and technology.
- It would be of interest to document how national and international seminars and symposia may have impacted research ideas and design between researchers in academia, industry, and government.
- Were techniques and instruments developed at the Hanford Site applied in non-governmental settings?
- How have Hanford researchers influenced R&D in non-governmental settings? What patents resulted from research at the Hanford Site?
- The origins of the U.S. Department of Energy's national laboratory concept arose out of compelling national need because of the nuclear weapons race. In that regard, the concept of a nationalized R&D industry was successful, concerns about environmental and human health notwithstanding. In the absence of any emergency situations, is a nationalized R&D force effective?
- Were the R&D contributions to nuclear energy, energy conservation, addressing environmental and nuclear legacy waste problems commensurate with their costs?
- Are there alternative ways to structure a nationalized R&D group? How does working in a national laboratory differ from other R&D settings such as industry and higher education?
- Hanford's R&D budgets have not been made available to the public. When they are, it would be useful to evaluate how much money was spent on basic research versus applied R&D. These figures would also prove useful in compiling a comparative data set of R&D spent at other nuclear sites around the country.
- The use of humans in radiation experiments financed by the U.S. Department of Energy and its predecessor agencies remains a topic of considerable interest and controversy. To the authors' knowledge, not one source document exists detailing all of the human experiments performed at the Hanford Site or by Hanford researchers off-site. It would be beneficial to compile these data and conduct oral interviews with both the researchers and the human subjects. Many scheduled follow-up studies proposed by the U.S. Department of Energy were never conducted or were cancelled. More work is needed to understand how researchers and decision makers rationalized the use of human subjects and the possible long-term health consequences they may have experienced.
- Radiation detection programs at other national laboratories have been linked to research on important military applications such as nuclear warfare, detecting radiation that could compromise the functioning of nuclear submarines and detecting radiation on the nuclear battlefield. Were any of the results of human radiation programs or radiation detection programs at the Hanford Site used in these other studies? If so, how were the results applied?
- It would be of value to document the popular views held by area residents about R&D at Hanford. Did their knowledge about particular experiments with animals and occupational injuries influence their attitudes toward atomic energy and government installations in general? Did these views affect the types of research performed?

Biological and Ecological Research Projects at the Hanford Site, 1951-1990

Year	Biological and Ecological Research Projects (As Indicated by Annual Reports)
1951 (GE 1951b)	• **Biological Sciences:** radioactive isotope studies on aquatic organisms; mean activity density of iodine-131 in jack rabbits; duck exposure to pile effluent; tritium and tritium oxide absorption and retention studies on humans, rats, and plants; uptake of radioactive strontium, yttrium, cesium, and iodine in plants; plutonium fixation in rats; biological effectiveness of radiation in bacteria; pile effluent water toxicity to fish and bean plants; iodine-131 studies on sheep
1952 (GE 1952b, 1953c)	• **Aquatic Biology:** pile effluent monitoring with Silver salmon, Chinook salmon, Rainbow trout; algae growth in pile effluent
	• **Metabolism Studies:** plutonium and tritium absorption and deposition in rats and rabbits; percutaneous absorption of tritium oxide in man; absorption and translocation of various fission products in plants; chronic effects of pile effluent on plants; effects of radiation from various isotopes on the growth of algae
	• **Toxicology:** radioiodine in sheep to study growth, birth defects, tumors, etc.; removal of plutonium in dogs via sirconium salts
	• **Biological Services:** biological monitoring of goats, plants, wildlife and waterfowl, sheep, rabbits, and cadaver ashes
	• **Biological Sciences:** removal of radiophosphorus contaminant from water in an aquarium community; extended retention of Rainbow trout in dilute reactor effluent; effect of reactor effluent on young Silver salmon; chronic effects of pile effluent water on cereal plants; metabolism of tritium gas in the rat; long-term retention of tritium in the rat and sheep; metabolism of hydrogen isotopes by algae, rat metabolism of deuterium and tritium; plutonium absorption studies of the rat; intravenous effects of plutonium on the dog; radioactive elements in "Jangle" soil; absorption and translocation of fission products by plants; radium analysis in humans
1953 (GE 1954a)	• **Biological Sciences:** abundance and radioactivity of crustacea in the Columbia River; reactor effluent monitoring with Chinook salmon, effects of reactor effluent on cereal plants; absorption and translocation of plutonium-239 and cerium-144 by plants; physical and chemical properties of soil types; mutagenic effects of phosphorous-32; effects of tritium oxide on Chorella pyrenoidosa; hydrogen isotope incorporation in growing algae; rat studies with deuterium and tritium, studies on the absorption of plutonium in rats; absorption and distribution of ruthenium in fowl; radium abundance in cadavers from the Pacific Northwest; comparison of CaEDTA with zirconium citrate in promoting excretion of plutonium from the dog; toxicity of iodine-131 in sheep, beta irradiation of the skin of the sheep
1954 (GE 1955a)	• **Biological Sciences:** effects of reactor effluent on aquatic and plant life; plant and animal absorption and metabolism of several fission products, plutonium, and tritium; effects of iodine-131 and external radiation administered to sheep; toxicity of radioactive particles; new or improved techniques of possible application to radiobiology
1955 (GE 1956b)	• **Biological Sciences:** efficacy of isotopic dilution; effects of production facility effluents on biota; studies on radioactive particles; plant and microbiological studies of several radioisotopes; animal metabolism studies of several radioelements; biological effects of iodine-131 in sheep and pigs
1956 (GE 1957a)	• **Biological Sciences:** skin irradiation and decontamination; studies utilizing radioactive particles; plant and microbiological studies; effects of production facility effluents on biota
1957 (GE 1958d)	• **Biological Sciences:** uptake, distribution, and turnover of radioelements in plants, animals, and communities; biological effects of ionizing radiations on certain organs and organisms; effects of reactor effluent on plant and animal life
1958 (GE 1959a)	• **Biological Sciences:** strontium and calcium metabolism uptake in plants and animals; cesium accumulation by aquatic plants and animals; observations of biological effects of iodine-131 in sheep and swine; studies on the retention and elimination of plutonium in swine; studies to determine a safe daily limit for ingestion of phosphorus and zinc; tungsten metabolism in rats; polonium distribution in yeast; radioactive particles inhalation studies; radiation effects and protection experiments on rats; effects of production facility effluents on biota
1959 (GE 1960a)	• **Biological Sciences:** strontium-calcium and cesium-rubidium-potassium studies on rats, ewes and suckling lambs, plants, and soil; strontium toxicity studies on miniature swine and rainbow trout; physiology and therapy studies on miniature swine and rabbits; iodine studies on sheep, swine, and leaves; phosphorus studies in mice and fish; zinc studies in rats and ram; radioactive particles in mice; biological effects of radiation in the rat and yeast; laboratory techniques; environmental studies on contamination in wildlife, Chinook salmon, marine mollusks, and C. Columnaris

Year	Biological and Ecological Research Projects (As Indicated by Annual Reports)
1960 (GE 1961a)	• **Biological Sciences:** strontium and calcium metabolism and toxicity studies in Rainbow trout, rats, miniature swine, large animals, and meal moths; iodine metabolism and toxicity studies in sheep and swine, milk, and thyroids; zinc, cesium, cerium, neptunium, and plutonium metabolism and toxicity studies in pregnant ewes, sheep, rats, and rams; inhalation studies of plutonium and cerium-144 oxygen-2; physiology, therapy, and radiation effects on trout, yeast, bean plants, barley plants, mice, and rats; hematology studies on miniature swine, Pitman-Moore swine, and Palouse and Hormel swine; environmental studies on fish, marine organisms, salmon, geese, grass fields, and C. Columnaris
1961 (GE 1962a)	• **Biological Sciences:** strontium and calcium toxicity and metabolism studies in swine and trout; radioiodine toxicity and metabolism studies in sheep and swine; heavy elements toxicity and metabolism studies in milk; inhalation studies; modification of radioelement deposition and radiation response studies; animal and cellular physiology studies in water absorption and ingestion, regenerating liver, Ephestia larvae, proliferating cells, and yeast; plant physiology studies on barley plants and others; ecology studies on waterfowl, salmon, Columnaris, beetles, natural plant communities, desert soil, and Arctic plants and animals
1962 (GE 1963a)	• **Biological Sciences:** toxicity and metabolism of radioelements (transuranium and rare Earth elements; radioiodine; strontium; other elements; inhalation studies); modification of radioelement deposition and radiation response; animal and cellular physiology; plant physiology • **Ecology:** reactor effluent monitoring; Chinook salmon spawning near Hanford; swimming performance of Chinook salmon reared in reactor effluent; incidence and virulence of Columnaris; plankton-water chemistry cycles in the Columbia; soil moisture depletion in a sagebrush and a greasewood community; radiocesium in northern Alaskan Eskimos and their foods; gamma-emitting radionuclides in Alaskan fish and plants
1963 (GE 1964a)	• **Biological Sciences:** toxicity and metabolism of radioelements (radioiodine studies; inhalation studies; transuranium and rare Earth elements studies; strontium studies; other elements studies); modification of radioelement deposition and radiation response (removal of internally deposited plutonium; excretion of plutonium-239 in the intestine; effects of plutonium in swine; influence of internal emitters; graft-host response in xenogenic chimeras; etc.); animal and cellular physiology – various studies; plant physiology (deposition of iodine-131 on vegetation; accumulation of iodide by barley plants; cesium toxicity in plants; removal of plutonium-239, tungsten-185 and lead-210 from plants) • **Ecology:** second survey of radioactivity in northern Alaskan Natives and their foods; incidence and virulence of Columnaris; Chinook salmon spawning near Hanford; swimming performance of Chinook salmon reared in reactor effluent; toxicity of industrial chemicals to fish; plankton-water chemistry cycles in the Columbia; vegetational analysis in a sagebrush stand; potassium and sodium cycling by greasewood and hopsage
1964 (PNL 1965a)	• **Biological Sciences:** inhalation studies on beagles and rats; effects of strontium in miniature swine and molting crayfish; radioiodine effects on sheep and cattle, sheep milk, and skin exposure; effects of cesium on sheep, trout, and feed and rumen contents; effects of transuranium and other rare earth elements on sheep, fatty livers, and miniature swine; studies on zinc-65, cadmium-115m, copper, and x-rays; modification of radioelement deposition and radiation response studies on dogs, rats, and cows-to-man via milk; animal and cellular physiology studies on rats, flour beetles, chimeras, Escherichia Coli, Neurospora, lung mitochondria; plant physiology; ecological studies among Alaskan Natives, animals, and forage plants; nuclide fallout studies on plants and browsing animals in Washington; annual forage yield and fallout deposition in a cheatgrass community; Columbia River plankton studies; Chinook salmon spawning; and reactor effluent monitoring
1965 (PNL 1966a, 1966d)	• **Biological Sciences:** radiation effects – general; toxicity of radioelements; inhalation studies; combating detrimental effects of radiation; molecular and cellular level studies; environmental radiation studies – terrestrial and aquatic • **Geology and Hydrology:** stratigraphy of the uppermost part of the basalt sequence; variations in the gravity field at Hanford; prediction of vertical infiltration as influenced by errors; horizontal infiltration of liquids into porous media; extension of solution capabilities by finite differences method; etc • **Mineral Chemistry:** montmorillonite exchange equilibria with strontium-sodium-cesium; self-diffusion of sodium in vermiculite; exchange of alkali metal cations on a natural stilbite; alkali metal cation equilibria with chabazite; unit cell size and cesium loading; cation exchange properties; dispersion characteristics of strontium; cesium and strontium distribution beneath liquid waste disposal sites • **Particles and Gases:** application and performance of a spinning disc aerosol generator; subisokinetic sampling of particles in an air stream; errors in sampling with nozzles; particle deposition and re-entrainment in long vertical conduits; an iodine-131 generator for field release studies; release of noble gases from aqueous solution
1966 (PNL 1967a)	• **Biological Sciences:** inhalation studies on dogs, rats, leukemia patients, and lung tissue; space nuclear systems studies on dog lungs, swine, and rats; detrimental effects of radiation studies on swine; molecular and cellular studies; environmental radiation studies (terrestrial) on Alaskan Natives, Hanford geese and gulls, native rodents, plants, soil, and potato tubers; environmental radiation studies (aquatic) on trout, salmon, bottom organisms, other vertebrates, freshwater mussels, adult Caddis flies, Fairy shrimp, and Fisherola Nutalli

Year	Biological and Ecological Research Projects (As Indicated by Annual Reports)
1967 (PNL 1968a)	• **Biological Sciences:** radiation effects studies; strontium-90 studies in miniature swine; inhalation studies (dogs, rats); space nuclear systems studies (dogs, rats, leaves); radioelement toxicity studies (sheep, swine, rats); radioelement removal studies (rats, dogs); molecular and cellular level studies (Neurospora Crassa, human testes, intestinal cells, fish)
	• **Terrestrial Ecology Studies:** Eskimo food chain; rodents; Darkling beetles; desert plants; local soil; greasewood; cheatgrass
	• **Aquatic Ecology Studies:** Chinook salmon; Steelhead trout; bottom organisms; Sockeye salmon; freshwater mussels
1968 (PNL 1970a)	• **Biological Sciences:** radiation effects studies (rats, miniature swine, rabbits, mice); strontium-90 studies in miniature swine (various studies); inhalation studies (dogs, rats, etc.); radioelement toxicity studies (rats, sheep, swine, crustaceans); radioelement removal studies (rats, dogs, etc); molecular and cellular level studies (salmon, trout, etc)
1969 (PNL 1967d, 1970d, 1970e)	• **Biological Sciences:** gastrointestinal irradiation studies; irradiating skin; radiation and human spermatogenesis; radiation effects in mice; structural effects of temperature in fish; metabolism and effects of radionuclides in the developing fetus and the young; effects of radioiodine in sheep and swine; effects of radiostrontium in swine; radiation induced leukemogenesis in swine; inhalation studies (rats, dogs, uranium miners); space nuclear systems studies; removal of internal emitters; blood irradiators; malonaldehyde as a radiolysis product of biological materials
	• **Environmental Studies:** ecological micrometeorology and climatology of the ALE Reserve; geology and paleoecology of the ALE Reserve; hydrologic cycling on the ALE Reserve; characteristics of Hanford project soils
	• **Terrestrial Ecology Studies:** plant and animal ecology; mechanisms of transport and uptake of radionuclides in Alaska; mathematical simulation of radionuclide transfer and food chain processes
	• **Aquatic Ecology:** characteristics of Columbia River; freshwater ecology of Rattlesnake Springs; rates and mechanisms of biogeochemical processes of the Columbia; effects of modification on the Columbia River ecosystems
	• **Geology/Hydrology:** Yakima basalt formation; incidence of earthquakes at Hanford; geophysical seismic exploration; bouguer gravity anomalies in the Hanford Res.; acceleration factors for flow of fluids through soils; flow system surfaces; transport analysis
	• **Mineral Chemistry:** dispersion characteristics of strontium; geochemical processes; strontium and cesium equilibrium distribution coefficients; cation exchange properties
	• **Particles and Aerosols:** compact high speed spinning disc aerosol generator; particle deposition and re-entrainment in vertical conduits; etc.
1970 (PNL 1971a, 1971b)	• **Biological Sciences:** gastrointestinal irradiation studies; radiation effects; metabolism and effects of radionuclides in the developing fetus and the young; effects of radiostrontium in miniature swine; radiation-induced leukemogenesis; inhalation studies; low-level plutonium inhalation studies in beagles; inhalation hazards to uranium miners; space nuclear systems studies; removal of internal emitters; structural effects of temperature in fish; blood irradiators
	• **Terrestrial Ecology:** wildfire, animal ecology studies; plant ecology studies; ecological micrometeorology and climatology of the ALE Reserve; characteristics of Hanford soils; geology and paleoecology of ALE Reserve; Hydrologic cycling on the ALE Reserve; Alaska studies; analysis of natural systems
	• **Aquatic Ecology:** ecological characteristics of the Columbia River; freshwater ecology of Rattlesnake Springs; rates and mechanisms of biogeochemical processes of the Columbia River; effects of modification on the Columbia River ecosystems; effects of modification of the Columbia River ecosystems; effects of short range particle irradiation on embryogenesis of marine Teleost fish; fixation and long-term accumulation of tritium in an experimental aquatic environment
1971 (PNL 1972a, 1972b)	• **Biological Sciences:** gastrointestinal irradiation studies; effects of irradiating skin; radiation effects on chylomicron synthesis and capillary permeability in mice; miniature swine as model for intracavitary irradiation; cellular regulatory mechanisms; metabolism and effects of radionuclides in the developing fetus and the young among rats; effects on radiostrontium in miniature swine; radiation-induced luekemogenesis; inhalation studies among experimental animals; low-level plutonium inhalation studies in beagles; inhalation hazards to uranium miners (biological studies); space nuclear system studies; removal of internal emitters; fine structural effects of temperature in fish; development and evaluation of blood irradiators; biological effects of intracorporeal radioisotope heat source
	• **Aquatic Ecology:** effects of thermal discharge on aquatic biota; effects of tritium on aquatic environs; factors affecting biogeochemical cycling; synergistic effects of temperature, pollutants, and disease in aquatic organisms
	• **Terrestrial Animal Ecology:** perturbed environs; irradiation effects; population dynamics and reproductive habits; environmental quality; Alaska studies; analysis of natural systems

Year	Biological and Ecological Research Projects (As Indicated by Annual Reports)
	• **Terrestrial Plant Ecology:** vegetational dynamics and characterization; nutrient and hydrological factors; influence of environ factors on arid soil respiration rate; ecological micrometeorology and climatology of the ALE Reserve
1972 (PNL 1973a, 1973b)	• **Biological Sciences:** toxicology of ingested, injected, and topically applied radionuclides (swine, rats); toxicology of inhaled plutonium and transplutonium elements (rats, dogs, etc.); inhalation hazards to uranium miners (beagles, hamster, rats, mice, mine atmosphere); evaluation of radionuclides in man; removal of radionuclides (rats, etc.); nuclear-powered prostheses and other human applications; mechanisms of radiation effects;
	• **Plutonium Studies:** interactions of plutonium aerosols with plant foliage; Nevada test site; behavior of waste radionuclides in soil-plant systems; methods for measuring plutonium complexation in soil and uptake by plants
	• **Marine Sciences:** thermal resistance; effects of short-range particle irradiation; geochemical ocean sections study (GEOSECS); physical and radiological chemistry of ocean solutions
	• **Analysis of Natural Systems:** sampling for radionuclides and other trace substances; bias in capture-recapture studies; sampling sagebrush on ALE; Antarctic seal study
	• **Characteristics of Hanford Soil and Aquatic Sediments:** influence of environmental factors on arid soil respiration rate; temperature and moisture; livestock grazing; mercury levels in particulate matter suspended in waters of Lower Columbia watershed
	• **Freshwater Ecology:** effects of thermal discharge on aquatic biota; effects of tritium on aquatic environs; factors affecting biogeochemical cycling; combined effects of temperature, pollutants, and disease in aquatic organisms
	• **Terrestrial Ecology:** terrestrial plant ecology studies; grasslands biome; terrestrial animal ecology; Alaska studies
1973 (PNL 1974a, 1974b)	• **Biological Sciences:** toxicology of ingested, injected, and topically applied radionuclides in rats, miniature swine, and mice; toxicology of inhaled plutonium and transplutonium elements in dogs, rats, and other rodents; inhalation hazards to uranium miners with rats, mice, and dogs; evaluation of radionuclides in man; removal of radionuclides in rats, miniature swine, and blood; mechanisms of radiation effects in N. Crassa, miniature swine, and mammalian cells
	• **Analysis of Natural Systems:** quantitative ecological aspects of nuclear power development; sampling for contaminants; modeling ecological systems; statistical and mathematical aspects of ecology
	• **Environmental Chemistry:** the potential for plutonium complexation in soil and uptake by plants; characteristics of Hanford soils and aquatic sediments; fate and effects of oil on marine coastal ecosystems; Hanford intercontractor support; radioecology of iodine-129; suspended particle interaction
	• **Freshwater Ecology:** effects of thermal discharge on aquatic biota and fish behavior and sensory physiology; combined effects on temperature, pollutants, disease; effects of tritium on aquatic environs; factors affecting biogeochemical cycling
	• **Marine Sciences:** bioenvironmental effects of effluents discharge from nuclear power plants to coastal waters; fate and effects of oil on marine coastal ecosystems; physical and radiological chemistry of ocean solutions; geochemical ocean study section (GEOSECS)
	• **Radiological Sciences:** radiological health research; iron-55 in human populations; radioanalytical procedures development; fate and effects of radionuclides in Alaska
	• **Terrestrial Ecology:** energy and water relations; mineral nutrients; manipulated ecosystems; consumer studies; primary producer studies; radiation studies; modeling
1974 (PNL 1975a, 1975b)	• **Biological Sciences:** low-level plutonium and transplutonium oxide inhalation studies in beagles; toxicology of inhaled plutonium and transplutonium elements in rodents; toxicology of inhaled beryllium compounds; development of aerosol exposure and analytical techniques; mechanisms of inhaled plutonium toxicity; toxicology of radionuclides in the fetal and juvenile mammal (rats); factors modifying radionuclide metabolism and effect in mice; toxicology of chronically fed strontium-90 in miniature swine; gut related studies of radionuclide toxicity (rats); acute toxicity of reactor catastrophe effluent; inhalation hazards to uranium miners (dogs, hamsters, rats, mice); evaluation of radionuclides in man (tissue samples, Richland school children); treatment for radionuclide incorporation (rats, dogs); chemical radiation protection; mechanisms of radiation effects (N. Crassa, miniature swine, leukemia virus)
	• **Analysis of Natural Systems:** sampling for contaminants; modeling ecological systems; statistical and mathematical aspects of ecology; quantitative ecology of impact evaluation

Year	Biological and Ecological Research Projects (As Indicated by Annual Reports)
	• **Environmental Chemistry:** potential for plutonium complexation in soil and uptake by plants; characteristics of Hanford soils and aquatic sediments; suspended particle interactions; Hanford intercontractor support • **Environmental Impact Assessments** • **Freshwater Sciences:** thermal effects on aquatic biota; combined effects of temperature and pollutants on fish; cold shock; Gable Mountain pond studies; salmon spawning studies; effects of modifications on aquatic ecosystems; ecological distribution and fate of plutonium and americium in processing waste ponds; effects of tritium on aquatic systems • **Marine Sciences:** non-nuclear effluents; thermal effects of effluent discharges; biological availability and effects of effluents; effects of low-level chronic irradiation; physical and radiological chemistry of ocean solutions; geochemical ocean sections study (GEOSECS); in situ pollutant studies; biogeochemistry of plutonium and americium in marine systems • **Radioanalytical Procedures Development** • **Terrestrial Ecology**
1975 (PNL 1976a, 1976b)	• **Biological Sciences:** development of aerosol exposure and analytical techniques; biological behavior of plutonium released with sodium; low-level plutonium and transplutonium oxide inhalation studies in beagles; low-level plutonium nitrate inhalation studies in beagles; toxicology of inhaled plutonium and transplutonium elements in rodents; disposition and biological effects of inhaled plutonium in miniature swine; inhalation hazards to uranium miners; inhalation hazards to coal miners (coal and diesel engine exhaust); toxicology of tritium and the noble gases; delayed effects of exposure to acid aerosols; toxicology of radionuclides in the fetal and juvenile mammals; factors modifying radionuclide metabolism and effects; gut-related studies of radionuclide toxicity; acute toxicity of reactor-catastrophe effluent; toxicology of chronically fed strontium-90 in miniature swine; biochemical indices of pollutant carcinogenicity; effects of pollutant metals in nutritionally deficient populations; effects of inhaled oil shale and spent shale particles in animals; evaluation of radionuclides in man; chemical radiation protection; treatment of radionuclide incorporation; treatment of toxic metal exposure associated with non-nuclear technologies; blood irradiation for medical applications; mechanisms of radiation effects; interaction of fossil-fuel derived toxic metals with biological membranes; biological effects of intracorporeal radioisotope heat sources; artificial heart program; alveolar clearance of inhaled metal oxides; effects of sulfur pollutants on lung physiology and biochemistry • **Analysis of Natural Systems:** sampling and modeling; population studies; statistical and mathematical aspects of ecology; quantitative ecology of impact evaluation; quantitative aspects of environmental plutonium studies • **Environmental Chemistry:** the potential for plutonium complexation in soil and uptake by plants; influence of soils and aquatic sediments on the chemical behavior, transport, and bioavailability of pollutants resulting from energy production; suspended particle interactions • **Freshwater Sciences:** effects of thermal discharges on aquatic biota; combined effects of waste heat and environmental factors acting in concert; effects of hydroelectric generation on riverine ecology; effects of water quality alterations on fish behavior; effects of behavior of fossil fuel elements in freshwater ecosystems Pacific Northwest energy related regional studies program; ecological distribution and fate of plutonium and americium in a processing pond; sublethal effects of tritium on aquatic systems; salmon spawning studies • **Marine Sciences:** effects of thermal discharges to coastal waters; non-nuclear effluents; bioavailability and impact of effluents on coastal ecosystems; effects of low-level chronic irradiation on embryonic development of marine fish and invertebrates; biogeochemistry of plutonium and americium in the marine environ; physical and radiological chemistry on ocean solutions; geochemical ocean sections study (GEOSECS) • **Radioanalytical Procedure Development** • **Terrestrial Ecology:** Hanford Reservation support services; weathering and aging of transuranics and radioecology of iodine-129; ecological micrometeorology and climatology of the ALE Reserve; terrestrial ecology studies; grassland biome; terrestrial animal ecology studies • **Shale Oil:** terrestrial effects of oil shale development; development of chemical reactions, stability and transport model of oil shale process wastes in soil • **Oil and Gas:** fate and effects of petroleum hydrocarbons in marine coastal ecosystems; long-term effects of hydrocarbons on selected ecosystems and associated organisms • **Nuclear Wastes (Fission):** suspended particle interactions and uptake in terrestrial plants; transuranic complexation in soil and uptake by plants; quantitative aspects of plutonium field studies; ecological distribution and fate of plutonium and

Year	Biological and Ecological Research Projects (As Indicated by Annual Reports)
	americium in a processing waste pond on the Hanford Reservation; use of analog elements to predict the equilibrium behavior of transuranic elements in the environ; weathering and aging of transuranics; biogeochemistry of plutonium and americium in the marine environ; radioecology of waste management areas (ROWMA); radioecology of uranium; radioecology of iodine-129 and technitium-99; influence of soils and aquatic sediments on the chemical behavior, transport, and bioavailability of pollutants
	• **Nuclear Fusion:** reviews for division of magnetic fusion energy; sublethal effects of tritium on aquatic systems; effects of low-level chronic irradiation on embryonic development of marine fish and invertebrates
	• **Solar Hydroelectric Generation:** effects of hydroelectric generation on riverine ecology
	• **Multitechnology Report:** analysis of natural systems; quantitative ecology of impact evaluation; geochemical ocean sections study (GEOSECS); chemistry of ocean solutions; in situ pollutant study of the New York bight; effects of water quality alterations on fish behavior; effects of thermal discharges on aquatic biota; combined effects of waste heat and environmental factors acting in concert; power plant heat and chemical effluent effects on selected marine and estuarine communities; immediate effects of anti-fouling biocides from coastal and offshore power plants; effects of thermal discharges to coastal waters; biological availability of energy related effluent material in the coastal ecosystem
	• **Hanford Support Services:** ARHCO studies
	• **Land Use and Coal Technology:** Hanford National Environmental Research Park (NERP) and ALE administration; terrestrial ecology; restoration of surface-mined lands; fossil fuel effluents in freshwater ecosystems
	• **Nuclear Wastes (Fission):** suspended particle interaction and uptake in terrestrial plants; transuranic complexation in soil and uptake by plants; quantitative aspects of plutonium field studies; ecological distribution and fate of plutonium and americium in a processing waste pond; use of analog elements to predict the equilibrium and behavior of transuranic elements in the environment; weathering and aging of transuranics; radioecology of uranium and waste management areas; transport and bioavailability of pollutants
	• **Nuclear Fusion:** biomagnetic effects; sublethal effects of tritium on aquatic systems; effects of low-level chronic irradiation on embryonic development of marine fish and invertebrates; effects of beryllium and lithium on aquatic systems; terstogenic effects of low-level magnetic fields
	• **Oil and Gas:** fate and effects of petroleum hydrocarbons in marine coastal ecosystems; effects of refinery wastes and oil from transfer facilities; long-term effects of hydrocarbons on selected ecosystems and associated organisms
	• **Hydroelectric Generation:** effects of hydroelectric generation on riverine ecology
1978 (PNL 1979a, 1979b)	• **Coal:** inhalation hazards to miners; lung toxicity; effects of pollutant metals; alveolar clearance of inhaled metal oxides; factors influencing cross-placental transfer and tetrogenicity of metallic pollutants; solvent refined coal biostudies
	• **Conservation:** mutagenic effects of electric fields
	• **Fission:** aerosol and analytical technology; inhaled plutonium oxide/nitrate in dogs; inhaled transuranics in rodents; toxicology of plutonium-sodium/sodium; inhalation hazards to uranium miners; toxicology of krypton-85; toxicology of thorium cycle nuclides; fetal and juvenile radiotoxicity; modifying radionuclide effects; gut-related studies of radionuclide toxicity; removal of deposited radionuclides; viral and radiation carcinogenesis; blood irradiators; cigarette smoke and plutonium
	• **Fusion:** biomagnetic effects
	• **Oil Shale:** late effects of oil shale pollution; mutagenicity of oil shale; development
	• **Multitechnology:** toxicology of inhaled acid aerosols; mobilization of deposited metals; metal-membrane interactions toxic effects of geothermal effluents
	• **National Environmental Research Park & Land Use:** various
	• **Synfuels:** various
	• **Oil and Gas:** fate and effects of petroleum hydrocarbons in marine ecosystems; long-term effects of hydrocarbons on ecosystems and organisms

Year	Biological and Ecological Research Projects (As Indicated by Annual Reports)
	• **Nuclear Wastes:** transuranic behavior in soils and plants; quantitative aspects of plutonium field studies; analog elements for transuranic chemistry; transuranic weathering in plants; radioecology of nuclear fuel cycles; environ behavior and effects of technetium and iodine • **Nuclear Fusion** • **Solar/Hydroelectric Generation** • **Satellite Power Systems** • **Multitechnology and Supporting Research** • **Alaskan Environmental Research**
1979 (PNL 1980a, 1980b)	• **Coal:** inhalation hazards to coal miners; effects of pollutant metals; alveolar clearance of inhaled metal oxides; factors influencing the cross-placental transfer and teratogenicity of metallic pollutants; solvent refined coal biostudies; perinatal effects of pollutants; mutagenicity of synfuels • **Conservation:** mutagenic effects of electric fields; electric field dosimetry; in vivo mutagenesis (electric fields); in vitro effects on electric fields • **Fission:** aerosol and analytical technology; inhaled plutonium oxide and plutonium nitrate in dogs; inhaled transuranics in rodents; cigarette smoke and plutonium; toxicology of plutonium-sodium; toxicology of sodium and lithium; inhalation hazards to uranium miners; toxicology of krypton-85; toxicology of thorium cycle nuclides; fetal and juvenile radiotoxicity; modifying radionuclide effects; gut-related studies of radionuclide toxicity; removal of deposited radionuclides; viral and radiation carcinogenesis; development of blood irradiators; transuranic behavior in soils and plants; quantitative aspects of transuranic field studies; analog elements for transuranic field studies; transuranic weathering in plants; radioecology of nuclear fuel cycles; environmental behavior and effects of technetium-99 and iodine-129 • **Fusion:** biological effects of magnetic fields; sublethal effects of tritium on aquatic systems; ecological effects of lithium and beryllium on important aquatic organisms and communities
1980 (PNL 1981a, 1981b)	• **Coal Studies:** inhalation hazards to coal miners; pollutant metals and malnutrition; alveolar clearance of inhaled metal oxides; factors influencing the cross placental transfer and teratogenicity of metallic pollutants • **Conservation:** mutagenic effects of electric fields; AC E-field dosimetry; in vivo mutagenesis; in vitro effects of electric fields • **Fission:** aerosol technology development; inhaled plutonium oxide/nitrate in dogs; inhaled transuranics in rodents; cigarette smoke and plutonium; toxicology of plutonium-sodium/sodium-lithium; inhalation hazards to uranium miners; toxicology of krypton-85; toxicity of thorium cycle nuclides; fetal and juvenile radiotoxicity; modifying radionuclide effects; gut-related studies of radionuclide toxicity; viral and radiation carcinogenesis; blood irradiators • **Fusion:** biological effects of magnetic fields • **Oil Shale:** health effects of synthetic fuels, terrestrial effects of development • **Multitechnology:** toxicology of inhaled acid aerosols; metal-membrane interactions; toxic effects of geothermal effluents • **National Environmental Research Park & Land Use:** terrestrial ecology; dynamics of wild populations; restoration of surface-mined lands; long-term ecological monitoring; application of long-term chemical biobarriers for U-tailings; re-vegetation • **Alaskan Resource Research:** ecological investigation of Alaskan resource development • **Synfuels:** ecological effects of coal conversion; toxicology • **Nuclear Waste (Fission):** transuranic behavior; analogs for transuranics; long-term plant availability of actinides; radioecology of nuclear fuel cycles; environmental behavior and effects of technetium and Iodine • **Marine Research Programs:** effects of energy systems effluents; bioavailability of effluent materials in coastal ecosystems; marine chemistry

Year	Biological and Ecological Research Projects (As Indicated by Annual Reports)
	• **Nuclear Fusion:** sublethal effects of tritium on aquatic systems; ecological effects of lithium and beryllium on aquatic communities; teratogenic effects of low-level magnetic fields • **Pumped Storage & Hydroelectric Development** • **Electric Field & Microwave Research**
1981 (PNL 1982a, 982b)	• **Coal:** mutagenicity of SRC materials; solvent refined coal biostudies; health effects of synthetic fuels; perinatal effects of SRC; teratology of SRC-II materials; inhalation hazards to miners; factors influencing the cross-placental transfer and teratogenicity of metallic pollutants • **Conservation:** mutagenic effects of electric fields; electric field dosimetry; in vivo mutagenesis of electric fields; in vitro effects of electric fields • **Fission:** aerosol technology development; inhaled plutonium oxide/nitrate and americium dioxide in dogs; inhaled transuranics in rodents; low-level plutonium-239 oxygen-2 lifespan studies; toxicology of plutonium-sodium; cigarette smoke and plutonium; toxicity of sodium and lithium; inhalation hazards to uranium miners; toxicology of krypton-85; toxicity of thorium cycle nuclides; fetal and juvenile radiotoxicity; gut-related radionuclide studies; statistical health effects studies; radioisotope customer list • **Fusion:** biological effects of magnetic fieldsMultitechnology: metal-membrane interactions; alveolar clearance of inhaled metal oxides • **Multitechnology:** metal-membrane interactions; alveolar clearance of inhaled metal oxides. • **National Environmental Research Park & Land Use:** terrestrial ecology; dynamics of wild populations; quantifying energy-related effects on mobile species; restoration of surface mined lands; long-term ecological monitoring; application of long-term chemical biobarriers for U-tailings; re-vegetation • **Alaskan Resource Research:** ecological investigation of Alaskan resource development • **Shale Oil:** terrestrial effects of development • **Synfuels:** ecological effects of coal conversion; coal liquefaction health and environmental effects • **Nuclear Waste (Fission):** transuranic element behavior; transuranic chemical species in ground water; quantitative aspects of transuranic and other radionuclide field studies; analog elements for transuranic chemistries; long-term plant availability of actinides; radioecology of nuclear fuel cycles; environmental behavior of technetium-99 and iodine-129 • **Marine Research:** trace metals metabolism in marine animals; bioavailability of energy-effluent materials in coastal ecosystems; marine chemistry of energy related pollutants • **Nuclear Fusion:** sublethal effects of tritium on aquatic systems ecological effects of lithium and beryllium on aquatic communities; teratogenic effects of low-level magnetic fields • **Pumped Storage & Hydroelectric Development:** effects of hydroelectric generation on riverine ecology • **Pathways Modeling, Assessment, and Hanford Project Support** • **Electric Field Research**
1982 (PNL 1983a, 1983b)	• **Coal Studies:** mutagenicity of SRC materials; solvent refined coal biostudies; health effects of synthetic fuels; perinatal effects of SRC; teratology of SRC-II materials; effect of pollutant metals • **Conservation:** teratology of guinea pigs exposed to electric fields; mutagenic effects of electric fields; electric field dosimetry; in vitro effects of electric fields • **Fission:** aerosol technology development; inhaled plutonium oxide/nitrate in dogs; inhaled transuranics in rodents; low-level plutonium-239 oxygen-2 lifespan studies; cigarette smoke and plutonium; toxicity of sodium and lithium; inhalation hazards to uranium miners; toxicology of kryton-85; toxicity of thorium cycle nuclides; modifying radionuclide effects; fetal and juvenile radiotoxicity; gut-related radionuclide studies; blood irradiators; health effects studies; radioisotope customer list • **Fusion:** biological effects of magnetic fields

Year	Biological and Ecological Research Projects (As Indicated by Annual Reports)
	• **Multitechnology:** metal-membrane interactions; alveolar clearance of inhaled metal oxides
	• **Terrestrial and Riverine Ecology:** terrestrial ecology; response of Columbia River fish to man-induced flow alterations; determination of long-term biological responses; ecological investigations in Alaska
	• **Marine Research:** entrance, behavior, and fate of trace metals in coastal waters; transport and bioavailability of trace metals in coastal ecosystems; biological effects of trace metals on marine animals
	• **Ecological Effects on Coal Conversion:** transport and fate; effects on aquatic and terrestrial systems; food chain transfer
	• **Solid Wastes (Mobilization, Fate, and Effects)**
	• **Radionuclide Fate and Effects:** environmental transport processes; radionuclide transport parameters; ecological interactions influencing remedial actions at waste sites; transuranic chemical species in groundwater; analogs for transuranic chemistries
	• **Statistical and Theoretical Research:** examining the fate of contaminants from energy developments; censuring wild and stray populations of large animals; designing studies to detect effects on mobile species
	• **Statistical and Quantitative Research:** developing methods to evaluate environmental behavior of radionuclides and population dynamics of large animals
1984 (PNL 1985a, 1985b)	• **Human Health Effects from Energy Generation:** statistical health effects study
	• **Carcinogenesis:** biostudies of complex mixtures; inhaled plutonium oxide/nitrate in dogs; low-level 239PuO2 lifespan studies; cigarette smoke and plutonium; toxicology of thorium cycle nuclides; inhalation hazards to uranium miners; fetal and juvenile radiotoxicity
	• **Mutagenesis:** mutagenicity of complex mixtures
	• **Systems Damage:** gut-related studies of radionuclides; teratology of complex mixtures; perinatal effects of complex mixtures; tissue dose in fossil fuel exposure; aerosol technology development; biological effects of magnetic fields
	• **General Life Sciences:** metal-membrane interactions
	• **Health and Environmental Risk Analysis:** complex mixtures-relative potency matrix
	• **Medical Applications of Nuclear Technology:** blood irradiators; radioisotope customer list
	• **Western Arid and Semi-Arid Ecosystems:** NERP; dynamics of arid land ecosystems
	• **Marine Sciences:** toxicity of sea-surface contaminants; metabolism of trace metals in marine animals; uptake and depuration of silver in mussels; partitioning coefficients (Kp) for metals in sediments and water column
	• **Mobilization, Fate, and Effects of Energy-Derived Chemicals:** hydrochemical phenomena controlling solute behavior; contaminant transfer through aquatic organisms
	• **Radionuclide Transport Research:** characterization of plant-produced ligands; metabolic behavior of technetium in plants; radionuclides transported by insects and fish
	• **Statistical and Quantitative Research on Wild Populations:** population studies of large animals; quantifying energy-related effects on mobile species
1985 (PNL 1986a, 1986b)	• **Human Health Effects:** statistical health effects study
	• **Health Effects of Radiation:** inhaled plutonium oxide/nitrate in dogs; oncogenes in radiation carcinogenesis; cigarette smoke and plutonium; low-level plutonium-239 oxygen-2 lifespan studies; inhalation hazards to uranium miners; toxicology of thorium cycle nuclides; aerosol technology development; fetal and juvenile radiotoxicity; gut-related studies of radionuclides
	• **Health Effects of Chemical Mixtures:** mutagenicity of complex mixtures; biostudies of complex mixtures; teratology of complex mixtures; perinatal effects of complex mixtures; health effects of complex mixtures; tissue dose in exposure to complex mixtures
	• **Medical Applications of Nuclear Technology:** blood irradiators; radioisotope customer list

Year	Biological and Ecological Research Projects (As Indicated by Annual Reports)
	• **Arid Land Sciences:** dynamics of arid land ecosystems; ecological studies at the NERP; biotic transport research • **Marine Sciences:** marine chemistry of energy-related contaminants; environmental stress and disease in marine animals; trace metal metabolism in marine animals • **Biogeochemical Phenomena in Indirect Exposure Pathways:** soil and plant processes controlling cation behavior in terrestrial systems; environmental behavior of inorganic anions; bioaccumulation and biochemical behavior of xenobiotic organic compounds in terrestrial systems; aquatic food chain transfer • **Subsurface Transport:** subsurface behavior of organic compounds and mixtures; hydrogeochemical phenomena controlling solute behavior • **Theoretical (Quantitative) Ecology:** dynamics of wild populations; quantifying energy-related impacts on mobile species; designs for environmental field studies
1986 (PNL 1987a, 1987b)	• **Human Health Effects:** statistical health effects study • **Health Effects of Radiation:** inhaled plutonium oxide/nitrate in dogs; oncogenes in radiation carcinogenesis; low-level plutonium-239 oxygen-2 lifespan studies; inhalation hazards to uranium miners; toxicology of thorium cycle nutrients; aerosol technology development; fetal and juvenile radiotoxicity; gut-related studies of radionuclides • **Health Effects of Chemical Mixtures:** mutagenicity of complex mixtures; complex mixtures biostudies; teratology of complex mixtures; perinatal effects of complex mixtures; tissue dose in complex mixture exposures • **Medical Applications of Nuclear Technology:** blood irradiators for medical applications development; radioisotope customer list • **Environmental Sciences:** detecting and managing change in terrestrial ecosystems; remote sensing; biotic transport processes in arid ecosystems; installation of precision-weighing lysimeters for evapotranspiration measurements at Hanford's ALE Reserve; biogeochemical phenomena; subsurface transport; marine sciences; theoretical ecology; exploratory research; university interactions; technology transfer
1987 (PNL 1988a, 1988b)	• **Human Health Effects Research:** statistical health effects study; iron stores and risk of cancer; interlaboratory toxicology data base; benefit-cost analysis of OHER research • **Health Effects Research in Biological Systems:** inhaled plutonium oxide/nitrate in dogs; inhalation hazards to uranium miners; mechanisms of radon injury; aerosol technology development; oncogenes in radiation carcinogenesis; molecular events during tumor initiation; fetal and juvenile radiotoxicity; molecular markers during development; molecular control of lung development; mutational studies in DNA targets • **Medical Applications of Nuclear Technology:** radioisotope customer list and shipment summary • **Environmental Sciences:** detecting and managing change in terrestrial ecosystems; biogeochemical phenomena; subsurface transport; marine sciences; theoretical ecology; exploratory research; university interactions; technology transfer; international relations
1988 (PNL 1989a, 1989b)	• **Human Health Effects Research:** statistical health effects study; iron stores and risk of cancer; interlaboratory toxicology knowledge base • **Health Effects Research in Biological Systems:** inhaled plutonium oxide/nitrate in dogs; low-level plutonium-239 oxygen-2 studies; inhalation hazards to uranium miners; mechanisms of radon injury; microdosimetry of radon daughters; growth factors in radiation carcinogenesis; oncogenes in radiation carcinogenesis; molecular events during tumor initiation; aerosol technology development; fetal and juvenile radiotoxicity; molecular markers during development; molecular control of lung development; mutation of DNA targets; synthesis of human genome information; genome graphics interface • **Medical Applications of Nuclear Technology:** radioisotope customer list • **Detecting and Managing Change in Terrestrial Ecosystems:** arid lands water balance; the effect of water and nitrogen perturbations on gas exchange in arid lands ecosystems; spatial variability in ecological factors influencing surface water distribution in the shrub-steppe; measuring landscape-level processes; floristics and climate in the deserts of Western United States

Year	Biological and Ecological Research Projects (As Indicated by Annual Reports)
	• **Biochemical Phenomena:** soil and plant processes controlling cation behavior in terrestrial systems; environmental behavior of inorganic anions • **Subsurface Science:** subsurface chemistry of organic chemical mixtures; subsurface behavior of organic fluids; radon transport modeling in soils; a finite element model of radon advection and diffusion in unsaturated cracked soils; subsurface microbial phenomena • **Marine Sciences:** environmental stress and disease in marine invertebrates • **Theoretical Ecology:** designs for environmental field studies; dynamics of wild populations; quantifying energy-related effects on mobile species • **Exploratory Research:** biogenic cycling of volatile nitrogen-containing species; remote thermal data for determining evapotranspiration in arid landscapes; comparison of techniques to elucidate adsorption phenomena in aqueous solutions; optimal measurement scales and experimental design for detecting and predicting global ecosystem change; variability and speciation of atmospheric mercury; environmental detection and fate of genetically engineered microorganisms; modeling salmon spawning habitat through remote sensing and geographic information system techniques
1989 (PNL 1990a, 1990b)	• **Human Health Research:** statistical health effects studies; iron stores and risks of cancer • **Biological Research:** inhaled plutonium oxide and plutonium nitrate in dogs; national radiobiology archives; low-level plutonium-239 oxygen-2 life-span studies; inhalation hazards to uranium miners; mechanisms of radon injury; microdosimetry of radon daughters; growth factors in radiation carcinogenesis; oncogenes in radiation-induced carcinogenesis; molecular events during tumor initiation; mutation of DNA targets; fetal and juvenile radiotoxicity; aerosol technology development • **General Life Sciences Research:** synthesis of human genome information; GnomeView, a graphical interface to the human genome • **Medical Applications:** radioisotope customer list • **Detecting and Managing Change in Terrestrial Ecosystems:** understanding terrestrial ecosystems; landscape ecological studies; arid lands water balance; a microcosm-level gas exchange study • **Subsurface Science:** subsurface microbial phenomena; groundwater dating at the SRS; subsurface chemistry of organic compounds and mixtures; chemistry/microbiology controlling chelated radionuclide transport; intermediate-scale subsurface transport of co-contaminants; subsurface flow physics of immiscible organic liquids; subsurface science program liaison; estimating steady date diffusive and advective fluxes from different soil types • **Marine Sciences:** cause and progression of leukemia in a marine invertebrate • **Theoretical Ecology:** dynamics of wild populations; designs for environmental field studies • **Exploratory Research:** biogenic cycling of volatile nitrogen-containing species; optimal measurement scales and experimental design for detecting and predicting global ecosystems change; a knowledge-based system for evaluating human disturbance and global climate change in arid regions; remote sensing of plant stress; fluxes of biogenic gases from the ocean surface; development of a passive diffusion sensor for the measurement of ambient mercury vapor concentrations
1990 PNL 1991a, 1991b)	• **Human Health Research:** statistical health effects studies; Tatum Dome cancer incidence study; iron stores and risk of cancer • **Biological Research:** inhaled plutonium oxide and plutonium nitrate in dogs; national radiobiology archives; low-level plutonium-239 oxygen-2 life-span studies; inhalation hazards to miners; mechanisms of radon injury; dosimetry of radon progeny; aerosol technology development; growth factors in radiation carcinogenesis; oncogenes in radiation-induced carcinogenesis; mutation of DNA targets; molecular events during tumor initiation; genotoxicity of inhaled energy effluents; fetal and juvenile radiotoxicity • **General Life Sciences Research:** GnomeView II, a graphical interface to the human genome; synthesis of human genome information • **Subsurface Science:** subsurface chemistry of organic chemical and organic-radionuclide mixtures; subsurface microbial phenomena; hydrogeologic and geochemical controls on microorganisms; pore water chemistry; chemistry/microbiology controlling chelated radionuclide transport; subsurface organic fluid flow research; intermediate-scale subsurface transport of co-contaminants

Year	Biological and Ecological Research Projects (As Indicated by Annual Reports)
	• **Terrestrial Science:** vegetation as a controlling influence in shrub-steppe water balance; landscape ecological studies; gas exchange dynamics and the effects of nitrogen and water; effects of soil water deficit and plant age; dynamics of soil inorganic nitrogen; the effects of fire on nitrous oxide flux; dominated old-fields; artificial intelligence in environmental research; environmental research parks; designs for environmental field studies
	• **Exploratory Research:** engineered microbes; contaminant transport modeling; Hollaender fellowship in contaminant transport; geostatistics; monitoring network design for risk-based groundwater remediation; remote sensing of plant stress; optimal measurement scales and experimental design for detecting and predicting global ecosystems change

PHYSICAL SCIENCE RESEARCH PROJECTS AT THE HANFORD SITE, 1951-1990

Year	Physical Science Research Projects (As Indicated by Annual Reports and with Asterisk by Quarterly Reports)
1951 (Mickelson 1951)	• **Radiological Records and Standards:** general; design standards; radiation monitoring activities; instrument calibrations
	• **Biophysics:** general; control laboratory and counting room; regional survey; synoptic meteorology; control services; experimental meteorology; geology; industrial hygiene; methods; radiochemical standards; soil science
1952 (GE 1952b)	• **Biophysics:** REDOX stack effluent gas and process water studies; determination of oil content in oil fog dispersion; radioactive isotopic build-up in 100-C reactor effluent; recirculated water analysis for 100-H; environmental contaminants originating from Hanford
	• **Synoptic/Experimental Meteorology:** weather forecasts; climatological statistics; oil-fog studies; air contaminant trajectory studies
	• **Geology/Hydrology:** drilling exploration; ground water contamination studies
	• **Industrial Hygiene/Medical:** particle exposure studies from 200 Areas, REDOX stack, 314 Building; non-radioactive health hazard studies (toxic metals, gases, or vapors)
	• **Instrument Development:** single channel gamma spectrometer using a sodium iodine acintillator; portable alpha survey instruments with scintillation detectors; C.P. meter; equipment for measuring the temperature below the river bottom; equipment for monitoring test wells; recording anemometer using an uninterrupted light beam; metal evaporator; etc.
	• **Soil Science:** absorption of cesium; studies of tank waste solutions penetrating the groundwater; accumulated sludge in the 105-H storage basin; polargraphic procedure for determining the total cation exchange capacities of soil
	• **Physics**
1953 (GE 1954c)	• **Radiological Engineering**
	• **Radiological Records and Standards**
	• **Biology Studies:** general; aquatic; metabolism and toxicology; biology controls
	• **Biophysics Section:** general; control laboratory and counting room; regional survey; synoptic meteorology; control services; experimental meteorology; earth sciences; industrial hygiene; instrument development; methods; physics; radiochemical standards
1954 (GE 1954d)	• **Radiological Engineering:** disposal of liquid wastes; disposal of process wastes to Columbia River; improvement of radiation protection design criteria; mechanical development
	• **Biology Program:** deduction of permissible human exposure by biological experimentation; effects of process effluents on aquatic organisms; effects of process effluent on plants and animals
	• **Biophysics Program:** monitoring methods; environmental studies and adequacy of waste disposal systems; instrumentation for radiation detection and measurement

Year	Physical Science Research Projects (As Indicated by Annual Reports and with Asterisk by Quarterly Reports)
1955 (Mickelson 1955)	• **Radiological Engineering:** disposal of liquid wastes from separations processes; disposal of process wastes to Columbia River – effluent dispersion patterns, ruptured fuel element waste disposal, downstream domestic usage of river water, reactor purge waste disposal, disposal of reactor experimental control solution, metal preparation area waste disposal; Columbia River studies
	• **Radiation Protection Design:** radiological disaster studies; radioactive particle hazard studies; hazards of possible process materials
	• **Mechanical Development**
	• **Biology Program:** deduction of permissible human exposure by biological experimentation – tritium absorption and metabolism, fission product absorption and metabolism, plutonium and absorption, respiratory metabolism and toxicity of radioactive particles, skin irradiation; effects of process effluents on aquatic organisms; effects of process effluent on plant and animal life – effect of reactor effluent on plants, absorption and translocation of radioelements on plants, toxicity of iodine-131 in animals
	• **Biophysics Program:** analytical methods; monitoring methods; counting methods; bioassay methods; special studies; geological studies; soil chemistry; soil physics; process waste disposal; diffusion and transport study; particle pick-up study; trajectory study (200 mile scale); atmospheric contamination problems; aerosol sampling; respirator testing; toxic materials
	• **Dosimetry:** gamma ray dosimetry; beta ray dosimetry; neutron dosimetry
	• **Instrumentation:** survey instrumentation; biological instrumentation; laboratory instrumentation; special purpose instrumentation
1956* (GE 1956c, 1956d, 1957b, Parker 1956)	• **Biology Program:** aquatic biology studies; experimental animal farm studies; metabolism studies; pharmacology and experimental therapeutics; plant nutrition and microbiology; minor problems research; biological analyses
	• **Biophysics:** chemical and monitoring methods; geological studies; meteorology studies; industrial hygiene studies
	• **Dosimetry Program:** gamma ray; beta ray; and neutron dosimetry; body monitor
	• **Instrumentation Program:** survey; biological; laboratory; and special purpose instrumentation
	• **Chemistry and Separations:** analytical chemistry; chemical effluents technology
1957* (GE 1957c, 1957d)	• **Biology Program:** aquatic biology; experimental animal farm; metabolism studies; pharmacology studies; plant nutrition and microbiology; biological analyses; radioecology
	• **Physics and Instrumentation:** meteorology; radiation dosimetry; instrumentation
	• **Chemistry and Separations Processes:** analytical chemistry; chemical effluents technology
1958* (GE 1958e, 1958f, 1958g, 1959b)	• **Biology Program:** aquatic biology; experimental animal farm; metabolism studies; pharmacology; plant nutrition and microbiology; radioecology
	• **Physics and Instrumentation:** meteorology; radiation dosimetry; instrumentation
	• **Chemistry and Separations Processes:** analytical chemistry; chemical effluents technology
1959 (GE 1960b)	• **Analytical and Counting Methods:** gamma ray spectrometric analysis; radiostrontium procedures with nitric acid; analysis of plutonium in urine; phosphorescence photons; determining radioactive ions in reactor effluent
	• **Bioassay Studies:** measurements of zinc
	• **Environmental Studies:** radioactivity in foods; measuring fallout radioisotopes on vegetation
	• **Reactor Studies:** the origin of effluent radioisotopes; sources of fission product radioisotopes in effluent; kinetics of arsenic retention; effects of temperature and flow velocity on the decontamination of reactor effluent
	• **Radiation Chemistry:** chemical protection from ionizing radiation; protolysis and hydrolysis reactions of erioglaucine in aqueous solutions; susceptibility of solutes to attack by free radicals in aqueous solution; electron-spin resonance studies; membrane electrode cell for the determination of dissolved oxygen

Year	Physical Science Research Projects (As Indicated by Annual Reports and with Asterisk by Quarterly Reports)
1960 (GE 1961b)	• **Atmospheric Physics:** diffusion and transport; tracer technology; dispersion measurements • **Radiological Physics:** radioactivity in Hanford employees; whole body counter; retention of radioisotopes; scanning counter; shadow shields; polyethylene long counter; radioactive neutron sources; Perlow neutron spectrometer; Van de Graaff accelerator helium ion source; positive ion accelerator target prep.; gamma ray source measurements; stopping power of aluminum for electrons; precision instrumentation for an electron; linearizing thermistor thermometer; data in calorimetry; liquid nitrogen level controller • **Radiological Chemistry:** analytical and counting methods – various; environmental studies – various; reactor studies – various; radiation chemistry – various • **Chemical Effluents Technology:** special geological studies – various; geochemical and geophysical research – various; effluents research and development – various; equipment and instrumentation – various • **Instrumentation:** contaminated air monitoring techniques; biology operation support instrumentation; dosimetry instrumentation; radiation detection development; general radiological monitor development; automatic monitoring and recording methods; special purpose instrument development
1961 (GE 1962b)	• **Atmospheric Physics:** diffusion and transport; deposition of airborne materials; vertical exchange coefficients and the vertical flux of matter; precipitation scavenging processes • **Radiological Physics:** shadow shield whole body counter; iodine-131 levels in human thyroids after nuclear fallout; factors influencing measurement of plutonium in wounds; retention of isotopes; detection of phosphorous-32 in vivo; precision long counter; radiation effects in BF3 tubes; neutron energy degradation; secular variation in neutron emission from plutonium; neutron moderator; target contamination; etc. • **Radiological Chemistry:** analytical and counting methods; radiation chemistry; environmental studies; reactor studies • **Chemical Effluents Technology:** soil chemistry, geochemistry; soil physics; geology; instrumental techniques; uranium oxidation and fission product volatility; micromeritics • **Instrumentation:** contaminated air monitoring techniques; biology support instrumentation; dosimetry instruments; radiation detection development; general radiological monitor and circuitry development; automatic monitoring and recording methods
1962 (GE 1963b)	• **Atmospheric Physics:** diffusion, deposition, and transport; precipitation scavenging processes; advances in tracer technology • **Radiological Physics:** whole body counters; cesium-137 in Eskimos; cesium-134 in Eskimos and fallout; cesium-137 urinary excretion by Eskimos; determination of phosphorous-32 in vivo; plutonium x-ray scintillation counters; pulse shape discrimination in fast neutron dosimetry; lithium sandwich fast neutron spectrometer; precision long counter; calibration of the polyethylene double moderator; cosmic-ray neutron measurements; film activation studies; calorimetric determination of antimony-124; heat output of plutonium; input and source impedance corrections in high precision voltmeter standardization • **Radiological Chemistry:** reactor studies; environmental studies; radiation chemistry • **Chemical Effluents Technology:** soil chemistry, geochemistry; soil physics; geology; radioisotopes as particles and volatiles; uranium oxidation and fission product volatility • **Instrumentation:** automatic monitoring and recording methods; general radiological monitoring and circuit development; dosimetry instrumentation; radiation detector development; contaminated air monitoring techniques
1963 (GE 1964b)	• **Atmospheric Physics:** prediction of environmental exposures; comparison of results of atmospheric diffusion; dispersion from elevated sources; precipitation scavenging processes; atmospheric tracer technology • **Radiological Physics:** radioactivity in Alaskan Eskimos; human uptake of cesium-137; determination of the avg. cesium-137 body burden of a population; scintillation counter for in vivo detection of plutonium; determination of phosphorous-32 in vivo; whole body counters; pulse shape discrimination; response of paired ion chambers; precision long counters; low energy fast neutrons; fast neutron medical facility & shielding; heat amplifier; theory on measurement on beams • **Radiological Chemistry:** radioanalytical models; reactor studies; environmental studies • **Chemical Effluents Technology:** soil chemistry, geochemistry; soil physics; geology; radioisotopes as particulates and volatiles

Year	Physical Science Research Projects (As Indicated by Annual Reports and with Asterisk by Quarterly Reports)
	• **Instrumentation:** automatic monitoring and recording methods; biology instrumentation; dosimetry instrumentation
1964 (PNL 1965b)	• **Atmospheric Physics:** model for diffusion and transport for a continuous release in stable atmosphere; measurements of dispersion; point source in an unstable atmosphere; atmospheric turbulence measurements; area dosage relationships; precipitation scavenging studies; sampler for recording concentration of airborne zinc sulfide; improvements in micrometeorological instrumentation
	• **Radiological Physics:** radioactivity measured in Alaskan Natives 1962-64; seasonal variations of cesium-137 in Alaskan Eskimos; Urinalysis for cesium-137; sodium-22 and iron-55 in people and food; radioactivity in an Eskimo; whole body counting; cesium-137 in Trappist monks; humans exposed to uranium mine atmospheres; etc.
	• **Radiological Chemistry:** radiation chemistry; radioanalytical methods; reactor studies; counting studies; environmental studies
	• **Chemical Effluents Technology:** soil chemistry and geochemistry; atmospheric radioactivity and fallout – aerosol studies
	• **Instrumentation:** radiological measurement methods and instrumentation; biological and biomedical instrumentation; dosimetry and atmospheric physics instrumentation
1965 (PNL 1966b, 1966c)	• **Atmospheric Sciences:** precipitation scavenging studies; Midas computer program for evaporation of falling drops and isentropic stream functions and trajectories; radio-telemetered meteorological and radiation observation network; instrumenting of the Queen Air aircraft for sampling of a zinc sulfide atmosphere tracer; prediction of exposures from an elevated source; spectral distribution of atmospheric turbulence
	• **Radiological Physics:** fast pulse averaging; time-resolved spectroscopy; specific ionization of monoenergetic protons in tissue-equivalent gas; grid-walled proportional counter; self-absorption in scintillation crystals; pulsed x-ray machine; local energy density and hit theory; long-term change of plutonium-beryllium neutron sources; local-energy-density distributions; calorimetric measurement of polonium-210; low energy fast neutron spectrometer; transition zone density; etc
	• **Radiation Chemistry:** determination of erythocyte size distribution; radiation induced hemolysis of erythrocytes; effect of chloride ion on the radiolysis of erioglaucine solutions; electron spin resonance detection instruments; cobalt-60 irradiator flow system for electron spin resonance long-lived organic radicals formed in irradiated aqueous solutions; reaction kinetics of free radicals
	• **Radiological Chemistry:** alpha-emitters in the Columbia; reduction of reactor effluent radionuclides; the effect of reduced alum addition on the removal of arsenic-75 at the KW reactor; ion exchange studies; seasonal variations in sediment deposition; use of zinc-65/cobalt-60 ratios to determine age and deposition rates; method for determination of phosphorous-32 and other radionuclides; behavior and transport of radionuclides in the Columbia; natural and artificial radionuclide concentration in farm produce; distribution and excretion of technetium in humans; lead-210 in Alaskan biological samples; etc.
1966 (PNL 1967b, 1967c)	• **Atmospheric Sciences:** lateral plume growth; diffusion studies; turbulence investigations; wake studies; instantaneous plume detection; precipitation scavenging studies; real-time sampling of airborne tracers with aircraft; meteorological analyses of PLUMBBOB nuclear tests; raindrop charge measuring systems; identification and measurement of hydrogen chloride gas in the atmosphere; krypton-85 as an atmospheric tracer; portable boom-type air sampler
	• **Radiological Chemistry:** tracers of atmospheric precipitation processes; atmospheric behavior of airborne radionuclides; krypton-85 as an atmospheric tracer; air sampling; gamma-ray spectrometer studies; trace elements in marine organisms and sediments; etc
	• **Radiological Physics:** instrumentation for time-resolved spectrometry; proton irradiation of gases; programmed distribution generator; whole-body counter design and calibration; film dosimeter performance criteria; etc
1967 (PNL 1968b, 1968c)	• **Radiological Chemistry:** trace elements absorption, content, and measurement; automatic real-time air monitoring; radionuclide distribution in salmon; use of photon spectrometry; liquid scintillation counting; etc.
	• **Radiological Physics:** hit theory; delta rays of heavy particles; microdosimetry; gas pressure control; etc.
	• **Earth Sciences:** landslide cause and significance; tectonic deformation and seismic activity; infrared imagery over surface water; palynology; geophysical evaluation of thick basalt sequences in the Pasco Basin; etc
1968 (PNL 1970b, 1970c)	• **Radiological Chemistry:** determination of trace levels of various elements in soft tissue; radionuclide distribution and concentration; radium and radiobarium measurement in seawater; iron-55 concentrations in sea organisms; transfer of airborne radionuclides; multielement analysis of aquatic organisms; etc.

Year	Physical Science Research Projects (As Indicated by Annual Reports and with Asterisk by Quarterly Reports)
	• **Radiation Chemistry:** rate constants of organic compounds; long-lived organic radicals formed by irradiation; flash photolysis
	• **Radiological Physics:** radiation dosimetry; hit theory and microdosimetry; energy absorption straggling, distribution, and spectra; pulse read ion chamber system; beam current regulator; fast pulse signal averaging; etc.
	• **Instrumentation:** solid state detectors; smoking control for canines; radon monitoring; tritium target scanning using surface contoured diodes; miniature high voltage supply; etc.
1969 (PNL 1970f, 1970g)	• **Atmospheric Sciences:** diffusion and transport; boundary layer studies; BOMEX studies; fallout studies; precipitation scavenging studies; radioisotopes as particles and volatiles; ecological micrometeorology and climatology
	• **Radiological Chemistry:** inhalation hazards to uranium miners; rates of deposition in Alaskan environ; physical and radiological chemistry of ocean solutions; geothermal ocean section study (GEOSECS); in-cloud scavenging; radioanalytical procedure development; radioactive fallout rates and mechanisms; iron-55 in the human population; radiation instrumentation
	• **Radiological Physics:** internal depositions of radionuclides in man; characterization of radiation-induced free radical reactions in aqueous systems; nuclear techniques in medical science; radiation dosimetry; radiation physics
1970 (PNL 1971c, 1971d)	• **Atmospheric Sciences:** precipitation scavenging studies; diffusion and turbulence studies; radioisotopes as particles and volatiles; ecological micrometeorology and climatology
	• **Radiological Chemistry:** inhalation studies; rates of deposition of airborne radionuclides in the Alaskan environment; physical and radiological chemistry of ocean solutions; geochemical oceans sections study (GEOSECS); in-cloud scavenging; radioanalytical procedure development; radioactive fallout rates and mechanisms; iron-55 and lead in human population of the world; radiation instrumentation; tracer studies in the Northeast Colorado Hail experiment program
	• **Radiological Physics:** internal depositions of radionuclides in man; characterization of radiation-induced free radical reactions in aqueous systems; nuclear techniques in medical science; radiation dosimetry; radiation physics
1971 (PNL 1972c, 1972d)	• **Atmospheric Sciences:** precipitation of scavenging tracers; raindrop trajectories; wind shear; precipitation washouts of various compounds; aerosol particle size distributions; isentropic trajectories; climatological measurements; radioactive particles; aircraft instrumentation; tracer studies; fallout rates and mechanisms; turbulence measurement studies; lab procedures
	• **Radiological Chemistry:** inhalation hazards to uranium miners; rates of deposition of airborne radionuclides in the Alaskan environment; physical and radiological chemistry of ocean solutions; geochemical oceans sections study (GEOSECS); in-cloud scavenging; radioanalytical procedure development; radioactive fallout rates and mechanisms; iron-55 and lead in the human population; radiation instrumentation; tracer studies in the METROMEX experiment; radionuclide decline in the Columbia River water
	• **Radiological Physics:** dosimetry of radionuclides in man; characterization of radiation-induced free radical reactions in aqueous systems; nuclear techniques in medical science; mechanisms of environmental exposure; radiological evaluation of post mortem tissue samples; radiation dosimetry; radiation physics
1972 (PNL 1973c, 1973d)	• **Atmospheric Sciences:** diffusion, deposition, and turbulence studies on various terrain and surface types; experimental and modeling efforts for deposition of particulates; resuspension of surface contamination; controlled tracer experiments; distribution of resuspended soil particles; removal of pollutants by wet scavenging processes; various scavenging experiments; METROMEX studies; global distribution of contaminants
	• **Radiological Chemistry:** inhalation hazards to uranium miners; fate and effects of radionuclides in Alaska; physical and radiological chemistry of ocean solutions; geochemical ocean sections study (GEOSECS); precipitation scavenging; radioanalytical procedure development; radioactive fallout rates and mechanisms; iron-55 in human populations; radiation instrumentation; tracer studies in the national hail research experiment; tracer studies in the metropolitan meteorological experiment; survey study of radionuclides and chemical pollutants; radionuclide decline in Columbia River water
	• **Radiological Physics:** characterization of radiation-induced free radical reactions; nuclear techniques in medical science; mechanisms of environmental exposure; evaluation of radionuclides in man; radiation dosimetry; radiation physics; radiation biophysics

Year	Physical Science Research Projects (As Indicated by Annual Reports and with Asterisk by Quarterly Reports)
1973 (PNL 1974c, 1974d)	• **Atmospheric Physics:** characterization of sources and ambient pollutants; transport, diffusion, and turbulence studies; atmospheric transformation processes; removal and resuspension processes (wet removal, dry deposition particles, and resuspension of particles); special studies (cooling tower and cooling pond atmospheric impact and ecological micrometeorology and climatology of the ALE Reserve)
	• **Instrumentation Technology:** radiation instrumentation – radiation chemistry; nuclear techniques in medical science
	• **Radiological and Chemical Physics:** radiation physics studies; radiation dosimetry; radiation biophysics
	• **Nuclear Medicine Technology and Other Health Applications:** nuclear powered prostheses; medical uses of isotopes
	• **Artificial Heart Program:** medical grade plutonium-238 from americium-241; recipient radiation exposure; population radiation exposure
1974 (PNL 1975c, 1975d)	• **Atmospheric Sciences:** survey of radioactivity and chemical pollutants studies; radioactive fallout rates and mechanisms studies; atmospheric aerosols and trace gases studies; regional transport of effluent studies; fallout phenomenology; Hanford regional meteorological studies; atmospheric diffusion, deposition, and transport phenomena; fundamental turbulence studies; tracer studies in the METROMEX experiment; precipitation scavenging studies; cooling tower and cooling pond atmospheric impact studies; radioisotopes as particles and volatiles; particle resuspension and translocation; ecological micrometeorology and climatology on the ALE Reserve
	• **Instrumentation Technology:** radiation instruments; in situ monitoring; activation analysis technology; chemical and molecular composition technology; environ pollution analysis; real-time measurement of airborne plutonium; nuclear techniques in medicine
	• **Radiological and Chemical Physics:** radiation physics studies; radiation dosimetry and radiation biophysics studies
	• **Analysis and Evaluation:** toxic substances in nuclear fuels; methodology for the assessment of social values
	• **Nuclear Medicine Technology & Other Health Applications:** nuclear powered prostheses; medical use of isotopes
	• **Artificial Heart Program**
1975 (PNL 1976c, 1976d)	• **Instrumentation Technology:** low-level radiological chemistry laboratory techniques; in situ monitoring; environmental pollution analysis; in situ pollutant measurements; trace constituent analysis by laser excitation; analytical techniques for measurement of technetium-99 in environmental samples; fast real-time measurement of extremely low-level airborne plutonium; environmental applications of holography; chemical reaction kinetics
	• **Radiological and Chemical Physics:** radiation physics; radiation dosimetry and radiation biophysics; dosimetry of particulate sources in the lung
	• **Health Protection:** dosimetry of internal emitters
	• **Analysis and Evaluation:** toxic substances in nuclear fuels; assessment of social values in thermal plant siting; considerations and methodologies for studies concerning the effect of low-level exposures to environmental contaminants
	• **Regional Studies:** Pacific Northwest regional assessment program
	• **Atmospheric Sciences:** survey of ambient airborne pollutants; radioactive fallout rates and mechanisms; atmospheric chemistry of coal conversion pollutants; chemical and physical properties of pollutants in the Colstrip power plant effluents; tracer studies – METROMEX; theoretical studies; atmospheric boundary layer studies; particle resuspension and translocation; radioisotopes as particles and volatiles; precipitation scavenging; aerosol and trace gas transformations; safety analysis and environmental effects studies; Pacific Northwest energy related regional assessment program; atmospheric impacts
1976 (PNL 1977c, 1977d)	• **Atmospheric Sciences:** surveys and weather modification; air flow and dispersion; deposition and resuspension; pollutant scavenging; pollutant transformations and interactions; composite models for atmospheric processes;
	• **Physics and Chemistry of Pollutant Interactions in the Environment:** reaction kinetics of combustion processes
	• **Radiological and Chemical Physics:** radiation dosimetry and radiation biophysics; dosimetry of particulate sources in the lung; radiation physics-basic collision cross sections, radioluminescence studies

Year	Physical Science Research Projects (As Indicated by Annual Reports and with Asterisk by Quarterly Reports)
	• **Characterization, Measurement, and Monitoring:** radiation instrumentation; environmental pollution analysis; oil shale and tar sand research; dosimetry of internal emitters; analytical techniques for technetium-99 measurement; trace constituent analysis by laser excitation; application to holography; real-time measurement of plutonium in air below MPC levels
1977 (PNL 1978c, 1978d)	• **Atmospheric Sciences:** precipitation scavenging in MAP3S; aerosols and trace gas transformations; MAP3S modeling studies; atmospheric boundary layer studies; particle resuspension and translocation; urban pollution characterization; transport and deposition; coal conversion pollutant chemistry; regional studies; air pollution dry deposition; comprehensive study of atmospheric diffusion prediction from meteorological parameters; alternate fuel cycle technologies; hemispheric pollution behavior studies of chemicals and radiation substances; fallout rates and mechanisms; meteorological effects of thermal energy releases (METER) • **Solar:** various studies • **Oil Shale:** dual tracer deposition; revised method of energy flux calculations; transport and dispersion in complex terrain; verification statistic for numerical models; etc. • **Coal:** formation of fly fish in coal combustion; mass spectrometric studies of gaseous effluents from in situ coal glassification • **Fission:** various studies • **Geothermal:** heavy-metal gaseous emissions from power plants; measurement of potentially toxic materials in geopressured geothermal fluids • **Multitechnology:** various studies
1978 (PNL 1979c, 1979d)	• **Atmospheric Sciences:** aerosol and trace gas transformations; precipitation scavenging in MAP3S; MAP3S modeling studies; regional studies; atmospheric boundary layer studies; coal conversion pollutant chemistry studies; particle resuspension and translocation; alternate fuel cycle technologies/thorium fuel cycle technologies; meteorological effects of thermal energy releases; fallout rates and mechanisms; air pollution dry deposition studies • **Radiation Physics:** initial interaction processes; track structure; energy transport; radiation dosimetry and radiation biophysics; microdosimetry of internal sources; dosimetry of internal emitters; real-time measurement of plutonium in air; analytical techniques for environmental sampling; radiation instrumentation • **Geothermal, Oil Shale & Coal Studies** • **Multitechnology:** environmental pollution analysis; pollutant characterization; applications of holography; certifies research materials
1979 (PNL 1980c, 1980d)	• **Atmospheric Sciences:** Project ASCOT; radiation measurements; tracer experiments; micrometeorological measurements; sulfur surface flux measurements; particle deposition studies; pollutant transformations and models; MAP3S studies; regional scale assessments; aircraft measurements/research; in-cloud conversion rates; wind and cloud studies; particle/gas deposition studies; deposition measurements; resuspension studies; airborne plutonium transport at Hanford; atmospheric radionuclide concentrations near Richland; Hanford dispersion model; research needs in the oil shale industry; atmospheric transport studies; Hanford turbidity • **Coal:** reaction kinetics of combustion products; solvent refined coal effluent/product characterization • **Radiation Physics:** initial interaction processes; track structure; energy transport; radiation dosimetry and radiation biophysics; microdosimetry and radiation biophysics; microdosimetry of internal sources; dosimetry of internal emitters; real-time measurement of plutonium in air at below-MPC levels; analytical techniques for measurement of technetium-99 in environmental samples; radiation instrumentation; magnetic field dosimetry development • **Geothermal:** heavy-metal and noxious-gas emission from geothermal resource development • **Oil Shale:** oil shale and tar sand research • **Multitechnology:** environmental pollutant characterization by direct-inlet mass spectrometry; trace analysis by laser excitation; direct coupled plasma emission spectroscopy; applications of holography
1980 (PNL 1981c, 1981d)	• **Atmospheric Sciences:** behavior of air pollution; types of pollutants emitted; transport and diffusion; physical and chemical transformations during transport; removal by wet and dry deposition processes; pollutant impacts on climate, bodies of water and organisms; deposition and resuspension studies of particulates; removal by deposition studies; oil shale studies

Year	Physical Science Research Projects (As Indicated by Annual Reports and with Asterisk by Quarterly Reports)
	• **Coal:** reaction kinetics of combustion products; modeling cellular effects of coal pollutants • **Fission/Radiation Physics:** initial interaction process; track structure; energy transport; radiation dosimetry and biophysics; microdosimetry of internal sources; dosimetry of internal emitters; radiological chemistry; magnetic field dosimeter development; transuranic chemical species in groundwater • **Geothermal:** heavy-metal and noxious-gas emission from geothermal resource development • **Oil Shale:** oil shale and tar sand research studies • **Multitechnology:** environmental pollutant characterization by direct inlet mass spectrometry; trace analysis by laser excitation; applications of holography
1981 (PNL 1982c, 1982d)	• **Atmospheric Sciences:** atmospheric diffusion in complex terrain; atmospheric boundary layer studies; pollutant transformation in the atmosphere; atmospheric carbon dioxide abundance; meteorological effects of thermal energy releases; theoretical studies and applications; air pollution dry deposition; particle resuspension and translocation; oil shale fugitive air emissions • **Coal:** reaction kinetics of combustion products; modeling cellular effects of coal pollutants • **Radiation Physics:** initial interaction process; energy transport; radiation dosimetry and biophysics; microdosimetry of internal sources; dosimetry of internal emitters; radiological chemistry; magnetic field dosimeter development • **Oil Shale:** oil shale and tar sand research studies • **Multitechnology:** environmental pollutant characterization by direct inlet mass spectrometry; trace analysis by laser excitation
1982 (PNL 1983c, 1983d)	• **Atmospheric Sciences:** pollutant characterization; boundary layer meteorology in complex terrain; dispersion, deposition, and resuspension of atmospheric pollutants; particle resuspension and translocation; theoretical studies and applications; oil shale fugitive air emissions • **Analytical Studies:** coal liquefaction product characterization; oil shale and tar sand research materials preparation and documentation • **Measurement and Dosimetry:** characterization of synfuels and combustion products; trace analysis by laser excitation; procedures development • **Radiation Physics:** initial interaction process; track structure; energy transport • **Dosimetry and Biophysics:** modeling and cellular studies; dosimetry; microdosimetry of internal emitters
1983 (PNL 1984c, 1984d)	• **Atmospheric Sciences:** transformation of energy related contaminants; atmospheric diffusion in complex terrain; atmospheric boundary layer studies; dry deposition; particle resuspension and translocation; theoretical studies and applications • **Analytical Studies:** chemical basis for the biological response to conversion materials; oil shale and tar sands research • **Measurement and Dosimetry:** characterization of synfuels and combustion products; lasers in analytical chemistry • **Radiation Physics:** initial interaction processes; track structure; radiation dosimetry; modeling and cellular studies; radiation biophysics; modeling cellular response to genetic damage; internal microdosimetry
1984 (PNL 1985c, 1985d)	• **Atmospheric Sciences:** atmospheric studies in complex terrain; atmosphere boundary layer studies; coupling/decoupling of synoptic and valley circulation; air pollution by dry deposition; oil shale fugitive air emissions and plume depletion; particle resuspension and translocation; theoretical studies and applications; processing of emissions by clouds and precipitation • **Analytical Studies:** chemical basis for the biological response to complex organic mixtures • **Measurement and Dosimetry:** supercritical fluid analytical methods; lasers in analytical chemistry • **Radiation Physics**: initial interaction processes; track structure; radiation dosimetry; modeling and cellular studies; radiation biophysics; modeling cellular response to genetic damage; internal microdosimetry

Year	Physical Science Research Projects (As Indicated by Annual Reports and with Asterisk by Quarterly Reports)
1985 (PNL 1986c)	• **Atmospheric Sciences:** atmospheric diffusion in complex terrain; coupling/decoupling of synoptic and valley circulation; atmospheric boundary layer studies; dry deposition and resuspension; processing of emissions by clouds and precipitation
1986 (PNL 1987c, 1987d)	• **Atmospheric Sciences:** atmospheric diffusion in complex terrain; coupling/decoupling of synoptic and valley circulation; atmospheric boundary layer studies; scavenging studies; Soviet accident fallout measurements • **Analytical Studies:** chemical basis for biological response to energy-related organics • **Measurement and Dosimetry:** supercritical fluid analysis methods; lasers in environmental research • **Radiation Biology:** initial interaction processes; track structure; radiation dosimetry; modeling and cellular studies; radiation biophysics; modeling cellular response to genetic damage
1987 (PNL 1988c, 1988d)	• **Atmospheric Sciences:** atmospheric studies in complex terrain; large-scale atmospheric transport and processing of emissions; measurement and analysis techniques; air-surface interaction • **Analytical Studies:** DNA adducts as indicators of health risks; Chernobyl database management • **Measurement and Dosimetry:** supercritical fluid analytical methods; lasers in analytical chemistry • **Physical Processes in Radiation Biology:** radiation physics – initial interaction processes, track structure; radiation dosimetry – dosimetry, modeling and cellular studies, radiation biophysics, modeling cellular response to genetic damage
1988 (PNL 1989c, 1989d)	• **Atmospheric Sciences:** atmospheric studies in complex terrain; large-scale atmospheric transport and processing of emissions; measurement and analysis techniques • **Analytical Studies:** Chernobyl database management • **Measurement and Dosimetry:** supercritical fluid analytical methods; lasers in environmental studies; laser measurements of lead-210; DNA adducts as indicators of health risks; biological effectiveness of radon particles • **Physical Processes in Radiation Biology:** radiation physics; radiation dosimetry; testing cell response models; radiation biophysics; modeling cellular response to genetic damage
1989 (PNL 1990c, 1990d)	• **Atmospheric Studies in Complex Terrain:** direct numerical simulation of turbulent boundary layer flows; modeling of slope flows; modeling and observations of a valley's heat budget; planning for Oak Ridge experiment; observations of thermally developed wind systems in mountainous terrain; atmospheric mass and heat budgets for a canyon land basin; use of a nonhydrostatic mesoscale model to simulate drainage flows in the presence of ambient winds • **Large-Scale Atmospheric Transport and Processing of Emissions:** precipitation chemistry observations; concentration and distribution of atmospheric peroxides over the Pacific; overview of the frontal boundary study; effect of uncertainty in RSM meteorological parameters on predicted pollutant wet deposition; MAP3S precipitation chemistry network; proposed test of two hypotheses related to concentration fluctuations • **Climate Change:** another Gremlin in the greenhouse (anthropogenic sulfur); preliminary analysis of deep convection in an ocean general circulation model; experiments on the influence of whitecaps on air/sea gas transport rates; second-generation model of greenhouse gas emissions; regional effects of climate change and carbon oxygen-2 fertilization on the natural and human environment • **Dosimetry Research:** Chernobyl database management; determination of radon exposure; DNA adducts as indicators of health risks; biological effectiveness of radon alpha particles; biological behavior and dosimetry • **Measurement Science:** capillary electrophoresis-mass spectrometry; lasers in environmental research • **Radiological and Chemical Physics:** radiation physics; radiation dosimetry; radiation biophysics; modeling cellular response to genetic damage
1990 (PNL 1991c, 1991d)	• **Atmospheric Research:** atmospheric diffusion in complex terrain; heat budgets of valley and basin atmospheres; atmospheric boundary layer studies; atmospheric chemistry program/PRECP; continuing development of the G-1 aircraft • **Climate Research:** atmospheric radiation measurement (ARM) program; carbon dioxide ocean research; progress report on the second generation model; resources analysis research project – MINK study

Year	Physical Science Research Projects (As Indicated by Annual Reports and with Asterisk by Quarterly Reports)
	• **Dosimetry Research:** Chernobyl database management; determination of radon exposure; DNA adducts as indicators of health risks; biological effectiveness of radon alpha particles • **Measurement Science:** capillary electrophoresis-mass spectrometry; lasers in environmental research • **Radiological and Chemical Physics:** radiation physics-various; radiation dosimetry-various; radiation biophysics-various; modeling cellular response to genetic damage-various

SECTION 8

SITE SECURITY

J.K. KEATING AND D.W. HARVEY
PACIFIC NORTHWEST NATIONAL LABORATORY

SECTION 8 - SITE SECURITY

Letter from President Franklin D. Roosevelt to Lieutenant General Leslie R. Groves, commander of the Manhattan Project, June 29, 1943: *"I am writing to you as the one who has charge of all the development and manufacturing aspects of this work [referring to the Manhattan Project]. I know that there are several groups of scientists working under your direction on various phases of the program. The fact that the outcome of their labors is of such great significance to the nation requires that this project be even more drastically guarded than other highly secret war development. As you know, I have therefore given directions that every precaution be taken to insure [sic] the security of your project."* - UPOA 1977, bk. 1, vol. 14, app. A-1

The purpose of security at the Hanford Site during the Manhattan Project was to prevent others from learning how to create an atomic bomb by "provid[ing] the secrecy and protection necessary to prevent all possible espionage, sabotage, damage, interference or other harmful effects which might endanger the successful completion of the Program" (UPOA 1977b, bk. 1, vol. 14, p.1.2). Throughout the Cold War, the purpose of the Safeguards and Security Program (as it was then called then) was to ensure "that appropriate protection is provided for government property and equipment to prevent the loss of classified information, and to prevent the loss, theft, diversion, or sabotage of SNM [special nuclear materials]" (DOE 1990 p.8-5).

As the first contractor at the Hanford Site, E.I. Du Pont de Nemours and Company (Du Pont) conceived and developed the security measures for the Manhattan Project. These measures were modified over the years in reaction to world events, increased knowledge of security, and subcontracting changes in the 1960s. The Atomic Energy Acts of 1946 and 1954, for instance, were pivotal in determining how information and materials were to be handled in relation to national security. In the background of all security measures was the need for secrecy during the Manhattan Project and the need to guard classified information during the Cold War.

The main security organization at the Hanford Site has been the Hanford Site Patrol from 1943 to the present with support from other organizations. Since the earliest days of the Manhattan Project, the Hanford Site Patrol has been the watchdog monitoring access to the Hanford Site. The major organizations teaming with the Hanford Site Patrol were the U.S. Army Corps of Engineers (1943-1947) and the Federal Bureau of Investigation (1943-present).

Security regulations established during the Manhattan Project contributed to the Hanford Site culture. Employees were surrounded daily in one way or another by aspects of security. They passed security slogan billboards on their daily ride to work. They presented their security identification badges to Hanford Site Patrol officers wherever they went. They had their personal belongings searched. They knew nothing about their work except what was necessary for their immediate assignment. They had to follow strict regulations regarding classified information and its accessibility. They had to be careful about what they discussed at work and at home. This was the environment employees accepted as part of life while living in a government town and working at the Hanford Site, and it continued to be part of the Hanford culture throughout the Cold War. With the change in mission, regulations have decreased, resulting in security becoming a less restrictive aspect of working at the Hanford Site.

Although less restrictive from an employee viewpoint, many of the security measures of the Cold War are still in practice today. Because numerous sources dealing with security issues at the Hanford Site are still classified, information about

CONTENTS

Divider Page Photo: Guard Tower, ca 1945 (Du Pont 1945a)

site security during the Cold War could not be fully developed in this section. For instance, it was very difficult to obtain any information concerning the practices of the Federal Bureau of Investigation at the Hanford Site.

HISTORICAL ELASTICITY

The security measures undertaken were numerous: background investigations of new employees, security education and information programs, classified areas and documents, security safeguards, and emergency preparedness. Each of these measures has required elasticity, stretching to support increased security needs and contracting when those needs no longer existed. Increased security needs that stretched the system were mainly caused by unexpected events or changes in the Hanford Site mission, funding, and international affairs.

One such unexpected event that stretched the security system during the Manhattan Project occurred when Japanese balloons, carrying incendiary and fragmentation bombs, appeared over the Hanford Site in 1945. Hanford security was put on alert and reinforcements from the Pasco Naval base were dispatched as backup to help intercept the balloons. Although security was not breached, the event did impact plant operations. One balloon, entangled with a transmission line of the Grand Coulee and Bonneville generating stations, caused a power surge on the lines connected to the reactors. Power to the reactors was knocked out only briefly because the automatic backup system temporarily restored power (Jones 1985, Sanger 1995).

Elasticity again came into play at the end of World War II. Secrecy was no longer as important as protecting classified information and nuclear materials. Increased attention was given to "strengthen[ing] security control from a physical and internal standpoint. This was accomplished gradually, and with a minimum of publicity by redistribution and augmentation of the exterior guard forces, higher clearance standards for old and new employees, a strict publicity control, and other similar measures based upon the principle of safeguarding that which the Project is known to possess" (UPOA 1977b, bk. 4, vol. 6, p. 11.1).

The elasticity of the security system was especially flexed when the Korean War started in 1950. Air patrolling increased to 24 hours per day (Rokkan 1997), and military troops were brought to the Hanford Site for protection against the threat of attack on the facilities. The U.S. Army established Camp Hanford and a network of anti-aircraft artillery and Nike missile sites. A description of these installations can be found in Section 9, Military Operations.

With the end of the Korean War, some security measures were relaxed (Ogletree 1977, Rokkan 1997). Routine air patrolling ended in 1964. The Hanford Site's main roads were opened for family and guest tours in 1967. The double badge system was replaced with a single badge indicating all pertinent access and clearance authorizations. Personnel and vehicle searches became less common. The Columbia River was opened to the public from the 300 Area to Ringold. However, the channel and the shoreline between the 300 Area and the island east of the area remained off limits. M-8 light-armored cars, vehicle rovers, and manned fence towers were no longer used.

The threat of international terrorism in the aftermath of the 1972 Olympic Games in Munich again triggered increased security. During this time, the Hanford Site Patrol increased its vigilance over all the major production areas. Officers patrolled the 300 and 400 Areas 24 hours a day and increased surveillance at the 100 Area river pump houses and the 200 Area tank farms. Detex clocks were installed at the Fast Flux Test Facility. Patrol officers used them to clock in at each of the locations in the Fast Flux Test Facility where they patrolled. "Razor wire fencing was installed, and several more patrol officers were assigned to guard the nuclear material (Ogletree 1977). The Hanford Site Patrol organized a Tactical Response Team in 1977 to better combat terrorism. Members of the Tactical Response Team were specially trained patrol officers with additional skills in special weapons, counter-terrorist tactics, crowd control, aerial surveillance, and forced entry (Thielman 1995).

During the late 1970s and into the 1980s, security measures were increased and upgraded at the PUREX Plant (although measures were appropriately reduced while it was not operating between 1972-1983), the Plutonium Finishing Plant, and the 100 N Reactor, which were the main operating plants at the time. Extra security fences with detection and alarm systems were constructed around the operation areas, and all personal possessions were searched at security checkpoints when anyone entered and left these areas. At the N Reactor, for example, several structures were added, such as rooftop guardhouses, river guard towers (see the Historic Property Inventory Form for the 181-NA Pump House Guard Tower in Appendix B on the Internet), and pill boxes in the hallways of the reactor building that armed security guards used for cover to watch for suspicious activity (DOE 1997a) (see Figure 2-8.1).

Figure 2-8.1. *Pill Box Used for Security Surveillance in the Hallway of the 100-N Reactor, 1986*

While certain measures were taken to increase security against terrorism, other security measures were relaxed in the 1970s and 1980s (Cadd 1998, DOE 1994c, Thielman 1995, Rokkan 1997). The entire stretch of the Columbia River through the Hanford Site was officially opened to the public in 1979 after being restricted since 1943. Air restriction over Hanford was modified to allow for more open air space. Magnetic cards were used to access buildings, which reduced the need for security personnel. The number of protected areas requiring high security was reduced when the 209-E Critical Mass Laboratory was closed and N Reactor and PUREX were placed on cold stand-by status in 1988 and 1990, respectively. The Hanford Site implemented the island concept in which only three areas were designated as highly protected areas: the Plutonium Finishing Plant, the Fast Flux Text Facility, and the K Basins.

Another area calling for elasticity concerned trespassers who, either innocently or maliciously, entered the Hanford Site. When innocent motorists, sightseers, photographers, and boaters unknowingly wandered onto the Hanford Site, the Hanford Patrol simply turned them around. Malicious trespassing was rare. These infractions included escaped criminals crashing through a barricade, poachers entering the Hanford Site from the Columbia River or on Rattlesnake Mountain to hunt, and pot hunters trespassing on the shoreline and islands to search for archaeological artifacts. In malicious trespassing cases, the Benton County Sheriff's Office or the Hanford Site Patrol issued citations or made arrests.

In the mid-1980s, on the day set aside to commemorate the bombing of Japan, demonstrators protested Hanford's continuing role in the creation of nuclear defense weapons. Some protestors were arrested for trespassing and inhibiting traffic (Curtis 1999). This demonstration, followed by the demise of the Cold War, signaled the change in the Hanford Site mission. With the mission changing from nuclear material production to environmental cleanup, it became apparent that the Safeguards and Security Program would need to be reevaluated. In May 1990, a team of contractor staff evaluated security and "discuss[ed] the proposed transition from a defense security posture to an industrial posture, with additional protection for remaining nuclear materials and classified information" and initiated the Security Transition Program at the Hanford Site (Thielman 1995, p.9).

This team gave way to a multi-contractor team led by the U.S. Department of Energy (DOE) that identified and implemented areas of change, such as consolidating materials in conjunction with consolidated security areas, reducing clearances, simplifying access to the Hanford Site, and using automation and technological security devices where possible (DOE 1994c, Thielman 1995). DOE saved time and money by drastically reducing the number of high level security clearances called Q and L (those already issued and to be issued) beginning July 1990. Other changes resulting from this effort were the closing of the 300 Area protected areas, consolidating materials requiring security measures, closing of most badge houses, and turning law enforcement over to the Benton County Sheriff's office (DOE 1994c, Thielman 1995) – all reflecting the elasticity that has been the mainstay of security at the Hanford Site.

ROLE OF SECRECY

The story of security at the Hanford Site involves the concept of secrecy, especially during the Manhattan Project when the United States was trying to develop nuclear weapons to end World War II. The shroud of secrecy played a major role in the daily lives of Hanford Site employees. Hanford Site employees were told little to nothing about the project on which they were working. They were told only that their work at the Hanford Site was going to help the United States win the war. Thus, secrecy and efficiency in their work was essential. Even the Selective Service, who deferred those working at the Hanford Site from military service, knew only that the Manhattan Project was related to a wartime effort (Van Arsdol 1992).

> ***"If the [Manhattan] Project was to be successful it was necessary to afford it maximum protection and to cloak it in the utmost secrecy. There were strong indications that the German government had already embarked upon a similar program. It was apparent that the successful completion of the American Project prior to the German might decide the course of the war and be a dominant factor in the post-war peace...It was to be expected that once the enemies of the United States learned of the Project, they would try every means within their power to sabotage it. The absolute necessity, then, of keeping all classified information concerning the Project from the enemy and of taking every precaution to secure it from attempts at sabotage was obvious."*** **- UPOA 1977b, bk. 1, vol. 14, p.1.1**

To keep employee spirits high while not knowing the Hanford Site's purpose, Colonel Franklin T. Matthias, one of Lieutenant General Groves' top assistants, attempted to boost morale in a speech made in 1944 to Hanford Site workers. One worker recalls:

> *"Thousands came. He wanted to get it across to everybody how important it was. Some people didn't seem as dedicated as they could have been. He made a pretty good speech. It gave us all a shot in the arm. When we left there we were ready to build a plant. He did say that it was impossible to tell us what we were doing because the enemy would like to know. We were not allowed, he said, to discuss it with each other, just like our foreman had told us. But he said I can tell you this much, that it's important and the enemy, Germany, is attempting to do the same thing we are, to build a plant like this. And whoever gets there first will win the war. And that was enough said. We didn't ask any further questions."* - Sanger 1995, p. 91

> **Before the mission of the Manhattan Project was publicly released, *"workers largely depended upon the word of the Army that secrecy was important, and the explanation had to be clearly made to them that, although they had no knowledge of the actual purpose of the [Manhattan] Project, they nevertheless had important information, through their knowledge of types of equipment used, materials handled, size of buildings, and number of people on the job. Grounds for protecting information were largely patriotism, loyalty to the fighting men, and the reasoning that the less publicity given the Project, the more difficult it would be for the enemy to acquire information about it and also, the greater would be the element of surprise."* - UPOA 1977b, bk. 1, vol. 14, p. 6.13**

Although prohibited from doing so, employees and Richland Village residents speculated about what was being produced at the Hanford Site since so little information was available to them. Some humorous theories included Pepsi-Cola, Kleenex, clothespins, fourth-term Roosevelt campaign buttons, and sandpaper – relating to the local dust storms and the abundance of sand. Other more serious theories included poison or nerve gas and solid rocket fuel to compete with the Germans (Sanger 1995, Loeb 1982). "All rumors regarding the purpose of the [Manhattan] Project, health hazards, or other rumors which might create unrest among the personnel were promptly traced to their source and disciplinary action was taken in the form of termination or reprimand" (UPOA 1977b, bk. 4, vol. 1, p. 6.2).

It has been reported that even the U.S. Army Corps of Engineers (as Matthias has said) "rigged up a cover story right at the beginning... We ended up calling Hanford a place to make RDX [a new explosive]. Nobody questioned it. Du Pont was known as an explosives maker" (Sanger 1995, p. 79). The objective was to try to keep people from speculating about the Hanford Site.

DECLARATION OF SECRECY

To keep people from talking, each employee received a security orientation in which they signed a Declaration of Secrecy form. They signed this form again when terminating their employment. In signing the form, the employee took a pledge of faith and allegiance to the United States, agreed not to disclose any classified information or materials to unauthorized individuals or to misuse the material, and understood that violation of the National Espionage Act or the Federal Sabotage Act was punishable by up to 10 years in prison or up to $10,000 in fines (UPOA 1977b). Information not to be disclosed to others included "references to the mission of the [Manhattan] Project, the size, scope, or nature of its operations, general problems worked on, materials and equipment used in restricted areas, connection and relationship to each other and to the [Manhattan Engineering] District of various projects and laboratories throughout the country, and professions or former connections of principal technical personnel" (UPOA 1977b, bk. 1, vol. 14, p. 6.12). Since1988, Hanford Site employees obtaining security clearances agree to and sign the Classified Information Nondisclosure Agreement, Standard Form 312, which serves the same general purpose as the original form.

COMPARTMENTALIZATION

The main way to stop information leaks at the Hanford Site was to limit the knowledge of each employee. This was called "compartmentalization of knowledge." *A Guide to Manhattan Project: The Official History and Documents* (UPOA 1977b, bk. 1, vol. 14, app. B-7, p. 4) states that "to minimize risk or compromise, employees of the [Manhattan] Project shall be organized into small working groups or teams so far as possible, each working on its own phase of the job and not being permitted to inspect or discuss the work being done by others." Therefore, the amount of information one person or group of persons obtained was limited. Employees were directed to focus only on their specific task and refrain from asking questions and unnecessarily discussing their jobs with fellow employees and even spouses or friends. Only a select group of employees knew the goal of the Manhattan Project (Loeb 1982, p. 26).

> ***"Compartmentalization of knowledge, to me, was the very heart of security. My rule was simple and not capable of misinterpretation—each man should know everything he needed to know to do his job and nothing else. Adherence to this rule not only provided an adequate measure of security, but it greatly improved over-all efficiency by making our people stick to their knitting. And it made it quite clear to all concerned that the project existed to produce a specific end product – not to enable individuals to satisfy their curiosity and to increase their scientific knowledge."*** **- Groves 1983, p. 140**

Compartmentalization was practiced in other areas of the Manhattan Project as well. The Corps required "equipment orders to commercial firms specify that an item not be manufactured and assembled at the same location. And when the production plants reached the point of start-up operations, plant managers received instructions to split up orders for raw materials among a number of suppliers so that the purpose for which they were being used could not be readily ascertained" (Jones 1985, p.269). In addition, blueprints were "broken down and distributed in such a way as to reveal as little as possible to any one individual about the overall character of the project" (Jones 1985, p.269). General construction specifications had to be relayed to the workers through hand-drawn sketches since only specific people had clearance to view official drawings (Sanger 1995).

The concept of compartmentalization, later commonly called "the need to know," continued through the end of the Manhattan Project and into the Cold War era. However, during the Cold War, compartmentalization was not focused on the creation of the bomb (by then a known fact), but on general national security and business sensitive information and activities.

CRITICS OF SECRECY

Some found the strict security regulations to be oppressive and excessive during the Manhattan Project. Because background investigations had to be so comprehensive they "result[ed] in delays in employment and hinder[ed] the already overworked recruitment drive" (Hageman 1945, p.73). "These security requirements, probably more than any other single subject, were a determining factor affecting practically every phase of the work performed under contracts supervised by the Manhattan District of the Corps of Engineers, and in many instances delayed, or increased the difficulty of, the work" (Du Pont 1945c, p.36).

Compartmentalization was a particular nuisance to scientists. Many felt that the restriction prohibiting them from exchanging data and discussing the Manhattan Project kept the technology from developing faster, thus causing "extended delays in achievement of scientific and technical objectives of the program". This was the view of Leo Szilard, a Manhattan Project scientist. He stated in 1946, "compartmentalization of information was the cause for failure to realize that light uranium [uranium-235] might be produced in quantities sufficient to make atomic bombs. We could have had it eighteen months earlier. We did not put two and two together because the two two's [sic] were in a different compartment" (Jones 1985, p.270). Others have commented that compartmentalization:

- *"puts you in the position of trying to do an extremely difficult job with three hands tied behind your back"* – 1943, Edward Condon, a physicist (Jones 1985, p. 271)
- *"resulted in a general loss of efficiency, required more supervision to obtain proper coordination, and, despite all attempts to counteract it, lowered morale and caused delays"* - Du Pont 1945c, p.38
- *"made it difficult to adequately impress workers with the importance and urgency of the project"* - Hageman 1945, p. 73
- resulted in *"bearing the seeds of unfortunate mistakes, duplication of research and ridiculous exaggerations. Inevitably, in some cases, scientists might waste time and effort solving a problem that had already been solved in another laboratory. They did not know it, however, because they had not been permitted to exchange information"* - Groueff 1967, p.43

Although, to take another look at it, Paul Loeb points out in his book *Nuclear Culture: Living and Working in the World's Largest Atomic Complex*:

> *"But if the security rules were frustrating, they removed from most Hanford workers the burden of judging the wisdom of what they were creating. Simply assuming that their efforts were necessary to win the war, they could immerse themselves in details. No one considered until much later (after the Hanford mission was revealed and the war was won) what other choices might have been possible. They knew only that their labors had brought to a close the war."* - Loeb 1982, p.26

BREACHES IN SECRECY

Although no security breaches are known to have occurred at the Hanford Site, the need for security intensified when it was thought security had been breached at the Radiation Laboratory of the University of California, Berkeley (Jones 1985, p. 263). Secret information was allegedly leaked from the Radiation Laboratory by Communist sympathizers working there in 1942. This activity seemed to have been funded by a "high official in the Soviet embassy in Washington [D.C.]" (Jones 1985, p. 264). In 1943, counter-intelligence agents discovered this breach in security. Employees found to be connected to the leaks were discharged or transferred to non-sensitive assignments. The counter-intelligence agents discovered and ended continued Soviet Union espionage activities at the Radiation Laboratory, Metallurgical Laboratory, and the Los Alamos Laboratory but not before crucial secret information about the reactor process, chemicals, and technical knowledge of the bomb reportedly had been leaked to Soviet agents (Jones 1985).

No specific espionage cases were identified at the Hanford Site. However, the Soviet Union's first plutonium production reactor built in 1948, called F-1, appears to be almost an exact copy of the Hanford Site's 305 Building Test Reactor built in 1943. Measurements of both were very similar in relation to power, uranium rod spacing, rod diameter, and amount of uranium in the reactors. Both reactors were constructed to evaluate the purity of graphite and uranium. The fact that the design details of the Hanford Site reactor were not released until after 1955, at least 7 years after the Soviet Union completed its reactor, adds to the notion of espionage although this remains unsubstantiated (Libby 1979, Rhodes 1995).

> ***"Soviet agents, masking as diplomatic and consular officials, turned to members of the Communist Party of the United States and to party sympathizers for assistance in penetrating American wartime institutions and projects. The Russians, making the plea that the Americans government was withholding important information and thus unnecessarily delaying Allied victory, recruited many native Communists and fellow-travelers to assist them in obtaining vital secretes about wartime activities."*** **- Jones 1985, p. 263**

SITE SECURITY ORGANIZATIONS

Overall management of security at the Hanford Site has always rested with the federal agency assigned responsibility for operating the site. The federal agencies managing the Hanford Site have been the U.S. Army Corps of Engineers (1942-1946), the Atomic Energy Commission (1947-1974), the Energy Research and Development Administration (1975-1976), and the U.S. Department of Energy (1977-present). Key players in the development and operation of security at the Hanford Site have been the subcontractors.

DEVELOPMENT OF THE SECURITY PROGRAM

In 1942, the Corps selected Du Pont as the prime contractor to create and operate the Hanford Site. In this role, Du Pont assumed the responsibility for protection and security of the Hanford Site. The first step Du Pont took in preparation for this responsibility was to gather a team from its Engineering, Explosives, and Patrol departments in Wilmington, Delaware where they had experience working at other war material production sites (Du Pont 1944-1946). This team developed a Plant Protection Program to satisfy Du Pont's security responsibilities and make "provisions for the establishment and operation of (a) an adequate plant patrol force, (b) effective personnel investigation measures, and (c) positive security regulations and practices for the safeguarding of military information" (Du Pont 1944-1946, sec.1/ Aug. 10, 1945, p.2). Colonel Matthias approved the Plant Protection Program in March 1943.

The overall security goals of the Plant Protection Program during the Manhattan Project were to "keep the Germans from learning anything about our efforts or our technical and scientific advancements, ensure a complete surprise when the bomb was first used in combat, [and] ...keep the Russians from learning of our discoveries and the details of our designs and processes" (Groves 1983, p.141).

In general terms, Du Pont's responsibilities were to "establish, maintain, and operate a protection program for the purpose of protecting the personnel and all property at [the Hanford Site] connected with the design, construction, and operation work by the [Du Pont] Company" (Du Pont 1944-1946, sec. 1/Aug. 10, 1945, Exhibit E, p.1). Details outlined in the Plant Protection Program included (Du Pont 1944-1946, sec. 1/Aug. 10, 1945, Exhibit E, p. 6):

- Establishing a qualified plant patrol force and ensuring patrol officers are provided appropriate training
- Operating effective administrative control and organization of the plant patrol
- Obtaining, constructing, and maintaining any equipment necessary for patrol officers to carry out duties (weapons, fences, towers, guard stations, administration buildings, communication links)
- Establishing patrol post orders and instructions

- Preventing any unauthorized removal of Manhattan Project information or equipment from the Hanford Site, any unauthorized access/entry of persons or vehicles to the site and specific site areas or buildings, and prohibited items from being brought on site
- Creating and operating a program to conduct background investigations on all employees, including fingerprints, photographs, medical exams, contacting personal references, obtaining Personnel Security Questionnaires, Data Cards, Espionage Act and Declaration of Citizenship papers on employees
- Preventing the inappropriate disclosure of Manhattan Project information by issuing safeguarding instructions and advice regarding classified areas and property (documents, equipment) and by providing appropriate facilities and storage for use in safeguarding information
- Creating blackout and air raid facilities

Du Pont had the "responsibility of protecting the physical project facilities against theft, sabotage, and espionage, and for assisting county, state and federal authorities in the enforcement of laws on the plant and in Richland Village" (Du Pont 1944-1946, sec. 3/Oct. 5, 1945, p.1). To accomplish this, Du Pont created a security force, the Hanford Site Patrol, which began its duties at the Hanford Site in 1943 when construction began (DOE 1997b, Du Pont 1944-1946). As construction was completed and operations began in 1944, the Hanford Site Patrol came under the new Protection Department, which was developed as an expansion of the Explosives Department (Du Pont 1944-1946). Around this same time, a Military Police Detachment was assigned to the site, and the Federal Bureau of Investigation also established a local branch office, which played a supportive role.

Although at the end of World War II the Atomic Energy Commission replaced the Corps and the General Electric Company became the prime site contractor, the Hanford Site Patrol remained as the main security organization. The U.S. Army's Camp Hanford enhanced site security from 1951-1960. Camp Hanford was an anti-aircraft artillery and Nike missile complex. See Section 9, Military Operations, for a description of this installation.

In 1965, the operation of the Hanford Site was diversified among several contractors (see Section 1, Construction History, for more information). Individual contractors developed their own accountability guidelines, including security. In general, the various contractors followed the Atomic Energy Commission's master guidelines. However, in some cases, contractor's interpretations had to be ironed out (Rokkan 1997).

Currently, the Safeguards and Security Program (as it is now called) basically still follows the original objectives of physically protecting facilities, property, and equipment; protecting and controlling special nuclear materials; preventing the unauthorized release of classified information; and protecting employees and the public (DOE 1990).

HANFORD SITE PATROL

The Hanford Site Patrol was established in mid-1943 to "prevent the loss, damage or destruction of Hanford Engineer Works' [Hanford Site's] property through theft or sabotage, protect the lives of the employees and residents of the [Manhattan] Project, enforce the laws of the United States and the State of Washington" (Du Pont 1945a, p. 189). By July 1944, the Hanford Site Patrol consisted of approximately 1300 officers whose responsibilities included patrolling construction areas and verifying clearances for access to the Hanford Site (Du Pont 1945a).

The responsibility to protect the Hanford Site was decentralized because of the distance between the various operations. "Operational direction [was] broken down by areas; each area reporting to a central headquarters. [This system] provided for the establishment of self-contained patrol forces in the 100, 200, and 300 Areas, the Hanford Camp [Hanford construction camp], and Richland" (Du Pont 1945a, p. 188) and eventually the 400 Area.

The types of infractions handled by the Hanford Site Patrol included insecure storage of classified materials, unauthorized entry, presence of prohibited items, improper use of keys, and, later, improper handling of classified materials in electronic communications. As a result of such activities, approximately 12 to 24 citations were issued per year. This number decreased as the number of employees and amount of classified material decreased. The disciplinary actions for these offenses depended on the severity and details of the infraction (Curtis 1999).

Patrol Headquarters buildings for which Historic Property Inventory Forms have been prepared are 1720-K for the 100 Areas, 2721-E for the 200 Areas, 3701-D and 3707-D for the 300 Area, and 4790 for the 400 Area (see the Historic Property Inventory Forms for these buildings in Appendix B on the Internet). The Patrol Headquarters building maintained communication through the Patrol Operations Center, which coordinated and relayed secure information in case of an emergency or security breach. The 3701-D Building also contained an Emergency Control Center in its basement for several years that could be isolated from outside air and, thereby, protect the emergency team inside from any airborne materials while they dealt with an emergency. The Patrol Operations Center has been located in the 200 East Area since 1982. It was previously located in the basement of the Federal Building in Richland. The Patrol Operations Center functioned as the main security operations communication headquarters for the Hanford Site. It was the main emergency center on the Hanford Site, as well as central dispatch, responsible for initiating emergency actions.

Patrol officers "were also sworn in as Auxiliary [Military Police] and, in emergency, would be subject to military control" (UPOA 1977b, bk. 4, vol. 6, p. 11.4). Hanford Site Patrol officers carried the highest security clearance on the Hanford Site (and still do today), the Q clearance, and were also held to the same requirement as other Hanford employees not to discuss their work or the Hanford Site with anyone (Beardsley 1998).

In 1952, the total number of Hanford Site Patrol officers had decreased to 595. It was again reduced to 149 officers by 1973. Reasons for the decrease were reactor and guard post closures, reduction of services, and budget constraints (Ogletree 1977). During the next 15 years, however, from 1973-1988, staffing increased because of continued tensions during the Cold War. By 1988, the number of patrol officers had increased to about 450 but was again reduced to less than 400 by 1990 after the threat of the Cold War had subsided (Walton 1999).

As of 1990, a force of Hanford Site Patrol officers was on duty at all times, 7 days a week, 24 hours per day operating as the primary security agency on the Hanford Site. By this time, the role of the Hanford Site Patrol was to provide protection service for the DOE and Site contractors by protecting certain nuclear materials, classified information, facilities, property, and personnel and enforcing federal, state, and local laws "through criminal investigation, traffic enforcement, access control, property inspections, and response/investigation of security incidents, and communication of emergency situations on and off the Hanford Site to Hanford personnel" (DOE 1990, p. 8-1).

Resources of the Hanford Site Patrol in 1990 included sufficient security staff for routine duties, an airborne Special Response Team, a criminal investigation unit, a canine (K-9) unit trained to detect explosives and drugs, a hostage-negotiation team, the Patrol Training Academy, standard and specialized alarm systems and weapons, and a vehicle fleet including sedans, four-wheel drives, helicopters, armored personnel carriers, and boats (DOE 1990).

Benton County Sheriff Deputy Status

Although law enforcement was under the jurisdiction of Benton County, the Benton County Sheriff deputized the Hanford Site Patrol. This status gave the patrol officers authority to administer state laws across the Hanford Site, an area too large for the county to cover adequately (Du Pont 1945a).

In their role as specially assigned Benton County deputies, the Hanford Site Patrol had to deal with drunk and disorderly persons, gambling, prostitution, fighting, homicides, and robberies (see Table 2-8.1). "During the construction period a clerk of the Kennewick, Washington, Justice of the Peace Court was stationed at

> ***"Because of the comparative magnitude of the Hanford Engineer Works [Hanford Site] and the number of persons employed during the construction period, the problem of law enforcement and jurisdiction was much greater than that encountered at the University of Chicago or at Clinton Laboratories. After conferences between State and Federal officials it was decided that concurrent jurisdiction [over the Hanford Site] would lead to confusion so it was decided to leave the administration and enforcement of State laws to the State and/or its subdivisions so long as such laws did not unduly interfere with the protection of the Project. Most of the Project lies within Benton County and as the county was not prepared to assume the entire burden the Benton County Sheriff deputized the Prime Contractor's [Du Pont] patrol officers."*** **- UPOA 1977b, bk. 4, vol. 1, pp. 7.1-7.2**

Table 2-8.1. Summary of Incidents the Hanford Site Patrol Dealt with in Their Role as Specially Assigned Benton County Deputies between March 15, 1943 and August 28, 1944

Incident	No.	Incident	No.	Incident	No.
Intoxication	3,156	Gambling	161	Sex Cases	69
Burglary	1,124	Trespassing	144	Draft Evasion	50
Petty Larceny	593	Drunken Driving	123	Un-Americanism	44
Assaults	522	Missing Persons	108	Natural Death	25
Public Nuisance	478	Auto Theft	105	Mob Demonstration	19
Grand Larceny	450	Juvenile Delinquency	97	Accidental Death	19
Wanted Elsewhere	217	Mental Cases	89	Violent Death	5
Vagrancy	197	Bootlegging	88	Suicide	4
Robbery	177				

Hanford and a jail was established as a branch of the Benton County Jail. From May 1943 to August 1944 a total of $72,986.24 was collected by Benton County as bail forfeitures and fines on the Project" (UPOA 1977b, bk. 4, vol. 1, p. 7.3). Although some lawbreakers were terminated, in general it was more important to return the employees to their jobs and keep them working on the project at hand (Sanger 1995). "Un-Americanism" was a catch-all category, which included possession of two-way radios and criticism or protest against President Roosevelt (Bubenzer in Sanger 1995, p. 94).

Patrol Officers Were Required to Maintain a Proper Appearance

Hanford Site patrol officers were trained not only on security enforcement matters but also on what was expected both on and off duty in proper and acceptable appearance, courtesy, etiquette, discipline, and speech. For example, below are some of the guidelines discussed in the Hanford Engineer Works Patrol Training Manual (GE 1958b, sect. VI, pp. 7-8):

- A lack of interest in what is being told one is a breach of courtesy. Do not allow your attention to wander while another person is speaking.
- Be strictly punctual at all engagements, whether business or social.
- Do not lean against a building or assume a loafing attitude while engaged in conversation, or otherwise.
- The Patrolman should refrain from making 'wise cracks' and sharp retorts regardless of the occasion.
- A pleasant facial expression is an asset. It is a smile (not a grin) that has been said to speak all languages.
- When you are in uniform, whether you are technically on or off duty, to the general public you are on duty. Be exemplary in your conduct and edifying in your conversation.
- Every Patrolman owes it to himself and to the profession as a whole to be scrupulously clean in person and dress.
- Cleanliness of person calls for frequent visits to the bathtub and showers; for a clean-shaven face; for the hair properly cut, clean, and well brushed; and for clean hands and fingernails. A sprinkling of dandruff on the collar and shoulders of a coat suggests slovenliness.
- Proper care of the teeth is essential not only because of appearance (this alone would compensate for the trouble), but for the preservation of health. Every precaution should be taken to guard against halitosis.
- It is an inexcusable breach of etiquette to sneeze or cough in the presence of others without first covering the mouth with the handkerchief.

The Hanford Site Patrol also handled traffic infractions. "The Highways on the Project could not be classed as public highways so the Commissioners of Benton County enacted a set of traffic regulations to be enforced by the special deputies on the Project. Other State laws pertaining to public health, sanitation, hunting, and fishing are administered and enforced throughout the area" (UPOA 1977b, bk. 4, vol. 1, pp. 7.1-7.2). The Washington State Patrol held the primary traffic enforcement jurisdiction for State Highway 240 in the southwestern part of the Hanford Site, and the Hanford Patrol, as Benton County deputies, functioned as a secondary enforcement.

In the early 1990s, security programs were cut back due to budget constraints and the relaxation of certain aspects of security at the Hanford Site. At this time, the Benton County Sheriff deputy status of the Hanford Site Patrol was terminated, but the federal status was maintained (Cameron 1998). In 1993, the Benton County Sheriffs Office took over the secondary traffic enforcement on Highway 240 and the primary traffic enforcement of roads on the Hanford Site.

Patrol Training Program

During the Manhattan Project, workers came with varied degrees of security enforcement knowledge. A detailed training program was organized in 1944 including "handling disorderly, violent, or injured persons," firearms and weapons, the Hanford Site Patrol systems and methods, and understanding the importance of the Hanford Site Patrol (Du Pont 1945a, p.195). In addition, workers were required to attend classes on such subjects as courtesy and discipline, Hanford Site geography, physical training, fire fighting, writing reports, and radio operation (Du Pont 1945a).

Figure 2-8.2. *Hanford Patrol Officers at Shooting Practice*

Throughout the Manhattan Project and Cold War era, patrol officers had hands-on training. They were trained in the use of 38-caliber revolvers, 45-caliber submachine guns, 30- and 50-caliber machine guns, 37-mm gas grenades and guns, smoke bombs, tear-gas guns, carbine, riot guns and other special weapons (Du Pont 1945c). A shooting range near the southern end of Gable Mountain was used for shooting practice (see Figure 2-8.2). Other firing ranges were located along Horn Rapids Road. Patrol officers were also trained in the operation of M-8 light armored tanks (see Figure 2-8.3) equipped with a turret 37-mm gun, a 30-caliber machine gun, and a 50-caliber machine gun (GE 1955b). All patrol personnel were eventually trained to work around and within radiation areas. They attended an initial 3-day course with an 8-hour refresher course every other year to obtain and keep their RAD (radiation absorbed dose) Worker Training current. Radiation training was necessary in case security situations on the Hanford Site required patrol officers to enter radioactive zones. This training is still required today, and Hanford Patrol training instructors have become certified as Radiation Worker Training instructors (Cameron 1998).

Figure 2-8.3. *Hanford Patrol Practice Exercise with Armored Tank Around 1949-1950*

Classes were initially taught at the training school building in the Hanford Site construction camp. Eventually, a training academy was established in buildings within the 1100 Area. In 1980, a new Patrol Training Academy complex was built on Horn Rapids Road where the firing ranges were located (Du Pont 1945a). This new Hanford Patrol Training Academy provided a range of training environments including classrooms, library resources, practice shoot houses, an exercise gym, and an obstacle course. In these environments, patrol officers have been taught defense and control tactics, crisis negotiation, safety and first aid, communication systems, and use of special equipment.

In 1977, the Hanford Site Patrol formed the first offensive, anti-terrorist team, called the Tactical Response Team, specifically trained in special weapons and tactics at a DOE facility. This team maintained roving patrol duties and was prepared to react to any emergency situations on the Hanford Site. The members of the Tactical Response Team were specially selected and underwent advanced and specialized training that included use of special advanced weapons, building entry and control, suspect apprehension, use of force, hostage rescue, crowd control, and physical fitness (see Figure 2-8.4) (Rokkan 1997, Cameron 1998). The Tactical Response Team was also trained in the operation of Fast Attack Vehicles (similar to dune buggies) and V-150 armored personnel carriers (DOE 1985). In the mid-1980s, a Special Response Team was developed as a supplement to the security provided by the Hanford Patrol and Tactical Response Team. The Special Response Team was centrally located in the 200 East Area and could quickly respond to any security need by using helicopters. The two teams were eventually combined.

Figure 2-8.4. *Special Response Team Practicing Building Entry, 1994*

COUNTER-INTELLIGENCE AND MILITARY POLICE

For the first year "internal security in the MED [Manhattan Engineer District] was supervised by War Department Counter Intelligence and was thus a responsibility of the Army The War Department's area of responsibility was to include all its civilian employees, as well as all civilians on military reservations or under military control" (Groves 1983, p.138).

In late 1943, changes in philosophy of the counter-intelligence operations of the War Department "made it impossible to rely any longer on the formerly very satisfactory centralized organization" (Groves 1983, p.139). The Corps was forced to create its own counter-intelligence operation.

The counter-intelligence operation maintained a liaison with the Du Pont security organization and "outside agencies: the Federal Bureau of Investigation, the Office of Naval Intelligence, Army Intelligence and Counter-Intelligence, State Police and, in most cases, local police and sheriff's offices in Washington, Idaho, Oregon, Montana, and Wyoming" with relation to prevention of sabotage and espionage of the Hanford Site (UPOA 1977b, bk. 4, vol. 6, p. 11.2)

During the Manhattan Project, typical counter-intelligence activities aimed at preventing sabotage and espionage toward the Hanford Site included a variety of methods such as monitoring conversations, censoring the media, conducting personal investigations, and editing outgoing mail. The focus was on safeguarding activities and personnel clearances and conducting investigations related to military but not civilian operations. Du Pont's security organization handled civilian operations. However, if Du Pont discovered questionable personnel, the case would be turned over to the Corps' counter-intelligence agents for investigation. The Federal Bureau of Investigation handled any espionage or sabotage cases outside the Hanford Site but could be a joint effort if the case appeared to be focused at Hanford (UPOA 1977b). For a detailed historical account of overall security and counter-intelligence at the Hanford Site, see *A Guide to Manhattan Project: Official History and Documents*, Book I, Volume 14, "Intelligence and Security" (UPOA 1977b) and *Manhattan: The Army and the Atomic Bomb*, Chapter XI, "Security" (Jones 1985).

The 744th Military Police Battalion, Company D, was assigned from Fort Lewis to the Hanford Site in July 1944 as a supplement to the Hanford Patrol. By the end of the month, this battalion was designated the Military Police Detachment No. 2. From 1943-1947, the Military Police patrolled the site perimeter and barricades set up around the Hanford Site, which at that time were the Richland, Prosser, and Yakima barricades (Ogletree 1977). They also secured the 213 Final Storage Magazine Building, the Hanford construction camp, the Midway Substation, and the Hanford Ferry and were used to guard classified shipments off site (Du Pont 1944-1946, Du Pont 1945a). In addition, two military police officers were assigned to patrol the Richland Village, mostly on weekends, to keep the peace (CREHST 1998). In

1947, the Military Police force was slowly phased out as the perimeter patrol, and their other security duties were transferred to the Hanford Site Patrol (Rokkan 1997). Eventually, the need for increased security capabilities lead to the development of Camp Hanford into which the troops from Fort Lewis were incorporated. At this time, the National Guard unit left the Hanford Site. For further information on the military at the Hanford Site, see Section 9, Military Operations.

FEDERAL BUREAU OF INVESTIGATION

The Federal Bureau of Investigation (FBI) established an office in Richland Village in the 1940s. The FBI had no criminal investigative jurisdiction at the Hanford Site. This duty was the responsibility of Benton County. Nor did the FBI maintain any security duties at the Hanford Site. However, since the Hanford Site was a federal project, the FBI received security clearances allowing them access to the entire Hanford Site and its buildings. The FBI had three main functions at the Hanford Site:

- Investigate federal crimes against the government, such as theft and fraud
- Investigate employee backgrounds for security clearance
- Investigate violations of the Atomic Energy Act, such as theft of classified materials or trespass in classified areas (Dellwo 1999, Parkhurst 1999)

Several permanently assigned agents maintained the office until 1961, when only one agent was permanently assigned. Supplemental agents were assigned during times of increased investigations. During the mid-1940s, normally 15-20 agents were present (see Figure 2-8.5). This office was part of the Washington State Resident Agency of the FBI and presumably was initially established to support the Hanford Site. However, the Richland office eventually was assigned to also cover all the southwest counties of the state in cases of violations of federal criminal statutes.

Figure 2-8.5. *Agents of the Federal Bureau of Investigation Assigned to Richland, February 10, 1948*

The Sheriff's office or the FBI, depending on the jurisdiction, was called in to handle any crimes committed on the Hanford Site. Crimes that required the involvement of the FBI most commonly included classified material and theft of government property even when the theft was minor or concerned an item that probably was just misplaced. A former FBI agent recalls that employees would take everything from tools and building materials to office supplies during the mid-Cold War period (Parkhurst 1999). An odd occurrence on the Hanford Site, as the same agent recalls, involved the increased use of office supplies during August, the month of lowest employee attendance. It was suspected that employees were taking extra office supplies home to stock up their children's school supplies before school started in September. However, petty theft such as this was a low priority during a time when the focus was to produce and protect plutonium. The attitude was that it would be more expensive for the FBI to catalog and track all the government supplies than it would be to purchase new supplies (Parkhurst 1999).

The FBI focused instead on investigating serious crimes. One unusual and unsolved case occurring during the early-mid Cold War era involved the disappearance of twelve copies of the same classified document over a weekend. The courier, who was supposed to deliver the registered documents from the 300 Area to the Federal Building on a Friday afternoon, never showed up. It turned out he was killed in an auto accident the next day in Idaho. The accident scene was searched, but the documents were never recovered. On another occasion, an employee ending employment at the Hanford Site decided to take a souvenir with him when he moved. He took a fuel element. When this was discovered, the FBI tracked down the individual and recovered the souvenir (Parkhurst 1999).

The FBI also investigated fraud against the government (misrepresentation or false facts reported to the government) and prosecuted some of the cases. Such investigation included payroll and cost misrepresentation (Parkhurst 1999).

In 1943, the FBI set up an office in Portland, Oregon, which was part of the Soviet Infiltration of the Radiation Laboratories Surveillance Program. Although not totally devoted to Hanford Site security, the program was set up because U.S. Army Intelligence in their top secret Operation Venona discovered a national security threat. The Soviet Union had implemented a major espionage endeavor directed at U.S. atomic weapons facilities (Dellwo 1999). Because the Soviets were considered allies at the time, the Portland FBI office tracked Soviets entering the country through the Portland area using undercover and secret operations. The FBI agents installed listening devices and wiretaps in the Portland area in vehicles, apartments, conference rooms, the Purchasing Commission office, and even a local communist party book store. The FBI monitored approximately 15-20 wiretaps in what was called "The Plant" where monitoring equipment was located.

The result of the monitoring was the FBI discovered that along with the Soviets coming to Portland for legitimate reasons under the Lend-Lease Program, many Soviet espionage agents also were entering the area. Some of these Soviets were trained chemists and physicists with an interest in the Hanford Site. A group of ten Soviets (some of whom were identified as having knowledge of nuclear science) requested a tour of the Columbia Basin region. Special undercover security precautions were taken to protect the secrecy of the Hanford Site. The tour vehicles were equipped with listening devices, and at least one undercover FBI agent accompanied the tour as a Columbia Basin Irrigation Project engineer. The tour was routed such that the Soviet agents were not near the Hanford Site, and when they asked what was done at the site, the reply was that the work was related to hydroelectric power. No Soviet espionage agents were ever discovered to have gained access to the Hanford Site (Dellwo 1999). For additional discussion of foreign espionage, see *Dark Sun: Making of the Hydrogen Bomb* (Rhodes 1995).

SECURITY MEASURES

Security measures were activities the Hanford Site Patrol and other organizations undertook to ensure secrecy in producing the atomic bomb during the Manhattan Project and to prevent the loss of classified information and loss, theft, diversion, or sabotage of nuclear materials during the Cold War.

PATROLLING

Figure 2-8.6. *Typical Guard Tower Around 1945*

Patrolling the Hanford Site was the most obvious of all the security measures. The foot and motor patrol routes covered the entry points along the fence and shore lines of the 100, 200, 300 and 700 Areas (GE n.d.(a)). At one point in 1945, a total of 225 patrol officers on rotation were manning towers (see Figure 2-8.6) along the 200 East Area fence lines before surveillance along these lines was done by vehicle (Sanger 1995). Patrol duties also included patrolling inside facilities and checking for unlocked filing cabinets containing classified materials (Beardsley 1998).

Initially, the Hanford Site Patrol was responsible for patrolling the perimeter of the Hanford Site. The Military Police took over that responsibility in August 1944, which included responsibility for the Midway Substation and the barricades around the perimeter. The Military Police were ordered not to hesitate to fire their weapons if an unauthorized individual tried to force through the barricades. This patrol duty also included a 24-hour-per-day patrol of the Rattlesnake Mountain area by vehicle (CREHST 1998, UPOA 1977b). In 1947, patrol of the perimeter boundary and operation of the perimeter barricades was reassigned to the Hanford Site Patrol.

Figure 2-8.7. *Hanford Boat Patrol Around 1949-1950*

The Columbia River became restricted to the public along the Hanford Site shoreline beginning February 1943, and the river remained off-limits until 1979 (Rokkan 1997). During that time, the river and shoreline were under surveillance by boat and air patrols (see Figures 2-8.7 and 2-8.8). Initially, the government took over the ferry at the Hanford town site in late 1943 and assigned its operation to Du Pont (Hanford Site Patrol) in association with the development of the River Patrol. Between 1944 and 1965, the River Patrol equipment and administration was transferred at various times between the government and Hanford Site Patrol. River patrolling was discontinued in the 1970s while the responsibility of the Hanford Site Patrol and then reinitiated in the early 1980s until Benton County took over the responsibility in the early 1990s (Du Pont 1945a, Ogletree 1977). Patrol officers used the 181-NA Pumphouse Guard Tower (see the Expanded Historic Property Inventory Form for the 181-NA Tower in Appendix B on the Internet) to monitor the river access to the 100-N Area and general area security.

Figure 2-8.8. *Hanford Air Patrol Around 1949-1950*

The Corps established a restricted air space over the Hanford Site during the Manhattan Project to prohibit unauthorized flights across the area. The Air Patrol was organized to enforce this restriction and protect the Hanford Site. The Corps operated the Air Patrol using L-5 and Cub aircraft. Later, civilian pilots working for the Atomic Energy Commission and then the U.S. Department of Energy took over the operation and covered seven different air patrol routes, each averaging 55 miles. These patrol routes supplemented the perimeter patrols (described above), and pilots could communicate any security infractions to each other. The pilots looked for and investigated suspicious activity and flew low enough to identify vehicles (Oregonian 1949). Most of their flight hours were dedicated to patrolling but also included transportation, photography, and spraying vegetation. The Air Patrol was scheduled as a daytime patrol but was on call 24 hours per day (Rokkan 1997).

When the Korean War started in 1950, the Atomic Energy Commission changed the Air Patrol to operate 24 hours per day for extra protection against possible air attacks. As a supplement, air reinforcements could be requested to dispatch fighter planes to the Hanford Site if necessary. In addition, the Atomic Energy Commission had connections with the Air Force's Othello Radar Station and the base at Moses Lake, both of which could be called to help track and intercept trespassers. The bases in Spokane, Portland, and McChord were also available for assistance. Trespassers were reported, chased, and forced to land so the incident could be investigated and possible fines issued by the Federal Aviation Administration (GE 1955b, Rokkan 1997). Routine patrolling of the Hanford Site by aircraft ended in 1964, but the

Air Patrol remained as an organization to monitor the restricted air space over the Hanford Site. The air restriction was modified in the mid-1970s to allow limited flights.

In 1983, the U.S. Department of Energy purchased two helicopters to enhance the Air Patrol's capabilities (see Figure 2-8.9). The helicopters provided aerial observation, investigation and pursuit, a 24-hour air patrol capability, aerial night vision (infrared), and quick transport of a Special Response Team to any area on Hanford Site in case of a security threat. A secondary duty of the helicopter fleet was to aid in the case of emergency or unusual occurrences outside of their main duties. Helicopters were discontinued, along with riverboat patrols, in the early 1990s (RHO 1986-1987, DOE 1985, Thielman 1995).

Figure 2-8.9. *Hanford Site Air Patrol Helicopters*

MEASURES RELATED TO SECRECY DURING THE MANHATTAN PROJECT

Secrecy was extremely important during the Manhattan Project because of an overriding fear the enemy would learn how to create the atomic bomb. Lieutenant General Groves said it was necessary to "minimize the likelihood of vital secrets falling into enemy hands" and the "free exchange of information had to be stopped, if we were to beat our opponents in the race for the first atomic bomb" (Groves 1983, p.140).

Because so many people with varied backgrounds were necessary for the Manhattan Project, it was probable that some potential spies and saboteurs were inadvertently hired. Manhattan Project officials assumed Germany, Japan and the Soviet Union "would learn of the atomic energy program and...use espionage to expand their knowledge of it and sabotage to destroy America's military advantage" (Jones 1985, p.260). Thus, the Corps bought time by using extensive counter-intelligence activities.

Conversation Monitoring

In 1943, agents from the Corps counter-intelligence organization began undercover work to identify any security breaches within the Hanford Site or the local community. The agents "occupied strategically located positions in the project offices, laboratories, and plants, set up listening posts, checked intensively into personal and other records of individuals under suspicion, and took other measures designed to solve espionage cases" (Jones 1985, p. 262). During personal espionage investigations, the Corps used special equipment, such as concealed listening and recording instruments and cameras with special telephoto lenses. Undercover agents would pose as regular members of the work force or community, such as painters, contractors, hotel clerks, tourists, electricians and even gamblers (Jones 1985). One agent had regular informants who would frequently eavesdrop on employees and residents. An agent recalls: "One story they picked up was that we [the Hanford Site] were making rockets, anti-personnel rockets. They told me who had said it. I told my office this person was starting rumors and they took care of it. The guy was young, and unmarried, and he got drafted real quick" (Sanger 1995, p. 140). However, as one former FBI agent remembers, he was amazed at the overall loyalty and trustworthiness of the American citizens with a few exceptions (Dellwo 1999). See Jones (1985) and Rhodes (1995) for further detail.

The counter-intelligence agents investigated unauthorized releases of classified information and in most cases "found that the information leaks...were the result of carelessness or ignorance on the part of the employee or individual with knowledge of the project. But because it was always possible such leaks were surface ramifications of much more dangerous espionage activity, all cases of careless handling of classified data received prompt and rigorous corrective action" (Jones 1985, p. 260). One former employee recalled a specific instance of sensitivity to careless talk: "When they spoke of radiation during that time they referred to it as 'activity.' I made the mistake in the hearing of one of our managers, I used the word radioactive, because I knew what I was talking about. But, oh my, I was taken into an office and security people told me that word is a No-No, NEVER say that again" (Sanger 1995, p. 176).

In July 1944, the counter-intelligence agents began monitoring telephone conversations as well (Du Pont 1944-1946). One agent recalls, "I was supposed to find out if anyone said anything that sounded wrong. I didn't know what I was listening for, except for anything that sounded covert or to find out if anyone were speculating about what was going on at the plant" Sanger 1995, p.135).

Code Words

Even though no public or private discussions of classified information was allowed, communication was necessary in many job situations between co-workers. Thus, they were advised to create code words for those classified matters which had to be discussed (UPOA 1977b). Code words between employees were never written down but simply memorized (Thayer 1996). This practice applied not only to construction workers but also to upper management who were forced to use code words as well. When Enrico Fermi and Arthur Compton (two experimental physicists and major figures involved in the Manhattan Project) visited the Hanford Site, they used the aliases Dr. Farmer and Mr. Comas so no one would link their names to the Manhattan Project and determine what was being produced at the Hanford Site (Groueff 1967).

The Hanford Site's technical employees used the following official code words, which were classified as secret (GE 1950a):

F Material	finished or purified graphite
Monster	shipping cask
MJ-1	Redox Process
MJ-2	234-5 Plutonium Finishing Plant
P-9	heavy water
P-10	Tritium
P-11	Critical Mass Program
P-12	Exponential Pile (Reactor) Program
RW	radiological warfare

Vendors and Purchasing

Individuals not on the Hanford Site employee list (such as vendors, consultants, and labor representatives) who required access to certain types of information or areas were cleared for those activities during the Manhattan Project (Du Pont 1945a). However, for security reasons, in most cases no information could be given to vendors regarding the plant product, which made the urgency and importance of materials difficult to communicate to vendors. In fact, it was mandatory for vendors who were manufacturing extremely important or critical pieces of equipment for the Hanford Site to have guards stationed in their shops to protect the equipment. To prevent a single vendor from obtaining too much classified information on one piece of equipment, several vendors would be used to manufacture different parts of that equipment (Du Pont 1945c).

Specialized procurement procedures were in place to ensure security of classified equipment orders. Purchase orders were designed to reveal no connection to Hanford's national defense mission. Only project numbers and purchase order numbers identified equipment, and some items were purchased using code names. No signatures of government employees appeared on any procurement paperwork unless it was used as the project's copy (Du Pont 1945c).

Media Censorship

Media censorship was critical because it would help ensure that "nothing [was] published that would in any way disclose vital information... [or] might attract attention to any phase of the project." These matters were kept out of "any magazine or newspaper that was likely to be read by an enemy agent or by anyone whose knowledge of scientific progress would enable him to guess what was going on" (Groves 1983, p.146).

Colonel Matthias and his staff went to the local and regional newspapers and radio stations to ask that they help keep the Manhattan Project quiet by not publishing or broadcasting reports or seeking any further information about the Hanford Site since it was related to an important war project. He also requested that any articles related to the Hanford Site not be published until he approved them. However, this was difficult for newspapers to do since they worried that their competitors would not hold back their stories. Special attention was given to the local community media because the government takeover of large amounts of land and the increase in population had raised questions and concerns in the community. Matthias encouraged cooperation by promising to provide reports of breaking news to newspapers that complied with his requests. In fact, the day the bomb was dropped he kept this promise and reported the event to the newspapers (Findlay and Hevly 1995, Sanger 1995, UPOA 1977b).

A larger effort than Matthias' local request was the Director of Censorship sending the Censorship Code notification to U.S. editors and broadcasters in 1943. This notification stated that no reports were to be released about "new or secret military weapons...experiments...involving: Production or utilization of atom smashing, atomic energy, atomic fission, atom splitting, or any of their equivalents. The use for military purpose of radium or radioactive materials, heavy water, high-voltage, discharge equipment, cyclotrons. The following elements or any of their components: Polonium, uranium, ytterbium, hafnium, protactinium, radium, thorium, deuterium" (UPOA 1977b, bk. 1, vol. 14, pp. 6.15-16).

Counter-intelligence agents were assigned to seek out any breach in the media that mentioned the Hanford Site or related Manhattan Project subjects. As one reviewer recalls: "Each of us was assigned so many periodicals and newspapers and we had to watch for words. One of the words was 'atom'" (Sanger 1995, p.140).

The newspaper Hanford employees read was the *Sage Sentinel,* which the Hanford Engineer Works Employees Association published from July 1943 to February 1945. This publication was for Hanford employees only and was not allowed to be taken off the Hanford Site even though the paper was under strict regulations not to publish any information revealing the purpose or progress of the Manhattan Project. Beginning in 1944, a City of Richland edition of the *Sage Sentinel* was published. However, it was edited to focus on world, national, and non-sensitive local news (Findlay and Hevly 1995, Du Pont 1945a).

MEASURES RELATED TO RICHLAND VILLAGE

No fence was installed around Richland Village although counter-intelligence agents closely watched its residents, and the Village Police kept a copy of a key to every house in town. Matthias asked Du Pont to keep the town clean and presentable since it would be open to outsiders and possibly a reflection on the Hanford Site (Findlay and Hevly 1995).

> ***"Once construction of the [Richland] village had ended, Groves and Matthias fully intended to build and patrol a fence around it, just as had been done at Oak Ridge and Los Alamos. They wanted to keep operations personnel 'under control for security reasons' and at the same time prevent access to Richland from outside the Project. But Du Pont objected to such controls over the townspeople and apparently succeeded in changing the Army's mind."*** **- Findlay and Hevly 1995, p. 39**

Richland Village was not fenced in but was still on federal property and thus "subject to regulations imposed by the Area Manager," as well as those of Washington State and Benton County (GE n.d.(b), p.3-4). The Hanford Site Patrol contained a separate patrol force located in Richland called the Village Police. They also held special Benton County deputy status and enforced the law within the City of Richland until 1958 when Richland became incorporated and was no longer a government-run city. Village Police responsibilities also included security of the 700 Area (see Figure 2-8.10), which was the Hanford Site's main administrative area (Du Pont n.d, Hunter 1998).

Although attempts may have been made to make Richland Village appear to be a normal town, the necessity for strict security still played a role. The Richland Village phone book was stamped as classified or restricted information. Richland residents' mail was examined to ensure no sensitive information was being communicated out of town through the mail. Telephone calls were tapped to listen for a breach of security or loose talk. Counter-intelligence agents disguised as ordinary residents of the town kept an eye on suspicious characters. Photography within Richland Village was permitted with the exception of the Administration Area. However, photos could not be commercially sold or published without the approval of the Area Manager. Residents owning firearms had to register them with the Hanford Site Patrol, and using personal radio transmitters required approval from the Security and Intelligence Office of the Atomic Energy Commission (Findlay and Hevly 1995, GE n.d.(b), Parkhurst 1999).

Figure 2-8.10. *Guard House to the West Entrance of the 700 Area Staffed by the Richland Village Police Around 1951-1952*

No one was allowed to live in Richland Village unless they were Hanford Site employees or family members. Background investigations were conducted to some degree on all the residents of the village (Parkhurst 1999). In one instance, when a vehicle with out-of-state license plates entered Richland, the Village Police followed the vehicle until the driver reached his destination and then questioned the driver to find out why he was in Richland Village (Rokkan 1997).

SECURITY EDUCATION PROGRAM

> *"The Security Education Program was designed to educate District personnel to become instinctively security-conscious. [Personnel were reminded not to] discuss nor [sic] to circulate news articles or rumors concerning specific types of project information. Appeal was made on the grounds of patriotism, loyalty to the fighting men and the desirability of preventing project information from reaching the enemy."* - UPOA 1977b, bk. 1, vol. 14, pp. S8-S9

The educational campaign to inform and remind employees of the importance of security was played out in films, posters, billboards, hanging mobiles, literature, specialized talks, telephone disks or stickers, and paycheck inserts (Du Pont 1944-1946, Du Pont 1945a, UPOA 1977b). For examples, see Figure 2-8.11 and Figure 2-12.3 in the Section 12, History of Workers. Other security slogans used on the Hanford Site were (Loeb 1982, p.33):

- *"Caution. Engage brain before starting mouth."*
- *"A secret can circle the globe without refueling."*
- *"Alcohol preserves almost anything except a secret."*

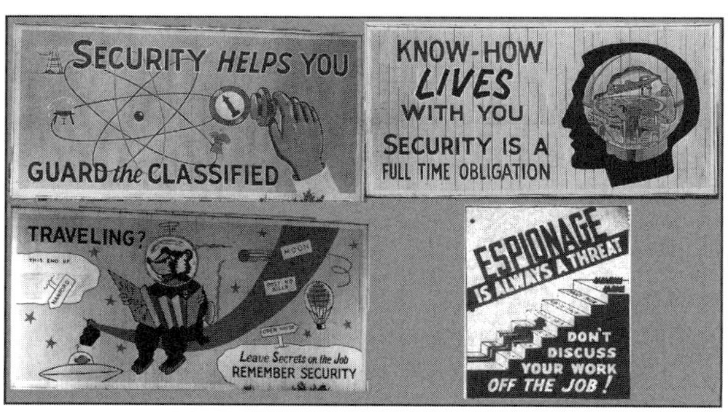

Figure 2-8.11. *Security Billboards in the Mid-1950s*

The security education campaign even extended into the lives of those living in Richland Village–to employees' family members, relatives, and others living in the community and supporting but not working at the Hanford Site. Security awareness talks and films were delivered at different community meetings to bring awareness to the townspeople. Signs with security slogans were posted around the village (Du Pont 1944-1946). Some Dupus Boomer (a local cartoon character) cartoons appeared in the Richland Villager newspaper during the late 1940s. They contained humorous security reminders to which residents working at the Hanford Site could relate (see Figure 2-8.12).

Security Bulletins were also used to keep the employees informed and reminded of security issues. These bulletins were distributed as early as February 1945 until January 1951. They covered issues such as safeguarding classified information, plant protection, explanations for recent security changes, proper security badge use, and prevention of espionage. Employees received the bulletins on an as-needed basis.

One of the paragraphs in Security Bulletin #3 illustrates the need for this form of security information. As World War II was ending and the world was becoming aware of the Hanford Site's mission, employees needed to understand that the world did not know everything about the Hanford Site and security must still be enforced:

Figure 2-8.12. *Dupus Boomer Cartoons Related to Hanford Site Security*

"Recent progress made by the United Nations atomic control committee does not lessen our responsibility. The opinion that the military application of atomic energy must be placed under absolute universal control increases our obligation at the present time. Collectively, and in many cases individually, we possess sufficient knowledge of process or operational secrets, which if exposed through violation of our Oath of Secrecy, would undermine or nullify our Government's advantage in its efforts to establish an effective control plan. The fact that our country must retain full control of the atomic bomb secret until controls are in actual operation is a somber reminder that our silence might well mean our salvation." - Farley 1946

Another paragraph in Security Bulletin #21 in 1948 illustrates the use of bulletins to emphasis certain security issues:

"Wives of project workers also come within this danger zone, not only because of the influence they might exert on their husbands, but because of the unfortunate habit some people have, of being less cautious about discussing their work in front of their wives than they are with social acquaintances or total strangers. Not being sworn to secrecy, as were their husbands, they may feel free to discuss all matters that they feel might be of mutual interest to members of their bridge club, social acquaintances or of more concern as a topic for group discussion. Espionage agents are always alert to such gossip, not so much for the value of the second hand information they might obtain, but as a means of identifying the originator of such revelations as a potential contact for further disclosures." - Farley 1948

CLASSIFIED MATERIALS CONTROL

Two types of materials were classified, documents and nuclear material, and access to them was controlled.

Documents

Much of the information created on the Hanford Site during the construction and operation phases of the Manhattan Project and Cold War eras, such as documents and blueprints, were (and are) considered sensitive. A classification program was created to handle the distribution, control, and destruction of such information. Management and key scientists determined the classification assignments. Nearly all information created was designated as classified (Gydesen 1999).

The main classification levels were Top Secret, Secret, Confidential, and Restricted. These classifications remained in use throughout the Manhattan Project and Cold War eras (Du Pont 1945a, UPOA 1977b, GE n.d.(b)). However, the definitions have been slightly modified over time, and the following categories have been added: Restricted Data, Formally Restricted Data, and National Security Information. To access classified information, the employee had to have the appropriate clearance and the "need to know." The definitions below are those used in the 1940s:

TOP SECRET classification referred to "certain secret documents, information, and material, the security aspect of which is paramount, and whose unauthorized disclosure would cause exceptionally grave danger to the nation" (UPOA 1977b, bk. 1, vol. 14, app. B-7, p. 2). This covered information such as production figures of end product and materials, military use of product, successful methods, plant locations, and set-ups.

SECRET classification referred to "information or features, the disclosure of which might endanger national security, cause serious injury to the interest or prestige of the nation or any governmental activity, or would be of great advantage to a foreign nation" (UPOA 1977b, bk. 1, vol. 14, app. B-7, p. 2). Manhattan Project information involving technical design, processes, construction methods, scope and size of the project, maps, and photographs was covered under this classification.

CONFIDENTIAL classification referred to "information or features...the disclosure of which might be prejudicial to the interest or prestige of the United States, a governmental activity, or an individual, or be of advantage to a foreign nation" (UPOA 1977b, bk. 1, vol. 14, app. B-7, p. 2). Information involving statistics on the materials and personnel required to construct the facility and the construction and production progress was covered under this classification.

RESTRICTED classification referred to "information or feature for official use only, or when disclosure should be limited for reasons of administrative privacy, or denied the general public" (UPOA 1977b, bk. 1, vol. 14, app. B-7, p. 3). Administrative and organization information was covered under this classification. This information was available to employees in general, such as through the security bulletins, but was not publicly available.

UNCLASSIFIED information included data that did not need safeguarding. Classified information was identified by a stamped mark denoting the classification level and category. The stamped marks were usually found on the top and bottom of each page of unbound documents, front and back of photographs, on the first few pages of securely fastened books or pamphlets, and under the scale on maps (UPOA 1977b). When mailed, the classified information (except Top Secret documents) was placed in double-sealed envelopes and sent by registered mail requiring a receipt (Du Pont 1945c).

Only authorized couriers were allowed to transport Top Secret documents. The couriers carried the documents inside two sealed envelopes under their shirts. The documents never left the courier's control until they were delivered.

A Son's Remembrance of His Father's Job as an Atomic Energy Commission Courier in the 1950s

Darrell House's father, as he remembers, was always armed due to his job. Darrell and his siblings knew their father was a courier and his job was very important. However, they were told not to discuss their father's job with anyone. If any one asked about their father's work or where he was during one of his deliveries, the children were to refer people to their mother who would field the questions. Darrell's father was out of town on courier deliveries for a combined total of 4-6 months per year. These trips would arise suddenly, sometimes scheduled only a day before the trip. Darrell's father would come home from work and ask his wife to pack enough clothes for x number of days. This was the only way his wife knew how long he would be gone. She eventually was able guess where he was heading depending on what clothes she was told to pack. One suit probably meant he was making a one day flight delivery to Washington D.C, and 5 or so days of clothes meant he was probably making a delivery to Los Alamos or elsewhere by train (House 1999).

Each employee who had access to classified information or documents was responsible "not to discuss such information, either orally or in writing, with anyone without first being assured of the other party's authority and need to receive such information" (Du Pont 1944-1946, sec. 2/Oct. 2, 1945, p. 13). To identify employees who were cleared to receive or transport classified materials, different types of authorization cards and material passes were issued during the Manhattan Project era:

- Authorization Cards contained the "name of the individual, the category of work for which he was 'cleared,' a description of his type of duties, reference to his supervisor, and personal data" (Du Pont 1945c, p.39). These cards identified the type of information the employee was authorized to see.
- Material and Package Passes were used to "control the movement of material and equipment between plant areas and off the plant" (Du Pont 1944-1946, sec. 2/Oct. 2, 1945, p. 9). These passes were issued only for approved transfers and only to authorized employees. In addition, Authorized Messenger passes were given to employees authorized to transport classified documents outside the barricades (GE 1955b).

Employees were also responsible for physically safeguarding the classified documents in their possession by following specific guidelines including preparation, mailing, handling, accountability, and disposal. A document accountability program had always been enforced at the Hanford Site, which inventoried the location and custodians of classified documents. Originally, teams of clerks personally inspected each employee's classified document holdings to verify the inventory. In the mid-1950s, the system used a punch card system in which a worker would feed a punch card to a computer that would print an individual's classified document inventory. This inventory was sent to the individual to verify, sign and return. Inventory audits were conducted to ensure that the inventory list detailed exactly what was in the individual's possession. Initially, individuals were required to verify inventories every 30 days; this was later changed to every 3 months, then every 6 months, and by the 1970s, every year (Gydesen 1999, Rokkan 1997). The document accountability program was discontinued in the early 1990s.

The storage of classified information had specific requirements. Information needed to be locked in fireproof cabinets or safes that had at least two separate combination locks or were secured by other reasonable measures during the Manhattan Project (Du Pont 1944-1946, UPOA 1977b). Hanford Site Patrol officers inspected offices each day in search of unlocked files containing classified materials. If found, the patrol officers investigated to determine if any information was missing (Beardsley 1998, Du Pont 1944-1946). During the Cold War era, classified information continued to be stored securely either within individual offices or storage vaults.

In 1944, the Classified Files Department developed two major holding locations for controlling classified information including storage, document tracking, and accountability: the 3706 Building in the 300 Area and the 712 Records, Printing, and Mail Office Building in the 700 Area (Du Pont 1944-1946, Gydesen 1999). In addition in 1975, the 3727 Classified Vault Storage Facility was built to securely store classified printed materials. This building was built with 1-foot thick walls and had a traditional lock and key entrance but with an additional secured entrance to the interior vault. That door was made of thick metal and had a combination lock. For more information, see the Historic Property Inventory Forms for the 712 and 3727 Facilities in Appendix B on the Internet.

To destroy classified information, either an authorized employee (in the presence of an authorized officer) or staff in the Classified Files Department shredded or burned the material. When the material was successfully destroyed, the witnessing officer signed a Certification of Destruction as well as a destruction report, if one was filed. The destruction report contained a description of the material, the location where it was destroyed, the date of destruction, and any other details pertaining to the destruction. This destruction procedure was followed for all classified documents, even worksheets (notes), drafts, carbon paper, etc. (UPOA 1977b).

The Operations Security program provided a process to evaluate and control technically unclassified information so that adversaries did not obtain bits and pieces of unclassified information that could help them identify classified information, projects, or activities. During the Manhattan Project and into the Cold War era, operations security was practiced at the Hanford Site to a certain degree through general site security regulations and classified information. However, no organized program addressed all the issues of operations security. An official and organized concept of operations security was developed as a result of problems during the Vietnam War. The United States instituted Operations Security in many government programs to avoid the release of unclassified information that could be beneficial to adversaries (Luczynski 1999).

The purpose of Operations Security at the Hanford Site has been described as "continual evaluation and reporting of routine operational activities to determine whether classified, sensitive/unclassified or company proprietary information is being inadvertently made available to an adversary or outside unfriendly element" (RHO 1978-1986, chap. 28, p. 1). Communication of information for general operations (such as procurement records, job announcements and resumes, telephone connections, drawings, and travel information) was examined as possible references for adversaries. Countermeasures were developed, recommended, and implemented to resolve any operations security issues (Luczynski 1999, RHO 1978-1986).

With the mission at the Hanford Site changing from nuclear material production to environmental cleanup, the U.S. Department of Energy has been declassifying many of the classified documents. See Chapter 3 for resources to finding information on the present status of formerly classified documents.

Nuclear Material

Classified and Radioactive Materials Passes were used to control "movement of process material or material which should not be inspected due to radioactivity, toxicity, possibility of contamination, security classification or similar reasons" (GE 1955b, sect. IX, pp. 2).

The most common movement of nuclear material on the Hanford Site involved transporting non-radiated fuel rods from the 300 Area to the 100 Areas reactors, irradiated fuel rods from the reactors to the 200 Areas separations plants, and plutonium from the 200 Areas to the storage vaults. In each case, the Hanford Site Plant Railroad transported the nuclear material in specialized rail cars escorted by armed security guards who were also sworn as Auxiliary Military Police. (For more detail, see the Expanded Historic Property Inventory Form for the Hanford Site Plant Railroad in Appendix B on the Internet.) Shipments of uranium billets (feed material for fuel manufacturing in the 300 Area) arrived at the Hanford Site from Chicago on rail cars highly protected by security guards, who on the trip back guarded shipments of classified metal scrap and turnings. The scrap and turnings were by-products of the production process at the Hanford Site that were shipped to other plants for reprocessing (UPOA 1977b, bk. 1, vol. 14, p. 5.3). The rail shipments came onto the Hanford Site through the Riverland Classification Yard (near the Midway Substation west of the Hanford Site) where the War Department turned over the rail cars and materials to Du Pont (Du Pont 1944-1946, UPOA 1977b).

> ***"A former worker at the Hanford Site explains the method of transferring plutonium from the 231-Z Plutonium Metallurgy Facility to the 213 Final Storage Magazine vaults in the side of Gable Mountain: "We would make up a little caravan of one car with the plutonium, one car ahead of us and one car behind us, with Army personnel with .45s and I think machine guns. The vault doors at the storage building required two combinations to open. As I recall, nobody was supposed to know both combinations."*** **- Sanger 1995, p.193**

Armed Hanford Site Patrol officers and military guards have always been assigned to escort classified material being moved on the Hanford Site, an assignment that remains essential today (Du Pont 1944-1946). The 213 vaults were also protected by a guard tower, mounted machine guns, a security fence, and roving patrol (CREHST 1998). For more information, see the Expanded Historic Property Inventory Form for the 231-Z Building in Appendix B on the Internet.

In 1970, the 2736-Z Primary Plutonium Storage Facility was constructed as the first building on the Hanford Site to be planned especially for the safe storage of plutonium and plutonium products and scrap. (For more information, see the Expanded Historic Property Inventory Form for the 2736-Z Facility in Appendix B on the Internet.) This building was constructed with important security and safety features in mind, such as the 6- to 8-inch thick, reinforced concrete walls, doors, and ceiling with vibration sensors; combination door locks and alarm system; and a detection, measurement, and alarm system called the Vault Safety and Inventory System that was installed in the mid-1980s. The Hanford Site Patrol always had controlled access to this building. Visitors were required to sign a log sheet when they entered and exited.

The most significant transfers of classified nuclear materials off site occurred when plutonium was taken to the plant at Los Alamos, New Mexico for use to produce the atomic bomb. Colonel Matthias and a Military Intelligence escort hand-carried the plutonium to Los Alamos the first time in February 1945. They drove to Portland, Oregon and then took a train to Los Angeles where they met a Los Alamos agent who had no idea that the "wooden box wrapped in brown paper" contained a flask of plutonium "suspended between shock absorbers" (Sanger 1995, p.195). Matthias urged the agent to request a locked compartment on the train to Los Alamos for security reasons. He reinforced the importance of protecting the package by stating that it cost $350 million dollars to make. The agent followed Matthias's advice (Groueff 1967).

As the amount of plutonium transferred to Los Alamos became larger, a more efficient transfer system was necessary. Military Intelligence controlled the delivery. . By mid-1945, they were using a convoy of Army olive-colored panel trucks with no windows or markings that were modified to hold 24 containers of plutonium. These trucks were typically used as Army ambulances and thus aroused little suspicion. Ten agents were assigned to each delivery convoy, which consisted of three trucks with one military police car in the lead and one at the rear. The agents were armed with shotguns, 38-caliber revolvers, and machine guns. The convoy vehicles were all in radio contact with each other, and they were instructed that in case of an accident they were to drive off the road up wind from the accident.

The convoy drove from the Gable Mountain vaults on the Hanford Site (where the trucks had been loaded by Hanford Site agents) approximately 700 miles to Fort Douglas in Salt Lake City, Utah to meet the Los Alamos convoy of agents. The agents responsible for transferring the plutonium from the Hanford Site convoy to the Los Alamos convoy dressed in protective clothing and wore "health badges" (dosimeters) (Sanger 1995, pp. 196-7). The transfer agents unloaded and loaded the convoys out of view of both sets of escort agents. At this point, the Hanford Site's role in the transfer was complete, and delivery to New Mexico became the responsibility of the Los Alamos escort agents (Groueff 1967). In two instances, rush deliveries of two containers were made to Los Alamos using C-47 airplanes (Sanger 1995).

> ***"Except for an officer riding in the ambulance, none of the drivers or soldiers knew what they were transporting. Matthias insisted on the convoy's taking different routes and never stopping at the same place to eat."*** **On occasion, he would send counter-intelligence agents to follow the convoy and make sure they were not developing routine activities that could be used to sabotage the convoy. - Groueff 1967, p. 311**

In August 1946, it became more acceptable to use rail transportation rather than truck convoys for these deliveries (UPOA 1977b). Security guards escorted the plutonium, which was transported in a vault in the middle section of a specialized rail car (Rokkan 1998).

CLASSIFIED AREAS AND CONTROL OVER ACCESS TO THEM

Access to the Hanford Site has always been controlled and limited to authorized persons wearing the appropriate badge. Not every badged employee could access every location or facility at the Hanford Site, and these restricted areas changed over time. For example, during construction "employees having access to 100 and 300 Areas were, as a general rule, not given access to 200 Areas and vice versa" (Du Pont 1944-1946, sec. 2/Oct. 2, 1945, p.15). Later on, employees were permitted access to all areas. However, specific buildings were still restricted to only authorized employees.

During the initial phases of construction, access to specific areas was constantly changing as construction progressed. During this period "certain sections and buildings in plant areas were restricted, and access to them was limited to persons who had been cleared to receive classified information and who had business to perform within the area" (Du Pont 1944-1946, sec. 2/ Oct. 2, 1945, p. 9). These initial areas included the 100 B Area, 100 D Area, 100 F Area, Test Pile (Reactor), Metal Fabrication, Cold Semi-Works (300 Area), 200 Area Canyon and Isolation Building, Temporary Construction Building, and 105 Warehouse. (Du Pont 1944-1946).

A former construction worker at the 100 B Area recalls an example of the changing access requirements: *"That thing [the B Reactor] was so secret, maybe a person would be cleared for only one side of it. Four sides square, and maybe a guy would be cleared for the intake side but he wouldn't be cleared for the exhaust side. And everybody was not cleared to go up in it. You get about halfway up...and they'd be an armed guard, and he would check you out and if your name was on the list you could go by. If it wasn't that was as far as you went. That was every day. Maybe you could go up Monday and you couldn't go Tuesday."* - Sanger 1995, p.136

Each area on the Hanford Site was under one of three access designations:

- **Controlled Area** where "control of entrance is desirable for general security, but within which all 'Restricted Data' is protected further by guarded repositories or confinement within 'exclusion' or 'limited' areas" (GE 1955b, sect. VIII, p. 1).
- **Limited Area** where "activities involving 'Restricted Data' are safeguarded, and within which freedom of movement of any individual does not in itself constitute access to 'Restricted Data' meaning that access is available to authorized personnel but their access to further information is restricted by guards or other internal controls" (GE 1955b, sect. VIII, p. 1). Limited areas required a specific clearance level to enter.
- **Exclusion Area** where "vital operations involving 'Restricted Data' [were] performed, and within which the freedom of movement of any individual of itself constitutes access to 'Restricted Data' meaning that just having access to this area would cause classified information to be available to the employee" (GE 1955b, sect. VIII, p. 1). Exclusion areas required the highest clearance level (Q) to enter (GE 1955b, GE 1958b).

In the mid-1980s, the Hanford Site was divided into five access categories, all of which required a security identification badge. However, badges were marked with different symbols to identify the employee's type of access clearance (see the Background Investigations and Clearances Sub-Section below):

- The definition of a **Controlled Area** basically stayed the same and covered the 700 and 100 Areas.
- The new **Security Area** covered controlled access including the area between the Wye and Yakima Barricades where the space was "subject to physical protection" (RHO 1986-1987, sect. 72-01.1, p. 3).
- The **Limited Area** became an "area enclosed by a physical barrier [and] subject to physical protection" such as the 100, 200, 300, and 400 Areas (RHO 1986-1987, sect. 72-01.1, p. 3).
- The **Protected Area** was normally located within a Limited Area and was also enclosed by a physical barrier. An example is the Plutonium Finishing Plant (PFP) (a protected area) within the 200 West Area (a limited area).
- The **Material Access** Area contained special nuclear materials located inside a Protected Area and was protected by a physical barrier (RHO 1986-1987).

Fences, Barricades, and Badge Houses

The Hanford Site was constructed with access control points at several locations. The main access control points were the perimeter barricades along the Hanford Site boundary, which was entirely enclosed by a three-strand barbed wire fence (see Figure 2-8.13). This fence carried "No Trespassing," "No Parking," and "No Photography" signs and warning signs at each end of Highway 240 as it entered the Hanford Site telling motorists to quickly leave the area if the light was flashing for threat of radiation (Rokkan 1997). The fence line displays only "No Trespassing" signs today.

The original perimeter barricades or badge houses, built in the early to mid-1940s, were the Richland/300 Area Barricade (see Figure 2-8.14), Prosser Barricade, and the Yakima Barricade. These barricades were located along the main access routes to the Hanford Site: Route 4 South near the 300 Area, Route 10 near the Yakima River, and Route 11A near its connection to Highway 24. The Richland and Prosser Barricades were consolidated and taken out of service with the construction of the Wye Barricade located at the intersection of Route 4 South and Route 2 South in 1959 (see Figure 2-8.13). The Yakima Barricade was replaced in 1982. (See the Historic Property Inventory Form for the Yakima Barricade, 604 Building, in Appendix B on the Internet for more information.)

The Hanford Site Patrol staffed these barricades and checked for the presence of security badges on individuals entering the Hanford Site. During the Manhattan Project when the Military Police staffed the perimeter barricades (badge houses) and employees were bussed on and off the Hanford Site, the Military Police stopped every bus at the barricade and a patrolman boarded, checked every individual for their security badge, and examined all personal items such as lunch boxes (CREHST 1998).

Secondary access control points were located at the main entrances to each of the construction areas, later operation areas. These were the area badge houses (also called guardhouses) and area fences. The Hanford Site Patrol, who checked the security badges of those entering the specific area for the appropriate clearance, staffed the area badge houses. It was also at this badge house where employees picked up their Area Badges and dosimeter (as described in Badges and Passes below). Examples of typical area badge houses are 1701-K and 3701-N (see the Historic Property Inventory Forms for these buildings in Appendix B on the Internet for more information).

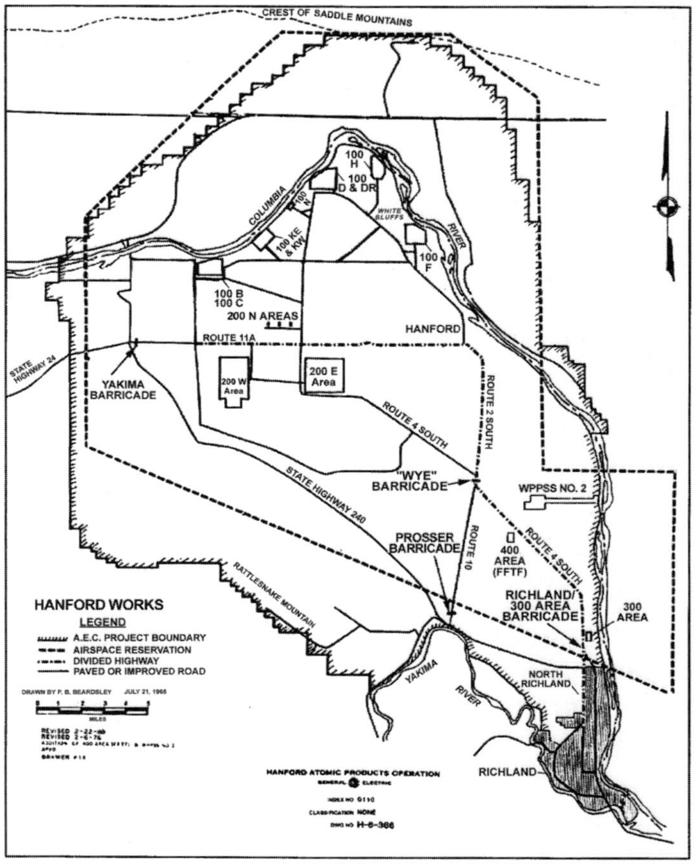

Figure 2-8.13. Map Showing the Security Barricades at the Hanford Site, Revised 1988

The security fences around the operation areas, such as the reactor and separation areas, were installed as certain facilities within the areas were completed and consisted mainly of tall, chain-link fences topped with strands of barbed wire. For example, when a river pump house was completed, a fence was erected all around it for protection. Then when another building such as a power house or office was completed, the fence was re-routed to also enclose those buildings until eventually the entire area was fenced off and accessible only through the area badge houses (Du Pont 1944-1946).

Figure 2-8.14. Access Control Point at the 300 Area Barricade, 1951

A third level of access control points was located at the entrances to specific protected areas (within the operation areas) requiring further access control and authorization to enter. These areas were surrounded by another chain-link security fence topped with strands of barbed and/or coils of razor wire and were accessed through a specific badge house. One example was the PUREX protected area, which was accessed through the 2701-AB PUREX Badge House and contained increased security measures such as metal detectors and x-ray machines (such as those at airports), special nuclear material detectors, key card access (issued for personnel accountability and interior area access controls), and alarm systems (RHO 1986-87). The 1112-N Badge House is another example. See the Historic Property Inventory Forms for the 2701-AB and 1112-N badge houses in Appendix B on the Internet for more information.

Badges and Passes

The employee badge system was one of the crucial access control tools supporting the Hanford Site security system. No one without a badge was allowed through any badge house. In addition, the badges indicated which areas the individual was authorized to enter. The major types of badges that have been worn on the Hanford Site are listed below (Du Pont 1944-1946):

- The **Photo-Identification Pass** was the general badge issued to every employee. It was used very generally to identify the employee and to gain access to the Administration Area (700 Area) and through the main Hanford Site perimeter barricades. An Employee's Temporary Tag Pass was sometimes used to temporarily identify the employee until a Photo Identification Pass could be issued. During construction before many areas were subject to more controlled access restrictions, the Photo Identification Pass (also called the Project Badge) was the only pass necessary on the Hanford Site. As the Hanford Site and its facilities developed, more and different areas, buildings, and sections of buildings were restricted according to the progress of construction and the security department discretion. This evolution led to the need for Area Badges to control access (Du Pont 1944-1946, Du Pont 1945a).
- The **Area Badge** was used as a secondary employee identification, but this badge allowed personnel access to the manufacturing areas to which the individual was authorized or cleared to enter because of their job. Instead of being carried by the employee at all times as was the case with the Photo-Identification Pass, this badge was issued and surrendered along with a personnel dosimeter or health instrument for radiation monitoring at the area badge house. To obtain an Area Badge, the employee had to call out his badge number to the patrolman at the area badge or guard house and show his Photo-Identification Pass (see Figure 2-8.15). The patrolman was responsible for obtaining a positive identification comparison between the two badges. Once the Area Badge was issued, the employee was allowed to enter the area. Infrequent, authorized visitors to an area would pick up an area clearance card upon entering the area. Employees were not allowed inside specific buildings in the area unless their Area Badge was marked by a specific symbol (see the Background Investigations and Clearances Sub-Section below) or unless the supervisor in charge issued them a red tag pass for that building. Upon leaving the area, the Area Badge and dosimeters were returned to the patrol officer and stored in that area's badge house. Monthly lists were sent to department heads showing employees who had Area Badges. They were reviewed and updated as necessary to control and monitor employee access (Du Pont 1944-1946).

Figure 2-8.15. *Hanford Patrol Officer Checking Photo Identification Pass and Issuing Gamma Pencil Dosimeter*

Patrol officers at the badge houses have always been required to visually verify that the individual passing through the checkpoint was the same person appearing in the photograph on their security badge. Badge photos had to be kept up-to-date with the person's appearance. For instance, a new picture was necessary if a beard was shaved or grown. If the visual comparison was not accurate enough, employees were advised to get a new photo taken (GE 1958b).

Background Investigations and Clearances

The personal and professional backgrounds of potential Hanford Site employees were (and continue to be today to a certain degree) investigated before they were employed. The goal of the personal background investigations in the early years of the Hanford Site was to make "every effort to find out before employing anyone whether there was anything in his background that would make him a possible source of danger, paying particular attention to his vulnerability to blackmail, arising from some prior indiscretion" (Groves 1983, p. 141). Generally, this basic goal held throughout the Cold War with additional attention added to ensure the applicant was in fact the person they claimed to be (Parkhurst 1999).

The investigation process began with employees filling out Personnel Security Questionnaires that detailed personal history, education (sometimes even including attendance and grades), employment history, organization memberships, foreign countries visited, and personal references (Du Pont 1944-1946). Attempts were made to verify this information and determine the character of the individual.

Originally, potential employees were investigated further if their job duties would require them to handle confidential or secret information or their Personnel Security Questionnaires revealed information that might lead to questionable loyalty, such as birthplace or military service in any country other than the United States, employment not fully detailed, and membership in questionable organizations. Sometimes the more thorough investigations went back as far as childhood. Even more thorough investigations were required for individuals who had oriental ancestry, were emigrants from enemy countries, had visited or previously lived in enemy countries since 1933, had family in an enemy country, had a criminal history of a nature that might endanger the Manhattan Project, or had served in the military in an enemy country (Groves 1983, UPOA 1977b).

Because the Hanford Site was on a fast track and required a large construction work force, occasionally investigators had to cut corners on background investigations and security clearances of employees to supply enough hands to Manhattan Project. "Faced with a continuing shortage of scientifically and technically trained personnel, project leaders early on had adopted the policy of weighing the degree of risk against the contributions an employee with security clearance problems could make in development of atomic weapons" (Jones 1985, p. 260). In addition, even when background investigations were not quite complete, employees were placed on non-classified assignments until cleared for other work (Groves 1983).

All employees were fingerprinted. The fingerprints of any individual matching arrest records at the FBI were further investigated. The arrest record was compared with what the individual had reported when applying. If any discrepancies existed, the individual was questioned. "Depending on his attitude when questioned, the seriousness of his arrest record, the quality of his work, his absentee record, and the need for men of his particular ability, he was either retained or discharged. No one was hired or kept on who had been convicted of rape, arson, or narcotics charges. Such persons were felt to be unreliable because of their demonstrated weakness in moral fiber and their liability to blackmail" (Groves 1983, p.145).

The standards for employment of the construction employees were less strict than that of the operations employees. So when construction came close to completion in the mid-1940s and employees would be free to transfer to the operation department, each construction employee was re-investigated to guarantee that they met the standards for the new position. Re-investigation included reviewing references, re-interviewing, and obtaining more personal history details. In some cases, the re-investigation determined that certain individuals who had worked on construction did not pass the requirements for operations work. They could be assigned jobs in non-classified areas with notes in their files that they were not to receive any area badges that gave clearance to enter the production areas. Even when current Du Pont employees from other sites where transferred to the Hanford Site, their personnel files were reviewed to ensure compliance with Hanford's security standards (Du Pont 1944-1946).

The same security requirements were applied to the hiring of subcontractors to work on the Hanford Site. "District contractors or subcontractors first had to be cleared before classified information could be made available to them. Company clearance[s were] designed principally to eliminate contractors with potentially subversive foreign affiliations and to prevent fraud and collusion, by disclosing the financial assets of the company, members of the directors, principle stockholders, etc. Clearance was based on...a credit report, and agency checks of key personnel" (UPOA 1977b, bk. 1, vol. 14, pp. 3.7-3.8).

Each Hanford Site employee was authorized or cleared to receive only certain types of information and enter only certain areas on the Hanford Site for security and secrecy reasons. To identify what clearances an employee was given, a clearance symbol system was established. The symbol that appeared on the employee's badge identified the clearance categories that the individual had been granted. The clearance symbols used were the Q, L, and WA.

The Q clearance was the highest clearance category and was originally indicated by a blue background on the badge. It cleared the individual for access to all types of classified information "as determined by job-related need-to-know" (RHO 1986-87, sect. 72-01, p. 1). The Q clearance also authorized the employee to transport classified documents and to have access to special nuclear materials and protected areas (RHO 1986-87, sect.72-01.1, p. 2). An L clearance was the next clearance category and was originally indicated by a yellow background on the badge. It also cleared the individual to receive classified information, but only at the confidential and restricted level and authorized the employee to have access to protected areas. The "WA" (work authorization) status was for employees who were not cleared to receive any classified information or enter any protected areas (RHO 1986-1987). Other badge symbols that were used previously to indicate access authorization to specific areas and buildings are shown in Table 2-8.2.

Table 2-8.2. *Badge Symbols for Secured Areas in the Mid-1980s*

Access Controlled Areas	Location	Badge Symbol
Limited Areas	100 Area	1
200 Areas		2
300 Areas		3
400 Areas		4
Protected Areas	K Storage Basin	K
100 Areas	N Reactor	N
	Vital Area (N Reactor/Control Room)	V
200 Areas	209-E Building	E
	PUREX Facility	X
	PFP	Z
300 Area	308, 324, and 325 Buildings	A
400 Area	Permanent Key Card Entry	P
	Temporary Key Card Entry	T
	Yard Access Only	Y
700 Area	Data Processing Center (Federal Bldg.)	D

During the Manhattan Project, many employees were required to hold Q clearances thus necessitating a large number of background investigations. The Federal Bureau of Investigation conducted many of the background investigations for Q clearances on Hanford Site applicants for the Atomic Energy Commission. The load of background investigations would frequently require the Richland Federal Bureau of Investigation office to bring in additional agents. The policy on these investigations was that all contacts were made personally. This meant they had to rely on other Federal Bureau of Investigation offices if an applicant ever lived, worked, went to school, etc., in another state. The Federal Bureau of Investigation agents in that state made the personal contacts and reported back to the Richland Federal Bureau of Investigation office. By 1971, the number of Q clearance background investigations had dropped significantly. However, the Federal Bureau of Investigation had begun to conduct reinvestigations on current employees in high security clearance positions to update their personal files (Parkhurst 1999).

During the late 1980s, all badges had a green (DOE standard) or gray (Hanford Site- specific) background, and the symbols were simplified. The letter clearance code was replaced with numbers, the higher the number, the more access allowed. The Q clearances were indicated by a 3, L clearances by a 2, and WA clearances by a 1. The Q/3 clearances had access to all types of information and areas as before. The L/2 clearances had access to limited and protected areas (K Basins, Plutonium Finishing Plant, Fast Flux Test Facility) and secret and confidential information. The WA/1 clearances had access to everything else not requiring a Q/3 or L/2 clearance. The only other symbol was Z clearance (referring to the building number of the Plutonium Finishing Plant, 234-5Z), which allowed access to the Plutonium Finishing Plant for those with at least a L/2 clearance. This information is available only on a secure, not public, website. For further information or verification of this information, contact Larry Wonch (Administrative Security at B&W Protection, Inc.).

Figure 2-8.16. *Subject to Search Billboard*

Figure 2-8.17. *Hanford Patrol Officer Inspecting a Vehicle at a Perimeter Barricade, 1957*

Safeguards

Security measures included safeguards, such as conducting searches, prohibiting certain items, establishing a lock and key system, securing computers, and installing electronic security devices.

Searches and Prohibited Items

As part of the security measures against sabotage and espionage at the Hanford Site, patrol officers conducted both body and vehicle searches to prevent contraband items from entering the Hanford Site and classified materials from leaving it.

Searches were more necessary and frequent during the Manhattan Project than during the later Cold War era. However, everything and everyone were still subject to search at any time (see Figure 2-8.16). In some restricted areas, every person and every vehicle were searched upon entry and exit (Du Pont 1944-1946, sec. 2/Oct. 2, 1945, p. 4). By the mid-1980s, locations requiring personal property searches during entry and exit were reduced to the following protected areas and facilities: 100 N Area, 100 K Basin, Plutonium Finishing Plant (234-5Z), Plutonium-Uranium Extraction Facility (202-A), Critical Mass Laboratory (209-E), Plutonium Fabrication Pilot Plant (308), and Fast Flux Test Facility (405). Routine and random searches were still performed across the rest of the Hanford Site (RHO 1986-1987). Personal searches included sporadic searches of the person's body and clothing and examination of bags, lunch pails, purses, etc. (GE 1958b). Vehicle searches involved examination of the interior (including glove compartment), trunk, under the hood, and under the vehicle body (chassis) (see Figure 2-8.17) (Du Pont 1944-1946, GE 1955b).

Items considered contraband or prohibited on the Hanford Site included firearms or weapons, cameras and recording or transmitting devices, illegal drugs and alcohol, explosives, and incendiary devices (RHO 1986-1987). Originally, the list also included binoculars, telescopes, film, and personal vehicles (Du Pont 1945a). Prohibited items were confiscated unless the individual had a Prohibited Articles Property Pass, which was issued for prohibited items approved for use on the Hanford Site because they were necessary to conduct specific work (RHO 1986-1987). Today, personal vehicle access is permitted on most of the Hanford Site with the exception of much of the 300 and 400 Areas and parts of the 100 and 200 Areas. The Hanford Site Patrol can periodically conduct vehicle and personal searches anywhere on the Hanford Site.

Lock and Key System

A lock and key system was started at the Hanford Site during the mid-1940s. "Each area [had] an independent key system, consisting of a single grand master with several master and sub-master keys subsidiary to it" (Du Pont n.d, p. 12) for buildings and equipment. "A group of these master type keys [were] the individual keys for (1) building door lock sets and padlocks; and (2) special purpose padlocks" (Du Pont n.d, p. 13). The Hanford Site Patrol has been responsible for the key system since October 1944 (Du Pont n.d.).

Magnetic (mag) Cards began to be used 1979 as a type of key to control access to buildings by authorized employees. The magnetic card was inserted into a reader at the entrance to a building. The reader scanned the card to identify the magnetic strip inside (Cadd 1998). If accepted, the door briefly unlocked for the employee to enter. Similarly, Key Cards were issued in some locations, such as PUREX, as employees passed through area badge houses. These cards were individually programmed to allow access to authorized locations within the area; for example, through certain doors and into certain rooms. OMNI Locks (key pads) were another device used at the Hanford Site to control access to restricted buildings or rooms. These locks used a security access code that had to be correctly punched into the keypad.

By the mid-1980s, two types of locks and keys existed: security and non-security. Upon official request, security keys were issued to employees with Q or L clearances for access to areas, buildings, etc., containing classified information or equipment when necessary to perform their work. Non-security keys to unclassified and non-sensitive areas were issued upon official request to any prime contractor employee. Keys were issued to other contractor employees when the contractor's security department submitted a request in writing and only for the areas, buildings, etc., associated with the contractor (RHO 1986-1987).

Computer Security

In the early days of computers, computer security consisted only of employees using passwords, backing up the information on their computers, and setting up lock and key systems to protect sensitive information. It was not until the U.S. Department of Energy issued Order 1360.2, Unclassified Computer Security for Sensitive Systems in 1980 that an official computer security plan was put into effect (Fluckiger 1998).

Computer security has been defined as "the protection resulting from all measures designed to prevent deliberate or inadvertent unauthorized disclosure, acquisition, manipulation, modification, or loss of information contained in a [computer] system" (RHO 1978-1986, chap. 21, p. 3). Computers were ranked as classified or unclassified. Classified computers were stored in fully enclosed rooms with security access alarms. Only authorized employees had access to the computer room and the computer system via access locks, identification-verifying systems, and password protection. Users of classified computers had to have the appropriate security clearance and special computer security training. Users of unclassified computers received computer security training as well. Even unclassified computers require a certain amount of security, such as the use of passwords. The Safeguards and Security personnel randomly examined unclassified computer systems to ensure they contained no classified information. (RHO 1978-1986).

Electronic Security Devices

Over the years, electronic security devices were incorporated into the Hanford Site security system. The use of electronic devices added state-of-the-art technology, efficiency, and effectiveness to security and allowed the Hanford Site Patrol to streamline its patrol routes and posts. The electronic devices cover two main areas, nuclear materials and intruders, and consist basically of detection systems, alarms, and complex computer systems. Because many of the electronic systems are presently in use, this discussion is limited to a few examples of either discontinued or non-sensitive systems.

Nuclear Material Control Systems: Until the 1960s, nuclear material was tracked manually on paper. During the 1960s, the General Electric Company used a punch card system to track nuclear material. Beginning in 1965 with contractor diversification at the Hanford Site, various systems were developed and used by individual contractors to track the material for which they were responsible. In 1967, the Atomic Energy Commission established the Nuclear Materials Management and Safeguards System. This system (applied at all the Atomic Energy Commission nuclear material sites and still in use) had a computerized database to track nuclear material at all the sites. The Hanford Site submitted nuclear material data each month to this database (Walker 1999).

From 1980-1992, the Westinghouse Hanford Company used the Safeguards Active Response Inventory System (SARIS) to track the inventory and location of nuclear material on the Hanford Site. It consisted of three separate computer systems: SARIS I located in the 308 Plutonium Fabrication Pilot Plant and the 309 Plutonium Recycle Test Reactor facility, SARIS II located in the 427 Fuels Materials and Examination Facility, and the Nuclear Material Safeguards Computer System located in the 234-5Z Plutonium Finishing Plant. At each of these locations, a nuclear material custodian logged in information on the movement and storage of the nuclear material. SARIS systems were later replaced with new and improved operating systems. See the Expanded Historic Property Inventory Forms in Appendix B on the Internet for more information on the buildings mentioned above.

The fourth computer system used by Westinghouse, the Vault Safety and Inventory System, was installed in the 2736-Z Primary Plutonium Storage Facility in 1985/1986 for "on-line, real-time monitoring of the items stored in vaults" (DOE 1990, p. 8-4). (For more information, see the Expanded Historic Property Inventory Forms for these buildings in Appendix B on the Internet.) This system was a state-of-the-art detection, measurement, annunciation, and alarm system. It could monitor the inventory of individual canisters containing plutonium by the unique electronic signature that each canister contained. This system is still in service.

The Plutonium Finishing Plant had an associated Central Alarm Station in the 2701-ZA Building (for more information, see the Historic Property Inventory Form for the 2701-ZA Building in Appendix B on the Internet). From this building, patrol officers watched the activities in the Plutonium Finishing Plant complex, including the shipping/receiving area, access gates, and storage vaults 24 hours per day. Alarms would alert the Hanford Site Patrol to breaches in security, and they would take appropriate action. This system is still in service.

Intruder Detection: During the early days at the Hanford Site, the only way to detect intruders was visually. Later, technological advances brought detection equipment to the Hanford Site. Many of these intruder detection systems are still in place and therefore cannot be discussed in detail. The types of detection systems that have been used on the Hanford Site include, but are not limited to, motion sensors of various types, surveillance cameras, night vision equipment, and metal detectors, such as those used at airports. The locations and use of these systems have changed over time as security requirements of specific areas have changed.

EMERGENCY PREPAREDNESS

The Hanford Site Emergency Preparedness Program was initiated in 1943 as the Project Emergency Plan to cover a range of emergency situations. Types of emergencies affecting site security include sabotage and hostage situations. For specifics about emergency situations such as natural disasters, fires, and medical emergencies, see Section 10, Worker Health and Safety. The original mission of the Project Emergency Plan was to provide "maximum protection to life and property during a period of emergency" (Du Pont 1945a, p. 1373).

The Project Emergency Plan during the Manhattan Project called for a response group, made up of a variety of staff members under a central coordinator, to report to the field manager for emergencies whether security related or not. This group received backup and support from a parallel response group consisting of the area engineer and the Hanford Site Patrol. The support group was prepared to contribute staff and equipment during emergencies. The Project Emergency Plan mainly consisted of warning notifications and procedures for emergency evacuation of personnel from the 100 Area, 200 Areas, and the Hanford construction camp. Evacuations were originally monitored and controlled within the Emergency Control Center located at the construction camp Patrol Headquarters. A secondary control center was located in Richland Village (Du Pont 1945a, GE 1958b).

Since the beginning of the Manhattan Project, practice evacuations were conducted regularly. The crash alarm emergency telephone system installed in the 100 and 200 Areas in late 1945 was tested during these practices (Du Pont 1944-1946). One scenario outlined emergency procedures in case of a night air raid on the Hanford Site. A total plant blackout was planned so that enemy aircraft would have difficulty sighting the targets (presumably the production facilities). As described in Security Bulletin #54 (Jaynes 1950), the Hanford Site Patrol would be responsible for notifying all key departments with duties to perform during this type of emergency. Employees would also be notified of the blackout. It was the employees' responsibility to remain indoors and turn off all exterior and interior lights. If it was necessary to use indoor lights, windows and doors would be covered with blackout curtains such that no light could escape. Any emergency outdoor lights used during a blackout were to be pointed downward. Passengers in vehicles (including both automobiles and trains) would stop the vehicle, turn off all lights, and enter the nearest building or, if not near a building, remain in the vehicle. The patrol officers would restrict vehicle traffic from entering or exiting the perimeter barricades and would be stationed at main road intersections on the Hanford Site to stop and hold traffic with their lights off until the blackout was canceled (Jaynes 1950).

With the Cold War, emergency preparedness concerns increased and the program's goal focused on protecting the Hanford Site employees first, the environment second, and the facilities last (Schuette 1998). The potential for sabotage of critical facilities and holding the world hostage with nuclear materials became the greatest concern during the Cold War era. By 1962, Radiological Emergency plans dictated the organization of Radiological Emergency Staff, personnel whose special skills and experience made them useful in responding to a radiological emergency. The Radiological Emergency Staff included plotters to track the radiological emissions and emergency status (levels of staff, equipment, and exposure), a message clerk to relay messages from the plotters to the radio room, runners to maintain team communication, a meteorologist to track the effect of the weather on the emergency, radiation survey team members, and a supervisor.

Although called the Radiological Emergency Staff, the members of this team were notified in case of emergencies caused by any release including biological, chemical, or radiological. When notified, the members reported immediately to the Plotting Room located in the basement of the last wing of the 703 Administration Building (for details see the Historic Property Inventory Form for 703 Building in Appendix B on the Internet). If the emergency occurred during off hours, the members were notified through the Richland Crash Alarm System, which was a "block of twenty ordinary residential telephones in member's [sic] homes, all of which could be connected into a conference call from the master phone at the Security Patrol Emergency Officer's desk in the basement of the 703 Building" (Heid 1962, no. 01.1, p.11). These phones would ring continuously, interrupt any busy phones while cutting off that connection, and deliver an emergency notification, instructions, and a call to duty.

Table 2-8.3. *Hanford Site Emergency Alarms*

	Meaning	Employee Action
Howler (ah-oo-gah)	Criticality	Run, don't walk, out of the building to staging area at least 100 feet away
Gong or Horn	Fire	Evacuate building and go to staging area
Wavering/Wailing Siren	Take Cover	Stay inside or seek shelter, do not leave building, close windows, doors, and ventilation systems
Steady Siren	Evacuation	Go to staging area after securing work station
Continuous Air Monitor	Airborne	Evacuate area immediately; stop work and exit building
or Ringing Bell and Flashing Red Light	Radioactivity	
Crash Alarm (telephone)	Emergency Instructions	Answer telephone crash alarm and listen for instructions

In the case of a radiological emergency, Hanford Site employees would be notified by the "Take Cover" and "Evacuation" Hanford Site Emergency Alarms (see Table 2-8.3). In the event of a radiological emergency evacuation, a caravan transported employees to staging areas located past the Yakima Barricade in the community of Whitstran, southwest of the Hanford Site (Heid 1962). Later, staging areas were developed in each Hanford Site area. In many cases, individual buildings had specific staging areas that were also used for general building emergencies such as fires.

The Emergency Preparedness Program was continually updated, modified, and streamlined to meet changing conditions. Training, testing, and practices were carried out throughout the year by individual contractors involved in the program, and once a year the Emergency Preparedness staff conducted a large-scale emergency exercise at the Hanford Site. In the event of an actual emergency, an incident command post would be set up in the field to handle the situation. Also, the Emergency Operation Center in the Federal Building in Richland would be staffed with appropriate personnel to support the command post and handle off-site public notification. Site security personnel stationed in the Emergency Operation Center would stay in contact with the Emergency Control Centers, which have been located in the 100-N, 200 East, 200 West, 300, and 400 Areas since the early 1980s (Schuette 1998).

The Emergency Operation Center (now called the Patrol Operations Center) is located in the 2721-E Building in the 200 East Area. Before 1983, the Patrol Operations Center was located in the basement of the Federal Building in Richland. Some specific examples of the Emergency Control Centers located at the Hanford Site over the years include one located in the 703 Building basement, one in the basement of the 3701-D Building that was a self-contained center with pressure-seal doors, and one located within the 6652-L Nike Missile Base Underground Missile Storage Facility where a war board would be assembled to track events in case of a nuclear accident or attack. See the Historic Property Inventory Forms in Appendix B on the Internet for the 3701-D and 6652-L facilities for more detail.

AREAS FOR FURTHER RESEARCH

Security at the Hanford Site has undergone many changes and challenges since the end of the Cold War. Security measures have been evaluated and reassessed as a result of the change in the Hanford Site mission. Security has become more relaxed in certain aspects and yet increased and improved in others. Along with this, the employee view of security sometimes focuses too much on the fact that the Cold War is over, not necessarily realizing that that does not change the security goals. Further research into the security changes and challenges caused by the end of the Cold War and new mission at the Hanford Site would be worthwhile.

Because many of the security measures of the Cold War are still in practice today, numerous sources dealing with security issues at the Hanford Site are still classified. When these documents are declassified, future researchers will be able to give a fuller account of site security during the Cold War than could be developed here. The following U.S. Department of Energy documents detail the transition effort:

- *Security Transition Program Office Progress Report* (Nousen 1994)
- *Security Transition Program Office (STPO): Technology Transfer of the STPO Process, Tools, and Techniques* (Hauth et al. 1994)
- Hanford Vision 2000 Team Security brochures:
  - *Security Transition: Changing the Face of Hanford* (DOE 1994c)
  - *Lessons from Hanford: Managing Change in the DOE Complex* (DOE 1994d)
  - *Hanford Security Transition: Tools for Change* (DOE 1994e)

A collection of unedited, unpublished, narrated video footage exists in the U.S. Department of Energy's Video Production department (currently run by Lockheed Martin Services, Inc.) in the Federal Building, Room 127. This collection includes historic and current footage of security operations, specifically, Tactical and Special Response Team exercises, special security vehicle use (helicopters, fast attack, tanks), patrol officer fitness training and requirements, K-9 Unit training, security control rooms, vehicle searches, check points, and examples and explanations of electronic security devices (x-ray machines, the Nuclear Material Tracking System, motion detectors). This footage, along with a supportive narrative and oral interviews, could be used to create a video history of security at the Hanford Site.

One of the sub-themes of Hanford Site security is the operation of the counter-intelligence organization. Further research into the detailed operations of counter-intelligence at the Hanford Site and in Richland Village would provide a view of an underground security force involved in security-related activities aimed at protecting national security. Although counter-intelligence operations are not unique to the Hanford Site, details of its operations during the Manhattan Project and Cold War would lend insight into counter-intelligence activities at a plutonium production facility and the relationship with national security.

Classification policies used by the Atomic Energy Commission and later the U.S. Department of Energy shielded both agencies from questions concerning releases of radioactivity, environmental degradation, and environmental health risks to workers and public. Questions concerning whether security practices were legitimate or valid could be a focus for future research.

Specific areas of study concerning the effect of site security on environmental and human health, labor unions, and research might be:

- How did the atmosphere of security during World War II affect what workers learned of the hazards of working at the Hanford Site? How did it influence the ways industrial injuries and diseases were handled in the medical and legal system?
- How did compartmentalization and secrecy affect the work place? How did these conditions impact future environmental and occupational health problems at the facility?
- How did the security and background checks shape community life in areas such as the environment, civil liberties, and civil rights? What issues were not raised at the Hanford Site for security reasons?
- What was the impact of the security system on the development and activities of labor unions? Did security regulations repress unions?
- How did the security system affect scientists and engineers. Did the security system restrict dissent among scientists?

SECTION 9

MILITARY OPERATIONS

D.W. HARVEY
PACIFIC NORTHWEST NATIONAL LABORATORY

SECTION 9 - MILITARY OPERATIONS

Military operations first came to the mid-Columbia region when the U.S. Naval Air Station in Pasco was completed in July 1942, 6 months before the Hanford Site was established (DOE 1995d). The U.S. Naval Air Station controlled 25 airfields, operated and maintained a base for naval aircraft units and aviation personnel, and conducted advanced training maneuvers. The only relationship between the Naval Air Station and the Hanford Site was, after the war, barracks from the U.S. Naval Air Station were barged up the Columbia River for use at the North Richland Construction Camp and the U.S. Army's Camp Hanford. Other military operations during World War II in the mid-Columbia region were at Fruitvale, a former airfield located 1 mile west of the Hanford Site's 300 Area, which was also used as a training range for bombing exercises.

While the Hanford Site played a critical national defense role, providing along with the Savannah River Site approximately two-thirds of the nation's weapons-grade plutonium between 1944-1964 (Albright et al. 1993), the Hanford Site was not considered a military installation. Even when the U.S. Army Corps of Engineers controlled the Hanford Site during the Manhattan Project (1943-1946), actual military operations were minimal. During the early post-World War II period, the United States recognized the need for substantial military protection of the Hanford Site because of the increased Cold War era tensions between the United States and the Soviet Union and the concurrent increase in plutonium production and expansion of Hanford Site facilities.

To provide military protection of the Hanford Site, the U.S. Army established Camp Hanford in 1951. Camp Hanford was a military post operated by the Army until 1961. Camp Hanford consisted of an extensive military compound in North Richland and various forward positions situated throughout the Hanford Site. The forward positions were the areas on the Hanford Site where the anti-aircraft artillery and later Nike missiles were located.

While Camp Hanford operated, the Atomic Energy Commission controlled the Hanford Site, and the Army occupied Atomic Energy Commission lands with a permit. The Atomic Energy Commission, a civilian agency, had taken over the Hanford Site from the Corps in January 1947. A comprehensive agreement between the Army and the Atomic Energy Commission, simply titled the "Army Agreement" (Contract No. DA-45-164-eng-1187) and dated March 1, 1951, provided the basic terms under which the Army would occupy, use, and develop, sometimes jointly, Atomic Energy Commission lands, structures, services, and utilities for Camp Hanford, both in the North Richland military compound and at the anti-aircraft artillery and later Nike missile positions. While the Army provided important military protection for the Hanford Site, the Atomic Energy Commission's authority was dominant unless the site came under hostile enemy attack.

The agreement between the Atomic Energy Commission and the Army was amended by several supplements. The later supplements provided for the restoration and return to the Atomic Energy Commission of various lands and facilities remaining under Army jurisdiction with the closure of Camp Hanford in 1961. The termination of a permanent military presence at the Hanford Site came about because of the rapidly changing military technology brought on by the international arms race. Soon after their introduction at the Hanford Site the 120 millimeter anti-aircraft guns became obsolete and were replaced with the more sophisticated Nike missiles.

CONTENTS

Divider Page Photo: Nike Missiles (NORAD photo)

The Nike missile systems subsequently became outdated with the development of the Intercontinental Ballistic Missiles in the United States and the Soviet Union.

With U.S. national defense relying on the nuclear triad of Intercontinental Ballistic Missiles, Trident submarines, and Strategic Air Command B-52s, protecting strategic defense facilities no longer required permanent military installations such as Camp Hanford. Security of the Hanford Site's plutonium production facilities for the rest of the Cold War became the responsibility of the Hanford Site Patrol.

Military Operations During the Manhattan Project

From the beginning, the Corps had utmost concern for security of the plutonium production facilities at the Hanford Site (see the discussion in Section 8, Site Security). In addition to the Hanford Site Patrol officers the Corps hired when construction began in 1943, the Corps also assigned a small force of military police and intelligence personnel to protect the Hanford Site. The following year, the military police were assigned to patrol the perimeter of the site. By the end of World War II, all patrol functions for the Hanford Site were turned over to the military police units (DOE 1995d).

Although the number of military police steadily increased after World War II, facility expansion as plutonium production increased brought pressure for a larger, more permanent military presence at the site. Troops from Fort Lewis arrived in 1948 on extended maneuvers to replace the military police. By 1949, a small group of National Guardsmen were assigned to the Hanford Site in addition to the troops from Fort Lewis, and at the same time the Atomic Energy Commission assigned guard duties to General Electric's security patrols, who were issued light armored tanks. The Army, however, viewed the contingent of National Guardsmen and the use of security patrols as inadequate to handle the increased security needs because they lacked anti-aircraft capabilities and sufficient manpower (Gerber 1992a).

The National Guardsmen and security patrols were stationed in North Richland, an area that underwent considerable growth during the late 1940s due to the expansion of the Hanford Site. By 1950-1951, approximately 25,000 people were living in North Richland, which included the first members of the Army's newly established Camp Hanford.

Establishment of Camp Hanford During the Cold War Era

The increasing strategic military importance of the Hanford Site and the growing number of personnel brought about a call for a permanent military installation in the area. In response, the U.S. Department of Defense established air defense systems at the Hanford Site and placed the U.S. Army in charge of operating those systems. The Army in turn established Camp Hanford.

The Atomic Energy Commission had constructed most of the structures in Camp Hanford's military compound beginning in 1947 when they built the North Richland Construction Camp as housing and offices during expansion of the Hanford Site. The U.S. Army Corps of Engineers was brought in to complete the construction of Camp Hanford, including structures in the compound and at the anti-aircraft artillery sites.

Camp Hanford's military compound in North Richland consisted of an administrative area (Camp Hanford Headquarters, which later became the Operations and Services Building for the Pacific Northwest National Laboratory), barracks (cantonment areas), the trailer park, "Bremerton" housing (prefabricated residential units from Puget Sound Naval Shipyard in Bremerton), a commercial district, hospital facilities, recreation centers, a school, power house, reservoir, pump house, an industrial area, and firing ranges (see Figure 2-9.1).

The final boundaries of the camp were set in 1955 when the Atomic Energy Commission transferred 1100 acres in North Richland to the Army

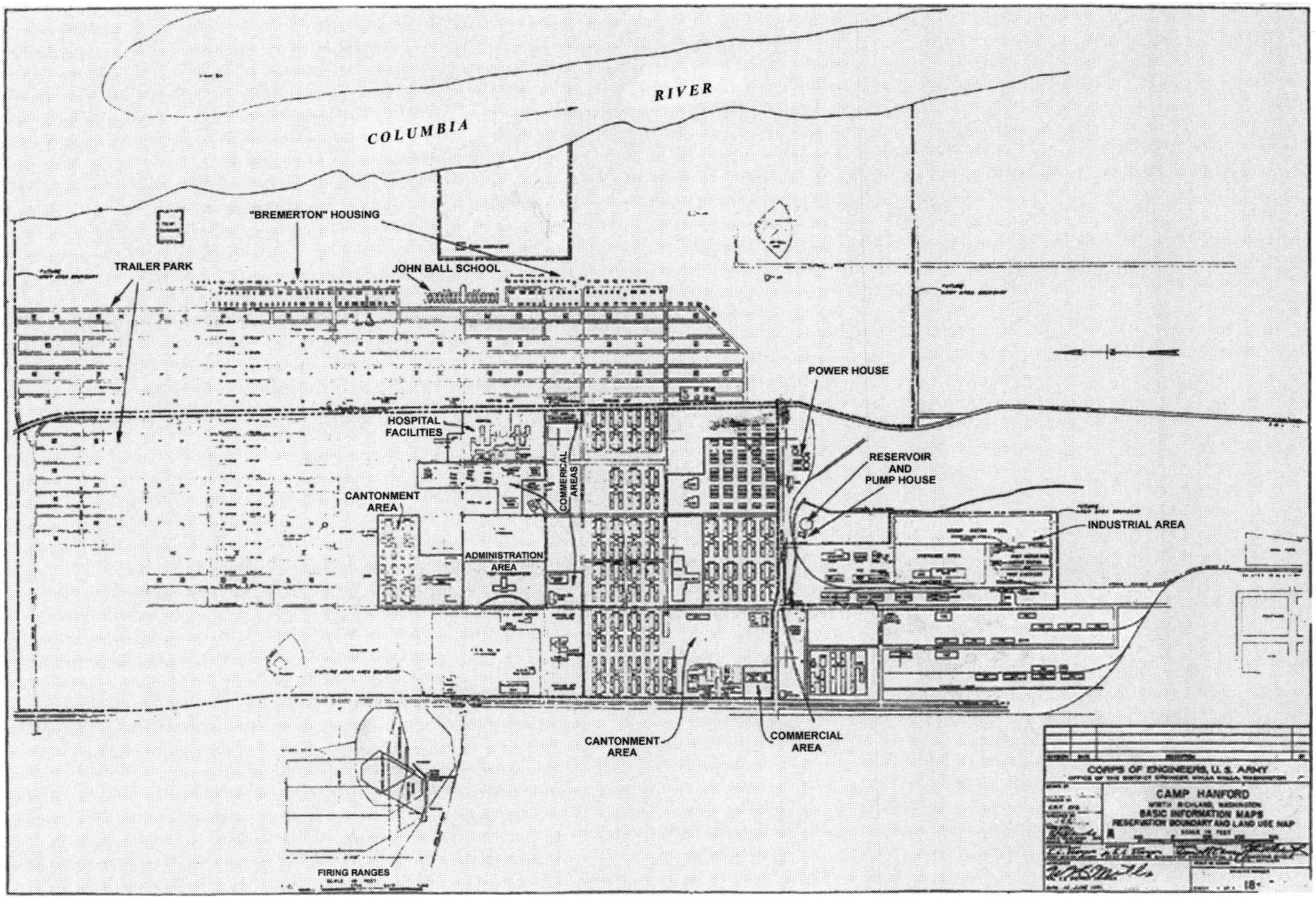

Figure 2-9.1. *Layout of Camp Hanford's Military Compound in North Richland, 1951*

(Franklin 1988), increasing its total area to 3700 acres. By 1955, Camp Hanford encompassed the former North Richland Construction Camp, including portions that had been the 3000 Area Camp used in World War II, most of the rest of the 3000 Area, and reportedly one of the largest trailer courts (2200 units) in the country at the time (*Tri-City Herald* 1961). Along with the land, the Army also acquired 68 barracks, 2 mess halls, 140 trailer lots, and 6 bathhouses.

The 1951 "Index Sheet to General and Detail Site Plan and Building Use Map" (available at the Library of the U.S. Department of Energy's Hanford Cultural Resources Laboratory) noted approximately 100 structures at Camp Hanford. The early housing consisted of two-story barracks obtained from the Pasco Naval Station, one-story barracks, and other buildings from the former construction camp at the Hanford townsite. The 1957 Index Sheet to General and Detail Site Plan and Building Use Map (available at the Library of the U.S. Department of Energy's Hanford Cultural Resources Laboratory) indicates 90 buildings plus approximately 140 "Bremerton" style duplexes. The buildings were arranged in a symmetrical layout with repetitive rows of barracks and other Army buildings of the 700-800 series of mobilization-type construction.

In 1950, the Sixth Army's 5th Anti-Aircraft Artillery Group and four attached gun battalions (the 83rd, 501st, 518th, and 519th) began to replace the National Guardsmen at the Hanford Site before the formal establishment of Camp Hanford. The 770th Gun Battalion, a Washington National Guard unit, was assigned to Camp Hanford in July 1951. The battalion returned to reserve status in 1952 when the 83rd Gun Battalion replaced it.

The first Army contingents to arrive at the Hanford Site to operate the air defense systems lived in temporary barracks until Camp Hanford was established in 1951. The initial Camp Hanford contingent "consisted of only eight officers and fourteen enlisted men plus civilians" (Army 1958, p. 1). At full strength, the contingent consisted of the four battalions, each responsible for four batteries (A, B, C, and D). Battery was the name used both for the sub-groups in the battalions and the locations where the artillery was installed. Each battery had 100 Army personnel assigned to it. In addition, another 400 soldiers were assigned to each of the four battalions along with 1500-3000 personnel including civilians who provided support at the Camp Hanford Headquarters in North Richland. Thus, roughly 5000 personnel were stationed at Camp Hanford at any one time, including civilian government employees and non-federal workers (Wellington 1998, Army 1961).

The 501st, 518th, and 519th battalions served at the Hanford Site from 1950-1958. The 83rd Gun Battalion, activated in 1950 as a 120 millimeter gun battalion, was assigned to Camp Hanford in 1952. The battalion was redesignated the 83rd Missile Battalion in 1954 in preparation for replacing the 120 millimeter guns with Nike Ajax missiles the following year. Beginning in late 1956, the 120 millimeter guns from the 501st, 518th, and 519th gun battalions were removed from Camp Hanford and reassigned to National Guard units (Aradcom Angus 1958). In 1958, the Nike Ajax missiles were removed from Camp Hanford and the 83rd Missile Battalion was deactivated.

The Hanford Site mission required tight security and the soldiers knew little of the uses of the facilities they were assigned to defend. They knew that the site helped manufacture atomic bombs but knew little of the other site missions. Frank Trent, Camp Hanford veteran, remarked, *"All I knew was we were there to defend it [the Hanford Site].... If an airplane flew over, and it was not identified, we were supposed to shoot it down."* - Mulick 1998, p. C1

Camp Hanford soldiers had to cope with the Hanford Site's intricate security and secrecy requirements and contend with harsh living conditions in the field. Knowing little of their mission or the nature of some of the facilities they were assigned to protect, the first anti-aircraft artillery personnel were transported to the forward positions that initially consisted of nothing more than sand and sagebrush. Initial assignments were to carve out roads and paths, set up 12- by 24-foot tents, and construct sandbagged gun emplacements. The batteries of each battalion rotated between the forward positions, where the anti-aircraft artillery (later the Nike missiles) were located, and the military compound in North Richland. But for "about [the first] two and one-half months, [the soldiers] were picked up each evening and driven back to North Richland to sleep" (Gerber 1992a).

Camp Hanford was *". . . never much more than a make-shift military base . . ."* (Mulick 1998, p. C3) according to some veterans. Conditions in the forward positions during the early years were described as primitive. *"Initially we lived in tents, hauled our water, dug latrines, ate food liberally sprinkled with sand, bathed in helmets and under home-made cold showers.... The equipment and vehicles were frequently obliterated by sand storms. The rutted roads scratched out of the desert made traveling slow, uncomfortable and caused maintenance to become a great problem"* (Army 1955, p. 9). Frank Trent, Camp Hanford veteran, *"...remembers burning shoe polish inside the tents and sleeping back-to-back during frigid winter nights just to keep warm.... soldiers also heated their squad tents by stealing heaters from railroad sheds"*. The soldiers had very little time off. *"Soldiers worked three straight weeks along the perimeter before working a week out of barracks in north Richland, where they'd perform maintenance and other upkeep tasks. Soldiers were granted their only off time during that week..."* - Mulick 1998, p. C3

Initial improvements to the forward positions were completed by spring of 1951 and included construction of revetments (walls) made of sandbags and wood planking to protect the gun emplacements, wire communications, roads, and temporary housing. By the end of the first 2 years, living conditions for the soldiers in the field had improved considerably. New latrines, mess halls, and sewage facilities were built, and water and power supplies were upgraded.

ANTI-AIRCRAFT ARTILLERY SITES

The primary mission of Camp Hanford was to establish air defense systems throughout the Hanford Site including the Riverland area, North Slope, and the area now known as the Fitzner/Eberhardt Arid Lands Ecology Reserve. This was accomplished by encircling the 100 and 200 Areas with sixteen anti-aircraft artillery batteries to protect the reactors and chemical separations plants (DOE 1995d) (see Figure 2-9.2). Position numbers, such as PSN 21, designated the batteries. Seven anti-aircraft artillery batteries were located on the North Slope, seven in the central area, and two in Riverland. No anti-aircraft artillery batteries were established in the area now known as the Fitzner/Eberhardt Arid Lands Ecology Reserve. Two base camps without gun emplacements were also located in the central area just south of the Columbia River. "Base Camp 130 and PSN 71, the D battery of the 83rd AAA gun Battalion, both immediately south of the Columbia river were abandoned by 1954, possibly because they were subject to periodic flooding..." (Maas 1993).

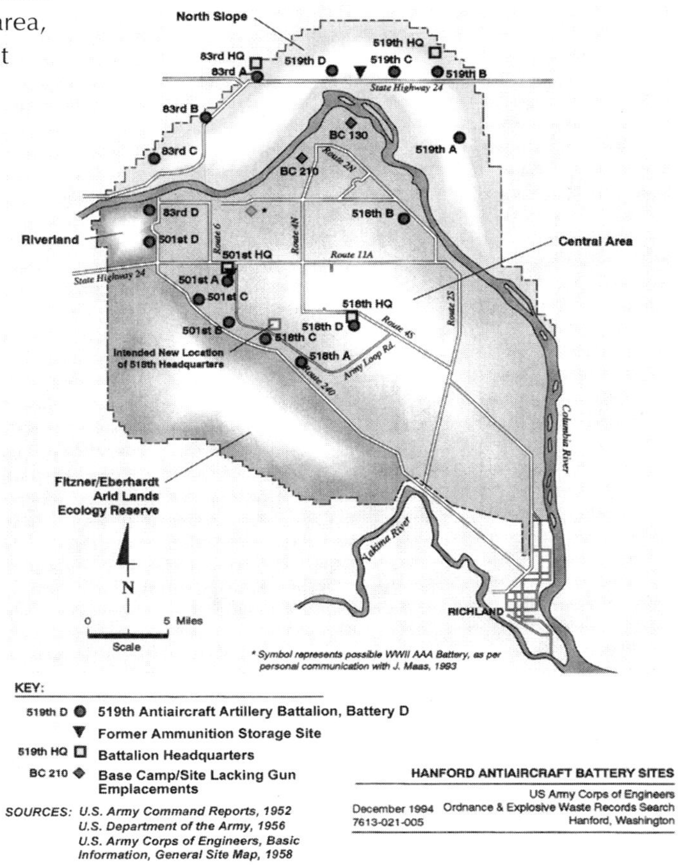

Figure 2-9.2. *Camp Hanford Anti-Aircraft Artillery Sites*

Typically, Camp Hanford's anti-aircraft artillery sites were each roughly 20 acres in size and contained any number of buildings (typically around 20), various utility distribution systems, roads, and sidewalks. Each site consisted of emplacements protected by revetments made of sandbags and wood planking, wooden structures, prefabricated metal buildings, and, later, permanent, concrete block structures. The most commonly constructed prefabricated buildings had aluminum walls and roofs with wooden or concrete floors set on concrete pier blocks.

The permanent structures included barracks, latrines, mess halls, craft shops, pump houses, motor pools, and radar facilities. Each site typically had a small arms range, water storage cistern, sanitary and sewage waste facilities. Pathways, sidewalks, roadways, and parking lots connected the structures. Some of the anti-aircraft artillery sites (later Nike missile sites) were larger and designated as battalion headquarters sites as well as administrative headquarters in the military compound. These headquarters sites had administrative facilities, recreation halls, and power plants.

The layout and design of the anti-aircraft artillery sites reflected a standard military arrangement of facilities separated by function. The four semi-circular, artillery emplacements with revetments were arranged in a square or rectangular plan separate from the residential and administrative facilities (see Figure 2-1.3 in Section 1).

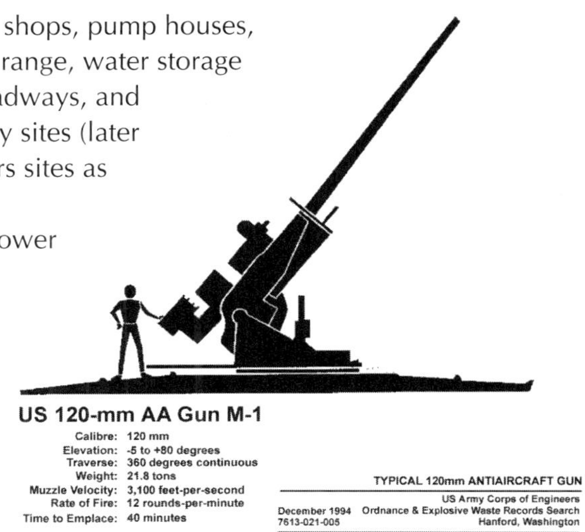

Figure 2-9.3. *120 Millimeter Anti-Aircraft Gun Used at Camp Hanford, 1950-1956*

The first Army defense weapons at the anti-aircraft artillery sites consisted of "64-120 mm antiaircraft guns, 64 - .50-caliber

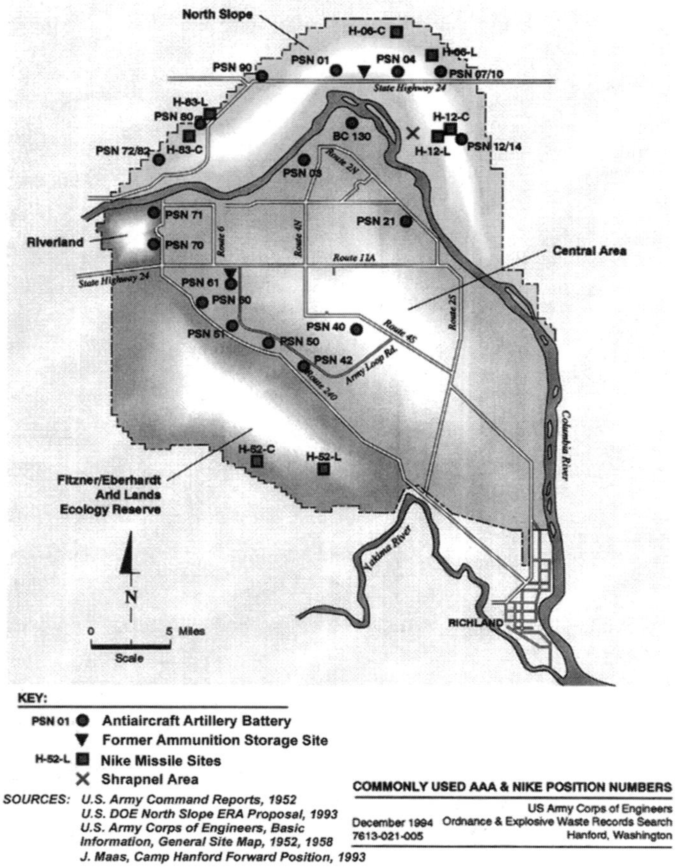

Figure 2-9.4. *Camp Hanford Anti-Aircraft Artillery and Nike Positions*

machine guns, .30-caliber machine guns, antitank rockets, hand grenades, and small arms such as pistols and rifles" (DOE 1995d, p. 26). Each battery was assigned four 120 millimeter anti-aircraft guns (see Figure 2-9.3). The 120 millimeter gun was "...designed to supplement the capabilities of the 90 mm gun and to provide an effective defense against high-altitude heavy bombers" (DOE 1995d, p. 28).

The most intact anti-aircraft artillery sites remaining on the Hanford Site are Positions 61, 51, 50, and 42 located along Army Loop Road, and Position 40 on 200 East Hill (see Figure 2-9.4). Position 61, a battalion headquarters facility, lacked gun emplacements and barracks. Today, these sites "consist of specially sparse debris-scatters associated with demolished buildings and their contents, concrete building pads, traces of former roads and paths, ammunition boxes, and the sandbagged [doughnut-shaped] revetments, wooden structures and other sandbagged/cobblestone structures associated with the former antiaircraft artillery sites" (Archaeological Site Form HCRC #95-600-005 on file at the Library of the U.S. Department of Energy's Hanford Cultural Resources Laboratory) (see Figure 2-9.5).

Figure 2-9.5. *Revetment at Anti-Aircraft Artillery Position 42, 1994*

Specific remains associated with the above sites include a cement obelisk at Position 40, a monument to the 518th Gun Battalion, Battery D. Painted on it is "D Battery 518th" (see Figure 2-9.6). At several of the positions a number of the partially collapsed emplacements include wooden tool sheds that feature painted tool outlines on the walls (see Figure 2-9.7). A concrete monument is at Position 50 with a painted inscription "518th AAA GUN BN (120 mm)" (see Figure 2-9.8). The opposite side of the monument has an inscribed description of the battalion's colors, coat of arms, and World War II service location (Archaeological Site Form HCRC #95-600-005 on file at the Library

Figure 2-9.6. *Obelisk at Anti-Aircraft Artillery Position 40 with Inscription of "D Battery 518th," 1994*

of the U.S. Department of Energy's Hanford Cultural Resources Laboratory).

Beginning in 1954, the Army supplemented the anti-aircraft artillery guns with Nike surface-to-air missiles. In late 1957 and early 1958, thirteen anti-aircraft artillery sites were phased out of service, and their associated structures and equipment were declared excess. The other three anti-aircraft artillery sites were retained and modified to support the three North Slope Nike sites.

Figure 2-9.7. *Interior of Tool Shed at the Anti-Aircraft Artillery Position 40, 1994*

NIKE MISSILE INSTALLATIONS

The Nike Ajax and Nike Hercules missile defense systems were developed in response to heightened Cold War tensions and the escalating international arms race. Nike missile systems were capable of engaging high speed aerial targets at greater ranges than conventional anti-aircraft artillery. From the early 1950s to the mid-1970s, Nike Ajax and Hercules missiles (see Figure 2-9.9) were deployed by the U.S. Army at 292 sites throughout the continental United States to protect major metropolitan areas and strategic military installations from enemy aerial attack.

Figure 2-9.8. *Concrete Monument to "518th AAA Gun Battalion (120mm)," at Position 50, 1994*

During the mid-1950s, Nike missiles gradually replaced most of the 120 millimeter anti-aircraft guns at the Hanford Site. The Nike Ajax was the first generation, pioneer surface-to-air Nike missile. The Nike Ajax systems employed liquid-fueled missiles armed with conventional warheads and "...used a ground-based guidance system with three integrated radar systems" (Bender 1998). First deployed nationally in 1954, the Nike Ajax missile (see Figure 2-9.10), 35 feet in length with a range of 10-12 miles, was a two-stage missile using a solid-fueled booster and a liquid-fueled sustainer motor to deliver high explosive warheads to a radar-determined intercept point with a targeted aircraft (McMaster et al. 1984).

The Nike Hercules missile (see Figure 2-9.11), 39 feet long with a range of 75-80 miles, was introduced in 1958 and gradually replaced the Ajax. "The Hercules was a two-stage missile which differed from Ajax in that the sustainer motor was solid-fueled, and the warheads were primarily nuclear" (McMaster et al. 1984, Summary Page). The Hercules missile was more powerful than the Ajax. It was faster, had more than seven times the range, protected more than nine times the area, and used conventional as well as nuclear warheads. "The advantage of the Hercules missile was its ability to attack high speed, high-flying aircraft formations at greater range with a single nuclear warhead" (DOE 1995d, p. 24).

Figure 2-9.9. *Nike Missiles Ajax, Hercules, and Zeus*

Between 1955 and 1958, the Army deployed Nike missiles at four locations on the Hanford Site. Whereas the position numbers for the anti-aircraft artillery were designated by "PSN," "H" designated those for the Nike missiles. Each of the four Nike missile sites had a control site and a launch site, which were indicated by "C" and "L" at the end of their position numbers. Three of the Nike installations were located on the Wahluke (North) Slope (H-06-C and H-06-L on Saddle Mountain,

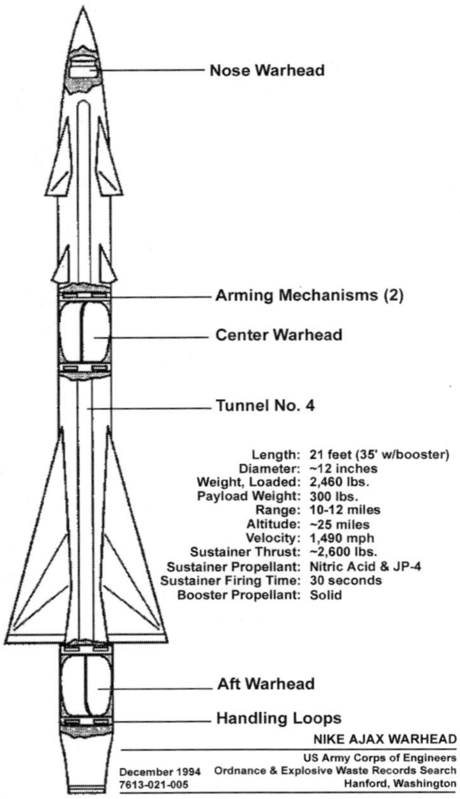

Figure 2-9.10. *Nike Ajax Missile*

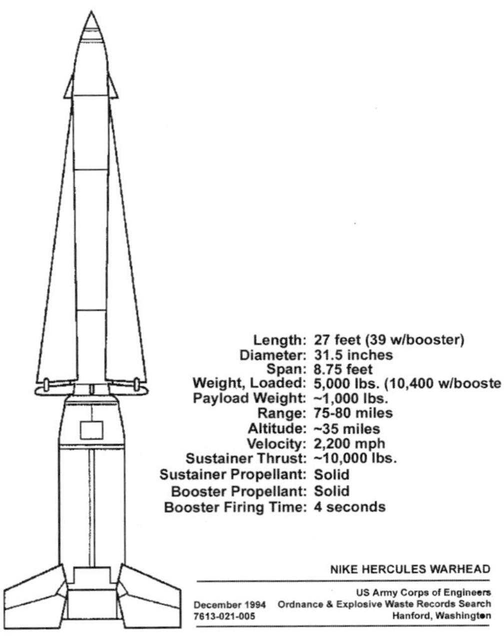

Figure 2-9.11. *Nike Hercules Missile*

H-12-C and L at Othello, and H-83-C and L at Priest Rapids) and one on the Fitzner/Eberhardt Arid Lands Ecology Reserve, H-52-C and L (see Figure 2-9.4).

> ***"The siting of Nike batteries was governed by many factors, including: regulations concerning minimum required distances separating populated areas from installations storing known amounts of explosives; maximum and minimum restrictions on the distance separating launch and control areas; the location of the control area on a topographic high or other area with an unobstructed radar view; and the arrangement of all batteries in a single defense into a coherent ring positioned at an optimum radius from the metropolitan or strategic center to be defended..."*** **- Binder 1992, p. 1**

All four batteries were of similar construction and layout, consisting of the control center, launch area, and associated barracks and administration buildings. Each Nike missile site had two launch sites, two underground missile storage magazines, twenty Ajax missiles, and eight launchers.

In 1958, H-06-L was modified to accommodate twelve Hercules nuclear missiles as well as retain the capability to launch Ajax missiles (DOE 1995d) (see Figure 2-9.12).

> *Differences in the technology of the Hercules missile and its related systems necessitated changes in the design and configuration of the sites that were equipped with this more advanced missile type. The changes made to the Nike Hercules launcher areas typically included: the addition of a separate warheading building; enhancements to the standby power generating facilities; a second (inner) layer of security fencing directly surrounding the underground missile storage and ground-level launch facilities; enhanced site security features and additional sentry stations; and, a kennel for sentry dogs. The need for increased site security reflected the fact that the Hercules missile – unlike its predecessor, the Ajax – could be armed with either conventional (high-explosive) or nuclear [W-31] warheads. Modifications to the control . . . areas of Hercules-equipped sites typically included the following elements: standby power generating system enhancements; . . . additional radar towers and improved radar systems; and, new buildings housing equipment related to those radar systems* (Bender 1998).

Whether a Nike site had Ajax or Hercules missiles, the battery control area and the launch area were located from 1000 to 6000 yards apart. "The minimum distance was determined by the maximum elevation tracking rate of the missile-tracking radar, and the maximum distance was determined by the maximum length of the cable available to connect the battery control trailer with the launcher control trailer" (Artillery School 1955, p. 3).

While the layout and type of structures within each area was site-specific, each installation had certain design and layout features and structures in common. The battery control area generally was located on 10-50 acres of land while the launch area was located on approximately 40-60 acres. Barracks were situated in either the launch area or the battery control area or both, as was the case at the Hanford Site. When located in the launch area, they were approximately one-quarter to one-half mile away from the actual launch sites (IT 1992).

Figure 2-9.12. *Former Nike Missile Launch Area (Position H-06-L) Located on the North Slope*

BATTERY CONTROL AREA

The battery control areas were normally placed on the highest possible point. They contained all the radar, guidance, electronic, and communications equipment needed to identify incoming targets, launch missiles, and direct and guide missiles in flight to intercept enemy aircraft. The areas included the acquisition radar, the target-tracking radar, the missile-tracking radar, the radar control trailer, the battery control trailer, and the maintenance and spare parts trailer.

Ground-based radar and guidance systems were a unique and defining feature of the Nike missile system. The ground-based radar and guidance systems were situated in mobile trailers and interconnected with communication cables. The radar units were either located atop metal towers or set on concrete foundation structures, piers, or pads with a variety of radar antennas (Bender 1998).

***"The [incoming] target was identified by the acquisition radar and then transferred to the target tracking radar, which tracked the target throughout its flight to the intercept point. To control flight and engage the warheads, Nike Ajax missiles used a guidance system coordinated with radar units on the ground. The Ajax was launched by remote control from a nearly vertical position..."* (see Figure 2-9.13). - McMaster et al.1984, p. 5-1**

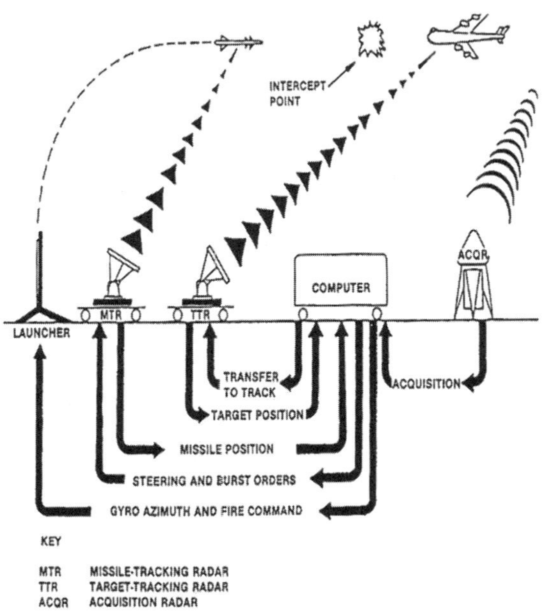

Figure 2-9.13. *Nike Ajax Missile System for Tracking and Intercepting Targets*

The most important trailers were the battery and radar control trailers. "The battery control trailer contained the acquisition radar cabinet assembly, the battery control console assembly, the computer assembly, an early warning plotting board, and an event recorder... The battery control console assembly contained the displays and controls required... The radar control trailer contained the target console assembly, the missile console assembly, the radar power cabinet assembly..." (McMaster et al.1984, pp. 5-7). The controls and displays necessary for the tactical operation of the battery were located in the battery control console. Two automatic plotting boards were mounted directly above the console. The larger board displayed the target position and the predicted point of interception. The smaller board plotted the altitude of the predicted interception point and provided the estimated time of flight to assist in plotting the interception (Artillery School 1955).

The Nike Ajax and Hercules missiles were guided by electronic beams and directed to their targets using the complex system of radar and tracking computers described above. Because of limitations in the tracking and guidance systems, only one missile could be in flight at a time. The Nike Ajax was armed with three high-explosive warheads that two arming mechanisms and five detonating cords activated after the computer in the battery control trailer issued burst orders.

> *"During the missile flight phase, steering commands are transmitted to the missile via the beam of the missile-tracking radar,... The missile burst command is automatically transmitted from the computer to the missile via the missile-tracking radar beam."* **- Artillery School 1955, p. 8**

LAUNCH AREA

The launch areas contained underground missile storage magazines and launch equipment, including buildings and structures used for testing and servicing the missiles. These included generator buildings, missile fueling and warheading facilities, missile assembly buildings, pump houses and other utilities, and administrative, housing, and recreation buildings. Most maintenance activities were conducted in the launch area.

The main function of the launch areas was to maintain the missiles in a combat-ready posture that required the assembly, storage, handling, and disposal of missile components and propellants as well as solvents, fluids, fuels and other materials required for a variety of support functions. All the operations necessary to make the missile flight-ready were conducted in the launch area.

> *"Missile maintenance was conducted in four locations: the magazine, above ground at the launcher, the fueling area, and the assembly building. Where the maintenance took place depended on the specific operation.... Major structural repairs required that the missile be defueled and returned to the assembly building."* **- LETC 1986, p. 30**

Missiles arrived at the launch site partially disassembled in thirteen individual components, unarmed, and defueled. On arrival, the missiles were taken to the Missile Assembly and Testing Building. Next, the missiles were transported to the Fueling and Warheading Station for fueling/ defueling and warheading. The process involved installing the two warheads, connecting the accessory power supply, and joining the missiles to the rocket boosters. For safety in case of an accidental explosion, the station was encircled by earth berms 8-10 feet high.

From the Fueling and Warheading Station, the missiles were taken to the underground missile storage facility, also called the magazine. "The [underground] magazine stored missiles and contained storage racks and a rail system used to deliver the missiles to the elevator. Once above ground, the missile was moved on rails to the launchers" (LETC 1986, p. 34). The launchers and transport rails were situated on top of the magazine. The missiles were brought up to the ground level on the elevator and manually pushed out the elevator door and into position on either side of the door. Firing procedures involved launching missiles from the elevator launcher and reloading until the magazine was exhausted (BOMARC/AF TALOS 1990) (see Figure 2-9.14).

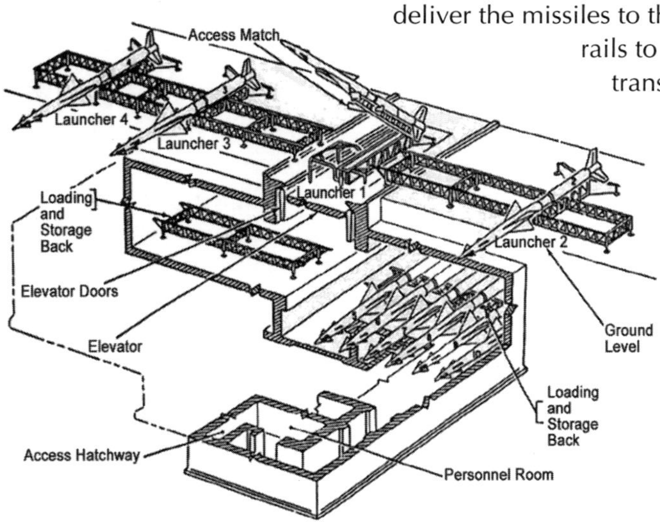

Figure 2-9.14. *Underground Storage Magazine and Launcher-Loader Assembly*

A single site usually had 12 launchers with 4 missiles at each launcher: 1 missile in position and 3 reloads for a total of 48 missiles per site. All the Nike missiles at the Hanford Site were rotated from underground storage to the ready line regularly as part of a maintenance program. The missiles were fueled and defueled with each rotation. Fueling and defueling was completed in the acid pit in the warheading area.

The launcher control trailer, located adjacent to the fueling and warheading area, contained the controls necessary to connect the trailer with the launcher designated to launch a missile. The trailer was also connected by cable to the battery control trailer located in the Battery Control Area.

Despite their widespread distribution throughout the United States and other parts of the world, Nike missiles were never used in actual warfare. Nike missiles were launched for training purposes only from remote desert sites in New Mexico and Texas. By 1960, the hardware for the Nike missile systems was transferred off site.

EXISTING EXAMPLE OF A NIKE MISSILE SITE: H-52-C AND H-52-L

Of the four Nike missile installations on the Hanford Site, only the launch and radar control site on the Fitzner/Eberhardt Arid Lands Ecology Reserve (H-52-C and H-52-L) remains relatively intact. The former battery control area, H-52-C, is located on top of Rattlesnake Mountain, while the former associated launching area, H-52-L (see Figure 2-9.15), is situated at the base of the mountain. The distance between the H-52 control and launch areas is approximately 3 miles.

The spatial arrangement of the buildings and structures in the launch and battery control areas was site-specific with no standard layout plans. The H-52 launch area was divided into two separate zones for functional and safety reasons (see Figure 2-9.16). One zone included the administrative, mess hall, residential, and recreational facilities. The other zone included the underground missile storage magazines, surface launcher-loader equipment, missile refueling and warheading area, missile assembly and testing building, launch control trailer, fuel storage sheds, pump house and generator facilities. While some Nike sites designated areas for housing separate from the battery control and launch areas, the H-52 installation had barracks in both areas (McMaster et al.1984).

Figure 2-9.15. *H-52-L Nike Missile Launch Site at the Fitzner/Eberhardt Arid Lands Ecology Reserve, 1994*

The underground missile storage facility, 6652-L (see Figure 2-9.17), for the H-52 launch area consisted of two underground storage magazines, elevator, and surface launcher-rail equipment. The two magazines had the storage capacity for twenty Ajax missiles and eight surface launchers. Each magazine contained a small personnel room with all the controls for raising and lowering the elevators and launchers, selecting the missile to be launched, and manually firing a missile if the signal cable from the control center failed. Adjacent to the personnel room was a large area and sub-floor for the storage of missiles and an elevator to carry the missiles to the surface. The magazines contained storage racks and a rail system used to deliver the missiles to the elevator. Once above ground, the missiles were moved on rails to one of four launcher-loader assemblies for firing.

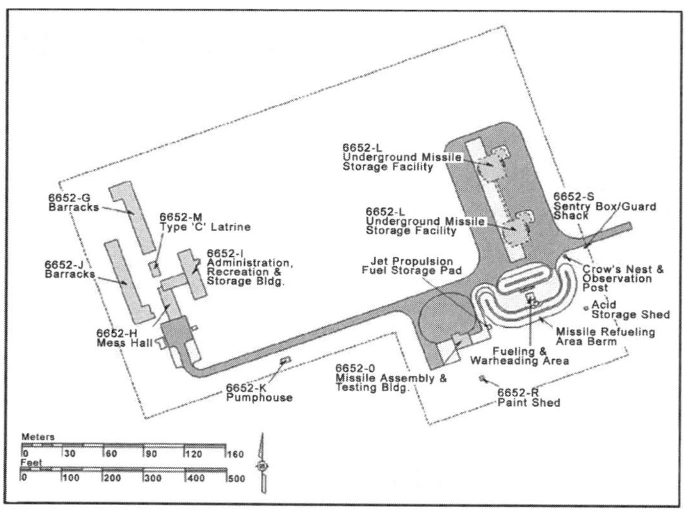

Figure 2-9.16. *Layout of the H-52-L Nike Missile Launch Area, 1998*

Following the deactivation of the Nike missile program, the south magazine was converted to an emergency relocation center for use in the event of a nuclear accident or if the Hanford Site came under attack. The north magazine remains relatively intact except for the removal of the missiles, elevators, launcher-loader assembly from both magazines, and concrete surfacing of the elevator doors.

Figure 2-9.17. *Underground Missile Storage Facility (6652-L) at the H-52-L Nike Missile Launch Site, 1995*

Figure 2-9.18. *Existing Barracks/Administration Building (6652-C) with "1st Missile Battalion, 52nd Artillery, Battery D" on the Wall at the H-52-C Battery Control Area, 1997*

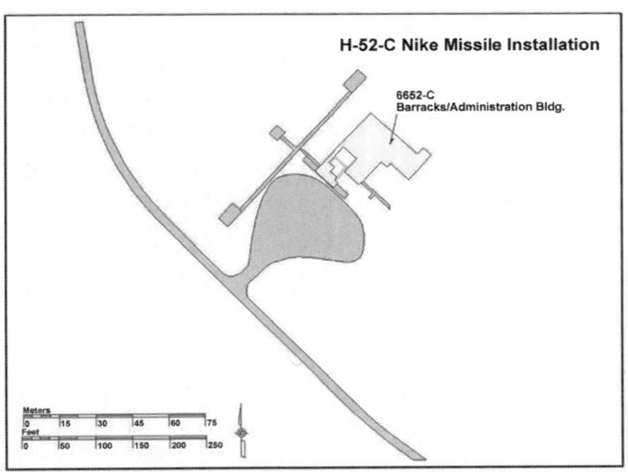

Figure 2-9.19. *Layout of the H-52-C Battery Control Area, 1998*

The Crows Nest/Observation Post and a guard shack are located at the entrance of the H-52 Nike installation. The Crows Nest was an elevated security observation post where guards could view the launch site and the surrounding area. The guard shack/sentry box is adjacent to the tower and located at the entrance to the facility. Since security was of the utmost importance at the Nike site, the sentry post was manned 24 hours a day.

Other H-52 launch area facilities that are extant include the barracks, mess hall, administration and recreation hall, pump house, latrine, paint shed, acid storage shed, and jet propulsion fuel storage pad (see Figure 2-9.16).

The H-52 battery control area combined the administrative functions and battalion officers' quarters in one building (6652-C). The target tracking radar, acquisition radar dome, and missile tracking radar equipment was situated on either concrete pads or concrete piers. Other radar equipment was housed in the Battery Control and Radar Control trailers. The trailers, interconnected with communication cables, were attached by a covered walkway to the barracks/administrative building. All that remains of the control center are the former administrative/barracks building (6652-C) (see Figure 2-9.18), concrete pads and piers, protective earth berms, walkways, and a parking lot (see Figure 2-9.19).

FIRING ACTIVITIES AND AMMUNITION STORAGE FACILITIES

During its tenure at Camp Hanford, the Sixth Army's 5th Anti-Aircraft Artillery Group conducted firing training at the nearby Yakima Firing Range using 120 millimeter guns. "Training was conducted approximately twice weekly with each of the battalions rotating from active status to training status. Prior to a training session at the firing center, only the guns from one battery of one battalion were moved to the Yakima Firing Center. The other battalions would then borrow the training guns during their firing at the Yakima Firing Center. This schedule permitted a sufficient contingent of men and guns to be on alert status at Camp Hanford at all times" (DOE 1995d, p. 15).

Three documented firing ranges included the Gable Mountain Firing Range, Hanford Firing Range, and the range west of the 300 Area, while small arms shooting ranges existed at a number of individual anti-aircraft artillery batteries (Maas 1993) (see Figure 2-9.20). Ammunition caches were stored at each of the batteries to provide small arms for men on assignment.

Each battery had an ammunition storage area that was constructed and located in whatever manner was functional to the mission. There was no apparent standard procedure for the construction and location of storage areas. For example, one battery had tents for storage, another had partially buried storage tents, and a third battery had completely dug-in and constructed ammunition dumps (DOE 1995d).

In addition, two central ammunition storage areas, sometimes referred to as "igloos," were established at the Hanford Site. One storage area with two igloos side by side was located on the North Slope between the 519th D (PSN 01) and C Batteries (PSN 04). The second storage area had two separate igloo clusters near the 501st A Battery (PSN 61) in the central area (see Figure 2-9.2). Additionally, small arms were stored and used at a number of locations on the Hanford Site, specifically at ammunition caches, anti-aircraft artillery batteries, and firing ranges.

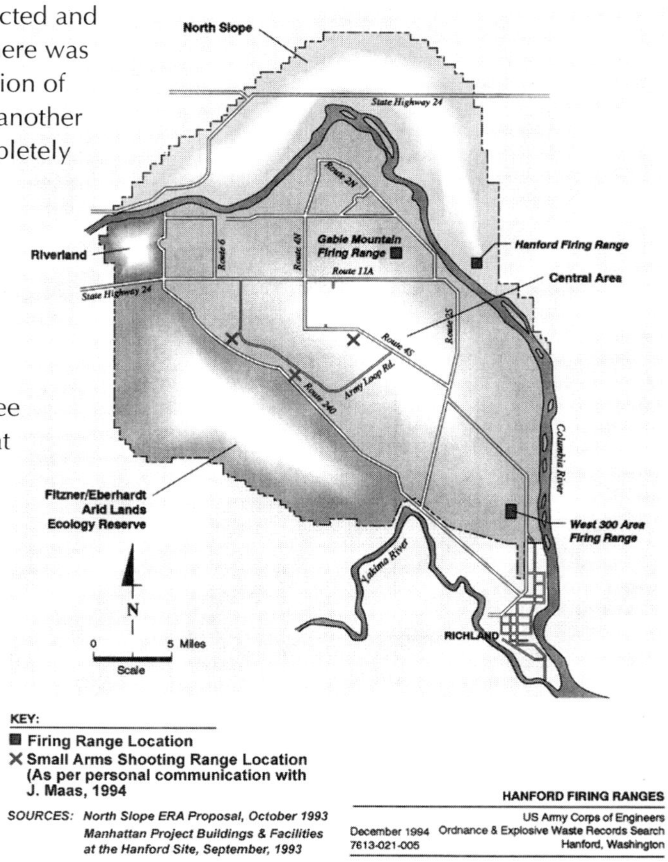

Figure 2-9.20. *Camp Hanford Firing Range Positions*

CAMP HANFORD INDUSTRIAL AREA

Camp Hanford's industrial facilities (see Figure 2-9.1) provided maintenance and repair of anti-aircraft artillery guns and the brigade motor pool (automobile maintenance and repair) and included craft workers/carpenter shops; metal fabrication shops; ordnance storage; warehouses for shipping and receiving dry goods; laundry; cleaners; photo and x-ray facilities; central steam plant; oil, fuel and paint storage depots; and other support services for the Camp's defense installations in the forward positions (anti-aircraft artillery and Nike installations) throughout the Hanford Site. A branch of the Hanford Site Plant Railroad terminated in the industrial area, and several spurs were situated adjacent to warehouse docks for the loading/unloading of supplies. By the late 1980s, nine buildings/structures remained in the industrial area associated with the Camp Hanford era.

CLOSURE OF CAMP HANFORD

By the late 1950s, the development of intercontinental ballistic missiles had rendered Nike missiles obsolete. On July 1, 1959, Camp Hanford became a sub-post of Fort Lewis. In October 1960, the Camp and its missile battalion and batteries were deactivated. On December 21, 1960, the Atomic Energy Commission terminated the land use permit for the former anti-aircraft artillery sites. The following year the Camp Hanford compound and structures were transferred to the Atomic Energy Commission. Camp Hanford officially closed on March 31, 1961.

The numerous barracks and other residential and recreational facilities located outside of the industrial zone, including the large trailer park, were built as temporary structures and were removed in the years following the closure of Camp Hanford. The industrial area buildings (former 3000 Area), concrete reservoir and pump house, roads, concrete pads, and walkways (now covered with grass) are the only extant remains of the Camp Hanford structures above ground.

Areas for Further Research

To provide military protection of the Hanford Site, the U.S. Army established Camp Hanford in 1951 and operated it until it closed in 1961. While the Army operated Camp Hanford, the Atomic Energy Commission controlled the Hanford Site. The Army occupied Atomic Energy Commission lands with a permit. The Atomic Energy Commission's authority was dominant unless the site came under hostile enemy attack.

Camp Hanford closed because rapidly changing military technology eliminated the need for the type of defense system Camp Hanford provided. Soon after their introduction at the Hanford Site the 120 millimeter anti-aircraft guns became obsolete and were replaced with the more sophisticated Nike missiles. The Nike missile systems subsequently became outdated with the development of the Intercontinental Ballistic Missiles in the United States and the Soviet Union—all this within a 10-year period.

Once the U.S. national defense relied on Intercontinental Ballistic Missiles, Trident submarines, and Strategic Air Command B-52s, Camp Hanford and other similar military installations were no longer needed to protect strategic defense facilities, such as the Hanford Site. The 10-year life of Camp Hanford lends itself to future researchers to study as a reflection of the world political situation and changing technology at the time. Some areas of study might be:

- Various national military crises affected military operations at the Hanford Site. . Future researchers may wish to investigate the results of heightened moments of crises on the military operations at the Hanford Site.
- What were the financial, logistical, political, and technical hurdles to overcome when within the 10-year period of Camp Hanford the military installed and then replaced the 120 millimeter anti-aircraft guns with Nike missiles and then removed the Nike missiles because they had become obsolete?
- How did the Army function under the authority of the Atomic Energy Commission, and what effect, if any, did their relationship have on both the military protection and production of plutonium at the Hanford Site?
- Another topic would be a detailed exploration of the daily work situation at Camp Hanford.

SECTION 10

WORKER HEALTH AND SAFETY

B.J. FREER
CH2M HILL HANFORD, INC.

SECTION 10 - WORKER HEALTH AND SAFETY

> *"...safety standards present social as well as technical problems"* **- Barton Hacker, science and technology historian (Hacker 1987, p. 9)**

The production of plutonium began at the Hanford Site in 1944. The Plutonium Finishing Plant and the Fast Flux Test Facility both shut down in 1989, leaving a legacy of worker health and safety concerns. Thus, the history of worker health and safety at the Hanford Site spans a period from Buck Rogers to Star Wars—from technology based in the 1930s to that of the 1980s. During this time, industrial radiation protection developed in conjunction with the nuclear industry.

When the Manhattan Project began, worker health and safety programs emerged from 1) industrial hygiene with its basis in early industrial experience and 2) medical, occupational, and scientific experiences with radium and x-rays. This is not to say that the Hanford Site worker health and safety programs emerged directly from these two areas. Rather, Hanford Site worker health and safety decision makers took what they could from these two areas, and, because of a lack of previous experience in industrial-scale plutonium production, they improvised the rest during the Manhattan Project and the early years of the Cold War.

During the Cold War, worker health and safety programs at the Hanford Site built on Manhattan Project experiences by applying lessons learned. In addition to Hanford Site experiences, worker health and safety professionals drew on knowledge gained at the University of Chicago's Metallurgical Laboratory from 1942-1943 and the Oak Ridge Reservation from 1943-1944 (Wirth 1951, p. 19ff). Lessons learned often resulted in new technologies and process modifications. These changes were often driven by increased demands for defense mission plutonium production. Worker health and safety programs at the Hanford Site during the Cold War also had access to safety data that had been accumulated in the work place and the results of applied research on specific processes and facilities. This information offered an opportunity to evaluate mistakes, accidents, and design flaws to improve policies, programs, and practices.

A major challenge of developing a worker health and safety program at the Hanford Site was addressing the unknown problems of producing plutonium. It was, therefore, necessary to develop health and safety practices through trial and error as well as through tried and tested industrial health and safety knowledge. This mixture of ad hoc and proven approaches was applied to the theoretical understanding of full-scale plutonium production at the time because there was no precedent.

Another major challenge was to incorporate radiation safety with general industrial safety and communicate the importance of complying with an emerging combination of safety practices without jeopardizing national security (Cantril and Parker 1951, p. 476ff). Worker health and safety dilemmas arose as disclosing information to protect workers from harm was weighed against the directive not to compromise the secret.

The Hanford Site's responses to these challenges are reflected in part in the historical records of fatalities and accidents at the Hanford Site. Hanford Site data from these records compare favorably with other industries (Strom 1997). The production of plutonium and cleanup of the site each bring their own challenges in the field of worker health and safety. Issues such as whether Hanford's worker population has an increased mortality level due to radiation exposure as well as questions about the relationship between

CONTENTS

Divider Page Photo: Geiger Counter (Neg. 10528-4)

occupational exposure to chemicals and illness are the subject of ongoing research and debate (see, for example, Mancuso et al. 1977, Gilbert et al. 1989, and Stewart and Kneale 1991) but not the subject of this section. The subject of this section is the history of worker health and safety at the Hanford Site in relation to the Manhattan Project and Cold War Era facilities.

> ***"Industrial safety, industrial hygiene, radiation protection, and fire are all different disciplines but actually they all overlap a little."*** **- William Taylor, Industrial Safety Engineer at the Hanford Site 1948-1994 (Taylor 1998b)**

DISCIPLINES OF WORKER HEALTH AND SAFETY

At the Hanford Site, the three main disciplines of industrial health and safety are: 1) industrial hygiene, 2) industrial safety and fire protection, and 3) radiation protection. Before the Manhattan Project, knowledge from these areas was not consolidated into one discipline or even connected by a set of related approaches. Throughout the Manhattan Project and Cold War, these three disciplines co-existed but remained separate at the Hanford Site.

Industrial hygiene is the "recognition, evaluation and control of those environmental factors or stresses, arising in or from the work place, which may cause sickness, impaired health and well-being, or significant discomfort and inefficiency among workers or among citizens of the community" (Corn 1983, p. 13). Radiation protection is the practice of monitoring and controlling ionizing radiation for the protection of workers. Industrial safety and fire protection assists workers in meeting the challenge of reducing their exposure to industrial hazards encountered in work environments. Industrial safety and fire protection often help workers find ways to accomplish their jobs while complying with the advice of industrial hygienists and radiation protection professionals.

While each of these fields developed separately at the Hanford Site, the urgency of the Manhattan Project demanded overlaps between them. In 1958, industrial hygienist Frank Patty (1958, p. v) noted that industrial hygienists "will find it increasingly necessary to acquire a working knowledge in radiation hygiene." According to Patty, radiation protection professionals "will find it increasingly necessary to recognize the health problems attending other physical forces - such as sound, heat, and light - [and] chemical contaminants." At the Hanford Site, the Manhattan Project created a combination of new challenges that health and safety professionals did not face before World War II. Despite overlaps in these fields, an integrated approach to worker health and safety was not developed at the Hanford Site. It is beyond the scope of this history to fully examine the results of such a lack of integration among the fields of worker health and safety, and future researchers may wish to explore this question.

According to managers responsible for worker health and safety at the Hanford Site during the Manhattan Project, many workers at that time possessed a basic knowledge of standard industrial hazards. Management believed this practical knowledge of workplace safety was sufficient for the new field of radiation protection to build on. The goal was to rapidly tailor workers' practical knowledge of industrial safety to basic radiation protection principles without compromising the Hanford Site's secret purpose. Basic principles of standard industrial safety, such as keeping work areas clean of debris, were thought to be similar to the basic safety principles of radiation protection. After completing training, workers were expected to be able to practice the fundamental rules of both without a great deal of trouble. On August 31, 1944, the head of radiation protection, Herbert M. Parker, wrote a letter to Assistant Plant Manager D. O. Notman that illustrates the point: "All operators [are] require[d] to know that they handle radioactive materials which emit alpha, beta and gamma radiation similar to the radiation from radium. The absorption of these types of radiation in matter can be explained on an elementary basis, and hence [also] the steps that each man must take at all times for his own protection and the protection of co-workers in addition to [using] the permanent protective devices. This is necessary and sufficient to enable men to do their jobs safely" (Parker 1944, p. 1).

At the end of World War II, the breakneck pace of the Manhattan Project gave way to the slow buildup of the Cold War. Brand new worker programs in radiation protection were in place, and they drew on proven industrial safety practices where it was useful. With the threat of war over, industrial hygiene, industrial safety and fire protection, and radiation protection continued to develop independently at the Hanford Site. Each program responded separately to the diverse array of worker health and safety challenges at the Hanford Site by defining the boundaries between them and ensuring their area of responsibility was fulfilled.

INDUSTRIAL HYGIENE

> ***"You can find radiation. I dare you to tell me what you are being exposed to in terms of chemicals; but you can get a meter and it will even tell you what isotope of radiation you are dealing with."* - Ed Beck, Industrial Hygienist at the Hanford Site 1989 to date (Beck 1998)**

Industrial hygiene is at the intersection of industry, safety, and medicine. Examples of areas where industry, safety, and medicine intersect are noise, heat, toxic chemicals, fumes, and dust. The role of industrial hygiene is to identify conditions that can lead to occupational disease and to analyze and control those conditions. This role is prevention of occupational disease, which differentiates industrial hygiene from other areas of worker health and safety. Industrial hygienists identify occupational maladies, such as repetitive motion disorder, before problems emerge in the workforce. An industrial safety and fire professional at the Hanford Site would often request an industrial hygienist to characterize a work site for hazards. The results of such hazards might be falling, tripping, electrocution, and suffocation. Industrial hygienists evaluated all types of work areas but in radiological work areas evaluated only non-radiological hazards.

Industrial hygiene is the most inclusive of the industrial health and safety fields at the Hanford Site because of its wide responsibilities. Reflecting on the nature of this work, one veteran industrial hygienist observed, "the work will never become monotonous or routine because the chemist, the engineer, and the physicist will keep introducing new and more or less hazardous materials and processes that require new developments for the evaluation and control of exposure attendant to their use" (Patty 1958, p. viii).

Historically, industrial hygiene developed expertise on worker safety from studies of World War I munitions factories. These studies stand out for their close attention to qualitative detail. For example, in their review of World War I research, Hamilton and Hardy focus on one case where "body cleanliness was more important than the removal of fumes. This seemed to be proved by the quite striking increase of cases of T.N.T. poisoning on hot summer days, when the skin was flushed and covered with sweat" (Hamilton and Hardy 1949, p. 3). As this anecdote indicates, knowledge often came as a result of sickness or accidents.

The scientists, Army officers, and Du Pont managers in charge of the Manhattan Project at the Hanford Site felt the production of plutonium posed tremendous risks. Just what these risks were from an industrial hygiene perspective was not clear at the outset of the Manhattan Project. However, the production schedule set by Lieutenant General Leslie R. Groves, commander of the Manhattan Project, was too rushed to diminish the level of uncertainty before pressing ahead with industrial plutonium production (Goldberg 1998, p. 61). In the face of substantial gaps in industrial hygiene knowledge of industrial plutonium production, the Manhattan Project went ahead as planned.

To the extent possible, Manhattan Project decision-makers at the Hanford Site based their general health and safety premises on applicable lessons already learned from an industrial hygiene perspective. Industrial hygienists amassed a record of applied research in the 1930s that was designed to improve industrial safety programs. In turn, Du Pont's industrial safety programs at munitions plants used industrial hygiene research. These programs were then used as templates for worker health and safety training plans at the Hanford Site. Du Pont's official history of the Manhattan Project at the Hanford Site attests to this: "In the original planning to provide a Safety Program for the Hanford Engineer Works [Hanford Site], *as a result of previous experience in safety training on other war construction projects,* it was agreed that Supervision would be responsible for the necessary safety education of the employee" (Du Pont 1945a, p. 173, italics added).

One area where previous industrial hygiene knowledge could be applied was hazard prevention analysis, which entailed an exhaustive evaluation of all known risks to a particular industrial process. Data from these evaluations were then incorporated into plans for design, construction, operation, and maintenance. At the Hanford Site, hazard prevention analysis took the form of the Job Hazard Breakdown. A 1958 manual, "H Reactor Operation: Job Hazard Breakdowns, All Buildings" contains a detailed list of potential job hazards for each task at the facility. Corresponding to this list of potential hazards is a list of "Safety Points." Workers were instructed to incorporate the safety points into their work routine to reduce the risk of injury from hazards. For example, the task, "Working in Pits," lists suffocation as a job hazard. Safety points for the hazard of suffocation included "Pit should be free of toxic gases before entering" and "If in doubt about pit contamination, use approved gas mask or self contained breathing apparatus." The final safety point for the job hazard of suffocation while working in pits notes: "Become familiar with methods of artificial respiration" (GE 1958a, p. H 5.72).

Frank Adley's account of how industrial hygiene was first applied at the Hanford Site: *"Dr. Herb Parker was responsible for health physics health matters and he was advised to get an industrial hygienist for a special problem he had. He was advised to contact Dr. Philip Drinker of the Industrial Hygiene Department at Harvard's School of Public Health in Boston. He did so and Drinker said he'd recommend Frank Adley, then in Seattle. Dr. Parker hired me on for this special problem. From this point on, industrial hygiene grew in size, having several professional hygienists and its own laboratory. Industrial hygiene handled all of Hanford's occupational health problems of an industrial hygiene nature."* - Adley 1998, p. 1

According to Frank Adley, the first industrial hygienist hired at the Hanford Site, "there was no industrial hygiene on the plant at startup" (Adley 1998, p. 1). As Adley recounts the Manhattan Project health and safety situation, "In the early production days at Hanford, radiological hazards were handled by Health Physics and those of a chemical nature were handled by Safety in [the] Chemical Processing Department."

When industrial hygiene started at the Hanford Site in the late 1940s, it was organized within the Medical Department. William D. Norwood, an industrial physician (see Figure 2-10.1), was the Superintendent of the Hanford Medical Department. P.A. Fuqua, also a physician, was the Assistant Director of the Medical Department (Wilson 1987, p. 1.5). The function of early programs in industrial hygiene was to monitor the workforce for known industrial diseases and to conduct assessments of work environments to promote healthy job sites. Other duties of the medical program included pre-employment and post-employment physical examinations for all workers (Cantril 1951). These practices continued at the Hanford Site with modifications.

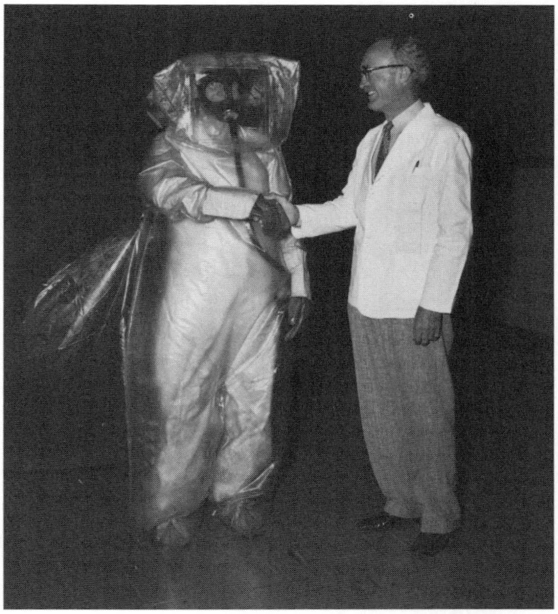

Figure 2-10.1. *William D. Norwood (on right) Shaking Hands with Worker*

INDUSTRIAL SAFETY AND FIRE PROTECTION

> *"Industrial safety in general is the mundane and the everyday stuff that is real easily forgotten, but more often than not it causes the accidents"* - Ed Beck, Industrial Hygienist at the Hanford Site 1989 to date (Beck 1998).

> *"The General Electric Company furnishes all safety equipment except safety shoes and common sense"* - General Electric Company, *Manual of Standard Practices* (GE 1954b).

While industrial safety and fire protection can be complex, "good housekeeping" is one of its basic messages. The "Master Safety Rules" the Atlantic Richfield Hanford Company posted at all workplaces at the Hanford Site in 1971 illustrate this point (see Figure 2-10.2). The first rule emphasizes using common sense while on the job, "Follow instructions. Don't take chances. If you don't know, ask."

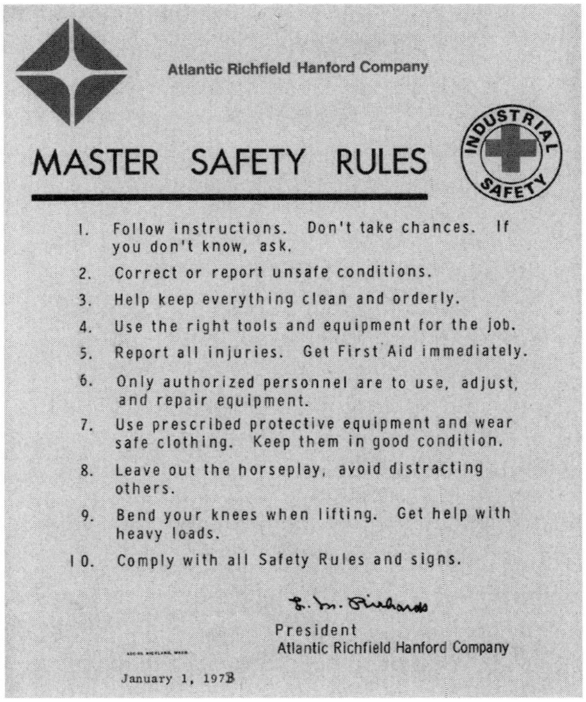
Atlantic Richfield Hanford Company

MASTER SAFETY RULES

INDUSTRIAL SAFETY

1. Follow instructions. Don't take chances. If you don't know, ask.
2. Correct or report unsafe conditions.
3. Help keep everything clean and orderly.
4. Use the right tools and equipment for the job.
5. Report all injuries. Get First Aid immediately.
6. Only authorized personnel are to use, adjust, and repair equipment.
7. Use prescribed protective equipment and wear safe clothing. Keep them in good condition.
8. Leave out the horseplay, avoid distracting others.
9. Bend your knees when lifting. Get help with heavy loads.
10. Comply with all Safety Rules and signs.

President
Atlantic Richfield Hanford Company

January 1, 1971

Figure 2-10.2. *Master Safety Rules, Atlantic Richfield Hanford Company, January 1, 1971*

Of course, industrial safety and fire protection involve much more than simply posting lists of safety rules. The industrial safety record at the Hanford Site is partly based on the extent to which the workers practiced "good housekeeping"—applying safe practices to specific work situations. H. Sudmann's 1992 paper on the causes of hand wounds at the Hanford Site's 200 Areas demonstrates that this emphasis is well placed: "The glamorless operations like housekeeping rarely receive the level of safety analysis or coverage as the production process does. Hands-on maintenance and cleanup of gloveboxes is [sic] still necessary and should be recognized as an accident waiting to happen. Of the 19 wound events, 12 (63%) resulted from hands-on cleanup or waste handling" (Sudmann 1992, p. 11).

The Hanford Site established industrial safety and fire protection standards to protect workers from avoidable accidents, such as falls and spills, and encourage them to properly operate and maintain mechanical equipment and systems. These standards were posted on the applicable equipment (see Figure 2-10.3). Beginning with the Manhattan Project and continuing through the Cold War, contractors employed safety engineers to carry out operational industrial safety.

The Hanford Site established the general framework for safety and fire protection on January 1, 1944, upon review of a comprehensive safety analysis conducted during the first few months of construction at the Hanford Site (Du Pont 1945a, p. 173). Safety personnel consisted of inspectors, often called safety engineers, who worked in the field with craft workers such as welders and carpenters. Safety inspectors reported to assistants who in turn reported to the Safety Superintendent. All safety personnel reported to the Central Safety Department, which reported to Operations. The Central Safety Department had inspectors assigned to each of the operations, such as fuel manufacturing, and those inspectors reported directly to the managers of their assigned operation. The logic of this reporting system formed the basis for operational industrial safety at the Hanford Site throughout the Cold War.

RADIATION PROTECTION

By the end of the 19th century, radiation safety was not foremost in the minds of the relatively few experimental scientists working with x-rays and the new found naturally occurring radioactive elements (Caufield 1989, p. 25). "Experimenters and physicians, laymen and physicists alike set up x-ray generating apparatus and proceeded about their labors with blithe lack of concern regarding potential dangers" (Kathren and Ziemer 1980, p. 1). Scientific fascination with documenting the strange properties of radioactivity, along with public curiosity about its uniqueness, were main features of the history of radiation in the late 19th and early 20th century (Caufield 1989, p. 29). Historian Barton C. Hacker observed, "As 1895 ended, Wilhelm Conrad Roentgen, professor of physics at the University of Wurzburg, published his discovery of x-rays and

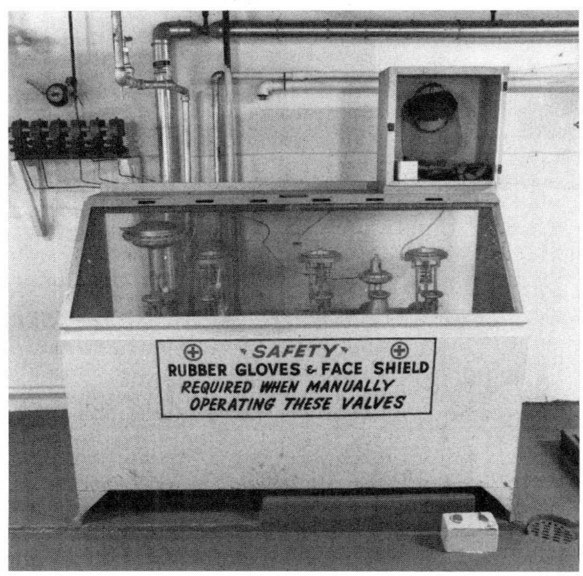

Figure 2-10.3. *Safety Sign and Protective Gear at Valve Station in 221-T Operating Gallery*

"In the early days there were no formal programs in health physics at universities, so none of us arrived at Hanford as health physicists, but we had some background. The first health physics classes at Hanford were held at Dr. Parker's home in Richland. There were about twelve original health physicists who took these first classes." **- Roger Hultgren, Operations and Radiation Protection at the Hanford Site 1944-1999 (Hultgren 1999a)**

explained how to produce them. He made headlines worldwide. Photographs like windows opened into living bodies amazed the public, but science was no less excited to learn that the physical world still held real surprises"(Hacker 1987, p. 10).

As a result of these discoveries and the slow realization that health effects were part of the new discovery, the "first decade saw proposed, if not widely adopted, almost every technique since used to prevent or control harmful effects of radiation" (Hacker 1987, p.10). The decade of the 1920s was marked by "the introduction of film badges for routine personnel monitoring, recognition of genetic effects of x-rays (for which Hermann Muller won the Nobel Prize in 1946), and the adoption of a unit for measuring radiation by the Second International Congress on Radiology in 1928" (Kathren and Ziemer 1980, p. 3). The Manhattan Project drew from this radiation protection information.

Radiation protection, also known by its Manhattan Project code name health physics, is concerned with the control, monitoring, protection, and evaluation of effects from radiation in a variety of settings. Large-scale industrial radiation protection came of age during the Manhattan Project, and the story of its subsequent development in the United States is, in many ways, tied to the Cold War. Comparing Manhattan Project radiation safety with pre-war experience, Hacker notes: "Although applied on a vastly larger scale, the trial-and-error methods devised over four decades to protect radiation workers mostly met the need. Radiation sources were shielded, careful work habits instilled, workplaces closely watched, workers screened to detect early signs of damage" (Hacker 1987, p. 4).

In 1942 as a first step in radiation protection on an industrial scale for the Manhattan Project, the U.S. Army Corps of Engineers and the University of Chicago established a special health physics division at the plutonium pilot plant (the Clinton Semiworks) at the Oak Ridge Reservation in Tennessee to study the health hazards posed by the plutonium production process. This health physics division developed monitoring instruments, standards for the measurement and control of radiation, and trained doctors and health physicists to work at the Hanford Site (Gerber 1992a, p. 28). Herbert M. Parker and Simeon T. Cantril were the principal architects of the health physics program at the Oak Ridge Reservation (Kathren et al. 1986, p. xvii). Parker was a noted British radiological physicist who had worked with Cantril at the Tumor Institute in Seattle since 1938. Parker and his senior assistant, the x-ray diffraction scientist, Carl C. Gamertsfelder, then headed a small group who arrived at the Hanford Site in the summer of 1944 to assume primary responsibility for radiological health and safety. Parker became Chief Supervisor of the Health Instruments Section on July 3, 1944, and Cantril became Assistant Supervisor in charge of the Industrial Medical organization on September 12, 1944 (Cantril 1945, p. 2).

Parker and Gamertsfelder's expertise fit well with the experience of E. I. Du Pont de Nemours. One of the reasons Du Pont had been selected as the primary contractor for the Hanford Site was because it "had emerged during the 1930s as a corporate leader in the still novel field of industrial safety, highly respected for efforts to protect workers in the notoriously risky manufacture of explosives" (Hacker 1987, p. 53).

When they arrived at the Hanford Site, Parker and Du Pont set about to establish radiation protection practices. Factors that played a part in the development of these practices were:

- Du Pont's emphasis on safety and experience in dealing with munitions production in World War I
- A large portion of the early craft workers (welders, carpenters, pipe-fitters) were unionized, which usually meant they had experience in safety practices and industrial hygiene

- Wartime secrecy prevented workers from receiving a full disclosure of the health and safety hazards. Information that might reveal the secret was not communicated to workers. A big part of the secret was radiation. Therefore, routine safety practices, industrial housekeeping, and sloganeering had to be stressed instead of information about radiation protection.

At the Hanford Site, radiation protection was first organized as the Health Instruments Section. The Health Instruments Section was within the Medical Department, and although the names of the organizations changed minimally over the years this organizational arrangement continued for three decades until the end of General Electric's tenure at the Hanford Site. Battelle Memorial Institute took over support of radiation protection duties in 1965. Herbert Parker was the first director of the Health Instruments Section, and he continued to influence the course of radiation protection at the Hanford Site through the early 1980s (Kathren et al. 1986).

In his history of dosimetry and its use in radiation protection programs at the Hanford Site, Robert Wilson notes that, "the original plans for Hanford included only a small 'hazards engineer' group assigned to the technical department for radiation control. This group's functions were envisioned as being similar to conventional production plant safety engineers or inspectors. However, the experience of du Pont staff and both Cantril and Parker at the Clinton [Oak Ridge Reservation] and University of Chicago Laboratories indicated that a much larger, technically oriented group was required" (Wilson 1987, p. 1.6).

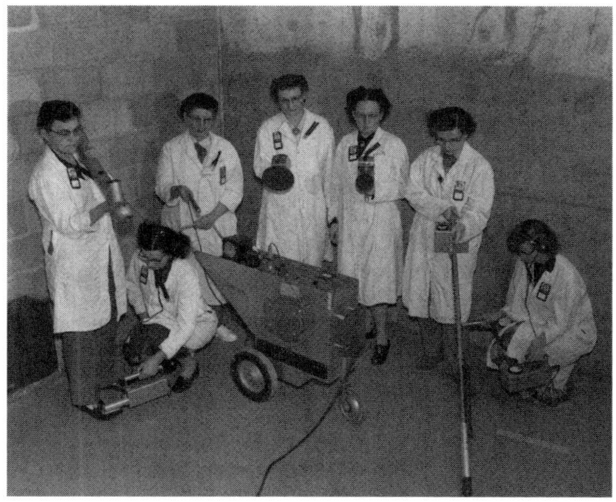

Figure 2-10.4. *Workers with Different Field Instruments for Detecting Radiation*

In the early days of the Manhattan Project, "the HI [Health Instruments] Section was divided into three groups: survey, personnel meters, and special studies. By August 1944 there were 27 members working in the section" (Wilson 1987, p. 1.5) although Parker had requested an initial staffing level of 138. Later, "under Parker's direction, it evolved quickly from a group concerned with shielding estimates to one concerned primarily with detection and measurement of radiation, and the translation of these data into necessary protective measures" (Kathren et al. 1986). Figure 2-10.4 shows a variety of field instruments for detecting radiation.

HEALTH AND SAFETY CHALLENGES

At its most basic level, the challenge of developing health and safety programs at the Hanford Site consisted of controlling known hazards and anticipating unknown problems that might be encountered during the production of plutonium. It was necessary, therefore, to proceed by trial and error in some areas while developing health and safety practices that applied tried and tested industrial health and safety knowledge to what was at the time a theoretical understanding of full-scale plutonium production.

Du Pont's lengthy experience in industrial safety for the manufacture of hazardous chemicals and munitions was retained in its corporate memory, embodied in the practical knowledge of many workers, and documented by industrial hygienists. The next step was to combine information on hazardous chemical production with knowledge of radiation gained at the University of Chicago's Metallurgical Laboratory (1942-1943) and the Oak Ridge Reservation (1943-1944) (Wirth 1951, p. 19ff). The resulting information was available to health and safety management at the Hanford Site to anticipate hazards stemming from the first full-scale attempt to produce plutonium.

Physicists working on the Manhattan Project expected that the large-scale production of plutonium would generate and release both known and unknown radiological hazards (Rhodes 1986). For this reason, and a host of others including secrecy, Lieutenant General Groves and the weapons designers of the plutonium-239 bomb decided to build the plutonium production facilities in a remote area. In constructing the plutonium production facilities, officials acted with great urgency and ruled out time to test scale models. Therefore, while some worker health and safety concerns that were unresolved at the time of construction became factors in site layout, other dangers were considered secondary to the mission of plutonium production. The huge buffer zones at the Hanford Site were there primarily to protect facilities from other facilities. The isolated location of the Hanford Site also exemplifies a Manhattan Project response to a perceived high level of known hazards as well as huge uncertainties. Hazards in this case relate to both engineering and health and safety concerns (Groves 1983). These concerns included the possibility of a massive explosion of reactors or a criticality at chemical separations facilities.

Another health and safety challenge was the unknown hazards that could result from producing a new element—plutonium. By the time of the Manhattan Project, the only experience with a similar element was "the thousand or so curies of radium that had up to that time been separated and used for radioluminous compounds, [which] resulted in many deaths in perhaps a few thousand occupationally exposed dial painters. At Hanford there was potential for exposure of tens of thousands of workers to the millions of curies"(Kathren et al. 1986, p. xviii). The radioluminous compounds referred to were from radium being used as "glow-in-the-dark" paint for dials on clocks, watches, and gauges. The number of deaths arose from dial painters tipping small brushes with their wetted lips and unintentionally ingesting radium. As the health effects began to emerge in the young female dial painters, "there were, of course, the usual protestations of innocence on the part of the company. It was not the paint but something else, even syphilis" (Stannard 1988, p. 24).

From this tragic episode in occupational health and safety history, standards had been implemented for work with radium, and protection of health and safety in Hanford Site plutonium production was based on knowledge derived from these experiences. How the tragedy of the young radium dial painters influenced worker health and safety on the Manhattan Project cannot be underestimated: "Had the radium cases not occurred, or had they begun to appear 15 or 20 years later than they did, or had the cases gone unnoticed, or had they come to the attention to less perceptive investigators, the story of the atomic energy program might have been a sad one. The radium martyrs provided the proof that even small quantities of radioactive substances can be lethal if strict standards of hygiene are not adhered to" (Eisenbud 1968, p. 6).

Although the production of plutonium at the Hanford Site was on a scale much larger than previous industrial experience with radium (Hacker 1987, Stone 1951), the hazards presented by plutonium and radium were comparable when scale was taken into account. For example, the soluble forms of radium and plutonium are similar in that they both have long half-lives compared with the human life span. However, at the Hanford Site, plutonium also existed in an insoluble form that had its own special characteristics. The half-life of radium-226 is 1602 years, and the half-life of plutonium-239 is 24,110 years (Stannard 1988, p. 3; Seaborg 1994, p. 380).

While using the experience with radium as a baseline, scientists sought to discover the properties peculiar to plutonium. The new isotope, plutonium-239, had never been analyzed until Glenn T. Seaborg transmuted a small amount from uranium in 1941. Scientists found that in addition to its inherent high toxicity, plutonium emits high energy alpha particles with particular hazards from ingestion or inhalation. Protection programs needed to prevent ingestion, inhalation, and skin absorption of these particles. In addition to plutonium, numerous other radioisotopes were formed during the plutonium production process. As with plutonium, scientists studied each new radioisotope and gauged its hazards (Hacker 1987).

Scientists conducted these studies by observing the health effects of these radioisotopes on animals. "It was soon apparent from experiments in rodents that Pu [plutonium] was about 15 times more acutely 'toxic' than radium-226 for acute (lethal) effects based on the radioactivity injected per unit body weight" (Voelz 1991, p. 682). Health and safety scientists from the Health Instruments Section at the Hanford Site drew on this research to establish tolerance levels for plutonium in water and air (Parker 1945). During the Cold War, concerns with toxicity and acute effects were augmented by research into dose and cancer-induction and incorporated into Hanford Site protection programs.

In an interview, the first Superintendent of the Hanford Medical Department, William Norwood, puts the early tolerance levels at the Hanford Site in perspective: "the worst possibility is plutonium. The amount allowed, total whole body was .04 microcuries, an amount equal to what you could put on the point of a very sharp pencil" (Sanger 1989, p. 150). Research on the metabolism of plutonium in humans informed detection programs by pointing out that bone and liver were the organs most likely to receive the highest concentrations from ingestion, while the lungs were an organ of concern when intake was through inhalation (Voelz 1991 pp. 688-690). As Stannard's (1988) review of the radiobiology literature indicates, a sizeable body of basic and applied research on the relationships between plutonium and health was conducted during the Cold War at the Hanford Site.

Another challenge was how to develop a method to determine whether workers had ingested plutonium. Norwood developed such a method and put it into practice by conducting applied occupational medical research on plutonium at the Hanford Site: "The people at greatest risk were working in the separations plants...Plutonium did show up, but well below the permissible limits" (Sanger 1989, p. 150). Norwood's research detected the presence and measured the amount of plutonium in the bodies of workers. Detection and measurement are not, in themselves, evidence of health effects. Norwood's research measured only what existed as a result of normal working conditions. He did not subject workers to experiments (DOE 1995c, pp. 5, 46-49). Future health effects researchers may well use Norwood's Hanford Site data as one source in their studies. However, Voelz (1991) notes it is crucial to understand that data on the detection and amount of radiation do not, in themselves, say anything about health effects. Even when this distinction is made, "Continued studies are also necessary, because long latent periods, probably over 30 years, are anticipated" (Voelz 1991, p. 692).

RESPONSIBILITIES FOR HEALTH AND SAFETY

> ***"The safety of all employees is a responsibility of the General Manager and each succeeding level of supervision"*** **- General Electric Company, *Manual of Standard Practices* (GE 1954b)**

On October 3, 1944, a meeting was held to define responsibilities for health and safety at the Hanford Site. The purpose of this meeting was to "clarify the position of the Medical Department and the HI [Health Instruments] group in relation to the responsibilities for radiation safety throughout the plant" (Norwood and Cantril 1944, p. 1). The Medical Department, the Health Instruments Section, a management official representing the Operating Department, and a management official representing the Maintenance Department attended the meeting.

On November 4, 1944, management at the Hanford Site received a statement of responsibilities, based on the items agreed to at the meeting. This memorandum contained two key points, 1) "Responsibility for maintaining radiation within safe limits in the operating areas rests with the area superintendent or those delegated by him," and 2) "The Medical Department, through representatives in the HI [Health Instruments] group, has the responsibility to management to insure [sic] that over-exposure does not occur" (Norwood and Cantril 1944, p. 1). Thus, the various operations organizations were responsible for day-to-day worker health and safety by limiting emissions and exposure. The Health Instruments Section advised management on strategies for avoiding worker over-exposure.

This document illustrates a principle that became a rule of thumb for worker health and safety at the Hanford Site: ultimate responsibility for health and safety rested with operations. Individual workers were responsible for their actions while on the job as in any other industrial setting. The rule of thumb was further clarified in a collection of papers published after the extreme secrecy of wartime ended, which the scientists who formulated health and safety programs for the Manhattan Project had written. Reflecting on the issue of responsibility for worker health and safety, the authors of one of these papers state, "From their experience the authors feel that the responsibility for protection against radiation, as against any other hazard, must rest with those who are charged with the primary program, not with the health service. Ultimately, those directing the work must assume the responsibility for the health of the workers who carry out their plans" (Cantril et al. 1951, p. 277).

In December 1954, the Hanford Site's main contractor, General Electric, issued the first edition of its *Manual of Standard Practices* (GE 1954b). This document contains a section titled "Safety and Health," which confirms that the rule of thumb set down during the Manhattan Project was carried through into the Cold War, meaning the bulk of routine health and safety activity was conducted in operational settings as a part of the job. Oversight of health and safety fell to separate groups and professions.

At the Hanford Site, this fundamental distinction meant that the health and safety groups served in a support capacity. "The responsibility of the health service is to give full cooperation and advice in the prevention and control of hazards and in the furtherance of safe working conditions" (Cantril et al. 1951, p. 277).

Although the statement of responsibilities document issued in November 1944 uses the specific term "radiation safety" (Norwood and Cantril 1944), a review of the historical record generally shows that the responsibilities formulated during this meeting applied to all phases of worker health and safety.

HEALTH AND SAFETY SUPPORT

> ***"We fought a fire in every building before there ever was one"*** **- Gordon Towne, Assistant Fire Chief, Hanford Fire Department, 1947-1978 (Towne 1998)**

The support fields were industrial hygiene, industrial safety and fire protection, and radiation protection. The primary role of the health and safety professionals was consultation rather than implementation. The health and safety professionals generally followed the three steps of any consultation process: 1) examine the situation, 2) write an evaluation and recommendations, and 3) follow-up when asked to return to evaluate mitigation efforts and compliance with recommendations.

The process of consultation at the Hanford Site may be illustrated by an example from an industrial hygiene report dated August 11, 1949 (Gill 1949). An industrial hygienist, Wallace E. Gill, examined the degreaser operation in the 105 Building at H Reactor on June 29 and July 13, 1949 and wrote a report based on an examination of the degreaser operation. Gill (1949) notes that an industrial hygiene study of this degreaser was "done at the request of the Medical Division" (Gill 1949, p. 2). The "Medical Division" referred to was the Industrial Medical Division, which was a part of the Medical Department. The primary reason for the request from the Industrial Medical Division was to evaluate the air concentration of a solvent used in the degreaser operation. Degreasing in this facility was defined as, "removing grease, oil, or wax from metal parts by immersing them in solvent vapor and/or boiling solvent, generally trichloroethylene or perchloroethylene" (Gill 1949, p. 5). The solvent used in this degreaser operation was trichloroethylene, which in 1949 had a specified maximum allowable concentration in air. Degreasing in the 105 H Building required operators to be near the large solvent tank used to clean metal parts. Therefore, the Industrial Medical Division initiated an investigation into the health aspects of this operation.

To examine the situation, Gill obtained access to the 105 H Building to take measurements of the workspace and make drawings of the degreaser. He observed workers using the degreaser and characterized this process. Finally, he took a series of air samples for velocity and trichloroethylene concentration. In this degreaser area, the highest air concentration found during tests on June 29, 1949 was 30 parts per million. The maximum allowable concentration was 100 parts per million.

Although the air concentration Gill found was under the limit, he listed ten recommendations "for the improvement of health and safety in these operations" (Gill 1949, p. 8). For example, recommendation number seven states, "Washing hands in the solvent should be forbidden" (Gill 1949, p. 11).

Anticipation and pre-planning were also aspects of health and safety support. One illustration of pre-planning was the Pre-Fire Plan program conducted by the Hanford Fire Department. This program commenced in the 1950s. In an interview, Towne discussed this form of health and safety planning:

Gordon Towne: *"The Pre-Fire Plans were detailed plans for every building on the complex. We would go through each building and write a plan as to how we would attack any fire at any given location in a building. This included a breakdown of all hazards, radiological aspects as well as chemicals... Pre-Fire Plan documents contained maps, written narrative, and diagrams. The Pre-Fire Plan for PUREX was 40 pages. They detailed everything pertaining to fire. We had full cooperation with building managers."*

Brian Freer: *"I liked what you said earlier, 'we fought a fire in every building before there ever was one.'"*

Gordon Towne: *"We would have sessions. Officers would get together, and brainstorm all the hazards we would think of in each building. We went through each of the buildings ourselves. The Fire Department closely followed the construction of each facility at Hanford. We knew them as they built them. We knew about plans and followed the construction all the way through. We received a number of AEC [Atomic Energy Commission] awards for this aspect of our fire safety. We actually took that idea and applied it throughout the AEC facilities. We went and explained the Pre-Fire Plan program, as it was developed at Hanford, to fellow firefighters at Oak Ridge, Livermore and other facilities."* - Towne 1998

Not all health and safety support work was conducted for one specific organization. In an interview, Towne recalled that fire prevention safety was the one aspect of Hanford Site safety that cross-cut operations, maintenance, construction, research and development, and administration areas.

In addition, the threat of fire extended beyond the built environment. According to Towne (1998), the natural environment at the Hanford Site accounted for a large portion of fire events historically and could easily transform the spark from a lightning strike into a threat to construction projects or reactor operations. Pre-planning then had to include the entire area of the Hanford Site.

Du Pont's history of the Hanford Site concurs with Towne's assessment: "One of the greatest concerns of the Fire Department was the thousands of acres covered with cheet grass, bunch grass, and sagebrush surrounding the plant areas and Hanford Camp. A potential fire hazard existed at all times" (Du Pont 1945a, p. 182). To illustrate the danger and unpredictability of "natural cover fires," as they were known to firefighters, consider that wind changes caused one fire on Rattlesnake Mountain to burn in four different directions in a 24-hour period. For Gordon Towne, the recurring hazards of natural cover fires are illustrated by a cluster of them he did not fight. When Towne was on vacation in July 1971, the Hanford Fire Department fought 64 natural cover fires in a 2-hour period (Towne 1998).

RELATIONSHIPS BETWEEN THE SUPPORT DISCIPLINES AND OPERATIONS

Authority is one issue that emerges when the responsibility for worker health and safety rests with operations. Another issue is compliance with recommendations. Together they lead to the question of what authority the health and safety support organizations had to enforce the recommendations they made to operations.

"If there was no fire we had recommendation authority only. If there was a fire in a building, then that building belonged to us."
- Gordon Towne, Assistant Fire Chief, Hanford Fire Department, 1947-1978 (Towne 1998)

From the time of the Manhattan Project to the early 1950s, radiation monitors reported to the Health Instruments Section, the radiation protection organization. Radiation monitors were the first line of defense from overexposure at the Hanford Site. They took readings from survey instruments and calculated worker exposure to radiation. Because the radiation monitors were not under the operations chain of command, they maintained their professional independence. However, by the mid 1950s most radiation monitors reported directly to the management of a particular facility rather than the Health Instruments Section. The Health Instruments Section still maintained a staff of radiation monitors primarily used for research purposes. With this alteration in the chain of command, the oversight and control function of radiation protection was less independent from the day-to-day concerns of operational activities. On the other hand, because the radiation protection function was responsible for the prevention of overexposure, the Health Instruments Section could now spend more time reviewing and investigating overexposure. Herbert Parker was one of the managers responsible for this change. He described the rationale behind it as reflecting "an attitude that ultimately each operator must take care of his own radiation monitoring, just as he protects himself by not walking under a ladder or by not handling acid without protective equipment" (Parker 1955, p. 903).

These changes meant that radiation monitors who commenced employment after the mid 1950s were less likely to have broad working experience in operational health physics at the Hanford Site. As a result, the work experience of radiation monitors became more and more facility specific. Radiation monitors who arrived after the mid 1950s also would have fewer opportunities for work experiences that would allow them to see how Hanford Site processes and functions were related to one another.

This division of work sometimes lead to confusion because of different lines of authority as a former industrial safety inspector, William Taylor, relates (see sidebar box). Lines of authority were more clear when it came to compliance issues. Towne (1998), relates:

> ***"The contractor, GE, had their own radiation monitor, and the support services group Hanford Laboratories, had their own monitor. One monitor is actually reporting to the operations manager and a second that would be reporting up the chain to the laboratories. In the beginning, the first ten times that happened to me or so it used to bother me. I'd be out there in the middle of night when the reactor was down and it was really hard to get time [for research].... But there was a good reason for it after I thought about this for a while. The guy I brought with me knew [how to monitor for my research], but he didn't necessarily know the reactor plant so they would give me somebody that knew the plant. So there was some advantage to having that plant based radiation monitor with me just to keep me out of trouble. So after a while I thought, yeah this is a good precaution, but it always added a couple of hours to the job."*** **- Taylor 1998b**

Gordon Towne: *"When we would get a request of that nature [on a compliance issue] we would respond as soon as possible, the evaluation was made and we made our recommendation to the operations manager. Normally it was corrected, or if there was a problem it was remedied."*

Brian Freer: *"Did you have a follow up after the recommendation was made?"*

Gordon Towne: *"Oh yes, always."*

Brian Freer: *"What kind of authority did the Fire Department have to enforce the recommendations?"*

Gordon Towne: *"If there was no fire we had recommendation authority only. If there was a fire in a building, then that building belonged to us."*

Brian Freer: *"Compliance with recommendations. Was it typically right away? Were there ever tough spots where compliance with fire safety recommendations was difficult? What might you do to facilitate that compliance?"*

Gordon Towne: *"Often it depended on money. If the money was in the budget, we got action. Suppose we recommended a sprinkler system for a building. They were expensive."*

Brian Freer: *"Were your recommendations factored into budgetary decisions?"*

Gordon Towne: *"I don't know how to answer that. Give you one example. We recommended that a sprinkler system be installed in all the power houses. Money wasn't in the budget. Then we had a fire in 184 East Powerhouse. Then all the Powerhouses were sprinklered. Where did all the money come from? I don't know. Money flowed after a major problem."*

Communication was an important ingredient in the relationship between operations and support groups. From the perspective of worker health and safety, at times this meant providing operations with information in a format that was directly relevant to operational objectives. An interview with Joe Soldat, health physicist at the Hanford Site from 1948-1995, illustrates this point (see sidebar box).

> ***"...when I was changing stack samples, if I noticed a high reading I would go talk to the operations manager in charge. One time I told an operations person, 'it looks to me like you have put out too much iodine.' He replied, 'is that bad?' I said, 'yep.' I explained to him the situation and he did whatever was necessary. When we got to REDOX, we had another innovation. We put the alarms from the count rate meter on the stack monitor in the operating office so that the guy in there could look at the recording chart. He didn't have to wait until somebody changed the sample to make adjustments." - Soldat 1999***

The Health Instruments Section at the Hanford Site put a great deal of effort into their calculations for acceptable operational activities and were responsible for informing operations of the importance of adhering to their guidelines.

> Joe Soldat recalls: *"I have a letter Art Keene [head of radiation protection in the 200 Areas] wrote to Koberg (Keene 1957). [Keene was responding to a letter that said], "we have a need for an extra push for plutonium. You said we can not release more than 10 curies a week. What if we did a special push and you allowed us to put out 30 curies in one week and we wouldn't do anything the rest of the time?" He wrote back, "No." Because the 10 curies per week was based on average long term meteorology, and if you put 30 in one week it might, at that particular time, put too much in the environment. Parker made it stick."* - Soldat 1999

An example of an instance in which the normal relationship between operations and support was abrogated is the Green Run, a secret military experiment conducted at the Hanford Site in December 1949 at the 221-T chemical separations facility. During the Green Run, operations suspended operational guidelines established by the Health Instruments Section on the authority of the U.S. military. This experiment became known as the Green Run because irradiated uranium fuel was still "green" when the chemical separations process began. "Green" meant the fuel had not been aged in a holding basin for the length of time recommended by the Health Instruments Section.

The Green Run set a benchmark for measuring environmental radiation levels emitted in the chemical separations process. At the Hanford Site, this test was accomplished by controlling the number of days spent fuel was cooled, bypassing filtration systems intended to capture emissions, carefully monitoring the chemical separations process, then measuring levels of airborne radioactivity. Exhaust stacks released a gaseous by-product, iodine 131, into the environment when the spent fuel was dissolved in acid. Airplanes traced the presence of iodine 131, as did stationary and mobile ground-based monitoring instruments as it moved through the atmosphere and subsequently deposited on the ground (Jenne and Healy 1950).

The nuclear arms race was behind this experiment because the United States wanted to know if the Soviet Union was neglecting to age their irradiated fuel. If the Soviet Union was not taking the same amount of time as the United States was to age the fuel, estimates of nuclear weapon stockpiles would have to be revised. Thus, the Green Run was an attempt to "test a methodology for monitoring the emerging Soviet nuclear program" (DOE 1995a, p. 216).

Although this test was not intended as a study of radiological effects on humans living in the pathway of this airborne release, "this test did result in the exposure of populated areas near the site" (DOE 1995a, p. 217). To put the Green Run release in historical context of other airborne radioactive releases, approximately 8,000 curies of iodine 131 were released into the atmosphere during this test: "Radioactivity released during the Green Run was, however, dwarfed by the amount of radioactive materials released during routine production activities at Hanford in the 1944-1948 period, before the installation of emission reduction equipment" (DOE 1995a, p. 217). Occupational and public health effects from the Green Run are still in litigation. This event was an aberration from operating limits already established, not an experiment to set standards.

> ***"The first impression that you get is that there are 5 radiation monitors for every industrial safety person at Hanford. But then when you take a look at the consequences to the public, to the city, to this area where we are sitting it is not too hard to rationalize. I didn't think they worked as hard as we did."*** **- William Taylor, Industrial Safety Engineer at the Hanford Site 1948-1994 (Taylor 1998b)**

RELATIONSHIPS AMONG THE SUPPORT DISCIPLINES

According to health and safety professionals who worked at the Hanford Site during the Cold War, the emphasis was on radiation protection. The historical emphasis on radiation protection, at times at the expense of other protection programs in areas such as chemical exposure, is supported by a recent report the National Economic Council issued at the request of President Clinton. An inter-agency panel of U.S. government departments (for example, Health and Human Services, Labor) reviewed occupational epidemiological studies pertaining to the Hanford Site and identified significant findings for this National Economic Council report, *The Link Between Exposure to Occupational Hazards and Illnesses in the Department of Energy Contractor Workforce* (NEC 2000, p. 2). The report notes: "there is a general lack of industrial hygiene monitoring data for chemical and physical hazards."

A 1992 historical overview, written from the standpoint of operational radiation protection at the Hanford Site, supports this assessment: "The previous emphasis on radiation safety, in many cases to the exclusion of industrial safety, no longer exists" (Sudmann 1992, p. 11). This does not imply that messages about standard industrial safety were not transmitted to individuals. To the contrary, the archival record indicates that the promotion of basic industrial safety was a highly visible, slogan-oriented, regular feature of an individual's work world at the Hanford Site from the Manhattan Project through the Cold War.

According to the remarks of the Closing Panel on "The Use of Risk Information in Making Clean-up Decisions" presented at the Second Annual Health of the Hanford Site Conference, the emphasis on radiation protection at the Hanford Site is a historic artifact (Takaro 1998). One reason for the emphasis was the perception of danger posed by radioactive materials being produced on an industrial scale for the first time. With the uncertainty of such factors in mind, management for worker health and safety at the Hanford Site decided early on to keep radiation exposure records on personnel, and they maintained an extensive radiation dosimetry record for work sites.

In contrast, large volumes of chemicals had been produced industrially for a number of years before the Manhattan Project. General familiarity with large-scale chemical usage in industry may, along with other reasons, have contributed to the fact that from 1943-1990, the Hanford Site did not maintain a database on chemical exposure to workers. On the issue of chemical exposure, the National Economic Council noted: "little is known about the actual levels of worker exposure to these materials. DOE does not maintain a centralized repository of industrial hygiene information on exposures to individual chemicals" (NEC 2000, p. 5). The National Economic Council report would be a good starting place for readers interested in exploring this topic further because it contains summaries of reports and a useful bibliography.

The recollections of Hanford Site health and safety professionals allow further exploration of radiation protection as it pertains to relationships between the disciplines. One health and safety professional, Les Cronk, had the unique perspective of working at the Hanford Site for many years, then transferring to a laboratory research setting at the Lawrence Livermore National Laboratory, and finally returning to the Hanford Site. He draws a distinction between industry and research in his interpretation of relationships between health and safety support at the Hanford Site (see sidebar box).

> ***"Hanford is contrasted with Livermore in that there is an emphasis on the radiological at Hanford. They are a lot different. At Hanford you have health physics personnel who are performing radiological surveys, taking air samples and such. Whereas in other facilities that person would be an 'all around' type person concerned with health and safety in general from a broader perspective."*** **- Cronk 1998**

The emphasis on radiation protection at the Hanford Site required a special force of radiation monitors and professionals more broadly trained in the field as a whole. Over time, a two-tier occupational structure in radiation protection emerged at the Hanford Site. However, routine monitoring required less formal training than did data analysis. With the unionization of radiation monitors, training in this technical field became geared more to vocational schools, community colleges, union halls, and contractors than to research universities. However, health physics became a field of study at several major universities, with professional certification from the American Academy of Health Physics becoming a reality in 1959.

"We didn't do any radiological work at all [as industrial hygienists]. I don't think it has ever been thought of or been allowed. Well, it's territory. The HPTs [Health Physics Technicians, formerly Radiation Monitors] continue to try to get into industrial hygiene at Hanford even today. Everywhere else in the real world, like at Commonwealth Edison and other places in the civilian nuclear reactor facilities, NRC [Nuclear Regulatory Commission] regulated nuclear facilities, or in research facilities that use radioisotopes which use RSOs [Radiation Safety Officer] and in these contexts the industrial hygiene professionals do as much radiological work as the RSOs. The industrial hygiene folks don't sign the papers, but they are smart enough to figure things out and use the survey instruments. Here at Hanford it is almost an Act of God to do a radiological survey no matter who you are." **- Beck 1998**

Herbert Parker noted that Hanford Site radiation monitoring was unique in its approach to the task of radiation protection. The reason for this, as he put it, was a "tendency in industry to incorporate the radiation protection function with overall safety and health functions." According to Parker, at the Hanford Site radiation protection was more closely related to operations than at other nuclear facilities: "protection requirements apply to practically every operating task, phase of design, and construction of improved facilities" (Parker 1955, p. 903).

WORKER PERSPECTIVES ON HEALTH AND SAFETY

Workers often faced the conflicting demands of ensuring health and safety while meeting the production requirements. Two sources relate the health and safety experiences of former Hanford Site workers in their own words: *Hanford and the Bomb: An Oral History of World War II* (Sanger 1989), which is an oral history focusing on the Manhattan Project, and the transcript of a February 3, 2000 public meeting the U.S. Department of Energy's Office of Environment, Safety, and Health sponsored (DOE 2000). According to Assistant Secretary of Environment, Safety, and Health, David Michaels, the purpose of this meeting was to listen to the health and safety experiences of current and former workers. The meeting was held at the Richland Federal Building. An estimated 600 people attended this meeting, and 50 people provided testimony. The time period discussed at the February 3, 2000 public meeting was from the start of operations to the present.

Orville F. Hill shared his perspective on work and radiation during the Manhattan Project:

> *"I wasn't really concerned about radiation in those days, but I had respect for it. When I look back, we did a lot of foolish things. We took risks, to get results rather than spend days and weeks designing equipment so that you could do everything without getting very close or taking any risks. We went ahead and did the work."* - Sanger 1989, p. 156

Former Hanford worker Jack Miller worked on the reactors. He recounts an experience removing spent fuel elements from one of Hanford's reactors:

> *"I can remember being on the rear face, wearing coveralls, cloth boots, cloth gloves, rubber boots and rubber pants, rubber hat and jacket and rubber glove and assault mask. It's difficult to talk into the communications mike [microphone] when you're wearing an assault mask, which is like an Army gas mask. I was trying to raise the front face, 'rear to front, rear to front,' no response. Then I tried 'rear to control,' there's always somebody*

there. No response. I got so damned disgusted. I started to slam that mike down and noticed I only had a foot of cord on the mike, one of the pieces of machinery coming by had cut the cord. I got my hands contaminated a number of times, because of a leaky glove. We would wash off with soap and water." - Sanger 1989, p. 143

Mac MacCready offered this perspective on precautions:

"In those days, we did a lot of precautions because we didn't know. All this process, you see, was designed on the basis of work done in the laboratory with a few micrograms of plutonium." - Sanger 1989, p. 147

Wakefield Wright had these recollections of his work life at Hanford:

"When we first came here, for years, they would come by and leave a bottle on the porch. And you were required to urinate in that bottle and then they would come and pick it up. We were tested once a month, during the war and afterward. We had yearly physical examinations. And, every day somebody would check your feet for radioactivity.... It was much easier after they dropped the bombs. Before the bombs, you would have to tell them, 'Put your confidence in me, because I am not letting you go somewhere where you would be hurt but at the same time I can't tell you what you are doing either.' Not many of them guessed. I couldn't say anything about radioactivity, not until after August 6." - Sanger 1989, p. 155

Retired Hanford worker George Foster spoke at the February 3, 2000 public meeting, (DOE 2000, pp. 73-79). He shared a story that illustrates how some of the occupational health and safety history of Hanford's Manhattan Project and Cold War era went undocumented. Foster arrived at the Hanford Site on February 2, 1948. He received a puncture wound with plutonium on December 29, 1952, but as he recounts below, he didn't know this at the time of the accident:

"I had on two pair of surgeon gloves, then a pair of canvas gloves, and then a big heavy rubber glove, working in that hood. And I couldn't quite reach the pipe. Well, when I pulled that up over there, why, I pricked my hand. I knew that I had ruptured the glove. And so I told them, 'I ruptured my glove.'

So they peeled that outer garment off, put it in a proper container, right on down to the last surgeon glove, and it was hot, over a million count again. They didn't know that [the puncture] had actually went into the skin, because there wasn't any blood or anything. They didn't even report it as an accident. I asked Dr. Brockman, she was the head of all the nurses — I mean, all the doctors and nurses, and I asked her about cutting it out. Oh, she talked back in her throat, oh, she says, it will just make another sore, just make another sore. So that was out. But anyway, I quit GE and went back out on construction." - DOE 2000

Foster goes on to recall a conversation with co-workers over a poker game. A co-worker said a woman at a Hanford laboratory was scared when she brought in a certain worker's urine sample because it was so "hot" with plutonium. No names were on the labels for bottled urine samples; the labels were numerically coded with employment identifiers. According to Foster,

"The lab worker said the number on the hot bottle was 6204. That was my number, 6 identified me as a pipefitter. Well, Bill Rice, the foreman, right away…called me in and wondered where I got it. And so at first I was at a loss. They told me I had plutonium. I was working in the 300 Area at the time. And they told me, they asked me if I knew where I got it. And he says, 'Well, there isn't any plutonium in the 300 Area. You must mean uranium, don't you?' And he says, 'No, plutonium.' We tore a hood out of a certain building in the 300 Area and [he] said we thought maybe you got it [there]. I told him, 'Oh, that's where I got the contamination, was with that wire brush.'" - DOE 2000

Later, Foster consulted with Hanford occupational medical doctors regarding his plutonium deposition.

"So they said they wanted that man [Foster] followed up with bottles...regardless of where that man went. [The doctors said I] should have been followed with bottles immediately [after the accident]. [Regarding his case, Foster overheard one Hanford medical professional say,] 'Well, we didn't have enough systems set up at that time.' [Another medical professional responded] 'Oh, yes, we did. That system was set up.' They [occupational medical doctors reviewing his case] were really chewing on each other in there." - DOE 2000

Another worker, Maffeo, arrived for work at the Hanford Site in 1949. He worked in the laboratories of the 200 East and 200 West Areas:

> *"The Hot War was over and the Cold War ensued. And there was a lot of pride out here, a lot of pride. You talk about safety. Yes. There was a lot of safety. And all the analytical procedures we had, we had a procedure on the left-hand side and safety on the step by step. If it had 20 steps, there would be 20 comments, beware of this, be careful of that, don't do this, don't do that. So there was a very, very —And yet there was problems. Always something popped up that would be wrong. And it was rectified. And there was a lot of pride in the people here. And I really thought that it was a magnificent effort. I mean, solid. And it wasn't just an individual effort. It was everybody that participated in this tremendous effort, because production, production, production was the key, and safety, safety, safety ran right alongside of it."* - DOE 2000, pp. 104-105

These first person accounts of worker health and safety suggest that occupational experiences were diverse and that workers remembering these experiences often do so in different voices and from heterogeneous vantage points. While some former Hanford workers highlight the plant's accomplishments, others point out the failures and accidents. As such, these remembrances also suggest not one single worker viewpoint typifies the Hanford occupational experience. Rather, these accounts may be a starting point for a closer look at other individual occupational experiences as a way of learning more about worker health and safety at the Hanford Site. Such personal accounts are important to the historic record of health and safety at the Hanford Site because the undocumented accident alongside the everyday experience demands the same attention and scrutiny as is given to statistical data. Through close attention to the detail of occupational experience, future historians will begin to shape a more complete picture of health and safety at the Hanford Site.

RESEARCH AND STANDARDS

Early in the 20th century, standards for radiation exposure were based on the concept of a tolerance dose. A tolerance dose was a fraction of the amount of radiation (x-rays in those days) that resulted in erythema (reddening of the skin). The tolerance dose, therefore, was a product of the directly observable, external physical effect of ionizing radiation and was defined as "the dose which an operator can, for a prolonged period of time, tolerate without ultimately suffering injury" (Arthur Mutscheller quoted in Hacker 1987, p. 15). After the Manhattan Project, the concept of tolerance was discarded in favor of "permissible radiation exposure" (AEC 1951c, p. 2103). The term "permissible radiation exposure" reflects a change in philosophy from earlier notions of "tolerance." For worker health and safety at the Hanford Site, the main difference between the two was that "permissible radiation exposure" assumed that any dose carried with it some degree of risk. In operational settings at the Hanford Site, a conservative limit, called an operational control, was set at a figure lower than the maximum permissible exposure. As of 1990, radiation protection philosophy was based on the concept of ALARA (as low as reasonably achievable). This philosophy attempted to achieve optimization between the costs of the detrimental effects of radiation exposure and the benefits derived from the use of radioactive materials. ALARA included technical as well as social and economic factors.

APPLIED RESEARCH

Scientists at the Hanford Site knew that information, practices, and policies regarding radiation health and safety had to be systematically evaluated through applied research. Because the Hanford Site was primarily a production facility, the applied research tended to be focused on real problems as they were discovered. As Findlay and Hevly (1995, p. 13) have observed, "this industrial story—less glamorous, perhaps, than the scientific story—has been slighted by historians, but it is every bit as important and as complicated." Health and safety professionals at the Hanford Site could inform themselves with basic research being conducted at the facility, at other sites associated with the nuclear defense mission, and at universities, but ultimately the identification and solution of problems was based on the constraints of operational requirements (Findlay and Hevly 1995, p. 72). For additional information, see Section 7, Research and Development.

Figure 2-10.5. *Worker Conducting Self-Survey of Thyroid*

The newly emerging field of health and safety in a plutonium production facility developed in tandem with day-to-day operations. At the Hanford Site, applied research was the most direct route of obtaining feedback on operational activities. Applied research used data from operational practices and routine monitoring: "their solutions tended to emphasize a high level of hands-on knowledge and direct problem-solving techniques" (Findlay and Hevly 1995, p. 52). One source of data from routine monitoring was equipment workers could use to survey their thyroid glands (see Figure 2-10.5). Such sources of information were used to keep the exposure levels of onsite workers beneath accepted limits and to prevent other industrial dangers from causing them harm (Selby and Soldat 1958).

Beginning in 1949, applied research at the Hanford Site began to include measurement of plutonium in the bodies of deceased workers. In an effort to correlate models of worker radiation dose with the actual amounts of plutonium absorbed into the body, tissue samples were taken from some deceased plutonium production workers with suspected high concentrations of internal contamination. A sample of non-plutonium workers, local residents, and residents from the northwest region were also autopsied to determine if any plutonium was present (see, for example, Newton et al. 1966). These tissue samples were taken at the time of autopsy and underwent laboratory analysis (Kathren 1995, p. 349). Before 1968, no informed consent was requested from the families of subjects specifically requesting permission to remove tissue and conduct analysis for plutonium. Between 1949 and 1968, family members appear to have authorized most of the autopsies, but the research to measure plutonium in tissues removed during autopsy was not commonly disclosed to those granting permission. Funds for this research most likely came from the budget of the Medical Department. However, other Hanford Site departments also may have contributed funds, although these are difficult to account for in the official records.

In 1968, the tissue analysis program became a voluntary program the Atomic Energy Commission formally recognized. In 1970, this program was named the United States Transuranium Registry. Through 1990, this program provided important feedback to verify models of radioisotope uptake into the human body based on the actual exposures of workers to plutonium and americium (Stannard 1988, p. 1469). This information was then factored into protection programs at the Hanford Site and elsewhere.

Another example of applied research on worker health and safety undertaken by Hanford Site staff is illustrated in an October 14, 1977 edition of Battelle's newsletter, *The Greenie*. This newsletter contains an article that profiled the Safety Analysis Section of the Energy System Department. The Associate Manager of that group, Ed Watson, called his environmental consequences team "tail-end Charlies" because they represented the last step in the process of protecting humans from "toxic materials at every stage of energy production" (Battelle 1977). Applied research this group conducted at the Hanford Site included sampling waste streams in various environmental media and checking the efficiency of emission filtration systems.

Applied research at the Hanford Site increased after the passage of the National Environmental Protection Act (NEPA). President Nixon signed NEPA into law on January 1, 1970. For these "tail-end Charlies," as Ed Watson noted, "NEPA is the biggest impetus for our business. It really spelled out the need for assessing the consequences of technologies to man and his environment. Now we have an overload of projects that touch on a variety of energy-generating systems" (Battelle 1977, p. 1).

Another applied research study was the whole body count of local school children that took place from the early 1960s to the mid-1970s (Honstead and Eichner 1972) (see Figure 2-10.6). This study was part of a larger worker surveillance program (Honstead 1968). A mobile whole body counter was used to detect and measure internally deposited radioactive material. For a physical and technical description of this mobile whole body counter, see, for example, Eichner 1969 or Brady and Swanberg 1965. Scientists studied the uptake measured in relation to the diet of the school children as a means of verifying environmental models (Wilson 1987, p. 4.15). The purpose of this program was to monitor the local population for fallout of nuclear weapons and to evaluate the Columbia River as a pathway for radionuclides to enter the human body through food intake. According to findings based on this study, the radiation found in the bodies of local children were trace amounts, enough to be detected but negligible when compared to public body burden limits. Thus, this particular study did not affect radiation standards for workers at the Hanford Site.

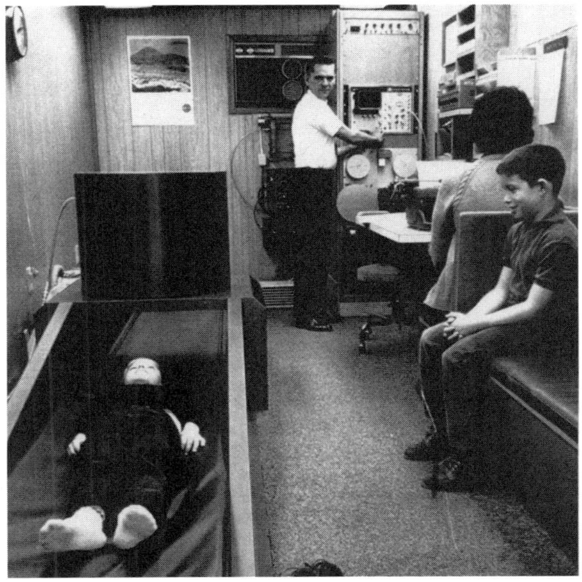

Figure 2-10.6. *Mobile Whole Body Counter Used for Monitoring School Children*

The literature on applied research efforts includes numerous other examples of studies conducted at the Hanford Site. For example, Parker "initiated expansion into a practical program in [radiobiology] linked to the hazards to humans and the environment that had arisen in their work" (Stannard 1988, p. 434). Commenting on this research, Stannard (1988, p. 435) notes, "the bulk of the work involved feeding contaminant [radio iodine] to sheep to tie as closely as possible to problems in the field" (see Figure 2-10.7). For additional information on the Biology Laboratory where the sheep were kept (108-F Building), see the Expanded Historic Property Inventory Form in Appendix B on the Internet. These studies consisted of both basic and applied research and contributed to basic scientific understanding of the relationship between radiation and health. However, the topics of experiments were often based on outstanding problems stemming from operations at the Hanford Site.

Figure 2-10.7. *Worker Feeding Sheep Radioactive Food at the 108-F Building*

EPIDEMIOLOGICAL RESEARCH ON HANFORD WORKERS

Worker health and safety at the Hanford Site has been the subject of ongoing epidemiological research since the 1960s. The primary question that scientists are interested in answering centers on whether or not a statistically significant relationship exists between worker radiation dose and increased mortality. Mancuso et al. (1977) published the first peer-reviewed article based on data from the Hanford Site in the open scientific literature. This study, "an investigation of mortality among Hanford workers," found "elevated mortality for multiple myeloma and cancers of the lung and pancreas" (Wilkinson 1991, p. 715). Gregg S. Wilkinson, editor of *Occupational Medicine's* "State of the Art Review on the Nuclear Energy Industry," goes on, "This was subsequently followed by another analysis conducted by Gilbert and Marks that disputed some of Mancuso's findings but confirmed the presence of excess multiple myeloma and pancreatic cancer mortality" (Wilkinson 1991, p. 715). According to Wilkinson, "An enormous controversy erupted...which continues to this day. The excess of multiple myeloma has continued, whereas the pancreatic cancer has diminished with additional follow-up" (Wilkinson 1991, p. 715). The recent National Economic Council report includes a table, "Summary of Statistically Significant Elevated Findings for Cancer at Department of Energy Sites," which lists cancer of the pancreas as a standardized mortality ratio study with statistically significant findings for specific organ site cancer. Cancers indicated by a statistically significant trend analysis for radiation versus indicated cancer include liver, pancreas, genital cancer, Hodgkin's lymphoma, and multiple myeloma (NEC 2000, p. 9).

The general reader should be aware that this is a complex issue specialists in the field of occupational health are currently debating. Readers may wish to consult overviews of the various papers to gain a sense of the differences of statistical interpretation assigned to the data sets. The National Economic Council report (NEC 2000), contains a useful overview of findings from occupational epidemiological studies. The Hanford Health Information Network website, http://www.doh.wa.gov/hanford/, contains, among many other useful health and safety-related publications, "Hanford-Related Studies" and "Hanford-Related Projects." "Hanford-Related Studies" provides an annotated list of health studies with a Hanford component. "Hanford-Related Projects" contains descriptions and contact information on projects, such as the Hanford Environmental Dose Reconstruction Project.

STANDARDS

Guidelines for radiation protection standards at the Hanford Site came from three main sources. The first source consisted of recommendations made by advisory committees such as the National Council on Radiation Protection and Measurements and the Federal Radiation Council. The recommendations they made did not carry a legal requirement for implementation at the Hanford Site. The second source of guidelines came from the federal agency responsible for the Hanford Site: the Atomic Energy Commission, the Energy Research and Development Agency, and the U.S. Department of Energy. The standards these agencies set typically followed standards of the National Council on Radiation Protection and Measurements and were legally binding and enforceable at the Hanford Site. The third source of radiation protection standards were set at the Hanford Site. These standards, known as operational controls, set lower working limits to ensure compliance with those set by the federal agency responsible for the Hanford Site.

Dade W. Moeller (1989), a health physicist, gives a historical perspective on the development of radiation protection standards in the 20th century. At the twenty-fifth annual meeting of the National Council on Radiation Protection and Measurements, Moeller discussed what he termed seven stages in the development of radiation protection standards. These stages show the trend toward lower limits and the transition from the concept of a tolerance dose to ALARA (as low as reasonably achievable):

Avoidance of Acute Effects (1900-1930)
Concern for Chronic Effects (1930-1950)
Concern for Genetic Effects (1950-1960)
Concern for Somatic Effects (Primarily Leukemia) (1960-1970)
Concern for Somatic Effects (Primarily Solid Tumors) (1970-present)
Application of a Risk-Based Approach (1980-present)
Approach to Professional Maturity (1990-present)

A historical overview of radiation protection guidelines enforceable at the Hanford Site is shown in Table 2-10.1. The general trend is toward lower worker exposure limits.

The Health Instruments Section established operational controls at the Hanford Site in conjunction with operational management. The purpose of operational controls was to maintain a conservative first line of protection against overexposure and ensure facilities operated within their safety envelope and in accordance with their design. Operational controls were always set lower than the maximum permissible limit.

The 1954 issue of the General Electric manual Radiation Protection Standards notes that operational controls were "substantially more conservative than official maximum permissible limits" (GE 1954q, p. 1). On October 12, 1967, the head of radiation protection in the 200 Areas, Art R. Keene, wrote a letter to Fred W. Albaugh to clarify the role of

Table 2-10.1. *Chronology of Radiation Standards: Occupational External Whole Body Dose Equivalent Limits, 1947-1977*

	National Council on Radiation Protection and Measurement (NCRP)		International Commission on Radiological Protection (ICRP)		1960 Federal Radiation Council (FRC)		Atomic Energy Commission (AEC) Energy Research and Development Administration (ERDA) Department of Energy (DOE)	
Year	**Criteria**	**Reference**	**Criteria**	**Reference**	**Criteria**	**Reference**	**Criteria**	**Reference**
1947	0.1 rad/day 0.5 rad/wk	NBS 1934 (Handbook 18)	0.2 rad/day 1.0 rad/wk				0.1 rad/day	NCRP NBS 1934 (Handbook 18)
1949	0.3 rad/wk	NBS 1949 (Handbook 42)	0.2 rad/day 1.0 rad/wk				0.1 rad/day	
1950	0.3 rad/wk		0.3 rad/wk				0.3 rad/wk 3.9 rad/13 wk	NBS 1950 (Handbook 47) AEC (DBM)
1951	0.3 rad/wk		0.3 rad/wk				3.0 rad/Series 2.0 rad/Series 3.9 rad/13 wks	AEC (LASL 1952 - WT 204) AEC (Ranger/ Greenhouse) AEC (DBM)
1952	0.3 rad/wk		0.3 rad/wk				3.9 rad/13 wks	AEC (DBM)
1954	3.0 rad/13 wks 0.3 rad/wk Max 15 rem/y	NBS 1954 (Handbook 59)	0.3 rad/wk				3.0 rad/13 wks 0.3 rad/wk Max 15 rem/y	NCRP NBS 1954 (Handbook 59)
1957	5 rem/y Ave 12 rem/y Max	Addendum NBS 1954 (Handbook 59)	0.3 rad/wk				3.0 rad/13 wks 0.3 rad/wk Max 15 rem/y	
1958	0.3 rem/wk Max 3 rem/13 wks 12 rem/y Max 5 (N-18) rem*	Addendum NBS 1954 (Handbook 59)	0.1 rem/wk 3.0 rem/13 wks 5 (N-18) rem*	ICRP 1959 (Report 1)			0.3 rem/wk 3.0 rem/13 wk 12 rem/y Max 5 (N-18) rem*	NCRP NBS 1954 (Handbook 59)
1959	0.3 rem/wk Max 3 rem/13 wks 12 rem/y Max 5 (N-18) rem*		0.1 rem/wk 3.0 rem/13 wks 5 (N-18) rem*				0.3 rem/wk 3.0 rem/13 wk 12 rem/y Max 5 (N-18) rem*	
1960	0.3 rem/wk Max 3 rem/13 wks 12 rem/y Max 5 (N-18) rem*		0.1 rem/wk 3.0 rem/13 wks 5 (N-18) rem*		3 rem/13 wk 5 rem/y Ave 12 rem/y Max 5 (N-18) rem*	FRC 1960 (Report 1)	3 rem/13 wk 5 rem/y Ave 5 (N-18) rem*	FRC 1960 (Report 1)
1965	0.3 rem/wk Max 3 rem/13 wks 12 rem/y Max 5 (N-18) rem*		3 rem/13 wk 5 rem/y Max	ICRP 1966 (Report 9)	3 rem/13 wk 5 rem/y Ave 12 rem/y Max 5 (N-18) rem*		3 rem/13 wk 5 rem/y Ave 5 (N-18) rem*	
1971	3 rem/13 wk 5 rem/y	NCRP 1971 (Report 39)	3 rem/13 wk 5 rem/y Max		3 rem/13 wk 5 rem/y Ave 12 rem/y Max 5 (N-18) rem*		3 rem/13 wk 5 rem/y Ave 5 (N-18) rem*	
1974	3 rem/13 wk 5 rem/y		3 rem/13 wk 5 rem/y Max		3 rem/13 wk 5 rem/y Ave 12 rem/y Max 5 (N-18) rem*		3 rem/13 wks 5 rem/y	NCRP 1971 (Report 39)
1977	3 rem/13 wk 5 rem/y		5 rem/y acceptable risk	ICRP 1977 (Report 26)	3 rem/13 wk 5 rem/y Ave 12 rem/y Max 5 (N-18) rem*		3 rem/13 wks 5 rem/y	

*N-18 is years of occupational exposure after age 18, where N is the worker's age in years – a guideline used historically to show maximum cumulative occupational radiation dose to an adult worker. This approach is no longer used in either the DOE or NRC occupational exposure regulations.

operational controls at the Hanford Site. In this letter, Keene wrote that operational controls "provided a rational tool for operating management to control their exposure in a manner which would assure [sic] compliance with the basic limits and still provide reasonable latitude throughout the month" (Keene 1967, p. 1; Hicks and Yesberger 1967).

In 1954, operational controls at the Hanford Site were based on a 20 percent safety factor for most types of radiation exposure. The same conservative approach was still evidenced in 1967. The operational control for radiation exposure was set at 4 rem. The Atomic Energy Commission maximum permissible limit was 5 rem, reflecting a safety factor of 20 percent. The 1951 REDOX technical manual provides a window into this topic from an operations perspective: "Because [of] the lack of full, exact knowledge, the permissible exposure limits at the Hanford Works [Hanford Site] are conservatively set and frequently reviewed in light of new information. The safety factors involved are not exactly known. In some cases they may be quite large; in others, just adequate" (AEC 1951c, p. 2103). Operational controls were the first line of administrative defense against overexposure throughout the Cold War.

In the case of external exposure, when a worker is removed from a radiation field the exposure stops. However, exposure from radioisotopes inside the body continues until the substance is removed, eliminated, or decays to stability. A body burden is the maximum permissible amount of a radioisotope allowed in a worker's body. At the Hanford Site, operational controls for body burdens have historically been set at one-fifth of the maximum permissible exposure limit.

At the Hanford Site, primary operational controls for preventing internal depositions were established by limiting the concentrations of radioisotopes in air and water. Secondary operational controls were maintained through the use of personnel protective gear such as filtered breathing masks. Thus, operational controls for internal exposure were different from controls for external exposure. However, they were often used in conjunction with one another.

Table 2-10.2. *Occupational Exposure Limits for Radioisotopes at the Hanford Site, 1954*

Maximum Permissible Amount of Radioisotope in Total Body and Maximum Permissible Concentration in Air and Water for Occupational Exposure at Hanford Atomic Products Operation[(a)]

Radioisotope (Body-soluble)	Critical Organ	Microcuries in Total Body	Microcuries per cc of Water	Microcuries per cc of Air
H^3 (HTO or T_2O)	Body	3000 (b)	2×10^{-1} (b)	1.2×10^{-5} (b)
Si^{31}	Body	76 (c)	2×10^{-1} (c)	3×10^{-5} (c)
C^{14} (CO_2)	Fat	250	3×10^{-3}	5×10^{-7} (bone)
P^{32}	Bone	10	2×10^{-4}	1×10^{-7}
S^{35}	Skin	100	5×10^{-3}	1×10^{-6}
$C1^{35}$	Body	200	2×10^{-3}	4×10^{-7}
A^{41}	Body	30	5×10^{-4}	5×10^{-7}
Fe^{55}	Blood	1000	4×10^{-3}	6×10^{-7}
Fe^{59}	Blood	11	1×10^{-4}	1.5×10^{-8}
Co^{60}	Liver	3	2×10^{-2}	9×10^{-6}
Ni^{59}	Liver	39	2.5×10^{-1}	2×10^{-5}
Cu^{64}	Liver	150	8×10^{-2}	6×10^{-6}
KR^{85} (4.4 hr)	Body	80 (c)	1×10^{-3} (c)	1×10^{-6} (c)
Sr^{89}	Bone	2	7×10^{-5}	2×10^{-8}
Sr^{90} (+ Y^{90})	Bone	1	8×10^{-7}	2×10^{-10}
Ru^{106} (+ Rh^{106})	Kidneys	4	1×10^{-3} (b)	3×10^{-8}
I^{131}	Thyroid	0.3	3×10^{-5}	3×10^{-9}
Xe^{135}	Body	100	1×10^{-3}	2×10^{-6}
Po^{210}	Spleen	0.02	3×10^{-5}	2×10^{-10}
Rn^{222} (+Daughters)	Body	---	2×10^{-6}	1×10^{-8}
Ra^{226} (+ 1/2 Daughters)	Bone	0.1	4×10^{-8}	8×10^{-12}
Th^{234}	Bone	120	3	6×10^{-7}
U^{233}	Bone	0.04	1.5×10^{-4}	1×10^{-10}
U^{235}	Bone	0.07 (c)	6×10^{-4} (c)	1×10^{-10} (c)
Pu^{239}	Bone	0.04	1.5×10^{-6}	2×10^{-12} (c)
Pu^{239} (insoluble)	Bone	0.008	---	2×10^{-12}
Am^{241}	Bone	0.056	1×10^{-4}	3×10^{-11}
Cm^{242}	Bone	0.05	9×10^{-4}	2×10^{-10}

(a) Recommendations of NBS 1953 (Handbook 52) unless otherwise indicated
(b) Recommendations of H. A. Kornberg
(c) Calculated by H. V. Clukey

Maximum permissible limits for air and water were first set down by Parker in "Tolerance Concentration of Product in Drinking Water" on August 4, 1945 (Parker 1945). The 1954 *Radiation Protection Standards* (GE 1954q, p. 3) manual lists occupational exposure limits for radioisotopes by microcuries per cubic centimeter of water and air (see Table 2-10.2).

PREVENTION AND PROTECTION

The prevention of health and safety problems was one of many factors decision-makers considered at the beginning of the Manhattan Project when the sites were selected. The primary goal during the Manhattan Project remained the production of a specified amount of plutonium by a set date. Manhattan Project management incorporated worker health and safety planning into the Hanford Site's wartime operations to the extent that it was feasible and did not jeopardize the ultimate goal.

For both the Manhattan Project and Cold War era, understanding the prevention and protection measures at the Hanford Site required understanding the types and effects of radiation. It is commonly thought that as the energy level of radiation increases, so does biological damage. This is not always the case. Consider cosmic rays, which have very high energy but most pass right through us without any interaction. Biological damage is instead a function of both the mass and energy of the impinging radiation. Massive particles such as alpha particles cause a great deal of damage in very small areas, whereas gamma rays may have only a few ionizations with the same unit volume. Ultimately, how damaging the type of radiation may be is a function of its ability to transfer energy to the medium it is traveling through.

Barrie Lambert (1991, p. 725) notes "it is generally accepted now that all exposure to ionizing radiation results in some damage to biologic tissue." However, he goes on, "the evidence for biologic effects" at extremely low levels is "scanty," or, in other words, at certain extremely low levels the damage caused may be within the ability of the body to heal itself. Repeated exposures to low levels of ionizing radiation are another matter. Repeated exposures are the main subject of ongoing discussions in the scientific literature regarding radiation dose and biological response. This type of radiation exposure is often referred to as chronic exposure. Acute exposures are larger doses of radiation that have been demonstrated to cause serious injury or even death at sufficiently high exposures. One related, and unresolved, scientific question pertains to the existence of a threshold level at or below which no biological harm occurs from ionizing radiation (Cronkite and Musolino 1996, Nussbaum 1998). For a scientific description of radiation, see the various publications on the Hanford Health Information Network website at http://www.doh.wa.gov/hanford.

RADIATION PROTECTION: SHIELDING, DISTANCE, AND TIME

The three main radiation protection principles are shielding, distance, and time. In earlier days, bronze foundries and blacksmith shops safeguarded artisans from molten metal with heat shields and by distancing workers from heat sources. While the tools of radiation detection and measurement have become more precise, the principles of shielding, distance, and time have remained constant. "Safety against sources of highly penetrating radiation is maintained, in general, by three methods. Massive

Example of Radiation Sources Requiring Protection

Workers in reactor buildings at the Hanford Site had to be protected from all sources of radiation. The reactors themselves were heavily shielded with a thermal shield and a biological shield. But all reactor internals and coolant piping became radioactive to a greater or lesser extent. Thus objects and materials to be monitored and protected against included the following:

- Discharged fuel slugs, which contained fission products and gave off alpha, beta, and gamma radiation in all energy ranges
- All pieces of equipment (fixtures, tubing, piping, fittings) removed from the reactor, which were generally highly radioactive
- Control rods used to control and stop the chain reaction
- Air and gases in the reactor
- Cooling water, which came in contact with the fuel elements and so contained minerals made radioactive by neutrons in the reactor
- Miscellaneous contaminated materials such as dust and moisture, which was irradiated
- Tools used to service contaminated or activated reactor components

shields are placed about the radiation source, or smaller movable shields may be erected about a temporary hazard. Where workers must handle radioactive material that emits penetrating radiation, they perform their tasks from a distance or they work only for brief periods in the radioactive area and thus limit total exposure, since the amount of radiation received is a factor not only of its intensity but also of duration of exposure" (AEC 1956, p. 19).

Shielding

The first principle in radiation protection is shielding. Shielding refers to physical barriers that block substances or energy from passing through. Lead, cement, earth and water are often used as materials in shielding against electromagnetic forms of ionizing radiation because of their high densities.

Reactors at the Hanford Site were surrounded by special layers of material to stop alpha and beta particles and gamma radiation as well as slow down and absorb neutrons. A concrete wall shielded the discharge face of the reactor. Once slugs of irradiated fuel were discharged, they fell into and were shielded by a deep pit of water. The chemical separations buildings at the Hanford Site are an example of facility design being driven by shielding requirements. As with the reactors, reinforced concrete walls separated operators from process equipment. Portable and temporary shielding was often used for maintenance and repair work in reactors and chemical separations facilities.

Shielding is not regulated. The regulations address maximum allowable exposure rates and accumulated dose. Shielding helps keep exposure rate and dose below allowable levels.

Distance

Maintaining distance from radiological materials is the second principle of protection against radiation. A basic rule of radiation physics is that radiation (except machine-generated radiation beams) is emitted equally in all directions. Radiation monitors, therefore, often use the inverse square law to calculate radiation intensity. "It states that the radiation dose rate from a small source drops off as the square of the distance from the source. That is the dose at 2, 3, 4, 5, etc. meters from a small source will be _, 1/9, 1/16, 1/25, etc. of the dose at 1 meter" (Frigerio 1968, p. 23-24). Radiation monitors use this rule of thumb to estimate radiation dose in specific areas.

At the Hanford Site, remote handling devices allowed workers to reduce exposures by maximizing their distance from a source. For instance after slugs were discharged into the pit of water on the back face of a reactor, they were handled with remote-controlled equipment.

Time

Limiting the time of exposure is the third principle in radiation protection. Time is a factor because radiation dose produces cumulative effects. Thus, when a radiation zone is established in an occupational setting, the amount of radiation present in that area is described in "exposure unit per hour." Hand-held radiation dose instruments used for monitoring work areas, such as the Cutie Pie (see Figure 2-10.8), give readouts in dose units per hour. The time allowed in a given radiation zone varied with the amount of radiation present. Workers in charge of reading the Cutie Pies were responsible for keeping track of the exact length of time allowed.

Figure 2-10.8. *Worker Taking a Dose-Rate Reading in a Radiation Zone with a Cutie Pie*

Dosimetry

Dosimetry refers to a measurement used to calculate a worker's exposure record. A program of personnel dosimetry was carried out at the Hanford Site during the Manhattan Project and throughout the Cold War. Two basic types of dosimetry are external and internal. External dosimetry measures radiation from sources outside the human body. Internal dosimetry measures radioisotopes that are deposited within the human body.

Three devices measured external dosimetry of workers at the Hanford Site: film badges, pocket ionization chambers, and later thermoluminescent dosimeters (TLDs). Workers were assigned these devices to track their individual exposures. Film badges, pocket ionization chambers, and TLDs were issued in the badge houses. The TLDs were first used at the Hanford Site in the 1970s and were calibrated to ensure they provided accurate exposure data. They were calibrated at the 3745 Standards Building and the 3746 Irradiation Physics Laboratory (for more information see the Expanded Historic Property Inventory Form for the 3746 Building in Appendix B on the Internet).

Film Badges

Workers had film badges clipped to outer clothing on the chests.. The film badges used Eastman-Kodak Nuclear Track Emulsion, Type A film. This method of dosimetry was used because ionizing radiation darkens photographic emulsions on film badges (Kathren 1974, p. 60). As Wilson (1987, p. 2.28) notes, "the film badge was considered the official record of exposure to external radiation on a long-term permanent basis."

Pocket Ionization Chambers

Pocket Ionization Chambers (PICs) "were the signal for investigating a potential unusual exposure and requesting special processing of the film badge to assist in determining the actual radiation exposure" (Wilson 1987, pp. 2.29-2.30). Workers often called these dosimetry devices "pencils" because of their shape. As with the film badge, Hanford Site workers wore pencil ionization chambers on the front of outer clothing. Two PICs were worn to reduce the chance of error in a PIC readout. The PICs were the dosimeters of choice whenever workers entered radiation fields sufficiently intense to warrant periodic evaluations of worker exposure as the work progressed. The extended processing times required for film badges and TLDs precluded their use under such circumstances.

Thermoluminescent Dosimeters

Dosimetry technology advanced in the 1960s, and personnel radiation protection programs at the Hanford Site used the new technology. . In 1972, the TLD replaced the film badge for radiation workers at the Hanford Site (Wilson 1987, p. 2.16). A former head of radiation instrumentation at the Hanford Site, Ron Kathren, explains how a TLD works: "Thermoluminescence is a property of all inorganic crystals. When exposed to ionizing radiation, some of the energy absorbed is stored via electrons...When the material is subsequently heated, these electrons return to the ground state, accompanied by the emission of a photon of visible light. The amount of light is proportional to the dose absorbed within the crystal" (Kathren 1974, p. 62). Table 2-10.3 compares photographic film and thermoluminescence.

Bioassay

Bioassay refers to measuring radiation in the human body from samples such as feces, tissue, breath, and urine. Instruments were designed to measure levels of radioactive materials in such samples. At the Hanford Site, if a bioassay reading was positive a technician calculated the percentage of the maximum permissible limit to which the worker had been exposed and estimated the total dose the worker would receive from the radioisotope uptake. This information was then related with previously recorded internal exposure (if any) and entered into a worker's ongoing exposure record.

Whole Body Counters

Whole body counters are instruments that measure the amount of internally deposited radiation without requiring a sample of body fluid or another component to be read in a laboratory (see Figure 2-10.9). At the Hanford Site, whole body counters were originally placed in large mobile trailers in the 1960s and used near operational areas. Later, a permanent whole body counting center was built in Richland. Workers were requested to go to this counting center for routine monitoring and evaluating potential internal exposure.

Table 2-10.3. Comparison of Two Personnel Dosimeters in Terms of the Ideal Dosimeter

Characteristic	Photographic Film	Thermoluminescence
Response		
Accuracy in the field	±25%	±10%
Energy dependence	High, factor of 30 at 30 keV	Low, ±25% from 0.1 to 1.25 MeV (LiF)
Angular dependence	High	Low
Range	10 mR to 10^3 R	10 mR to 10^5 R[a]
Mixed radiations	Good	Fair
Cost	Low	Low
Small size and convenient shape	Fair	Very good
Ease of handling and ruggedness	Fair	Very good
Adaptability	Poor	Excellent
Longevity of wearing interval	1 month	Indefinite
Shelf life	1 year	Indefinite
Resistance to environment	Poor	Excellent
Readout (includes rapidity, capability for subsequent readout, or reuse)	Fair	Good

(a) May be extended to 10^6 or more by proper dosimeter selection. Two methods were used to measure internal exposure: bioassays and whole body counters.

Some whole body counters have large crystals that translate the energy emitted from internally deposited radioactive materials into electric signals. These signals are then translated into information about the energy and relative abundance of the radionuclide present. When a positive reading occurs, a technician calculates a radiation dose for the worker. This technique is applicable only to the assay of gamma-emitting isotopes.

A wound counter was also developed to take readings of alpha radiation contamination that deposited in cuts and punctures in the skin.

RADIATION PROTECTION INSTRUMENTS

In addition to the instruments used to measure radiation for the calculation of a worker's exposure record, Hanford personnel protection programs used a variety of survey instruments to measure the presence of radiation and to set exposure and dose rates. Three categories of survey instruments were: detection, exposure rate, and dose rate.

Detection instruments are designed to signal the presence of radioactive materials and display their data in relative units such as counts per minute. They are used to detect the presence or confirm the absence of radiation. A Geiger Counter is an example of a hand-held detection instrument used to survey workers for contamination before entering and exiting radiation zones (see Figure 2-10.10)

Exposure rate instruments measure ionization in the air. They are used to establish regulatory boundaries such as the boundary of a radiation zone.

Figure 2-10.9. *Whole Body Counter Equipment*

Dose rate instruments compensate for the effect of the radiation energy on its ability to produce a dose in tissue. They display results in some unit of dose per unit time (such as, millirem/hour). They are essential tools in radiation work because they establish the level of hazard present on a job site. The information dose rate instruments provide allows work plans to be timed accurately so worker exposure is kept as low as possible. The Cutie Pie was the mainstay dose-rate meter at the Hanford Site and was used inside radiation zones to monitor worker dose rates (see Figure 2-10.8).

Hanford procedures required workers to be monitored whenever they exited a radiation zone. The environment around the radiation zone was also monitored.

Exit Surveys

Exiting radiation zones called for special procedures. A piece of yellow and magenta tape on the ground identified the entry to radiation zones. Before workers crossed over the tape to exit a radiation area, a radiation protection technician surveyed the surface of their bodies and hands with a detection instrument. Next, the technician passed the probe attached to the instrument slowly across the bottom of shoes. The worker being surveyed lifted up one foot at a time and stepped across the line of tape into a clean area after the bottom of the shoe was cleared (see Figure 2-10.10).

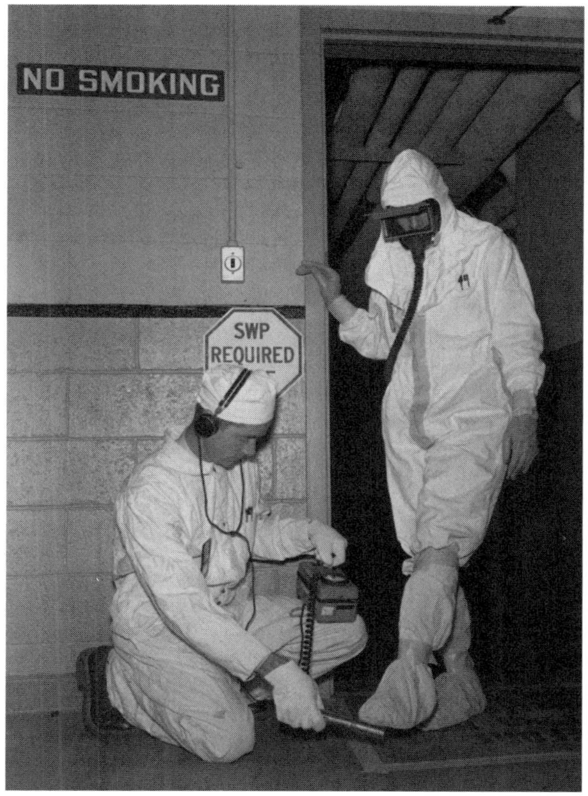

Figure 2-10.10. *Worker Being Surveyed with a Geiger Counter upon Exiting a Radiation Zone*

At some locations at the Hanford Site, workers surveyed themselves when leaving a radiation control area instead of having a technician do the survey. The self-detection instruments were often designed to take readings of the hands and feet. When a self-survey was called for, workers placed their hands and feet inside ports of the machine for a few seconds. The machine then instructed the worker to exit the radiation zone or re-survey (see Figure 2-10.11).

Figure 2-10.11. *Portal Instrument Used for Self-Survey When Exiting a Radiation Zone*

Continuous Air Monitors

Workplace air monitoring instruments included mobile units with audible and visual alarms. These units processed information and activated an alarm if a radiation level exceeded its pre-set limit. Other air monitoring instruments pulled air through a filter that had to be removed and taken to a laboratory for analysis. The monitors with filters were better for obtaining relatively accurate estimates of airborne concentrations. Emission stacks at chemical separations facilities were equipped with stack monitors to provide a reading of radiation levels in the exhaust stream.

Environmental Monitoring

Radiation monitoring also included a comprehensive environmental surveillance program. Surveys were used for off-site monitoring whereby samples were taken from continuous air monitors stationed throughout the Pacific Northwest (for more information, see the Historic Property Inventory Forms for the 2614-E-2 and 2614-W-2 Air Monitoring Stations in Appendix B on the Internet). Vegetation samples and water samples taken from the Columbia River downstream from the Hanford Site and in the Pacific Ocean were part of this data collection program. Farm products were also monitored. Samples were drawn from representative farms in areas surrounding the Hanford Site.

One of the main isotopes measured for was iodine-131. Iodine-131 was released to the atmosphere as an off-gas during the process of extracting plutonium. Because of the air-ground-vegetation-animal-human pathway of iodine-131, milk and other agricultural products including poultry, beef, and produce were monitored for its presence (DOE 1995a, p. 91).

Iodine-131 was also an isotope of interest because of its ability to preferentially collect in the thyroid gland, thus possibly predisposing some individuals to increased risk of thyroid health problems. Readers interested in the results of these ongoing surveys should refer to annual editions of the Hanford Site's Environmental Report and the Hanford Environmental Dose Reconstruction project publications (see References Cited and Bibliography for Further Information).

MEDICAL EXAMINATIONS

Monitoring for radiation exposure at the Hanford Site was part of a larger occupational program of medical examinations. Prospective workers at the Hanford Site underwent pre-employment and post-employment physical examinations. Medical evaluations also were an ongoing series of routine interval examinations.

> ***"Industrial medicine and [the] employee both gain most when the prevention of occupational disease is the main objective of their joint efforts. For this reason a program of frequent medical examinations was emphasized. The purposes of these examinations were:***
>
> 1. ***To permit the employee to discuss with the plant physician any symptoms or problems related to health, and particularly those which may have had an occupational origin or may have contributed to absenteeism***
> 2. ***To detect general as well as special abnormalities that may have arisen because of the work or working conditions. The aim was to try and detect the earliest possible sign of change resulting from occupation and, by frequent reexaminations, to follow the course of any such abnormality. Treatment, change of work, or further investigation depended upon the specific findings and the causative agent.***
> 3. ***To detect any working condition in any plant that may have produced deleterious effects upon the health of the workers so that such conditions might be corrected."* - Cantril 1951, p. 295**

At a conference presentation in December 1954, Herbert Parker, placed the role of occupational medicine at the Hanford Site in a wider industrial context. His thoughts were based, in part, on confidence in the Hanford Site worker safety program: "Annual physical examinations and periodic blood counts have their place, but not significantly more so in the atomic energy industry than in other industries" (Parker 1955, p. 903).

SAFETY PROCEDURES

The managers responsible for the work being undertaken determined the safety procedures. They consulted with support services at the Hanford Site, typically the industrial safety and fire protection groups, to develop these procedures. Safety procedures were posted at all job sites and included in manuals available for workers to review. For example, safety rules are documented in General Electric's *Manual of Standard Practices* published in 1954 (GE 1954b).

At the Hanford Site, radiological safety procedures were first expressed to workers through the Special Hazards Bulletin that the Special Hazards Committee prepared. This Committee was a branch of the Central Safety Committee (Cantril 1951, p. 304). As Wilson (1987, p. 1.10) notes, "the SHBs [Special Hazards Bulletins]had official sanction at Hanford. The first bulletins were published in 1945. Three other bulletins were subsequently published to provide guidance in handling situations that required control and special consideration because of radiological conditions." Table 2-10.4 lists some of the Special Hazards Bulletins issued in the 1950s.

Table 2-10.4. *Examples of Special Hazards Bulletins Issued in the 1950s*

SHB No.	Radiation Protection Standard Subject	Issue Date
1	Responsibility for Work in Radiation Zones	August 1, 1952
2	Procedure at Time of a Radiation Incident	August 1, 1952
3	Radioactive Solid Waste Disposal	December 15, 1954
4	Radiation Zone Injuries	August 1, 1952
5	Shipment of Radioactive Materials	August 1, 1952
6	Radiation Incident Investigation	May 1, 1953
7	Release of Equipment from Radiation Zone Status	August 1, 1952
8	Emergency in a Radiation Zone	August 1, 1952

In addition to the information the Special Hazards Bulletins disseminated, the Special Work Permit was designed to control exposure to radiation hazards. The Special Work Permit established procedures for and communicated information about radiation safety. "This control form was completed prior to starting any radiation work in order to define the task, identify locations, and specify the type of special protective measures that would be required for the duration of the permit" (Wilson 1987, p. 1.10). Many former Hanford Site workers recall the daily routine of changing into and out of "SWP" clothing such as coveralls and gloves. Special Work Permits sometimes also required workers to wear additional equipment such as respirators with special filters (see Figures 2-10.12 and 2-10.13).

Figure 2-10.12. *Special Work Permit Clothing Arranged on Floor*

Figure 2-10.13. *Workers in Special Work Permit Clothing*

EMERGENCY PREPAREDNESS

Emergency Preparedness was the plan to handle a range of emergency situations. Types of emergencies affecting health and safety include natural disasters, fires, and medical emergencies. The procedures planned for any emergency situation along with specifics concerning such situations as sabotage and the taking of hostages are discussed in Section 8, Site Security.

> ***"Even though all practical measures are taken to minimize the probability of a serious radiation event occurring at the HAPO [Hanford Atomic Products Operations - now Hanford Site], there will always remain the possibility that such an event can occur. The intent of this Manual of Emergency Radiological Plans is to provide improved capability for coping with such an event. It is not intended that these plans and procedures be considered as inflexible courses of action which cannot be deviated from, but rather that they serve as guides to supplement the use of good judgement. Any emergency action should be directed toward the goal of minimizing personnel exposures."*** **- Heid 1962, preface**

One of the key emergency situations for which the Hanford Site prepared was radiological accidents. The Emergency Radiological Plans and Procedures manuals were for large-scale radiological accidents. The Pacific Northwest National Laboratory maintains a historical collection of these manuals. See the sidebar box for an example of the type of information in these manuals.

Training and Communication

The Atomic Energy Acts of 1946 and 1954 formalized national security policies pertaining to atomic energy. Thus, federal law influenced programs designed to protect worker health and safety at the Hanford Site. The Hanford Site established a hazard-disclosure policy that was based on national security policies regarding classified information about nuclear weapons. The national security policies dictated what subjects were only available to workers on a need-to-know basis as security clearance determined. In summarizing a November 1944 report issued by Cantril, *Responsibilities of Health Group for Radiation Safety,* Wilson (1987, p. 1.7) observed, "those in charge of the work had the responsibility to inform their workers, within the limits of security, of the nature of the hazards they encountered." Another document has a slight variation, "within the limits of the security clearance of the employee and in line with the hazards-disclosure program" (Cantril 1951, p. 303).

Figure 2-10.14. *Billboards Displaying Safety Slogans*

A major concern was that secrecy inhibited communication about radiation hazards. Most workers could not even be told that radiation existed at the Hanford Site. Uranium was referred to as "base metal," and plutonium as "product." Concerns about secrecy also stemmed from project leaders' fears that an already tight labor supply (Cantril 1951, p. 292) might become tighter if previously undisclosed aspects of the job were suddenly emphasized. After the Hanford Site completed its initial mission at the end of World War II, it became evident that management's fears had some basis in fact. "Once employees learned what they had been making, they inundated Du Pont's Medical Department with questions about the dangers of the workplace" (Findlay and Hevly 1995, p. 53). Cantril and Parker wrote a document entitled *Status of Health and Protection at the Hanford Engineer Works* "shortly after the dropping of atomic bombs on Japan...to reassure [the workforce] concerning their safety on the job and to emphasize the fact that great care had been taken to safeguard their welfare" (Stone 1951, p. 476).

What could be communicated was stressed heavily in safety slogans and posters. According to Hacker (1987, p. 56), "training and indoctrination had become a growing part of the [Manhattan Project] Health Division plan. Constant reminders of the hazards and directions on how to cope with them helped workers protect themselves. They also helped avoid stalling the project." Du Pont and later General Electric continually reminded workers of the need for safety. Billboards displayed safety slogans that often related to a specific time of year, such as Valentines Day (see Figure 2-10.14).

Articles in the *Hanford Engineer Works News* also reminded workers of the need for safety, although a review of these articles suggests that the plant newspaper was not used as a forum to educate workers on the relationship between radiation safety and industrial safety. Few articles on safety link radiation and safety or even mention radiation. Most articles could have applied to non-radiological industries, such as the article, "Learn Safety

Instructions and Follow Them" or the industrial cartoon character Safety Sam exhorting fellow workers to "Report or Eliminate Hazardous Conditions Promptly" (GE 1947a, p. 4).

All communications campaigns portrayed safety as something to be excited about, and workers were given incentives to be involved. Contests were held, such as the 100 B Safety Contest (see Figures 2-10.15 and 2-10.16).

Introductory safety training took place when workers began employment or started a new job. Training consisted of a review of technical and operations manuals. One such manual is the General Electric *Manual of Standard Practices* with a section on "Safety Program-Safety Orientation." The manual states, "New General Electric employees receive their first formal contact with the safety program from the Employee and Public Relations Department. In addition, the area Safety Engineers will arrange for safety orientation of new employees within one week after arrival in production area. The employee's immediate supervisor is responsible for instructing new personnel on departmental and building safety rules" (GE 1954b, p. 2.1). A review of such manuals was often followed by formal class instruction and on-the-job-training.

MEDICAL DEPARTMENT

William Norwood played a central role in organizing the Hanford Site Medical services and facilities. Du Pont hired Norwood in February 1943. From February 1943 to March 1944, Norwood divided his time between the Metallurgical Laboratory in Chicago and the Oak Ridge Reservation. His primary mission during this time was to develop an industrial medical program for the Hanford Site and to assemble qualified personnel to take on responsibilities in the new program. Norwood arrived at the Hanford Site in March 1944 as Superintendent of the Medical Department. He held this position into the 1960s.

Figure 2-10.15. *100-B Reactor Safety Contest: How Many Salt Tablets?*

Norwood organized the Medical Department into divisions and sections. The two divisions were Village Medical and Industrial Medical. The Village Medical Division handled the health care needs of people living in the town of Richland. The Industrial Medical Division provided occupational medical services to operating personnel at the Hanford Site. The Industrial Medical Division contained two sections: Plant Medical and Health Instruments. The Plant Medical Section's responsibility was the health and safety of the workers. The main functions of the Plant Medical Section included (Cantril 1945, p. 2):

- Examine workers to detect over-tolerance amounts of radioactive substances either on or in the body, which might lead to later injury
- Conduct tests that give early warning of injury
- Care for workers' plant traumatic injuries and emergencies
- Care for workers' minor illnesses at First Aid Stations

Figure 2-10.16. *Prizes Awarded for a Safety Contest in 1953*

To staff the Plant Medical Section, J.A. Quigley, Hanford's first industrial physician, arrived in December 1943 (Cantril 1945, p. 4). Quigley's arrival coincided with the initial manufacturing of fuel in the 300 Area. A Supervisor of Nurses arrived in July 1944, who set up procedures and gave specialized training to industrial nurses. Three industrial physicians were on staff at the Hanford plants through September 1945 (Norwood and Fuqua 1945), located as follows:

- One in the 200 Areas
- One in the 100 B and 100 D Areas with one day a week at the 100 F Area
- One in the 300 Area two days a week, in the 100 F Area one day, and the remainder at Kadlec Hospital to assist with exams

As of August 1945, the remainder of the Plant Medical Section staff consisted of:

- Six physicians
- Twenty-six registered nurses
- Twenty-six laboratory and x-ray technicians
- Four medical helpers
- One Receptionist

This staffing arrangement remained throughout the mid-1960s. The Hanford Environmental Health Foundation began providing occupational health services for the Hanford Site in the mid-1960s. From that time onward, industrial physicians began practicing in administrative settings in north Richland.

Kadlec Hospital opened on March 15, 1944. Pre-employment physical examinations for the Plant Medical Section were conducted there throughout the 1950s and 1960s.

The Hanford Site contained seven First Aid Stations. The main functions of these facilities included:

- Care for industrial injuries
- Examining center for the Plant Medical physician
- Care for minor illnesses, thus aiding in keeping the employee on the job

Each First Aid Station was equipped with:

- Treatment room
- Recovery room with bed
- Clinical laboratory
- Physician's office
- Toilet facilities

Special equipment included:

- Resuscitator
- Diathermy
- Small trunk packed with emergency supplies for use in event of area evacuation
- Two ambulances always ready at the Richland Station
- One or two ambulances always on call at the Patrol Headquarters in each area

From the start of operations in 1944 until August 1, 1945, the First Aid Stations performed 5,768 treatments and 17,908 re-treatments.

HANFORD FIRE DEPARTMENT

Whereas shielding, distance, time, monitoring, safety procedures, training, and communication were preventive measures, the Hanford Fire Department was one of the established protection measures. Since 1943, the Hanford Fire Department has been responsible for fire safety at the Hanford Site. During 1943 and 1944, the Hanford Fire Department built eight fire stations: Richland, White Bluffs, Central Shops, 100-B, 200 West, 200 East, 100-D, and 100-F, in that order. The available equipment the fire department used to fight fires consisted of several 1000-gallon tank trucks, several 60-750-gallon pumpers, one gas truck, one foam truck, and several personnel vehicles. The Inspection Division, developed in 1943, detected and corrected fire hazards. This included educating Hanford workers about hazards and the use of fire extinguishers. The first fire breaks along roads, railroad tracks, and utility lines were created during 1943-1944 and totaled approximately 1260 miles. Since none of the vehicles were capable of off-road fire response, grass fire vehicles were created out of Army Weapon Carriers that were equipped with 287-gallon water tanks and pumping units (Du Pont 1945a).

As the Hanford Site developed and began operation, the recognition of operational and industrial fire hazards lead to increased fire response capabilities, equipment, and further advancement of the Hanford Fire Department. It developed into a multi-dimensional emergency response force dedicated to the protection of life and property at the Hanford Site.

The fire department's duties not only included fire fighting and prevention, but also several other emergency response activities. The fire department was the lead within the Hanford Site's Incident Command Agency that "works closely with facility management to assure [sic] that emergencies are mitigated in an effective and efficient manner" (HFD n.d., p. 3). This agency ensured that appropriate organization and staffing were set up to handle each emergency and that all involved understood the situation. The department was also trained and equipped to respond to medical emergencies with basic life-support service, ambulance transportation, and extraction tools such as the "Jaws of Life."

The Hanford Fire Department's capabilities also included special situations involving hazardous materials. For these situations, firefighters were trained and equipped to respond to and control radiological and chemical events, such as spills, and provide medical care or decontamination of contaminated or exposed individuals. The specialized hazardous material response vehicle contained the necessary equipment and information to assess and mitigate such events. It included protective clothing, weather data, an unknown substance categorization kit, and a computer system aiding in emergency operation recommendations (safe evacuation distances, plume forecasting, correct protective clothing for particular situations). The Hanford Fire Department was also trained to rescue individuals from above- or below-ground spaces, confined spaces, contaminated spaces, water, and collapsed structures (HFD n.d., DOE 1998d). Although mainly dedicated to the Hanford Site, the Hanford Fire Department also assisted local communities and other fire districts when needed. Hanford Fire Department firefighter training is still ongoing and includes structural and wilderness firefighting.

The types of fire response vehicles assigned to each station were based on the needs of the station's response area. For example, a 5,000-gallon truck was assigned to the 100/600 Area station (609 Building) because of the probability of its responding to natural cover or grass fires. The first response hazardous material vehicle was assigned to the 200 Area station (609-A Building) because of the nature of the surrounding facilities and probability of hazardous materials response calls occurring there (Rice 1998).

Hanford Site structures were equipped with fire alarm and/or fire suppression systems that signaled the 200 Area fire station when triggered. The Hanford Fire Department regularly tested and maintained these systems (HFD n.d.). Hanford Site buildings were required to post a fire emergency exit plan so workers would know how to exit the building and where the staging areas were outside the building.

None of the original eight fire stations exist today. The Hanford Fire Department force is currently located in four fire stations (Brown 1994, DOE 1998d): the 100/600 Area fire station built in 1961, the 200 Area fire station built in 1965, the 300 Area fire station built in 1964, and the 400 Area fire station (4704-S Building) built in 1973. These fire stations are staffed 24 hours per day, 7 days a week by rotating firefighting platoons. The platoons consist of one officer, four fire fighters, and one paramedic or emergency medical technician. For more information on the 300 Area fire station (3709-A Building), see the Historic Property Inventory Form in Appendix B on the Internet.

RADIOLOGICAL DECONTAMINATION

Along with the Hanford Fire Department, the Hanford Site set up radiological decontamination procedures and facilities as protection measures. If a radiological accident occurred, each facility had decontamination procedures in place. An example of general radiological decontamination guidelines is detailed in Norwood (1975, p. 208-212). Historically, procedures for personnel decontamination stipulate that decontamination should occur in an area close to where the exposure took place. "Unless the skin is broken, most decontamination is done by the radiation monitors who receive special training in this field" (Norwood 1975, p. 207).

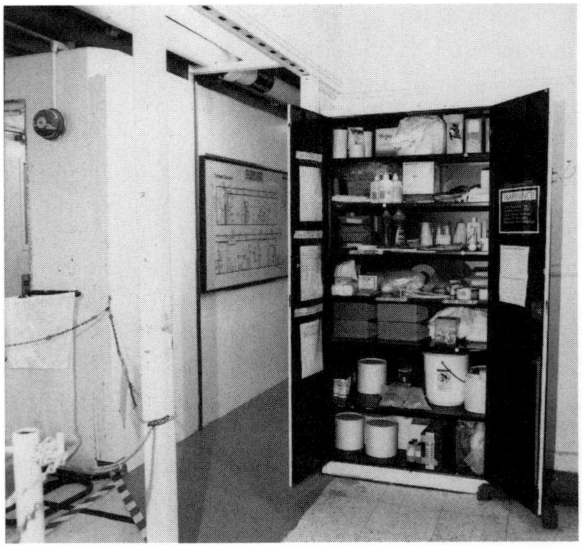

Figure 2-10.17. *T-Plant Decontamination Supply Cabinet*

Personnel decontamination refers to a process of removing radioactivity from clothing and the body. A worker who was contaminated with radioactivity was not to leave the Hanford Site until decontamination had been completed. Between 1943 and 1990, facilities that posed risks of exposure to radiation had decontamination stations. These stations included showers and sinks with storage cabinets positioned nearby for convenient access to supplies for washing away surface contamination from the skin and hair (see Figure 2-10.17). Norwood describes the procedure for transporting workers who were injured and emitting radioactivity from their bodies:

> *"After removal of an injured contaminated employee from the zone of exposure, operating personnel have been instructed to measure the emanation and, if high, to remove all clothing and wash the patient thoroughly with water. This might be a lifesaving procedure. Then patient would then be wrapped in blankets. Morphine would be quickly given for pain. Gross hemorrhage or other emergencies would be treated by the occupational nurse or physician to combat major shock."* - Norwood 1964, p. 474

> ***"We settled for grease pits which are available in area garages for use on instant notice. These are usually within a hundred yards of our area medical facilities. This is a cement pit used for greasing cars. The patient would be placed on the cement floor adjacent to the pit, while the doctor and associates stand in the pit. The pit wall will protect their bodies. Portable lead glass screens and lead bricks are stored in the garages and could be quickly assembled to protect the heads and necks of the members of the medical team and radiation monitor, while performing emergency treatment."*** **- Norwood 1964, p. 476**

Before the Emergency Decontamination Facility was completed in 1967 (see the Expanded Historic Property Inventory Form for the 748 Emergency Decontamination Building in Appendix B on the Internet), facilities for providing emergency medical care of radiating patients were "made available in each of the operating areas" (Norwood 1964, p. 475). Grease pits were decided on as the site for medical care of this special situation.

Six years before the Emergency Decontamination Facility was completed, an uncontrolled, self-sustaining criticality called a "nuclear excursion" occurred at the SL-1 Reactor in Idaho Falls in January 1961. Three fatalities resulted. Very high dose rates were documented in the facility and in the area nearby. The experience of radiological response teams and accident analysis caused health and safety professionals at the Hanford Site to re-assess their ability to treat a medical situation involving radiation. As a result of the assessment of the accident in Idaho Falls, the Emergency Decontamination Facility (748 Building) was built at the Hanford Site to safeguard a medical team and provide a space for treatment of a medical injury involving radiological contamination (Norwood and Quigley 1968, p. 608).

The historic significance of the Emergency Decontamination Facility (748 Building) is augmented by its role as the treatment facility and temporary residence for Harold R. McCluskey (1912-1987), a Hanford worker involved in the 1976 accident at 242-Z Waste Treatment Facility. At 2:55 a.m. on August 30, 1976, McCluskey was working the graveyard shift as a nuclear operator recovering americium. The process equipment consisted of ion-exchange columns housed inside glove boxes. Ion-exchange columns function like home water softeners and are designed to remove a desired metallic element, in this case americium, from a waste stream. Before August 30, 1976, this facility had been on standby for 4 months due to a strike, which started April 30, and ongoing labor-management negotiations. Once the strike ended, the americium recovery equipment in the 242-Z Facility was restarted. A short time after startup, McCluskey noticed an unusual sound and dark brown smoke coming from inside the glove box. He placed a small step ladder near the portal to investigate and soon thereafter the ion-exchange column exploded. "He was externally contaminated with ^{241}Am [americium-241], sustained a substantial internal deposition of this isotope, was burned with concentrated nitric acid and injured by flying debris about the face and neck" (Breitenstein 1983, p. 855).

After receiving first aid at the 242-Z Facility, at 5:14 a.m. McCluskey arrived by ambulance at the Emergency Decontamination Facility approximately 25 miles away. A team commenced medical treatment and decontamination of McCluskey. It soon became apparent to team members that McCluskey would be unable to be transferred to the main hospital because radiological contamination continued even after repeated decontamination. Medical treatment and decontamination were restricted to the Emergency Decontamination Facility and so was McCluskey. The Emergency Decontamination Facility became a temporary home to McCluskey and he remained confined there for 78 days. On the 79th day, McCluskey, his wife, and dog moved into a mobile trailer parked along the north wall of the Emergency Decontamination Facility. After 103 days, McCluskey visited his home in Prosser, and 47 days later moved home permanently (Breitenstein 1983, pp. 860-861). McCluskey was known to have remarked, "that was the longest graveyard shift I've ever worked."

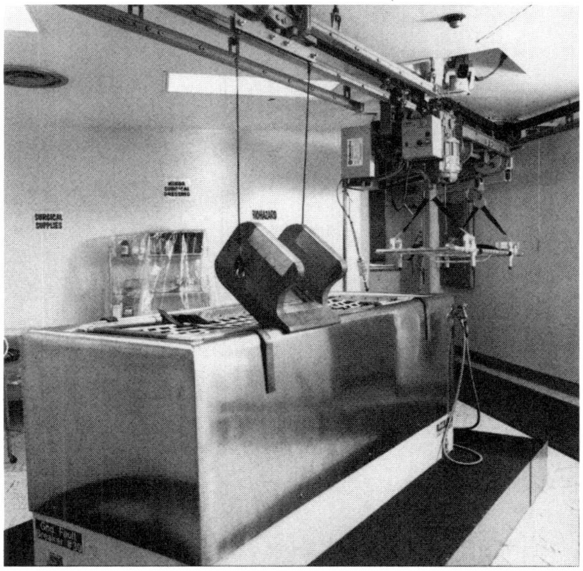

Figure 2-10.18. *Emergency Decontamination Equipment in the 748 Building*

The Emergency Decontamination Facility was originally designed for short-term care of patients in special situations. Although the McCluskey experience is an exception, the Emergency Decontamination Facility remains operational for the temporary care of patients injured from radiation. The main goal of the facility is to prepare patients for transfer to the appropriate medical facility. The Emergency Decontamination Facility is staffed only during emergencies. It is owned by the U.S. Department of Energy and presently operated by the Hanford Environmental Health Foundation. Figure 2-10.18 shows decontamination equipment in the Emergency Decontamination Facility, and the Expanded Historic Property Inventory Form in Appendix B on the Internet gives further information about the facility.

> ***"In 1972 the unusual occurrence reporting (UOR) system was initiated by the U.S Atomic Energy Commission, Richland Operations Office. Between 1972 and 1986, 1,742 UORs were issued for events that occurred in chemical processing facilities in the Hanford 200 Areas. In 1981 the U.S. Department of Energy (DOE), Richland Field Office (RL) directed that events of less consequence than an UOR also be documented. The Off Normal Report (ON) was initiated, and between 1981 and 1986 over 1,000 reports were issued. These reports and investigations are available to the public at the DOE Richland Operations Public Reading Room in Richland, Washington." -*** **Sudmann 1992, p. 1**

RESULTS OF PREVENTION AND PROTECTION EFFORTS

The results of prevention and protection efforts at the Hanford Site are reflected in the fatality, accident, and property damage/loss rates. According to official definitions applicable to the Hanford Site, accidents were reported according to two categories: "A" and "B" (AEC 1975, pp. 147-150). Type A category accidents included fatalities, damage of $100,000 or more, or excessive over exposure to radiation. Type B accidents were below this category but above certain thresholds with regard to the severity of injury and included property losses above $25,000. Other accident categories have emerged over time. The following fatality, accident, and property damage/loss rates from 1943-1990 found in Strom (1997) and Warner (1993) provide only the Type A and B category accident rates.

FATALITIES

No occupational fatalities through acute exposure to radiation as a result of Hanford operations have been officially documented. See the references cited in the above section, Epidemiological Research on Hanford Workers for estimates of the number of fatalities caused through chronic occupational exposure to radionuclides and chemicals resulting from Hanford operations. Epidemiological data cannot determine the cause of individual fatalities. In the categories of construction and operations, a total of 45 fatalities on the project have been documented for the period 1943-1990. Of the total fatalities, 36 or 80 percent occurred before 1965. Through 1990, the most recent fatality involved a non-project salvage operator who died as a result of injuries when a bunker collapsed at the 184 Building in July 1981. Dan Strom observes that fatalities at the Hanford Site have, "a 20 year average rate of about 2.2 per 100,000, or 1 per 45,000 worker years" (Strom 1997, p. 337).

COMPARISON WITH NATIONAL STATISTICS

The injury rates at the Hanford Site as well as rates for other accidents have been below the national averages for industry as reported by the National Safety Council. As Strom (1997, p. 333) points out in his review of the data, "The highest fatal accident rate at Hanford, 55 per 100,000, occurred in 1944. For comparison, the 1976 National Safety Council fatal accident in the construction sector was 56.8 per 100,000." Hanford's worst year, 1944, was better than an average year for industry in the 1970s. In 1976, the Hanford Site fatality rate per 100,000 was 2.2. This compares with rates of 11.1 for the government sector, 31.3 for the transport and public utilities sector for the same year (Strom 1997, p. 338). Table 2-10.5 compares the fatality rates for the Hanford Site and other industries for 1976, 1991, and 1994.

Table 2-10.5. *Comparison of Fatal Accident Rates between the Hanford Site and Other Industries*

Fatal Accident Rates	Mean Rate 1976[a] (10^{-5} y^{-1})	Mean Rate 1991[b] (10^{-5} y^{-1})	Mean Rate 1994[c] (10^{-5} y^{-1})
All groups	14.2	9.0	4.1
Hanford	2.2[d]	1.3[e]	1.3[e]
Trade	6.4	4.0	1.6
Manufacture	8.9	10.1	3.8
Service	8.6	4.0	1.5
Government	11.1	9.0	2.8
Transport and public utilities	31.3	22.0	12.1
Construction	56.8	31.0	14.7
Mines and quarries	62.5	43.0	26.7
Agriculture (1973-80)	54.1	44.0	26.2

(a) National Safety Council. *Accident Facts.* Chicago: National Safety Council; 1976 ed.; 1977.
(b) National Safety Council. *Accident Facts.* Chicago: National Safety Council; 1991 ed.; 1992.
(c) National Safety Council. *Accident Facts.* Itasca, Illinois: National Safety Council; 1994 ed.; 1995.
(d) Twenty-year average, 1976-1995.
(e) Ten-year average, 1986-1995.

From 1943-1990, the 1944 Hanford Site fatality rate of 55 per 100,000 is more than double the next highest figure of 23 per 100,000 set in 1943, both of which were construction years. Figure 2-10.19 compares the fatalities that occurred on Atomic Energy Commission projects with those that occurred nationwide and provides a breakdown of the causes of the Atomic Energy Commission fatalities. Figure 2-10.20 compares the severity rates for injuries that occurred on Atomic Energy Commission projects with those that occurred nationwide.

INTERNAL DEPOSITION ACCIDENTS - RADIOLOGICAL

According to statistics radiological health and safety support kept at the Hanford Site, eight occurrences have resulted in workers receiving internal depositions exceeding standards. The maximum internal deposition of 170 body burdens resulted from the ion exchange accident that occurred August 1976. No internal exposures exceeding DOE standards have been documented between January 1985 and 1990. In January 1985, a worker in the 234-5Z Plutonium Finishing Plant received 1.6 times the maximum permissible body burden from a puncture wound.

WHOLE BODY PENETRATING EXPOSURE ACCIDENTS - RADIOLOGICAL

A total of seventeen workers have received occupational, whole body penetrating radiation exposure that exceeded 5 rem per year or 3 rem per quarter from operations at the Hanford Site. Of the seventeen exposures, six exceeded 10 rem. The maximum dose was 110 rem whole body penetrating radiation exposure. No whole body over exposures have occurred at the Hanford Site since December 1977.

EXTREMITY EXPOSURE ACCIDENTS - RADIOLOGICAL

Extremity exposure is exposure to hands, arms, legs or feet. This is the most common type of exposure. A total of twelve workers have received exposures to their hands, which exceeded the level of 75 rem per year or 25 rem per quarter. A highly localized maximum dose was 80,000 rem to a portion of the hand of a laboratory chemist who was operating an x-ray diffraction unit. Only one over exposure to an extremity has occurred since 1972. That was in June 1983 as a result of a technician putting his hand in front of an x-ray beam.

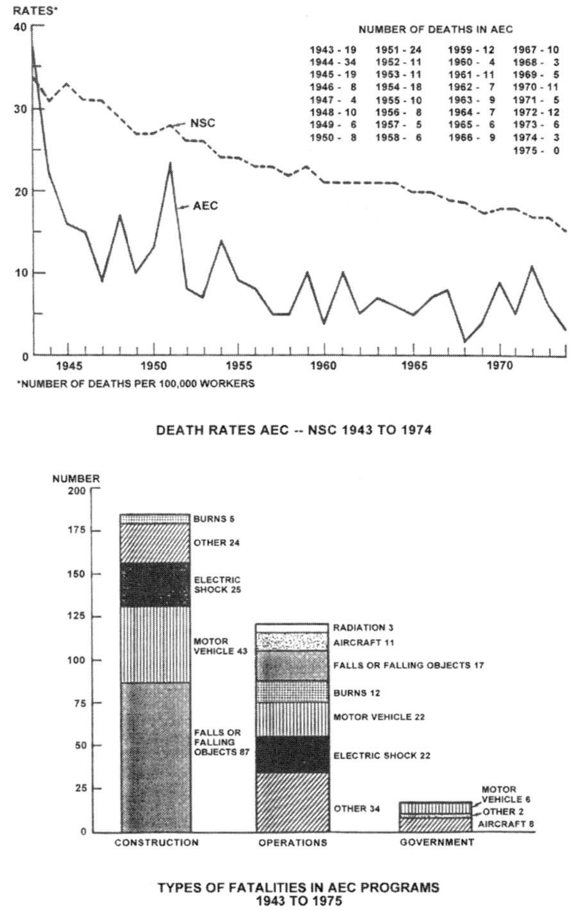

Figure 2-10.19. *Comparison of Fatalities at Atomic Energy Commission Facilities with Those Reported by the National Safety Council, 1943-1974*

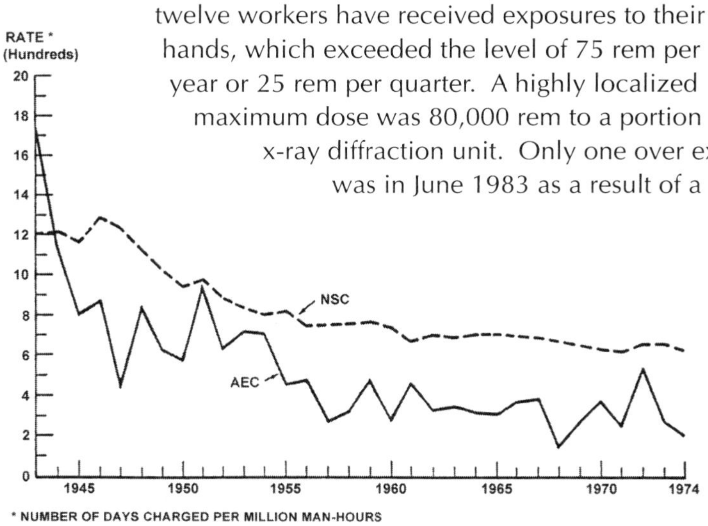

Figure 2-10.20. *Comparison of Injury Severity Rates at Atomic Energy Commission Facilities with Those Reported by the National Safety Council, 1943-1974*

CRITICALITY ACCIDENTS - RADIOLOGICAL

A criticality accident is an event where fissionable material undergoes nuclear fission in an uncontrolled fashion. This causes a sudden release of energy in the form of particulate radiation, electromagnetic radiation, and heat. Two criticality accidents occurred during operations at the Hanford Site: one on November 16, 1951 at the Critical Mass Laboratory (209-E Building) and the other on April 7, 1962 at the RECUPLEX Facility. A historical accident review document notes the following points of interest in the Critical Mass Laboratory accident:

"The Hanford Homogeneous Reactor Experiment was an experimental program carried out in the Critical Mass Laboratory. Its purpose was to investigate the physics of plutonium solutions to determine safety limitations for process facilities. During one such experiment a partially full spherical assembly containing plutonium nitrate solution as fuel was accidentally brought to a prompt critical condition by too-rapid withdrawal of the safety rod...Analysis indicated that a total energy release of about 3MWs (about 8×10^{16} fissions) had occurred...There was no physical damage to the facility...Radiation doses to personnel in the control room, which was located 200 feet away as a protection against such occurrence, measured up to 600 mrem [millirem]." - DOE 1991a, pp.2-7

During the 1962 RECUPLEX criticality accident, three workers received whole body penetrating radiation exposures exceeding standards. A process review history provides this description of the accident:

"A nuclear excursion occurred in a general purpose transfer tank (the K-9 vessel) in the Recouplex [sic] plutonium waste recovery facility of the 234-5 Z building. At the time of the incident, Recouplex [sic] was not in operation. It had been shut down for a cleanup which was essentially complete. Two employees who were present observed a blue flash and heard a noise like an electric arc. Evacuation was immediate, and all emergency procedures were executed properly. Exposures for the three most exposed individuals were 110, 43, and 1.43 rem." - DOE 1991a, pp. 2-13

TRANSPORTATION ACCIDENTS

Six significant accidents have involved the transportation of radioactive material. However, in only one of those accidents was any radioactive material (uranium oxide) released to the environment. This accident occurred April 1956.

PROPERTY DAMAGE OR LOSS ACCIDENTS

Property damage or loss exceeding $50,000 has occurred forty times. The maximum loss of $1,174,000 resulted from damage to a hot process pipe trench when a 32-ton cover block fell during remote crane operations in the B Plant in February 1980.

AREAS FOR FURTHER RESEARCH

The history of health and safety at the Hanford Site is broad in scope, and researchers are in the early phases of writing about the subject. Issues such as whether the Hanford Site's worker population has an increased mortality level due to radiation exposure as well as questions about the relationship between occupational exposure to chemicals and illness are the subject of ongoing research and debate (see, for example, Mancuso et al. 1977, Gilbert et al. 1989, and Stewart and Kneale 1991) but are not the subject of this section. This section discusses the programs established to monitor worker health and safety and the environment in relation to the Hanford Site facilities. The long-term effectiveness of these programs and the effects of working with or living near radiation and chemicals are only briefly discussed. Even within the limited scope of this book, however, many avenues for further research are still available:

- What was the effect of security on health and safety? Did restrictions on communication cause health and safety problems?
- Are health and safety programs more effective when managed by operations or when divorced from conflict of interest and managed by separate health and safety organizations? What lessons have been learned from radiobiology research conducted at the Hanford Site?
- Despite overlaps in industrial hygiene, industrial safety and fire protection, and radiation protection, an integrated approach to worker health and safety was not developed at the Hanford Site during the Manhattan Project and Cold War Era. What effect did this lack of integration have on worker health and safety?
- Taken as a whole, what do the results of health studies that use Hanford Site data tell us about the relationship between radioactivity and human health?
- What does the occupational safety record tell us about the relative safety of these industries?

- What might be learned about the Cold War by comparing health and safety aspects of plutonium production facilities in the United States with those in the former Soviet Union?
- Why does ALARA (as low as reasonably achievable) refer only to radiological exposure at the Hanford Site?

Numerous health studies and projects have been conducted to determine the effect on people's health of the operations at the Hanford Site. Future researchers may wish to follow up on the results of these studies as a way of measuring the success of the prevention and protection efforts. The seven health studies and projects listed below, each of which published a series of reports, use Hanford Site data. Some of the data researchers used in these studies and projects included occupational external dosimetry records, social security records, county death certificates, airborne radioactive release information, and medical case history combined with medical evaluation. Some of the studies focused on occupational health, others on public health, and the Epidemiological Evaluation of Childhood Leukemia and Paternal Exposure to Ionizing Radiation combined the two approaches.

- The Hanford Environmental Dose Reconstruction (HEDR) Project covers the time period 1944-1972. The aim of this project was to determine the amount and identify the pathways of past radioactive releases from the Hanford Site. This information was then used in the Hanford Thyroid Disease Study.
- The Hanford Thyroid Disease Study was carried out by the Fred Hutchinson Cancer Research Center in Seattle, Washington. The Center collected data from the years 1944-1957 to examine whether thyroid disease is associated with residence in the vicinity of the Hanford Site. An act of Congress in 1988 authorized this study and the HEDR Project and they were funded by the Centers for Disease Control and Prevention.
- The Epidemiological Evaluation of Childhood Leukemia and Parental Exposure to Ionizing Radiation was a study pertaining to the years 1957-1991. This study examined whether an association exists between parental preconception radiation exposure and childhood cancer.

- The R-11 Survey used data from graduating high school senior classes and fourteen representative towns in the vicinity of the Hanford Site to examine whether or not an association exists between radiation exposure and health effects. This survey was funded in conjunction with a legal case involving the Hanford Downwinders Coalition.
- Research for the Hanford Mortality Study began in 1965. The Atomic Energy Commission initially sponsored this study. . Thomas Mancuso was the original director of this project. The research was designed to identify correlations between mortality rates and occupational radiation exposure (Stewart and Kneale 1991, Gilbert 1991).
- The Hanford Worker Study covers the time period 1944-1986. It used the same basic data sets as the Hanford Mortality Study but received private funding. The results of the Hanford Worker Study were published in March 1993 in the *American Journal of Industrial Medicine.*
- The Comprehensive Epidemiological Data Resource is a U.S. Department of Energy program. It is intended to be a public-use database of epidemiological information for health research.

Additional information on the relationship between health and radioactivity is maintained by the Hanford Health Information Archives, which is funded by the U.S. Department of Energy and located at Gonzaga University's Foley Center Library.

SECTION 11

TRANSPORTATION COMMUNICATION SITE SUPPORT

J.K. KEATING AND D.W. HARVEY
PACIFIC NORTHWEST NATIONAL LABORATORY

SECTION 11 - TRANSPORTATION, COMMUNICATION, AND SITE SUPPORT

Many of the siting criteria the Manhattan Engineer District of the U.S. Army Corps of Engineers established for the nation's first plutonium production facility at the Hanford Site related to the available infrastructure: transportation, water, and power.

The Hanford area already contained functioning roads and railroad tracks for local and regional transportation of people and goods. To produce plutonium, the Hanford Site required convenient access to allow construction activities to begin without first having to build roads and railroad tracks. A viable road system was necessary to transport employees, construction vehicles, and other supplies across the Hanford Site. Because of the widespread location of construction areas and the volume and types of materials involved, railroads were also a necessary method of transportation.

"Approximately 30,000 gpm [gallons per minute] of the purest water attainable [was] required for process cooling in each of the 100 Areas, in addition to the 5000 gpm of filtered water in each 200 Area. Additional quantities also [were] required for boiler make-up, condensing purposes, and for sanitary and miscellaneous services. In all, the average requirement for each 100 Area [varied] from 50,000 to 60,000 gpm with emergency demands up to 90,000 gpm." **- Du Pont 1945c, p. 139**

To produce plutonium, the Hanford Site required a reliable water supply that was relatively pure, cold, close to the project area so only a minimum pumping head would be required, and could provide the necessary volume of water (HAI 1987). The Columbia River, which flowed through the Hanford area, provided the necessary water volume, reliability, purity, temperature (averaging between 50-60 degrees Fahrenheit), and proximity to potential process areas requiring minimum pumping head.

A reliable and nearby power source was also needed to avoid building long transmission lines, which would have been difficult due to the wartime scarcity of metal. The power source had to be dependable and able to supply at least 100,000 kilowatts. This capacity was necessary to power the water pumps to cool the reactors, operate machinery, and supply water for workers at the Hanford Site (HAI 1987).

The Bonneville Power Administration was already established in the area and producing electricity year round from the Bonneville and Grand Coulee Dams. It also operated the Midway Substation located northwest of and adjacent to the Hanford Site. The Bonneville Power Administration and the Pacific Power and Light Company already operated power lines that serviced the nearby towns within the proposed project area (Du Pont 1945c). Existing Bonneville Power Administration operational facilities meant additional generating equipment did not have to be installed for year-round water availability and no storage of extra emergency supplies of coal or oil were needed for power production (HAI 1987). This convenient and capable power source provided the electricity the Hanford Site would need with relatively minor construction of additional transmission lines and substations until growth requirements dictated otherwise.

The Hanford area met the infrastructure siting criteria better than the other candidate areas on the

CONTENTS

Divider Page Photo: Hanford Site Buses, October 12, 1962 (Neg. 30321-4, Box PH1005)

West Coast the Corps was considering. (For more information on site selection criteria, see Section 1, Construction History.) The components within the existing infrastructure, although relatively minimal in relation to what the site would require, offered a base from which to begin construction.

After World War II, the Hanford Site experienced several expansions beginning in the late 1940s and continuing through the 1960s in response to increased Cold War tensions. International events associated with the intensification of the Cold War during the late 1940s increased allocations to national defense and expanded America's nuclear weapons program. The schism between the United States and the Soviet Union "led to a dramatic increase in plutonium production and construction of ancillary buildings and structures at all [Atomic Energy Commission] facilities, including Hanford" (DOE 1997b, p. 5.7).

Between the late 1940s and mid-1960s, the Atomic Energy Commission (which took over management of the Hanford Site from the Corps in 1947) built six new reactors (DR, H, C, KW, KE, and N). Also during this time period, the U.S. Department of Defense installed anti-aircraft artillery sites and Nike missile sites (see Section 9, Military Operations). The expansion of the plutonium production facilities and military installations at the Hanford Site resulted in an increased need for and subsequent expansion of transportation, communication, and other support facilities. Table 2-11.1 shows the structures built during the Cold War era that contributed to the expansion of infrastructure facilities.

By the 1980s, growth still continued but as the result of a changing Hanford Site mission. Environmental restoration and research and development activities became the main focus of Hanford Site projects and programs. In support of the new mission, the infrastructure expanded.

Table 2-11.1. *Hanford Site Structures Contributing to Infrastructure Growth during the Cold War Era*

Where	When	What
100 Areas	late 1940s-1960s	Post-War Reactors (DR, H, C, KW, KE, and N)
	1952	Experimental Animal Farm and Aquatic Biology Lab at 100-F Area
	1963-1964	Electrical Generating Plant in 100-N Area
200 Areas	late 1940s	Plutonium Finishing Plant, C Plant Hot Semiworks
	late 1940s-1980s	High-Level Waste Storage Tanks (Tank Farms)
	1950s	Reduction Oxidation Chemical Processing Plant (REDOX), U-Plant, Plutonium Uranium Extraction Plant (PUREX)
	1970-1990	Modular Office Units
300 Area	late 1940s	Analytical Chemistry Laboratory
	1950s	Laboratories: Radiochemistry, Physics and Metallurgy, Radiometallurgy, Biophysics, Metal Fabrication Development, Reactor Technology; Chemical Materials Engineering
	1960s	Fuels Manufacturing (Coextrusion Shop), Plutonium Fuels Pilot Plant, Plutonium Recycle Test Reactor, Liquid Metal Fast Breeder Reactor, Chemical Engineering, Shielded Materials, Sodium Test Facilities, Fast Reactor Thermal Engineering, Core Segment Development
	1970s	High-Temperature Sodium Facility, Life Sciences Laboratory
400 Area	late 1970s	Fast Flux Test Facility
600 Area	1950s	Antiaircraft Artillery and Nike Missile Positions
	1970s-1980s	Washington Public Power Supply System
1100 Area	early 1950s	Consolidated Transportation Facility, Southern Connection Railroad
3000 Area	1950s	Camp Hanford (North Richland)

> ***"The isolation of the [Hanford] site and the distance separating the various manufacturing areas made the establishment of a complete transportation network a necessity. An extensive system of roads and railroads was designed and constructed to expedite transfer of the enormous quantities of construction equipment and materials to the various areas as well as to provide adequate transportation facilities in case of an emergency in one or more of the areas during plant operations."* - UPOA 1977a, Book 4, Vol. 3, p. 7.4**

Transportation Plan

The Corps planned the Hanford Site to be spread out with miles of space between each reactor in the 100 Areas, between the operations in the 100, 200 and 300 Areas, and between those operations areas and the Richland townsite. With so much space required, efficient transportation to and from these areas during the construction and operation eras was an important component for success of the project. Figure 2-11.1 shows the roads and railroads on the Hanford Site in 1973.

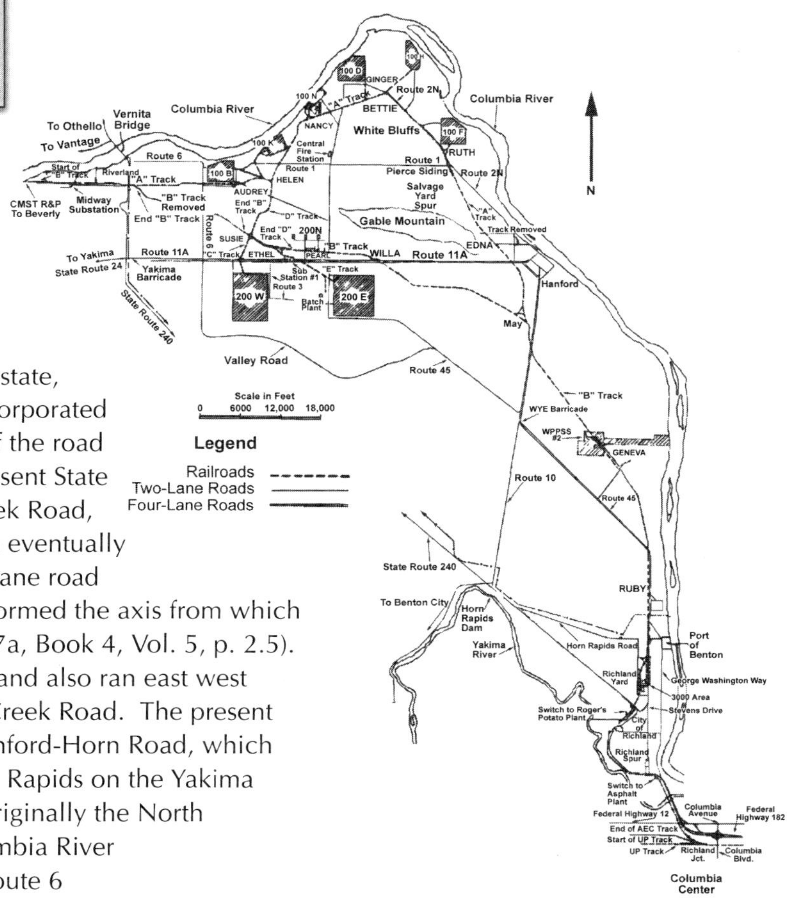

Figure 2-11.1. *Roads and Railroads at the Hanford Site, 1973*

Roads

A viable road system was necessary to transport employees, construction vehicles, and other supplies across the Hanford Site. When the Corps assumed ownership of the Hanford Site, a total of 570 miles of roads were present. They consisted of state, county, township, and private roads. The Corps incorporated as many existing roads as possible into the design of the road grid for the Hanford Site (see Table 2-11.2). The present State Highway 11-A was originally the Hanford-Cold Creek Road, which ran east-west across the project area and was eventually rebuilt along the same route as a double separated-lane road (Du Pont 1945a). The Hanford-Cold Creek Road "formed the axis from which all Project roads were later developed" (UPOA 1977a, Book 4, Vol. 5, p. 2.5). The present Route 1 was originally the Allard Road and also ran east west across the project area, north of the Hanford-Cold Creek Road. The present Route 10 and Route 2 South were originally the Hanford-Horn Road, which ran north-south through the Hanford Site from Horn Rapids on the Yakima River to the town of Hanford. Route 2 North was originally the North End Cut-off Road and generally paralleled the Columbia River between the towns of Hanford and White Bluffs. Route 6 and Highway 24 were originally the Upper River Road, which ran west from the west end of the Allard Road to a point near the Vernita Bridge (the Route 6 portion) and then south to the Hanford-Cold Creek Road. The north-south section of Route 6 was originally a branch road that ran between the east end of the Upper River Road and the Hanford-Cold Creek Road.

Table 2-11.2. *Hanford Site Use of Pre-1943 Roads*

Pre-1943 Roads	Adapted by the Hanford Site
Hanford-Cold Creek Road	Route 11-A (between Yakima Barricade and Route 2 South)
Allard Road	Route 1 (north of Hanford Cold-Creek Road)
Hanford-Horn Road	Route 10 and Route 2 South (between Yakima River's Horn Rapids and Route 11-A)
North End Cut-off Road	Route 2 North (between Route 11-A and 100-D Area)
Upper River Road and branch to the south	Route 6 (between Vernita Bridge Area and 100-B Area) and State Highway 24 (section between Vernita Bridge and Route 11-A)

The majority of the new Hanford Site roads were constructed of asphalt with a 2-inch-thick bituminous mat surface over a 3 or 5-inch base of crushed rock or stabilized aggregate. In addition, a network of graded and graveled roads existed within and between towns, substations, and along transmission lines. These ranged from 62 miles of black-topped roads, 107 miles of gravel-surfaced roads, 16 miles of graded and drained roads, and 385 miles of unimproved roads and trails (Du Pont 1945f).

The Corps constructed approximately 95 miles of temporary roads to accommodate the heavy road traffic during construction of the Hanford Site. The subcontractors, Guerin Brothers and Myer Bros.-N.M. Ball and Sons, Excavating Contractors, constructed the majority of these roads with the exception of those built in the 700 Area and Richland Village by Smith-Hoffman and Wright and some secondary roads built by the prime contractor at the Hanford Site, E.I. Du Pont de Nemours and Company. Some of these roads were constructed along grids that could later be used as permanent road locations (Du Pont 1945f, Du Pont 1945a). These "temporary roads were constructed of compacted sand and gravel eight to twelve inches thick, varying from 16 feet to 50 feet in width" (UPOA 1977a, Book 4, Vol. 5, p. 5.28). Sand and gravel used in road construction and stabilization was obtained from building excavations or local borrow pits (Du Pont 1945f, p. 25).

Many of the temporary roads were constructed in the Hanford construction camp and paved with a plant-mix bituminous surface. The Hanford construction camp was the only place bituminous surfacing was used for temporary roads (Du Pont 1945a, p. 484).

The Corps based the design of the road system on the estimated peak construction-phase traffic requirements and traffic requirements during operation in case of an emergency evacuation of the site. All main roads were improved to support the traffic load of the project by being reshaped, widened, and resurfaced with asphalt and gravel. New major roads included Route 3 between the 200 East and West Areas, Army Loop Road (also called the Valley Road) running along the south and west side of the 200 Area from Route 4 South and Route 11-A (Hanford-Cold Creek Road), and the access road to Rattlesnake Mountain (Du Pont 1945a, UPOA 1977a).

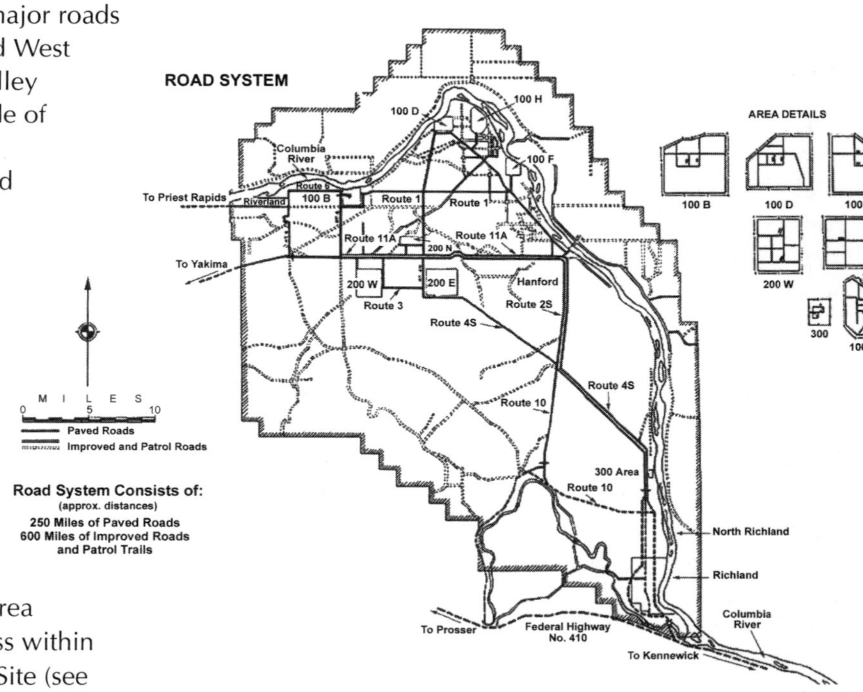

Figure 2-11.2. *Roads at the Hanford Site Around 1950*

For the Manhattan Project, the Corps designated the Hanford Site roads as Inter-Area or Intra-Area roads. Inter-Area (sitewide) roads provided access within and across the Hanford Site (see Figure 2-11.2). The main north-south and east-west Inter-Area roads had two 20-foot lanes divided by a 10-foot strip. Primary interconnecting roads and area approach roads were 30 feet wide. Intra-Area (those within an area) roads provided access within individual operation areas on the Hanford Site (see Figure 2-11.2). Du Pont's Wilmington Design Division created the Hanford Site's intra-area road layout. These roads varied in width from 12 to 20 feet (Du Pont 1945f). The majority of roads constructed within the process areas were placed at right angles to each other and ran north-south or east-west. A total of 56.64 miles of roads and 6.82 miles of walks were constructed inside the 100 (B, D, F), 200 (East, West, North), 600, 300, and 700 Areas by the end of construction in 1945 (Du Pont 1945a).

In the 1950s-1970s of the Cold War era, the Atomic Energy Commission constructed additional operation areas, such as the remaining 100 Areas (C, D, H, K, N) and 400 Area. These new construction activities required the expansion of the Hanford Site roadway grid (see Figure 2-11.3). In addition, maintenance, repairs, and resurfacing of old roads (all roads were resealed every 4-7 years) and construction of new roads were undertaken to bring the Hanford Site road system into compliance with Washington State Highway Department standards (GE 1950b). By 1957, the approximate total miles of roads was 868, including 268 miles of primary roads, 175 miles of secondary roads, and 425 miles of trails and dirt roads (GE n.d (a)).

To accommodate the increase in traffic in the 1970s and 1980s, major access routes, such as Route 4 South and Route 10, have been reconstructed. Approximately 32 miles of the Hanford Site inter-area roads are now four-lane. All of the 122 miles of intra-area roads are two lane (DOE 1990).

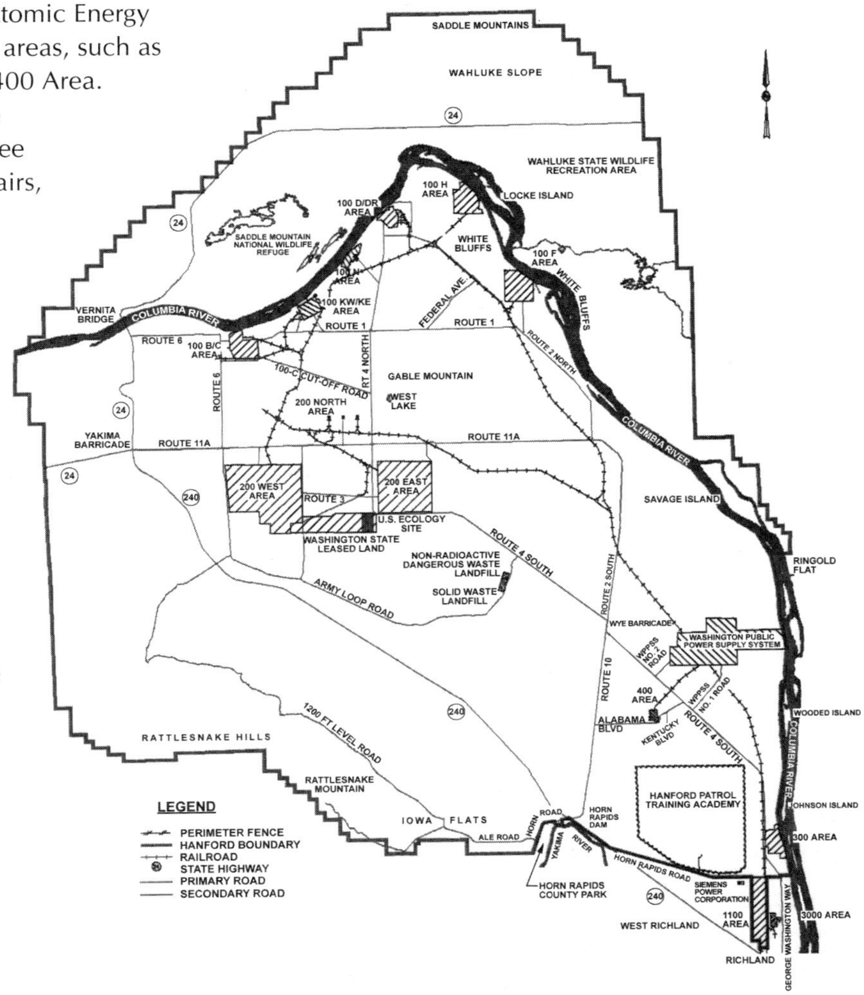

Figure 2-11.3. *Roads and Railroads at the Hanford Site, 1994*

HANFORD SITE PLANT RAILROAD

The Priest Rapids branch line of the Chicago, Milwaukee, St. Paul and Pacific Railroad had been constructed in 1913 between Beverly Junction and the towns of White Bluffs and Hanford, a total of 46 miles. This railroad branch followed the southern side of the Columbia River and serviced the fruit growers and ranchers in the area. Approximately 25 miles of this branch line was located in the northern section of what was to be the Hanford Site's reactor areas (100 Areas). The Corps took over this section of the railroad in April 1943 for use on the Hanford Site.

The Corps began immediately to improve and extend the existing line to support the needs of the construction activities and to speed up delivery and distribution of construction materials. The Guy F. Atkinson Company handled the railroad modifications, construction of temporary and permanent railroad track, and construction of the Riverland Classification Yard. During the expansion of the existing railroad line, material shortages and wartime priorities meant that mostly salvaged materials were used. A total of 35 miles of standard gauge, temporary rail, weighing 65-85 pounds per yard, were laid to the construction material plants and storage areas, such as the ballast and concrete aggregate plants (Du Pont 1945f). The Hanford Site used these tracks for transportation of equipment and materials across the site during construction.

It was also necessary to build branch lines to other construction areas of the Hanford Site. Some of these same tracks were used during operation of the site's production areas. In the 200 East and West Areas, 4200 feet and 5000 feet (respectively) of standard gauge railroad track were converted from temporary construction period track to permanent operation track (Gerber 1993c).

By the end of World War II, the Hanford Site had 158 total miles of railroad tracks. The railroad track connected all the major operation areas including the Richland Village and the Administration Area (see Figure 2-11.4). The railroad system included improved track of the Chicago, Milwaukee, St. Paul and Pacific Railroad line and temporary track at the Riverland Yard and between the former original townsites of White Bluffs and Hanford (Hageman 1945).

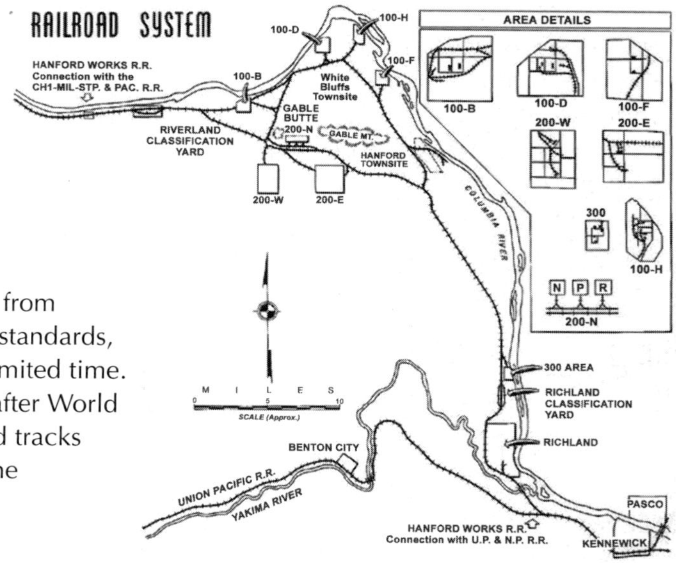

Figure 2-11.4. *Hanford Site Railroad System Around 1950*

The Corps had constructed the site tracks during World War II from what was available, which was often inferior materials of low standards, with the assumption that they would only be in service for a limited time. With the decision to continue production at the Hanford Site after World War II, it was necessary to replace the temporary, low-standard tracks with new treated ties, tie plates, and rails as well as improve the grades, alignments, and create safe roadbeds and tracks (GE 1950b). Morrison and Knudsen Construction Company did this work between 1948 and 1949 (Brands 1949). As the Hanford Site grew with the addition of more reactors and operation areas, the tracks were extended to tie these areas together with the railroad system. After a 1979 inspection, the Federal Railroad Administration recommended replacing rail that was laid before 1936 or weighed less than 90 pounds per yard with heavier, updated rail. During the 1980s, rail weighing 115 pounds per yard was installed on the main lines in the 100-K, 100-N, 200, and 300 Areas as well as 5 miles of the southern connection lines south of the Hanford Site (Haakenson 1998).

RIVERLAND CLASSIFICATION YARD

The Riverland Classification Yard was constructed in 1943 on the northwestern edge of the Hanford Site because that was where the tracks of the Priest Rapids branch entered the site. The Hanford Site Transportation Department assumed control over and classified incoming rail cars before they were sent across the site (DOE 1995d, p. 7). "Classification" refers to the process of tracking incoming rolling stock by recording when each rail car arrived onsite, when the car was delivered to its destination, what cargo was unloaded, and the date the car was to leave the site. "Declassification" was the process in reverse for outgoing rail cars, including when the rail car left and what cargo it was carrying (Haakenson 1998).

The Riverland Classification Yard had a capacity of 225 cars, contained 7 tracks, and was where the majority of the Hanford Site's railroad maintenance and operations facilities were initially located. To supply water to the steam locomotives, the Riverland Yard had a 25,000-gallon, elevated storage tank for treated water. At an additional 25,000-gallon, elevated storage tank, a centrifugal pump pulled the water from a deep well into the water storage tank (Du Pont 1945a). The only railroad access to the Hanford Site was through the Riverland Yard until the Richland Classification Yard and Southern Connection line were built in 1950. By 1964, the Richland Yard had taken over all the rail transportation responsibilities and the Riverland Yard was decommissioned.

RICHLAND CLASSIFICATION YARD - SOUTHERN CONNECTION

The Corps had constructed the Riverland Classification Yard in the northwest corner of the Hanford Site because the only rail access to the Hanford Site in the 1940s was via the Priest Rapids branch, which ran through the northern section of the site. Because the section of track from Beverly to the Riverland Yard was considered "extremely vulnerable to sabotage, natural rock slides, and damage from floods" (GE 1951a, p. 1), the Corps had considered installing a railroad junction south of Richland but to expedite construction of the Hanford Site decided against it. Another problem was the

Riverland Yard was the only connection to the Hanford Site so "freight traffic from the south and southwest, or originating from any point in the United States and transported by carriers other than the Milwaukee, could not make reasonable connections with either the Hanford Works [Hanford Site] railroad system or the Milwaukee system for economical interchange" (GE 1951a, p. 1). Thus after World War II, it was decided an additional access route and classification yard were necessary.

It was not until 1949 that plans for the Southern Connection and the Richland Classification Yard in the 1100 Area materialized when construction and growth of the Hanford Site increased in response to the Cold War. J. A. Terteling and Sons, Inc. won the contract for the new railroad, which was completed in 1950. The length of the Southern Connection is approximately 10 miles. It travels between the Richland Yard on the southern end of the Hanford Site, along the west edge of the City of Richland, and across the Yakima River to the Richland "Y" area where it connects with the Union Pacific Railroad main line near Columbia Center (GE 1951a).

This project also included the construction of the Yakima River railroad bridge crossing. The Yakima River railroad bridge was constructed in spans, using bridge sections salvaged from older bridges in the Pacific Northwest. Two of the spans used had originally been built in 1899 and three other spans in 1910.

Once completed, the southern connection became the main access port from which supplies entered the Hanford Site. The Richland Classification Yard eventually expanded and took over the role from the Riverland Yard as the maintenance shop for rolling stock as well as other site vehicles.

PROCESS AND SERVICE TRACKS

Railroad tracks were constructed either as process tracks or as service tracks. Process tracks required higher standards because they would carry special loads. For instance, the tracks that carried the irradiated fuel rods from the 100 Areas reactors to the 200 Areas chemical separations plants were process tracks and were constructed of heavier rail and on a wider roadbed than service tracks. Process track rail weighed a minimum of 80 pounds per yard, and the roadbeds were 28 feet wide, whereas service track rail weighed 65-85 pounds per yard, and the roadbeds were 18 feet wide. There are approximately 40 miles of process track. All other tracks were considered service tracks and were used to transport construction materials, coal, chemicals, fuel, equipment and other materials.

MAINTENANCE

Maintenance of railroad tracks and components has been an important ongoing task because railroad operations involved the transportation of special nuclear materials, chemicals, and supplies between areas. In the 1950s, railroad tracks were maintained comparable to common standards of the era. The process track for special loads was maintained comparable to mainline track levels of common carrier railroads. The service track for general freight was maintained comparable to branch line levels of common carrier railroads, while the remaining track for construction and industrial uses was maintained at safe and economical levels (GE 1950b). Maintenance included activities such as adding ballast and oil to the railroad bed, removing sand, re-aligning ties, and replacing materials. During the Manhattan Project and early Cold War era, maintenance of the rolling stock took place in the Riverland Classification Yard (Gerber 1993c).

The Hanford Site Plant Railroad operates between Richland Junction *"where it joins the Union Pacific commercial tracks and ranges to the abandoned Chicago, Milwaukee, St. Paul and Pacific tracks near the Vernita Bridge at the north boundary of the Hanford Site. Approximately 105 miles of the system are considered in service to active facilities across the site. Roughly 22 miles of track are in standby or out-of-service condition."* - DOE 1990, p. 10-1

1980S HANFORD SITE PLANT RAILROAD

Even while new railroad spurs were being laid into the 100-N Area and the 400 Area as those areas were being developed, the old spurs servicing the other 100 Areas were deactivated along with the reactors. As the coal fired powerhouses on the Hanford Site were replaced with diesel run systems, much of the railroad system was released from the responsibility of transporting coal (Taylor 1998a). However, the Hanford Site Plant Railroad still operates in a limited capacity to service the current Hanford Site missions. For more information on the Hanford Site Plant Railroad, see the Expanded Historic Property Inventory Form in Appendix B on the Internet.

TRANSPORTATION EQUIPMENT

MANHATTAN PROJECT

As important as the roads and rails at the Hanford Site was the equipment that ran on them. The Corps established the Area Engineer's Transportation Department in 1943 to manage the transportation equipment used on the Hanford Site. This equipment consisted of the railroad rolling stock, ferry, buses, cars, trucks, and construction vehicles.

Rolling stock was scarce during the Manhattan Project due to wartime needs. Rail cars and locomotives were leased from the Milwaukee Road and other companies. Additional rolling stock was transferred from the U.S. Army supply. Most of the rolling stock was standard scale but in poor condition. The rolling stock fleet included 120-ton and 80-ton diesel electric locomotives, cabooses, gondolas, and specialized cars such as transformer, box, dump, cask, flat, and tank cars (GE 1950b, UPOA 1977a). Custom rolling stock was necessary for transportation of special materials such as irradiated uranium, which was carried by "well-bottomed flat cars with special buckets and shielded casks" (Du Pont 1945c, p. 180). A couple of 40-ton furnace cars and some 90-ton underslung cask cars were fabricated or modified on the Hanford Site. The cask cars carried irradiated fuel rods from the reactors to the chemical separations plants (see the Expanded Historic Property Inventory Form for the Hanford Site Plant Railroad in Appendix B on the Internet).

During construction in 1943, the Milwaukee Railroad was delivering 600 carloads of building materials and equipment per week to the Riverland Classification Yard. Material from aggregate pits on the Hanford Site was carried to construction areas across the site at a total of 50 cars per day. By the 1950s, the majority of movement on the railroad tracks was for delivery of coal across the site and transportation of irradiated uranium fuel rods (for more information see the Expanded Historic Property Inventory Form for the Hanford Site Plant Railroad in Appendix B on the Internet).

The Hanford ferry that operated along the Columbia River out of the town of Hanford had been servicing the towns and surrounding region when the Corps took it over and Du Pont operated it during construction of the project. The Corps acquired all the ferry equipment in the takeover, including tugboats. Under the Corps, the ferry continued to connect State Highway 11-A (running east-west from the Yakima Barricade to the original Hanford townsite) and U.S. Highway No. 395 (along the northeast side of the river) for vehicle and passenger traffic crossing the Columbia River (Du Pont 1945f). The Corps had also acquired the Richmond Ferry operating near the Vernita Bridge and used the equipment (a barge-type ferry and small boat) at the Hanford Ferry location (Du Pont 1945a).

The Hanford bus service shuttled workers *"between the Richland residential districts and the Area Bus Terminal; non-stop service between the bus terminal and specific work areas; and shuttle service within the 100 and 200 Manufacturing Exclusion Areas. Except for privately owned or other Government vehicles the system [was] the sole means by which operating personnel [could] reach the Manufacturing Areas."* - GE 1950b, p. 9

In 1944, the Corps faced the problem of how to transport 45,000 workers to work locations scattered across 670 square miles. The problem was complicated by the fact that during World War II tires and gasoline were rationed. The Corps decided to establish a free bus service. The Transportation Corps of the Army provided the buses (see Figure 2-11.5). Peak bus use was in September 1944 with 904 buses in service. This system was the "largest bus operation in the world for any given territory or city" at this time (UPOA 1977a, p. 9.5). During construction of the Hanford Site between 1943-1945, approximately 340,000,000 bus service passenger-miles were driven. By 1950, the monthly average alone was approximately 4,500,000 of bus service passenger-miles (Hageman 1945, GE 1950b, UPOA 1977a).

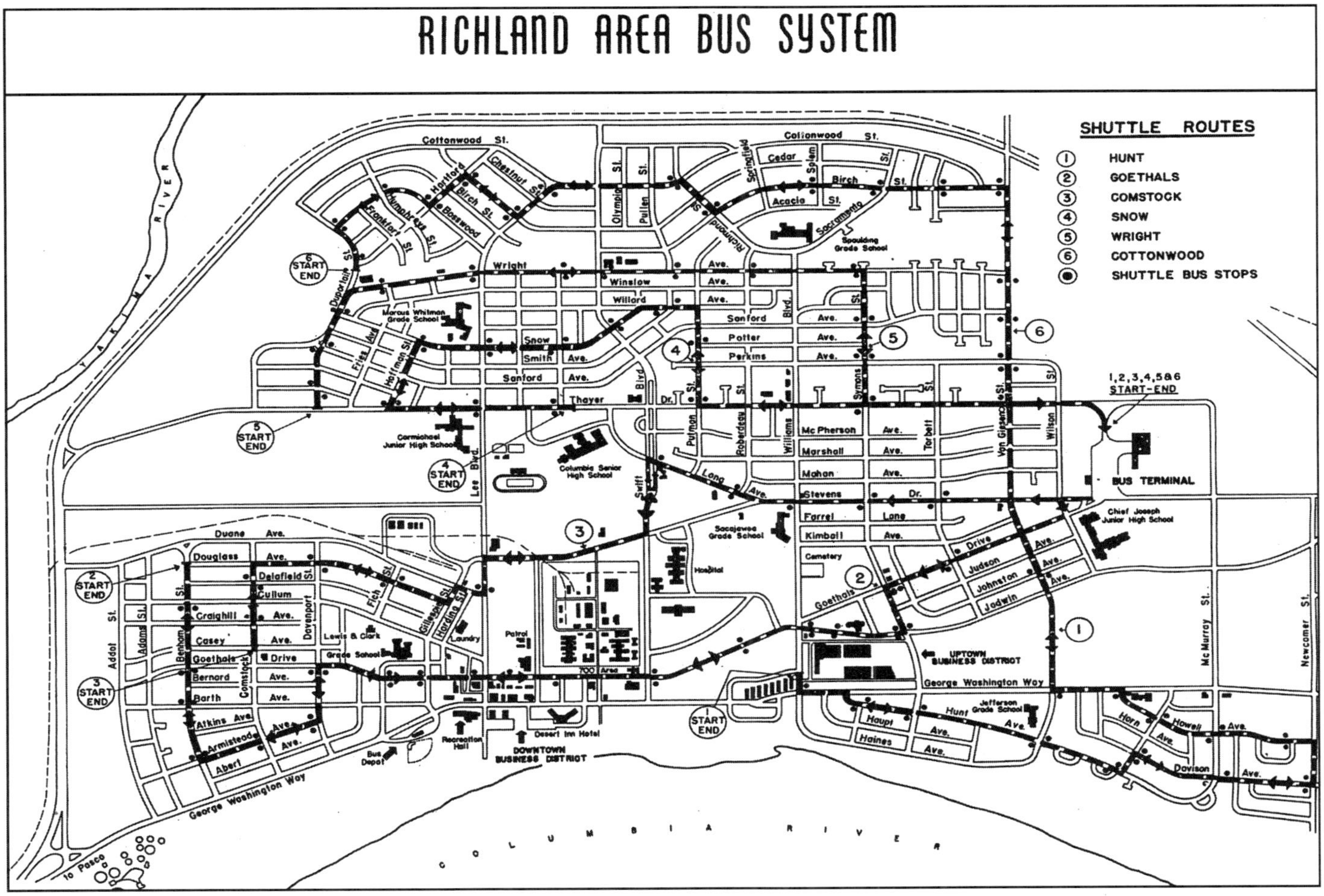

Figure 2-11.5. Map of the Richland Area Bus System Around 1950

The Corps also established a public bus service for Richland. It was a separate bus service from the one used by employees for work. The Richland public bus service had a fleet of 37 buses and carried approximately 47,000 passengers each month (GE 1950b).

In the mid 1940s, the fleet of transportation vehicles at the Hanford Site included approximately 1800 road vehicles (sedans, carryalls, trucks), 900 buses, 2600 construction vehicles (dump trucks, tractors), 50 emergency vehicles (fire, ambulance), and 500 railroad cars and locomotives (Matthias n.d.). Other types of vehicles consisted of construction equipment such as cranes, bulldozers, graders, scrapers, and tractors; and road maintenance equipment such as snowplows, dump truck sanders, and tractors. A towing fleet of vehicles provided emergency service for the Hanford Site vehicles and equipment (GE 1950b).

COLD WAR ERA

After World War II, the 1100 Area was established along Stevens Way in north Richland as the support services center for the Hanford Site containing general stores, shipping, receiving, transportation, maintenance, and warehouse facilities. During World War II, these services had been dispersed among the various areas on the site (see Figure 2-11.6) and in Richland (GE 1950b, Matthias n.d.).

The 1100 Area expanded during the 1950s with the increase in shipping, receiving, vehicle repair, and maintenance activities in response to Hanford Site growth. The 1171 Main Shop Building, completed in 1954 as the Transportation Maintenance Shop, consolidated the Hanford Site transportation and maintenance shops across the site into one facility. The 1171 Building was built to service the 2000 various vehicles used on the Hanford Site at the time. The facility consisted of a body repair and paint shop, a high-bay area for locomotives and larger equipment repair, a few maintenance shop areas, a vehicle wash facility, and an automotive parts storage area (for more information, see the Expanded Historic Property Inventory Form for the 1171 Building in Appendix B on the Internet).

Figure 2-11.6. *100-H Area Garage and Service Station Around 1950*

Another facility with vehicle maintenance and repair services during the 1950s and 1960s was the Army's Camp Hanford Brigade Motor Pool facility located in the 3000 Area. The Army had established the Motor Pool to maintain its vehicles supporting the anti-aircraft artillery sites and Nike missile installations. Located in the 1226 Building in the industrial area of Camp Hanford in North Richland, the Motor Pool continued to provide vehicle maintenance services for the Hanford Site until 1971 even though Camp Hanford closed in 1961 when the Nike missile sites were shut down. With the Army's departure from North Richland, the 1226 Building came under the control of the Atomic Energy Commission. A branch of the Hanford Site Plant Railroad terminated in the 3000 Area, and several spurs extended to industrial facilities and warehouses, including the 1226 Building, for the purpose of loading and unloading supplies.

HANFORD AIRPORT

The Corps built the Hanford Airport to support the need for airplane security surveillance of the large area encompassing the Hanford Site. Air surveillance continued through 1964. The Hanford Airport also gave key construction managers quick access to major regional airports and provided air-express delivery of items critical for constructing the Hanford Site. The first airstrip on the Hanford Site was located in the Hanford construction camp at the former original Hanford townsite. The airstrip consisted of a 30-foot wide, 2000-foot long, single lane landing strip constructed of blacktop.

Two elements led to the need for a new airstrip. As the camp grew larger and closer to the airstrip, proper take-off and landing clearances could not be obtained. At the same time, the number of Air Transport Command air-express shipments ordered for the Hanford Site was increasing. The original airport was abandoned and a new Hanford airport was constructed by 1945 at a location between the Hanford construction camp and the south end of Gable Mountain. This airport consisted of two landing strips of bituminous material, each 200 feet wide, one 2400 feet long and the other 4000 feet. The airport included two wooden hangers, one hutment, and refueling capabilities for army planes that consisted of electric gasoline pumps and underground storage tanks. The airport was enclosed by a barbed wire fence (Du Pont 1945a). In 1944, the "Day's Pay" B-17 bomber (see Figure 2-12.4 in the Worker History Section) made its ceremonial visit to the Hanford Airport. "Day's Pay" was purchased for the U.S. Air Force (at the time called the U.S. Army Air Corps) by the Hanford Site employees who each donated a day's pay to the cause (Thayer 1996). This airport was abandoned, probably around 1964 when the air patrol security surveillance was terminated.

The Richland Airport, built in 1944 west of the City of Richland, was used as the headquarters for the Hanford air patrol employees and air fleet. According to a retired air patrol pilot, the fleet consisted of a twin-engine Beechcraft, a Stearman spray plane, and two Super Cub planes. In 1960, the Port of Benton acquired this airport from the Atomic Energy Commission, and it was opened to private air traffic (POB 1998).

WATER UTILITIES

When the Hanford Site was constructed, existing water supply services consisted of domestic water systems and wells; town water systems in Richland, Hanford, and White Bluffs; and irrigation districts with associated powerhouses, pumping stations and irrigation ditches (Du Pont 1945f, Parker 1979, Harris 1972). These systems produced inadequate volumes of water needed for construction and operation of the Hanford Site.

Figure 2-11.7. Typical Floating Pump House on the Columbia River Around 1945

The water supply for the Hanford Site needed to cool the reactors, refrigerate, feed boilers, condense steam, put out fires, provide drinking water and sanitation services (Hageman 1945). The existing systems were updated with new equipment (storage tanks, booster pumps) and improved by drilling more wells and adding new pipelines to increase the capacity.

In 1943, the Corps acquired the water system that had serviced the town of White Bluffs. This system was expanded by combining existing wells with a new well and connecting them to a new system with pipelines, a booster station, and an elevated storage tank to provide water service in the area. The Corps also acquired the water system that had serviced the town of Hanford for use at the Hanford construction camp. However, this system alone was not sufficient for the camp's requirements. The Corps constructed a new system for the camp and operated it in addition to the original system. The new system included wells, pumps, storage tanks, and water mains (Du Pont 1945a).

Figure 2-11.8. McGee Well, Ground Water Storage Tank, and Booster Around 1945

The 100 Areas were supplied with Columbia River water through pumping stations and 108 miles of temporary water lines (Du Pont 1945f). The first water pumped from the Columbia River was to the 100-B Area concrete plant via a river pump house (Du Pont 1945a). This temporary system also included a chlorinator station, wooden storage tank, and a piping system. The 100-D and 100-F Areas obtained the same type of system, although their pump houses were located on floating rafts in the Columbia River (see Figure 2-11.7) (Du Pont 1945a). Each of the 100 Areas reactors was eventually supplied with a Columbia River pump house and water lines, which subsequently supplied all the water needed on the Hanford Site for use as sanitary, fire, and process water.

A well at the McGee Ranch supplied water to the 200 Areas and Central Shops during construction. The McGee Ranch had been built in 1928 in the northwest corner of what is now the Hanford Site (see Figure 2-11.8). The Corps took over the McGee well and water lines in 1943 and connected them to the booster station constructed in the 200 East Area. The Allard/Coyote Pumping Station had a line connected with the McGee line near the Central Shops area. Together, the McGee-Allard/Coyote Water System comprised 93,000 linear feet of pipe (44,800 feet of which was wooden), five storage tanks providing several functions (general storage, emergency water storage, elimination of natural gas in water, and reduction of pipe line pressure), and seven booster pumps (gasoline and electric-driven) (Du Pont 1945a).

Once the 100 Areas were equipped with river water pump houses and the 200 Areas were equipped with reservoirs and filter plants, water lines between the two areas completed the export water system in 1944. The export water system (see Figure 2-11.9) was designed to supply water from the Columbia River to the 100-B, 100-D and 200 Areas. The export water system also supplied water to various locations in the 100 and 600 Areas, and the 251-W Electrical Substation (DOE 1994b). "About 232 miles of pipe [was] installed for steam and water services of which more than half were required for construction purposes. This is exclusive of the large quantities of larger sized water pipe and special stainless steel process piping" (Hageman 1945, p. 70).

The export water lines (still in existence today) connecting the 100 and 200 Areas were constructed of reinforced concrete, measure 24-42 inches in diameter, and total approximately 25 miles (DOE 1994b). Concrete, which was already in use elsewhere in the United States and performing well, was chosen as the material best suited for the lines. In addition, the use of concrete eliminated using valuable steel and saved the cost of transporting materials because the pipes were constructed on the Hanford Site from existing aggregate materials (Du Pont 1945c). The 100-B Area contains the primary water pumping facility from which export water is distributed to other areas. The 100-D Area contains the emergency backup water pumping facility.

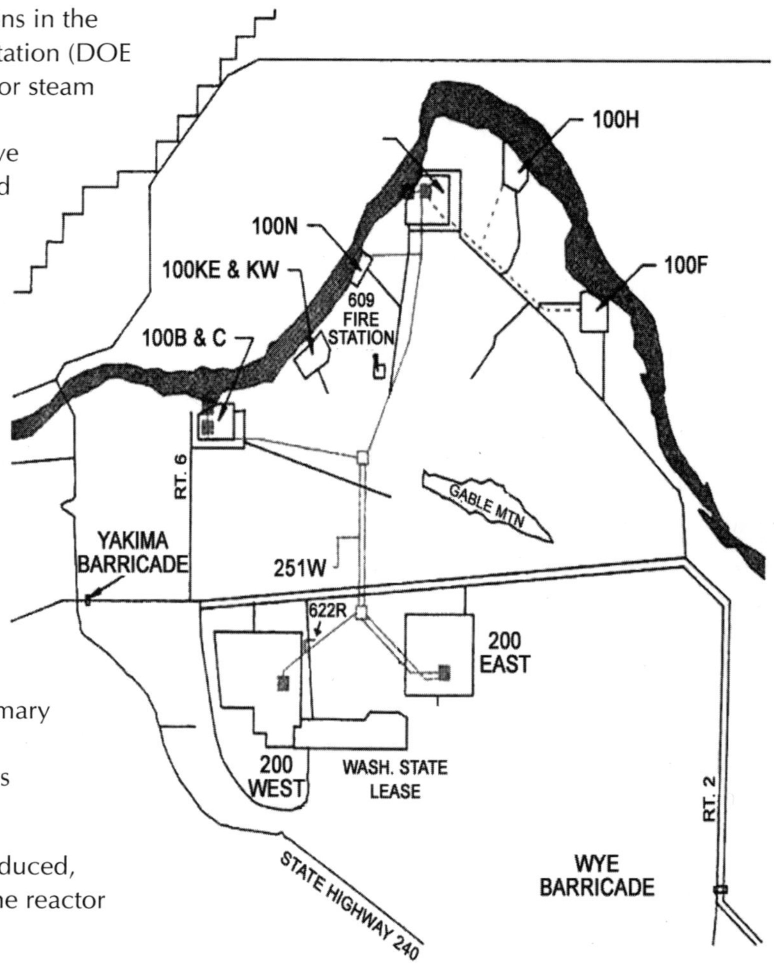

Figure 2-11.9. *Export Water System at the Hanford Site Around 1990*

Until 1967, when reactor operations were being reduced, the water system was maintained to comply with the reactor operation standards. During the 1970s, the system was repaired and maintained on an as-needed basis only until 1978 when the pumps, valves, screens, and buildings were rehabilitated. The reinforced concrete export lines between the 100 and 200 Areas were inspected in 1984 during valve modifications and determined to still be "in a highly reliable and very good condition" (DOE 1994b, p. 2-2). Today, the export water system is still in operation to supply the 200 Areas with raw water. However, as the 100 Areas and their individual water systems have been deactivated, ties to the export lines have been made to continue to supply water to the 100 Areas for limited usage needs.

WATER FLOW ROUTE

In each 100 Area, water was pumped from the Columbia River by motor-driven pumps (steam-driven pumps for backup) in the 181 River Pump Houses (see Figure 2-11.10) through concrete water lines into the two reservoirs within the 182 Reservoir and Pump House facility, one of which was for emergency storage (see Figure 2-11.11). The 182 Reservoir and Pump House facility then distributed water to the 183 Filter Plant/Chemical Treatment Building for purification (see Figure 2-11.12).

Originally it was thought that the 185 Deaeration Plants would be needed at this step to remove gases from the water, however, that turned out to be unnecessary. The 186 Demineralization Plants were also believed to be necessary for further purification. Only one 186 Demineralization Plant was built. It was determined to be unnecessary and so was never used.

Water from the 183 Building was pumped to clearwells for later use in the reactors. Two 300,000-gallon, above-ground water storage tanks at each area were also filled with an emergency water supply (see Figure 2-11.13). If both steam and electrical power were to fail, gravity would flow water from the tanks through the reactor to keep it cool for at least a short time (UPOA 1977a, Hageman 1945).

The 183 Pump Houses in the 100-B and 100-D areas also pumped water to the 200 Areas via the export water line. In the 200 Areas, water from the export water line is received at the 282 Reservoir and Pump House and pumped to the 283 Filter Plant for treatment and storage in clearwells until used as sanitary water within the 200 Areas line (see Figure 2-11.14).

Figure 2-11.10. *181-D River Pump House from Aerial View Looking South Around 1945*

SITE WATER SYSTEMS

As the 100 Area reactors were shut down, the only water needed was for the minimal activities still ongoing and in case of fire. The 100-B Area's original system was deactivated with the exception of the export water system. A small hydro-chlorinator system was brought in to supply the 100-B Area with the limited amount of water it needed. The 100-D Area also added a hydro-chlorinator system to provide backup for the export water system, water for its own minimal needs, and water for the 100-F and 100-H Areas. When the 100-F and 100-H Areas water systems were deactivated, they were networked to the 100-D Area's system for their limited water needs. The 100-K Area water system was constructed in 1955 with two identical systems (KE and KW) to service the 100-K Area. The 100-KW system was deactivated, and the 100-KE system was maintained to service the limited needs in the 100-K Area and its treatment plant (Kelly 1998, Rohl 1998).

Figure 2-11.11. *182-D Reservoir and Pump House from Aerial View Looking North Around 1945*

Because the N Reactor, constructed in 1964, had a closed-loop cooling system requiring much less water than the once-through reactors, the 100-N Area's water system contained a smaller river-water pumping facility than that of the other 100 Areas. However, it maintained the same basic pump houses, filtration plant, and clearwell facilities. The 100-N Area's water system was unique in that it also provided water to generate steam as the fuel source for producing electricity at the Hanford Generating Plant. The 100-N River Water Pumping Station was deactivated in 1990, leaving the water line originally designed to provide emergency cooling water for the 100-N Area as the primary water source (DOE 1994b, 1997a).

Figure 2-11.12. *183-D Filter Plant from Aerial View Looking Northwest Around 1945*

The water supply for 200 North was independent of the Hanford Site water system. It contained two drilled wells and water lines running from the wells to each of the 212 buildings (the only buildings needing water in 200 North) and on to the 251-N Electrical Substation. A separate line ran from the wells and connected to the export water lines running from the 100 Areas to the 200 East and West Areas (Du Pont 1945a, Gerber 1993c). The 251-W Electrical Substation, built in

Figure 2-11.13. *Typical Water Storage Tank at the Hanford Site Around 1945*

Figure 2-11.14. *282-E Reservoir and Pump House, 283-E Filter Building, and 284-E Powerhouse from Aerial View Looking Southeast, 1994*

1944, contained its own sanitary water system. The export water system for 200 East and 200 West, also constructed in 1944, supplied water for the chemical separations process and also for sanitary, fire, and other process needs (DOE 1994b). The export water system remains today as the water supply for these areas.

During construction, the 300 Area received water temporarily from small Columbia River pumping facilities and existing irrigation lines. The 300 Area's water supply system eventually developed into a system independent of the rest of the Hanford Site and serviced "research, laboratory, engineering development, fuel fabrication, and support facilities" (DOE 1994b, p. 1-7). The 300 Area's water system contains a river water pump house, treatment plant and storage tanks. Emergency or backup water can be obtained from the City of Richland. The water is chlorinated before it is distributed to the 300 Area buildings for various uses such as processing, heating, cooling, laboratory use, and sanitary use.

The 400 Area water system was constructed in 1978 to service the Fast Flux Test Facility and support buildings. The water is pumped from three deep-water wells (DOE 1994b). Each well can individually support the pumping capacity for the total system flow so the other two wells are used for backup and emergency needs. Water is pumped to three above ground tanks, chlorinated, and distributed for use as sanitary, fire, and process water. The distribution pumping station for the 400 Area is the 481 Primary Pump House, which contains two sanitary water pumps. The 481-A Secondary Pump House is for emergency backup.

In the 600 Area, each of the following was built with its own water system: 609 Fire Station, 622 Meteorological Complex, 6652-C Rattlesnake Mountain Observatory, 6652-K ALE field laboratories, Patrol Training Academy, and the Yakima Barricade systems. Each individual system was designed to service specific buildings at a specific location. These systems contain packaged water treatment plants, developed springs, pressurized tanks, and wells (DOE 1994b).

The water supply for the 700 and 1100 Areas during the Manhattan Project was obtained from individual wells in the Richland area. The water was chlorinated before it was distributed for sanitary, fire, laboratory, heating, cooling, and drinking uses. This system contained a 25,000-gallon storage tank (Du Pont 1945f, 1945a)

SANITARY WATER

Sanitary water is the term used to define raw water that has been purified or chlorinated for consumption by humans. The raw water was purified in treatment plants located within each of the areas. The resulting sanitary water was then stored in clearwells or tanks or piped directly to the buildings and facilities for use.

When construction of the Hanford Site first began, the Corps brought sanitary drinking water to the site in water trucks and barrels. By the end of 1943, however, the water lines designated for sanitary water use were chlorinated and ready for distribution (UPOA 1977a).

Sanitary water was supplied through 15 individual systems on the Hanford Site within the 100, 200, 300, 400, and 600 Areas and mainly serviced process plants, laboratories, and office buildings (see Table 2-11.3). As of 1990, ten of the systems obtained raw water from the Columbia River with river water pump houses, and the other five used underground deep-water wells (DOE 1994b).

Table 2-11.3. *Hanford Site Export Water Systems*

System	Water Source and Distribution Facility	Capacity (gallons per minute)
100-B Area	Columbia River – Pump House	70,5000
100-D Area	Columbia River – Pump House	54,000
100-K Area	Columbia River – Pump House	200,320
100-N Area	Columbia River via Export System	420,000
200 East	Columbia River via Export System	2,600
200 West	Columbia River via Export System	2,600
251-West	Packaged Water System	30
300 Area	Columbia River – Pump House	3,200
400 Area	Three Deep Water Wells	220 each
609 Fire Station	Packaged Water System	125
622-R Meteorological Complex	Packaged Water System	30
16652-C Rattlesnake Mt. Observatory	Rattlesnake Mountain Spring Well	<30
6652-K ALE Field Station	Deep Water Well	approx. 60
Patrol Training Academy	Deep Water Well	<30
Yakima Barricade	Deep Water Well	<30

WATER FOR FIRE CONTROL

Water for fire control was provided via raw and sanitary storage distribution systems and stored in elevated storage tanks at each of four headquarters buildings. The Corps established the four headquarters buildings in the 100, 200 and 300 Areas. In the 200 East and West areas, an extra reserve water tank for fire control was located at the T, U, and B process areas. The 300 Area system had a series of storage tanks and five electrically driven water pumps.

PROCESS WATER

Process water is the term used to define water that has been purified to an acceptable level for use in the manufacturing process. Large quantities of process water were needed in the 100, 200, and 300 areas to cool the reactor cores, create steam condensate for power, and flush away waste from the chemical separation process. The most economical way to provide a dependable, continuous water supply to the operations areas was to pump water stored in one area to the other area. The Corps built the water storage system in the 100 Areas and pumped it from there to the 200 Areas. An equally dependable power supply was needed to power the water pumps. The 230-kilovolt distribution loop out of the Midway Substation powered motor-driven pumps, which were backed up by steam-driven pumps in case of an emergency. In case power to both types of pumps failed, elevated emergency water tanks would supply water to the reactors for at least a short time. Within the 200 East and West areas, process water supply lines ran from the 282 Reservoir and Pump House to the 221, 224, and 231 buildings for use in the chemical separations process (Gerber 1993c).

ELECTRICAL POWER UTILITIES

Electrical power was as critical as water for the Hanford Site to function. The Corps estimated that an 86,000-kilovolt-amperes, uninterrupted power system at 95 percent power would be required. This estimate led to the need for major additions to and upgrades of the existing system.

When the Hanford Site was established, the Bonneville Power Administration was producing electricity year round. The Bonneville Power Administration operated a double transmission line of 230-kilovolts each that ran between the

Bonneville and Grand Coulee Dams on the Columbia River and crossed the northwest section of the Hanford Site. The Bonneville Power Administration also operated a substation at Midway near the northwestern boundary of the Hanford Site. From the Midway Substation, a 115-kilovolt line extended east to the original Hanford townsite. Although the Bonneville Dam had no surplus power available at the time of the construction of the Hanford Site, the Grand Coulee Dam was able to produce surplus power immediately with the installation of more generators (Du Pont 1945c).

Existing transmission lines and substations provided a basic network for the construction and initial operations of the Hanford Site. The Pacific Power and Light Company operated a 66-kilovolt transmission line in the area that serviced nearby towns and connected to the Priest Rapids Irrigation District power lines at the Allard/Coyote Rapids Substation. The Bonneville 115-kilovolt line between Walla Walla and Midway connected with the substation of the town of Hanford (UPOA 1977a). The 66-kilovolt transmission lines were powered by the 115-kilovolt line connected to the Hanford Substation and by a branch line of the 115-kilovolt Bonneville line running between Midway and Walla Walla (Du Pont 1945c).

Early construction power was provided through portable gasoline-driven generators and the existing substations at Hanford, Richland, White Bluffs, and Allard/Coyote Rapids until additional substations were constructed (UPOA 1977a). The White Bluffs Substation and distribution system serviced the project shops and storage buildings, pipe fabrication plant, cement block plant and nearby wells. The Hanford Site used existing power and lighting connections during renovations of existing homes and buildings (Du Pont 1945a).

EARLY ELECTRICAL POWER

Temporary substations were constructed in the 100, 200, 700 (Hanford Administration), and 3000 (Camp Hanford) areas with capacities of 3000-5000 kilovolt-amperes for use in those areas. The Corps acquired the 66-kilovolt transmission lines and substations the Pacific Power and Light Company and the Priest Rapids Irrigation District operated. Nineteen miles of branch lines were attached to the existing 66-kilovolt transmission lines to temporarily supply power to the 100 and 200 Areas. Another 19 miles of 6.9-kilovolt distribution line were also installed to service the shops, wells, and other installations. Permanent, primary substations built in the 300, 700, and 1100 Areas provided construction power for those areas (Du Pont 1945a).

The main power supply for the Hanford Site was routed through the Midway Substation, located between and powered by the Bonneville and Grand Coulee Hydro-Electric Plants. This substation was reconfigured with "an additional 50,000-kilovolt-ampere, 230-kilovolt to 115-kilovolt bank," by the Corps to satisfy the Hanford Site's permanent electric power needs (UPOA 1977a, Book 4, Vol. 3, p. 702). By 1944, the Midway Substation was the major power source for the Hanford Site (PRC 1992).

All the existing Bonneville Power Administration transmission lines remained intact within the newly established Hanford Site. An agreement was developed that directed the Bonneville Power Administration to continue to maintain the lines (PRC 1992). However, installation of additional lines was still required. The Bonneville Power Administration built 52 miles of new 230-kilovolt transmission lines (Du Pont 1945c, UPOA 1977a).

The Corps acquired the 53.6-mile portion of the 66-kilovolt Yakima-Hanford-Pasco transmission line that existed within the Hanford Site between the Midway Substation and Richland from the Pacific Power and Light Company. Only the reconditioned part of the line between Hanford and Pasco was used for permanent power distribution. The Corps also acquired the Allard/Coyote Rapids and White Bluffs substations and the Hanford Substation along with 8.5 miles of 115-kilovolt transmission line (UPOA 1977a).

TRANSMISSION LINES

By 1953, after numerous additions and upgrades, the electrical system consisted of 230-kilovolt, 115-kilovolt, 66-kilovolt, and 13.8-kilovolt transmission lines that tied all the major operation areas to the Midway Substation and to each other (GE n.d.(a)) (see Figure 2-11.15).

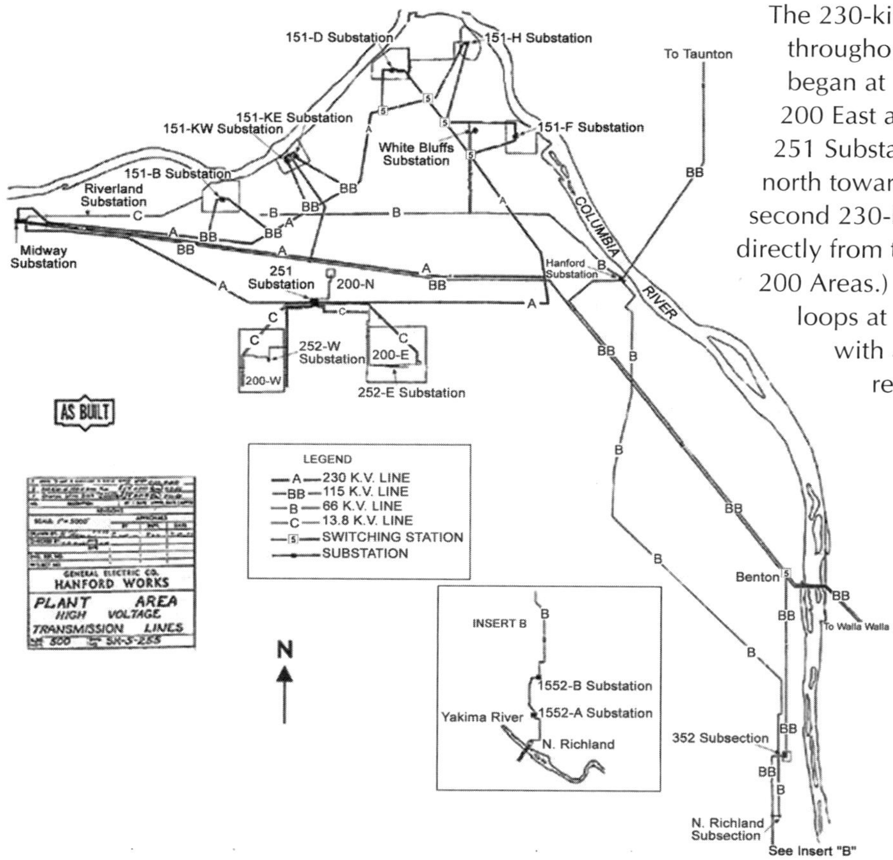

Figure 2-11.15. *High Voltage Transmission Lines at the Hanford Site, 1953*

The 230-kilovolt lines were placed in loops throughout the 100 and 200 Areas. The major loop began at the Midway substation and extended to the 200 East and West Areas where it tied in with the 251 Substation. The line then continued east and north toward the 100 Areas. (At this location, a second 230-kilovolt line was connected that extended directly from the Midway Substation, bypassing the 200 Areas.) Switching stations directed branch lines or loops at each of the 100 reactor areas to connect with a 151 Substation. Each branch line reconnected with the major 230-kilovolt line at another switching station and the entire line went back to the Midway Substation to complete the loop. In general, the 230-kilovolt grid extended through substations in each area where the power was reduced to varying voltages to service different needs of the specific areas (GE n.d.(a)).

One of the two main 115-kilovolt lines traveled at an east/west axis across the Hanford Site from the Midway Substation to a point west of the Hanford Substation. Here it turned and went to the Benton Switch Station where it divided. One branch crossed the Columbia River to Walla Walla and a second branch headed south. The southern branch extended through the 352 Substation in the 300 Area and into Richland to connect with the 751 and 1151 Substations. Power distributed from the two latter substations was through 13.8-kilovolt and 2.5-kilovolt lines. A second 115-kilovolt line traveled across the Columbia River from Taunton and through the Hanford Substation. From here the line extended south to parallel the other 115-kilovolt line until a point near the Benton Switch Substation where it crossed the river again and headed to Walla Walla (GE n.d.(a)).

The one main 66-kilovolt transmission line primarily operated between the 100 Areas and North Richland. The line originated at the Midway Substation and extended east through the northern section of the Hanford Site. At a point near the 100-F Area, a branch line extended directly north and tied in with the White Bluffs Substation. The main line continued along the Columbia River, passing through the Hanford Substation. The line then extended south to connect to the 115-kilovolt line at a point in North Richland (GE n.d.(a)).

The 13.8-kilovolt lines were located mainly within the 100 and 200 Areas. A 13.8-kilovolt line existed between the Midway Substation and the 151-B Substation in the 100-B Area. It passed through the Riverland Yard substation, powering the yard with a 2.3-kilovolt line. In addition, lines ran between all the other 100 Areas (F, H, D, K and eventually N) for emergency backup in case the 230-kilovolt loop lines in the 100 Areas should fail (Davie 1998). In the 200 Areas, the 13.8-kilovolt lines looped between the 251 Substation and the 252 East and West Substations (GE n.d.(a)).

As the Hanford Site grew, additional transmission lines were installed to support the new developments. The 100-N Area, established in the 1960s, was connected to the 13.8-kilovolt and 230-kilovolt transmission lines via its 151 Substation. The 400 Area Fast Flux Test Facility, developed in the late 1970s, had two 451 substations connected to two 115-kilovolt lines: one line from the Benton Switching Station and one from the White Bluffs Substation. A 13.8-kilovolt tie line from the 300 Area was also connected to the 400 Area (DOE 1990) (see Figure 2-11.16).

SUBSTATIONS

The Midway Substation was originally the main electrical distribution station. In addition, the Hanford Substation and the Richland Substation, which existed before the Manhattan Project, were also used in the electrical grid for the Hanford Site. However, major upgrades, including additional substations, have been made to the electrical system to accommodate changing requirements brought on by growth of the Hanford Site.

The 100, 200, 300 and 700 Areas were all constructed with primary substations (see Figure 2-11.17). The primary substations 151 and 251 in the 100 and 200 areas, respectively, were powered by 230-kilovolt transmission lines and associated loops, whereas the primary substations 351-A and 351-B in the 300 and 700 areas, respectively, were powered by 66-kilovolt transmission lines (Du Pont 1945f). In addition, secondary substations and distribution stations were constructed in the 100 and 200 Areas as steps in the electrical distribution system. Two substations were constructed in the 400 Area in the 1980s. The 451-A Substation serves the Fast Flux Test Facility and associated buildings, while the 451-B Substation serves the Fuels and Materials Examination Facility and associated buildings.

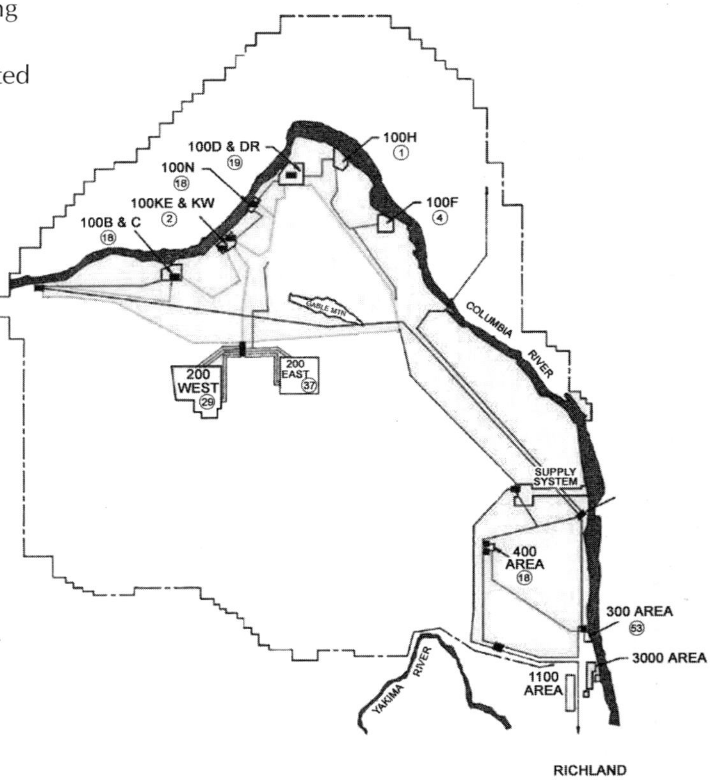

***Figure 2-11.16.** Hanford Site Electrical System Around 1990*

***Figure 2-11.17.** 151-D Primary Substation from Aerial View Looking Northwest Around 1945*

Additional substations were also built throughout the 600 Area. The Benton Switch Substation was built in 1948. A new White Bluffs Substation was built in 1976 west of the 300 Area. The H.J. Ashe Substation, also built in 1976, is located on the east side of the Hanford Site. In 1986, the H.J. Ashe Substation took over the electrical maintenance and operations responsibilities from the Midway Substation, which had been an electrical maintenance headquarters (PRC 1992).

Between 1979 and 1985, the Supervisory, Control, and Data Acquisition System was installed at the 251-W Substation. This system "monitors, provides status, and [has] alarms to allow the dispatcher to change electrical routings remotely through seven primary substations...and six secondary substations" within the 100, 200, and 300 Areas (DOE 1990, p. 3-1).

> ***"Under the agreement between the [Washington Public Power] Supply System and the [Atomic Energy Commission], the government paid all costs associated with producing the steam; the Washington Public Power Supply System bore all costs associated with transporting the steam from the 109-N Building, powering the turbines in the 185-N Building, and transporting the electricity to the Bonneville Power Administration grid."* - DOE 1997a, p. 2-9**

HANFORD GENERATING PLANT

In 1963, President John F. Kennedy broke ground for the Hanford Generating Plant, which would generate electricity from steam the N Reactor produced. The N Reactor was completed in 1964 as a dual-purpose reactor to produce plutonium and use the heat generated to produce electricity in the Hanford Generating Plant.

During nuclear reactions within the N Reactor core, the generated heat was absorbed by water pumped through the process tubes. The heated water was then sent to the 109-N Building's steam generators where the heat was transferred to a secondary system, causing water to boil and produce steam. The water was then recirculated to the reactor core. Some of the steam produced was used within 100-N Area, however, the majority was transferred to the 185-N Export Powerhouse turbine, also known as the Hanford Generating Plant, where it fed two turbine generators producing approximately 860 million watts of electricity. This power was delivered to the 155-N Export Power Switchyard and on to the Bonneville Power Administration grid providing power to the Pacific Northwest. Because of occasional N Reactor outages, the electrical power the Hanford Generating Plant produced was not dependable on a daily basis without extra assistance. The Bonneville Power Administration provided that assistance by supplying power during outages.

Because N Reactor was built with a dual purpose, it was planned that if the production of weapons-grade plutonium was eliminated, the reactor could still be operated economically to produce steam for the Hanford Generating Plant. In 1973, this plan was realized when plutonium production at the N Reactor was halted, and the N Reactor's only function was to generate steam. However in 1982, N Reactor's dual purpose was reinstated to include weapons-grade plutonium production. In 1986, N Reactor was put on standby status for safety concerns and inspection. During this time no electricity or plutonium was produced, and in 1988 N Reactor was placed on cold-standby status. Eventually in 1991, the U.S. Department of Energy ordered the final shutdown of N Reactor, ensuring the end of producing power through the N Reactor and the Hanford Generating Plant. For more information on the Hanford Generating Plant, see the Reactor Operations Section and the Historic Property Inventory Form for the Hanford Generating Plant (185-N) in Appendix B on the Internet.

STEAM PRODUCTION

The production of steam on the Hanford Site was necessary as a backup power supply in case of power failure. It was also used for cooking, heating facilities and water. During construction of the Hanford Site, steam was used in some of the construction processes, such as heating bituminous materials to build roads (Du Pont 1945a).

Figure 2-11.18. *Typical Overhead Steam Lines Around 1945*

The steam production system consists of powerhouses and a network of steam lines. Temporary steam lines were first installed and totaled approximately 39 miles, the majority of which were in the Hanford construction camp area. These steam lines had an average operating pressure of 70 pounds per square inch and ran above ground on wooden pipe supports (see Figure 2-11.18), except in the 1100 Area and in parts of the 200 Areas where the lines were underground (Du Pont 1945a). The steam lines were made from seamless steel pipe with welded and flanged joints, except for the lines in construction areas that were made of screwed steel pipe.

Wherever temporary steam was needed on the Hanford Site, portable, single-unit boilers were used. In the case of the Hanford construction camp, railroad locomotive boilers were used to make steam for the powerhouse (Thayer 1996). Within the Hanford construction camp, Central Shops, and Camp Hanford, a total of 24 semi-permanent powerhouses were constructed containing hand-fired boilers. However, an estimated maximum of over 331,000 pounds per hour of steam would be required for each reactor. Thus, additional powerhouses were installed throughout the 100, 200, 300, and 700 Areas (Du Pont 1945c). The powerhouses supplied power to steam turbine pumps for the secondary reactor coolant system in the 100 Areas and supplied heat to facilities throughout the Hanford Site where needed.

Figure 2-11.19. *184-B Powerhouse from Aerial View Looking Northeast Around 1945*

In the 100 Areas, a powerhouse was built in each of the B, D and F reactor areas during the Manhattan Project (see Figure 2-11.19). They each contained four coal boilers with capacities of 100,000 pounds per hour and were able to "generate steam at 225 pounds per square inch pressure and at about 50 degrees superheat" (Du Pont 1945c, p. 167). Powerhouses were built at each additional reactor area (H, K, and N) as they were constructed.

The 100-KE and KW reactors operated with a dual-purpose capability to provide heat to their work areas. Heat exchangers transferred the heat in the waste water from the reactor's graphite block to an ethylene glycol water solution. This solution then transferred the heat into the air ducts of the 100-K Reactor buildings (Carlisle and Zenzen 1996).

The 100-N Area steam system was unique because of its production of electricity at the Hanford Generating Plant. Steam production was terminated in the 100-N Area in conjunction with N Reactor going to standby status. Oil-fired boilers (one with a capacity of 575,000 pounds per hour and two with a capacity of 200,000 pounds per hour) supplied standby power (DOE 1997a).

The 200 East and West Areas each have one powerhouse (for more information see Figure 2-11.14 and the Historic Property Inventory Form for the 284-E Powerhouse in Appendix B on the Internet). The original boilers were designed to operate at 225 pounds per square inch pressure (Du Pont 1945c). Steam served industrial equipment and process needs in the 200 Area (Rockwell n.d.) These two powerhouses are connected by a 4.5-mile, two-directional, steam tie line built in 1989.

Fuel elements were fabricated in the 300 Area. An electrical power failure in the 300 Area did not pose the serious threat to the production of fuel elements that it did to the production and separation of plutonium. Nevertheless, the 384 Powerhouse was built in 1943 to supply steam to the 300 Area for facility heat, compressed air, humidification, and emergency electrical power via coal-fired boilers and later oil-fired boilers (Du Pont 1945c, Gerber 1993c). Steam was also used to produce fuel elements. New and more efficient package boilers have been installed in the 200 and 300 Areas to replace the old powerhouses (Taylor 1998a).

SEWER SYSTEMS

Because the existing sewage systems in the 1940s did not provide the capacity or locations needed by the Hanford Site, additional sewage lines, tanks and treatment plants were constructed. Service began with temporary installations during construction. As construction was completed, many installations were deactivated because of decreased usage and permanent installations were placed into service. The Hanford Site created two types of sewer systems: sanitary and process.

SANITARY SEWERS

Sanitary sewers handle general sanitary waste. During the Manhattan Project, the greatest demand for sanitary sewer systems was where the largest number of workers was located—in the Hanford construction camp, 3000 Area Camp, and the Central Shops. In addition to using existing building sewer systems and improved former sewer systems at the original Hanford townsite, the Hanford construction camp contained many septic tank and drain field systems and three gravity-flow sanitary sewage treatment systems that chlorinated, flocculated, and settled sewage before it was disposed into the Columbia River. The 3000 Area Camp and the Central Shops area each contained a main sewage line connected to individual buildings. The main sewage line drained into a large septic tank and then to settling basins (Du Pont 1945a).

During construction, "33 gravity flow independent sanitary sewage system[s] were provided with individual septic tanks and drain fields" within the 100, 200, 300, and 600 Areas (Du Pont 1945f, p. 29). Where the Corps was reusing existing buildings that had plumbing, the systems were either connected to a more permanent sewer system or the existing system was upgraded with new septic tanks and drain fields. Indoor sanitary sewers only existed within office buildings in the 100, 200 and 300 Areas that had individual systems with underground septic tanks connected to drain fields (Du Pont 1945a). Currently, the Hanford Site contains 85 operational septic tanks and drain field installations and a Central Sewer System consisting of a lagoon and aeration facility located near the 100-N Area.

The sanitary sewer system contains sewer lines constructed of concrete and vitrified clay and spigot pipe (Du Pont 1945f). Sewer lines were connected to septic tanks, except in the 700 and 1100 Areas where they were connected to the Sanitary Sewer Disposal Plant in Richland Village. Septic tanks were rectangular and had anywhere from 8-inch to 1-foot-thick walls and floors depending on the number of people served each day. The tanks drained to the soil through drain fields built around the tanks (Gerber 1993c). The reactor areas each contained three to five septic tanks, whereas the 200 and 300 Areas required additional tanks due to the greater number of employees in those areas.

PROCESS SEWERS

Process sewers handled the removal of process waste (also known as effluent) from the reactors and other operations buildings. A minimum of one process sewer was located in each of the 100, 200, 300, and 600 Areas. The effluent in the 100 Areas was neutralized and if necessary transferred through effluent pipelines to a retention basin (see Figure 2-11.20) where it was cooled and then deposited in the Columbia River via outfall sewer lines or open trenches. Process facilities in the 200 East and West Areas each contained three separate process sewer systems. Effluent from the 221, 222, 224, and 291 buildings was carried to underground composite storage tanks in one system. Process cooling water from the 221 and 224 buildings was carried to cooling water retention basins in a second system, which emptied into open trenches or ponds. Chemical waste water, building drainage, steam condensate, and acidic and non-acid effluent were carried in the third system directly to open drainage ditches. The 300 Area effluent lines drained directly to two process ponds until 1974 after which process trenches were used (Du Pont 1945f).

Figure 2-11.20. *107-F Retention Basin from Aerial View Looking Northwest Around 1945*

Process sewer lines were of two types: acid proof and non-acid proof. Acid proof lines were mainly vitrified clay and earthenware pipe with acid proof joints. However, cast iron, wrought iron, steel, stainless steel, and concrete pipes were also used for specialized purposes. Non-acid proof pipes were constructed of vitrified clay and concrete with concrete joints (Du Pont 1945f).

For further information on the handling of waste at the Hanford Site, see Section 6, Waste Management.

TELEPHONE COMMUNICATION SYSTEM

The existing telephone companies providing service in the area were the Columbia River Telephone Company, Kennewick Valley Telephone Company, Oregon-Washington Telephone Company, Eltopia Telephone Company, and Mesa Telephone Company. The Columbia River Telephone Company serviced White Bluffs and Hanford. The Kennewick Valley Telephone Company serviced an area that included the town of Richland. These companies, however, had insufficient experience, technical capability, and equipment to support the telephone system needs of the Hanford Site.

MANHATTAN PROJECT

For the Manhattan Project, the Corps acquired the Columbia River Telephone Company and the portion of the Kennewick Valley Telephone Company system servicing Richland. The Columbia River, Kennewick Valley, and Oregon-Washington Telephone companies made new connections to their existing systems for use during construction after which the Corps acquired these installations (Du Pont 1945a).

> **The Signal Corps determined that *"seven permanent manually-operated telephone exchanges would be required for handling the five operating areas, the experimental (300) Area, and the Richland Village; in addition to one manually-operated temporary exchange for the Construction Camp at Hanford."* - Du Pont 1945c, p. 163**

The Prime Contractor at the Hanford Site (Du Pont), American Telephone and Telegraph Company, and the Chief Signal Officer of the Corps designed the communication system for the Hanford Site. The Corps' Signal Corps Division of the Ninth Service Command was in charge of expanding the existing systems, using the Pacific Telephone and Telegraph Company to do the work. The telephone system was designed "in accordance with War Department regulations governing the installation of [a] Government-owned telephone system" (UPOA 1977a, Book 4, Vol. 3, p. 7.6). Three individual exchanges were installed in the 100 Areas, and two exchanges were installed in the 200 Areas. The main exchange (the 702 Central Telephone Exchange Building) was constructed in the 700 Area. It contained all the central telephone switching equipment and personnel to provide connections to Hanford Site facilities.

During construction, the Signal Corps also installed temporary telephone lines that tied into the existing system during construction. Four temporary switchboards (Richland, Central Shops/Cold Creek, Pasco, Hanford) were set up initially and provided 1100 lines. In the construction of these exchanges, 75 miles of trunk cable and 3000 miles of local circuits were installed. Modifications were made to some existing lines to increase the service. For instance, a "repeater station on the main truck line north of the 300 Area was installed to improve transmission" (UPOA 1977a, Book 4, Vol. 3, p. 5.33). However, lines were also removed if they interfered with project construction.

Three permanent exchanges (Olive, Pine, and Maple) were built in 1944 and contained manually operated switchboards within the 1720 Patrol Headquarters buildings. The 200 East and West Areas also had individual permanent exchanges, servicing the Riverland Classification Yard, 200-North Area, and 622 Meteorological Complex. The 700 and 1100 Areas had a single system with four types of service: official and administrative, private residence, commercial stores, and emergency/service/ administrative (Du Pont 1945a).

The telephone service consisted of five systems. The first system was the Inter-Area System, which connected the various area telephone exchanges across the Hanford Site. From a main switchboard in the 702 Central Telephone Exchange Building, connections were made to each area's patrol headquarters, the supervisors' offices in the 200 Areas, and to the 300 Area's 3706 Radiochemistry Laboratory Building "with inter-connecting trunk lines between the 100 and 200 Process Areas" (Du Pont 1945f, p. 31). The second system was the Intra-Area System with connections within each area. The third system was the Teletalk/Inter-Communication System existing within each area's major buildings and between process buildings. The fourth system was the Power Dispatch System that connected each primary substation magneto

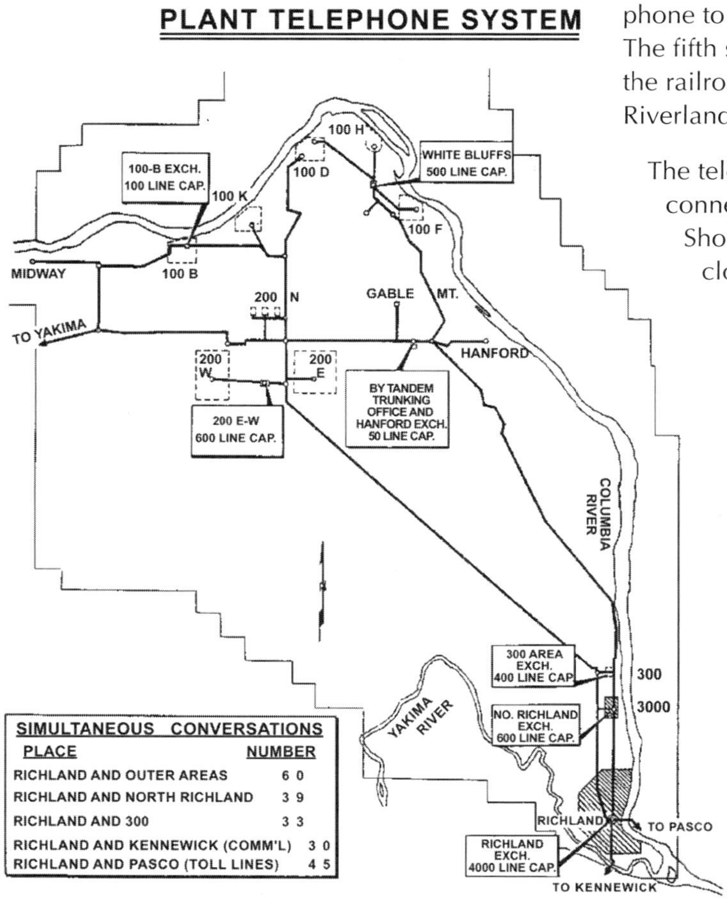

Figure 2-11.21. *Telephone System at the Hanford Site in the 1950s*

phone to the Chief Power Dispatcher in the 251-N Substation. The fifth system was the Railroad Dispatch System that connected the railroad dispatcher stations to the Chief Dispatcher at the Riverland Classification Yard (Du Pont 1945f, 1945a).

The telephone network, when completed in 1944, serviced connections between all offices in the area. When the Central Shops/Cold Creek Exchange and the Hanford Exchange were closed, the remaining security telephones were routed to the Hanford Fire Station.

Cold War Era

At the end of 1946, the telephone systems were converted to dial operation to increase speed of service and decrease cost. To make the unattended (no operator) lines safe from failure, they were tested remotely and emergency manual switchboards were installed (UPOA 1977a).

By 1957, the Hanford Site was using seven telephone exchanges throughout the 100, 200, 300, 600, and 700 Areas with a total of 6520 lines in service (see Figure 2-11.21) (GE n.d.(a)). The network expanded and improved with the growth and technological needs of the Hanford Site. Two telephone switching equipment exchanges supported the Hanford Site: North Richland exchange and the 200 Area exchange. These exchanges could access the public Switched Network, Federal Telephone System, and operator assistance.

Areas for Further Research

The Corps selected the Hanford area to produce plutonium because the area already had the key infrastructure components needed to begin construction: transportation, water, and power. Between initial construction and the height of the Cold War, the infrastructure systems at the Hanford Site expanded exponentially. The Corps, their successors, and contractors transformed agrarian support systems to an industrial scale system of roads, railroads, and the equipment that traveled on them, water and power utilities, sewer and communications systems. Each of these systems offers areas for further research. Specific areas of interest future researchers might explore are

- Day-to-day problems the Corps and Du Pont faced in scaling up an agricultural infrastructure system to industrial scale in less than a 2-year period
- How the Corps and Du Pont improvised substitutions for materials normally needed when expanding an infrastructure but not readily available during World War II
- Beyond what is already described in this book, what effect the need for secrecy and site security had on the Hanford Site infrastructure systems
- Whether in the course of expanding the water and power systems during the Manhattan Project and Cold War Era, the engineers designed energy and water conservation into the expanded systems

SECTION 12

HISTORY OF WORKERS AT THE HANFORD SITE

D.C. STAPP
PACIFIC NORTHWEST NATIONAL LABORATORY

SECTION 12 - HISTORY OF WORKERS AT THE HANFORD SITE

Life as a worker at the Hanford Site differed according to the particular era. What it was like to be a Hanford worker depended on whether the worker was one of the first to hire on in 1943 when the Manhattan Project was gearing up for World War II, was hired on in 1990 as the mission changed from production for the Cold War to environmental cleanup, or hired on in any of the other years during the Manhattan Project and Cold War, 1943-1990. So far as work at the Hanford Site is concerned, that period is divided into three eras:

- 1943 to 1945, when workers designed and constructed the Hanford Site
- 1944 to 1957, when workers operated and expanded the Hanford Site
- 1958 to 1990, when workers operated and diversified the Hanford Site

In November 1942, the government decided to pursue full-scale construction of plutonium production facilities that ultimately were located at the Hanford Site (Gosling 1994, p. 15). Although construction would not begin until March 1943, E.I. Du Pont de Nemours and Company, the contractor in charge, knew that obtaining the workforce would be a major task and thus began almost immediately to seek employees for the Hanford Site. After recruiting the workforce, Du Pont built two reactors, a separations plant, and support structures, which were operating by the end of December 1944.

This accomplishment was amazing considering that design decisions had to be made for untried technologies, workers had to be brought to an area with no existing housing and limited infrastructure, and the operating facilities had to be constructed. In addition, these activities had to be conducted with the utmost secrecy and security while worker health and safety, union, and minority issues were taken into consideration.

Much has been made of the strategy the U.S. Army Corps of Engineers used to design the plutonium production facility at the Hanford Site. The strategy, a tripartite scheme, involved the

- Theoretical expertise of physicists and scientists located at various centers but primarily in the Metallurgical Laboratory at the University of Chicago
- Construction and operations experience of E. I. Du Pont de Nemeurs and Company in Wilmington, Delaware
- Decision making and funding of the Manhattan Project by the U.S. Army Corps of Engineers

The synergy resulting from the interactions between the "hard-driving, thoroughly competent, industrial minded, scientific engineers and executives at Wilmington and the highly intelligent and theoretically inclined scientists at Chicago" (Groves 1983, p. 79), along with the ability of the Corps to manage the uncertainties associated with the reactors and separations technologies is often credited as a major key to the success of the Manhattan Project. Examples of the synergy are the decision to pursue the bismuth phosphate process over the lanthanum fluoride process and the decision to add extra fuel loading channels in the reactors. The former decision guaranteed that at least some plutonium could be extracted. The latter decision ensured that the reactors could sustain a reaction.

CONTENTS

Divider Page Photo: Workers Passing through Security Gate, 1944-1945 (Fermi and Samra 1995, p. 33)

"...we considered building a large deionizing plant for the second reactor so that at least one of the three piles [reactors] would certainly be operable at all times. I was discussing the advisability of this with G. M. Read of Du Pont, one night at Hanford, when Dr. Hillberry came into the room. I asked him for his views, and he replied that he did not think we would need the deionizing plant, but if we did, we could not do without it. I turned to Read and said, 'Go ahead and build it.' Hillberry then asked what it would cost and I told him that it would be somewhere between six and ten million dollars. He replied, 'I'm glad I didn't know that when I gave my opinion.' Such quick decisions were not too frequent and they were always preceded by as much research, study, and thought as could be devoted to them without delaying the completion of the project. Nevertheless, there were many decisions that had to be made when the unknown factors far outweighed the known. We built only the one deionizing plant and fortunately never found any need for it...." - Groves 1983, p. 81

The Corps had a basic philosophy it followed in making Manhattan Project decisions. The philosophy can be characterized as "if it is necessary, do it." When the path to an atomic bomb was clouded by the uncertainty associated with both gaseous diffusion technology and reactor technology, Lieutenant General Leslie R. Groves, commander of the Manhattan Project, decided to pursue both. The justification was that the United States could not fail, for, if we did, Germany might have the atomic bomb before us. This philosophy extended to all decisions the Corps made for the Manhattan Project.

The Du Pont designers and managers were under intense pressure to make technical, cost, and schedule decisions in the face of constant uncertainty, knowing full well the price of failure. To make decisions in the face of uncertainty, Du Pont had a tradition of using the "flexible" approach (Carlisle and Zenzen 1996, p. 109) of component sub-systems that could be changed as necessary. When building the Hanford Site, Du Pont focused on sub-systems for which design certainty existed and allowed time for sub-components that needed more research. Hounshell (1992, pp. 236-261) notes this

"...Du Pont was under contract to design and build this equipment following the basic data provided by the group in Chicago. So therefore when the group at Chicago gave us what to us was official data, we simply had to follow it. Now it happened several times during the design and construction of this place that they recalculated things, see, this was all very new, new physics. They were breaking ground all the time. They recalculated the situation a few times for certain design changes. Well, we proceeded to make them. Then, finally, as time was going, for one reason or another, Du Pont felt we should put a stop to this if possible, and the chief engineer, Slim Read, said he wasn't going to accept any more changes.

Of coarse [sic], that is a position you can take but you can't hold it. So when the Chicago people came through with another change, we had to make it, even though we said we would not make any more. And that is exactly what happened, they came through with another change, and unless we had made it we had every reason to suppose the plant wouldn't work. That was the change to add some extra tubes. It seems the plant was designed to have a certain driving power in terms of capacity of the plant to permit the fission reaction. If anything were to happen so that the capacity was diminished, then of coarse [sic] the plant wouldn't run. We had decided that we keep a 10 percent safety factor on this driving power, 10 percent excess 'k' if you want the jargon.

We made the change Chicago specified, but we insisted there still be a 10 percent excess 'k' as a safety factor, and that's what saved the day, because lo and behold when they tried to operate the place some unexpected things happened and unless we had this excess it wouldn't have run. We were glad we had the excess, and the excess was provided in the design by changing certain dimensions and setting the number of tubes of canned uranium in the pile [reactor]. The net of it was that more tubes were made available than the design called for, to the extent of 10 percent, and that let us operate even in the face of this unexpected phenomenon that occurred after startup...." - George Graves, the Du Pont engineer credited with making the conservative decision to add more tubes to the reactor design, quoted in Sanger 1995, pp. 57-58 organizational model is the one that enabled Du Pont to develop nylon in the 1930s. Along with its flexible component sub-systems, Du Pont had a tradition of designing conservatively, which crept into many of the basic design decisions at the Hanford Site and was often a source of friction with academics.

The friction among the three groups, especially between the academics and Du Pont, is well documented by historical writers and by the participants themselves. We only describe it briefly here to illustrate the atmosphere within which the early workers constructed and operated the Hanford Site.

WORKERS WHO CONSTRUCTED THE HANFORD SITE: 1943-1945

Finding skilled and unskilled laborers for the construction and organization of the Hanford Site in the numbers needed during wartime was a challenge. The military had already taken a large portion of the labor pool for the fighting forces. In addition, the military's demand for weapons and general supplies further reduced the available labor supply. The Hanford Site's chief local competition for labor was the Boeing Company in Seattle and Kaiser Aluminum in Vancouver (Goldberg 1998, p. 57). Du Pont, therefore, recruited workers nationwide. An additional challenge to recruiting workers was the remote location of the Hanford Site and lack of services. All manner of trades were recruited, from pipe fitters to machinists to secretaries, cooks, butchers, and photographers. Recruiters concentrated on men who were not liable to be drafted, such as those aged 38 or over with children, or men who had physical or occupational draft deferments (Sanger 1995, p. 68).

> ***"Du Pont's recruiters, numbering 153 at the peak, fanned out across 47 states [in addition to the State of Washington], Alaska and Canada and interviewed a total of 262,040 applicants. This effort was initiated, with the participation of the War Manpower Commission and the U.S. Employment Service, at the beginning of the project because it was then recognized that the State of Washington provided an insufficient worker pool, given the war projects then in progress in western Washington."***
> **- Thayer 1996, p. 93**

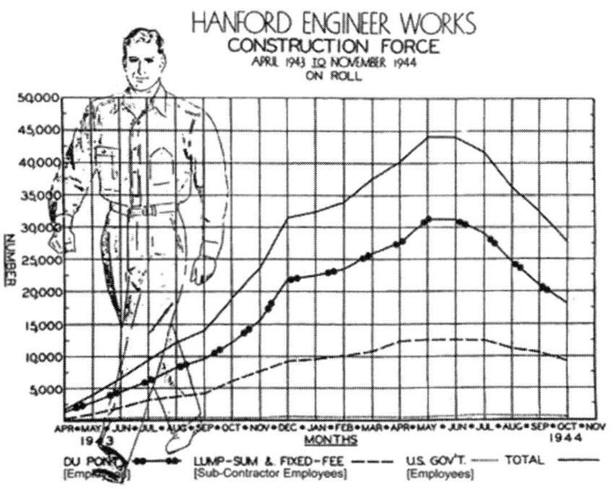

Figure 2-12.1. *Number of Construction Workers, 1943-1944*

Nevertheless with the Corps's assistance, Du Pont found construction workers. Construction at the Hanford Site began in March 1943. By May 1944, the labor force constructing the Hanford Site had peaked at approximately 45,000 (see Figure 2-12.1). Findlay and Hevly (1995, pp. 16-17) report that a severe labor shortage (up to 50 percent) existed for the first 14 months of construction. Between 1943 and 1945, it took 260,000 interviews to find 94,307 people to hire. After overcoming the initial labor shortage, the Hanford Site still experienced shortages of certain types of construction workers such as pipe fitters. To obtain the needed skills, Du Pont and the Corps used various creative strategies to obtain qualified workers.

One strategy cited for drawing workers to the Hanford Site was the high rate of pay. Many people were attracted to the Hanford Site for this reason alone. Rates at the Hanford Site averaged $8 per day for unskilled laborers, where $3 to $4 per day was typical for laborers in other parts of the nation. Skilled labor, pipe fitters, and electricians earned $15 per day compared to the typical $10 per day elsewhere.

Another strategy for dealing with the labor shortage was to increase the work week. Ten-hour days were the norm, and employees worked 5-1/2 to 6 days per week. A delicate balance existed between the pain of long hours and the reward of large paychecks enhanced by overtime pay (Findlay and Hevly 1995, p. 19). Many also worked Sundays.

> *"We worked enough people overtime on Sunday so that what the rest of them had to do on Monday got off to a good start. They used the day off to catch up on a lot of little things that would be in the way next week. They were buying time. The Army said, 'Don't hesitate to use money to buy time. Every day may be significant.'"* - Walter O. Simon, a chemist and Hanford's first operations manager with Du Pont, quoted in Thayer 1996, p. 93

Example of Advertisement for Construction Workers
(Van Arsdol 1992, p. 18)

War Workers Needed on
Southeastern Washington
Construction Project

Transportation Advanced
Attractive Scale of Wages

- 54-59 Hours Per Week
- Immediate Living Facilities for Employed Persons Only
- Applicants Must Possess Statement of Availability
- Company Representative Interviewing Applicants 8:30 a.m. to 4:30 p.m.

U.S. Employment Service War Manpower Commission
Portland, Oregon

Once Du Pont obtained the workers to build the Hanford Site, retaining them became the challenge. Thayer (1996, p. 93) reports a turnover rate of 4.2 hires for every 1.0 worker, which was acknowledged as high. Thayer goes on to indicate the nature of work was such that the turnover rate did not impede progress as long as replacement workers were found. Because of persistent problems in recruiting and maintaining an adequate workforce, living conditions were made more tolerable with amenities, such as recreational facilities, churches, a gymnasium, and a library. In addition, workers were allowed to bring their wives and children. Four reasons commonly cited for the high turnover are isolation of the site, miserable living conditions, requirement to maintain secrecy, and the sense that the Hanford Site was not vital to the war effort.

"At Texarkana, I was working at a concrete plant, making $33.33 a week, that was good money then, working as a common laborer. I worked there a couple of years, and when I left there I heard about this job in McAlester, Oklahoma, in about '42, a naval air station. That job kinda went down, and we heard about this job in Washington state. I had an uncle who come out here, and he wrote back and said they was paying a dollar a hour. I say, 'What?' My brother say 'A dollar an hour?' I say to my brother, 'You going?' My brother was working on the railroad. He say, 'I don't know. Them jobs don't last long.' I say, 'Man, I'm going, you do what you want. I got enough money for you to go.' We got together, my brother, Vanis, myself and another boy. We came out to Hanford, in the late summer of '43. We went to work on labor day." **- Willie Daniels quoted in Sanger 1995, p. 114**

We are all good American people
Who are trying to do our part
But to feel that you're blindly working
Will finally break anyone's heart.

There is something about really knowing
The thing that you're trying to do,
A sense of accomplishment maybe,
That warms one's heart through and through.

And that is the reason I'm leaving.
It's not that I hate the dust;
Its not that my bunk mates snore too loud,
Or my lunches contain some crust.

It's just that I want to be doing
A thing I can see and know
Be it a ship or airplane or what not.
The thing is I just want to know.

- Lois May Lyon, head of employee placement at the Hanford Site in 1944, quoted in Van Arsdol 1992, p. 19

Labor shortages prompted the recruitment of union workers. Using the war as justification, the Corps called on unions to accept responsibility for providing a quota of workers at the Hanford Site each month (Hales 1997, pp. 168-171). Thus, organized labor was an important source of labor in building the Hanford Site, although, as Colonel Franklin T. Matthias (one of Lieutenant General Groves top assistants) recalled, no formal agreement was signed with any of the unions involved at the Hanford Site.

"We didn't try to keep organizers away, we accepted the idea of union organization. We did not have a project contract. We agreed at the beginning with the labor leaders in the region that we would follow standard union practices. We set the basic rate at $1.05, the average between Seattle and Spokane. That was for common labor. Electricians and pipe fitters got more." **- Matthias quoted in Sanger 1995, p. 79**

Union leaders, such as the Teamsters leader Dave Beck in Seattle, had trouble pressuring Hanford management on union issues because of the appeal from the Corps to support the war effort and the fact that the unions could not supply the numbers of skilled labor anyway because of labor shortages. Still, the fact that about one-quarter of Matthias' construction journal entries were union related indicates that union matters required constant attention. Jurisdictional disputes between various unions and between categories of jobs caused delays in construction schedules. The Building Trades Council of Pasco could not furnish enough workers, and shortages existed in some crafts while others had surpluses. Machinists were often asked to perform other duties such as millwright work. This, of course, brought immediate protests that had to be resolved. In the summer of 1944, Matthias and Groves met with Martin Durkin, President of the United Association of Steamfitters and Plumbers, to appeal to him on the grounds of "the country's most vital war project" and to issue instructions to release jurisdictional claims and help fill 1000 jobs (Goldberg 1998, pp. 59-60; Thayer 1996, pp. 89-93). When Harry Thayer asked Matthias how much time was lost in labor disputes, Matthias replied, "Very little" (Thayer 1996, p. 175).

"I dealt with the unions, and Du Pont and I did a good job of batting problems back and forth and confusing the labor leaders. I did that also with the colonel in Washington D.C. who was head of labor relations. We kicked problems back and forth. I would say, that bastard in Washington doesn't know what he is doing. He would do the same thing, and blame me for all the problems. It worked great. We kept kicking them around until they disappeared." **- Matthias quoted in Sanger 1995, p. 79**

The unions were strong in the 1940s, and many Hanford workers, especially skilled workers, were union members. Thus in training, skills, organization, and work habits, the Hanford Site was characteristic of the major large-scale engineering projects of the era. Without this experienced workforce, which existed nationwide, the project may never have been able to proceed as rapidly as it did.

LIVING CONDITIONS

When the Corps arrived to begin constructing the Hanford Site, little available housing existed in the general area for the workers needed to construct and operate the plant. The 800 or so residents of White Bluffs, Hanford, and the former Richland townsite also had to move and find new housing. The former and new residents found living quarters among the following options (Findlay and Hevly 1995, pp. 24-25):

- Construction workers, as many as 45,000, moved into the Hanford construction camp located where the town of Hanford had been, near the reactor construction areas. Workers initially lived in tents until the camp was constructed. At peak occupancy, it included 131 barracks for nearly 25,000 men, 64 barracks for 4,350 women, 880 hutments for 10,000 men, and 3,600 trailer lots (see Figure 2-1.1 in the Construction History Section). In addition to housing, the workers had mess halls, recreational facilities, and various services. Segregated quarters existed for Americans of African descent.

- Key supervisors and essential office workers occupied much of the new government-owned village of Richland. The government constructed two-bedroom, single-story duplexes (for example, B house); three-bedroom, two-story houses (for example, F house) (see Figure 2-12.2); apartment buildings; prefabs; and dormitories for male (such as the K unit) and female (such as the J unit) workers. The commercial and residential buildings were all government-owned and maintained. Only about twenty pre-1943 buildings were incorporated into the new design (Kubik 1994, pp. 33-47).

- Mechanics, common laborers, and their families tended to live in Pasco and Kennewick located to the south of the Hanford Site. Pasco would grow from 3,913 to 8,500 by the mid-1940s, Kennewick from 1,918 to 7,500, Richland from 0 to 7000 in March 1943 and 15,000 by 1945. African Americans were restricted to living in east Pasco. The few workers and their families of Hispanic descent also lived in Pasco.

The construction crew worked long hours 6 days a week but still found time for recreation. Finding ways to keep the "troops" entertained was thought to be key to keeping them working at the Hanford Site. Dances, movies, and baseball games all became important, authorized activities. Bars in the camp were well patronized, and unauthorized activities such as gambling and prostitution existed.

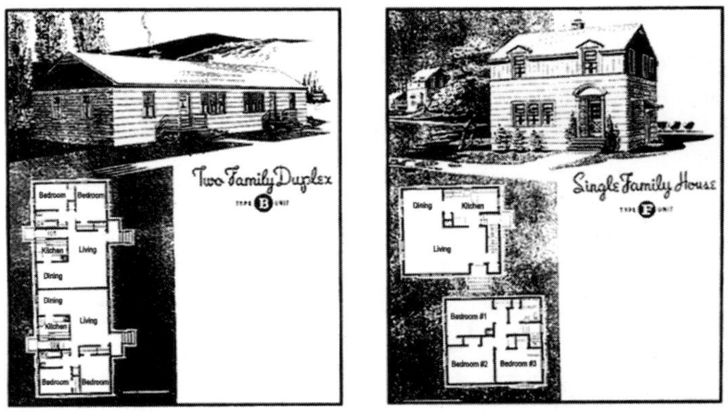

Figure 2-12.2. *Two Styles of Housing Built in Richland, Washington during the 1940s for Workers at the Hanford Site*

The presence of a few thousand single women in a construction camp of over 40,000 men, mostly single, presented its own problems. The women's barracks had housemothers and were fenced and patrolled for security and protection. Groves noted the women in the Hanford construction camp suffered from their own morale problems, as any group would "in an isolated area under rugged conditions, with few of the amenities of normal life" (Groves 1983, p. 90). Groves cites Mrs. Maris, Supervisor of Women's Activities, as doing an excellent job of helping the women cope. She in turn credited Du Pont administrators at the Hanford Site with helping her do what was necessary to keep living conditions bearable for the women (Groves 1983, p. 91). To recruit and retain workers, Hanford Site officials allowed married men to bring their wives and children with them. Measures taken to improve life for female workers and female family members at the Hanford Site included development of a nursery and community center, schools, formation of local Red Cross and Girl Scout organizations, introduction of a women's apparel store, initiation of 5:00 p.m. bus service to Pasco, and arrangements for women to attend dances with the Navy Air Force contingent in Pasco.

More than one story exists about the adverse living conditions created by the large construction project. Winds were especially notable, made worse by the loose sand so prevalent in the area that led to horrific sandstorms. A local historian noted: "Residents nicknamed the winds 'termination winds.' The morning after a storm of high winds and blowing dust, there would be a long line of workers at the employment office seeking to draw their termination pay and leave the area" (Kubik 1994, p. 44).

The sheer isolation of the project was also untenable to many. The adverse living conditions, housing shortages, and general lack of amenities resulted in the substantial turnover rate.

Secrecy and Security

Few of the construction workers at the Hanford Site knew they were building a plant to produce materials for a new type of weapon. They were told that the work was related to the war effort but were required to avoid any discussions about their work. This prohibition applied to discussion with co-workers as well as family members. Many stories abounded about what was being manufactured at the Hanford Site. Some of these included Eisenhower re-election buttons, explosives, and mining rare ore. Stories abound about workers who talked about their work being discharged immediately. Much speculation existed about the nature of work going on at the Hanford Site. Even after production began, only a few actually knew before the official announcement was made.

No incidents of espionage are known to have occurred at the Hanford Site during World War II, although records from those conducting investigations have not been released. Richard Rhodes (1995, pp. 38-39) indicates that the design of the first Soviet nuclear reactor matched the test reactor built at the Hanford Site in 1944, suggesting that some Soviet espionage was successful.

Perhaps the biggest challenge for security forces at the Hanford Site during construction was controlling the rowdy construction crews. Incidences recorded during peak construction are summarized in Table 2-8.1 in Section 8, the Site Security Section.

HEALTH AND SAFETY

Initial health and safety issues were construction related. The Hanford Site was a massive construction project involving many workers under a lot of pressure to keep to the schedule. However, the work was generally typical of large construction projects at the time. Engineering projects of this nature had begun in the 1930s with such undertakings as Grand Coulee Dam. Thus, a trained work force did exist. Thayer (1996, p. 93) reports a rate of 206,000 worker hours per major injury, which was about 2.5 times safer than for the general construction industry at the time.

URGENCY AND COMMITMENT

Although most construction workers did not realize the Hanford Site would produce materials for the atomic bomb, they did at least know their work was related to the war effort. Construction workers were regularly reminded of the urgency of their work and told every day it could spell the difference between victory and defeat.

> ***"The pipe fitters once sat out for an hour and a half one Monday morning...I took the stage in the recreation building –it held a thousand people –and they had loudspeakers to talk to themselves. Apparently they had a date with a union guy –an official –to be there to help them. There was one guy on the stage that I knew –a young fellow that was a pipe fitter, and I said, 'Give me a chance at the microphone.' And he said, 'Good, these guys sure need something.'***
>
> ***So I gave them a little talk, and I said, 'You know, you guys are breaking your promise. Your union agreed that we'd have no strikes. Now far more important than that, you are having a strike and not working on a project that is of tremendous importance to this country. Now, I thought you appreciated that, but apparently there are some of you that are moving this strike, and they're working for the Germans, and if I knew who it was that was fomenting this thing, I'd arrange to have them sent back to Germany, where they belong...'***
>
> ***If they'd had guns they would have shot me! And after I got them quieted down I said, 'Look, take it easy. I'm not calling you traitors, but some of you are acting like it. Now how about going back to work and doing what you promised, and what we need badly. I'll have the buses at the door in ten minutes.' And then they cheered and that was the end of it. They all went back to work, and their union leaders came as they were rushing out the doors, and they wouldn't go back in to listen to them. And I said, 'I'll set up a deal, and we'll discuss all these problems you have and we'll do something about it....'"*** **- Matthias quoted in Thayer 1996, pp. 175-176**

Other reminders came in the form of billboards, pamphlets, speeches, and cartoons (see Figure 2-12.3). Such reminders did not mean the workers were not committed to the war effort. Hanford workers and their families purchased U.S. savings bonds and made financial sacrifices as was common throughout the country. An example of the level of commitment is the "Days Pay" story (see sidebar box and Figure 2-12.4).

Figure 2-12.3. *Examples of Motivational and Security Signs at the Hanford Site during the 1940s*

> **Major Hanford Event: Days Pay**
>
> ***"The Hanford crafts [unions] were solidly patriotic. On their own initiative they asked everyone for a day's pay and raised $162,000 in seven weeks, enough to buy a four-engine B-17 bomber for the Air Corps. The bomber, named the 'Day's Pay,' flew in from Boeing Field in Seattle to the Hanford air strip in July of 1944 for a memorable ceremony to present the bomber to the Fourth Air Force. Matthias remarked, 'The principal talk at the ceremony was by Major Grashin, one of the officers who made the 'Death March' on Bataan. His talk was inspiring and he spared his audience none in his descriptions.' Concerning the overall effect of buying the B-17, Matthias said, 'This activity, conceived by the workmen and handled by them, coupled with the talk by Major Grashin was the most effective single morale builder during the job and did much to develop an attitude of teamwork and desire to help the war than any other thing.'"*** **- Thayer 1996, p. 99**

Figure 2-12.4. *Ceremony at the Hanford Site Celebrating the Arrival of Day's Pay*

Opportunities for Females and Minorities

As was common during World War II, women and minorities helped solve labor shortages in almost all job categories. The wartime draft exaggerated this condition as it took a disproportionate number of white male workers, leaving the civilian workplace increasingly female and "colored" (Hales 1997, p. 191). At the Hanford Site, only whites and African Americans were actively recruited. Hispanic Americans were discouraged, even as labor shortages were in dire straits, because it would require building a third segregation of camp facilities.

In general, work at the Hanford Site conformed to gender and ethnic social and occupational stratification of the 1940s. Women and minorities were hired to do work deemed too menial and undignified for white men (Hales 1997, p. 219). The major job categories noted for women were nurses, file clerks, stenographers, secretaries, and other service and administrative roles. They also filled some of the more technical and scientific roles where not enough men were available. In the mess hall at the Hanford construction camp, white women served the food while African American women cleaned up afterwards. African Americans comprised the major minority group, generally holding laborer and other unskilled positions.

While state law did not mandate absolute racial segregation, as at the Oak Ridge Reservation and other installations, managers at the Hanford Site opted to segregate and prevent any possible racial problems. Living quarters such as the Hanford construction camp, trailer camp, and barracks were all segregated by sex and race. Additionally, mess halls and taverns were segregated by race (Sanger 1995, pp. 68-69). According to Sanger (1995, p. 68), of those living at the Hanford Site in 1944, 65% were white men, 25% were white women, 9% were "colored" men, and 1% were "colored" women.

Scholars have written little about the role of women at the Hanford Site, and even less about minorities, male or female. The noted exception is *Their Day in the Sun: Women of the Manhattan Project* (Howes and Herzenberg 1999), which documents the experiences, accomplishments, and contributions of women. One of the most well known female physicists of the Manhattan Project, Leona Woods Marshall, worked at the Hanford Site assisting her husband, John Marshall, also a physicist, in overseeing the operation and construction of the plutonium production reactors. Jane Hall, also a physicist, worked on health physics and the development of radiation detection instrumentation (Howes and Herzenberg 1999, p. 129). In January 1945, Hanford's Laboratory Division began recruiting women, primarily from the local area, to fill technician positions in the chemical-analysis laboratories. Women like Marge Nordman DeGooyer came to the Hanford Site as news of jobs with good wages spread around the country. They were required to be between the ages of twenty-one and forty, have high school educations, pleasing personalities, and be alert and intelligent. After brief instruction on security and safety, they were given aptitude tests and received the essential training to complete their work.

Other women who worked at the Hanford Site included members of the Women's Army Corps (WACs). The WACs were assigned to work at all three of the Manhattan Project sites between 1943-1945. The Hanford Site had a small group of WACS, from sixteen to twenty-four, all assigned to production work. Buena Maris was personally recruited by Groves from her position as Dean of Women at Oregon State College in September 1943. Maris was made Supervisor of Women's activities and was the only woman who attended weekly senior staff meetings. Her primary mission was to deal with problems that led to high female turnover rates. Maris hired housemothers for each women's barracks, organized a library and Girl Scout Troop, started a Red Cross chapter, supported beauty contests and shopping trips, and even managed to get gravel walkways replaced with asphalt. The Girl Scout troop and library she founded are still in operation today (Gerber 1993d, Howes and Herzenberg 1999).

WORKERS WHO OPERATED AND EXPANDED THE HANFORD SITE: 1944-1957

This section explores the lives of Hanford workers once the production complex started operating. The section is divided into three parts:

- World War II Operations
- Post World War II Operations
- Living Conditions.

WORLD WAR II OPERATIONS

As construction of the production facilities neared completion, it was time to transition the workforce to operations. Many of the construction crew moved on to new construction projects, but some stayed to fill the multitude of jobs required to run the Hanford Site. No nuclear industry existed at this time from which to draw workers. As a result, Du Pont had to find experienced chemists, metallurgists, electrical and mechanical engineers. In February 1943, Du Pont planned for an operating force of 4000. Based on this estimate, 507 were to be above the rank of foreman, and 116 were ranked as supervisors or above (Pehrson 1943, p. 5).

One strategy Du Pont used to obtain the operations staff was to transfer them from other Du Pont facilities. Findlay and Hevly (1995, pp. 41-42) report that Du Pont moved 1532 employees to the Hanford Site between February 1944 and January 1945. Many of these individuals had worked at Du Pont ordnance plants in Colorado, Illinois, Tennessee, and Utah. Most were white, 30-40 years old, and generally had a science and engineering background.

To prepare for operations, workers went through a rigorous training program. Reactor control operators traveled to the Oak Ridge Reservation in Tennessee to train on the air-cooled prototype reactor that was operating and producing small quantities of plutonium. Other training, such as for crane operators in the separations plants, occurred onsite.

In many respects, operating the Hanford Site was similar to operating other industrial plants, such as chemical and water plants, so little training was needed for many jobs. The major differences, according to Raymond Genereaux (design project manager for the chemical separations plants at the Hanford Site) were radiation and critical mass (Sanger 1995, pp. 58-65). Radiation meant protecting workers in ways industry had not done previously. Workers often had to perform work remotely using periscopes, through leaded glass or using sound only. Critical mass also required workers to design instrumentation and equipment that was smaller and had minute margins of error.

Health and Safety

Because operations at the Hanford Site posed the same type of health and safety risk as operating any industrial plant, the same type of health and safety measures were implemented. In addition, measures to protect against radiation were also required. This was accomplished with a variety of shielding materials, limiting time of exposure, and increasing distance.

"...greater awareness of Hanford's mission might have driven labor away from the site by revealing the risks associated with plutonium production. [Lieutenant] General Groves generally feared letting the workers learn too much about the invisible dangers surrounding Hanford. As the time for starting up the first reactors neared, Du Pont executives in Wilmington ordered their managers at Hanford to practice a 'complete evacuation' of the 100 and 200 areas as well as the construction camp. Groves and Matthias refused to allow such a potentially alarming drill, and permitted instead only a limited evacuation of the 100 B and 100 D areas. In his diary, Matthias explained that a wholesale drill 'might be disastrous to the project as it might cause a large number of people to leave if their fears for safety were increased. It would also be sufficiently upsetting so that we could expect a serious effect both on security and manpower facilities to finish the job.' Vetoing Du Pont's request, the Army determined that, for the sake of efficiency and secrecy, construction workers ought not to learn about all of Hanford's unsafe working conditions. [Lieutenant] General Groves and Colonel Matthias, protecting 'the best interests of the United States,' would take no steps to alert employees to the many risks they faced by working at Hanford." **- Findlay and Hevly 1995, p. 20**

Shielding Equivalents Used at the Hanford Site
(Sanger 1995, p. 62)

One foot of lead equaled:
- One yard of steel
- One fathom of concrete (6 feet)
- One rod of earth (16.5 ft)
- One mile of air

Dosimeters and clothing provided other protective measures for the workers. The health and safety risk posed to workers, especially during start-up, were apparently understood by those in charge.

The radiation in the separations plants created a logistical problem for workers, especially those working with the large cranes. Raymond Genereaux explained what it was like training crane operators to use a periscope so they could see behind the 5-foot wall of concrete:

> *"Oh, they thought that [using the periscope] would be terrible. The first thing you know they were getting the biggest kick out of looking through the periscope, with this great big hook going down and hooking on to a wheelbarrow or something down there in the cell area, picking it up and gently putting it down. They had a ball. They were fantastic. The crane operators learned in ten hours or so. Once they got onto the idea, they could do it. They were like kids."* - Sanger 1995, p. 63

Genereaux also provides insight into how radiation concerns affected the separations plants:

> *"We installed the equipment remotely, even at first when there was no radiation, because we didn't want workmen in the cells. That would violate the whole principle. You would eventually have to replace equipment and you could not put anybody in the cells then. So you had to know how to do it ahead of time. Sometimes during construction, somebody would want to do things faster and not use the mockup building. We had some battles and we had fun, but we were all friends."* - Sanger 1995, pp. 63-64

The critical mass issue concerned making sure that the quantities of fissile material being stored were below the amount required to sustain a chain reaction. As plutonium began to be produced, separated, and concentrated, it became important to ensure that storage of the material was always below critical mass. Both radiation and critical mass concerns became major drivers for procedures that Hanford workers had to follow.

Technical Dilemmas

The Hanford Site entered the startup phase in spring 1944, beginning with the fuel manufacturing operation, followed by the B Reactor, and then the separations plants. The fuel manufacturing process provided the biggest technical dilemma for the workers. Walter Simon explains:

> *"Now, one of the most difficult problems we had before plutonium production began was making the uranium slugs [fuel elements]. The uranium was held in an aluminum can, a slug, about eight inches long and an inch and five-eighths in diameter. The can had to fit very tightly with no air space or bubbles. They couldn't leak because if water got into the uranium it destroyed the ability to react. So the concept was that the scientific people would find out how to do this and give us instruction. They found out how to design it but they never made a slug in the laboratory that didn't leak.*
>
> *Well, the summer of '44 was coming along and the reactor was shaping up and there would be nothing to put into it. We had a production superintendent at Hanford named Earl Swensson, who was a real dyed-in-the-wool production man. This was a case where one man did sell an idea. He said you know they'll never make one of these in the lab, even if they work on it for 10 years. It's a statistical matter. Why don't we make a thousand a day, we'll examine each one and test them all, and the poor ones we'll strip the aluminum can off and save the uranium and the next day we'll make another thousand. The first day a thousand failed, but there were maybe 10 better than the others and we tried to figure out why these 10 were better. The next day maybe they had 18 that were better. And they kept doing this and lo and behold after about three weeks there was one perfect can.... They made five good ones the next day, and 10 the next and after a while out of a thousand they were making 500 and then 600 a day that were all right. That's how they did it. It was a little terrifying because if we didn't have them it would stop the whole thing. The reactor would be ready on September 15 and we would have nothing to put into it."* - Walter O. Simon quoted in Sanger 1995, p. 154

Whereas the fuel manufacturing process provided the biggest technical challenge to the workers, the B Reactor startup presented the most dramatic. The B Reactor startup problem has been retold many times (for details see Section 3, Reactor Operations) and is a testimony to the value of Du Pont's practical abilities and the physicists' knowledge.

The separations plants startup proceeded without any major problems. That is especially impressive given that scientists and engineers moved from producing micrograms of plutonium in a laboratory setting to kilograms in separations plants, which represented a billion-fold scale-up (Gosling 1994, p. 27).

Secrecy and Security

To those in the know, the startup phase for the plant and subsequent operations must have been engrossing, considering that it affected the immediate race for the bomb and the long-term race for the scientific frontier. But few people knew much about these aspects of the job. Even those who knew about the mission knew little more than what their particular jobs entailed. By limiting knowledge, any one person could not reveal too much information to the enemy. Groves' philosophy toward communication was termed "compartmentalization," which meant "each man should know everything he needed to know to do his job and nothing else" (Groves 1983, p. 140).

In addition to internal security, external security was also a prime concern. External threats were primarily related to espionage and generally managed by Hanford security and undercover agents from the Federal Bureau of Investigation. Checkpoints placed at strategic locations prevented anyone from getting onsite without proper identification (see Figure 2-12.5).

One external threat was the launching of Japanese balloon bombs. Following a United States raid on Tokyo after the Japanese bombed Pearl Harbor, the Japanese produced about nine thousand rice paper balloon bombs. Easterly winds carried them to the North American mainland, including some in the Hanford area. One of the balloons tangled in a power line that stretched between Bonneville and Grand Coulee dams resulting in a power outage at the Hanford Site. Although the power outage was shorter than a minute, it took 3 days to restore the reactors to full power (Jones 1985, Sanger 1995). For more information, see also McDowell (1993) and the *Tri-City Herald* (1995).

Figure 2-12.5. *Female Workers Proceeding through Security Gates after Showing Badge, 1944-1945*

> The Japanese balloon bombs "*were a real worry. Everybody at the plant was always looking up because of the statistical chance of one falling. I remember seeing 40 at one time going over. The Navy planes at Pasco chased them regularly but they had poor luck. Matthias and I went over to prod them about getting better protection. They never managed to shoot one down, although a number came down on the project, but away from buildings. The bombs never went off, but the balloons blew around and the military were a great sight trying to round these up without being blown up themselves.*" - Walter O. Simon quoted in Sanger 1995, p. 155

> ***"I couldn't sleep. I lay watching the sky grow lighter, the river and hills coming into view. I thought of the thousands and thousands and thousands of years in which men have lived so briefly and have struggled so tirelessly to conquer the elements and their secrets and how well the secrets have been kept. I thought of the close race between success and failure on this discovery. And I couldn't help but believe that God, wearying of this long and tortuous war, had finally, reluctantly, given us this terrible weapon with which to end it...***
>
> ***I wish we could keep this discovery here. But to whomever it is entrusted, is entrusted the future of the world. I hope that the keepers of the secret will use it only for the good of mankind and never for vain or selfish purposes either for themselves or the people they represent."*** **- Reva Baumgarten Matthias, the wife of Colonel Matthias quoted in Van Arsdol 1992, pp. 75-76**

When operations commenced, the push was on to produce as much plutonium as soon as possible. Small amounts were sent regularly to Los Alamos, initially to support research and then to make three bombs: the first for the Trinity test explosion at Alamogordo, New Mexico, the second for Nagasaki, and the third to be available if needed.

World War II ended on August 14, 1945. Only 8 days prior with the bombing of Hiroshima, the secret of Hanford's wartime mission was revealed to the workers and the rest of the world (see Figure 2-12.6). At last Hanford workers and their families learned how they had been instrumental in helping to win the war. After the end of World War II, the full impact of atomic weaponry started to be realized by some at the Hanford Site just as it did around the world.

POST WORLD WAR II OPERATIONS

With the secret out and the war over, a brief period of uncertainty ensued. Would the Hanford Site close? Would it be time to move on? Production continued. Plutonium separations activities increased, in fact, to maintain a nuclear stockpile, but the number of contractor staff dropped to 5,000 by the end of 1945, and the work week was reduced from 48 hours to 40 hours (Findlay and Hevly 1995, p. 55). However by 1946, the situation became clear. America had a new opponent, the Soviet Union, and needed nuclear weapons to maintain its superiority. Over the next 9 years (1947-1955), five more reactors and three

EXTRA The Villager EXTRA

IT'S ATOMIC BOMBS

President Truman Releases Secret of Hanford Product

Information Is Made Public This Morning

News Spreads Slowly, Surprises Everyone Here

Jubilation And Satisfaction Follows Revelation Of Product Manufactured Here

Development Of Bomb Traced

Begun In 1939; Plant Expanded In June, 1942

SPECIAL—Today President Truman, in an offical White House release, broke the biggest secret of World War II—and perhaps the greatest secret of any war—when he informed Americans that the U. S. Army Air Forces had released on the Japanese an Atomic bomb containing more power than 20,000 tons of TNT—and that the Hanford Engineer Works is one of three plants in the country manufacturing the new bombs.

The bomb, made in the Hanford Engineer Works, in Oak Ridge, near Knoxville, Tennessee, and an unnamed installation near Santa Fe, New Mexico, produces more than 2,000 times the blast of the largest bomb ever used before.

How Much Damage?

Richland Gets Ready To Entertain Press

THE HANFORD ENGINEERING WORKS Bring V-J DAY CLOSER!

The bombs blast even the landscape out of sight. Nothing is impossible. The first bomb was dropped on Hiroshima a few hours ago. Observers report that the explosion was thousands of times greater than an earthquake and may change the course of civilization.

Atomic power was released against the Japs in answer to their refusal to the ultimatum issued last week. Source of the power is said to be coal, oil and power produced by the great dams in the Northwest and Tennessee.

In making the announcement, President Truman said that the bomb has added new and revolutionary increase in destruction on the Japanese. Mr. Truman went on that "it is an Atomic bomb. A harnessing of the basic power of the universe, the force from which the sun draws its power."

What is Atomic Bomb?

Figure 2-12.6. *News Article Telling Workers at the Hanford Site What They Had Been Producing*

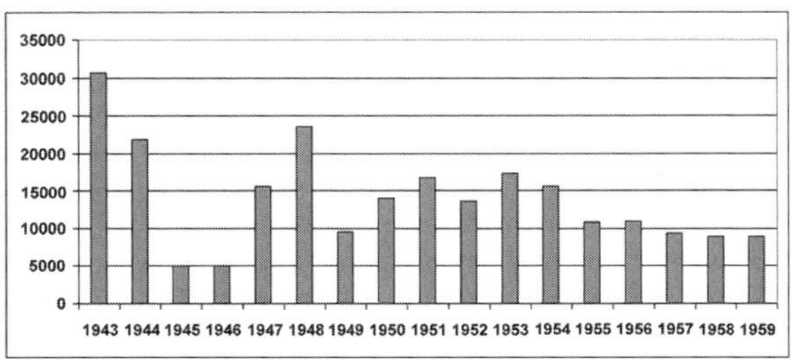

Figure 2-12.7. *Employment at the Hanford Site, 1943-1959*

new chemical separations facilities would be built at the Hanford Site requiring many additional workers (see Figure 2-12.7). An indicator of the number and type of scientists and engineers working at the Hanford Site during the post-war period is shown in Table 2-12.1.

Table 2-12.1. *Number and Type of Scientists and Engineers in 1947*

Chemical Engineers	104	Civil Engineers	36
Chemists	91	Physicists	28
Mechanical Engineers	65	Metallurgists	9
Electrical Engineers	52		

The life of the worker following World War II was affected by a

- Shift in organizational culture
- Change in nature of the war being fought and quantity of plutonium needed
- Addition of onsite research and development expertise
- Increase in power of organized labor
- Awareness of health and safety
- Community alignment with the atomic mission

Shift in Organizational Culture

Major changes occurred on two organizational fronts following the war. First was the transition from military control to public control. On August 1, 1946, President Truman signed the Atomic Energy Act, which placed the control of nuclear weapons and technology under the U.S. Atomic Energy Commission. On January 1, 1947, the Atomic Energy Commission took over control of the Hanford Site from the Corps. The shift from a military to a civilian organization changed the way decisions were made at the Hanford Site and the way site contractors were managed. Instead of Groves and his staff, a commission of five people appointed by the President ran the Atomic Energy Commission and made the major decisions affecting the Hanford Site. Four committees were established. The General Advisory Committee was composed of nine scientists who advised the Atomic Energy Commission on scientific and technical matters. Recognizing the need for a separate division and special advisory committee for the life sciences, the Advisory Committee for Biology and Medicine was formed. The Military Liaison Committee provided expertise in military-related matters. The Atomic Energy Commission also established the Advisory Committee on Reactor Safety to provide recommendations related to reactor safety. To maintain congressional oversight of the Atomic Energy Commission, Congress established the Joint Committee on Atomic Energy.

The second major change was the change in site contractors. On September 1, 1946, General Electric Company took over management of the Hanford Site from Du Pont. This was significant because, as Carlisle and Zenzen (1996) point out, these two companies had drastically different corporate cultures and approaches to project management. For the next 10 years, General Electric maintained the organization Du Pont had established. On September 1, 1956, General Electric reorganized. Instead of the Manufacturing Department controlling all production activities (production of fuel elements, production of plutonium, etc.) and the Engineering Department controlling all research, development, and architectural services, the Hanford Site was reorganized according to steps in the plutonium production process:

- Fuels Preparation Department responsible for manufacturing the fuel elements
- Irradiations Processing Department responsible for irradiating the fuel
- Chemical Processing Department responsible for separating out the plutonium
- Hanford Laboratories Operations responsible for helping the operations departments solve technical problems as well as managing the Biology and Medicine program, plutonium recycling program, and miscellaneous minor research and development programs
- Relations and Utilities Operations responsible for general services such as patrol of the Hanford Site, operation of the Hanford community, electrical, telephone, transportation, and other services not economical to decentralize
- Construction Engineering Operations responsible for construction activities

The number of individuals working for the government, Du Pont, General Electric, and construction firms between 1943 and 1957 is presented in Table 2-12.2. The totals are different from those shown in Figure 2-12.7 because the two data sets do not account for the same groups of employees.

Table 2-12.2. *Hanford Site Personnel, 1943-1957*

		Government Admin			Operations					Construction Contractors		
Date		Corps/ AEC	Military	Total	Production	R&D	Arch. Eng. Mgt. Ser.	Community (Inc. Hosp.)	Total	Major Sub-Cont.	Other	Total
1943	07-03	143		143						4,271	1,840	6,111
	09-30	247		247						9,648	4,181	13,829
	12-27	254		254						21,517	8,942	30,459
1944	03-11	280		280						25,778	11,086	36,864
	06-29	310		310					1,384	30,413	12,606	43,019
	09-28	325	175	500					3,983	21,253	10,702	31,955
	12-31	291	175	466					5,289	10,958	5,248	16,206
1945	03-31	276	224	500					6,099	334	0	334
	06-30	247	268	515					5,476	29	0	29
	09-30	224	225	449					4,662	0	79	79
	12-31	231	243	474					4,659	0	92	92
1946	03-31	232	316	548					4,662	0	403	403
	06-30	274	259	533					4,303	0	260	260
	09-30	300	290	590	3,317	23		850	4,190	0	335	335
	12-31	332	270	602					4,479	0	141	141
1947	03-31	295	262	557					4,821	0	215	215
	06-30	247	12	259					5,295	0	302	302
	09-30	254	3	257					6,347	1,305	857	2,162
	12-31	309	1	310	5,708	71	520	1,050	7,349	5,538	2,368	7,906
1948	06-30	345		345					8,658	9,704	6,495	16,199
	12-31	343		343	5,775	299	1,438	1,115	8,627	8,397	6,274	14,671
1949	06-30	330		330					7,403	2,705	2,565	5,270
	12-31	338		338	5,486	396	635	884	7,401	792	994	1,786
1950	06-30	347		347					7,758	3,080	1,563	4,643
	12-31	341		341	5,809	451	804	822	7,886	3,953	1,855	5,808

		Government Admin			Operations					Construction Contractors		
Date		Corps/ AEC	Military	Total	Production	R&D	Arch. Eng. Mgt. Ser.	Community (Inc. Hosp.)	Total	Major Sub-Cont.	Other	Total
1951	06-30	346		346					8,650	4,145	2,635	6,780
	12-31	358		358	6,677	675	1,055	700	9,107	5,391	1,916	7,307
1952	03-31	437		437					8,910	4,578	2,228	6,806
	06-30	452		452					8,907	3,286	1,983	5,269
	09-30	444		444					8,795	2,524	1,647	4,171
	12-31	455		455	6,695	682	780	641	8,798	3,369	1,029	4,397
1953	03-31	449		449	6,507	759	812	630	8,708	4,686	1,297	5,983
	06-30	462		462	6,455	760	844	631	8,690	6,536	942	7,478
	09-30	453		453	6,325	723	890	607	8,545	7,158	1,338	8,496
	12-31	444		444	6,353	680	882	611	8,526	7,705	1,238	8,943
1954	03-31	414		414	6,389	731	870	605	8,595	8,296	1,889	10,185
	06-30	397		397	6,506	721	851	613	8,691	7,974	2,005	9,979
	09-30	380		380	6,592	742	762	629	8,725	7,865	624	8,489
	12-31	362		362	6,838	679	762	614	8,893	5,417	937	6,354
1955	03-31	346		346	7,225	631	633	612	9,101	1,945	480	2,425
	06-30	336		336	7,456	648	611	604	9,319	1,226	184	1,410
	09-30	324		324	7,461	684	534	585	9,264			1,376
	12-31	320		320	7,624	680	521	584	9,409			1,207
1956	03-31	318		318	7,682	638	499	573	9,392			337
	06-30	304		304	7,707	680	484	547	9,418			909
	09-30	297		297	7,526	767	49;8	393	9,184			1,568
	12-31	295		295	7,468	749	625	361	9,203			1,428
1957	03-31	294		294	7,401	744	591	348	9,084			1,344
	06-30	289		289	7,355	755	582	351	9,043			963

Change in Nature of the War Being Fought and Quality of Plutonium

During World War II, Hanford workers knew that whatever it was they were doing they needed to accomplish it as fast as possible to win the war. The Government's announcement on August 12, 1945, was the first time the majority of workers knew what they were making at the Hanford Site. While the compartmentalization concept was still in place limiting workers' holistic understanding, at least they now knew what they were making.

Along with this change in understanding came a change in enemy and nature of the battle. Replacing the fascist regimes of Germany and Japan almost immediately was the Soviet Union, Stalin, and the threat of communism. The nature of the battle went from open warfare to a cold war in which the concept of the nuclear stockpile appeared. It was no longer a race for the bomb. Instead, it was a race to possess the most bombs. During the early years, the United States was in a comfortable position of strength as the only country to possess an atomic bomb. That changed on September 23, 1949, when President Truman announced the Soviet Union had exploded an atomic bomb during the previous month.

The fear of communism and gradual decay in Soviet Union and U.S. relations led to expansion at the Hanford Site for the next decade. While the urgency to produce may have been less than during World War II, an air of urgency still remained. The military established production goals, and the Atomic Energy Commission pressured contractors to meet these goals. Anything that might cause the goals to be missed was cause for concern. The production mentality intensified during this period and lasted throughout the Cold War.

The Soviet threat affected Hanford workers in various other ways as well. As the Soviet threat materialized, emphasis on security re-emerged. Espionage and anti-communism slogans became prominent. Cartoons and signs regularly reminded staff to keep their lips sealed about their work. The threat of enemy attack also became a greater possibility than it was during World War II with the nuclear arms race necessitating anti-aircraft artillery sites to be established surrounding the Hanford Site. These were later replaced by Nike missile bases.

Addition of Onsite Research and Development Expertise

Various departments at the Hanford Site developed research capability soon after operations began to encounter problems such as graphite expansion and fuel ruptures. The desire for greater efficiency and safer operations stimulated other research efforts.

One major area of research concerned health effects from radiation. Even before construction of the Hanford Site was complete, research on the impact of reactor effluents on fish was sponsored at the University of Washington Applied Fisheries Laboratory (Becker 1990). By 1945, this research was conducted at the Hanford Site, which was the anchor for what was to become one of the major radiobiological research centers in the nation (Stannard 1988). The fisheries work quickly expanded into research on plants and animals (for more information, see Section 7, Research and Development). President Eisenhower's Atoms for Peace initiative in 1954 brought additional research projects because the Hanford Site was one of the few centers of nuclear expertise in the country if not the world.

By the mid-1950s, research and development at the Hanford Site had grown to the extent that it made sense to unite the researchers under one umbrella. When General Electric reorganized in 1956, it created the Hanford Laboratories under the direction of Herbert M. Parker. Parker was Hanford's chief health physicist and supervised the Health Instruments Section during the Manhattan Project. By 1958, 1200 staff were working at the Hanford Laboratories, most of whom were assigned to the following departments:

- Reactor Fuels Research and Development – 257
- Chemical Research and Development Operation – 221
- Laboratory Auxiliaries Operation – 220
- Radiation Protection Operation – 159
- Biology Operation – 77

Working in a research and development environment was significantly different from working in an operations environment. In the latter, work life could be relatively mundane except when breakdowns and malfunctions required attention. Research staff were generally of a different type, often possessing advanced degrees and an academic as opposed to an applied science orientation. Parker had cited in 1948 a report from an Atomic Energy Commission board that identified the problems of attracting good researchers to an operating site such as the Hanford Site.

"Freedom of enquiry [sic] vs. urgency of assignment. One of the major satisfactions of an investigator in a university or an enlightened research institute is his freedom to choose his research problems. The makers of Atomic Energy Commission policy will realize that the motives of wartime service must now undergo a change because scientific workers will find programmatic research or imposed assignments unsatisfying unless sweetened with freedom to follow their own choice of problem for part of their working time, and also the development of new ideas by these workers may prove of inestimable importance." **- Board of Review of Biological and Medical Research in a letter to Mr. David E. Lilienthal cited in Parker 1948b**

The experience gained by building, operating, and refining nuclear production facilities made the Hanford Site one of the few centers for nuclear expertise in the country. The emerging leaders in the nuclear technology field were General Electric (the prime contractor at the Hanford Site), Combustion Engineering, and Westinghouse. As these and other companies gained experience in this emerging field and explored the idea of industrial applications, Hanford staff became recognized for their expertise. When President Eisenhower's Atoms for Peace initiative appeared in 1954, new opportunities for Hanford staff developed as they reached outside the Hanford Site to share information with their counterparts. This trend created a new breed of Hanford worker, one more dedicated to the nuclear field than to a particular cause.

The concept of developing better ways to operate at the Hanford Site was expected of all workers. The General Electric Company instituted an employee suggestion system when it arrived at the Hanford Site. Individuals providing suggestions for better ways of doing things were given cash awards. Such awards were a common feature in the plant newspaper. The dollar value of these innovations, both to the employee and the government, is shown in Table 2-12.3.

Table 2-12.3. *Results of General Electric Suggestion System, 1947-1957*

Years	Awards	Savings
1947	$485	$12,746
1948	$2,065	$58,973
1949	$4,110	$84,838
1950	$7,605	$98,673
1951	$8,860	$94,239
1952	$13,885	$129,828
1953	$13,100	$108,390
1954 (6 mo)	$7,160	$58,263
1954/55	$20,584	$59,089
1955/56	$22,770	$149,901
1956/57	$22,789	$214,556

Increase in Power of Organized Labor

Organized labor's role changed following World War II. During World War II, labor conflicts were minimized for several reasons. The Buildings Trade Council of Pasco was a dominant labor group at the Hanford Site as was the Western Conference of Teamsters, headquartered in Seattle. In general, the unions supported a no-strike pledge and set aside jurisdictional boundaries (in other words, when shortages existed, machinists let carpenters work as millwrights, plumbers permitted boilermakers to work at plumbing, etc.) largely because of their appreciation for the critical nature of the work being undertaken.

Table 2-12.4. *Labor Unions Representing Hanford Workers*

	Craft	Name	Local Number
General			
1.	Asbestos Workers	International Association of Heat and Frost Insulators and Asbestos Workers	No. 82
2.	Boilermakers	International Brotherhood of Boilermakers, Blacksmiths, Iron Shipbuilders and Helpers of America	No. 104 and No. 541 (Welders)
3.	Bricklayers	Bricklayers, Masons and Plasterers International Union of America	No. 7
4.	Carpenters	United Brotherhood of Carpenters and Joiners of America	No. 1849
5.	Cement Masons	Operative Plasterers and Cement Finishers International Association	No. 478
6.	Electricians (Wiremen)	International Brotherhood of Electrical Workers	No. 112
7.	Iron Workers (Structural)	International Assoc. of Bridge Structural and Ornamental Ironworkers	No. 14
8.	Laborers (General)	International HOD Carriers, Building and Common Laborers Union of America	No. 348
9.	Linemen	International Brotherhood of Electrical Workers	No. 77
10.	Machinist	International Association of Machinists	No. 1743
11.	Millwrights	International Brotherhood of Carpenters and Joiners of America	No. 1699
12.	Operating Engineers (Tractor)	International Union of Operating Engineers	No. 370
13.	Painters (Spray)	Brotherhood of Painters, Decorators and Paperhangers of America	No. 427
14.	Plumbers	United Association of Journeymen Plumbers and Apprentices of the Plumbing and Pipefitting Industry of the United States and Canada	No. 598
15.	Roofers	United Slate, Tile and Composition Roofers, Damp and Waterproof Workers Association	No. 54
16.	Sheet Metal Workers	Sheet Metal Workers International Association	No. 42
17.	Teamsters Flat Bed, Misc. 1-4 Ton	International Brotherhood of Teamsters, Chauffeurs, Warehousemen and Helpers of America	No. 839
Other			
1.	GE Maintenance, Operators, Hospital Workers, Janitors and Firemen	Hanford Atomic Council Metal Trades	
2.	J. A. Jones Office Workers	Office Employees International Union	No. 100
3.	J. A. Jones Draftsmen and Surveyors	Technical Engineers and Architects Association	No. 17
4.	Independent	National Federation of Federal Employees	No. 788
5.	Guards	Hanford Guards Union	No. 21

Following the war, organized labor's presence at the Hanford Site strengthened. In 1947, the Atomic Energy Commission signed an agreement with the American Federation of Labor Building and Construction Trades Unions. The American Federation of Labor affiliates representing Hanford workers as well as two unions that were not affiliated, the National Federation of Federal Employees and the Hanford Guards Union, are shown in Table 2-12.4. In 1949, operating personnel were allowed to unionize and voted to be represented by the American Metal Trades Council of the American Federation of Labor (Findlay and Hevly 1995, pp. 73-75). This led to the birth of the Hanford Atomic Metals Trade Council or HAMTC as it is still known today. Despite the increasing power of the unions, the Atomic Energy Commission retained its authority to approve wage rates and benefits.

Between 1947 and 1955, approximately 80 work stoppages involving unskilled laborers occurred, most lasting less than 1 week. Dispute over jurisdiction (which union had the right to do the work) was the most commonly cited reason for striking. Nearly 3 percent of work time was lost as a result of the work stoppages (GE n.d. (a), p. 80).

Awareness of Health and Safety

With the announcement of Hanford's role in atomic materials production came the awareness that workers were being subjected to radioactivity, something few knew anything about. As discussed in Section10, Worker Health and Safety, the Hanford Site monitored human and environmental health, but how much of this was understood by workers and their families is uncertain. Workers in the reactor and separations areas were told about the hazards. Others were not told. Concerns about health and safety became more common during the 1950s.

Getting "crapped up" (becoming contaminated with radioactive material) and sent to the first aid station to shower off was fairly routine and not considered a big deal by many workers. Workers stationed in radioactive areas were required to take bioassays (provide urine and fecal specimens) regularly and leave specimens on their front porch for pickup to measure possible radiation uptake. The number of radiation incidents recorded between 1944 and 1957 are shown in Table 2-12.5.

Table 2-12.5. *Radiation Incidents, 1944-1957*

	Class I[a]		Class II[b]	
Year	No. Incidents	No. People	No. Incidents	No. People
1944	0	0	3	5
1945	35	100	6	11
1946	39	84	4	5
1947	27	46	2	2
1948	38	51	2	5
1949	36	53	0	0
1950	20	35	5	8
1951	25	70	13	53
1952	71	125	12	23
1953	69	199	26	43
1954	76	232	20	20
1955	147	329	15	25
1956	88	160	13	13
1957 (7 months)	11	56	5	5

(a) Class I incidents are those where the weekly radiation limit is not exceeded but a potential exists. Class I incident rate reduction is largely due to a different definition of Class I incidents in 1956.

(b) Class II incidents are over-exposures exceeding 300 M.R.E.P. per week. No lost time resulted from radiation overexposure. REP is an obsolete unit of absorbed dose in tissue. It was originally derived from roentgen equivalent physical and is approximately equivalent to the rem. M means milli REP.

The general impression of many former workers is they felt at the time that adequate measures were taken to protect them. This was evidenced on a 1997 trip to C Reactor when an artifact assessment inspection took place. Several former workers helping to conduct the assessment were surprised at the more stringent protective measures now required to enter radiation zones. As more knowledge about radiation emerged, greater advancements in protective measures were made (for more information, see Section 10, Worker Health and Safety).

Public concern about the hazard of nuclear testing, atmospheric nuclear fallout, and siting proposed nuclear facilities mounted in the 1950s. This led to increasing local concern for the effect of the Hanford Site on the environment. The Columbia River Advisory Group, which had been formed by the Atomic Energy Commission in 1949 and consisted of

> ***"Sometimes I just knew I'd done everything perfectly and was feeling pretty elated when old 'Pluto' [the name given to one of the hand instruments that sniffed out contamination] would give me the word that somehow or another the hot stuff had escaped once more and I was in for a session of decontamination.***
>
> ***Wash it, scrub it, paint it, check it, and sometimes dispose of it—I never cleaned so much stuff that already looked clean. I remembered the first day when I saw a girl scrubbing one particular area on the floor. It looked so clean I wondered why she didn't go on to another part of the floor—but no, she stayed in that one spot, over and over. I was sure she or I was ready for the mental hospital.***
>
> ***When we couldn't get our hand count down, we dunked them in a solution of 'goop.' Usually only one dunking of that was necessary. It was supposed to take off a very thin layer of skin. Sometimes it took three or four dunkings and the bus would have to wait for us to the consternation of the other passengers.***
>
> ***One night several men came on the bus, late, and literally glowing with cleanliness with hair that stood straight up like it had just been washed, which it had. They tried to appear nonchalant but everyone knew they had really been in a hot spot without knowing it or perhaps had a spill. They took a ribbing that time.***
>
> ***So we made jokes about the coveralls which weren't always our size and put up with the daily bumpy bus rides and our confidence grew as we found ourselves continuing in a state of good health. We were checked once a month. The rumors flew on that score, but the blood tests and other tests belied the rumors."*** **- Yvette Berry, former worker, quoted in Van Arsdol 1992, p. 68**

water pollution and public health officials from the U.S. Public Health Service and Oregon and Washington, became increasingly alarmed about the high levels of radioactivity being released in the reactor effluent. Such concerns led to more and more radiobiological studies being conducted at the Hanford Site (for more information see Section 7, Research and Development and Section 10, Worker Health and Safety).

Community Alignment with Atomic Mission

Whereas the community had been aligned with the effort to win the war, during this period the local community began to align itself with the atomic mission. This attachment was a reflection of the individuals and their families who worked at the Hanford Site along with others in the community who either depended on the Hanford Site or were otherwise proud of their community's association with it. Thus, symbols of the atomic era began to appear (for example, "Richland, the Atomic Bustin Village of the West"). The Richland High School adopted the Hanford Bombers logo in the late 1940s. Organizations and businesses began taking atomic monikers as their theme in the 1950s, such as the Atomic City Kiwanis, Atomic City Toastmasters, and Atomic Lanes Bowling Alley.

LIVING CONDITIONS

When plutonium production operations began, the Corps shut down the Hanford construction camp. Many of the new operations staff moved into Richland, doubling the population to 15,000 by early 1945. The importance of keeping workers satisfied with living conditions was clearly recognized by Du Pont and its architect-engineer, G. Albin Pehrson.

> ***"...there would undoubtedly be a psychological hazard in a too-cramped plan. Although city dwellers are confined to narrow lots and restricted views, these are an accepted part of their environment. In the desert, where space is the key characteristic of the view, a cramped village of cramped houses would be out of character, a palatable and conscious discord."*** **- G. Albin Pehrson, Du Pont architect-engineer, discussing the design needs of Richland in Pehrson 1943, p. 17**

Community Snapshot - May 1, 1953

"RICHLAND, WASHINGTON – home of permanent operating employees at Hanford Works – is situated at the confluence of the Yakima and Columbia Rivers, comprises ten square miles of land owned by the United States of America, and is operated by General Electric Company, as agent of the United States Atomic Energy Commission.

It has a population of 27,000 people at present, which is an increase of 5,000 since 1950. The Atomic Energy Commission presently employs 450 people whose annual going rate of payroll is $2,200,000. Eighty percent of these employees now live in Richland. General Electric Company presently employs 8,665 people whose annual going rate of payroll is $46,500,000. Eighty-two percent of these employees now live in Richland.

Richland School District #400 comprises a high school, two junior high schools, and seven elementary schools... There are 6,060 housing units owned by the Government and there are 1,000 housing units under construction by private capital; 860 dormitory rooms that will accommodate 1,095 people. All houses, both Government-owned and privately-owned, are connected to sewer and water mains and there are 5,277 residential telephone connections. There are two banks here whose combined deposits currently total $16,500,000. There are 8,000 automobiles registered by private owners in Richland.

There are seventeen churches of which three are Government-owned and fourteen have been constructed by private capital. Nineteen clubs and organizations occupy space in sixteen Government-owned quarters and three privately owned buildings.

At present, there are two major shopping districts located about a mile apart, and five neighborhood shopping centers --generally consisting of a food and a drugstore and a service station... Modern hospital facilities are available in Richland and medical and dental care is provided by twenty-seven doctors and dentists engaged in private practice.

Richland, at present, has a population that is equal to the combined populations of its near-by sister cities of Kennewick and Pasco. Recent newspaper releases by the Hanford Office of the Atomic Energy Commission indicate that Richland may climb to a population of 30,000 in the not too distant future.

At present there is a construction camp known as North Richland five miles north of Richland, which is owned by the Government and operated under private contract. The present population of the Camp is 5,745 people; the number of construction employees as of May 1, 1953, is 5,802 whose current annual going rate of payroll is $31,916,560. The peek [sic] employment for the current expansion program in progress at Hanford is expected to reach 10,000 sometime in 1954, and then decline as construction is completed. There is also a Class "A" Army Camp at North Richland, but personnel and payroll data are not available.

There are thirteen businesses in North Richland whose annual sales in 1952 amounted to $2,718,159. Housing in North Richland for the civilian population is in Government-owned barracks, a large government-owned trailer camp where many people live in their privately-owned trailers, and in 200 Government-owned temporary houses." - A copy of this community snapshot is filed in the Pehrson 1943 document in the Reference Section of the Richland Public Library.

As Findlay and Hevly note (1995, p. 39), good housing would help employees cope with the inhospitable new country into which they had been thrust. They cite an example of Du Pont's concern for its employees' living conditions, namely the fight to prevent fencing in Richland. While the Corps would still patrol the village and the residents' outgoing letters would still be read, Du Pont apparently drew the line at fencing in the village as was done at the Oak Ridge Reservation in Tennessee and Los Alamos, New Mexico. Du Pont ran the government-owned town, but the Corps continued to play a role and the two organizations did not always agree. The Army hired Du Pont to run and police the town as part of their contract. When Du Pont acknowledged that they wished to do so without Army interference, Matthias resisted, citing corporate rigidity as a main factor. Tensions mounted between the Army and Du Pont in the fall of 1944 when

Du Pont accused the Army of interjecting too much control over Richland Village affairs (Findlay and Hevly 1995, pp. 43-45).

The more than 50 social organizations, 13 churches, a weekly paper named *The Villager*, and the Villagers, Inc. (an umbrella organization of the various community groups) formed a cohesive community (Hales 1997, p. 235). Residency in Richland, beyond a small group of construction workers, was restricted to certain classes of workers (Findlay and Hevly 1995, pp. 35-37). For example, laborers, janitors, and other manual workers were excluded from residing in Richland. When one did qualify to live in Richland, the type of government house a worker could rent depended on one's job classification:

- Operator to Foreman Classification: Types A and B and Prefabs
- Assistant and Shift Supervisor Classifications: Types F and H
- Chief Supervisors and Superiors: Types D, E, and G

The 1946-1947 expansion at the Hanford Site caused another housing shortage and more houses were constructed in Richland. As a result, a new construction camp arose in North Richland (in the area around the present Battelle complex) that had a population of 13,000 by 1948. As additional housing was constructed, school teachers and truck drivers also were able to obtain housing in Richland (Findlay and Hevly 1995, pp. 40-41). By 1950, Richland had grown to nearly 22,000, an increase of about 50 percent in 4 years. Along with the new housing, the Atomic Energy Commission constructed another commercial district known as the Uptown Shopping Center.

However, living conditions continued to be intolerable. The winds continued to wreak havoc, compounded by the dust problems created by stripping off vegetation during construction. Over time, as lawns were seeded, parks planted, and other disturbed areas stabilized, the dust problem subsided.

WORKERS WHO OPERATED AND DIVERSIFIED THE HANFORD SITE: 1958-1990

Worker life at the Hanford Site continued changing throughout this next Cold War period. Two factors were responsible for many of the changes: 1) the nuclear stockpile reached sufficient levels, leading to a stable period of bomb production as opposed to the earlier periods of rapid production and 2) increasing concern over human and environmental health risks posed by the nuclear complex.

EFFECT OF AN ADEQUATE NUCLEAR STOCKPILE ON WORKERS

From day one at the Hanford Site, military requirements for plutonium had led Hanford staff to push hard. First was the drive for bombs to end World War II, then came the demand for bombs for weapons testing to refine the capability, and finally came the increasing demand for larger and larger stockpiles of nuclear weaponry. Finally by the early 1960s, the nation's Cold War planners grew comfortable with the plutonium supplies, and projections for future production began to decrease. The Limited Test Ban Treaty (signed in 1963 by Great Britain, the Soviet Union, and the United States to ban nuclear tests in the atmosphere, outer space, and under water) was in place, and the nation possessed an enormous amount of plutonium.

As a result, the production capability at the Hanford Site was no longer needed. After almost two decades of annual production increases, the Hanford Site embarked on a plan in 1963 that led to the shutdown of eight of the nine reactors by 1971. The impact was two-fold:

- First, employment decreased. The elimination of some jobs was deferred as nuclear facilities had to be defueled and deactivated, but slowly the numbers of workers within the production groups declined.
- Second, employment diversified. To help offset the economic impact of closing the reactors, the Atomic Energy Commission initiated diversification efforts to help maintain employment levels.

The Employment Roller Coaster

Beginning in the 1960s, the Hanford roller coaster of economic boom and bust intensified. The second major down cycle began in 1963 (see Figure 2-12.8) (the first being immediately after the war in 1945-1946) with the announcement that three reactors would be closing. This down cycle continued throughout the decade as more closures were announced. Finally in early 1971, the government closed the KE Reactor and gave notice that N Reactor would also close.

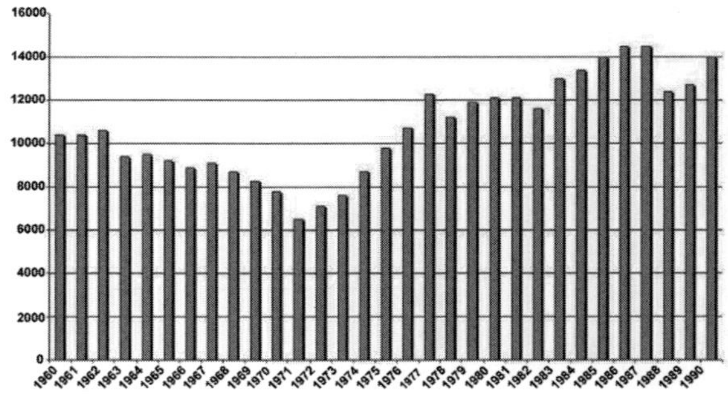

Figure 2-12.8. *Employment at the Hanford Site, 1960-1990*

Despite the shutdown of the reactors, the 1970s brought an end to the down cycle. Construction began on the Fast Flux Test Facility, a prototype breeder reactor in 1970 (for more information, see Section 7, Research and Development). An agreement that kept the N Reactor operating for electricity production only was forged in 1972, preventing closure of that facility. Construction of the Washington Public Power Supply System (Supply System) nuclear power plants Numbers 2, 1, and 4, commenced in 1972, 1975, and 1978, respectively. In 1976, an evaluation of the Hanford Site for long-term storage of nuclear waste began.

By the mid 1980s, however, the bust cycle began again. Cost overruns and failed projections on electricity needs in the Northwest led to cancellation of two of the Supply System plants (Numbers 1 and 4). As a result of the Reagan Administration defense buildup, major improvements in the production facilities at the Hanford Site helped offset some of the decline, but that was short lived. In 1983, Congress canceled the breeder reactor program, jeopardizing the future of the Fast Flux Test Facility. In 1987, following the explosion of the Chernobyl reactor in the Soviet Union, the N Reactor was initially shut down for safety upgrades. Stockpiling of plutonium and citizen concerns about health and environmental issues led to the decision not to restart N Reactor, obviating the need for any further production-related work at the one remaining separations plant, PUREX (Plutonium-Uranium Extraction Plant). The Hanford Site's production era came to a close.

Diversification Efforts

Diversifying the Tri-City economy away from its primary dependence on plutonium production funding emerged as a theme at the Hanford Site in the early 1960s to help offset employment cutbacks. This theme continued throughout the remainder of the Cold War. The Atomic Energy Commission tried various mechanisms in concert with local community leaders to bring new nuclear missions in particular and new industries in general to the Tri-Cities (Findlay and Hevly 1995, pp. 215-300). These efforts had roots in the 1950s as Hanford staff increasingly worked on projects not pertaining to the Hanford Site. Some of this work came through the Atomic Energy Commission as it built new facilities around the country at Savannah River, Rocky Flats, Idaho, and elsewhere. In some cases, other work opportunities for staff came through the General Electric Company, which was pursuing commercialization of nuclear technology at its San Jose facility.

Another symbol of diversification appeared when Congressional and local leaders launched a vigorous campaign to build a steam plant to generate electricity from N Reactor heat. This battle, termed the last major eruption of the public versus private power controversy (Dyke 1989, p. 161), was a catalyst for local involvement. During this time, local economic interests formed the Tri-City Nuclear Industrial Council (later named the Tri-City Industrial Development Council) to seek out opportunities. Also during this time, the *Tri-City Herald* emerged as a strong supporter for new programs at the Hanford Site (Pugnetti 1975).

A major diversification effort came in 1964 when the Atomic Energy Commission moved from a single-contractor to a multi-contractor approach. Several new companies replaced the General Electric Company, which had been the main site contractor since replacing Du Pont in 1946 (see chart at end of Chapter 1 for list of contractors). The new contractors committed to promoting diversification by helping to create jobs that would not be dependent on the Atomic Energy Commission. These activities met with limited success.

> ***"Hanford is also a major research and development center of the Atomic Energy Commission with an annual R&D budget in excess of $30 million. About one-third of this R&D effort is in support of Hanford production operations and the remainder is directed toward a variety of peaceful applications of atomic energy. About three-fourths of this work is conducted within the Hanford laboratories organization and the remainder by research and engineering groups directly attached to production organizations." -*** **Holsted and Albaugh 1964a, p. 10**

Hanford's research and development capability continued to attract new work for staff. The Atomic Energy Commission elevated the status of the Hanford Laboratories in 1964, establishing it as a national laboratory (now called Pacific Northwest National Laboratory) and awarding Battelle Memorial Institute the contract to run it. Included in this contract was an innovative contracting mechanism that also permitted Battelle to use government facilities for private work. Diversification efforts provided new opportunities for Hanford Site workers. By building on the unique experience gained at the Hanford Site, many staff were able to become involved in new nuclear applications.

EFFECT OF HUMAN AND ENVIRONMENTAL SAFETY CONCERNS ON WORKERS

During the 1960s and 1970s, the issue of radioactivity and human and environmental health intensified. The stimulus for much of this was the rapid growth of the commercial nuclear industry. As public opinion increasingly opposed nuclear technology, anti-nuclear activists eventually turned their attention to the nuclear production complex, in particular the risk associated with the stored and newly generated wastes associated with the facilities (Wasserman and Solomon 1982, Del Tredici 1987). The increasing awareness, knowledge, and concern for radiation and its potential impact on workers, surrounding populations, and the environment had three major effects on Hanford workers: 1) increase in regulations and procedures, 2) increase in governmental oversight, and 3) development of a negative image of the Hanford Site to many outsiders.

Increasing Regulations, Orders, and Procedures

As safety concerns increased, the number of regulations, orders, and procedures requiring compliance grew in geometric proportion. This added to the bureaucracy as new divisions were created and others reorganized. For example, the Division of Biology and Medicine evolved into the Environmental Safety and Health Division. These changes were designed to balance power within the agency, which previously had been held mainly by the production division. To implement the various directives, the government had to add staff to ensure compliance. Over time, the system became bogged down with organizations that had overlapping responsibilities, requirements that conflicted, and audits that seemingly never ended. The impact on management and workers was significant. Many workers believed that the requirements to follow new work procedures actually reduced safety because they took away their ability to use common sense. Managers often complained that the time they spent with auditors and addressing their findings hurt their ability to get their work done. Nevertheless, the new requirements were successful in providing more balance between production and safety goals.

Increasing Governmental Oversight

The Hanford Site and the rest of the production complex are government-owned, contractor-operated facilities. The licensing and regulatory functions concern only commercial nuclear enterprises. The government maintains technically qualified staff at the facility to oversee the work of the contractors. Achieving the optimal balance of government to contractor staff is difficult. Without enough government staff, the contractor might abuse the system. With too much oversight, the contractor's progress might be impeded. At the Hanford Site, the number of federal employees is 3-5 percent of the contractor staff. Typically, it has been hard for the government to maintain a well-qualified, experienced staff because it lacks the incentive systems of private industry.

In 1975, the licensing and regulatory functions of the Atomic Energy Commission were separated and placed into a new agency, the Nuclear Regulatory Commission. The remaining functions were placed into the Energy Research and Development Administration, which was reorganized into the U.S. Department of Energy (DOE) in 1977.

In general, the increasing attention on the production complex resulting from concern about nuclear risk pressured the government to exercise more control on its contractors. By the 1960s, however, the bureaucracy was becoming unmanageable and simply adding staff to supervise, inspect, and monitor contractors only increased the bureaucracy. Adding DOE orders to implement regulations required writing numerous procedures that required audits and reports, all of which had questionable benefit. The effect on worker life was pronounced.

An example cited by former workers was the contrast of building N Reactor with building the Fast Flux Test Facility. By the 1970s, when the Fast Flux Test Facility was being designed and built, the government brought many Nuclear Navy people, such as Milt Shaw (a protégé of Admiral Rickover), into the Atomic Energy Commission and successor organizations to oversee nuclear activities. The problem was the Nuclear Navy culture was very different from the production complex culture. Operating a nuclear reactor on a submarine was different from operating a nuclear reactor in the desert. The endless reviews and questioning of contractor recommendations made the experience frustrating for many.

Increasing Negative Image of the Hanford Site

On July 23, 1971, the Supreme Court of the United States ruled in the Calvert Cliffs decision that the Atomic Energy Commission had violated the National Environmental Policy Act of 1969 with regard to the environmental impact of proposed nuclear plants (Fehner and Holl 1994, p. 15). This marked the threshold of formal public involvement at Atomic Energy sites such as the Hanford Site. Through the 1970s and 1980s, various environmental impact statements were initiated that focused outside attention on the Hanford Site and provided a forum for those concerned about Hanford operations.

The Nuclear Waste Policy Act of 1982 identified the Hanford Site as a possible repository for high-level radioactive waste. Although the Hanford Site was not chosen, the Nuclear Waste Policy Act mandated public information and community hearings and also stated that "affected tribes" be included in the public involvement process. Hanford land has been used historically by the Confederated Tribes of the Umatilla Indian Reservation, the Yakama Indian Nation, and the Nez Perce Tribe. The government also consulted with the Wanapum Band, a group indigenous to the area and located at Priest Rapids just upstream from the Hanford Site. In general, the Tribes took a hard line on the need to protect and restore the environment and questioned projects that conflicted with their traditional values.

By the mid 1980s, two of Hanford's most powerful advocates, Washington State Senators Warren G. Magnuson and Henry M. Jackson, left office. Although one of their successors, Slade Gorton, continued to lobby for Hanford interests, growing doubts and fears about nuclear energy and nuclear waste could be found nationwide, and Hanford's image, especially on the west side of the state, was suffering. During the same period, several citizen advocacy groups formed including the Hanford Education Action League, which defined its main goal as citizen advocacy and public access to information. They also challenged nuclear weapons policies, supported the permanent closure of PUREX, and promoted the safe disposal of waste. Many who lived downwind of the Hanford Site (known as "Downwinders") received publicity as they pursued claims that their health had been impacted by past radioactive emissions. Whistle blowers, employees who had been punished by employers for raising operational concerns, also received much publicity that contributed to the Hanford Site's negative image (D'Antonio 1993). This reached a peak in 1986 when Brock Adams defeated Slade Gorton's re-election bid by campaigning against Gorton's record of too strongly supporting Hanford's "bomb factory" (D'Antonio 1993, pp. 189-190; Findlay and Hevly 1995, pp. 298-300).

The negative publicity also affected Hanford workers, their families, and their community. Some questioned the severity of possible health ramifications from working at or near the Hanford Site. Some fought back, organizing a local support group called the Hanford Family to defend the Hanford Site. Formed in 1986 by Mike Fox, a Hanford engineer, the

Hanford Family educated the public about nuclear processes, challenged bad press, and campaigned for continued use of Hanford's reactors and work force. Other Hanford workers formed the group World Citizens for Peace to consider the morality of Hanford operations (D'Antonio 1993). Later, as the plutonium production era came to a close and waste management became the biggest priority, the former negative imagery was transformed as plutonium production workers were retrained as environmental technicians to clean up past messes (Noonan 2000, pp. 34, 38).

MISCELLANEOUS CHANGES

The period between 1960 and 1990 brought with it an assortment of other changes that bear on worker history.

Living Conditions

In 1958, the Atomic Energy Commission transferred ownership of Richland to the residents after 11 years of efforts to privatize the town. Occupants were allowed to buy the properties they had been renting, a city council was elected, and the various governmental services were transferred to the city. This had all been made possible by the Atomic Energy Community Act of 1955, which enabled residents of the three atomic cities (Hanford, Los Alamos, and Oak Ridge) to assume the obligations of local self government (Kubik 1994, pp. 62-72). After 15 years of Richland being a company town that the government and its contractors owned and operated, residents could finally live in a non-company town (Findlay and Hevly 1995, pp. 8-113).

Major Event – President Kennedy Visits the Hanford Site

In October 1963, President John F. Kennedy came to the Hanford Site for the groundbreaking of the Hanford Generating Plan at the N Reactor (see Figure 2-12.9). It was the first time that the public could enter the Hanford Site, and an estimated 37,000 people did just that. Faced with sweltering temperatures, the people heard the President praise the accomplishments of the past and paint a rosy picture for the future. This was the biggest event ever held in the Tri-Cities and was a major morale booster for Hanford employees and their families. In retrospect, the event is a symbol for the opening of the site that occurred in the coming years.

Changes in Security

President Kennedy's visit and speech to 37,000 Tri-Citians in 1963 marked the start of a gradual relaxation of security requirements at the Hanford Site. Within the next few years, the Columbia River was opened to recreationists, except for the shore bordering the Hanford Site. Workers still had to go through checkpoints, vehicles were still subject to searches, the Hanford Site Patrol still guarded the site, and lockdowns (gates closed and all personnel searched before exiting the facility) still occurred regularly, but the threat of espionage had peaked with the anti-communism movement of the 1950s.

Figure 2-12.9. President John F. Kennedy at Ground Breaking of the Hanford Generating Plant at N Reactor, October 1963

Security requirements started tightening again in the 1970s with the advent of skyjacking and bombings. Concern centered in two major areas: the threat of theft of special nuclear materials and the threat of causing a major nuclear accident. By the 1980s, the government ordered several security upgrades. These included fences with razor wire such as those surrounding the N Reactor and PUREX, pill boxes in hallways for security guards, patrol towers for armed guards, and generally increased patrolling, including the use of helicopters. Various efforts were funded to identify weaknesses in defenses that terrorists could exploit. Although these anti-terrorist measures were clearly seen and understood by the workers, they do not seem to have felt particularly at risk. For more detail on security, see Section 8, Site Security.

Educated Workforce

By the 1960s, an influx of a new generation of nuclear workers began. Major universities had nuclear engineering departments and were producing a steady stream of nuclear engineers, who with their advanced degrees came to work at the Hanford Laboratories. Before long, the Tri-Cities were being recognized as having "...an unusually high percentage of professional and technical people, because of the highly technical character of the Hanford production and research and development activities" (Holsted and Albaugh 1964a, p. 9). Throughout the period, the Hanford Site maintained its place as second only to the Seattle area for having the largest concentration of PhDs in the State of Washington.

1986 – A Year of Change

The year 1986 brought with it a series of events that forever changed the U.S. Department of Energy:

- January 1986 – The space shuttle Challenger exploded, killing all aboard. This event sent shock waves through everyone involved in technology, sending the message that things can go awry.
- February 1986 – Because of concerns from Downwinders and other environmental groups, DOE responded with the release of 19,000 pages of historical records documenting past releases of chemicals and radionuclides to the surrounding environment. This release of documents generated much attention toward and adverse publicity about the Hanford Site and signified the start of long-term investigations concerning potential health effects.
- March 1986 – Public involvement in the Defense Waste Environmental Impact Statement process began. This provided public forums for groups concerned about the effect of the Hanford Site on human health and the environment.
- April 1986 – The Chernobyl Nuclear Plant exploded in Ukraine. Reports circulated almost immediately that a nuclear plant with a similar design was operating at the Hanford Site, which brought interest by the media.
- Summer 1986 – Whistle blower issues regarding illegal waste disposal, the removal of radioactivity signs before Governor Booth Gardener's visit to the Hanford Site, and routine nuclear engineering standards violations provided more negative publicity (D'Antonio 1993).
- Fall 1986 – Westinghouse is named winner of the Hanford Site Consolidation contracts, signifying a return to the concept of a single operating contractor for the Hanford Site.

Because of similarities between the Chernobyl plant and the Hanford Site's N Reactor (both were graphite moderated, water-cooled reactors), the press became intensely interested in the Hanford Site following the explosion at Chernobyl. Faced with external demands for the closure of the Hanford Site's last production facility, the response from workers was predictable. The N Reactor was safe and should keep operating. To verify or disprove that belief, numerous internal and external task forces and commissions reviewed the technical data over the next 2 years to determine if a similar explosion could occur at the Hanford Site. Most were unanimous that a similar explosion could not occur, but the N Reactor was shut down anyway because the end of the Cold War eliminated the need for more plutonium. Although the Chernobyl accident did not cause the shutdown of the N Reactor, public interest and the resulting DOE response marked the start of opening the DOE complex to the public.

Worker Complacency

During one of the N Reactor reviews following the Chernobyl explosion, the topic of worker complacency arose. The following quote from a member of the Roddis Panel (commissioned by the Secretary of Energy to evaluate the safety of Hanford's N Reactor) in 1986 indicates that complacency might have been a problem at the Hanford Site but goes on to explain the prevalence of complacency in the nuclear industry in general:

> *"I regret to say that there was also evidence of complacency at the plant, at the lower levels. Admittedly I have had only a short visit, so some of the observations may have simply been aberrations from the norm, but there were an uncomfortable number of small but significant deviations from familiar safety practice. The sense that came through was one of people doing a job in a context of safety rules whose significance had somehow been lost, and which were therefore not deemed very important. This was somewhat puzzling, since the upper management seemed well focused; yet the plant workers must somehow reflect the values passed on from above.*

This is not as defamatory as it may sound. It is virtually impossible to maintain the necessary sense of insecurity and concern for the occurrence of improbable events, when everyday experience seems to belie the need. It requires constant attention to maintain a taut ship. Pilots have a similar problem, as do the Air Force personnel who man the missile silos, to say nothing of the vast majority of drivers. And it is still a problem in the nuclear industry, despite the Kemeny Commission report. Few humans will perform to the best of their ability, year after year, with only inspirational speeches to motivate them." - Lewis 1986, p. 4

Change in Mission from Production to Cleanup

By 1988, it was clear that Hanford's production era had come to a close. The Hanford Site's only remaining production reactor would remain closed, thereby obviating the need for any further fuel manufacturing or separations work. Within months, DOE declared that the Hanford Site had a new mission: cleanup. Hanford's job would be to clean up the contamination resulting from over 40 years of manufacturing plutonium. Preliminary estimates suggested the cleanup might require 30 years and $26 billion to accomplish.

Faced with cleanup legislation and motivated groups calling for action, DOE, the U.S. Environmental Protection Agency, and the Washington State Department of Ecology entered into closed door meetings to establish cleanup schedules and commitments. The result was a federal facility compliance agreement that identified milestones and a 30-year schedule, known as the Tri-Party Agreement (Ecology 1994).

The change in mission presented a new challenge for DOE. Environmental cleanup differs from nuclear production in significant ways. One primary difference is the uncertainty surrounding cleanup projects. Much of the cleanup work involves characterizing the contamination through sampling, assessing the risk contamination presents to human health and the environment, evaluating appropriate cleanup technologies, and doing remedial work. Such work was challenging enough in the industrial sector where contamination was primarily chemical. At the Hanford Site, where much of the contamination was radiological, the challenges were monumental. Because most of the staff in the late 1980s came from a nuclear production background, they were often ill-equipped for the new type of work required (EMO 1992).

AREAS FOR FURTHER RESEARCH

Throughout the Manhattan Project and Cold War Era, the types of occupations for which workers were hired reflect the changing mission of the Hanford Site. The changing mission affected how workers were recruited, their working and in the early days their living conditions, the number of workers on the site at any one time, their salaries, and the issues confronting them. This book on facilities only briefly touches on these topics.

Much remains to be discovered about work life at the Hanford Site. The combination of production of nuclear materials, new technology, radioactivity, and secrecy make work at the Hanford Site a topic of interest that should continue to attract researchers for a long time. Two general areas that could benefit from additional research are 1) seeking information that is presently lacking and 2) finding answers to historical questions concerning workers at the Hanford Site that could not be addressed within the scope of this book on facilities.

ADDITIONAL INFORMATION

To enrich what is presently known about work life at the Hanford Site, researchers need to seek more first-person accounts, photographs/films/videos, and memorabilia.

First Person Accounts

Some first-person accounts of work life at the Hanford Site, also referred to as oral histories, do exist. Accounts collected by David Sanger (1995) focused on personal remembrances by key individuals regarding the design, construction, and operation of the Hanford Site during the Manhattan Project. Ted Van Arsdol (1992), a local newspaperman, collected stories and incorporated them in his history of the Hanford Site. Interactive interviews such as those conducted by Thayer (1996) and Stannard during the 1980s (Stannard 1988) provide good information because interviewers had particular questions and pursued the answers. Similar interactive interviews were conducted in the mid-1990s by Darby Stapp concerning the N Reactor (DOE 1997a, Volume 3) and Thomas Marceau and Michele Gerber. The Marceau-Gerber interviews are housed at the Columbia River Exhibition of History, Science, and Technology but have not yet been approved for publication.

Despite the efforts of these researchers, much remains to be done. Given the advancing ages of Hanford Manhattan Project and Cold War workers, an active oral history program should commence immediately to complete previous efforts and conduct additional interviews with key informants. Beyond simply collecting reminiscences, a valuable effort in its own right, such a program must develop a research agenda to guide the interview process so that the types of questions identified in this volume can be pursued. The questions provided below are a start. Other sections in this book have similar additional research directions that may be applicable to oral historical work.

Photographs, Films, and Videos

One of the principal goals of worker history research is to enable the audience to understand what it was like to work at the Hanford Site. Words alone cannot accomplish this goal. Fortunately, thousands of photographs and many films exist that illustrate work life. Unfortunately, the photographs and films are not catalogued and in some cases have not even been located. With the rapid dismantling of the site, it is imperative that these visual history treasures be located, catalogued, and preserved.

Memorabilia

Memorabilia also provide information about work life at the Hanford Site. Unfortunately, such materials are rare. Some signs, calendars, mementos, etc., have survived and are being preserved. Others can be reconstructed from photographs and films, but more memorabilia are needed.

One particularly rich item, the Hanford Site newspaper (*GE News* from 1946-1964 and *Hanford Project News* from 1965-1980) does exist in its entirety. The 30+ years of weekly newspapers contain much information about work life at the Hanford Site. Included are descriptions of workers, events that occurred onsite, and methods management used to change worker behavior. These papers should be catalogued and indexed quickly so that future researchers can locate information relevant to their research.

Newspapers Published for Hanford Workers

The Sage Sentinel:	1944-1946
GE News:	1946-1964
Hanford Project News:	1965-1980

HISTORICAL QUESTIONS CONCERNING WORKERS

Our understanding of worker life at the Hanford Site will continue to improve as additional research is conducted. The following questions are some warranting further work:

- Is Gerber (1992a) correct that Hanford managers were irresponsible when they subjected workers, surrounding populations, and the environment to radioactive and chemical wastes? Or should Gerber have considered the Cold War context in which these decision makers had to work as Kathren (1993) suggests? More work is needed to understand how decision makers faced tradeoffs regarding the need to fulfill a mission and the need to do so safely. The notions of acceptable risk and what is safe are relevant here. The following comment was made by a member of the Roddis Panel, commissioned by the Secretary of Energy to evaluate the safety of Hanford's N Reactor, which had certain design similarities to the Soviet Union's Chernobyl nuclear plant. The Secretary wanted an independent assessment to ensure that a similar explosion could not occur at the Hanford Site:

  "Let me say at the outset that your specific wish [your letter of May 2 to Dr. Roddis] to be 'absolutely certain... that there is no possibility of an accident similar to the one that occurred in the Soviet Union' cannot be fulfilled, simply because risk is a matter of degree, and the complete avoidance of risk is a fantasy. I will stress this point throughout, because it seems to be imperfectly understood within the Department. We

often heard people opine that the reactor was 'safe' and that is a meaningless observation. Decision making involves either overt or subliminal balancing of risks and other costs against benefits. Just as there are (allegedly) no free lunches, there are no risk-free technologies. The risk is real, and no risk is worth accepting without commensurate benefit, while great risks are acceptable in return for great benefits. The benefits associated with operation of the N Reactor lie in its provision of special nuclear materials for defense programs, and the assessment of the value of these benefits lies without our charter...."
- Lewis 1986, p. 1

The Hanford Environmental Dose Reconstruction studies conducted by the Centers for Disease Control and Prevention and other future studies of worker health will help address these questions.

- In 1994, the U.S. Department of Energy (DOE) conducted interviews for the Advisory Committee on Human Radiation Experiments (ACHRE). Among the many questions to investigate would be what was the effect of the experiments on the workers at the Hanford Site? Did the results of the experiments improve safety and health practices for the workers? The ACHRE Archive is at George Washington University, and the website is http://www.gwu.edu/~nsarchiv/radiation/index.html.

- During the 1980s as public concern over the nuclear industry heightened, safety concerns also were directed toward the DOE production complex. Charges such as the following were made that production complex staff had not kept pace with training and operational practices in the nuclear industry, suggesting that safety was being compromised at DOE facilities:

  "The mishaps with the P Reactor [a reactor at the Savannah River Site where several safety-related incidents occurred in the 1980s] showed just how far practices in the defense reactor program have diverged from those in the commercial nuclear industry. The accident at Three Mile Island taught utilities that the unexpected could occur and led to major improvements in operating practices and training, instrumentation, accident analysis, and maintenance. But the veil of secrecy surrounding the weapons program prevented these changes from becoming part of the DOE knowledge base...

  This isolation was illustrated last year, when a well-known nuclear safety expert met with nuclear-safety staff of a DOE contractor. The expert wondered why he did not recognize any of the employees from national and international meetings. The staff members told him that they did not attend those meetings because they could learn nothing from them. The employees seemed unaware that a gap had grown between their practices and those of their colleagues in the commercial nuclear industry." - Ahearne 1989, p. 27

  Did a similar gap exist at the Hanford Site between Hanford nuclear workers and industrial nuclear workers? If so, in what areas did the Hanford Site differ, and why did the Hanford Site not keep up with Industry standards? Budget restrictions? Arrogance? The subject of maintaining contact with state-of-the-art industrial advancements and improvements has important implications for government-sponsored facilities.

- To better understand the impact of the Hanford Site on the nuclear industry, research is needed to determine the extent of worker migration from the Hanford Site to other nuclear settings. In particular, Findlay and Hevly (1995) and others have suggested that many of the Hanford Site's best people transferred to General Electric's commercial nuclear operation when General Electric left the Hanford Site in 1964. Is that true? Did these individuals go on to make major contributions to the nuclear industry? Research focused on the extent to which the Hanford Site staff reached out to industry at conferences and through publication would also shed light on the question of the impact of the Hanford Site on the nuclear industry.

- In the 1970s, a major migration of staff from the nuclear division of the U.S. Navy to the Hanford Site occurred. The two organizations had drastically different approaches to operating procedures and conflicts emerged. Research is needed to understand how the interactions between these two distinct corporate cultures evolved.

- Comparative research is needed to explore similarities and differences among the nuclear production sites, between the production sites and the nuclear industry, and between the production complex and the military complex. Insights gained from the way the managers of these various entities organized their workers to develop new technology and manage complex and dangerous technology might have value to future large-scale governmental and private sector technology projects.

- How have workers at the Hanford Site negotiated their roles as atomic weapons and energy employees with their personal lives? Were they afforded opportunities to partake in the forums of public activism or public participation? If so, how? If not, was it because of fear of redress by fellow employees or local residents?

- A number of authors have examined the impact of the Hanford Site on the Tri-Cities (see, for example, Deutschmann 1952, Findlay and Hevly 1995, Findlay 1996, Fleischer 1974; Noonan 2000). However, the social implications of nuclear communities need to be undertaken on a more comparative scale. How did workers shape or define life at the Hanford Site and in Richland versus social life at Los Alamos and the Oak Ridge Reservation? Did the conglomeration of people from all over the country create a unique mix of new social customs or invented traditions? Furthermore, the experiences of African Americans who worked at the Hanford Site during construction and operation are missing. Future studies need to address and document worker history for both minorities and women.

- Concerning the larger history of work and labor relations in the United States, where does the Hanford Site fit in? What were the conflicts over union negotiations? How do worker experiences differ from other Du Pont and General Electric operations throughout the country?

- Most recently reversing a position held for decades, DOE has admitted that many workers who built America's nuclear weapons products became ill because of exposure to radiation. In light of this, is it wrong to conclude that workers correctly remember feeling safe but that their trust was misplaced? How may this announcement affect the morale of workers at DOE installations today?

CHAPTER 3

GUIDE TO RESOURCES

S.D. CANNON
PACIFIC NORTHWEST NATIONAL LABORATORY

CONTENTS

Photo: Interior of the Hanford Technical Library (Neg. 25819-4)

CHAPTER 3 - GUIDE TO RESOURCES

The authors of the Hanford Site historic district book surveyed a broad range of resources in writing their respective sections. Sources they used are cited in the text and listed in the References Cited section of this book. Sources they did not use are listed in the section, "Bibliography for Further Information." With both listings available, the interested reader may pursue a topic at whatever level of detail they desire.

This guide to resources offers a springboard to future researchers by describing the collections of information about the Hanford Site. The focus is on the repositories at the Hanford Site and in the Tri-Cities area. Contractors for the U.S. Department of Energy (DOE) operate many of the repositories described. Repositories of information about the Hanford Site not included in this guide are those such as the Washington State Historical Society, the U.S. Army Corps of Engineers, and former contractors, for example, Du Pont de Nemours (whose archives are housed at the Hagley Museum and Library in Wilmington, Delaware), General Electric, and Westinghouse companies.

Not all collections described are directly accessible by public researchers. Those denoted in the "Publicly Accessible Collections" sub-sections have staff available to help public researchers. Those in the "Other Collections" sub-sections are not accessible to public researchers. You can request materials needed from these "Other Collections" through the Public Requests Service (Tel. 509-376-1418, E-Mail public_requests_hanford_docs@rl.gov). For a duplication fee, the Public Requests Service provides information that has been cleared for public release. You can request information that has not been cleared from the Freedom of Information Act Officer (U.S. Department of Energy, P.O. Box 550, Mailstop A7-75, Richland, Washington 99352, Tel. 509-376-6288). DOE staff and their contractors should contact the "Other Collections" directly for information about access. In addition to this guide, the Internet has information on how to access Hanford Site information (http://www.hanford.gov/boards/openness/factsheets.htm).

The types of resources for which collections are identified in this chapter are:

- Documents
- Newspaper Articles
- Photographs
- Drawings, Maps, and Plans
- Videotapes and Audiotapes
- Artifacts
- Automated Resources

DOCUMENTS

Documents about the Hanford Site are numerous. The collections described below represent the majority of the Hanford Site documents. Not all collections described are directly accessible by public researchers. Those denoted in the "Publicly Accessible Collections" sub-section have staff available to help public researchers. Those in the "Other Collections" sub-section are not accessible to public researchers. You can request materials needed from these "Other Collections" through the Public Requests Service (Tel. 509-376-1418, E-Mail public_requests_hanford_docs@rl.gov). For a duplication fee, the Public Requests Service provides information that has been cleared for public release. You can request information that has not been cleared from the Freedom of Information Act Officer (U.S. Department of Energy, P.O. Box 550, Mailstop A7-75, Richland, Washington 99352, Tel. 509-376-6288). DOE staff and their contractors should contact the "Other Collections" directly for information about access. In addition to this guide, the Internet has information on how to access Hanford Site information (http://www.hanford.gov/boards/openness/factsheets.htm).

PUBLICLY ACCESSIBLE COLLECTIONS

Hanford Technical Library

Location:	Consolidated Information Center, Washington State University at Tri-Cities Campus, 2770 University Drive, Room 101, Richland, Washington
Mailing Address:	Pacific Northwest National Laboratory, P.O. Box 999, Mailstop P8-55, Richland, Washington 99352
Telephone:	(509) 372-7430
E-Mail:	pnl.techlib@pnl.gov
Website:	http://www.pnl.gov/tech_lib/home.html
Fax:	(509) 372-7431
Hours:	Monday - Thursday 8 a.m. - 9 p.m.
	Friday 8 a.m. - 7 p.m.
	Saturday 10 a.m. - 4 p.m.
	Monday - Friday 8 a.m. - 5 p.m. (Technical Reports collection)

The Pacific Northwest National Laboratory (PNNL) operates the Hanford Technical Library to provide technical information support to PNNL and Hanford Site staff. The public is welcome to use the library's resources but may not check out materials. Self-service photocopiers, microform reader- printers, and computers are available to the public for research purposes.

The library provides PNNL and Hanford Site staff with many resources and services, many directly at the desktop. PNNL and Hanford Site staff should refer to the library's internal web site or contact the library for additional information.

The library contains books, journals, industry standards, technical reports, and reference materials in print and microform as well as access to a variety of electronic resources. Many of the library's resources are related to the history of the Hanford Site, such as:

- Nuclear Science Abstracts
- Energy Research Abstracts
- Large collection of historical Hanford generated reports

Public Reading Room of the U.S. Department of Energy

Location:	Consolidated Information Center, Washington State University, Tri-Cities Campus, 2770 University Drive, Room 101L, Richland, Washington
Mailing Address:	P.O. Box 999, Mailstop H2-53, Richland, Washington 99352
Contact:	Terri Traub, Tel. (509) 372-7443
E-Mail:	doe.reading.room@pnl.gov
Website:	http://reading-room.pnl.gov
Fax:	(509) 372-7444
Hours:	Monday - Friday 10:00 a.m. - 5:00 p.m.

The Pacific Northwest National Laboratory operates the Hanford Public Reading Room to provide the public with access to information and documents about DOE and Hanford activities. The collection contains primarily paper copies of technical reports. Materials are for use in the Reading Room but copies can be made for a fee. A catalog of the collection is searchable over the Internet at http://rrcatalog.pnl.gov and, where available, contains links to electronic versions of reports. The Reading Room also provides the public with Internet access to DOE and Hanford related web sites.

Administrative Record Public Access Room

Location: 2440 Stevens Center Place, Room 1101, Richland, Washington
Mailing Address: P.O. Box 950, Mailstop H6-08, Richland, Washington 99352
Contact: Debra A. Isom, Tel. (509) 376-2530
E-Mail: Debra_A_Debbi_Isom@rl.gov
Website: http://www2.hanford.gov/arpir
Fax: (509) 376-4989
Hours: Monday - every other Friday 9-11:30 a.m. and 1-3:30 p.m.

The Administrative Record Public Access Room houses documents referred to in the *Hanford Federal Facility Agreement and Consent Order* (Tri-Party Agreement) as administrative records. Administrative records are documents and information relied on to arrive at a final decision for remedial action or hazardous waste management. An administrative record is established for each operable unit and Treatment, Storage, and Disposal group and contains all documents with information considered in arriving at a record of decision or permit. All documents contained in the administrative record files are publicly available.

Public Requests Service

Contact: Anne L. Weaver, Tel. (509) 376-1418
E-Mail: public_requests_hanford_docs@rl.gov

The Public Requests Service allows individuals to obtain copies of a variety of cleared documents generated at the Hanford Site, including reports, speeches, procedures, videos, and photographs. The Service searches a number of databases, charges $0.10/page for most paper documents and duplication costs for other forms of information. If the Service is unable to provide a requested document, the requester may be referred to the Freedom of Information Act process.

Freedom of Information Act Requests

Mailing Address: Freedom of Information Act Officer, U.S. Department of Energy, P.O. Box 550, Mailstop A7-75, Richland, Washington 99352

Telephone: (509) 376-6216

When information is not otherwise publicly available, anyone can request documents through the DOE Freedom of Information Officer at the Hanford Site. Requests must indicate that the request is being made under the *Freedom of Information Act*, identify the information being sought so an employee familiar with the subject can locate it in a reasonable amount of time, and must express a willingness to pay costs for search time and duplication of documents. A public, non-commercial requester can receive the first 100 pages and 2 hours of search time at no cost. Search and retrieval of classified documents requested through the *Freedom of Information Act* is no different than for unclassified documents. Classified documents, however, undergo a declassification review. A requester does not need to know where records are stored to make a request.

OTHER COLLECTIONS

You can request materials needed from these "Other Collections" through the Public Requests Service (Tel. 509-376-1418, E-Mail public_requests_hanford_docs@rl.gov). For a duplication fee, the Public Requests Service provides information, which has been cleared for public release. You can request information that has not been cleared from the Freedom of Information Act Officer (U.S. Department of Energy, P.O. Box 550, Mailstop A7-75, Richland, Washington 99352, Tel. 509-376-6288). DOE staff and their contractors should contact the "Other Collections" directly for information about access.

Contractor Central Files

Contractor Central Files houses record copies of unclassified documents the U.S. Department of Energy and its contractors have generated – both internal and external publications. Documents may be reviewed before ordering a copy. The Contractor Central Files also houses controlled unclassified documents, those that have been assigned a document control number and for which distribution is tracked. Controlled unclassified documents may be obtained by calling the Controlled Document Management Office, Federal Building, Room 386, Tel. 509-376-6831. Need-to-know is required for access to both unclassified and unclassified controlled documents.

Classified Document Control Center

The Classified Document Control Center is the repository for classified and declassified historical documents that are either Project Hanford Management Contract (PHMC) documents or DOE documents in the custody of PHMC. The documents are filed alpha/numerically by document number, but database searches may be performed using author name, key words, subject, or title. An appropriate security clearance and the need-to-know are required for access to classified matter.

Classified Matter Control Center

The Classified Matter Control Center is a repository for classified and declassified historical documents that are either PNNL) documents or DOE documents in the custody of PNNL. The documents are filed numerically by document number, but database searches may be performed using author name, key words, subject, or title. An appropriate security clearance and the need-to-know are required for access to classified matter.

Hanford Site Records Holding Area

Official records of the DOE Richland Operations Office and its contractors that must be retained because of their record value are retired to the Hanford Site Records Holding Area (RHA). When retired, records are packed into record storage boxes and sent to RHA. Information on the contents of each box is entered into an automated data system called the RHA-MIS, and the box is shelved in the 712 Building Records Storage Vault for a period of time. The length of time that a given class of records must be maintained in storage is specified by the applicable DOE Records Schedule, or in the more generic General Records Schedule published by the National Archives and Records Administration (NARA). If lengthy record retention is required by the applicable records schedule, the box is shipped to the Federal Records Center in Seattle, Washington, which is operated by NARA, or to other record storage facilities provided by DOE. The total volume of records in storage now exceeds 80,000 ft^3. Other than the small percentage of records that have been identified for permanent (historical) retention, all other records are normally destroyed at the end of their retention periods.

Records in storage may be retrieved by DOE or by the contractor that retired them. To access official records in retirement, DOE Richland Operations Office, Office of River Protection, and contractor staff should make the request through their own records management group. Others should make the request by contacting the contractor:

DOE Records Officer	Gail Splett	376-8274
Records Holding Manager	George Vondruska	376-5087

Environmental Restoration Contractor Resource Center

The Environmental Restoration Contractor Resource Center contains documents from various Hanford Site contractors from the 1940s-1990s. The collection consists primarily of documents used for research on waste site characterization and remediation along with some documents and reference materials originating from the U.S. Environmental Protection Agency, state, and other federal agencies.

CH2M HILL, Hanford, Inc. Geosciences Library

The Geosciences Library houses numerous technical reports and other geologic, geochemical, hydrologic, and geophysical documents from various DOE contractors at the Hanford Site from the 1950s-1990s. The collection relates primarily to Hanford Site wells, vadose zone, groundwater monitoring and remediation, and environmental monitoring and surveillance activities for buildings, facilities, and waste sites in the 100, 200, 300, 400, and 1100 areas.

NEWSPAPER ARTICLES

Local and regional newspapers have published many articles written about the Hanford Site and its facilities over the past 50 years. This section describes newspaper collections of interest to Hanford Site researchers. These newspaper collections are publicly available.

RICHLAND VILLAGER

Location:	Richland Public Library, 955 Northgate Drive, Richland, Washington	
Contact:	Kathy Knudsen, Tel. (509) 942-7457	
E-Mail:	referenc@richland.lib.wa.us	
Website:	www.richland.lib.wa.us	
Fax:	(509) 942-7447	
Hours (Winter):	Monday -Friday	9:30 a.m. - 9 p.m.
	Saturday	9:30 a.m. - 5 p.m.
	Sunday	1 - 5 p.m.
Hours (Summer):	Monday -Thursday	9:30 a.m. - 9 p.m.
	Friday -Saturday	9:30 a.m. - 5 p.m.
	Sunday	Closed

The *Richland Villager* newspaper was a weekly newspaper published for Hanford Site employees and their families from November 1, 1945 through March 2, 1950. The Richland Public Library has copies of the *Richland Villager* newspaper on microfilm from 1945-1950 and some but not all of the hard copies. The successor to the *Richland Villager* was the *Columbia Basin News*, which the Richland Public Library also has on file. You may ask reference questions on the Richland Library website by selecting "reference" – "select topic" – "information" and then sending an E-Mail with your reference question.

HANFORD PLANT NEWSPAPERS

Location:	Columbia River Exhibition of History, Science, and Technology, 95 Lee Boulevard, Richland, Washington 99352	
Contact:	Connie Estep, Curator, Tel. (509) 943-9000	
E-Mail:	cestep@crehst.org	
Website:	http://www.crehst.org	
Fax:	(509) 943-1770	
Hours:	Monday - Saturday	10 a.m. - 5 p.m.
	Sunday	noon - 5 p.m.

The Columbia River Exhibition of History, Science, and Technology (CREHST) has bound volumes of the Hanford Site weekly newspaper from 1947-1982. The newspaper is variously titled: *Hanford GE News, Hanford Project News*, and *Rockwell News*. Access to holdings not on exhibit is by advance appointment. Please allow a minimum of 2 weeks for access to holdings not on exhibit.

COMMUNITY NEWSPAPERS

Location:	Mid-Columbia Library, Kennewick Branch, 1620 S. Union, Kennewick, Washington	
Telephone:	(509) 783-7878	
E-Mail	galaxy@mcl-lib.org	
Website:	www.mcl-lib.org	
Fax:	(509) 735-2063	
Hours:	Monday, Tuesday, Wednesday, Thursday	9 a.m. - 9 p.m.
	Friday - Saturday	9 a.m. - 5 p.m.
	Sunday	1 - 5 p.m.

The *Tri-City Herald* newspaper has published many articles on the Hanford Site. All issues (1948-current) are recorded on microfilm and are housed at the Mid-Columbia Library, Kennewick Branch. The library also has copies of the *Kennewick Courier* from September 9, 1943 to February 4, 1950. Other local libraries have copies of the *Tri-City Herald* as well.

REGIONAL NEWSPAPERS

The Spokesman-Review in Spokane, Washington has published many articles on the Hanford Site. A computer database is available for keyword searches for articles dating back to July 1994. Hard copy files of newspaper clippings dating back to the 1920s are categorized by subject. Search requests may be directed to the Library Research Line (Tel. 800-789-0029, ext. 5576). Search requests for photographs should be directed to the Photo Research Line (Tel. 800-789-0029, ext. 5416). For both extensions, leave a message and someone will return your call. The *Spokane-Review* charges for their research service. The rates are available on request.

The Seattle Times and *Seattle Post-Intelligencer* have also published numerous articles about the Hanford Site. An automated search service like that of the *Spokesman-Review* is not available. However, the Seattle Public Library has all the *Seattle Times* and *Seattle Post-Intelligencer* newspapers on file and available through the Magazine and Newspaper Department (Tel. 206-386-4680) for in-person research using CD-ROMs and microfilm. You can request the microfilm at your local library through an inter-library loan. The Business Technology Department (Tel. 206-386-4645) has a card file that indexes Hanford-related articles from 1945-1990s. The Humanities Department (Tel. 206-386-4625) also holds a card file index on the Hanford Site dating from August 1945 to April 1992 that includes articles from both newspapers.

PHOTOGRAPHS

Hanford Site photographs are stored at numerous locations. Below are the locations with the largest collections. Not all collections described are directly accessible by public researchers. The one denoted in the "Publicly Accessible Collection" sub-section has staff available to help public researchers. Those in the "Other Collections" sub-section are not accessible to public researchers. You can request materials needed from these "Other Collections" through the Public Requests Service (Tel. 509-376-1418, E-Mail public_requests_hanford_docs@rl.gov). For a duplication fee, the Public Requests Service provides information that has been cleared for public release. You can request information that has not been cleared from the Freedom of Information Act Officer (U.S. Department of Energy, P.O. Box 550, Mailstop A7-75, Richland, Washington 99352, Tel. 509-376-6288). DOE staff and their contractors should contact the "Other Collections" directly for information about access. In addition to this guide, the Internet has information on how to access Hanford Site information (http://www.hanford.gov/boards/openness/factsheets.htm).

Publicly Accessible Collection

Columbia River Exhibition of History, Science, and Technology

Location:	95 Lee Boulevard, Richland, Washington 99352	
Contact:	Connie Estep, Curator, Tel. (509) 943-9000	
E-Mail:	cestep@crehst.org	
Website:	http://www.crehst.org	
Fax:	(509) 943-1770	
Hours:	Monday - Saturday	10 a.m. - 5 p.m.
	Sunday	noon - 5 p.m.

The Columbia River Exhibition of History, Science, and Technology (CREHST) has an extensive collection of historic photographs of the Hanford Site. CREHST's total collection numbers approximately 2500 photographs, including many photographs of the Hanford Site and City of Richland from 1943-1945 with original negative numbers. CREHST has two indices to this group of photographs. In addition, CREHST has over 300 photographs with negative numbers, which are in the *Hanford News* volumes (1947-1982) stored at CREHST. CREHST also has a small collection of prints of the former Hanford and White Bluffs town sites from around 1910-1940. The museum will be creating a database of its photograph collections in the near future for retrieval by subject headings. As groups of photographs are entered in the database, they will be scanned and made available on the Internet. Copies of photographs are available. Please allow a minimum of 2 weeks.

Other Collections

You can request materials needed from these "Other Collections" through the Public Requests Service (Tel. 509-376-1418, E-Mail public_requests_hanford_docs@rl.gov). For a duplication fee, the Public Requests Service provides information that has been cleared for public release. You can request information that has not been cleared from the Freedom of Information Act Officer (U.S. Department of Energy, P.O. Box 550, Mailstop A7-75, Richland, Washington 99352, Tel. 509-376-6288). DOE staff and their contractors should contact the "Other Collections" directly for information about access.

Hanford Site Photography Archive

The Hanford Site Photography Archive provides complete photographic services and is the official, centralized repository of Hanford Site photographs. Its holdings include negatives of photographs taken at the Hanford Site since 1944. The Archive has a database and catalogs of its holdings. If the negative number of the desired photograph is known, an order may be placed with the Public Requests Service for a print. Negatives are not loaned.

Hanford Site Records Holding Area

The Hanford Site Records Holding Area has a collection of negatives dating from 1943 to the early 1990s that have been transferred from the Hanford Site Photography Archive and from various Hanford Site contractors. Roughly 50,000-60,000 negatives are stored at this repository.

A collection originally housed by CH2M HILL was retired to the Records Holding Area in early 2000. This collection has around 75 binders (approximately 2000 photographs) of prints primarily of site cleanup and building demolition activities dating from the mid-1980s through the mid-1990s.

Hanford Cultural Resources Laboratory

The Hanford Cultural Resources Laboratory collection that PNNL operates consists primarily of prints (approximately 200) and slides (several thousand) of cultural resource projects on the Hanford Site dating from the mid-1980s. The collection also includes aerial photographs of the Hanford Site (some dating to before 1943), two binders of photographs of the former Hanford and White Bluffs townsites, and Herbert Krieger's 1926 photographs of the Wanapum Indians.

Bechtel Hanford, Inc.

Environmental Information Systems at Bechtel Hanford, Inc. maintains photographs and negatives (approximately 2000) related to waste management, cultural resources, and geologic site projects. This collection includes five map drawers of aerial photographs of the Hanford Site as well as one file drawer of negatives of U.S. Geographic Survey aerial photographs from 1948. Their collection also includes approximately 2000 photographs that have been scanned into the Waste Information Data System.

Bechtel Hanford, Inc. maintains a collection of approximately 1500 photographs of the B Plant from the 1950s-1990s. This collection covers construction and deactivation, operations, cranes, equipment, interiors and exteriors of buildings. Some of the photographs are organized by subject, such as canyon crane, cask loading, cell construction, HEPA filters, hot pipe trench, and sand filter building. In addition, Bechtel Hanford, Inc. maintains a collection of 1950s-1990s photographs of PUREX construction, operations, and equipment.

CH2M HILL, Hanford, Inc.

The CH2M HILL collection consists of approximately 100 photographs of construction at the Hanford Site from the early 1940s through the construction of N Reactor in the 1960s.

Fast Flux Test Facility

The Fast Flux Test Facility maintains approximately 1500 photographs of the facility and the 400 Area dating from 1970-1990s. The photographs at this repository are mostly color prints.

Fuel Fabrication Facility for N Reactor Fuel

The collection in the Fuel Fabrication Facility includes approximately 1000 photographs, primarily of the Fuel Fabrication Facility complex buildings from 1979-1990s, with most dating from the mid-to-late 1980s. Some exceptions in this collection included two binders of the construction of the Plutonium Recycle Critical Facility and photographs of the 313 Building dating from 1969.

Plutonium Finishing Plant

The Plutonium Finishing Plant maintains a collection of approximately 1500 photographs, primarily from the 1950s-1990s of interiors and exteriors of the 234-5Z Building, including instrumentation and equipment, shielded hoods, glove boxes, sludge stabilization, and non-destructive assay equipment. The collection also includes a small collection of World War II era photographs and some photographs from 1948-1949 of the construction of the Plutonium Finishing Plant.

T Plant

The photographic collection at T Plant consists of approximately 1500 photographs. This collection includes photographs of T Plant facilities, employees, interiors and exteriors of the T Plant Canyon Building, PUREX, decontamination of the centrifuge, decontamination activities, and aerial photographs of the rail yard.

DRAWINGS, MAPS, AND PLANS

Drawings, maps, and plans for the Hanford Site are numerous and in a variety of locations. Below are the collections containing the largest numbers of drawings, maps, and plans. The collections described are not publicly accessible. You can request materials needed from these "Other Collections" through the Public Requests Service (Tel. 509-376-1418, E-Mail public_requests_hanford_docs@rl.gov). For a duplication fee, the Public Requests Service provides information that has been cleared for public release. You can request information that has not been cleared from the Freedom of Information Act Officer (U.S. Department of Energy, P.O. Box 550, Mailstop A7-75, Richland, Washington 99352,

Tel. 509-376-6288). DOE staff and their contractors should contact these collections directly for information about access. In addition to this guide, the Internet has information on how to access Hanford Site information (http://www.hanford.gov/boards/openness/factsheets.htm).

HANFORD CULTURAL RESOURCES LABORATORY

The Hanford Cultural Resources Laboratory has a collection of historic and current maps of the Hanford Site, some of which are U.S. Geological Survey maps.

U.S. DEPARTMENT OF ENERGY

The U.S. Department of Energy, Richland Operations Office, has all original Hanford Site engineering drawings. Over 300,000 engineering drawings have been created, dating back to the early 1940s. The drawings cover all phases and aspects of the Hanford Site.

VIDEOTAPES AND AUDIOTAPES

Numerous Hanford Site videotapes and audiotapes exist. The following list describes the main collections and their locations. DOE contractors have collections of their own videotapes and audiotapes as well. Only the Columbia River Exhibition of History, Science, and Technology collection is publicly accessible.

PUBLICLY ACCESSIBLE COLLECTION

Columbia River Exhibition of History, Science, and Technology

Location:	95 Lee Boulevard, Richland, Washington 99352	
Contact:	Connie Estep, Curator, Tel. (509) 943-9000	
E-Mail:	cestep@crehst.org	
Website:	http://www.crehst.org	
Fax:	(509) 943-1770	
Hours:	Monday – Saturday	10 a.m. – 5 p.m.
	Sunday	noon – 5 p.m.

The Columbia River Exhibition of History, Science, and Technology (CREHST) has several series of oral history tapes, including 1998 interviews with former Camp Hanford residents, 1995-1996 interviews described below, the 1991 Old Timer's Topical Colloquium II, and 1990 interviews with former B Reactor staff. CREHST is just beginning to transcribe these tapes. Researchers may listen to tapes of these oral history sessions by making an appointment during office hours with CREHST a minimum of 2 weeks in advance.

Members of the Hanford Site Historic Buildings Mitigation Project interviewed several groups of former employees who volunteered to share their reminiscences about the Hanford Site. The interviews were conducted weekly from December 14, 1995 to January 25, 1996. Attempts were made to record all the interviews. However, the recording equipment did not always work. Transcripts of the tapes were anticipated but have not been made. Conversations recorded on these tapes covered a broad range of topics such as working under tight security, specific accomplishments, challenges, entertainment, etc. Different aspects of the same topic might be explored each week, depending on the group's membership. Nevertheless, the central focus of "What was it like to work at Hanford?" always remained the same.

OTHER COLLECTIONS

You can request materials needed from this "Other Collection" through the Public Requests Service (Tel. 509-376-1418, E-Mail public_requests_hanford_docs@rl.gov). For a duplication fee, the Public Requests Service provides information that has been cleared for public release. You can request information that has not been cleared from the Freedom of

Information Act Officer (U.S. Department of Energy, P.O. Box 550, Mailstop A7-75, Richland, Washington 99352, Tel. 509-376-6288). DOE staff and their contractors should contact the "Other Collections" directly for information about access.

HANFORD SITE VIDEO ARCHIVE

The Hanford Site Video Archive is the official, centralized repository of Hanford produced and other videotapes and audiotapes depicting the Hanford Site and its operations. An automated index to videotape holdings exists.

ARTIFACTS

Objects that illustrate the history of the Hanford Site are considered to be special visual images of the past. Artifacts such as historic signs, instruments, tools and equipment for various professions, facility models, uniforms, and furniture (to name a few) are collected by professionals to provide a sense of life in the past. The *Hanford Curation Strategy: Manhattan Project and Cold War Era Artifacts and Records* describes the methods used to select and preserve artifacts from the Hanford Site. Listed below is the contact information for locating the artifacts that have been preserved. This location is publicly accessible.

COLUMBIA RIVER EXHIBITION OF HISTORY, SCIENCE, AND TECHNOLOGY

Location:	95 Lee Boulevard, Richland, Washington 99352	
Contact:	Connie Estep, Curator, Tel. (509) 943-9000	
E-Mail:	cestep@crehst.org	
Website:	http://www.crehst.org	
Fax:	(509) 943-1770	
Hours:	Monday - Saturday	10 a.m. - 5 p.m.
	Sunday	noon - 5 p.m.

The Columbia River Exhibition of History, Science, and Technology (CREHST) is a nonprofit educational museum and science center that manages the Hanford Site Manhattan Project/Cold War Era collection of artifacts. The artifacts CREHST manages are in a variety of locations. A few are continuously on display, but the majority are either stored in warehouses or are still in their original buildings. Call CREHST for information concerning particular types of artifacts. Access to holdings not on exhibit is by advance appointment. Please allow a minimum of 2 weeks.

AUTOMATED RESOURCES

Numerous automated resources are available for accessing information about the Hanford Site. Not all automated resources described are directly accessible by public researchers. Those denoted in the "Publicly Accessible Collections" sub-section have staff available to help public researchers. The one in the "Other Collection" sub-section is not accessible to public researchers. You can request materials needed from the "Other Collection" through the Public Requests Service (Tel. 509-376-1418, E-Mail public_requests_hanford_docs@rl.gov). For a duplication fee, the Public Requests Service provides information that has been cleared for public release. You can request information that has not been cleared from the Freedom of Information Act Officer (U.S. Department of Energy, P.O. Box 550, Mailstop A7-75, Richland, Washington 99352, Tel. 509-376-6288). DOE staff and their contractors should contact the "Other Location" directly for information about access. In addition to this guide, the Internet has information on how to access Hanford Site information (http://www.hanford.gov/boards/openness/factsheets.htm).

PUBLICLY ACCESSIBLE COLLECTIONS

Hanford Declassified Document Retrieval System

Website: http://www2.hanford.gov/declass/

The Hanford Declassified Document Retrieval System is the Internet site, along with OpenNet (http://www.osti.gov/opennet/), where declassified documents are made available. The intent is to provide easy and timely access to recently declassified Hanford documents, including documents declassified in response to Freedom of Information Act requests. This database is updated regularly as more documents become available.

Hanford Geographic Information System

Website: http://www.bhi-erc.com/projects/p_m/eis/hgis/hgis.htm
Location: 3350 George Washington Way, Mailstop H0-20, Richland, Washington 99352
Contact: Linda A. Dietz, Tel. 372-9378
E-Mail: ladietz@bhi-erc.com
Fax: (509) 375-4644

The Hanford Geographic Information System (HGIS) provides online Hanford Site geographical data. HGIS has detailed mapping of all areas at the Hanford Site, showing the location of buildings and other structures, underground lines, waste sites, test wells, fences, streets and highways, utilities, railroads, rivers, ponds, and lakes. High-quality Hanford Site maps can be generated from HGIS data (see http://www.bhi-erc.com/mojomap/hanford_MF.htm).

U.S. Department of Energy Information Bridge

Website: http://www.osti.gov/bridge/
Mailing Address: U.S. Government Printing Office, Federal Depository Library Program, 732 North Capitol Street NW, Washington, D.C. 20401
Telephone: (888) 293-6498
E-Mail: gpoaccess@gpo.gov
Prerequisites: Internet Access. System functionality requires a 486 Pentium PC, MAC or UNIX work station capable of running an Internet browser, Netscape 3.01 or greater or Microsoft 4.0 or greater and your browser must be JavaScript enabled and accept cookies.

The U.S. Department of Energy Information Bridge is a public website that provides free access to the full text and bibliographic records of DOE research and development reports in physics, chemistry, materials, biology, environmental sciences, energy technologies, engineering, computer and information science, renewable energy, and other topics. This system includes a growing collection of over 2 million searchable pages in over 27,000 reports DOE and the DOE contractor research and development community have produced and released to the public since January 1996.

Unique functionality includes full-text searching of each page of the entire collection. Users can electronically access, locate, search and download reports, select easy or advanced search options and limit search queries to specific data fields combined with Boolean operators to increase search precision. Reports can be viewed in GIF, TIF, or PDF format and can be downloaded in PDF or native formats. Access to the DOE Information Bridge is available without passwords or registration restrictions.

U.S. Department of Energy Information Locator

Website: http://www.directives.doe.gov/
E-Mail: explorer@lanl.gov

The U.S. Department of Energy Information Locator is a two-stage information database. The initial Locator allows a full text search of DOE records that have been submitted to the Information Locator. The Complex Search Form allows searches by field.

U.S. Department of Energy OpenNet Database

Website: http://www.osti.gov/opennet/
Contact: Douglas Zimmerman, Tel. (301) 903-4864
E-Mail: Doug.Zimmerman@hq.doe.gov

OpenNet is an electronic method of searching for documents, particularly those DOE has declassified. Many are available in full text form on the Internet, and all are available as paper copy. Other materials for which you can search on OpenNet are historical records and weapons films. The Office of Science and Technical Information, an office of DOE, operates OpenNet.

Waste Information Data System

Website: http://www.bhi-erc.com/projects/p_m/eis/wids/wids.htm
Location: 3350 George Washington Way, Mailstop H0-20, Richland, Washington 99352
Contact: Linda A. Dietz, Tel. 372-9378
E-Mail: ladietz@bhi-erc.com
Fax: (509) 375-4644

The Waste Information Data System (WIDS) is part of the Environmental Sites Database and offers Hanford personnel an online list and description of all Hanford waste sites. WIDS shows the official site name and alias, its location, description, history, dimensions, waste description and volume, process history, and other data, as applicable.

OTHER COLLECTION

You can request materials needed from this "Other Collection" through the Public Requests Service (Tel. 509-376-1418, E-Mail public_requests_hanford_docs@rl.gov). For a duplication fee, the Public Requests Service provides information that has been cleared for public release. You can request information that has not been cleared from the Freedom of Information Act Officer (U.S. Department of Energy, P.O. Box 550, Mailstop A7-75, Richland, Washington 99352, Tel. 509-376-6288). DOE staff and their contractors should contact the "Other Collection" directly for information about access.

Hanford Environmental Information System

The Hanford Environmental Information System (HEIS) offers online data from analytical sampling results related to characterization, monitoring, and cleanup. HEIS includes air sampling, biota, groundwater, soil, and mixed media; for example, analyses of Hanford waste sites, fauna, flora, and groundwater wells. The potential contaminants of concern analyzed include radiological and chemical.

CHAPTER 4

RECOMMENDATIONS FOR FUTURE USES

T.E. MARCEAU
BECHTEL HANFORD, INC.

CONTENTS

Photo: Department of Energy Transportation Maintenance Building Now Used by the Livingston Rebuild Center for Locomotive Repairs and Railroad Worker Training.

CHAPTER 4 - RECOMMENDATIONS FOR FUTURE USES

The purpose of this book is to preserve in words, diagrams, and photographs the structures at the Hanford Site because the majority of them have been or will be demolished. However, because of the national and international significance of the Hanford Site, certain properties should be left in place to illustrate the scope and scale of the various missions that developed the atomic energy program and supported national security. This chapter describes those properties proposed to be left in place, the public process that arrived at the recommendations, and introduces a vision of heritage tourism at the Hanford Site.

Whereas Chapter 2 describes the various missions, in particular the process for producing plutonium, as a way of preserving the Hanford Site in book form, this chapter recommends which structures should be preserved in their actual form. The highest form of preservation is retention in place. Properties that are occupied in association with ongoing functions are maintained and serviced so as to remain in operable condition.

> **Questions Posed by the Advisory Council on Historic Preservation**
>
> ***"Is it enough to see an interpretive display about the development of the atomic bomb? Would the opportunity to view an actual 1950s nuclear reactor control room or the workspaces where Fat Man and Little Boy were assembled enhance public understanding of this complicated period of American history?"*** **- ACHP 1991, p. 24**

Fortunately, the Hanford Site still retains the potential for in-place public interpretation. This is despite the fact that with the focus now on cleanup, the majority of Manhattan Project and Cold War Era buildings and structure no longer make a contribution to the Hanford mission and are scheduled for new uses or demolition. Having outlived their original usefulness, however, makes some of the properties that played key roles in the production of plutonium prime candidates for preservation. For example, given the international significance of the 105-B Reactor and its association with the initial development of atomic energy, this facility should be retained as an interpretive center. Likewise, the 100-K Area Flocculation Basins and Clearwells were not designed to accommodate the rearing of salmon, sturgeon and other fish species, but the operation of these pools as a hatchery by the Yakama Indian Nation is a perfect example of adaptive use that provides both economic benefit and returns a natural resource to the Columbia River.

Such examples of alternate use offer the opportunity for heritage tourism. Heritage tourism, defined as the visitation of sites of historic or cultural importance, would allow members of the local, national, and international communities to literally walk behind the fences and experience the facilities first-hand. Additionally, the identification, collection, and preservation of significant objects relating to Hanford's history (for example, tools, instruments, machinery, clothing) are key components of any public interpretation initiative. While it is not within the scope of this chapter to finalize the details of how site visitation would be accomplished or how visitors would be accommodated, it is appropriate both to advance the concept and to identify the buildings and infrastructure that would allow for it.

OBJECTIVES OF PRESERVATION IN PLACE

The decision to retain a facility in place must be based on an explicit understanding of what its preservation will accomplish. As scholarly and public interest in the Manhattan Project and Cold War era continues to grow, attention to the Hanford Site's contribution to those periods will increase. Preservation of certain structures at the Hanford Site would contribute to public education and interpretation. Information of interest to researchers and the public will include topics such as the history of the Hanford Site, nuclear technology and accomplishments, environmental issues, and social and cultural aspects of life at the Hanford Site. In addition to public education and interpretation, economic development is a second area in which Hanford facilities could make a continuing contribution to the local community.

However, the decision to retain a facility in place must be tempered by the long-term commitment of resources required to maintain that facility in a safe condition, particularly if the facility will be open to the public. In addition to funding, other factors that affect in-place preservation at the Hanford Site include environmental regulations that require removal of the structure, radiological or chemical contamination, disintegration of the structure, and limited or no educational or interpretive value.

PUBLIC PARTICIPATION

To help determine which structures should be retained as interpretive centers, museums, industrial or manufacturing facilities, or other private enterprises, the U.S. Department of Energy (DOE) invited the public to a series of meetings called Issues Exchanges. This was in keeping with the National Historic Preservation Act, which advocates that Federal agencies "seek and consider the views of the public when taking steps to identify historic properties, evaluate effects, and develop alternatives" (36 CFR 800 1986, 800.1[c][2][iv]).

The first Issues Exchange was held on October 29, 1997. To acquaint those in attendance with the overall goals and to ensure participation on an equal footing, pertinent sections of the *Hanford Site Manhattan Project and Cold War Era Historic District Treatment Plan* (Marceau 1998) were reviewed. The need for public involvement to identify buildings to retain in place was the central point of the presentation. On December 2, 1997 a second Issues Exchange was held to determine how the public could participate as decision-makers. As a result of this meeting, a Federal/Public working group was formed. Additional working sessions were held on January 27, February 11, March 4, and April 2, 1998.

RECOMMENDATIONS FOR PUBLIC EDUCATION AND INTERPRETATION

The Federal/Public working group decided that public education and interpretation would best be served by heritage tourism. To address the goal of preservation in place through heritage tourism, the working group envisioned a tour of the Hanford Site as it might take place some time after the year 2010. The working group reasoned that a full appreciation of the Hanford Site could only be obtained by ensuring that selected structures across the site be maintained in place so as to present to the visitor the true scale of the industrial landscape.

So visitors can understand the purpose for which the structures were built, the working group designated the thematic approach outlined in the *Hanford Site Manhattan Project and Cold War Era Historic District Treatment Plan* (Marceau 1998) as the organizing principle for determining the selection of buildings that would constitute the tour. Consideration was given to selecting buildings that would interpret the process for producing plutonium at the Hanford Site:

- First, preparing the uranium for the reactors, known as fuel manufacturing
- Second, irradiating the uranium in the reactors to produce plutonium, known as reactor operations
- Third, separating the plutonium from the base uranium, known as chemical separations
- Fourth, refining and then converting the plutonium for use in atomic weapons, known as plutonium finishing

Figure 4.1 shows these processes and the key structures where the processes took place. Interpretation would be linked to these processes, and the tour would be structured in the same order as the processes.

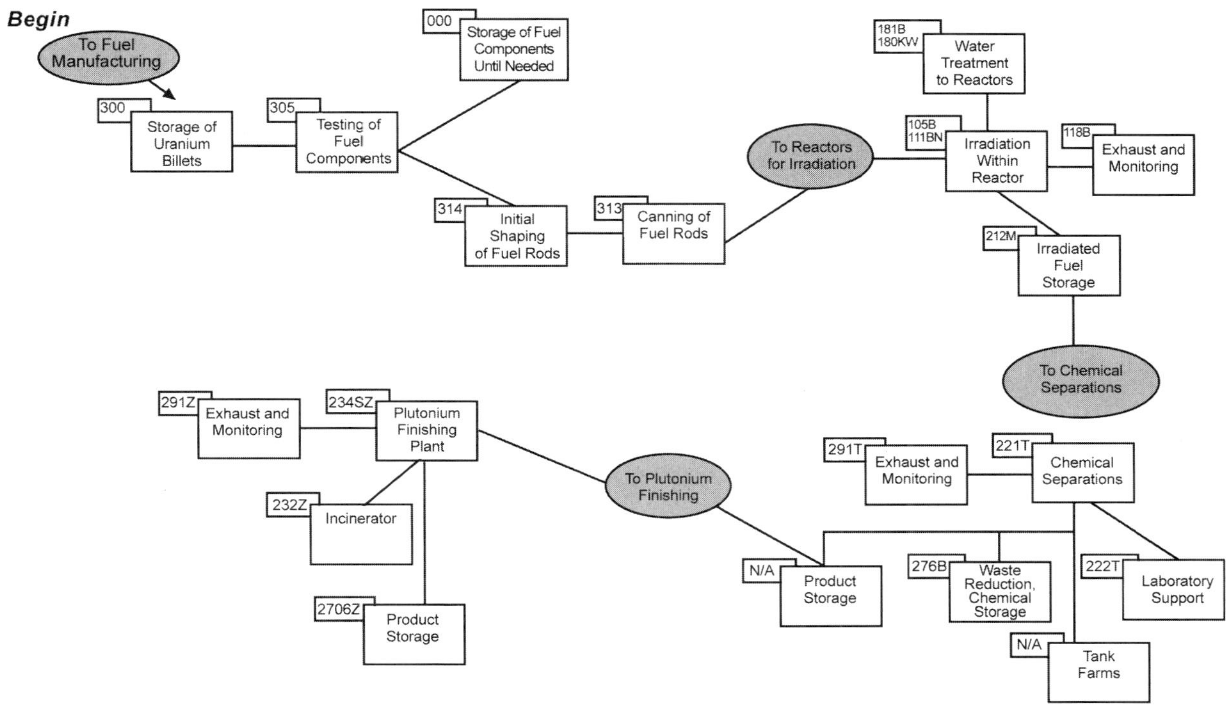

Figure 4.1. *Plutonium Production Process and Key Structures for Those Processes at the Hanford Site*

HERITAGE FACILITIES

Fuel Manufacturing (300 Area)

The first step in the production of plutonium was the milling of uranium fuel elements (commonly referred to as "fuel slugs") within the 300 Area. This process began with the arrival of purified uranium billets from offsite sources. These billets, cylindrical bars of metallic uranium tetrafluride (UF_4), were stored in specially designed buildings to ensure both security and accountability. Samples of these billets were tested at random to ensure quality. Billets of suitable quality were first extruded into rough uranium cores, then fabricated into fuel elements. Fabrication involved machining to proper size, grinding, cleaning, cladding, canning or jacketing as appropriate, and testing (see Chapter 2, Section 2 for a detailed discussion). Following fabrication, the fuel elements were sent to the 100 Area for irradiation within the reactors. The working group selected the following buildings as those necessary to tell the story of fuel manufacturing:

303-A - Fresh Metal Storage Building: The 303-A Building was built in 1943 to store uranium billets (known as "fresh metal" because of the form in which they were received). It also functioned as a holding area where unirradiated fuel elements could be inspected and tested for correct measurements and component assembly to ensure not only that fabrication had been successful but also that the fuel elements would not fail under irradiation. "Dummy" fuel elements (metallic spacers used to arrange the fuel elements within the reactor core) were brought back from the 100 Area to the 303-A Building to be cleaned, re-measured, and inspected for reuse. The 303-A Building is still used as a storage space for uranium billets received before the closure of N Reactor. The building is a one story, one room, concrete and concrete block structure. No architectural modifications are visible.

314 - Metallurgical Engineering Laboratory: The initial steps in fuel fabrication consisted of heating raw uranium billets in a furnace, then passing the heated billets through an extrusion press to produce fuel elements of specific dimensions. Both steps were conducted in an inert gas atmosphere within the 314 Building. After extrusion, hydrogen gas was removed ("outgassed") from the fuel elements, which were then straightened and sent to the 313 Building for machining and jacketing. These operations began in 1944. The 314 Building continued to heat, extrude, and straighten uranium fuel elements, provide autoclave and radiographic testing of canned elements, and assist in uranium scrap processing operations until 1971 when the last of the single-pass reactors closed. Modifications following that date allowed the Pacific Northwest National Laboratory to conduct a variety of research projects within the facility.

313 - Metal Fuels Fabrication Facility: The mission of the 313 Metal Fuels Fabrication Facility was to machine bare uranium fuel elements to desired dimensions for irradiation in Hanford's reactors, to jacket or can the fuel elements, and test the jackets for proper bonding and sealing (see the Historic American Engineering Record for the 313 Building in Appendix B on the Internet). Machining operations began in the 313 Building in December 1943. Canning operations began March 1944. The finished fuel elements were first irradiated in the 105-B Reactor in September 1944. In 1949, lithium-aluminum alloy fuel targets were fabricated. Following irradiation in the 105-B Reactor, these targets produced the tritium used in the first hydrogen weapons test. Over its operational lifetime, the 313 Building fabricated fuel for all of Hanford's single-pass reactors. This facility also produced some components of the fuel elements used in the 105-N Reactor. In addition to fabrication, researchers developed and continuously improved methods for canning fuel elements within the 313 Building. Lastly, uranium scrap recovery processes, recovery of uranium from waste acids, and recovery of chemical wastes were all undertaken within the 313 Building. The 313 Building has been emptied and is awaiting demolition.

305 - Test Pile (Reactor)/Hot Cell Verification Building: Built in 1944, the 305 Test Pile (Reactor) was the first operating reactor on the Hanford Site. "The low power (about 50 watts), graphite-moderated pile was originally used as a quality assurance tool to test samples of graphite, uranium, aluminum tubes and canning materials (for fuel rod [element] jackets), and other materials to be used in the original production reactors (B, D, and F Reactors)" (see the Expanded Historic Property Inventory Form for the 305 Building in Appendix B on the Internet). When analyzed against samples of known purity, the reactivity of uranium samples tested in the 305 Building could be used to predict the efficiency of the sample under full-scale production conditions. Because the graphite used in the production reactors had to be extremely pure, tests conducted on graphite samples within the 305 Building were critical to the success of reactor operations. An addition was made to the building in the late 1970s "when most of the 305 Building was converted to the Hot Cell Verification Facility, a non-radioactive ('cold') prototype and all purpose support facility for the Fuel Manufacture and Examination Facility in the 400 Area" (see the Expanded Historic Property Inventory Form for the 305 Building in Appendix B on the Internet). In light of this new mission for the 305 Building, uranium was removed from the test pile in 1974-1975. The pile itself was removed and buried in 1977-1978. Further modifications followed in 1985 and the early 1990s when the 305 Building was first adapted to support Rockwell's Process Facility Modifications Program and later the 100-K Spent Nuclear Fuel Basins Program, respectively.

Reactor Operations (100 Area)

Uranium fuel elements were irradiated within one of nine reactors over the productive lifetime of the Hanford Site. Given the heat generated by fissioning, cooling of the reactor cores was an essential element in the production of plutonium. So too was monitoring the process itself to ensure minimal impact on the health of workers and the environment. The working group selected the following buildings as those necessary to tell the story of fuel irradiation within the reactors:

181-B - River Pump House: "The basic functions of the 181 Buildings were to draw water from the Columbia River for reactor cooling, emergency and fire water storage, and for sanitary water purposes" (see the Historic Property Inventory Form for the 181-B River Pump House in Appendix B on the Internet). The 181-B River Pump House provided these services for the B Reactor and the 100-B Area beginning in 1944. Following enlargement in 1951, it served the C Reactor as well. Water was also pumped from the 181-B Building to the 200 Areas for use in the chemical separations process and to fulfill emergency and sanitary needs. The 181-B Building currently provides essential water services only to the 200 Areas.

183-KW - Filter Plant: Before water could be used in the reactors, it had to be filtered and chemically treated. The 183-KW Building was designed to filter and chemically treat the Columbia River water before its introduction into the KW Reactor. Treatment reduced or eliminated film on or corrosion of the process tubes within the reactor. In addition to measuring raw water, operations within the 183-KW facility included "chlorination of raw water, addition of coagulants to raw water, pH correction of filtered water, addition of corrosion inhibitor to process water, and influent and effluent control" (see the Historic Property Inventory Form for the 183-KW Filter Plant in Appendix B on the Internet). The 183-KW Building was a multi-component, multi-operational complex that could operate as a single plant, two plants, or six plants. The facility started operating in 1955 and was deactivated in 1971. However, as with 183-KE, some of the flocculation and sedimentation basins have recently been proposed for use in fish rearing.

1116-N - Control Room Simulator: The 1116-N Building is a mobile office building containing the N Reactor Control Room Simulator. The design and instrumentation of the simulator's interior is an exact copy of the layout of the control room in the 105-N Building. Constructed in 1982, the simulator was an "essential part of the [operations and safety upgrade] training program and was a manifestation of the trend toward greater safety and environmental protection" in the 1980s (see the Historic Property Inventory Form for the 1116-N Control Room Simulator in Appendix B on the Internet). As part of ongoing training, modifications to the equipment within the simulator were made as changes were introduced to the actual control room in the N Reactor. The Control Room Simulator, while in excellent condition, is no longer used.

105-B - B Reactor: Constructed in 1943-1944, the 105-B Reactor began operations on September 26, 1944 as the world's first full-scale, plutonium production reactor. It was the first of Hanford's original three reactors to attain full power and, hence, represents the beginning of the nuclear age. Plutonium produced in the B Reactor was used in the world's first nuclear explosion on July 16, 1945 at Alamogordo, New Mexico and thus represents the beginning of the era of nuclear weaponry. B Reactor also produced the plutonium used in the Fat Man bomb exploded over Nagasaki, Japan on August 9, 1945 and, therefore, represents the culmination of the Manhattan Project. These first-of-a-kind achievements distinguish the B Reactor and caused it to be listed on the National Register of Historic Places in 1992. Due to the sustained efforts of the B Reactor Museum Association, B Reactor has been set aside for retention in place as a public museum.

116-B - B Reactor Exhaust Stack: Integral to the operation of the B Reactor, the 116-B Stack vented air from the confinement zone of the reactor to the atmosphere. Originally linked directly to the 105-B Building by concrete ducts, alterations in the 1950s diverted air first through the 117-B Exhaust Air Filter Building and then through the 116-B Stack. This diversion was undertaken as a means of diminishing exposure to waste gases created and released during reactor operations. The 116-B Stack still stands beside the B Reactor and is listed as a contributing structure on the National Register nomination.

212-N - Lag Storage Building: Built in 1945, the 212-N Building provided underwater storage of irradiated fuel elements from the 100 Areas until Chemical Separations was able to process them. The reinforced concrete storage basin was 20 feet deep. This building is locked and radiological warning signs are posted due to residual contamination.

Chemical Separations (200 Area)

Once cooled, the irradiated fuel elements were transported to the 200 Area where plutonium and other derived isotopes were separated from the base uranium. Multiple chemical baths were required for this process, and continued quality control was essential. Again, as with irradiation monitoring, monitoring for worker health and impact to the environment was integrated into the process. Waste reduction and waste control were other key elements of this step. The working group selected the following buildings as those necessary to tell the story of chemical separations:

222-T - Process Control Laboratory: The 222-T Building was the first of three identical chemical separations support laboratories to be completed in 1944. B-Plant and U-Plant likewise each had a 222 support facility. "Because all work in the radioactively 'hot' T Plant had to be conducted remotely, the 222-T Building provided a very necessary laboratory setting in which samples drawn from the chemical separations process could be radiochemically analyzed to make sure the process was working correctly" (see the Expanded Historic Property Inventory Form for the 222-T Building in Appendix B on the Internet). The facility was first used to train instrument technicians. More important, however, staff in the Process Control Laboratory not only created, tested, and improved tools for remote handling of hazardous materials, they also assisted in developing "methodical, standardized counting and measuring techniques" and the equipment necessary to ensure uniform analytical procedures across the U.S. Department of Energy national complex (see the Expanded Historic Property Inventory Form for the 222-T Building in Appendix B on the Internet). This facility now supports administrative functions.

221-T - T Plant: The 221-T Plant was constructed between 1943-1944 as the first radiochemical processing facility at the Hanford Site and the first full-size radiochemical processing plant in the world (see the Historic American Engineering Record for the 221-T Plant in Appendix B on the Internet). T Plant began processing irradiated fuel elements from the B Reactor on the night of December 26, 1944. The nuclear materials for the world's first and third atomic explosions were processed in T Plant. The importance of T Plant diminished with the startup of the more efficient REDOX Plant in 1952 and diminished again with the startup of the even more efficient PUREX Plant in 1956. T Plant discontinued service as a chemical processing facility within 3 months of PUREX's activation. In 1957, the mission at T Plant changed dramatically when it was designated as the decontamination facility for equipment on the Hanford Site, a role it still fulfills today.

291-T - T Plant Exhaust Stack: The 291-T Stack "served to propel the off-gases from T Plant dissolver cells...upward and outward and to add a sufficient volume of diluting air so that gases could be dispersed in what was thought to be a safe manner" (see the Historic Property Inventory Form for the 291-T Stack in Appendix B on the Internet). Meteorological monitoring, begun as early as 1943, and vegetation monitoring initiated in 1944-1945 indicated that improvements were required. Sand filters were added in 1947. Silver reactor filters replaced these in 1950. Together with the 221-B Building, these buildings "became the first radiochemical processing facilities in the world to have their air emissions filtered in such a manner. The silver filtration technology soon became the world standard and has been used by the nuclear industry and nuclear defense sites ever since" (see the Historic Property Inventory Form for the 291-T Stack in Appendix B on the Internet). The 291-T Stack played a pioneering role in meteorological studies and air filtration techniques. The stack remains standing awaiting its final disposition.

276-B - Organic Make Up/Solvent Storage: The 276-B Building was constructed in 1965 "as part of the mid-1960s Waste Management Program transition of B Plant from a chemical separations plant into a waste reduction and storage facility" for processing waste streams from REDOX and PUREX (see the Historic Property Inventory Form for the 276-B Building in Appendix B on the Internet). The organic solvent solutions used in waste reduction operations within B Plant were made in the 276-B Building. These solutions were then pumped into the 221-B Building to extract strontium 89/90 from radioactive wastes being stored in the 200 Area single-shell tanks. With the recent deactivation of B Plant, this facility has no current function.

Plutonium Finishing (Z Plant-200 West)

Plutonium produced on the Hanford Site was first shipped to Los Alamos as plutonium nitrate paste, and later to Rocky Flats as metallic plutonium. The final processing of plutonium nitrate to metallic plutonium was accomplished in the Plutonium Finishing Plant. Here, as elsewhere along the process chain, waste control and monitoring were key concerns. The working group selected the following buildings as those necessary to tell the story of plutonium finishing:

234-5Z - Plutonium Finishing Plant: The 234-5Z Building, constructed in 1948, was a plutonium conversion and fabrication facility wherein plutonium nitrate was first converted to metallic plutonium and then made into weapons pits. Operations began on July 5, 1949. Initial operations relied on manual manipulation of plutonium mixtures within glove boxes (Rubber Glove Line). On March 18, 1952, remote, mechanized processing commenced with startup of the Remote Mechanical A Line, which "although partially dismantled...still stands today as the best and only representative of the earliest technology in remote, mechanized final plutonium processing" (see the Expanded Historic Property

Inventory Form for the 234-5Z Building in Appendix B on the Internet). The A Line was modified and expanded several times over its operational lifetime. In the mid-1960s, the C Line was added in response to increased demands for finished plutonium. Before the decision in 1965 to concentrate weapons fabrication work at Rocky Flats and Savannah River, operations within the 234-5Z Building had begun to diversify to include the "production of 'tailor-made' blends of Pu [plutonium] oxide that served as experimental fuel for the emerging nuclear power industry" (see the Expanded Historic Property Inventory Form for the 234-5Z Building in Appendix B on the Internet). At this same time, the 234-5Z Building was designated as the site where plutonium nitrate recovered from commercial reactor spent fuels would be reprocessed (Gerber 1996, p. 5-4). All operations ceased in 1989 when the 234-5Z Building was shut down.

291-Z - Z Plant Air Filter and Exhaust Stack: Constructed in 1948-1949, the primary function of the 291-Z Building was to draw gaseous effluents from the 234-5Z Building and exhaust them into the atmosphere. Before their release, effluents were passed through filters and treated to reduce human and environmental exposure to the waste gases released during plant operations. Between 1962-1964, additional connections were also made to the 232-Z, 236-Z, and 242-Z Buildings. The 291-Z Stack continues "to circulate and exhaust air from PFP complex facilities" (see the Historic Property Inventory Form for the 291-Z Building in Appendix B on the Internet).

232-Z - Waste Incinerator Facility: In 1958, the Hanford Site Chemical Processing Department proposed the construction of an incinerator (to be known as the Contaminated Waste Recovery Process Facility) "to recover plutonium from miscellaneous solid wastes produced by the adjacent Plutonium Finishing Plant (PFP), the Plutonium Isolation Facility [Building 231] and the REDOX and PUREX plants" (see the Historic American Engineering Record for the 232-Z Facility in Appendix B on the Internet). Operational in 1962, the 232-Z Facility was the first of its kind, the archetype for plutonium-recovery incinerators. Not only was plutonium recovered through this process, but incineration also served as a means to limit additional sub-surface contamination by significantly reducing the amount of contaminated wastes that had to be buried. "Following the 1970 rebuilding and re-start, the 232-Z incinerator operated for approximately three years, when it was shut down as part of an overall PFP review of 'interfaces with the environment'" (see the Historic American Engineering Record for the 232-Z Facility in Appendix B on the Internet).

2736-Z - Primary Plutonium Storage Facility: The 2736-Z Building "was constructed during 1970-1971 as the first building on the Hanford Site to be designed specifically for the storage and safe keeping of plutonium (Pu) products and scrap" (see the Expanded Historic Property Inventory Form for the 2736-Z Building in Appendix B on the Internet). Construction of this facility was an outgrowth of increased safety and security concerns expressed during the late 1960s. In 1972, the Hanford Site was designated as the Central Scrap Management Organization for the Atomic Energy Commission. The 2736-Z Building figured prominently in this decision. Between 1984-1987, a "state-of-the-art detection, measurement, annunciation and alarm system for Pu [plutonium] inventory protection" was installed which eventually attracted the notice of the International Atomic Energy Agency (see the Expanded Historic Property Inventory Form for the 2736-Z Building in Appendix B on the Internet). In 1994, the 2736-Z Building became a part of the International Safeguards Program. Plutonium is still stored in vaults in this building.

SITE VISITATION

The working group realized that a number of the structures selected for preservation are highly contaminated and may never be available for public access. Others among the facilities described above may be demolished because of existing Tri-Party Agreement milestones that may not be alterable. For those that are not demolished or irreversibly contaminated, the cost of stabilization and long-term maintenance will depend on the current condition of the building, its size and complexity, and its intended use in public education and interpretation; for instance, visiting the interior or only viewing the exterior.

Assuming that the issues associated with cost are resolved and assuming further that issues are resolved surrounding the safety and health concerns of opening a former plutonium production complex to the general public, how might visitation proceed? Considering transportation first, three media are available: water, road, and rail.

The Columbia River runs some 50 miles through the Hanford Site. Its waters, islands, and shorelines support abundant wildlife, and the views change with each bend in its course. Taking advantage of this, Columbia River Journeys and Hanford Reach Columbia River Tours both offer guided tours of the river. Thus commercial access via the river is already in place.

Over its operational lifetime, 268 miles of primary paved roads and 105 miles of railroad track were laid out across the Hanford Site. The U.S. Department of Energy currently offers road tours of the Hanford Site on an as-needed basis. Tours are conducted in large (47-passenger) or small (30-passenger) buses as required. Tours are confined to the major paved roads, and no exiting from the bus is allowed during the tour. Thus, visitor access to the interior of the Hanford Site is already in place.

> ***"The second segment consists of the single tracks beyond the Washington Public Power Supply System which loop throughout the Hanford Site with spur lines into each area. This segment should be preserved as a means to help educate the public on the cultural resources within the Hanford Site. For example, the railroad would be an excellent means of transporting the visiting public out to the 105-B Reactor. The route would allow for the interpretation of the Hanford story, from the geological events responsible for the formation of the valley through the development of the Manhattan Project and Cold War Era facilities. Since the railroad, in its present state, is fully operational, no appreciable work would be involved in its maintenance and preservation. The use of the present rail system also minimizes the impacts to the ecology of the area, since its removal would probably do more damage than leaving it in place. Continued preservation efforts would be minimal due to the favorable climate (i.e., the Port of Big Pasco has 100 year old rail still in use."*** **- T.H. Gronewald, President, Washington State Historical Railroad Society in Gronewald 1996**

The Washington State Historical Railroad Society has collected and continues to acquire rolling stock unique to or characteristic of the Hanford Site. Their goal is to exhibit this stock at the B Reactor Museum following its opening. A second goal long held by the Society is to use the rail system to transport visitors across the Hanford Site. Working through the Columbia River Exhibition of History, Science, and Technology, one scenario discussed in a number of public forums calls for the visiting public to buy a tour pass that would allow them to use any or all of these modes of transportation.

Having secured transportation, what will the visitor see? The Manhattan Project and Cold War Era resources comprise but one of many landscapes that together make up the Hanford Site. To ensure a broad audience, tourism should embrace all of the Hanford Site's heritage. The geological history of the mid-Columbia Basin is contained within the mountains, ridges, valleys, soils, and rocks now bounded by the Hanford Site. Grasses, plants, and brush are distributed across this natural landscape and support a wide variety of birds, insects, and mammals. These resources should be interpreted together with the river and its resources. All the forgoing constitute the cultural resources of the Native American community. Working cooperatively with them, aspects of their subsistence, legendary and religious heritage that may be shared with non-native peoples could be interpreted. The White Bluffs-Hanford Heritage group has personal knowledge of the agricultural history of the region. Their membership is a source of information from which to interpret the historic-archeological resources left in place following forced abandonment of the towns of Hanford, White Bluffs, and the surrounding areas in 1943. Finally, the Hanford Retirees Association and the B Reactor Museum Association could be called on to script the story of the Manhattan Project and Cold War years. There is no lack of things to see and experience.

Actions that need to be taken to allow for tourism include, but are not limited to, the following:

- Studies to characterize the radiological and hazardous materials of facilities that will be opened to the public
- Market analysis to identify the potential visitor and the cost of visitation
- Creation of a detailed Hanford Site Visitation Plan
- Development of design criteria to ensure uniformity in interpretive signs, exhibits, and displays

RECOMMENDATIONS FOR ADAPTIVE USE AND ECONOMIC DEVELOPMENT

While tourism may support part of the costs associated with opening the Hanford Site for visitation, additional funding sources for in-place preservation need to be found. The cost savings associated with not demolishing structures represent one potential source. More promising alternatives are adaptive use and economic development.

On August 20, 1998, the Office of the Inspector General issued an audit report entitled, *The U.S. Department of Energy's Facility Reuse at the Rocky Flats Environmental Technology Site* (OIG 1998). The report criticized the Rocky Flats Field Office for "scheduling" demolition of all site facilities, "without having formally analyzed the reuse potential of those facilities," and speculated that, "If reuse was feasible, the community surrounding the site could potentially realize significant economic benefits and DOE could avoid spending millions of dollars in demolition costs" (OIG 1998). In short, the Inspector General was calling Rocky Flats to task for failure to fully meet its responsibilities under the National Historic Preservation Act.

> ***"Prior to acquiring, constructing, or leasing buildings for purposes of carrying out agency responsibilities, each Federal agency shall use, to the maximum extent feasible, historic properties available to the agency."*** **- National Historic Preservation Act, 16 USC 470 (110)(a)(1)**

The working group recommends that the DOE, Richland Operations Office continue to examine its future facility needs to determine which, if any, facilities may be retained in support of future missions following site cleanup. This would be particularly true for facilities that could function for continued materials testing, process development, applied research and development, or other uses. This examination would not only be in keeping with the intent of the National Historic Preservation Act but would also eliminate any similar admonishment by the Inspector General that might be directed at the Hanford Site.

> *"...any Federal agency...shall, to the extent practicable, establish and implement alternatives for historic properties, including adaptive use, that are not needed for current or projected agency purposes, and may lease an historic property owned by the agency to any person or organization...if the agency head determines that the lease...will adequately insure [sic] the preservation of the historic property."* - 16 USC 470 (111)(a)

The 300 Area has been identified under a number of scenarios developed for the Hanford Site *Comprehensive Land Use Plan* (DOE 1996d) as a location for "Industrial and Commercial" development. To this end, DOE has begun to actively market this area of the Hanford Site through its Site Infrastructure Division. For example, Richland Specialty Extrusions is using the Sutton press (once used to extrude uranium fuel elements) to make aluminum baseball bats among other products in the 313 Building under a lease with DOE. Leases or property transfers have been accomplished for the 306-E, 308, and 318 Buildings. The

Table 4.1. *300 Area Buildings Recommended to Be Retained for Adaptive Use and Economic Development*

	Public Education/ Interpretation	Adaptive Use	Economic Development
305-B Engineering Development Laboratory Annex	√	√	
306-W Materials Development Laboratory		√	√
309/309-E Plutonium Recycle Test Reactor	√		
325/325-A Radiochemistry Laboratory/Cerium Recovery	√		
326 Physics and Metallurgy Laboratory	√		√
327 Post Irradiation Test Laboratory			√
329 Biophysics Laboratory	√		√
337 Technical Management Facility			√
337-B High Temperature Sodium Facility			√
3709-A Fire Station			√
3745-A Electron Accelerator Building		√	
3745-B Positive Ion Accelerator Building			√

314 Building is under discussion. Working on the principle that historic properties associated with ongoing functions will be retained in place, the working group identified an additional twelve buildings that should be advanced as preservation candidates. Table 4.1 lists these facilities.

DOE has made advances in preserving locations outside the 300 Area as well. For instance, the Port of Benton has acquired properties in the 400 Area, the 3000 Area, and more recently the 1100 Area. The Port then leases these properties to small businesses and startup companies to assist local economic development. The leasing of a portion of the 1171 Building to Livingston Rebuild Center for locomotive repairs and railroad worker training is a prime example of adaptive use coupled with local economic development. Here the Port coupled private economic pursuit with historic preservation by including a restriction in the deed that ensures retention of the significant historic features of the facility. The working group strongly supports DOE's plan to take an aggressive role in championing the use of its historic buildings as a means of promoting economic diversification in the Tri-Cities.

KEY DATES

KEY DATES OF EVENTS AT THE HANFORD SITE 1943-1990

Owner/Prime Contractor

Date	Event
Feb 25, 1941	Glenn Seaborg at the University of California Berkeley discovers plutonium-239 more likely to fission than uranium-235
U.S. Corps of Engineers Manhattan Engineer District	
Aug 1942	The U.S. Corps of Engineers is assigned to be in charge of the Manhattan Project
Nov 12, 1942	U.S. Government decides to build plutonium production reactors and separation facilities at the Hanford Site
Dec 2, 1942	Enrico Fermi achieves first nuclear reaction in the Chicago Pile-1 Reactor at the University of Chicago's Staff Field
U.S. Corps of Engineers Manhattan Engineer District E.I. Du Pont de Nemours and Company	
Dec 21, 1942	Du Pont receives contract to construct and manage the world's first plutonium production facility
MANHATTAN PROJECT 1943-1946	
1943	Test reactor in Building 305
Mar 1943	Construction of Hanford Engineer Works (present Hanford Site) begins
Nov 4, 1943	Air cooled pilot-scale graphite reactor completed at Oak Ridge Reservation, Tennessee
Dec 1943	Fuel manufacturing begins in 313 Building
Mar 20, 1944	First experimental canning operations start in the 313 Building
Sep 26, 1944	Startup of B Reactor, world's first full-scale plutonium production reactor
Dec 17, 1944	Startup of D Reactor
Dec 26, 1944	Startup of T Plant (bismuth phosphate separation process)
1944-1945	Installation of 64 underground, single-shell waste tanks
1945	U Plant (221-U Building) and its bulk reduction facility (224-U Building) built but do not go on line; 221-U instead is used for training, decontamination, and repair

Owner/Prime Contractor **Date**	**Event**
1945	Original radiochemistry laboratory completed (3706 Building)
Jan 16, 1945	Plutonium finishing operations begin in 231-Z Building
Feb 25, 1945	Startup of F Reactor
Apr 13, 1945	Startup of B Plant (bismuth phosphate separation process)
Apr 1945	Construction of Hanford Engineer Works completed
May 7, 1945	Germany Surrenders
Jul 16, 1945	First atomic explosion, Trinity test at Alamogordo, New Mexico
Aug 6, 1945	First atomic bomb (Little Boy) made with uranium-235, produced from the gaseous diffusion plant at the Oak Ridge Reservation in Tennessee, dropped on Hiroshima, Japan
Aug 9, 1945	Atomic bomb (Fat Man) made with Hanford Site plutonium, dropped on Nagasaki, Japan
Aug 14, 1945	Japan Surrenders
Mar 5, 1946	Winston Churchill gives his Iron Curtain speech saying a Cold War is well under way
Aug 1, 1946	Atomic Energy Act passed
U.S. Corps of Engineers Manhattan Engineer District General Electric Company	
Sep 1, 1946	General Electric Company becomes the prime site contractor of the newly named Hanford Works; the Atomic Energy Commission does not formally approve the contract until 1947 when the Atomic Energy Commission takes charge of the Hanford Site
COLD WAR ERA 1947-1990	
1947-1949	Installation of 42 underground, single-shell waste tanks
Atomic Energy Commission General Electric Company	
Jan 1, 1947	The Atomic Energy Commission takes charge of the former Manhattan Project sites; the site is renamed the Hanford Works
Mar 12, 1947	Truman Doctrine
Jun 23, 1948	Berlin blocked and air lift results

Owner/Prime Contractor **Date**	**Event**
1949	Startup of C Plant (pilot for reduction and oxidation separation process - REDOX)
Jul 5, 1949	Startup of Rubber Glove (RG) Line in the newly completed Plutonium Finishing Plant (Z Plant)
Aug 29, 1949	Soviet Union detonates its first atomic bomb
Oct 29, 1949	Startup of H Reactor
1950-1952	Korean War Expansion
1950-1952	Installation of 18 underground, single-shell waste tanks
Oct 3, 1950	Startup of DR Reactor
1951	Startup of 242-T and 242-B Evaporators to process T and B Plant radioactive high-level waste stored in tank farms to reduce volume by boiling off the liquid
Mar 28, 1951	Camp Hanford established
1952	Startup of Physical Constants Test Reactor
1952	Startup of S Plant (reduction and oxidation separation process - REDOX)
1952	Startup of U Plant to recover uranium from T and B Plant radioactive high-level waste
1952	Startup of Experimental Animal Farm and Aquatic Biology Laboratory
Mar 18, 1952	Startup of Remote Mechanical A Line in the (Z Plant)
Oct 31, 1952	First hydrogen bomb tested at the Pacific Proving Grounds
Nov 18, 1952	Startup of C Reactor
1953-1955	Eisenhower Expansion
1953-1955	Installation of 21 underground, single-shell waste tanks
Jul 27, 1953	Korean War Armistice signed
1954	Startup of Thermal Test Reactor
Aug 30, 1954	Atomic Energy Act (allows for more private, commercial atomic applications)
Dec 1954	Startup of KW Reactor
Feb 1955	Startup of KE Reactor

Owner/Prime Contractor

Date	**Event**
1955	Installation of Nike missile sites
1956-1964	Major Defense Production
1956	Moving all chemical processing work from the 231-Z Building to the Plutonium Finishing Plant
1956	Shutdown of B Plant
1956	Startup of Uranium Trioxide Plant, which was modified to convert uranium to uranium-trioxide
Jan 1956	Startup of A Plant (plutonium-uranium extraction process - PUREX)
Mar 20, 1956	Shutdown of T Plant
1957-1958	700 Area no longer a restricted area; fence removed and homes and commercial property sold to private owners
Oct 4, 1957	Soviet Union launches Sputnik
1958	Shutdown of U Plant
1960	Startup of the Plutonium Recycle Test Reactor (309 Building)
1961	Shutdown of Nike missile sites
1961	Shutdown of Camp Hanford
Oct 12-28, 1962	Cuban missile crisis
1963-1964	Electrical Generating Plant in 100-N Area
1963-1964	Installation of four underground, single-shell waste tanks
Aug 5, 1963	Britain, the United States, and the Soviet Union sign the Limited Test Ban Treaty to outlaw nuclear tests in the atmosphere, under water, and in outer space
Dec 31, 1963	Initial critical startup (deadline date) of N Reactor
Mar 1964	Actual startup of operations of N Reactor, which is the first dual-purpose reactor operated in United States
May 1964	Startup of the Plutonium Reclamation Facility
Dec 31, 1964	Shutdown of DR Reactor
Jan 1, 1965	U.S. Testing Company (assumes responsibility for bioassays, film badges, environmental samples)

Owner/Prime Contractor Date	Event
Jan 4, 1965	Battelle Memorial Institute assumes responsibility for management of the Hanford Site laboratories, renaming them the Pacific Northwest Laboratory
Mar 1, 1965	ITT Federal Support Services, Inc. assumes responsibility for support services
Apr 21, 1965	Shutdown of H Reactor
Jun 25, 1965	Shutdown of F Reactor
Aug 1, 1965	Hanford Environmental Health Foundation assumes responsibility for health services.
Nov 1, 1965	Douglas-United Nuclear, Inc. (joint venture between McDonnell Douglas Aircraft Company and United Nuclear Corporation) assumes responsibility for reactors and fuels
Jan 1, 1966	ISOCHEM, Inc. assumes responsibility for chemical processing and plutonium finishing
Jun 26, 1967	Shutdown of D Reactor
Jun 30, 1967	Construction completed on the High Temperature Lattice Test Reactor, 318 Building
Atomic Energy Commission Atlantic Richfield Hanford Company	
Sep 1, 1967	Atlantic Richfield Hanford Company becomes prime operator and assumes responsibility for chemical processing
Dec 1967	Shutdown of S Plant (REDOX)
1968-1988	Installation of 28 underground, double-shell waste tanks
Feb 12, 1968	Shutdown of B Reactor
Apr 25, 1969	Shutdown of C Reactor
Feb 1, 1970	Shutdown of KW Reactor
Jul 1, 1970	Westinghouse Hanford Company contracts to build and operate the Fast Flux Test Facility in the 400 Area and to operate the Hanford Engineering Development Laboratory
Jan 28, 1971	Shutdown of KE Reactor
1972-1982	Mostly fuel-grade plutonium produced
1972	Shutdown of the Uranium Trioxide Plant

Owner/Prime Contractor **Date**	**Event**
1973-1976	Startup of two new evaporator plants, 242-S and 242-A
Oct 11, 1974	President Ford signs the Energy Reorganization Act abolishing the Atomic Energy Commission and establishing the Energy Research and Development Administration
Energy Research and Development Agency Atlantic Richfield Hanford Company	
Jan 19, 1975	Energy Research and Development Administration activated; site renamed the Hanford Reservation
U.S. Department of Energy Rockwell Hanford Operations	
Jul 1, 1977	Rockwell Hanford Operations becomes prime operator and assumes responsibility for reprocessing operations
Oct 1, 1977	U.S. Department of Energy replaces the Energy Research and Development Administration; site renamed the Hanford Site
Feb 9, 1980	Startup of the Fast Flux Test Facility
1983-1987	Only weapon-grade plutonium produced
Nov 4, 1983	Restart of the Uranium Trioxide Plant
Jan 6, 1987	Shutdown of N Reactor
Mar 1, 1987	ICF Kaiser Hanford Company assumes responsibility for architecture and engineering
U.S. Department of Energy Westinghouse Hanford Company	
Jun 29, 1987	Westinghouse Hanford Company awarded the consolidated management and operations contract, which includes reactor operations, fuel manufacturing, chemical engineering, and waste management activities
1988	Shutdown of A Plant (PUREX)
May 1989	Shutdown of Plutonium Finishing Plant
Oct 1989	Cold War ends

LIST OF FIGURES

LIST OF FIGURES

CHAPTER 1 - HISTORIC OVERVIEW

CHAPTER 2 - PLUTONIUM PRODUCTION FACILITIES

SECTION 1 - CONSTRUCTION HISTORY

SECTION 2 - FUEL MANUFACTURING

SECTION 3 - REACTOR OPERATIONS

SECTION 4 - CHEMICAL SEPARATIONS

SECTION 5 - PLUTONIUM FINISHING

SECTION 6 - WASTE MANAGEMENT

Section 7 - Research and Development

Section 8 - Site Security

SECTION 9 - MILITARY OPERATIONS

SECTION 10 - WORKER HEALTH AND SAFETY

SECTION 11 - TRANSPORTATION, COMMUNICATION, AND SITE SUPPORT

Section 12 - History of Workers

Chapter 3 - Guide to Resources

Chapter 4 - Recommendations For Future Uses of Structures

LIST OF TABLES

LIST OF TABLES

CHAPTER 1 - HISTORIC OVERVIEW

CHAPTER 2 - PLUTONIUM PRODUCTION FACILITIES

SECTION 1 - CONSTRUCTION HISTORY

SECTION 2 - FUEL MANUFACTURING

SECTION 3 - REACTOR OPERATIONS

SECTION 4 - CHEMICAL SEPARATIONS

SECTION 5 - PLUTONIUM FINISHING

SECTION 6 - WASTE MANAGEMENT

SECTION 7 - RESEARCH AND DEVELOPMENT

SECTION 8 - SITE SECURITY

SECTION 9 - MILITARY OPERATIONS

SECTION 10 - WORKER HEALTH AND SAFETY

SECTION 11 - TRANSPORTATION, COMMUNICATION, AND SITE SUPPORT

SECTION 12 - HISTORY OF WORKERS

CHAPTER 3 - GUIDE TO RESOURCES
CHAPTER 4 - RECOMMENDATIONS FOR FUTURE USES OF STRUCTURES

REFERENCES CITED

REFERENCES CITED

36 CFR 800. 1986. Advisory Council on Historic Preservation, "Protection of Historic Properties: Regulations of the Advisory Council on Historic Preservation Governing the Section 106 Review Process." U.S. Code of Federal Regulations.

40 CFR 61. 1997. "National Emission Standards for Hazardous Air Pollutants." U.S. Code of Federal Regulations.

48 FR 44716. September 29, 1983. "Archeology and Historic Preservation; Secretary of the Interior's Standards and Guidelines." *Federal Register.*

15 USC 2601 et seq. (as amended). October 11, 1976. *Toxic Substances Control Act.* Public Law 94-469.

16 USC 470 et seq. (as amended). 1992. *National Historic Preservation Act of 1966.* Public Law 102-575 Title XL.

33 USC 1251 et seq. (as amended). December 27, 1977. *Clean Water Act of 1977.* Public Law 95-217.

33 USC 1251 et seq. (as amended). June 30, 1948. *Federal Water Pollution Control Act.* Chapter 758, 62 Stat. 1155.

33 USC 1401 et seq. (as amended). October 23, 1972. *Marine Protection, Research, and Sanctuaries Act of 1972.* Public Law 92-532.

42 USC 10101 et seq. (as amended). January 7, 1983. *Nuclear Waste Policy Act of 1982.* Public Law 97-425.

42 USC 2011 et seq. (as amended). August 1, 1946. *Atomic Energy Act of 1946.* Chapter 724, 60 Stat. 755.

42 USC 300f et seq. (as amended). December 16, 1974. *Safe Drinking Water Act of 1974.* Public Law 93-523.

42 USC 4321 et seq. (as amended). January 1, 1970. *National Environmental Policy Act of 1969.* Public Law 91-190.

42 USC 6901 et seq. (as amended). October 31, 1976. *Resource Conservation and Recovery Act of 1976.* Public Law 94-580.

42 USC 7401 et seq. (as amended). July 14, 1955. *Clean Air Act.* Chapter 360, 69 Stat. 322.

42 USC 9601 et seq. (as amended). December 11, 1980. *Comprehensive Environmental Response, Compensation, and Liability Act of 1980.* Public Law 96-510.

ACHP - Advisory Council on Historic Preservation. 1991. *Balancing Historic Preservation Needs with the Operations of Highly Technical or Scientific Facilities.* Washington, D.C.

Adley, F. 1998. Letter from F. Adley (retired from Hanford Environmental Health Foundation) to C. Bonadie (Hanford Environmental Health Foundation), November 3, 1998, Richland, Washington. Copy on file at the Library of the U.S. Department of Energy's Hanford Cultural Resources Laboratory, Richland, Washington.

AEC - U.S. Atomic Energy Commission. 1951a. *Redox Technical Manual.* HW-18700, Richland, Washington.

AEC - U.S. Atomic Energy Commission. 1951b. *Uranium Recovery Technical Manual.* HW-19887, Richland, Washington.

AEC - U.S. Atomic Energy Commission. 1951c. *Control of Radiation Hazards in the Atomic Energy Program.* Washington, D.C.

AEC - U.S. Atomic Energy Commission. 1956. *Radiation Safety in Atomic Energy Activities.* Staff Report to the Atomic Energy Commission, Twenty-First Semiannual Report, Washington, D.C.

AEC - U.S. Atomic Energy Commission. 1960. *The Future Role of the Atomic Energy Commission Laboratories, Volume I: Analysis and Conclusions.* Washington, D.C.

AEC - U.S. Atomic Energy Commission. 1965. *Redox Training Manual.* Richland, Washington.

AEC - U.S. Atomic Energy Commission. 1973. *Roads and Railroads, Hanford Works.* Drawing No. H-7-618, Rev. 4, Richland, Washington.

AEC - U.S. Atomic Energy Commission. 1975. *Operational Accidents and Radiation Exposure within the Atomic Energy Commission.* WASH 1192, Washington, D.C.

Ager, G.O., and L.M. Knights. 1963. *Fire Fighting Criteria for 234-5.* HW-76269 Rev, General Electric Company, Richland, Washington.

Agnew, S.F. 1997. *Hanford Tank Chemical and Radionuclide Inventories: HDW Model Rev. 4.* LA-UR-96-3860, Los Alamos National Laboratory, Los Alamos, New Mexico.

Ahearne, J.F. 1989. "Fixing the Nation's Nuclear-Weapons Plant." *Technology Review,* July 1989, pp. 24-29.

Albright, D., F. Berhout, and W. Walker. 1993. *World Inventory of Plutonium and Highly Enriched Uranium, 1992.* Oxford University Press, New York.

American Nuclear Society. 1992. *Controlled Nuclear Chain Reaction: The First 50 Years.* American Nuclear Society, La Grange Park, Illinois.

Anderson, J.D. 1990. *A History of the 200 Area Tank Farms.* WHC-MR-0132, Westinghouse Hanford Company, Richland, Washington.

Anderson, J.D. 1996. *The History of the 200 Area Burial Ground Facilities.* WHC-EP-0912, Westinghouse Hanford Company, Richland, Washington.

Anderson, K.J. 1972. *Prototype Mining Equipment, Z-9 Enclosed Trench.* ARH-2377, Atlantic Richfield Hanford Company, Richland, Washington.

Anderson, L.E., B.J. Kelman, and R.J. Weigel. 1984. *Interaction of Biological Systems with Static and ELF Electric and Magnetic Fields: 23rd Hanford Life Sciences Symposium.* U.S. Department of Energy, Pacific Northwest Laboratory, and the Electric Power Research Institute, Richland, Washington.

Anonymous. 1996. "The Way We Were." *Hanford Reach,* August 12, 1996, Richland, Washington.

Aradcom Argus. 1958. *Army Air Defense Command.* Colorado Springs, Colorado.

Architecture. 1995. "Corrugated Metal Catches On." Vol. 84, No. 8.

ARHCO - Atlantic Richfield Hanford Company. 1967a. *Z Plant Report, Task I-II and Recovery Operations, September 5, 1967 through December 31, 1967.* ARH-11 RD, Richland, Washington.

ARHCO - Atlantic Richfield Hanford Company. 1967b. *Monthly Report, 200 Areas Operation, September 1967.* ARH-59 DEL, Richland, Washington.

ARHCO - Atlantic Richfield Hanford Company. 1968a. *Monthly Report, 200 Areas Operation, September 1968.* ARH-308 DEL, Richland, Washington.

ARHCO - Atlantic Richfield Hanford Company. 1968b. *Z Plant Report, Task I-II & Recovery Operations, January 1, 1968 through December 31, 1968.* ARH-294 RD, Richland, Washington.

ARHCO - Atlantic Richfield Hanford Company. 1969a. *Monthly Report, 200 Areas Operation, March 1969.* ARH-1102 DEL, Richland, Washington.

ARHCO - Atlantic Richfield Hanford Company. 1969b. *Z Plant Report, Task I-II & Recovery Operations, 1969.* ARH-1048 RD, Richland, Washington.

ARHCO - Atlantic Richfield Hanford Company. 1976a. *Atlantic Richfield Hanford Company Monthly Report, March 1976.* ARH-LD-215-B, Richland, Washington.

ARHCO - Atlantic Richfield Hanford Company. 1976b. *Atlantic Richfield Hanford Company Monthly Report, July 1976.* ARH-LD-219-B, Richland, Washington.

ARHCO - Atlantic Richfield Hanford Company. 1977. *Atlantic Richfield Hanford Company Monthly Report, February 1977.* ARH-LD-226-B, Richland, Washington.

Army - U.S. Department of Army. 1955. *83rd Antiaircraft Artillery Missile Battalion, Camp Hanford, Washington: Information for personnel assigned to the 83rd AAA Missile Battalion.* Carlisle, Pennsylvania.

Army - U.S. Department of Army. 1958. "Last Year Sparked by Many Events." *Tumbleweed.* Vol. 2, No. 6, March 26, Camp Hanford, Washington.

Army - U.S. Department of Army. 1961. "Final Souvenir Edition." *Tumbleweed.* Vol.5, No. 1, January 25, Camp Hanford, Washington.

Artillery School - Artillery and Guided Missile School. 1955. *Organization Procedures and Drill for NIKE 1 Units.* ST-44-161-3D, Antiaircraft and Guided Missiles Branch, Fort Bliss, Texas.

ASME - American Society of Mechanical Engineers. 1952. *Uranium, Plutonium and Industry: A Summary of the U.S. Atomic Energy Program.* New York.

Ashton, L. 1999. "Towns Wasted by War Effort." *The San Diego Union-Tribune,* February 21, 1999, San Diego, California.

Backman, G. 1999. Interview of George Backman (Personnel Protection Supervisor for Chemical Separations at the Hanford Site, 1948-1988) by Brian Freer (CH2M HILL Hanford, Inc.), April 9, 1999. Copy of notes on file at the Library of the U.S. Department of Energy's Hanford Cultural Resources Laboratory, Richland, Washington.

Bailey, R.W., and M.S. Gerber. 1997. *PUREX/UO3 Facilities Deactivation Lessons Learned History.* HNF-SP-1147, Rev. 2, Project Hanford Management Contractor, Richland, Washington.

Bair, W.J. 1965. "Inhaled Radioactive Particles and Gases." *Health Physics* 10(12):861-1259.

Bair, W.J. 1988. *25th Annual Life Sciences Symposium: Radiation Protection, A Look to the Future.* Pergamon Press, New York.

Baker, R.G. 1957. *Project Proposal, Revision 2: Improved Task I and Task II Facilities, 234-5.* HW-52395, General Electric Company, Richland, Washington.

Baker, S.M., J.L. Devary, R.P. Elmore, R.F. Lorang, A.J. Rossi, and M.D. Freshley. 1988. *U1/U2 Uranium Plume Characterization, Remedial Action Review and Recommendation.* WHC-EP-0133, Westinghouse Hanford Company, Richland, Washington.

Ballinger, M.Y., and R.B. Hall. 1991. *A History of Major Hanford Facilities and Processes Involving Radioactive Material.* PNL-6964 HEDR, Pacific Northwest Laboratory, Richland, Washington.

Ballou, J.E. 1974. "Radiation and the Lymphatic System." *Proceedings of the 14th Annual Hanford Biology Symposium.* Atomic Energy Commission and Pacific Northwest Laboratory, Oak Ridge, Tennessee.

Balogh, B. 1991. *Chain Reaction: Expert Debate and Public Participation in American Commercial Nuclear Power, 1945-1975.* University Press, Cambridge.

Barthel, D. 1996. *Historic Preservation: Collective Memory and Historical Identity.* Rutgers University Press, New Brunswick, New Jersey.

Barton, G.B. 1950a. *RE: Analysis of a Standard Pu Solution by the Control Laboratories in 231 and 234-5 Buildings.* HW-18319, General Electric Company, Richland, Washington.

Barton, G.B. 1950b. *Analysis of a Standard Pu Solution by the Control Laboratories in 231 and 234-5 Buildings.* HW-19545, General Electric Company, Richland, Washington.

Battelle. 1977. "Tale End Charlies." *The Greenie,* XIV (40).

Battey, L.J. 1962. "Justification for Project CGC-912, Waste Treatment Facility, Z Plant." Memorandum to G.P. Kesel, May 28, 1962, General Electric Company, Richland, Washington.

Beardsley, P. 1998. Interview of Paul Beardsley (patrol officer at the Hanford Site from 1943-1945 and 1946-1977) by Joy Keating (Pacific Northwest National Laboratory) on October 21, 1998. Copy of notes on file at the Library of the U.S. Department of Energy's Hanford Cultural Resources Laboratory, Richland, Washington.

Beaver, W. 1990. "Nuclear Power Goes On-Line: A History of Shippingport." *In Economics and Economic History,* No. 105, Greenwood Press, Westport, Connecticut.

Bebbington, W.P. 1990. *History of Du Pont at the Savannah River Plant.* E.I. Du Pont de Nemours and Company, Wilmington, Delaware.

Beck, E. 1998. Interview of Ed Beck (Industrial Hygienist, 1989 to date) by Brian Freer (CH2M HILL Hanford, Inc.), November 17, 1998. Copy of notes on file at the U.S. Department of Energy's Hanford Cultural Resources Laboratory, Richland, Washington.

Becker, C.D. 1990. *Aquatic Bioenvironmental Studies: The Hanford Experience 1944-84.* Studies in Environmental Science, Vol. 39, Elsevier, New York.

Bell, R.S. 1950. *A History of Startup and Operations of the 234-5 Facility During the Year 1949.* HW-15953, General Electric Company, Richland, Washington.

Bell, R.S. 1951a. *S Division Monthly Report, 234-5 Operations, April 1951.* HW-21097 DEL, General Electric Company, Richland, Washington.

Bell, R.S. 1951b. *S Division Monthly Report, 234-5 Operations, June 1951.* HW-21563 DEL, General Electric Company, Richland, Washington.

Bell, R.S. 1952. *Synopsis of Scheduling for the Completion of the RMA Line.* HW-23232, General Electric Company, Richland, Washington.

Bender, D.E., ed. 1998. *Nike History Online.* March edition. [Online] Available URL: http://alpha.fdu.edu/~bender/nikenws.html

Benoliel, R.W. 1952. *Service of Production Bombs, Plutonium Purification and Fabrication, 234-5 Building.* HW-24153, General Electric Company, Richland, Washington.

Berube, N.A. 1977a. "Z Plant Process Preventive Maintenance Program." Memo to R.B. Gelman, D.G. Harlow, and R.L. Hibbard, March 9, 1977, Atlantic Richfield Hanford Company, Richland, Washington.

Berube, N.A. 1977b. "Oxide Line Preventive Maintenance Program." Memo to R.B. Gelman, D.G. Harlow, and R.L. Hibbard, March 24, 1977, Atlantic Richfield Hanford Company, Richland, Washington.

Berube, N.A. 1977c. "Solvent Extraction and Miscellaneous Treatment Process Preventive Maintenance Program." Memo to R.B. Gelman, D.G. Harlow, and R.L. Hibbard, March 25, 1977, Atlantic Richfield Hanford Company, Richland, Washington.

Berube, N.A. 1977d. "Preventive Maintenance Program for Z Plant Service Equipment." Memo to R.B. Gelman, D.G. Harlow, and R.L. Hibbard, May 2, 1977, Atlantic Richfield Hanford Company, Richland, Washington.

Bierman, S.R. 1964. *Pulsed Neutron Source Measurements of the Z-9 Crib.* HW-80913, General Electric Company, Richland, Washington.

Binder, M.S. 1992. *Literature and Field Evaluation of Potential Contamination at North Slope Nike Batteries, Hanford Defense, Washington.* Prepared for IT Corporation, Task Order 1-92-19, Richland, Washington.

Blasewitz, A.G. 1951. *RECUPLEX Feasibility Report.* HW-22596, General Electric Company, Richland, Washington.

Bodnar, J. 1992. *Remaking America: Public Memory, Commemoration, and Patriotism in the Twentieth Century.* Princeton University Press, Princeton, New Jersey.

Boehm, G.A.W., and A. Groner. 1986. *The Battelle Story: Science in the Service of Mankind.* Battelle Memorial Institute, Columbus, Ohio.

BOMARC/AF TALOS. 1990. *NIKE Quick Look III.* Fort Worth, Texas. Copy on file at the Library of the U.S. Department of Energy's Hanford Cultural Resources Laboratory, Richland, Washington.

Bonham, K., L.R. Donaldson, R.F. Foster, A.D. Welander, and A.H. Seymour. 1948. "The Effect of X-Ray on Mortality, Weight, Length, and Counts of Erythrocytes and Hematopoietic Cells in Fingerling Chinook Salmon, *Oncorhynchus tshawytscha* Walbaum." *Growth* 12:101-121.

Borgeson, M.E. 1966. *Soil Mining and Leaching Facilities, 216-Z-9 Waste Disposal Cavern, Conceptual Equipment Engineering Study.* ISO-448, Isochem, Inc., Richland, Washington.

Borgeson, M.E. 1967. *Budget Design Criteria, 216Z-9 Cavern Soil Mining and Leaching Facilities.* ISO-866, Isochem, Inc., Richland, Washington.

Bowles, K.C. 1976. *The Summary of Tests on "A" Billets with Respect to OD Unbonds Between Zircoloy-2 Clad and Uranium.* UNI-579, United Nuclear Industries, Inc., Richland, Washington.

Bracchini, M.A. 1997. "The History and Ethics Behind the Manhattan Project." University of Texas, Austin. [Online] Available URL: http://www.me.utexas.edu/~uer/manhattan/index.html

Braden, D.E. 1958. *Design Scope: Reduction of Radiation Exposure, RMA Line, 234-5 Building.* HW-55865, General Electric Company, Richland, Washington.

Braden, D.E. 1963. *Design Scope of the Waste Treatment Facility - Z Plant, Project CGC-912, Revision Number 1.* HW-76055, General Electric Company, Richland, Washington.

Braden, D.E. 1965. *Plutonium Blending Facility - Z-Plant, FY-1967 Budget Facility Description.* RL-SEP-368, General Electric Company, Richland, Washington.

Braden, D.E., C.J. Berglund, S.G. Smolen, and R.L. Stevenson. 1962. *Hazards Evaluation, Project CGC-979, RMC Button Line Filtrate Handling Facility, Z Plant.* HW-75134, General Electric Company, Richland, Washington.

Braden, D.E., C.L. Brown, B.F. Judson, and C.W. Nilson. 1963. *Chemical Processing Department, Hazards Evaluation Task Force, Waste Treatment Facility - Design Scope Revision No. 1.* HW-78577, General Electric Company, Richland, Washington.

Braden, D.E., and R.A. Ciccarelli. 1963. *Americium Recovery Facility, Design Scope.* HW-79369, General Electric Company, Richland, Washington.

Brady, D.N., and F. Swanberg. 1965. "The Hanford Mobile Whole-Body Counter." *Health Physics* 11:1221-1224.

Brands, J.W. 1949. "Recommendation Report No. 129: Railroad and Highway Maintenance." Memorandum to R. T. Cooke (Superintendent, Transportation Division), December 19, 1949, E.R. 4372, Hanford Works Control Division, Richland, Washington.

Breitenstein, B.D. 1983. "Medical Management and Chelation Therapy." *Health Physics* 45(4):855-866.

Brown, A.T., S.W. Dodd, J.A. Glasscock, E.M. Kuykendall, and J.M. Parr. 1980. *Plutonium Retrieval System PRE Operations Manual.* RHO-CD-1080, Rockwell Hanford Operations, Richland, Washington.

Brown, C.L. 1953. *Process Test MS-235-5, Reduction of Task III Cooling Time, RM Line.* HW-26776, General Electric Company, Richland, Washington.

Brown, D.J. 1994. "Milestone Report, LLP-94-011, Inspect FY 1996 Landlord Line Item (Central Fire Station)." Memorandum from D.J. Brown (Westinghouse Hanford Company) to C.C. Chamberlin-Dow, June 24, 1994. Copy on file at the Library of the U.S. Department of Energy's Hanford Cultural Resources Laboratory, Richland, Washington.

Bruns, L.E., R.Y. Lyon, and W.F. Unzicker. 1963. *Z Plant Weekly Report: RECUPLEX, Incinerator & Task I-II, January through December 1962.* HW-72224 RD, General Electric Company, Richland, Washington.

Burley, H.H. 1956. *Fuel Element Technical Manual.* HW-4000, General Electric Company, Richland, Washington.

Burnside, R.J., D.R. Skeen, S.H. Finfrock, A.H. Robinson, and H. Toffer. 1986. "Flux Flattening in a Graphite Moderated Core." *Transactions of the American Nuclear Society,* 53:463-464.

Bush, S.H. 1965. *Irradiation Effects in Cladding and Structural Materials.* Rowman & Littlefield, New York.

Bustad, L.K. 1964. "Biology of Radioiodine." *Proceedings of the Hanford Symposium on the Biology of Radioiodine.* Pergamon Press, Oxford.

Bustad, L.K., and R.O. McClellan. 1966. *Swine in Biomedical Research.* Atomic Energy Commission and Battelle Memorial Institute, Richland, Washington.

Cadd, T. 1998. Telephone call from Joy Keating (Pacific Northwest National Laboratory) to Tedd Cadd (Security Operations, Pacific Northwest National Laboratory), November 2, 1998. Copy of notes on file at the Library of the U.S. Department of Energy's Hanford Cultural Resources Laboratory, Richland, Washington.

Cahow, D.L., K.O. Fein, and G.L. Hanson. 1979a. *Safety Assessment Document - Z Plant RMA Oxide Line.* RHO-CD-454, Rockwell Hanford Operations, Richland, Washington.

Cahow, D.L., K.O. Fein, and G.L. Hanson. 1979b. *Safety Assessment Document - Z Plant RMA Oxide Line.* RHO-CD-454 Rev. 1, Rockwell Hanford Operations, Richland, Washington.

Cameron, B. 1998. Telephone calls from Joy Keating (Pacific Northwest National Laboratory) to Bruce Cameron (Patrol Training, B&W Protec, Inc.), October 6, 1998, November 2, 1998, and March 25, 1999. Copy of notes on file at the Library of the U.S. Department of Energy's Hanford Cultural Resources Laboratory, Richland, Washington.

Campbell, H.E. 1948/1989. "Photo from the Files." *Grapevine Magazine*, June 1989.

Cantril, S.T. 1945. "Plant Medical Section—Activities to August 1, 1945." *Operation of Hanford Engineer Works — Medical Department Plant and Village Medical,* HAN-73214, Book 6, E. I. Du Pont De Nemours and Co., Inc., Richland, Washington.

Cantril, S.T. 1951. "Industrial Medical Program—Hanford Engineer Works." In *Industrial Medicine on the Plutonium Project,* R.S. Stone (ed), McGraw-Hill, New York.

Cantril, S.T., and H.M. Parker. 1951. "Status of Health and Protection at the Hanford Engineer Works." *Industrial Medicine on the Plutonium Project,* R.S. Stone (ed.), McGraw-Hill, New York.

Cantril, S.T., J.J. Nickson, J.E. Wirth, and R.S. Stone. 1951. "Requirements of an Adequate Health Service in Relation to Atomic Research and Industrial Development." In *Industrial Medicine on the Plutonium Project,* R. S. Stone (ed.), McGraw-Hill, New York.

Carlisle, R.P. 2000. "Comment on Two Sections: Fuel Manufacturing and Plutonium Finishing." E-Mail message from Rodney Carlisle (Professor of History, Rutgers University, and author of *Supplying the Nuclear Arsenal: American Production Reactors, 1942-1992*) to Darby Stapp (Pacific Northwest National Laboratory), January 17, 2000. Copy on file at the Library of the U.S. Department of Energy's Hanford Cultural Resources Laboratory, Richland, Washington.

Carlisle, R.P., and J. Zenzen. 1996. *Supplying the Nuclear Arsenal: American Production Reactors, 1942-1992.* The Johns Hopkins University Press, Baltimore, Maryland.

Carlson, C.M., and R. Lyon. 1996. *Last Line of Defense: Nike Missile Sites in Illinois.* C. Whitacre (ed.), U.S. National Park Service, Denver, Colorado.

Carpenter, R.W. 1993. *100 D Area Technical Baseline Report.* WHC-SD-EN-TI-1181, Westinghouse Hanford Company, Richland, Washington.

Carpenter, R.W. 1994a. *100 B Area Technical Baseline Report.* WHC-SD-EN-TI-220, Westinghouse Hanford Company, Richland, Washington.

Carpenter, R.W. 1994b. *100 K Area Technical Baseline Report.* WHC-SD-EN-TI-239, Westinghouse Hanford Company, Richland, Washington.

Carr, P.J. 1958. *Completion Report, Expansion of 200 Area Facilities.* HW-24800-105, General Electric Company, Richland, Washington.

Cartmell, E. 1959. *Chemistry for Engineers.* Butterworths Scientific Publications, London.

Caufield, C. 1989. *Multiple Exposures: Chronicles of the Radiation Age.* University of Chicago, Chicago.

CEC - Committee on Energy and Commerce. 1986. *American Nuclear Guinea Pigs: Three Decades of Radiation Experiments on U.S. Citizens.* U.S. House of Representatives, 99th Congress, 2nd Session, U.S. Government Printing Office, Washington, D.C.

Chapman, V.R., and W.N. Mobley. 1952. *Production Test 234-5-1-MS: Evaluation of Use of Filter Paper in Conjunction with Filter Boats.* HW-25131, General Electric Company, Richland, Washington.

Chatters, J.C. 1989. *Hanford Cultural Resources Management Plan.* PNL-6942, Pacific Northwest Laboratory, Richland, Washington.

Chiaramonte, G.R. 1972. *Design Criteria, Disposal of Z Plant Wastes.* ARH-2335, Atlantic Richfield Hanford Company, Richland, Washington.

Christensen, E.I., and W.J. Maraman. 1969. *Plutonium Processing at the Los Alamos Scientific Laboratory.* LA-3542, Los Alamos Scientific Laboratory, Los Alamos, New Mexico.

Ciccarelli, R.A. 1958. *26" House Vacuum System - 234-5 Building.* HW-58025, General Electric Company, Richland, Washington.

Ciccarelli, R.A. 1959a. *Radiography Facility - Design Basis - Project CG-811.* HW-58258, General Electric Company, Richland, Washington.

Ciccarelli, R.A. 1959b. *Shielding - RMC Button Line, Project CG-734.* HW-59836, General Electric Company, Richland, Washington.

Clarke, W.J., E.B. Howard, and P.L. Hackett. 1968. "Myeloproliferative Disorders of Animals and Man." *Proceedings of the Atomic Energy Commission Symposium Series.* Oak Ridge, Tennessee.

Clemens, W.H. 1988. *Manufacturing Process Specifications.* WHC-CM-5-20, Westinghouse Hanford Company, Richland, Washington.

Clukey, H.V. 1957. *The Hanford Atomic Project and Columbia River Pollution.* HW-54243-Rev, General Electric Company, Richland, Washington.

Cochrane, R.B. 1977. *Functional Design Criteria, Z Plant Fire Protection, Project B-184.* RHO-CD-92, Rockwell Hanford Operations, Richland, Washington.

Cockeram, D.J. 1977. *Rockwell Hanford Operations Monthly Report - July 1977.* RHO-LD-10-B, Rockwell Hanford Operations, Richland, Washington.

Cockeram, D.J. 1978. *Rockwell Hanford Operations Monthly Report - January 1978.* RHO-LD-10-B Jan 78, Rockwell Hanford Operations, Richland, Washington.

Collins, P.E. 1952. *Improvement Program - 234-5 RMA Line.* HW-25340, General Electric Company, Richland, Washington.

Compton, A.H. 1956. *Atomic Quest, A Personal Narrative.* Oxford University Press, Oxford.

Coney, W.B. n.d. *Preservation of Historic Concrete: Problems and General Approaches.* Preservation Briefs #15, U.S. Department of the Interior, National Park Service Rocky Mountain Regional Office, Denver, Colorado.

Congress - U.S. Congress. 1991. *Complex Cleanup. The Environmental Legacy of Nuclear Weapons Production.* OTA-0-O-484, U.S. Government Printing Office, Washington, D.C.

Cooper, V.R. 1951. *Progress Report for September, 1951.* HW-22205, General Electric Company, Richland, Washington.

Corbell, D.W. 1975. "Panel Replacement Design." Memo to D.A. Armstrong, June 4, 1975, Atlantic Richfield Hanford Company, Richland, Washington.

Corbell, D.W. 1978. *Project Quality Requirements Plan for Project B-169, Criticality Alarm Systems.* RHO-CD-36-B-169, Rockwell Hanford Operations, Richland, Washington.

Corbell, D.W. 1979. *Project Quality Requirements Plan for Project B-195, Materials Management Information System, 234-5 Z Building, 200 West Area.* RHO-CD-36-B-195, Rockwell Hanford Operations, Richland, Washington.

Corn, J.K. 1983. "Historical Review of Industrial Hygiene." In *Industrial Hygiene—The Future,* J.K. Corn (ed.), Annals of the American Conference of Governmental Industrial Hygienists, Vol. 5, Cincinnati, Ohio.

Cote, S.L. 1994. *100-N Area Technical Baseline Report.* WHC-SD-EN-TI-251, Westinghouse Hanford Company, Richland, Washington.

Cowley, W.L. 1978. *Functional Design Criteria, Z-Plant Ventilation and Services Upgrade, Project B-178.* RHO-CD-622 1, Rockwell Hanford Operations, Richland, Washington.

Crease, R.P. 1999. *Making Physics, A Biography of Brookhaven National Laboratory, 1946-1972.* University of Chicago Press, Chicago, Illinois.

CREHST - Columbia River Exhibition of History, Science and Technology. 1998. Oral History Tapes: Camp Hanford Interview: Frank Krul and Bill Battershell. May 16, 1998, Richland, Washington.

Crigler, J.W. 2001. Telephone call to John Crigler (External Affairs, Bechtel Hanford, Inc.) from Sandra Cannon (Pacific Northwest National Laboratory), December 4-5, 2001. Copy of notes on file at the Library of the U.S. Department of Energy's Hanford Cultural Resources Laboratory, Richland, Washington.

Crocker, R.L. 1984. *RMA History I - Startup Criteria - SAR.* SD-CP-TI-036, Rockwell Hanford Operations, Richland, Washington.

Cronk, L. 1998. Interview of Les Cronk (Health and Safety professional at the Hanford Site, 1954-1997) by Brian Freer (CH2M HILL Hanford, Inc.), December 11, 1998. Copy of notes on file at the Library of the U.S. Department of Energy's Hanford Cultural Resources Laboratory, Richland, Washington.

Cronkite, E.P., and S.V. Musolino. 1996. "The Linear No-Threshold Model: Is it Still Valid for the Prediction of Dose-Effects and Risks from Low Level Radiation Exposure?" *Health Physics,* 70(6):775-776.

Cross, F.T. 1992. "Indoor Radon and Lung Cancer, Reality or Myth?" *29th Hanford Symposium on Health and the Environment,* Battelle Press, Columbus, Ohio.

Curtis, G. 1999. "Statistical Data Request." Electronic mail message from Greg Curtis (B&W Protec, Inc.) to Joy Keating (Pacific Northwest National Laboratory), March 1, 1999. Copy on file at the Library of the U.S. Department of Energy's Hanford Cultural Resources Laboratory, Richland, Washington.

D'Antonio, M. 1993. *Atomic Harvest: Hanford and the Lethal Toll of America's Nuclear Arsenal.* Crown Publishers, Inc., New York.

Dahlen, P.A. 1946. *Review of Blistered Slug Investigation.* HW-7-3737, E.I. Du Pont de Nemours and Company, Richland, Washington.

Davie, K. 1998. Telephone call from Joy Keating (Pacific Northwest National Laboratory) to Kenneth Davie (DynCorp Engineer, Site Utilities), July 14, 1998, Richland, Washington. Copy of notes on file at the Library of the U.S. Department of Energy's Hanford Cultural Resources Laboratory, Richland, Washington.

DeFord, D.H. 1991. *Z Aggregate Area Management Study Technical Baseline Report.* WHC-IP-0776, Westinghouse Hanford Company, Richland, Washington.

DeFord, D.H. 1992. *Semi-Works Aggregate Area Management Study, Technical Baseline Report.* WHC-SD-EN-019, Westinghouse Hanford Company, Richland, Washington.

DeFord, D.H. 1993. *100-F Reactor Site Technical Baseline Report Including Operable Units 100-FR-1 and 100-FR-2.* WHC-SD-EN-TI-169, Westinghouse Hanford Company, Richland, Washington.

DeFord, D.H. 1994. *300-FF-2 Operable Unit Technical Baseline Report.* BHI-00012, Bechtel Hanford, Inc., Richland, Washington.

DeFord, D.H. 1996. *100-IU-6 Operable Unit Technical Baseline Report.* BHI-00146, Bechtel Hanford, Inc., Richland, Washington.

DeFord, D.H., and R.W. Carpenter. 1995. *S Plant Aggregate Area Management Study Technical Baseline Report.* BHI-00176, Bechtel Hanford, Inc., Richland, Washington.

DeFord, D.H., and M.W. Einan. 1995. *100-H Technical Baseline Report.* BHI-00127, Bechtel Hanford, Inc., Richland, Washington.

DeJong, R. 1995. Interview of Rudy DeJong (Construction Foreman at B Reactor with Du Pont 1943-1946) by Thomas E. Marceau (Bechtel Hanford, Inc.), December 14, 1995. Tape on file at the Columbia River Exhibitions of History, Science, and Technology, Richland, Washington.

Dellwo, R. 1999. Interview of Robert (Bob) Dellwo (former agent with the Federal Bureau of Investigation from 1942-1948) by Joy Keating (Pacific Northwest National Laboratory), January 19, 1999. Copy of notes on file at the Library of the U.S. Department of Energy's Hanford Cultural Resources Laboratory, Richland, Washington.

DeLong, C.W., R.C. Thompson, and H.A. Kornberg. 1954. "Percutaneous Absorption of Tritium Oxide." *American Journal of Roentgenology, Radium Therapy and Nuclear Medicine,* Vol. 71, No 6, pp. 1038-1045.

Del Tredici, R. 1987. *At Work in the Fields of the Bomb.* Harper and Row, New York.

DeNeal, D.L. 1965. *Historical Events, Reactors and Fuels Fabrication.* RL-REA-2247, U.S. Atomic Energy Commission, Richland, Washington.

DeNeal, D.L. 1970. *Historical Events - Single Pass Reactors and Fuels Fabrication.* DUN-6888, Douglas United Nuclear Inc., Richland, Washington.

Deutschmann, P.J. 1952. *Federal City: A Study of the Administration of Richland, Washington, Atomic Energy Commission Community.* Masters Thesis, University of Oregon. The Richland Public Library has a copy on microfilm.

Dickeman, R.L. 1962. Letter from R.L. Dickeman (General Electric Company) to J.T. Stringer March 22, 1962. Copy on file in the Multi-Prod-Tr Prod. Co. Product File in the Federal Records Center, Seattle, Washington.

DOE - U.S. Department of Energy. 1980. *Final Environmental Impact Statement: Management of Commercially Generated Radioactive Waste.* DOE/EIS-0046F, Washington, D.C.

DOE - U.S. Department of Energy. 1981a. *Study of Anticipated Impact on DOE Programs from Proposed Reductions to the External Occupational Radiation Exposure Limit.* DOE/EV-0045 (Rev. 2-81), Washington, D.C.

DOE - U.S. Department of Energy. 1981b. *Monitored Retrievable Storage Concept: A Review of Its Status and Analysis of Its Impact on the Waste Management System.* DOE/NE-0019, Washington, D.C.

DOE - U.S. Department of Energy. 1983. *Monitored Retrievable Storage Proposal Research and Development Report.* DOE/S-0021, Washington, D.C.

DOE - U.S. Department of Energy. 1985. Hanford Patrol Flight Operations. Video Tape No. 2185/M, September 28, 1985, Lockheed Martin Services, Inc., Video Productions. Available in the Federal Building, Room 127, Richland, Washington.

DOE - U.S. Department of Energy. 1989. *Decommissioning of Eight Surplus Production Reactors at the Hanford Site, Richland, Washington.* DOE EIS-0119D, Draft Environmental Impact Statement, Washington, D.C.

DOE - U.S. Department of Energy. 1990. *Hanford Site Infrastructure Plan.* DOE/RL-89-31, Richland, Washington.

DOE - U.S. Department of Energy. 1991a. *Hanford Process Review.* DOE/AD-0015, Washington, D.C.

DOE - U.S. Department of Energy. 1991b. *Hanford 300 Area Development Plan.* DOE/RL-91-09, Richland, Washington.

DOE - U.S. Department of Energy. 1992a. *Environmental Restoration and Waste Management (EM) Program, an Introduction.* DOE/EM-0013P, Washington, D.C.

DOE - U.S. Department of Energy. 1992b. *Remedial Investigation/Feasibility Study Work Plan for the 100-FR-1 Operable Unit, Hanford Site, Richland, Washington.* DOE/RL-90-33, Richland, Washington.

DOE - U.S. Department of Energy. 1992c. *S Plant Aggregate Area Management Study Report.* DOE/RL-91-60, Richland, Washington.

DOE - U.S. Department of Energy. 1992d. *T Plant Source Aggregate Area Management Study Report.* DOE/RL-91-61, Richland, Washington.

DOE - U.S. Department of Energy. 1993. *Risk Management Study for the Retired Hanford Site Facilities.* WHC-EP-0619, Vol. 3, Richland, Washington.

DOE - U.S. Department of Energy. 1994a. *Characterization of Decontamination and Decommissioning Wastes Expected from the Major Processing Facilities in the 200 Areas.* WHC-EP-0787, Richland, Washington.

DOE - U.S. Department of Energy. 1994b. *Hanford Site Water Systems Plan.* RL-WSP-94-01, Richland, Washington.

DOE - U.S. Department of Energy. 1994c. "Security Transition: Changing the Face of Hanford." Brochure: *Hanford Vision 2000 Team Security,* RL-D94-034, Richland, Washington. Copy available in the U.S. Department of Energy Richland Operations Office Public Reading Room.

DOE - U.S. Department of Energy. 1994d. "Lessons from Hanford: Managing Change in the DOE Complex." Brochure: *Hanford Vision 2000 Team Security*, RL-D94-047, Richland, Washington. Copy available in the U.S. Department of Energy Richland Operations Office Public Reading Room.

DOE - U.S. Department of Energy. 1994e. "Hanford Security Transition: Tools for Change." Brochure: *Hanford Vision 2000 Team Security*, RL-D95-003, Richland, Washington. Copy available in the U.S. Department of Energy Richland Operations Office Public Reading Room.

DOE - U.S. Department of Energy. 1995a. *Human Radiation Experiments: The Department of Energy Roadmap to the Story and the Records.* DOE EH-0445, Washington, D.C.

DOE - U.S. Department of Energy. 1995b. "First Phase of Hanford Environmental Restoration Facility (ERDF) Nears Completion Ahead of Schedule and Under Budget." Hanford News Release, August 31, 1995, Richland, Washington.

DOE - U.S. Department of Energy. 1995c. *Human Radiation Experiments Associated with the U.S. Department of Energy and Its Predecessors.* DOE EH-0491, Washington, D.C.

DOE - U.S. Deaprtment of Energy. 1995d. *Ordnance and Explosive Waste Records Search Report.* DOE/RL-94-07, Rev. 0, Richland, Washington.

DOE - U.S. Department of Energy. 1995e. *Closing the Circle on the Splitting of the Atom, The Environmental Legacy of Nuclear Weapons Production in the United States and What the Department of Energy is Doing About It*, DOE/EM-0266, Washington, D.C. [Online] Available URL: http://www.lastinglegacy.net/thestory/text/close/closetoc.htm

DOE - U.S. Department of Energy. 1996a. *Programmatic Agreement Among the U.S. Department of Energy-Richland Operations Office, the Advisory Council on Historic Preservation, and the Washington State Historic Preservation Office for the Maintenance, Deactivation, Alteration, and Demolition of the Built Environment on the Hanford Site, Washington.* DOE/RL-96-77, Rev. 0, Richland, Washington.

DOE - U.S. Department of Energy. 1996b. *Plutonium: The First 50 Years, United States Plutonium Production, Acquisition, and Utilization from 1944 through 1994.* DOE/DP-0137, Washington, D.C. [Online] Available URL: http://www.osti.gov/html/osti/opennet/document/pu50yrs/pu50y.html

DOE - U.S. Department of Energy. 1996c. *300 Area Waste Acid Treatment System Closure Plan.* DOE/RL-99-11, Rev. 1, Richland, Washington.

DOE - U.S. Department of Energy. 1996d. *Draft Hanford Remedial Action Environmental Impact Statement and Comprehensive Land Use Plan.* DOE EIS-0222D, Richland, Washington.

DOE - U.S. Department of Energy. 1996e. *Hanford Site Solid Waste Landfill Interim Closure Plan.* DOE/RL-90-38, Richland, Washington.

DOE - U.S. Department of Energy. 1996g. *Plutonium Finishing Plant Stabilization: Final Environmental Impact Statement.* DOE/EIS-0244-F, Richland, Washington.

DOE - U.S. Department of Energy. 1997a. *N Reactor Comprehensive Treatment Report, Hanford Site, Washington.* DOE/RL-96-91, Draft A, Richland, Washington.

DOE - U.S. Department of Energy. 1997b. *National Register of Historic Places Multiple Property Documentation Form – Historic, Archaeological and Traditional Cultural Properties of the Hanford Site, Washington.* DOE/RL-97-02, Richland, Washington.

DOE - U.S. Department of Energy. 1997c. *Hanford Site Environmental Management Specifications: Section 4.2.2 Waste Management Project.* DOE RL-97-55, Rev. 1, Richland, Washington.

DOE - U.S. Department of Energy. 1997d. Visitor Orientation Pamphlet. RL-W94-054, Richland, Washington. Copy available in the U.S. Department of Energy Richland Operations Office Public Reading Room.

DOE - U.S. Department of Energy. 1997e. *Linking Legacies: Connecting the Cold War Nuclear Weapons Production Processes to Their Environmental Consequences.* DOE/EM-0319, Washington, D.C.

DOE - U.S. Department of Energy. 1998a. *Screening Assessment and Requirements for a Comprehensive Assessment, Columbia River Comprehensive Impact Assessment.* DOE/RL-96-16, Rev. 1, Richland, Washington.

DOE - U.S. Department of Energy. 1998b. *200 Areas Remedial Investigation/Feasibility Study Implementation Plan- Environmental Restoration Program.* DOE/RL-98-28, Draft B, Richland, Washington.

DOE - U.S. Department of Energy. 1998c. *Hanford Cultural Resources Management Plan: A Preservation and Protection Strategy.* DOE/RL-98-10, Richland, Washington.

DOE - U.S. Department of Energy. 1998d. *Hanford Fire Department Home Page.* Richland, Washington. [Online] Available URL: http://www.hanford.gov/fire/index.htm

DOE - U.S. Department of Energy. 1998e. "Radioactive Waste Tank Expert Panel to Meet Late June, Hanford Site Panel Will Review and Analyze Draft Contamination Report." Press Release, RL-OEA 98-059, June 8, 1998, Richland, Washington.

DOE - U.S. Department of Energy. 1998f. *Environmental Assessment Relating to the Decommissioning of the Strontium Semi-Works Facility.* DOE EA 0259, Richland, Washington.

DOE - U.S. Department of Energy. 2000. *U.S. Department of Energy Environment, Safety, and Health Public Meeting on February 3, 2000 at 6:30 p.m.*, Richland Washington. [Online] Available URL: http://tis.eh.doe.gov/advocacy/meetings/hanford/000203hanford.pdf

DOE - U.S. Department of Energy. 2000a. *United States Nuclear Tests July 1945 through September 1992.* DOE/NV-209-REV 15, Las Vegas, Nevada. [Online] Available URL: http://www.nv.doe.gov/news&pubs/publication/historyreports/pdfs/DOENV209 REV 15.pdf

DOE - U.S. Department of Energy. 2001. *Radioactive Waste Management Manual.* DOE M 435.1-1, approved 7-09-99, change 1: 6-19-10, Washington, D.C.

Donnell, D. 1948. *Dupus Boomer Cartoons.* Collection of cartoons published from 1947-1948 in the Richland Villager, Kennewick, Washington.

Dorian, J.J., and V.R. Richards. 1978. *Radiological Characterization of the Retired 100 Areas.* UNI-946, United Nuclear Industries, Richland, Washington.

DOW - U.S. Department of War. n.d. *Background Information on the Hanford Engineers Works.* Washington, D.C.

Dowis, W.J. 1986. *Work at Hanford 1943-1983.* Manuscript at U.S. Department of Energy Public Reading Room, Richland, Washington.

Draney, W.G., and D.J. Hill. 1976. *OSHA Type Inspection of Z Plant.* ARH-CD-637, Atlantic Richfield Hanford Company, Richland, Washington.

Drucker, H., and R.E. Wildung. 1975. "Biological Implications of Metals in the Environment." *Proceedings of the 15th Annual Hanford Life Sciences Symposium.* Energy Research and Development Agency and Pacific Northwest Laboratory, Oak Ridge, Tennessee.

Du Pont - E.I. Du Pont de Nemours and Company. n.d. *Hanford Engineer Works Patrol History 1944-45.* Copy on file in the Library of the U.S. Department of Energy's Hanford Cultural Resources Laboratory, Richland, Washington.

Du Pont - E.I. Du Pont de Nemours and Company. 1944-1946. Memoranda for the File, Protection Department. HAN-73214, Book 9, Richland, Washington. Copy available in the U.S. Department of Energy Richland Operations Office Public Reading Room.

Du Pont - E.I. Du Pont de Nemours and Company. 1944. *Hanford Operating Procedure for Coating and Canning Heavy Metal Slugs - Memorandum.* HAN-73214, Book 10, Part I, Richland, Washington.

Du Pont - E. I. Du Pont de Nemours and Company. 1945a. *Construction of Hanford Engineer Works: History of the Project.* HAN-10970, Vols. 1-4, Wilmington, Delaware.

Du Pont - E.I. Du Pont de Nemours and Company. 1945b. *Hanford Engineering Works, History of the Project, Vol. II.* HAN-10970, Richland, Washington.

Du Pont - E.I. Du Pont de Nemours and Company. 1945c. *Design and Procurement History of Hanford Engineer Works and Clinton Semi-Works.* IN-6263, Volume 2, Wilmington, Delaware. Copy available in the U.S. Department of Energy Richland Operations Office Public Reading Room.

Du Pont - E.I. Du Pont de Nemours and Company. 1945d. *HEW Technical Manual, Section A - Metal Preparation.* HW-10475A, Wilmington, Delaware.

Du Pont - E.I. Du Pont de Nemours and Company. 1945e. *Design and Construction History of Hanford Engineer Works and Clinton Semi-Works.* IN-06263, Vols. 1 and 2, Wilmington, Delaware.

Du Pont - E.I. Du Pont de Nemours and Company. 1945f. *Completion Report Manhattan District, Hanford Engineer Works, TNX Plant.* War Construction Division, Wilmington, Delaware.

Du Pont - E.I. Du Pont de Nemours and Company. 1945g. *History of Operations, 1 January 1944 to 20 March 1945.* OUT-1462, Richland, Washington.

Du Pont - E.I. Du Pont de Nemours and Company. 1946a. *Operation of Hanford Engineer Works. Introduction.* HAN-73214, Richland, Washington.

Du Pont - E.I. Du Pont de Nemours and Company. 1946b. *Operation of Hanford Engineer Works. Book 4.* HAN-73214, Richland, Washington.

Du Pont - E.I. Du Pont de Nemours and Company. 1946c. *Operation of Hanford Engineer Works. Book 7.* HAN-73214-DEL, Richland, Washington.

Du Pont - E.I. Du Pont de Nemours and Company. 1946d. *Operation of Hanford Engineer Work. Book 11.* HAN-73214-DEL, Richland, Washington.

Du Pont - E.I. Du Pont de Nemours and Company. 1946e. *Operation of Hanford Engineer Works. Book 12.* HAN-73214, Richland, Washington.

Du Pont - E.I. Du Pont de Nemours and Company. 1946f. *Operation of Hanford Engineer Works. Book 14.* HAN-73214, Richland, Washington.

Du Pont - E.I. Du Pont de Nemours and Company. 1946g. *Operation of Hanford Engineer Works. Book 15.* HAN-73214, Richland, Washington.

Dyke, R.W. 1989. *Mr. Atomic Energy: Congressman Chet Holifield and Atomic Energy Affairs, 1945-1974.* Greenwood Press, New York.

Ecology - Washington State Department of Ecology, U.S. Environmental Protection Agency, and U.S. Department of Energy. 1994. *Hanford Federal Facility Agreement and Consent Order.* Document No. 89-10, Rev. 3, The Tri-Party Agreement, Ecology, Olympia, Washington.

Eichner, F.N. 1969. *Whole-Body Counter Laboratory Mobile Unit B: Description and Operation.* BNWL-1154, Pacific Northwest Laboratory, Richland, Washington.

Eisenbud, M. 1968. "Radionuclides in the Environment." In *Diagnosis and Treatment of Deposited Radionuclides,* H.A. Kornberg and W.D. Norwood (eds.), Excerpta Medica Foundation, Monographs in Nuclear Medicine and Biology No. 2, Printed in The Netherlands by Reidel, Dordrecht.

Elderkin, C.E., and J.J Fuquay. 1965. *Recent Hanford Diffusion Results Significant for Estimating Pollution Potential.* BNWL-SA-429, Pacific Northwest Laboratory, Richland, Washington.

Elgert, O.J. 1968. *Appraisal of Plutonium Finishing Plant Operations.* CPD-179, U.S. Atomic Energy Commission, Richland, Washington.

EMO - Environmental Management Operations. 1992. *Schedule Optimization Study, Hanford RI/FS Program, Volume 2: Final Report.* EMO-1080, Vol. 2, Richland, Washington.

Endow, N., and L.F. Kendall. 1951. *Analytical Procedures for the Plutonium Metal Fabrication Process.* HW-21457, General Electric Company, Richland, Washington.

ERDA - U.S. Energy Research and Development Administration. 1976a. *Investigation of the Chemical Explosion of an Ion Exchange Resin Column and Resulting Americium Contamination of Personnel in the 242-Z Building, August 30, 1976.* BNWI-1007 DEL, Richland, Washington.

ERDA - U.S. Energy Research and Development Administration. 1976b. "Safety of NonReactor Nuclear Facilities," *ERDA Manual, Chapter 0531.* Richland, Washington.

Facer, J.F., and B. Weidenbaum. 1951. *The Recovery of Plutonium from Metal Wastes.* HW-22136, General Electric Company, Richland, Washington.

Farley, T.B. 1946. "G.E. Security Bulletin #3 – Responsibility." December 11, 1946, General Electric Company, Richland, Washington. Copy available in the U.S. Department of Energy Richland Operations Office Public Reading Room.

Farley, T.B. 1948. "G.E. Security Bulletin #21 – It Could Happen Here." March 22, 1948, General Electric Company, Richland, Washington. Copy available in the U.S. Department of Energy Richland Operations Office Public Reading Room.

Fecht, J.B., J.D. Anderson, R.D. Carter, D.W. Jeppson, J.C. Willi, and R.H. Wilson. 1973. *Hazards Review - Plutonium Effluent Cleanup - Z Plant - Project HCP-669 - Phase II.* ARH-2902, Atlantic Richfield Hanford Company, Richland, Washington.

Fehner, T.R., and J.M. Holl. 1994. *Department of Energy 1977-1994: A Summary History.* DOE/HR-0098, U.S. Department of Energy, Washington, D.C.

Felt, R.E. 1974. "Incinerator Operations Analysis." Memo to G.A. Nicholson, November 8, 1974, Atlantic Richfield Hanford Company, Richland, Washington.

Fermi, L. 1957. *Atoms for the World: United States Participation in the Conference on the Peaceful Uses of Atomic Energy.* University of Chicago Press, Chicago, Illinois.

Fermi, R., and E. Samra. 1995. *Picturing the Bomb: Photographs from the Secret World of the Manhattan Project.* Harry N. Abrams, Inc., New York.

Fillmore, J.W. 1965. "Z-9 Crib." Memo to S.G. Smolen, May 25, 1965, General Electric Company, Richland, Washington.

Fillmore, J.W., E.M. Johnston, J.F. Kane, and W.L. Lawton, Jr. 1964. *Hazards Evaluation: Electrical and Hydraulic Mezzanine - RMA Line - 234-5 Building.* HW-81360, General Electric Company, Richland, Washington.

Finch, L.W. 1957. *Planning Meeting Minutes: Replacement of Space (CWS) Filters - Z Plant.* General Electric Company, Richland, Washington.

Findlay, J.M. 1996. "Atomic Frontier Days: Richland, Washington and the Modern American West." *Journal of the West* 34 (3):32-41.

Findlay, J.M., and B. Hevly. 1995. *Nuclear Technologies and Nuclear Communities: A History of Hanford and the Tri-Cities, 1943-1993.* The Hanford History Project, Center for the Study of the Pacific Northwest, University of Washington, Seattle, Washington.

Fitzner, R.E., and R.H. Gray. 1991. "The Status, Distribution and Ecology of Wildlife on the U.S. DOE Hanford Site: A Historical Overview of Research Activities." *Environmental Monitoring and Assessment* 18:173-202.

Fleischer, C.C. 1974. *The Tri-City Nuclear Industrial Council and the Economic Diversification of the Tri-Cities, Washington, 1963-1974.* Masters Thesis, Washington State University. A copy is in the Washington State University Library collection at the Tri-Cities Campus.

Fluckiger J.D. 1998. Telephone call from Joy Keating (Pacific Northwest National Laboratory) to J.D. Fluckiger (Systems Engineer, Pacific Northwest National Laboratory), November 23, 1998. Copy of notes on file at the Library of the U.S. Department of Energy's Hanford Cultural Resources Laboratory, Richland, Washington.

FMRC - Factory Mutual Research Corporation. 1969. *Glovebox Window Materials: A Glovebox Fire Safety Application.* TID- 24896, Norwood, Massachusetts.

Foote, K.E. 1997. *Shadowed Ground: America's Landscapes of Violence and Tragedy.* University of Texas Press, Austin.

Ford, D. 1982. *The Cult of the Atom: The Secret Papers of the Atomic Energy Commission.* Simon and Schuster, New York.

Foster, R.F. 1957. *Recommended Limit on Addition of Dichromate to the Columbia River.* HW-49713, General Electric Company, Richland, Washington.

Foster, R.F. 1962. *Evaluation of Radiological Conditions in the Vicinity of Hanford for 1961.* HW-71999, General Electric Company, Richland, Washington.

Franklin - Franklin County Historical Society. 1988. "Transfer of North Richland Trailer Camp." *Franklin Flyer.* Pasco, Washington.

FRC - Federal Radiation Council. 1960. "Background Material for the Development of Radiation Protection Standards - Staff Report." Report No. 1, Washington, D.C.

Frigerio, N. A. 1968. *Your Body and Radiation.* Understanding the Atom Series, U.S. Atomic Energy Commission, Oak Ridge, Tennessee.

Fritz, J.R. 1955. *Design and Development Study, Task II Replacement, RM Line, 234-5 Building, D.O. 100991.* HW-37152, General Electric Company, Richland, Washington.

Fritz, J.R. 1957. *Vacuum System Requirements for RMA and RMC Lines.* General Electric Company, Richland, Washington.

Fritz, J.R. 1958. *Specifications for RMC Button Line Installation in 234-5 Building.* HWS-6250, General Electric Company, Richland, Washington.

Fritz, J.R. 1960. *RMC Button Line.* ATP-2101, General Electric Company, Richland, Washington.

Fritz, J.R., C.F. Yuenger, and J.L. Weeks. 1956. *Improvements to Task I and Task II, 234-5 Building.* HW-44105, General Electric Company, Richland, Washington.

Gaines, M. 1970. *Atomic Energy.* Grosset and Dunlap, New York.

Gallant, T.R. 1977a. *Functional Design Criteria, Safeguards and Security Upgrading.* RHO-CD-96, Rockwell Hanford Operations, Richland, Washington.

Gallant, T.R. 1977b. *Functional Design Criteria, Safeguards and Security Upgrading - Z Plant Calorimeter.* RHO-CD-96 Sup 1, Rockwell Hanford Operations, Richland, Washington.

Gallant, T.R. 1977c. *Conceptual Design Report, Access Road - 231-Z Safeguards and Security Upgrading.* RHO-CD-130, Rockwell Hanford Operations, Richland, Washington.

Gamertsfelder, C.C. 1963. *Plans and Hazards Analysis for the First Hanford* I^{131} *Field Release Test.* HW-78312, General Electric Company, Richland, Washington.

GAO - General Accounting Office. 1991. *Operation of Monitored Retrievable Storage Facility is Unlikely by 1998.* GAO/RCED-91-194, Washington, D.C.

Garfield, J.S., and H.J. Clausen. 1978. *Conceptual Design Report, Z Plant Calorimeter, Project B-158.* RHO-CD-233, Rockwell Hanford Operations, Richland, Washington.

GE - General Electric Company. n.d (a). *Gėneral Electric Company Hanford Atomic Products Operation Data Book.* HW-40154, Richland, Washington.

GE - General Electric Company. n.d.(b). *Hanford Engineer Works: Security Handbook.* Copy available in the U.S. Department of Energy Richland Operations Office Public Reading Room.

GE - General Electric Company [sic - Du Pont]. 1944. *Hanford Technical Manual, Section C.* HW-10475-C DEL, Richland, Washington.

GE - General Electric Company. 1947a. "Report Conditions Promptly." *Hanford Engineer Works News,* April 25, 1947.

GE - General Electric Company. 1947b. *Hanford Engineer Works Monthly Report, July 1947.* HEW-7283 DEL, Richland, Washington.

GE - General Electric Company. 1947c. *Hanford Engineer Works Monthly Report, August 1947.* HEW-7504 DEL, Richland, Washington.

GE - General Electric Company. 1947d. *Hanford Works Monthly Report, November 1947.* HW-8267 DEL, Richland, Washington.

GE - General Electric Company. 1949a. *Hanford Works Monthly Report, May 1949.* HW-13561 DEL, Richland, Washington.

GE - General Electric Company. 1949b. *Hanford Works Monthly Report, July 1949.* HW-14043 DEL, Richland, Washington.

GE - General Electric Company. 1949c. *Hanford Works Monthly Report, August 1949.* HW-14338 DEL, Richland, Washington.

GE - General Electric Company. 1949d. *Hanford Works Monthly Report, September 1949.* HW-14596 DEL, Richland, Washington.

GE - General Electric Company. 1949e. *Hanford Works Monthly Report, October 1949.* HW-14916 DEL, Richland, Washington.

GE - General Electric Company. 1949f. *Hanford Works Monthly Report, November 1949.* HW-15267 DEL, Richland, Washington.

GE - General Electric Company. 1950a. *Hanford Codes and Jargon.* HW-18223, Richland, Washington.

GE - General Electric Company. 1950b. *Hanford Works Transportation Division Operation.* HW-22059, Richland, Washington.

GE - General Electric Company. 1950c. *Hanford Works Monthly Report, December 1949.* HW-15550 DEL, Richland, Washington.

GE - General Electric Company. 1950d. *Hanford Works Monthly Report, October 1950.* HW-19325 DEL, Richland, Washington.

GE - General Electric Company. 1951a. *Design and Construction History – Project C-185 Railroad Connection South of Richland: Construction, Relocation, and Rehabilitation.* 712 Building, Box-035732, Richland, Washington.

GE - General Electric Company. 1951b. *Biology Research Annual Report.* HW-25021, Richland, Washington.

GE - General Electric Company. 1951c. *Health Instruments Divisions Research and Development Programs for the Advisory Committee on Biology and Medicine.* HW-19816-DEL, Richland, Washington.

GE - General Electric Company. 1951d. *Hanford Works Monthly Report, December 1950.* HW-19842 DEL, Richland, Washington.

GE - General Electric Company. 1951e. *Hanford Works Monthly Report, February 1951.* HW-20438 DEL, Richland, Washington.

GE - General Electric Company. 1951f. *Hanford Works Monthly Report, March 1951.* HW-20671 DEL, Richland, Washington.

GE - General Electric Company. 1951g. *Hanford Works Monthly Report, May 1951.* HW-21260 DEL, Richland, Washington.

GE - General Electric Company. 1951h. *Hanford Works Monthly Report, August 1951.* HW-22075 DEL, Richland, Washington.

GE - General Electric Company. 1952a. *Design and Construction History - Project C-165-A, Pile Area "H."* Hanford Works, Richland, Washington.

GE - General Electric Company. 1952b. *Radiological Sciences Department Annual Report 1952.* HW-27814, Richland, Washington.

GE - General Electric Company. 1952c. *Hanford Works Monthly Report, January 1952.* HW-23437 DEL, Richland, Washington.

GE - General Electric Company. 1952d. *Hanford Works Monthly Report for February 1952.* HW-23638 DEL, Richland, Washington.

GE - General Electric Company. 1952e. *Hanford Works Monthly Report, March 1952.* HW-23982 DEL, Richland, Washington.

GE - General Electric Company. 1952f. *Hanford Works Monthly Report for April 1952.* HW-24337 DEL, Richland, Washington.

GE - General Electric Company. 1952g. *Journal, Project C-198 Phase I.* CEV-6041, Richland, Washington.

GE - General Electric Company. 1952h. *Hanford Works Monthly Report for May 1952.* HW-24605 DEL, Richland, Washington.

GE - General Electric Company. 1952i. *Hanford Works Monthly Report for June 1952.* HW-24928 DEL, Richland, Washington.

GE - General Electric Company. 1952j. *Journal, Project C-198.* CEV-8015, Richland, Washington.

GE - General Electric Company. 1952k. *Hanford Works Monthly Report for August 1952.* HW-25533 DEL, Richland, Washington.

GE - General Electric Company. 1952l. *Hanford Works Monthly Report for September 1952.* HW-25781 DEL, Richland, Washington.

GE - General Electric Company. 1952m. *Hanford Works Monthly Report for October 1952.* HW-26047 DEL, Richland, Washington.

GE - General Electric Company. 1952n. *Hanford Works Monthly Report for November 1952.* HW-26376 DEL, Richland, Washington.

GE - General Electric Company. 1953a. *Manufacturing Department Radiation Incident Investigation No. 65.* HW-28021, Richland, Washington.

GE - General Electric Company. 1953b. *1952 at Hanford Works.* HW-26705-DEL, Richland, Washington.

GE - General Electric Company. 1953c. *Biology Research Annual Report 1952.* HW-28636, Richland, Washington.

GE - General Electric Company. 1953d. *Hanford Works Monthly Report for December 1952.* HW-26720 DEL, Richland, Washington.

GE - General Electric Company. 1953e. *Hanford Works Monthly Report for January 1953.* HW-26946 DEL, Richland, Washington.

GE - General Electric Company. 1953f. *Hanford Works Monthly Report for February 1953.* HW-27288 DEL, Richland, Washington.

GE - General Electric Company. 1953g. *Hanford Works Monthly Report for March 1953.* HW-27624 DEL, Richland, Washington.

GE - General Electric Company. 1953h. *Hanford Works Monthly Report for April 1953.* HW-27932 DEL, Richland, Washington.

GE - General Electric Company. 1953i. *Hanford Works Monthly Report for May 1953.* HW-28267 DEL, Richland, Washington.

GE - General Electric Company. 1953j. *Design and Construction History, Project C-198, 234-5 Building Program, Phases II and III.* HW-24800-36, Richland, Washington.

GE - General Electric Company. 1953k. *Hanford Works Monthly Report for August 1953.* HW-29229 DEL, Richland, Washington.

GE - General Electric Company. 1953l. *Hanford Works Monthly Report for September 1953.* HW-29513 DEL, Richland, Washington.

GE - General Electric Company. 1953m. *Hanford Works Monthly Report for October 1953.* HW-29794 DEL, Richland, Washington.

GE - General Electric Company. 1954a. *Biology Research Annual Report 1953.* HW-30437, Richland, Washington.

GE - General Electric Company. 1954b. *Manual of Standard Practices.* HW-45674, Richland, Washington.

GE - General Electric Company. 1954c. *Radiological Sciences Department Annual Report 1953.* HW-30464, Richland, Washington.

GE - General Electric Company. 1954d. *Radiological Sciences Annual Report Research and Development Fiscal Year 1954.* HW-33128, Richland, Washington.

GE - General Electric Company. 1954e. *Hanford Works Monthly Report for December 1953.* HW-30423 DEL, Richland, Washington.

GE - General Electric Company. 1954f. *Hanford Works Monthly Report for January 1954.* HW-30724 DEL, Richland, Washington.

GE - General Electric Company. 1954g. *1953 at Hanford.* HW-30000 DEL, Richland, Washington.

GE - General Electric Company. 1954h. *Hanford Works Monthly Report for February 1954.* HW-31006 DEL, Richland, Washington.

GE - General Electric Company. 1954i. *Hanford Works Monthly Report for March 1954.* HW-31267 DEL, Richland, Washington.

GE - General Electric Company. 1954j. *Design and Construction History, Project C-413 CG-413 Expansion of 234-5 Building Capacity.* HW-24800-90, Richland, Washington.

GE - General Electric Company. 1954k. *Hanford Works Monthly Report for April 1954.* HW-31734 DEL, Richland, Washington.

GE - General Electric Company. 1954l. *Hanford Works Monthly Report for May 1954.* HW-31964 DEL, Richland, Washington.

GE - General Electric Company. 1954m. *Hanford Works Monthly Report for July 1954.* HW-32624 DEL, Richland, Washington.

GE - General Electric Company. 1954n. *Hanford Works Monthly Report for August 1954.* HW-32889 DEL, Richland, Washington.

GE - General Electric Company. 1954o. *Hanford Works Monthly Report for October 1954.* HW-33585 DEL, Richland, Washington.

GE - General Electric Company. 1954p. *Hanford Works Monthly Report for November 1954.* HW-33962 DEL, Richland, Washington.

GE - General Electric Company. 1954q. *Radiation Protection Standards.* HW-25457 Rev., Richland, Washington.

GE - General Electric Company. 1954r. *Hanford Works Monthly Report for March 1954.* HW-31267 DEL, Richland, Washington.

GE - General Electric Company. 1954s. *Hanford Works Monthly Report for June 1954.* HW-32317 DEL, Richland, Washington.

GE - General Electric Company. 1955a. *Biology Research Annual Report 1954.* HW-35917, Richland, Washington.

GE - General Electric Company. 1955b. *Regulations and General Instructions Governing Hanford Works Security Patrolmen.* Richland, Washington.

GE - General Electric Company. 1955c. *Radiation Monitoring Reactor Section Procedures.* HW-45674, Richland, Washington.

GE - General Electric Company. 1955d. *Hanford Works Monthly Report for January 1955.* HW-34631 DEL, Richland, Washington.

GE - General Electric Company. 1955e. *Hanford Works Monthly Report for March 1955.* HW-35891 DEL, Richland, Washington.

GE - General Electric Company. 1955f. *Hanford Works Monthly Report for April 1955.* HW-36440 DEL, Richland, Washington.

GE - General Electric Company. 1955g. *Hanford Works Monthly Report for June 1955.* HW-37658 DEL, Richland, Washington.

GE - General Electric Company. 1955h. *Hanford Works Monthly Report for July 1955.* HW-38375 DEL, Richland, Washington.

GE - General Electric Company. 1955i. *Hanford Works Monthly Report for October 1955.* HW-39751 DEL, Richland, Washington.

GE - General Electric Company. 1955j. *Hanford Works Monthly Report for November 1955.* HW-40182 DEL, Richland, Washington.

GE - General Electric Company. 1955k. *Hanford Works Monthly Report for December 1954.* HW-34147 DEL, Richland, Washington.

GE - General Electric Company. 1955l. *Hanford Works Monthly Report for May 1955.* HW-36928 DEL, Richland, Washington.

GE - General Electric Company. 1956a. *1955 at Hanford Works.* HW-39900, Richland, Washington.

GE - General Electric Company. 1956b. *Biology Research Annual Report 1955.* HW-41500, Richland, Washington.

GE - General Electric Company. 1956c. *Quarterly Progress Report Research and Development Activities, April-June, 1956.* HW-44041, Richland, Washington.

GE - General Electric Company. 1956d. *Quarterly Progress Report Research and Development Activities in the Field of Radiological Sciences, July-September, 1956.* HW-46333, Richland, Washington.

GE - General Electric Company. 1956e. *Monthly Report: Hanford Atomic Products Operation for January 1956.* HW-41205 DEL, Richland, Washington.

GE - General Electric Company. 1956f. *Monthly Report: Hanford Atomic Products Operation for February 1956.* HW-41702 DEL, Richland, Washington.

GE - General Electric Company. 1956g. *Monthly Report: Hanford Atomic Products Operation for March 1956.* HW-42219 DEL, Richland, Washington.

GE - General Electric Company. 1956h. *Monthly Report: Hanford Atomic Products Operation for April 1956.* HW-42626 DEL, Richland, Washington.

GE - General Electric Company. 1956i. *Monthly Report: Hanford Atomic Products Operation for May 1956.* HW-43137 DEL, Richland, Washington.

GE - General Electric Company. 1956j. *Monthly Report: Hanford Atomic Products Operation for June 1956.* HW-43938 DEL, Richland, Washington.

GE - General Electric Company. 1956k. *Monthly Report: Hanford Atomic Products Operation for July 1956.* HW-44580 DEL, Richland, Washington.

GE - General Electric Company. 1956l. *Monthly Report: Hanford Atomic Products Operation for August 1956.* HW-45115 DEL, Richland, Washington.

GE - General Electric Company. 1957a. *Biology Research Annual Report 1956.* HW-47500, Richland, Washington.

GE - General Electric Company. 1957b. *Quarterly Progress Report Research and Development Activities in the Field of Radiological Sciences, October-December, 1956.* HW-48044, Richland, Washington.

GE - General Electric Company. 1957c. *Quarterly Progress Report Research and Development Activities in the Field of Radiological Sciences, January-March, 1957.* HW-49793, Richland, Washington.

GE - General Electric Company. 1957d. *Quarterly Progress Report Research and Development Activities in the Field of Radiological Sciences, April-June, 1957.* HW-51321, Richland, Washington.

GE - General Electric Company. 1957e. *Chemical Processing Department Monthly Report for December 1956.* HW-47675 DEL, Richland, Washington.

GE - General Electric Company. 1957f. *Chemical Processing Department Monthly Report for January 1957.* HW-48132 DEL, Richland, Washington.

GE - General Electric Company. 1957g. *Chemical Processing Department Monthly Report for July 1957.* HW-51802 DEL, Richland, Washington.

GE - General Electric Company. 1957h. *Chemical Processing Department Monthly Report for December 1957.* HW-54319 DEL, Richland, Washington.

GE - General Electric Company. 1958a. *H Reactor Operation, Power Operation, Job Hazard Breakdowns, All Buildings.* Richland, Washington. Copy available in the U.S. Department of Energy Richland Operations Office Public Reading Room.

GE - General Electric Company. 1958b. *Regulations and General Instructions Governing Hanford Works Security Patrolmen.* With 1961 and 1964 revisions, Richland, Washington.

GE - General Electric Company. 1958c. *Chemical Processing Department Monthly Report for February 1958.* HW-55215 DEL, Richland, Washington.

GE - General Electric Company. 1958d. *Hanford Biology Research Annual Report for 1957.* HW-53500, Richland, Washington.

GE - General Electric Company. 1958e. *Quarterly Progress Report Research and Development Activities in the Field of Radiological Sciences, January-March, 1958.* HW-55586, Richland, Washington.

GE - General Electric Company. 1958f. *Quarterly Progress Report Research and Development Activities in the Field of Radiological Sciences, April-June, 1958.* HW-56928, Richland, Washington.

GE - General Electric Company. 1958g. *Quarterly Progress Report Research and Development Activities in the Field of Radiological Sciences, July-September, 1958.* HW-57908, Richland, Washington.

GE - General Electric Company. 1959a. *Hanford Biology Research Annual Report for 1958.* HW-59500, Richland, Washington.

GE - General Electric Company. 1959b. *Quarterly Progress Report Research and Development Activities in the Field of Radiological Sciences, October-December, 1958.* HW-58833, Richland, Washington.

GE - General Electric Company. 1959c. *Chemical Processing Department Monthly Report for March 1959.* HW-59849 DEL, Richland, Washington.

GE - General Electric Company. 1959d. *Chemical Processing Department Monthly Report for July 1959.* HW-61366 DEL, Richland, Washington.

GE - General Electric Company. 1960a. *Hanford Biology Research Annual Report for 1959.* HW-65500, Richland, Washington.

GE - General Electric Company. 1960b. *Radiological Chemistry Operation Semiannual Report, July-December, 1959.* HW-63824, Richland, Washington.

GE - General Electric Company. 1960c. *Chemical Processing Department Monthly Report for December 1959.* HW-63313 DEL, Richland, Washington.

GE - General Electric Company. 1960d. *Chemical Processing Department Monthly Report for January 1960.* HW-63706 DEL, Richland, Washington.

GE - General Electric Company. 1960e. *1959 Annual Report, Hanford Atomic Products Operation.* HW-64000 DEL, Richland, Washington.

GE - General Electric Company. 1960f. *Chemical Processing Department Monthly Report for February 1960.* HW-64089 DEL, Richland, Washington.

GE - General Electric Company. 1960g. *Chemical Processing Department Monthly Report for April 1960.* HW-64991 DEL, Richland, Washington.

GE - General Electric Company. 1960h. *Chemical Processing Department Monthly Report for September 1960.* HW-66958 DEL, Richland, Washington.

GE - General Electric Company. 1960i. *Chemical Processing Department Monthly Report for October 1960.* HW-67252 DEL, Richland, Washington.

GE - General Electric Company. 1960j. *Chemical Processing Department Monthly Report for November 1960.* HW-67459 DEL, Richland, Washington.

GE - General Electric Company. 1961a. *Hanford Biology Research Annual Report for 1960.* HW-69500, Richland, Washington.

GE - General Electric Company. 1961b. *Research and Development Activities in the Radiological Sciences, Physical Sciences Portion, January-December, 1960.* HW-70050, Richland, Washington.

GE - General Electric Company. 1961c. *Chemical Processing Department Monthly Report for December 1960.* HW-67985 DEL, Richland, Washington.

GE - General Electric Company. 1961d. *Chemical Processing Department Monthly Report for January 1961.* HW-68345 DEL, Richland, Washington.

GE - General Electric Company. 1961e. *Chemical Processing Department Monthly Report for February 1961.* HW-68700 DEL, Richland, Washington.

GE - General Electric Company. 1961f. *Chemical Processing Department Monthly Report for March 1961.* HW-69051 DEL, Richland, Washington.

GE - General Electric Company. 1961g. *Chemical Processing Department Monthly Report for April 1961.* HW-69443 DEL, Richland, Washington.

GE - General Electric Company. 1962a. *Hanford Biology Annual Report for 1961.* HW-72500, Richland, Washington.

GE - General Electric Company. 1962b. *Research and Development Activities in the Radiological Sciences, Physical Sciences Portion, January-December, 1961.* HW-73337, Richland, Washington.

GE - General Electric Company. 1962c. *Chemical Processing Department Monthly Report for February 1962.* HW-72890 DEL, Richland, Washington.

GE - General Electric Company. 1962d. *Chemical Processing Department Monthly Report for May 1962.* HW-73884 DEL, Richland, Washington.

GE - General Electric Company. 1963a. *Hanford Biology Research Annual Report for 1962.* HW-76000, Richland, Washington.

GE - General Electric Company. 1963b. *Hanford Radiological Sciences Research and Development Annual Report for 1962.* HW-77609, Richland, Washington.

GE - General Electric Company. 1963c. *Chemical Processing Department Monthly Report for August 1963.* HW-78817 DEL, Richland, Washington.

GE - General Electric Company. 1963d. *Hazards Summary Report. Volume 3 – Description of the 100-B, 100-C, 100-D, 100-DR, 100-F, and 100-H Production Reactor Plants.* HW-74094, Vol. 3, Richland, Washington.

GE - General Electric Company. 1963e. *Hazards Summary Report. Volume 3 – Description of the 100-KE and the 100-KW Production Reactor Plants.* HW-74095, Vol. 3, Richland, Washington.

GE - General Electric Company. 1964a. *Hanford Biology Research Annual Report for 1963.* HW-80500, Richland, Washington.

GE - General Electric Company. 1964b. *Hanford Radiological Sciences Research and Development Annual Report for 1963.* HW-81746, Richland, Washington.

GE - General Electric Company. 1964c. *Chemical Processing Department Monthly Report for December 1963.* HW-80243 DEL, Richland, Washington.

GE - General Electric Company. 1965a. *Chemical Processing Department Monthly Report for June 1965.* RL-SEP-618 DEL, Richland, Washington.

GE - General Electric Company. 1965b. *234-5 Building Diversification Survey.* RL-SEP-737, Richland, Washington.

GE - General Electric Company. 1965c. *Chemical Processing Department Monthly Report for October 1965.* RL-SEP-837 DEL, Richland, Washington.

GE - General Electric Company. 1966. *Chemical Processing Department Monthly Report for December 1965.* RL-SEP-913 DEL, Richland, Washington.

Gephart, R.E. 1999. "A Short History of Plutonium Production and Nuclear Waste Generation, Storage, and Release at the Hanford Site." PNNL-SA-32152, Pacific Northwest National Laboratory, Richland, Washington.

Gephart, R.E., and R.E. Lundgren. 1997. *Hanford Tank Cleanup: A Guide to Understanding the Technical Issues.* PNNL-10733, Third edition, Pacific Northwest National Laboratory, Richland, Washington.

Gerber, M.S. 1991. *Historical Genesis of Hanford Site Wastes.* WHC-SA-1224-FP, Westinghouse Hanford Company, Richland, Washington.

Gerber, M.S. 1992a. *On the Home Front. The Cold War Legacy of the Hanford Nuclear Site.* University of Nebraska Press, Lincoln, Nebraska.

Gerber, M.S. 1992b. *Legend and Legacy: Fifty Years of Defense Production at the Hanford Site.* WHC-MR-0293, Rev. 2, Westinghouse Hanford Company, Richland, Washington.

Gerber, M.S. 1992c. *Past Practices Technical Characterization Study - 300 Area - Hanford Site.* WHC-MR-0388, Westinghouse Hanford Company, Richland, Washington.

Gerber, M.S. 1993a. *Summary of the 100-B/C Reactor Operations and Resultant Wastes, Hanford Site.* WHC-SD-EN-RPT-004, Rev. 0, Westinghouse Hanford Company, Richland, Washington.

Gerber, M.S. 1993b. *Multiple Missions: The 300 Area in the Hanford Site History.* WHC-MR-0440, Westinghouse Hanford Company, Richland, Washington.

Gerber, M.S. 1993c. *Manhattan Project Buildings and Facilities at the Hanford Site: A Construction History.* WHC-MR-0425, Westinghouse Hanford Company. Richland, Washington.

Gerber, M.S. 1993d. *The Hanford Site: An Anthology of Early Histories.* WHC-MR-0435, Westinghouse Hanford Company, Richland, Washington.

Gerber, M.S. 1993e. "Lessons of History at the Hanford Site." *Universe*, Vol. 6, No. 2, pp. 13-19, Washington State University, Pullman, Washington.

Gerber, M.S. 1993f. *A Brief History of the PUREX and UO3 Facilities.* WHC-MR-0437, Westinghouse Hanford Company, Richland, Washington.

Gerber, M.S. 1994a. *Dramatic Change at T Plant.* WHC-MR-0452, Westinghouse Hanford Company, Richland, Washington.

Gerber, M.S. 1994b. "B Reactor Startup 50 Year Ago Today." *Hanford Reach*, September 26, 1994, Richland, Washington.

Gerber, M.S. 1994c. *A Brief History of the T Plant Facility at the Hanford Site.* WHC-MR-0452, Add. 1, U.S. Department of Energy, Richland, Washington.

Gerber, M.S. 1995. "Hanford's Historic Reactors." *Columbia*, Vol. 9 (1):31-36.

Gerber, M.S. 1996. *The Plutonium Production Story at the Hanford Site: Processes and Facility History.* WHC-MR-0521, Rev. 0, Westinghouse Hanford Company, Richland, Washington.

Gerber, M.S. 1997. *History and Stabilization of the Plutonium Finishing Plant PFP Complex, Hanford Site.* HNF-EP-0924, U.S. Department of Energy, Richland, Washington.

Gerber, M.S. 1999. "Film Discovery Highlights Hanford's Early History." *Hanford Reach*, September 13, 1999, Richland, Washington.

Gerber, M.S. 2000. "Site Fire Precautions Have Long History." *Hanford Reach*, July 24, 2000, Richland, Washington.

Gerber, M.S., D.W. Harvey, and J.G. Longenecker. 1997. "The Manhattan Project and Cold War Eras, Plutonium Production at the Hanford Site, December 1942-1990." In *National Register of Historic Places Multiple Property Documentation Form – Historic, Archaeological and Traditional Cultural Properties of the Hanford Site, Washington*, DOE/RL-97-02, Rev. 0, U.S. Department of Energy, Richland, Washington.

Gifford, A.T. 1965. "Plutonium from Nuclear Fuel Services." Letter to J.H. Warren, April 9, 1965, U.S. Atomic Energy Commission, Richland, Washington.

Gilbert, E.S. 1991. "Studies of Workers Exposed to Low Doses of External Radiation." *Occupational Medicine* 6(4):665-680.

Gilbert, E.S., and S. Marks. 1979. "Analysis of the Mortality of Workers in a Nuclear Facility." *Radiation Research* 79:122-148.

Gilbert, E.S., and L.E. Sever. 1987. *Draft Presentation Studies of Congenital Malformations and the Hanford Site.* PNL-10469-225, Pacific Northwest Laboratory, Richland, Washington.

Gilbert, E.S., G.R. Petersen, and J.A. Buchanan. 1989. "Mortality of Workers at the Hanford Site: 1945-1981." *Health Physics* 56:11-25.

Gill, W.E. 1949. *Report of a Study of the Health Aspects of Degreaser Operation in the Airlock of the 105 Building in the 100 H Area.* HW 14140, U.S. Atomic Energy Commission, Richland, Washington.

Gimpel, R.F., G.L. Gurwell, and R.W. Szempruch. 1975a. *Analysis of Compliance of Room 139, 234-5 Z Building, with Plutonium Storage Criteria.* ARH-CD-385, Atlantic Richfield Hanford Company, Richland, Washington.

Gimpel, R.F., G.L. Gurwell, and R.W. Szempruch. 1975b. *Analysis of Compliance of Room 145, 234-5 Z Building, with Plutonium Storage Criteria.* ARH-CD-386, Atlantic Richfield Hanford Company, Richland, Washington.

Gimpel, R.F., G.L. Gurwell, and R.W. Szempruch. 1975c. *Analysis of Compliance of Room 174, 234-5 Z Building, with Plutonium Storage Criteria.* ARH-CD-387, Atlantic Richfield Hanford Company, Richland, Washington.

Gimpel, R.F., G.L. Gurwell, and R.W. Szempruch. 1975d. *Analysis of Compliance of Room 175, 234-5 Z Building, with Plutonium Storage Criteria.* ARH-CD-388, Atlantic Richfield Hanford Company, Richland, Washington.

Gimpel, R.F., G.L. Gurwell, and R.W. Szempruch. 1975e. *Analysis of Compliance of Room 192-B, 234-5 Z Building, with Plutonium Storage Criteria.* ARH-CD-391, Atlantic Richfield Hanford Company, Richland, Washington.

Gimpel, R.F., G.L. Gurwell, and R.W. Szempruch. 1975f. *Analysis of Compliance of Room 192-C, 234-5 Z Building, with Plutonium Storage Criteria.* ARH-CD-392, Atlantic Richfield Hanford Company, Richland, Washington.

Gimpel, R.F., G.L. Gurwell, and R.W. Szempruch. 1975g. *Analysis of Compliance of Room 194, 234-5 Z Building, with Plutonium Storage Criteria.* ARH-CD-393, Atlantic Richfield Hanford Company, Richland, Washington.

Gimpel, R.F., G.L. Gurwell, and R.W. Szempruch. 1975h. *Analysis of Compliance of Room 225, 234-5 Z Building, with Plutonium Storage Criteria.* ARH-CD-394, Atlantic Richfield Hanford Company, Richland, Washington.

Glasscock, J.A., A.R. Heath, and E.M. Kuykendall. 1982. *Plutonium Retrieval System PRE User Manual.* RHO-MA-236, Rockwell Hanford Operation, Richland, Washington.

Glover, G.A. 1978. *Work Plan for Depressurization and Balancing of 234-5 Building.* RHO-CD-329, Rockwell Hanford Operations, Richland, Washington.

Goldberg, S. 1998. "General Groves and the Atomic West: The Making and the Meaning of Hanford." *The Atomic West,* B. Hevly and J.M. Findlay (eds.), Center for the Study of the Pacific Northwest in association with University of Washington Press, Seattle.

Goldschmidt, B. 1982. *The Atomic Complex: A Worldwide Political History of Nuclear Energy.* American Nuclear Society, LaGrange Park, Illinois.

Gosling, F.G. 1994. *The Manhattan Project: Making the Atomic Bomb.* DOE/HR-0096, U.S. Department of Energy, Washington, D.C.

Gosling, F.G. 1999. *The Manhattan Project: Making the Atomic Bomb.* DOE/MA-0001-01/99, U.S. Department of Energy, Washington, D.C. [Online] Available URL: http://www.osti.gov/bridge/product.biblio.jsp?osti_id=303853

Graf, W.A., G.R. Kiel, C.W. Nilson, and R.L. Stevenson. 1962. *Project CGC-912, Waste Treatment Facility, Z Plant, Preliminary Hazards Review.* General Electric Company, Richland, Washington.

Graves, A.W., and L.E. Kusler. 1978. *Z Plant Decontamination and Decommissioning Engineering Plan.* RHO-CD-420, Rockwell Hanford Operations, Richland, Washington.

Gray, R.H. 1985. *Health and Environmental Research on Complex Organic Mixtures: 24th Hanford Life Sciences Symposium.* U.S. Department of Energy and Pacific Northwest Laboratory, Richland, Washington.

Gray, R.H. 1990. *Environmental Monitoring, Restoration, and Assessment: What Have We Learned?* 8th Hanford Symposium on Health and the Environment, Pacific Northwest Laboratory, Richland, Washington.

Gray, R.H., and C.D. Becker. 1993. "Environmental Cleanup: The Challenge at the Hanford Site, Washington, USA." *Environmental Management,* Vol. 17, No. 4, pp. 461-475.

Gronewald, T.H. 1996. Electronic mail message to T.E. Marceau (Bechtel Hanford Incorporated) from T.H. Gronewald (President, Washington State Historical Railroad Society) regarding Issues to be addressed in the Hanford Cultural Resources Management Plan, July 31, 1996.

Gross, C.N., and A.B. Greninger. 1949. *234-5 Remote Mechanical Line Design Bases and Schedules.* HW-14231, General Electric Company, Richland, Washington.

Groswith, C.T. 1949. *Peroxide Process - 234-5 Building.* HW-14710, General Electric Company, Richland, Washington.

Groueff, S. 1967. *Manhattan Project: The Untold Story of the Making of the Atomic Bomb.* Little, Brown, and Company, Boston, Massachusetts.

Groves. L.R. 1983. *Now it Can be Told: The Story of the Manhattan Project.* Da Capo Press, New York.

Gydesen, S. 1999. Telephone call from Joy Keating (Pacific Northwest National Laboratory) to Shirley Gydesen (Safeguards and Security, Pacific Northwest National Laboratory), May 6, 1999. Copy of notes on file at the Library of the U.S. Department of Energy's Hanford Cultural Resources Laboratory, Richland, Washington.

Haakenson, J.D. 1998. Telephone call from Joy Keating and Dave Harvey (Pacific Northwest National Laboratory) to John D. Haakenson (DynCorp Plant Railroad Operations Manager), July 20, 1998 and August 11, 1998. Copy of notes on file at the Library of the U.S. Department of Energy's Hanford Cultural Resources Laboratory, Richland, Washington.

Hacker, B.C. 1987. *The Dragon's Tail: Radiation Safety in the Manhattan Project, 1942-1946.* University of California, Berkeley.

Hacker, B.C. 1994. *Elements of Controversy: The Atomic Energy Commission and Radiation Safety in Nuclear Weapons Testing 1947-1974.* University of California Press, Berkeley.

Hackett, P.L. 1949. *Biology Seminar: The Present Status of Polonium Tolerance Estimation.* HW-12572, General Electric Company, Richland, Washington.

Hageman, R.C. 1945. *Hanford: Threshold of an Era.* Available at U.S. Department of Energy Public Reading Room, Richland, Washington.

HAI - History Associates Incorporated. 1987. *History of the Production Complex: The Methods of Site Selection.* DOE/NV/10594—H1 UC2, prepared for the U.S. Department of Energy, Rockville, Maryland.

Hales, P.B. 1997. *Atomic Spaces: Living on the Manhattan Project.* University of Illinois Press, Chicago.

Hall, J.C. 1972. *Budget Facility Study: Liquid Organic Waste Collection and Incineration.* ARH-2450, Atlantic Richfield Hanford Company, Richland, Washington.

Hamilton, A., and H.L. Hardy. 1949. *Industrial Toxicology.* P.B. Hoeber Publisher, New York.

Hampton, J.C. 1973. "The Cell Cycle in Malignancy and Immunity." *Proceedings of the 13th Annual Hanford Biology Symposium,* Atomic Energy Commission and Pacific Northwest Laboratory, Oak Ridge, Tennessee.

Hanson, G.L. 1971. *Radioactive Effluent Reduction from 200 Area Facilities.* ARH-1972, Atlantic Richfield Hanford Company, Richland, Washington.

Hanson, G.L. 1972. *Radioactive Effluent Reduction from 200 Area Facilities, Phase II.* ARH-2419, Atlantic Richfield Hanford Company, Richland, Washington.

Hanson, G.L. 1977. *Final Safety Analysis Report, Contaminated Soil Removal Facility, 216-Z-9 Enclosed Trench Project HCP-687.* ARH-LD-124, Add 1, Atlantic Richfield Hanford Company, Richland, Washington.

Hanson, G.L. 1978. *Addendum to ARH-LD-124: Final Safety Analysis Report, Contaminated Soil Removal Facility, 216-Z-9 Enclosed Trench Project HCP-687.* RHO-LD-62, Rev. 1, Rockwell Hanford Operation, Richland, Washington.

Harmon, K.M. 1957. *Monitoring of Direct Chemical Assay of Fabricated Parts.* HW-47989, General Electric Company, Richland, Washington.

Harmon, K.M., and W.H. Reas. 1957. *Conversion Chemistry of Plutonium Nitrate.* HW-49597A, General Electric Company, Richland, Washington.

Harris, M.P. 1972. *Goodbye, White Bluffs.* Franklin Press, Yakima, Washington.

Hauth, J.T., C.R.J. Forslund, and J.A. Underwood. 1994. *Security Transition Program Office (STPO): Technology Transfer of the STPO Process, Tools, and Techniques.* DOE/RL-94-124, U.S. Department of Energy, Richland, Washington.

Hayes, D.F. 1956. *A Summary of Accidents and Incidents Involving Radiation in Atomic Energy Activities, June 1945 through December 1955.* TID-5360, U.S. Atomic Energy Commission, Washington, D.C.

HEDR - Hanford Environmental Dose Reconstruction Project Technical Steering Panel. 1994a. *Summary: Radiation Dose Estimates from Hanford Radioactive Material Releases to the Air and the Columbia River.* Washington State Department of Ecology, Olympia, Washington.

HEDR - Hanford Environmental Dose Reconstruction Project Technical Steering Panel. 1994b. *Representative Hanford Radiation Dose Estimates.* Washington State Department of Ecology, Olympia, Washington.

Heeb, C.M. 1994. *Radionuclide Releases to the Atmosphere from Hanford Operations, 1944-1972.* PNWD-2222 HEDR, Battelle, Pacific Northwest Laboratories, Richland, Washington. For updated release numbers starting August 1951, see http://www.pnl.gov/eshs/cap/hra/pub/hedr.html.

Heeb, C.M., and D.J. Bates. 1994. *Radionuclide Releases to the Columbia River from Hanford Operations, 1944-1971.* PNWD-2223 HEDR, Battelle, Pacific Northwest Laboratories, Richland, Washington.

Heid, K.R. 1962. *Emergency Radiological Plans and Procedures.* HW 70935, U.S. Atomic Energy Commission, Richland, Washington.

Heid, K.R., B.D. Breitenstein, H.E. Palmer, B.J. McMurray, and N. Wald. 1979. *The 1976 Hanford Americium Accident.* TID-28938, Pacific Northwest Laboratory, Richland, Washington.

Helgeson, G.L. 1953. *Radiation Studies for Task III Design.* HW-28918, General Electric Company, Richland, Washington.

Helgeson, G.L. 1956a. *Further Notes on the Surface Dosage of Plutonium Metal.* HW-41369, General Electric Company, Richland, Washington.

Helgeson, G.L. 1956b. *Past Exposure Problems in Room 242, 234-5 Building.* HW-43784, General Electric Company, Richland, Washington.

Heriford, A. 1995. Interview of Annette Heriford (pre-Hanford Site resident, Co-Chair, White Bluffs-Hanford Association) by Thomas E. Marceau (Bechtel Hanford, Inc.), December 14, 1995. Tape on file at the Columbia River Exhibition of History, Science, and Technology, Richland, Washington.

Heubach, E.C. 1979a. *Conceptual Design Report - Z Plant Ventilation and Services Upgrade, Project B-178.* RHO-CD-490 1, Rockwell Hanford Operations, Richland, Washington.

Heubach, E.C. 1979b. *Functional Design Criteria, Z-Plant Ventilation and Services Upgrade, Project B-178.* RHO-CD-622 FDC Rev 1, Rockwell Hanford Operations, Richland, Washington.

Heubach, E.C. 1979c. *Conceptual Design Report, Z-Plant Ventilation and Services Upgrade, Project B-178.* RHO-CD-622 CDR, Rockwell Hanford Operations, Richland, Washington.

Hewlett, R.G., and F. Duncan. 1969. *Atomic Shield: A History of the United States Atomic Energy Commission.* WASH-1215, U.S. Atomic Energy Commission, Washington, D.C.

Hewlett, R.G., and O.E. Anderson, Jr. 1990. *The New World: A History of the United States Atomic Energy Commission, Volume I, 1939-1946.* University of California Press, Berkeley, California.

Hewlett, R.G., and J.M. Holl. 1989. *Atoms for Peace and War, 1953-1961: Eisenhower and the Atomic Energy Commission.* University of California Press, Berkeley, California.

HFD - Hanford Fire Department. n.d. *Advice from the Hanford Fire Marshal.* Brochure by the U.S. Department of Energy, Richland, Washington. Brochure available in the lobby of the 3790 Central Badging Office.

HGIS - Hanford Geographic Information System. 2001. *Environmental Information Systems Maps.* Bechtel Hanford, Inc. [Online] Available URL: http://www.bhi-erc.com/dm/hgis/hgis.htm

Hicks, H.G., and G.R. Yesberger. 1967. *Report on Radiation Protection Standards, Procedures, Practices, and Exposure Records.* U.S. Department of Energy, Richland, Washington.

Hildreth, N.T. 1950. *Report for Week Ending 4-28-50, Vacuum Problem in 234-5.* HW-17763, General Electric Company, Richland, Washington.

Hill, O.F. 1955. *Symposium on Iodine Problem.* HW-39073, General Electric Company, Richland, Washington.

Hill, O.F., and F.J. Leitz. 1952. *Quarterly Report, October through December 1951, Chemical Research Section.* HW-23160, General Electric Company, Richland, Washington.

Hill, O.F., and V.R. Cooper. 1958. "Scale-Up Problems in the Plutonium Separations Program." *Industrial and Engineering Chemistry* 20(4).

Hinds, N.R., and L.E. Rogers. 1991. *Ecological Perspectives of Land Use History: The Arid Lands Ecology (ALE) Reserve.* PNL-7750, Pacific Northwest Laboratory, Richland, Washington.

Hinkson, A.H. 1977. *Criticality Safety Review - Z Plant.* RHO-CD-37, Rockwell Hanford Operations, Richland, Washington.

Holsted, P.G., and F.W. Albaugh. 1964a. *Hanford as an Experimental Reactor Site.* U.S. Atomic Energy Commission, Richland, Washington.

Holsted, P.G., and F.W. Albaugh. 1964b. "Hanford Capabilities." Brochure prepared by the General Electric Company for the U.S. Atomic Energy Commission, Richland, Washington.

Honstead, J.F. 1967. *Program for Evaluating Environmental Radiation Dose to Children.* BNWL-SA-1288, Battelle, Pacific Northwest Laboratories, Richland, Washington.

Honstead, J.F. 1968. "Radionuclide Burden-Diet Relationships Near a Nuclear Facility." In *Diagnosis and Treatment of Deposited Radionuclides,* H.A. Kornberg and W.D. Norwood (eds.), Excerpta Medica Foundation, Amsterdam, Netherlands.

Honstead, J.F., and F.N. Eichner 1972. *Dietary and Body Burden Data and Dose Estimates for Local School Children and Teenagers.* Y80054, Pacific Northwest Laboratory, Richland, Washington.

Hopkins, Jr., H.H., and A.C. Jones. 1978. *Z Plant Decommissioning Study.* RHO-CD-251, Rockwell Hanford Operations, Richland, Washington.

Hounshell, D.A. 1992. "Du Pont and the Management of Large-Scale Research and Development." In *Big Science: The Growth of Large-Scale Research,* P. Galison and B. Hevly (eds.), Stanford University Press, Stanford, California.

House, D. 1999. Telephone call from Joy Keating (Pacific Northwest National Laboratory) to Darrell House (son of Atomic Energy Commission courier), January 28, 1999. Richland, Washington. Copy of notes on file at the Library of the U.S. Department of Energy's Hanford Cultural Resources Laboratory, Richland, Washington.

Howell, W.P., J.L. Kenoyer, M.L. Kress, K.L. Swinth, C.D. Corbit, L.V. Zuerner, D.M. Fleming, and H.W. DeHaven. 1989. *A Historical Review of Portable Health Physics Instruments and Their Use in Radiation Protection Programs at Hanford, 1944-1988.* PNL-6980, Pacific Northwest Laboratory, Richland, Washington.

Howes, R.H., and C.L. Herzenberg. 1999. *Their Day in the Sun: Women of the Manhattan Project.* Temple University Press, Philadelphia, Pennsylvania.

Hultgren, R. 1999a. Interview of Roger Hultgren (Chemical Separations Manager at the Hanford Site, 1944-1989) by Brian Freer (CH2M HILL Hanford, Inc.), April 8, 1999. Copy of notes on file at the Library of the U.S. Department of Energy's Hanford Cultural Resources Laboratory, Richland, Washington.

Hultgren, R. 1999b. Interview of Roger Hultgren (Chemical Separations Manager at the Hanford Site, 1944-1989) by Brian Freer (CH2M HILL Hanford, Inc.), May 27, 1999. Copy of notes on file at the Library of the U.S. Department of Energy's Hanford Cultural Resources Laboratory, Richland, Washington.

Hungate, F.P. 1965. "Radiation and Terrestrial Ecosystems." *Health Physics* 11(12):1255-1675.

Hunter, D. 1998. Telephone calls from Joy Keating (Pacific Northwest National Laboratory) to Dick Hunter (former patrol officer at the Hanford Site from 1948-1952, Hanford employee until 1988), October 15 and 23, 1998. Copy of notes on file at the Library of the U.S. Department of Energy's Hanford Cultural Resources Laboratory, Richland, Washington.

IC - International Conference on the Peaceful Uses of Atomic Energy. 1956. *Proceedings of the International Conference on the Peaceful Uses of Atomic Energy Held in Geneva 8 August to 20 August, 1955.* United Nations, New York.

IC - International Conference on the Peaceful Uses of Atomic Energy. 1958. *Peaceful Uses of Atomic Energy.* United Nations, Geneva, Switzerland.

ICF KH - ICF Kaiser Hanford Company. 1994. *Hanford Site Map.* ZMWN0021.DWG, Richland, Washington.

ICRP - International Commission on Radiological Protection. 1959 *Recommendations of the International Commission on Radiological Protection.* Report No. 1, Pergamon Press, Oxford.

ICRP - International Commission on Radiological Protection. 1966. *Recommendations of the International Commission on Radiological Protection.* Report No. 9, Pergamon Press, Oxford.

ICRP - International Commission on Radiological Protection. 1977. *Recommendations of the International Commission on Radiological Protection.* Report No. 26 (Annals of the ICRP 1(3)), Pergamon Press, Oxford.

Ingalls, W.P. 1952. *Remote Mechanical Operating Methods Vs. Rubber Glove Operating Methods for the 234-5 Process.* HW-25728, General Electric Company, Richland, Washington.

Ingalls, W.P. 1953. *Project CG-551, Addendum to Basic Process and Equipment Requirements for Modification of RMA Tasks II & III - 234-5 Building Expansion.* HW-29041, General Electric Company, Richland, Washington.

Ingalls, W.P. 1967. "Recovery of Pu From Z-9 Crib, Budget Item for FY 69." Memo to R.E. Olson, March 15, 1967, Isochem, Inc., Richland, Washington.

Isochem, Inc. 1967. *Chemical Processing Division Monthly Report for August 1967.* ISO-714 DEL, Richland, Washington.

IT - International Technology Corporation. 1992. *NIKE Missile Site Proposed Characterization Methodologies - Hanford Site North Slope Era.* Prepared for the Westinghouse Hanford Company, Richland, Washington.

Jaynes, R.E. 1950. G.E. Security Bulletin #54. Emergency Blackouts – Plant Areas. September 7, 1950, General Electric Company, Richland, Washington. Copy available in the U.S. Department of Energy Richland Operations Office Public Reading Room.

Jech, J.J. 1999. Interview of John Jech (Health Physicist at the Hanford Site, 1950-1988) by Brian Freer (CH2M HILL Hanford, Inc.), April 9, 1999. Copy of notes on file at the Library of the U.S. Department of Energy's Hanford Cultural Resources Laboratory, Richland, Washington.

Jenne, D.E., and J.W. Healy. 1950. *Dissolving of Twenty Day Metal at Hanford.* HW-17381-DEL, General Electric Company, Richland, Washington.

Jensen, M.A. 1999. "Hanford Reactor Operating History." Electronic mail message from Mark A. Jensen (Certified Reactor Operator at the Hanford Site, 1980-present), to Michelle Skinner (U.S. Department of Energy), September 23, 1999. Copy on file at the Library of the U.S. Department of Energy's Hanford Cultural Resources Laboratory, Richland, Washington.

Johnson, R.W., and M.G. Schene (eds.) 1987. *Cultural Resources Management.* Robert E. Krieger Publishing Company, Malabar, Florida.

Jones - J.A. Jones Construction Company. 1973. *Purchase Order: Gamma Radiation Indicating System.* JAJ-11193, Richland Washington.

Jones, V.C. 1985. *Manhattan: The Army and the Atomic Bomb.* Center of Military History, Washington, D.C.

Judson, B.F. 1956. *Semi-Works Operation of the RECUPLEX Facility, Volume I.* HW-44776, Vols. I and II, General Electric Company, Richland, Washington.

Kathren, R.L. 1974. "Plant and Personnel Radiation Monitoring." In *Nuclear Power Reactor Instrumentation Systems Handbook*, J.M. Harrer and J.G. Beckerley (eds.), Vol. 2, TID-25952-P2, U.S. Atomic Energy Commission, Washington, D.C.

Kathren, R.L. 1993. "Review of On the Home Front: The Cold War Legacy of the Hanford Nuclear Site." *Health Physics* 64 (5):553-554.

Kathren, R.L. 1995. "The United States Transuranium and Uranium Registeries: 1968-1993." *Radiation Protection Dosimetry*, 60(4):349-354.

Kathren, R.L., and P.L. Ziemer (eds.). 1980. *Health Physics: A Backward Glance.* Thirteen Original Papers on the History of Radiation Protection, Pergamon Press, New York.

Kathren, R.L., R.W. Baalman, and W.J. Bair (eds.). 1986. *Herbert M. Parker: Publications and Other Contributions to Radiological and Health Physics.* Battelle Press, Richland, Washington.

Kattner, W.T. 1949. *300 Area Scrap Reclamation: Summary of Discussion on July 7, 1949.* HW-14209, General Electric Company, Richland, Washington.

Keene, A.R. 1957. "I-131 Control." Letter from Art R. Keene (Head of Radiation Protection, General Electric Company) to D.R. Koberg (General Electric Company), April 1, 1957, Richland, Washington. Copy on file at the Library of the U.S. Department of Energy's Hanford Cultural Resources Laboratory, Richland, Washington.

Keene, A.R. 1967. "Maximum Permissible Limits-Operational Controls and Working Limits." Letter from Art R. Keene (Head of Radiation Protection, General Electric) to Fred W. Albaugh (General Electric Company), October 12, 1967, Richland, Washington. Copy on file at the Library of the U.S. Department of Energy's Hanford Cultural Resources Laboratory, Richland, Washington.

Kelly, L. 1998. Telephone call from Joy Keating (Pacific Northwest National Laboratory) to Lynn Kelly (DynCorp Engineer, Site Utilities), July 14, 1998, Richland, Washington. Copy of notes on file at the Library of the U.S. Department of Energy's Hanford Cultural Resources Laboratory, Richland, Washington.

Kendall, L.F. 1949. *Assay, Specific Gravity, and Iron Methods for 234-5 Process Solutions.* HW-13145, General Electric Company, Richland, Washington.

Kendall, W.R. 1964. *Plutonium Buy-Back Facility - Z Plant, FY-1966 Budget Facility Description.* HW-81542, General Electric Company, Richland, Washington.

Kennedy, J.F. 1963. "Radio and Television Address to the American People, July 26, 1963." [Online] Available URL: http://www.clw.org/pub/clw/coalition/ken0763.htm

Kerr, W.B. 1952. *Final Report: Production Test - 234-5: Plant Process Evaluation, Precipitation of Plutonium (IV) Oxalate.* HW-26450, General Electric Company, Richland, Washington.

Ketzlach, N. 1955a. *A Critical Mass Study of Hood No. 5 and the Filter Boat Cleaning and Testing Hood in 234-5 Building.* HW-38029, General Electric Company, Richland, Washington.

Ketzlach, N. 1955b. *Nuclear Safety Requirements of 234-5 Building Control Laboratory.* HW-38904, General Electric Company, Richland, Washington.

King, T.F., P.P. Hickman, and G. Berg. 1977. *Anthropology in Historic Preservation: Caring for Culture's Clutter.* Academic Press, New York.

Kirk, G. 1997. *337 and 337-B Buildings: Potential for National Register of Historic Places Eligibility.* PNNL-SA-29461, Pacific Northwest National Laboratory, Richland, Washington.

Kleinman, D. L. 1995. *Politics on the Endless Frontier: Postwar Research Policy in the United States.* Duke University Press, Durham, North Carolina.

Knight, R. 1999a. Interview of Russ Knight (Chemical Separations Technician at the Hanford Site, 1945-1989) by Brian Freer (CH2M HILL Hanford, Inc.), April 8, 1999. Copy of notes on file at the Library of the U.S. Department of Energy's Hanford Cultural Resources Laboratory, Richland, Washington.

Knight, R. 1999b. Interview of Russ Knight (Chemical Separations Technician at the Hanford Site, 1945-1989) by Brian Freer (CH2M HILL Hanford, Inc.), May 27, 1999. Copy of notes on file at the Library of the U.S. Department of Energy's Hanford Cultural Resources Laboratory, Richland, Washington.

Knights, L.M. 1966. *Critical Mass Control Specification, Removal of Plutonium from 9-B Hood Exhaust Duct, 234-5 Building.* ISO-311, Isochem, Inc., Richland, Washington.

Knights, L.M. 1968. "Dry Powder Evaluation Facility." Memo to W.P. Ingalls, May 31, 1968, Atlantic Richfield Hanford Company, Richland, Washington.

Knights, L.M. 1969. *The Continuous Incineration of Plutonium-Bearing Scrap.* ARH-SA-39, Atlantic Richfield Hanford Company, Richland, Washington.

Kornberg, H.A. 1964. *Progress in Biological Research: A Summary of Hanford Laboratories Achievements in These Programs Under General Electric 1956-1964.* HW-83613, General Electric Company, Richland, Washington.

Kornberg, H.A., and W.D. Norwood. 1968. *Diagnosis and Treatment of Deposited Radionuclides.* Excerpta Medica Foundation, Monographs on Nuclear Medicine and Biology No. 2.

Krieg, R.H. 1996. Telephone call from Ian Lindsay (Pacific Northwest National Laboratory) to R.H. Krieg (Pacific Northwest National Laboratory), August 3, 1996. Copy of notes on file at the Library of the U.S. Department of Energy's Hanford Cultural Resources Laboratory, Richland, Washington.

Krigens, A.G., and R.B. Laws. 1968. *Design Criteria, Continuous Solvent Extraction of Plutonium and Americium Waste Treatment Facility, 242-Z Building.* ARH-406, Atlantic Richfield Hanford Company, Richland, Washington.

Kubik, B. 1997. *Building 313: Fuel Fabrication Building.* DOE-0256, Westinghouse Hanford Company for the U.S. Department of Energy, Richland, Washington.

Kubik, B.J. 1994. *Richland.* City of Richland, Washington.

Kuklinski, G.B., and R.E. Olson. 1969. *Plutonium for Recovery from Scrap at Atlantic Richfield Hanford Company.* ARH-SA-26, Atlantic Richfield Hanford Company, Richland, Washington.

Lambert, B. E. 1991. "The Adequacy of Current Occupational Standards for Protecting the Health of Nuclear Workers." *Occupational Medicine (State of the Art Reviews)*, 6(4):725-740.

Lane, W.A. 1965. *Historical Events of the Chemical Processing Department.* RL-SEP-855 DEL, General Electric Company, Richland, Washington.

Lang, D. 1948. *Early Tales of the Atomic Age.* Doubleday and Co, Inc., Garden City, New York.

LANL - Los Alamos National Laboratory. 1995. "Radiation Protection and the Human Radiation Experiments." *Los Alamos Science*, Vol. 23.

Lanouette, W. 1999. "Hiroshima: World Culture as the Event 'War.'" Paper presented at a UNESCO Conference on World Heritage of the 20th Century "Modernity and Barbarism," The Bauhaus, Dessau, Germany.

LaRiviere, J.R. 1959. *Project Proposal, Revision 1, Vacuum System Improvements, 234-5.* HW-59156, General Electric Company, Richland, Washington.

Larson, D.E. 1996. *Hanford High-Level Waste Vitrification Program at the Pacific Northwest National Laboratory: Technology Development – Annotated Bibliography.* PNNL-10955, Pacific Northwest National Laboratory, Richland, Washington.

LASL - Los Alamos Scientific Laboratory. 1952. *Operation Ranger Nevada Test Site, Jan - Feb 1951 Volume 5 Program Reports - Operational.* WT-204, University of California, Los Alamos, New Mexico.

Laws, R.B. 1973. *Criticality Overflow Drains at Z Plant.* ARH-2406, Atlantic Richfield Hanford Company, Richland, Washington.

Lee, H.A. 1972a. *A Proposed Method of Fire Protection for the E-3 and E-4 Exhaust Systems in the 234-5 Building.* ARH-2611, Atlantic Richfield Hanford Company, Richland, Washington.

Lee, H.A. 1972b. *Design Criteria: Fire Protection for the E-3 and E-4 Exhaust Systems in the 234-5 Building.* ARH-2677, Atlantic Richfield Hanford Company, Richland, Washington.

LETC - Law Engineering Testing Company. 1986. *Investigation of Former NIKE Missile Sites for Potential Toxic and Hazardous Waste Contamination.* Vol. 1, for the U.S. Department of Army, Atlanta, Georgia.

Lewis, H.W. 1986. Letter dated October 1, 1986 to the Secretary of Energy, John. S. Herrington. Accession No. 3514, included in the Roddis Panel Reports on N Reactor Safety, Hanford Public Reading Room, Richland, Washington.

Lewis, M. 1996. Telephone call from Ian Lindsay (Pacific Northwest National Laboratory) to Milton Lewis (retired director of N Reactor programs), June 28, 1996, Richland, Washington. Copy on file at the Library of the U.S. Department of Energy's Hanford Cultural Resources Laboratory, Richland, Washington.

Libby, L.M. 1979. *The Uranium People.* Crane Russak and Company and Charles Scribner's Sons, New York.

Linenthal, E.T. 1995. *Preserving Memory: The Struggle to Create America's Holocaust Museum.* Penguin Books, New York.

Loeb, P. 1982. *Nuclear Culture: Living and Working in the World's Largest Atomic Complex.* Coward, McCann & Geoghegen, Inc., New York.

Low, A.J. 1967. "Plutonium Recovery From Scrap." In *Chemical Engineering,* (74)11:132-134.

Lowenthal, D. 1985. *The Past is a Foreign Country.* Cambridge University Press, Cambridge.

Luczynski, G. 1999. "OPSEC History." Electronic mail message from Gene Luczynski (Safeguards and Security Services, Pacific Northwest National Laboratory) to Darby Stapp (Pacific Northwest National Laboratory), March 29, 1999. Copy on file at the Library of the U.S. Department of Energy's Hanford Cultural Resources Laboratory, Richland, Washington.

Ludowise, J.D. 1980. *Plutonium Reclamation Facility, Engineering Technical Reference.* RHO-MA-246, Rockwell Hanford Operations, Richland, Washington.

Maas, J. 1993. *List of Designations on Camp Hanford General Site Maps, Camp Hanford Forward Positions for January 1952-February 1959.* U.S. Army Corps of Engineers, Seattle, Washington.

Mahaffey, J.A. 1989. "Modeling for Scaling to Man: Biology, Dosimetry, and Response." *Proceedings of the 26th Hanford Life Sciences Symposium,* Pergamon Press, New York.

Mahlum, D.D. 1977. "Developmental Toxicology of Energy-Related Pollutants." *Proceedings of the 17th Annual Hanford Biology Symposium,* U.S. Department of Energy and Pacific Northwest Laboratory, Oak Ridge, Tennessee.

Mahlum, D.D., R.H. Gray, and W.D. Felix. 1980. "Coal Conversion and the Environment: Chemical, Biological, and Ecological Considerations." *Proceedings of the 20th Annual Hanford Life Sciences Symposium,* U.S. Department of Energy and Pacific Northwest Laboratory, Oak Ridge, Tennessee.

Malody, C.W. 1976. "Analysis of Z-9 Soil." Letter to O.J. Elgert, November 5, 1976, Atlantic Richfield Hanford Company, Richland, Washington.

Mancuso, T.F., A. Stewart, and G. Kneale. 1977. "Radiation Exposures of Hanford Workers Dying from Cancer and other Causes." *Health Physics,* 33:369-385.

Marceau, T.E. 1998. *Hanford Site Manhattan Project and Cold War Era Historic District Treatment Plan.* DOE/RL-97-56, Rev. 1, Bechtel Hanford, Inc. for the U.S. Department of Energy, Richland, Washington. [Online] Available URL: http://www.hanford.gov/docs/rl97-56/appa.htm

Marshall, L.D. 1977. *Conceptual Design Report, 241-Z Waste Sump Facility Replacement.* ARH-CD-921 RD, Atlantic Richfield Hanford Company, Richland, Washington.

Matthias, F.T. n.d. *Constructing the Hanford Atom Factory.* Hanford Engineer Works, Richland, Washington.

Matthias, F.T. 1945. *Journals and Notes, 1943-1945.* Unpublished diary, U.S. Department of Energy Public Reading Room, Richland, Washington.

Maxfield, H.L. 1979. *Handbook, 200 Area Waste Sites.* RHO-CD-673, Rockwell Hanford Company, Richland, Washington.

Mayancsik, B.A. 1988. *400 Area Facilities Catalog.* WHC-IP-0269, Westinghouse Hanford Company, Richland, Washington.

McDowell, J. 1993. "The Year They Firebombed the West." *American Forests* (May/June): pp. 22, 23, 55.

McKee, D.W. 1959. *Inspection Facility - Design Criteria, Project CG-811.* HW-60676, General Electric Company, Richland, Washington.

McKnight, J.P. 1972a. *Design Criteria, HC-9B Fluorinator Off-Gas Jet Recycling System.* ARH-2328 1, Atlantic Richfield Hanford Company, Richland, Washington.

McKnight, J.P. 1972b. *Design Criteria Supplement, HC-9B Fluorinator Off-Gas Jet Recycling System and Disposal of Z Plant Wastes.* ARH-2517 1, Atlantic Richfield Hanford Company, Richland, Washington.

McMaster, B.N., J.B. Sosebee, W.G. Fraser, K.C. Govro, C.F. Jones, S.A. Grainger, and K.A. Civitarese. 1984. *Historical Overview of the Nike Missile System.* DRXTH-AS-IA-83016, Environmental Science & Engineering, Inc. for the U.S. Army Toxic & Hazardous Materials Agency, Aberdeen Proving Grounds, Maryland.

Meacham, J. E., R. J. Cash, D. R. Dickinson, F. R. Reich, J. M. Grigsby, A. K. Postma, and M. A. Lilga. 1996. *Assessment of the Potential for Ferrocyanide Propagating Reaction Accidents.* WHC-SD-WM-SARR-038, Rev. 1, Westinghouse Hanford Company, Richland, Washington.

Meitner, L., and O.R. Frisch. 1939. "Disintegration of Uranium by Neutrons: a New Type of Nuclear Reaction." *Nature,* vol. 143, February 11, 1939, pp. 239-40.

Menzel, D.B. 1970. "Pollution and Lung Biochemistry." *Proceedings of the 10th Annual Hanford Biology Symposium,* Archives of Internal Medicine Symposia, Vol. 9, American Medical Association, Chicago, Illinois.

Mickelson, M.L. 1951. *Annual Report of the Radiological Sciences Department.* HW-25709, General Electric Company, Richland, Washington.

Mickelson, M.L. 1955. *Radiological Sciences Department Annual Report on Research and Development Fiscal Year 1955.* HW-39558, General Electric Company, Richland, Washington.

Miller, R.L., and R.K. Wahlen. 1987. *Estimates of Solid Waste Buried in 100 Area Burial Grounds.* WHC-EP-0087, Westinghouse Hanford Company, Richland, Washington.

Mitchell, K.C. 1999. *Rhetoric to Reality, A Cold War Context Statement for the Pantex Plant.* Mason and Hager Corporation, Amarillo, Texas.

Mobley, W.N. 1957. "Status of Fire Protection Activities - 234-5 Building." Memo to W.K. MacCready, April 19, 1957, General Electric Company, Richland, Washington.

Moeller, D.W. 1989. "History and Perspective on the Development of Radiation Protection Standards." *Radiation Protection Today-The NCRP at Sixty Years,* Proceedings of the Twenty-Fifth Annual Meeting of the National Council on Radiation Protection and Measurements, Proceedings No. 11, National Council on Radiation Protection and Measurements, Washington, D.C.

Moeller, D.W. 1992. *Environmental Health.* Harvard University Press, Cambridge, Massachusetts.

Moore, R.L. 1992. *As I Knew Him: An Autobiography.* Privately Published, Richland, Washington. Copy available in the Hanford Technical Library.

Moorman, L. 1960. *Project Unitization Report: RMC Button Line - 234-5 Building.* General Electric Company, Richland, Washington.

Mulick, C. 1998. "Camp Hanford - Site Wasn't Easy for Young Soldiers Who Served as the Last Line of Defense." *Tri-City Herald,* March 8, 1998, Kennewick, Washington.

Murphy, J.G. 1976. *PRE Report for January 1976.* ARH-CD-542 1, Atlantic Richfield Hanford Company, Richland, Washington.

Murphy, J.G. 1978. *Nuclear Materials Control, Nuclear Materials Accounting Manual.* RHO-MA-142, Rockwell Hanford Operations, Richland, Washington.

National Safety Council. 1977. *Accident Facts.* National Safety Council; 1976 ed., Chicago.

National Safety Council. 1992. *Accident Facts.* National Safety Council; 1991 ed., Chicago.

National Safety Council. 1995. *Accident Facts.* National Safety Council; 1994 ed., Itasca, Illinois.

NBS - National Bureau of Standards. 1934. "Radium Protection for Amounts Up to 300 Milligrams." Handbook No. 18, Washington, D.C.

NBS - National Bureau of Standards. 1949. "Safe Handling of Radioactive Isotopes." Handbook No. 42, Washington, D.C.

NBS - National Bureau of Standards. 1950. "Recommendations of the International Commission on Radiological Protection." Handbook No. 47, Washington, D.C.

NBS - National Bureau of Standards. 1953. "Maximum Permissible Amounts of Radioisotopes in the Human Body and Maximum Permissible Concentrations in Air and Water." Handbook No. 52, Washington, D.C.

NBS - National Bureau of Standards. 1954. "Permissible Dose from External Sources of Ionizing Radiation." Handbook No. 59, Washington, D.C.

NCRP - National Council on Radiation Protection and Measurements. 1971. "Basic Radiation Protection Criteria - Recommendations." Report No. 39, Washington, D.C.

NEC - National Economic Council. 2000. *The Link Between Exposure to Occupational Hazards and Illnesses in the Department of Energy Contractor Workforce.* Inter-Agency Working Groups, Washington, D.C.

Nemzek, T.A. 1972. *Plutonium Effluent Cleanup, Plutonium Finishing Plant Project No. HCP-669.* Directive No. HW-611, U.S. Atomic Energy Commission, Richland, Washington.

Newton, C.E., Jr., K.R. Heid, H.V. Larson, and I.C. Nelson. 1966. *Tissue Sampling for Plutonium through an Autopsy Program.* BNWL-SA-918, Battelle, Pacific Northwest Laboratories, Richland, Washington.

Nichols, K.D. 1987. *The Road to Trinity: A Personal Account of How Nuclear Policies Were Made.* William Morrow and Co., New York.

Nickola, P.W. 1977. *The Hanford 67-Series: A Volume of Atmospheric Field Diffusion Measurements.* PNL-2433, Pacific Northwest Laboratory, Richland, Washington.

Nickola, P.W., J.V. Ramsdell, C.S. Glantz, and R.E. Kerns. 1983. *Hanford Atmospheric Dispersion Data: 1960 Through June 1967.* PNL-4814, Pacific Northwest Laboratory, Richland, Washington.

Nightingale, R.E. 1962. *Nuclear Graphite.* Academic Press, New York.

Noonan, C.F. 2000. *Federal City Revisited: Atomic Energy and Community Identity in Richland, Washington.* Masters Thesis, Ball State University. Copy in the Richland Public Library and U.S. Department of Energy Public Reading Room.

Norwood, W.D. 1964. "Planning for Care of Injured Radiating Patients." *Archives of Environmental Health,* 8:474-479.

Norwood, W.D. 1975. *Health Protection of Radiation Workers.* Charles C. Thomas, Springfield, Illinois.

Norwood, W.D., and S.T. Cantril. 1944. Memorandum from W.D. Norwood and S.T. Cantril (E.I. Du Pont de Nemours and Company), November 4, 1944, HW 7-854, Richland, Washington.

Norwood, W.D., and P.A. Fuqua. 1945. "Medical Department Experience: Organization, Plant, and Village Medical July through December 1945." *Operation of Hanford Engineer Works — Medical Department Plant and Village Medical,* HAN-73214, Book 6, E. I. Du Pont De Nemours and Co., Inc. Richland, Washington.

Norwood, W.D., and E.J. Quigley. 1968. "Experimental Radiosurgery for Care of Injured Radiating Patients." In *Diagnosis and Treatment of Deposited Radionuclides,* H.A. Kornberg and W.D. Norwood (eds.), Excerpta Medical Foundation. Amsterdam, Netherlands.

Nousen, P.A. 1994. *Security Transition Program Office Progress Report.* DOE/RL-94-114, U.S. Department of Energy, Richland, Washington.

NPS - National Park Service. 1990. *How to Apply the National Register Criteria for Evaluation.* National Register Bulletin 15, Washington, D.C.

Nuclear Waste News. 1996a. "Hanford: DOE May Stabilize, Dispose On-Site 280 Kg of Plutonium-Bearing Waste." May 9, 1996, Business Publishers, Inc., Silver Spring, Maryland.

Nuclear Waste News. 1996b. "Hanford: ROD Schedule Calls for Stabilization of Plutonium Finishing Plant by 2004." July 3, 1996, Business Publishers, Inc., Silver Spring, Maryland.

Nussbaum, R.H. 1998. "The Linear No-Threshold Dose-Effect Relation: Is It Relevant to Radiation Protection Regulation?" *Medical Physics,* 25(3):291-299.

O'Farrell, T.P. 1973. "Project ALE: A Natural Desert Community." *Pacific Search Journal of Natural Science in the Pacific Northwest* (July).

O'Neill, D. 1995. *The Firecracker Boys.* St. Martin's Press, New York.

O'Neill, D. 1998. "Alaska and the Firecracker Boys: The Story of Project Chariot." In *The Atomic West,* B. Hevly and J.M. Findlay (eds.), University of Washington Press, Seattle.

Ogletree, L.R. 1977. "Hanford Patrol Historical Data." Letter from L.R. Ogletree (Manager, Safeguards and Security, Rockwell Hanford Operations) to K.H. Jackson (Director, Safeguards and Security Division, U.S. Energy Research and Development Administration, Richland, Washington), August 24, 1977, Tracking No. 9725-R-1, Richland, Washington. Copy on file at the Library of the U.S. Department of Energy's Hanford Cultural Resources Laboratory, Richland, Washington.

OIG - Office of the Inspector General. 1998. *The U.S. Department of Energy's Facility Reuse at the Rocky Flats Environmental Technology Site.* DOE/IG-0425, Washington, D.C.

Olson, P.A. 1948. *Fish and Fish Problems on the Hanford Reservation.* HW-11642, General Electric Company, Richland, Washington.

Olson, R.E. 1974. "Replacement of Plexiglass Hood Panels." Memo to D.W. Corbell and R.L. Hibbard, February 15, 1974, Atlantic Richfield Hanford Company, Richland, Washington.

Oregonian, The. 1949. "Planes, Cars Mount Warlike Guard over Hanford Atom Plant." September 11, 1949. Photographers: Allan deLay of the Oregonian and Robley Johnson of the Atomic Energy Commission.

Palmer, J.H. 1957. *Cambridge Special "G" Type Space Filters, Moisture and Burning Tests.* General Electric Company, Richland, Washington.

Park, J.F., and R.A. Pelroy (eds.). 1989. "Multilevel Health Effects Research: From Molecules to Man." *Proceedings of the 27th Hanford Symposium on Health and the Environment,* Battelle Press, Columbus, Ohio.

Parker, H.M. 1944. Letter from H.M. Parker (E.I. Du Pont de Nemours and Company) to D.O. Notman, 31 August 1944, HW-7-588, Du Pont Company, Richland, Washington. Copy on file at the Library of the U.S. Department of Energy's Hanford Cultural Resources Laboratory, Richland, Washington.

Parker, H.M. 1945. *Tolerance Concentration of Product in Drinking Water.* HW 7-2119, E.I. Du Pont de Nemours and Company, Richland, Washington.

Parker, H.M. 1948a. *Speculations on Long-Range Waste Disposal Hazards.* HW-8674, General Electric Company, Richland, Washington.

Parker H.M. 1948b. *Present Status of the Hanford Works Biology Program, and its Proposed Future Development.* HW-8527, Hanford Works, Richland, Washington.

Parker, H.M. 1955. "Radiation Protection in the Atomic Energy Industry." *Radiology,* 65(6): 903-911.

Parker, H.M. 1956. *Radiological Sciences Department Quarterly Progress Report on Research and Development Activities, January-March, 1956.* HW-42403, General Electric Company, Richland, Washington.

Parker, M.B. 1979. *Tales of Richland, White Bluffs and Hanford 1805-1943.* Ye Galleon Press, Fairfield, Washington.

Parkhurst, C. 1999. Telephone call from Joy Keating (Pacific Northwest National Laboratory) to Clem (Clement) Parkhurst (former agent with the Federal Bureau of Investigation stationed in Richland, Washington from 1950-1971), January 14, 1999. Copy of notes on file at the Library of the U.S. Department of Energy's Hanford Cultural Resources Laboratory, Richland, Washington.

Patton, D.R. 1964. *Bill of Materials: Paging System - RMC Button Line.* General Electric Company, Richland, Washington.

Patty, F. 1958. *Industrial Hygiene and Toxicology.* Interscience Publishers, New York.

Pearce, D.W. 1967. *Research and Development in Environmental Sciences and Resources Management.* BNWL-453, Battelle, Pacific Northwest Laboratories, Richland, Washington.

Pehrson, G.A. 1943. *Report on the Hanford Engineer Works Village.* Office of G. Albin Pehrson, Architect Engineer, Spokane, Washington.

Peterson, M.L. 2000. "Comments on Fuel Manufacturing." E-mail message from Merrell L. Peterson (Autoclave/Lead Chemical Operator 1981-1988) to Ellen L. Prendergast (U.S. Department of Energy-Richland), February 5, 2000. Copy on file at the Library of the U.S. Department of Energy's Hanford Cultural Resources Laboratory, Richland, Washington.

Phillips, R.D. 1978. "Biological Effects of Extremely Low Frequency Electromagnetic Fields." *Proceedings of the 18th Annual Hanford Life Sciences Symposium,* U.S. Department of Energy, Pacific Northwest Laboratory, and the Electric Power Research Institute, Oak Ridge, Tennessee.

Pinto, A.P. 1978. *Conceptual Design Report, Project B-137, 241-Z Sump Improvements.* RHO-CD-513, Rockwell Hanford Operations, Richland, Washington.

Pinto, A.P. 1979. *Functional Design Criteria, Revision 1, Z Plant Fire Protection, Project B-184.* RHO-CD-92 Rev. 1, Rockwell Hanford Operations, Richland, Washington.

Plock, E.V. 1949. "Property Record for Project C-198." General Electric Company, Richland, Washington.

PNL - Pacific Northwest Laboratory. 1965a. *Hanford Biology Research Annual Report for 1964.* BNWL-122, Richland, Washington.

PNL - Pacific Northwest Laboratory. 1965b. *Hanford Radiological Sciences Research and Development Annual Report for 1964.* BNWL-36, Richland, Washington.

PNL - Pacific Northwest Laboratory. 1966a. *Pacific Northwest Laboratory Annual Report for 1965 in the Biological Sciences.* BNWL-280, Richland, Washington.

PNL - Pacific Northwest Laboratory. 1966b. *Pacific Northwest Laboratory Annual Report for 1965 in the Physical Sciences, Volume 1: Atmospheric Sciences.* BNWL-235, Richland, Washington.

PNL - Pacific Northwest Laboratory. 1966c. *Pacific Northwest Laboratory Annual Report for 1965 in the Physical Sciences, Volume 2: Radiological Sciences.* BNWL-235, Richland, Washington.

PNL - Pacific Northwest Laboratory. 1966d. *Pacific Northwest Laboratory Annual Report for 1965 in the Physical Sciences, Volume 3: Earth Sciences.* BNWL-235, Richland, Washington.

PNL - Pacific Northwest Laboratory. 1967a. *Pacific Northwest Laboratory Annual Report for 1966 to the USAEC Division of Biology and Medicine, Volume 1: Biological Sciences.* BNWL-480, Richland, Washington.

PNL - Pacific Northwest Laboratory. 1967b. *Pacific Northwest Laboratory Annual Report for 1966 to the USAEC Division of Biology and Medicine, Volume 2: Physical Sciences, Part 1: Atmospheric Sciences.* BNWL-481 Part 1, Richland, Washington.

PNL - Pacific Northwest Laboratory. 1967c. *Pacific Northwest Laboratory Annual Report for 1966 to the USAEC Division of Biology and Medicine, Volume 2: Physical Sciences, Part 2: Radiological Sciences.* BNWL-481 Part 2, Richland, Washington.

PNL - Pacific Northwest Laboratory. 1967d. *Pacific Northwest Laboratory Annual Report for 1966 to the USAEC Division of Biology and Medicine, Volume 2: Physical Sciences, Part 3: Earth Sciences.* BNWL-481 Part 3, Richland, Washington.

PNL - Pacific Northwest Laboratory. 1968a. *Pacific Northwest Laboratory Annual Report for 1967 to the USAEC Division of Biology and Medicine, Volume 1: Biological Sciences.* BNWL-714, Richland, Washington.

PNL - Pacific Northwest Laboratory. 1968b. *Pacific Northwest Laboratory Annual Report for 1967 to the USAEC Division of Biology and Medicine, Volume 2: Physical Sciences, Part 2: Radiological Sciences.* BNWL-715 Part 2, Richland, Washington.

PNL - Pacific Northwest Laboratory. 1968c. *Pacific Northwest Laboratory Annual Report for 1967 to the USAEC Division of Biology and Medicine, Volume 2: Physical Sciences, Part 4: Earth Sciences.* BNWL-715 Part 4, Richland, Washington.

PNL - Pacific Northwest Laboratory. 1970a. *Pacific Northwest Laboratory Annual Report for 1968 to the USAEC Division of Biology and Medicine, Volume 1: Life Sciences, Part 1: Biological Sciences.* BNWL-1050 Part 1, Richland, Washington.

PNL - Pacific Northwest Laboratory. 1970b. *Pacific Northwest Laboratory Annual Report for 1968 to the USAEC Division of Biology and Medicine, Volume 2: Physical Sciences, Part 2: Radiological Sciences.* BNWL-1051 Part 2, Richland, Washington.

PNL - Pacific Northwest Laboratory. 1970c. *Pacific Northwest Laboratory Annual Report for 1968 to the USAEC Division of Biology and Medicine, Volume 2: Physical Sciences, Part 3: Instrumentation.* BNWL-1051 Part 3, Richland, Washington.

PNL - Pacific Northwest Laboratory. 1970d. *Pacific Northwest Laboratory Annual Report for 1969 to the USAEC Division of Biology and Medicine, Volume 1: Life Sciences, Part 1: Biological Sciences.* BNWL-1306 Part 1, Richland, Washington.

PNL - Pacific Northwest Laboratory. 1970e. *Pacific Northwest Laboratory Annual Report for 1969 to the USAEC Division of Biology and Medicine, Volume 1: Life Sciences, Part 2: Ecological Sciences.* BNWL-1306 Part 2, Richland, Washington.

PNL - Pacific Northwest Laboratory. 1970f. *Pacific Northwest Laboratory Annual Report for 1969 to the USAEC Division of Biology and Medicine, Volume 2: Physical Sciences, Part 1: Atmospheric Sciences.* BNWL-1307 Part 1, Richland, Washington.

PNL - Pacific Northwest Laboratory. 1970g. *Pacific Northwest Laboratory Annual Report for 1969 to the USAEC Division of Biology and Medicine, Volume 2: Physical Sciences, Part 2: Radiological Sciences.* BNWL-1307 Part 2, Richland, Washington.

PNL - Pacific Northwest Laboratory. 1971a. *Pacific Northwest Laboratory Annual Report for 1970 to the USAEC Division of Biology and Medicine, Volume 1: Life Sciences, Part 1: Biological Sciences.* BNWL-1550 Part 1, Richland, Washington.

PNL - Pacific Northwest Laboratory. 1971b. *Pacific Northwest Laboratory Annual Report for 1970 to the USAEC Division of Biology and Medicine, Volume 1: Life Sciences, Part 2: Ecological Sciences.* BNWL-1550 Part 2, Richland, Washington.

PNL - Pacific Northwest Laboratory. 1971c. *Pacific Northwest Laboratory Annual Report for 1970 to the USAEC Division of Biology and Medicine, Volume 2: Physical Sciences, Part 1: Atmospheric Sciences.* BNWL-1551 Part 1, Richland, Washington.

PNL - Pacific Northwest Laboratory. 1971d. *Pacific Northwest Laboratory Annual Report for 1970 to the USAEC Division of Biology and Medicine, Volume 2: Physical Sciences, Part 2: Radiological Sciences.* BNWL-1551 Part 2, Richland, Washington.

PNL - Pacific Northwest Laboratory. 1972a. *Pacific Northwest Laboratory Annual Report for 1971 to the USAEC Division of Biology and Medicine, Volume 1: Life Sciences, Part 1: Biological Sciences.* BNWL-1650 Part 1, Richland, Washington.

PNL - Pacific Northwest Laboratory. 1972b. *Pacific Northwest Laboratory Annual Report for 1971 to the USAEC Division of Biology and Medicine, Volume 1: Life Sciences, Part 2: Ecological Sciences.* BNWL-1650 Part 2, Richland, Washington.

PNL - Pacific Northwest Laboratory. 1972c. *Pacific Northwest Laboratory Annual Report for 1971 to the USAEC Division of Biology and Medicine, Volume 2: Physical Sciences, Part 1: Atmospheric Sciences.* BNWL-1651 Part 1, Richland, Washington.

PNL - Pacific Northwest Laboratory. 1972d. *Pacific Northwest Laboratory Annual Report for 1971 to the USAEC Division of Biology and Medicine, Volume 2: Physical Sciences, Part 2: Radiological Sciences.* BNWL-1651 Part 2, Richland, Washington.

PNL - Pacific Northwest Laboratory. 1973a. *Pacific Northwest Laboratory Annual Report for 1972 to the USAEC Division of Biomedical and Environmental Research, Volume 1: Life Sciences, Part 1: Biological Sciences.* BNWL-1750 Part 1, Richland, Washington.

PNL - Pacific Northwest Laboratory. 1973b. *Pacific Northwest Laboratory Annual Report for 1972 to the USAEC Division of Biomedical and Environmental Research, Volume 1: Life Sciences, Part 2: Ecological Sciences.* BNWL-1750 Part 2, Richland, Washington.

PNL - Pacific Northwest Laboratory. 1973c. *Pacific Northwest Laboratory Annual Report for 1972 to the USAEC Division of Biomedical and Environmental Research, Volume 2: Physical Sciences, Part 1: Atmospheric Sciences.* BNWL-1751 Part 1, Richland, Washington.

PNL - Pacific Northwest Laboratory. 1973d. *Pacific Northwest Laboratory Annual Report for 1972 to the USAEC Division of Biomedical and Environmental Research, Volume 2: Physical Sciences, Part 2: Radiological Sciences.* BNWL-1751 Part 2, Richland, Washington.

PNL - Pacific Northwest Laboratory. 1974a. *Pacific Northwest Laboratory Annual Report for 1973 to the USAEC Division of Biomedical and Environmental Research, Part 1: Biological Sciences.* BNWL-1850 Part 1, Richland, Washington.

PNL - Pacific Northwest Laboratory. 1974b. *Pacific Northwest Laboratory Annual Report for 1973 to the USAEC Division of Biomedical and Environmental Research, Part 2: Ecological Sciences.* BNWL-1850 Part 2, Richland, Washington.

PNL - Pacific Northwest Laboratory. 1974c. *Pacific Northwest Laboratory Annual Report for 1973 to the USAEC Division of Biomedical and Environmental Research, Part 3: Atmospheric Sciences.* BNWL-1850 Part 3, Richland, Washington.

PNL - Pacific Northwest Laboratory. 1974d. *Pacific Northwest Laboratory Annual Report for 1973 to the USAEC Division of Biomedical and Environmental Research, Part 4: Physical and Analytical Sciences.* BNWL-1850 Part 4, Richland, Washington.

PNL - Pacific Northwest Laboratory. 1975a. *Pacific Northwest Laboratory Annual Report for 1974 to the USAEC Division of Biomedical and Environmental Research, Part 1: Biomedical Sciences.* BNWL-1950 Part 1, Richland, Washington.

PNL - Pacific Northwest Laboratory. 1975b. *Pacific Northwest Laboratory Annual Report for 1974 to the USAEC Division of Biomedical and Environmental Research, Part 2: Ecological Sciences.* BNWL-1950 Part 2, Richland, Washington.

PNL - Pacific Northwest Laboratory. 1975c. *Pacific Northwest Laboratory Annual Report for 1974 to the USAEC Division of Biomedical and Environmental Research, Part 3: Atmospheric Sciences.* BNWL-1950 Part 3, Richland, Washington.

PNL - Pacific Northwest Laboratory. 1975d. *Pacific Northwest Laboratory Annual Report for 1974 to the USAEC Division of Biomedical and Environmental Research, Part 4: Physical and Analytical Sciences.* BNWL-1950 Part 4, Richland, Washington.

PNL - Pacific Northwest Laboratory. 1976a. *Pacific Northwest Laboratory Annual Report for 1975 to the ERDA Division of Biomedical and Environmental Research, Part 1: Biomedical Sciences.* BNWL-2000 Part 1, Richland, Washington.

PNL - Pacific Northwest Laboratory. 1976b. *Pacific Northwest Laboratory Annual Report for 1975 to the ERDA Division of Biomedical and Environmental Research, Part 2: Ecological Sciences.* BNWL-2000 Part 2, Richland, Washington.

PNL - Pacific Northwest Laboratory. 1976c. *Pacific Northwest Laboratory Annual Report for 1975 to the ERDA Division of Biomedical and Environmental Research, Part 3: Atmospheric Sciences.* BNWL-2000 Part 3, Richland, Washington.

PNL - Pacific Northwest Laboratory. 1976d. *Pacific Northwest Laboratory Annual Report for 1975 to the ERDA Division of Biomedical and Environmental Research, Part 4: Physical and Analytical Sciences and Analysis and Assessment.* BNWL-2000 Part 4, Richland, Washington.

PNL - Pacific Northwest Laboratory. 1977a. *Pacific Northwest Laboratory Annual Report for 1976 to the ERDA Assistant Administrator for Environment and Safety, Part 1: Biomedical Sciences.* BNWL-2100 Part 1, Richland, Washington.

PNL - Pacific Northwest Laboratory. 1977b. *Pacific Northwest Laboratory Annual Report for 1976 to the ERDA Assistant Administrator for Environment and Safety, Part 2: Ecological Sciences.* BNWL-2100 Part 2, Richland, Washington.

PNL - Pacific Northwest Laboratory. 1977c. *Pacific Northwest Laboratory Annual Report for 1976 to the ERDA Assistant Administrator for Environment and Safety, Part 3: Atmospheric Sciences.* BNWL-2100 Part 3, Richland, Washington.

PNL - Pacific Northwest Laboratory. 1977d. *Pacific Northwest Laboratory Annual Report for 1976 to the ERDA Assistant Administrator for Environment and Safety, Part 4: Physical and Technical Programs.* BNWL-2100 Part 4, Richland, Washington.

PNL - Pacific Northwest Laboratory. 1978a. *Pacific Northwest Laboratory Annual Report for 1977 to the DOE Assistant Secretary for Environment, Part 1: Biomedical Sciences.* PNL-2500 Part 1, Richland, Washington.

PNL - Pacific Northwest Laboratory. 1978b. *Pacific Northwest Laboratory Annual Report for 1977 to the DOE Assistant Secretary for Environment, Part 2: Ecological Sciences.* PNL-2500 Part 2, Richland, Washington.

PNL - Pacific Northwest Laboratory. 1978c. *Pacific Northwest Laboratory Annual Report for 1977 to the DOE Assistant Secretary for Environment, Part 3: Atmospheric Sciences.* PNL-2500 Part 3, Richland, Washington.

PNL - Pacific Northwest Laboratory. 1978d. *Pacific Northwest Laboratory Annual Report for 1977 to the DOE Assistant Secretary for Environment, Part 4: Physical Sciences.* PNL-2500 Part 4, Richland, Washington.

PNL - Pacific Northwest Laboratory. 1979a. *Pacific Northwest Laboratory Annual Report for 1978 to the DOE Assistant Secretary for Environment, Part 1: Biomedical Sciences.* PNL-2850 Part 1, Richland, Washington.

PNL - Pacific Northwest Laboratory. 1979b. *Pacific Northwest Laboratory Annual Report for 1978 to the DOE Assistant Secretary for Environment, Part 2: Ecological Sciences.* PNL-2850 Part 2, Richland, Washington.

PNL - Pacific Northwest Laboratory. 1979c. *Pacific Northwest Laboratory Annual Report for 1978 to the DOE Assistant Secretary for Environment, Part 3: Atmospheric Sciences.* PNL-2850 Part 3, Richland, Washington.

PNL - Pacific Northwest Laboratory. 1979d. *Pacific Northwest Laboratory Annual Report for 1978 to the DOE Assistant Secretary for Environment, Part 4: Physical Sciences.* PNL-2850 Part 4, Richland, Washington.

PNL - Pacific Northwest Laboratory. 1980a. *Pacific Northwest Laboratory Annual Report for 1979 to the DOE Assistant Secretary for Environment, Part 1: Biomedical Sciences.* PNL-3300 Part 1, Richland, Washington.

PNL - Pacific Northwest Laboratory. 1980b. *Pacific Northwest Laboratory Annual Report for 1979 to the DOE Assistant Secretary for Environment, Part 2: Ecological Sciences.* PNL-3300 Part 2, Richland, Washington.

PNL - Pacific Northwest Laboratory. 1980c. *Pacific Northwest Laboratory Annual Report for 1979 to the DOE Assistant Secretary for Environment, Part 3: Atmospheric Sciences.* PNL-3300 Part 3, Richland, Washington.

PNL - Pacific Northwest Laboratory. 1980d. *Pacific Northwest Laboratory Annual Report for 1979 to the DOE Assistant Secretary for Environment, Part 4: Physical Sciences.* PNL-3300 Part 4, Richland, Washington.

PNL - Pacific Northwest Laboratory. 1981a. *Pacific Northwest Laboratory Annual Report for 1980 to the DOE Assistant Secretary for Environment, Part 1: Biomedical Sciences.* PNL-3700 Part 1, Richland, Washington.

PNL - Pacific Northwest Laboratory. 1981b. *Pacific Northwest Laboratory Annual Report for 1980 to the DOE Assistant Secretary for Environment, Part 2: Ecological Sciences.* PNL-3700 Part 2, Richland, Washington.

PNL - Pacific Northwest Laboratory. 1981c. *Pacific Northwest Laboratory Annual Report for 1980 to the DOE Assistant Secretary for Environment, Part 3: Atmospheric Sciences.* PNL-3700 Part 3, Richland, Washington.

PNL - Pacific Northwest Laboratory. 1981d. *Pacific Northwest Laboratory Annual Report for 1980 to the DOE Assistant Secretary for Environment, Part 4: Physical Sciences.* PNL-3700 Part 4, Richland, Washington.

PNL - Pacific Northwest Laboratory. 1982a. *Pacific Northwest Laboratory Annual Report for 1981 to the DOE Office of Energy Research, Part 1: Biomedical Sciences.* PNL-4100 Part 1, Richland, Washington.

PNL - Pacific Northwest Laboratory. 1982b. *Pacific Northwest Laboratory Annual Report for 1981 to the DOE Office of Energy Research, Part 2: Ecological Sciences.* PNL-4100 Part 2, Richland, Washington.

PNL - Pacific Northwest Laboratory. 1982c. *Pacific Northwest Laboratory Annual Report for 1981 to the DOE Office of Energy Research, Part 3: Atmospheric Sciences.* PNL-4100 Part 3, Richland, Washington.

PNL - Pacific Northwest Laboratory. 1982d. *Pacific Northwest Laboratory Annual Report for 1981 to the DOE Office of Energy Research, Part 4: Physical Sciences.* PNL-4100 Part 4, Richland, Washington.

PNL - Pacific Northwest Laboratory. 1982e. *Pacific Northwest Laboratory Annual Report for 1981 to the DOE Office of Energy Research, Part 5: Environmental and Occupational Protection, Assessment, and Engineering.* PNL-4100 Part 5, Richland, Washington.

PNL - Pacific Northwest Laboratory. 1983a. *Pacific Northwest Laboratory Annual Report for 1982 to the DOE Office of Energy Research, Part 1: Biomedical Sciences.* PNL-4600 Part 1, Richland, Washington.

PNL - Pacific Northwest Laboratory. 1983b. *Pacific Northwest Laboratory Annual Report for 1982 to the DOE Office of Energy Research, Part 2: Environmental Sciences.* PNL-4600 Part 2, Richland, Washington.

PNL - Pacific Northwest Laboratory. 1983c. *Pacific Northwest Laboratory Annual Report for 1982 to the DOE Office of Energy Research, Part 3: Atmospheric Sciences.* PNL-4600 Part 3, Richland, Washington.

PNL - Pacific Northwest Laboratory. 1983d. *Pacific Northwest Laboratory Annual Report for 1982 to the DOE Office of Energy Research, Part 4: Physical Sciences.* PNL-4600 Part 4, Richland, Washington.

PNL - Pacific Northwest Laboratory. 1984a. *Pacific Northwest Laboratory Annual Report for 1983 to the DOE Office of Energy Research, Part 1: Biomedical Sciences.* PNL-5000 Part 1, Richland, Washington.

PNL - Pacific Northwest Laboratory. 1984b. *Pacific Northwest Laboratory Annual Report for 1983 to the DOE Office of Energy Research, Part 2: Ecological Sciences.* PNL-5000 Part 2, Richland, Washington.

PNL - Pacific Northwest Laboratory. 1984c. *Pacific Northwest Laboratory Annual Report for 1983 to the DOE Office of Energy Research, Part 3: Atmospheric Sciences.* PNL-5000 Part 3, Richland, Washington.

PNL - Pacific Northwest Laboratory. 1984d. *Pacific Northwest Laboratory Annual Report for 1983 to the DOE Office of Energy Research, Part 4: Physical Sciences.* PNL-5000 Part 4, Richland, Washington.

PNL - Pacific Northwest Laboratory. 1985a. *Pacific Northwest Laboratory Annual Report for 1984 to the DOE Office of Energy Research, Part 1: Biomedical Sciences.* PNL-5500 Part 1, Richland, Washington.

PNL - Pacific Northwest Laboratory. 1985b. *Pacific Northwest Laboratory Annual Report for 1984 to the DOE Office of Energy Research, Part 2: Ecological Sciences.* PNL-5500 Part 2, Richland, Washington.

PNL - Pacific Northwest Laboratory. 1985c. *Pacific Northwest Laboratory Annual Report for 1984 to the DOE Office of Energy Research, Part 3: Atmospheric Sciences.* PNL-5500 Part 3, Richland, Washington.

PNL - Pacific Northwest Laboratory. 1985d. *Pacific Northwest Laboratory Annual Report for 1984 to the DOE Office of Energy Research, Part 4: Physical Sciences.* PNL-5500 Part 4, Richland, Washington.

PNL - Pacific Northwest Laboratory. 1986a. *Pacific Northwest Laboratory Annual Report for 1985 to the DOE Office of Energy Research, Part 1: Biomedical Sciences.* PNL-5750 Part 1, Richland, Washington.

PNL - Pacific Northwest Laboratory. 1986b. *Pacific Northwest Laboratory Annual Report for 1985 to the DOE Office of Energy Research, Part 2: Environmental Sciences.* PNL-5750 Part 2, Richland, Washington.

PNL - Pacific Northwest Laboratory. 1986c. *Pacific Northwest Laboratory Annual Report for 1985 to the DOE Office of Energy Research, Part 3: Atmospheric Sciences.* PNL-5750 Part 3, Richland, Washington.

PNL - Pacific Northwest Laboratory. 1987a. *Pacific Northwest Laboratory Annual Report for 1986 to the DOE Office of Energy Research, Part 1: Biomedical Sciences.* PNL-6100 Part 1, Richland, Washington.

PNL - Pacific Northwest Laboratory. 1987b. *Pacific Northwest Laboratory Annual Report for 1986 to the DOE Office of Energy Research, Part 2: Environmental Sciences.* PNL-6100 Part 2, Richland, Washington.

PNL - Pacific Northwest Laboratory. 1987c. *Pacific Northwest Laboratory Annual Report for 1986 to the DOE Office of Energy Research, Part 3: Atmospheric Sciences.* PNL-6100 Part 3, Richland, Washington.

PNL - Pacific Northwest Laboratory. 1987d. *Pacific Northwest Laboratory Annual Report for 1986 to the DOE Office of Energy Research, Part 4: Physical Sciences.* PNL-6100 Part 4, Richland, Washington.

PNL - Pacific Northwest Laboratory. 1988a. *Pacific Northwest Laboratory Annual Report for 1987 to the DOE Office of Energy Research, Part 1: Biomedical Sciences.* PNL-6500 Part 1, Richland, Washington.

PNL - Pacific Northwest Laboratory. 1988b. *Pacific Northwest Laboratory Annual Report for 1987 to the DOE Office of Energy Research, Part 2: Environmental Sciences.* PNL-6500 Part 2, Richland, Washington.

PNL - Pacific Northwest Laboratory. 1988c. *Pacific Northwest Laboratory Annual Report for 1987 to the DOE Office of Energy Research, Part 3: Atmospheric Sciences.* PNL-6500 Part 3, Richland, Washington.

PNL - Pacific Northwest Laboratory. 1988d. *Pacific Northwest Laboratory Annual Report for 1987 to the DOE Office of Energy Research, Part 4: Physical Sciences.* PNL-6500 Part 4, Richland, Washington.

PNL - Pacific Northwest Laboratory. 1989a. *Pacific Northwest Laboratory Annual Report for 1988 to the DOE Office of Energy Research, Part 1: Biomedical Sciences.* PNL-6800 Part 1, Richland, Washington.

PNL - Pacific Northwest Laboratory. 1989b. *Pacific Northwest Laboratory Annual Report for 1988 to the DOE Office of Energy Research, Part 2: Environmental Sciences.* PNL-6800 Part 2, Richland, Washington.

PNL - Pacific Northwest Laboratory. 1989c. *Pacific Northwest Laboratory Annual Report for 1988 to the DOE Office of Energy Research, Part 3: Atmospheric Sciences.* PNL-6800 Part 3, Richland, Washington.

PNL - Pacific Northwest Laboratory. 1989d. *Pacific Northwest Laboratory Annual Report for 1988 to the DOE Office of Energy Research, Part 4: Physical Sciences.* PNL-6800 Part 4, Richland, Washington.

PNL - Pacific Northwest Laboratory. 1990a. *Pacific Northwest Laboratory Annual Report for 1989 to the DOE Office of Energy Research, Part 1: Biomedical Sciences.* PNL-7200 Part 1, Richland, Washington.

PNL - Pacific Northwest Laboratory. 1990b. *Pacific Northwest Laboratory Annual Report for 1989 to the DOE Office of Energy Research, Part 2: Environmental Sciences.* PNL-7200 Part 2, Richland, Washington.

PNL - Pacific Northwest Laboratory. 1990c. *Pacific Northwest Laboratory Annual Report for 1989 to the DOE Office of Energy Research, Part 3: Atmospheric Sciences.* PNL-7200 Part 3, Richland, Washington.

PNL - Pacific Northwest Laboratory. 1990d. *Pacific Northwest Laboratory Annual Report for 1989 to the DOE Office of Energy Research, Part 4: Physical Sciences.* PNL-7200 Part 4, Richland, Washington.

PNL - Pacific Northwest Laboratory. 1991a. *Pacific Northwest Laboratory Annual Report for 1990 to the DOE Office of Energy Research, Part 1: Biomedical Sciences.* PNL-7600 Part 1, Richland, Washington.

PNL - Pacific Northwest Laboratory. 1991b. *Pacific Northwest Laboratory Annual Report for 1990 to the DOE Office of Energy Research, Part 2: Environmental Sciences.* PNL-7600 Part 2, Richland, Washington.

PNL - Pacific Northwest Laboratory. 1991c. *Pacific Northwest Laboratory Annual Report for 1990 to the DOE Office of Energy Research, Part 3: Atmospheric and Climate Research.* PNL-7600 Part 3, Richland, Washington.

PNL - Pacific Northwest Laboratory. 1991d. *Pacific Northwest Laboratory Annual Report for 1990 to the DOE Office of Energy Research, Part 4: Physical Sciences.* PNL-7600 Part 4, Richland, Washington.

PNL - Pacific Northwest Laboratory. 1993. *Arid Lands Ecology (ALE) Facility Management Plan.* PNL-8506, Richland, Washington.

PNNL - Pacific Northwest National Laboratory. 1999. *Our History.* [Online] Available URL: http://www.pnl.gov/glance/history.html

POB - Port of Benton. 1998. *Richland Airport.* Richland, Washington. [Online] Available URL: http://www.port-of-benton.com/web/r_air.html

Pope, D. 1998. "Antinuclear Activism in the Pacific Northwest: WPPSS and Its Enemies." In *The Atomic West,* B. Hevly, and J.M. Findlay (eds.), University of Washington Press, Seattle.

PRC - PRC Environmental Management, Inc. 1992. *RCRA Facility Assessment – Final Report.* Prepared for the U.S. Environmental Protection Agency, Washington, D.C.

Prudich, T. 1947. *Discussion of Questions Concerning Specification Letters 234-1, 235-1, 234-2 and 234-3.* HW-08106, General Electric Company, Richland, Washington.

Public Law 101-510. 1990. *National Defense Authorization Act for Fiscal Year 1991.* Section 3137, "Safety Measures for Waste Tanks at Hanford Nuclear Reservation," 104 Stat. 1833 (1990).

Pugnetti, F.T. 1975. *Tiger by the Tail: Twenty-Five Years with the Stormy Tri-City Herald. Tri-City Herald,* Pasco, Washington.

Purcell, R.H. 1966. *Plutonium Recycle Test Reactor Experience, 1962-1966.* BNWL-SA-557, Battelle, Pacific Northwest Laboratories, Richland, Washington.

Quinn, F.J., Jr. 1952a. *Report: 234-5 Building RM Line Equipment Tests - Task III Mixer.* HW-24084, General Electric Company, Richland, Washington.

Quinn, F.J., Jr. 1952b. *Report: 234-5 Building RM Line Equipment Tests - Task III Furnace Calibration.* HW-24165, General Electric Company, Richland, Washington.

Ramsdell, J.V., C.S. Glantz, and R.E. Kerns. 1985. "Hanford Atmospheric Dispersion Data 1959-1974." *Atmospheric Environment* 19(1):83-86.

Reas, W.H. 1956. *Americium-Curium Correction.* HW-41274, General Electric Company, Richland, Washington.

Reberger, D.W. 1977. *Z Plant Routine Sampling Schedule for Ceramic Grade Plutonium Dioxide - Fast Flux Test facility Cores 3 and 4.* RHO-CD-67, Rockwell Hanford Operations, Richland, Washington.

Reinker, P.H. 1964. "Plutonium Waste Incineration." Letter to A.T. Gifford, June 11, 1964, General Electric Company, Richland, Washington.

Relander, C. 1956. *Drummers and Dreamers.* Caxton Printers, Caldwell, Idaho.

Remaize, J. 1996. Telephone call from Ian Lindsay (Pacific Northwest National Laboratory) to John Remaize (former fuels production manager in the 333 Building), June 28, 1996, Richland, Washington. Copy of notes on file at the Library of the U.S. Department of Energy's Hanford Cultural Resources Laboratory, Richland, Washington.

RHO - Rockwell Hanford Operations. 1978-1986. *Security Manual.* RHO-MA-101, Richland, Washington.

RHO - Rockwell Hanford Operations. 1979a. *Rockwell Hanford Operations Monthly Report, January 1979.* RHO-PB-SR-10-B-CP-Jan 79, Richland, Washington.

RHO - Rockwell Hanford Operations. 1979b. *General Radio-Chemical Training Manual.* RHO-MA-235, U.S. Department of Energy, Richland, Washington.

RHO - Rockwell Hanford Operations. 1979c. *B Plant Operations Training Manual.* RHO-MA-226, U.S. Department of Energy, Richland, Washington.

RHO - Rockwell Hanford Operations. 1984. *Program Business Management, Rockwell Hanford Operations Chemical Processing Monthly Report, July 1984.* RHO-PB-SR-10-B-CP-July 84, Richland, Washington.

RHO - Rockwell Hanford Operations. 1985a. *Chemical Processing Programs, Monthly Status Report, February 1985.* RHO-PB-SR-10-B-CP-Feb 85, Richland, Washington.

RHO - Rockwell Hanford Operations. 1985b. *Chemical Processing Programs, Monthly Status Report, July 1985.* RHO-PB-SR-10-B-CP-July 85, Richland, Washington.

RHO - Rockwell Hanford Operations. 1986-1987. "Security." *Standard Requirements and Procedures,*Volume 7, RHO-GM-MA-2, Richland, Washington.

RHO - Rockwell Hanford Operations. 1986a. *Chemical Processing Programs, Monthly Status Report, January 1986.* RHO-PB-SR-10-B-CP-Jan 86, Richland, Washington.

RHO - Rockwell Hanford Operations. 1986b. *Chemical Processing Programs, Monthly Status Report, July 1986.* RHO-PB-SR-10-B-CP-Jul 86, Richland, Washington.

RHO - Rockwell Hanford Operations. 1986c. *Chemical Processing Programs, Monthly Status Report, December 1986.* RHO-PB-SR-10-B-CP-Dec 86, Richland, Washington.

Rhodes, R. 1986. *The Making of the Atomic Bomb.* Simon and Schuster, New York.

Rhodes, R. 1995. *Dark Sun, the Making of the Hydrogen Bomb.* Simon and Schuster, New York.

Rice, M. 1998. Telephone Communication to Martha Rice (Hanford Fire Department) from Joy Keating (Pacific Northwest National Laboratory), December 15, 1998, Richland, Washington. Copy of notes on file at the Library of the U.S. Department of Energy's Hanford Cultural Resources Laboratory, Richland, Washington.

Rickard, W.H. 1972. "Rattlesnake Hills Research Natural Area." In *Federal Research Natural Areas in Oregon and Washington. A Guidebook for Scientists and Educators,* J.F. Franklin et al. (eds.), U.S. Department of Agriculture, Portland, Oregon.

Rickard, W.H. 1979. *Terrestrial Ecology Studies in the Vicinity of Washington Public Power Supply System, Nuclear Projects 1 & 4: Progress Report for 1978.* Pacific Northwest Laboratory, Richland, Washington.

Rickard, W.H. 1988. *Shrub-Steppe: Balance and Change in a Semi-Arid Terrestrial Ecosystem.* Elsevier, New York.

Riddell, F.A. 1948. *Smohalla's Village.* Document on File with the Department of Anthropology, Washington State University, Pullman, Washington.

Roberts, R.E. 1958. *History of Airborne Contamination and Control - 200 Areas.* HW-55569 RD, General Electric Company, Richland, Washington.

Roblyer, S.P. 1994. *Plutonium and Tritium Produced in the Hanford Site Production Reactors.* WHC-SD-CP-RPT-014, Westinghouse Hanford Company, Richland, Washington.

Rodgers, L.H. 1986. *1985 Remote Mechanical C RMC Line Campaign Report.* RHO-RE-EV-85, Rockwell Hanford Operations, Richland, Washington.

Roemer, J.J. 1970. *Hazards Review, Dry Glove Box for Analysis of Plutonium Oxide, 234-5 Building.* ARH-1853, Atlantic Richfield Hanford Company, Richland, Washington.

Roesch, W.C. 1964. *Progress in Radiological Physics: A Summary of Hanford Laboratories Achievements in These Programs Under General Electric,* 1947-1964. HW-83605, General Electric Company, Richland, Washington.

Rohay, V.J. 1997. *Conceptual Model for Solid Waste Landfill.* BHI-01063, Bechtel Hanford Inc., Richland, Washington.

Rohl, D. 1998. Telephone call from Joy Keating (Pacific Northwest National Laboratory) to Douglas Rohl (DynCorp Principle Engineer, Site Utilities), July 14 and 20, 1998, Richland, Washington. Copy of notes on file at the Library of the U.S. Department of Energy's Hanford Cultural Resources Laboratory, Richland, Washington.

Rohrmann, C.A. 1956. "Process Engineering at the Hanford Separation Plants." *Nucleonics* 14(6), pp. 66-68.

Rokkan, B.J. 1997. *Planes to Planes.* Copy on file at the East Benton County Historical Museum, Kennewick, Washington.

Rokkan, B.J. 1998. Interview of Bill (Rocky) Rokkan (former pilot and member of the Security Division with the Atomic Energy Commission and U.S. Department of Energy from 1948-1981), October 20, 1998. Copy of notes on file at the Library of the U.S. Department of Energy's Hanford Cultural Resources Laboratory, Richland, Washington.

Rydell, R.W. 1985. "Architectural Frontiers: An Introduction." *Pacific Historical Review,* Vol. 54, No. 4, pp. 397-403.

Sackschewsky, M.R., D.S. Landeen, G.I. Baird, W.H. Rickard, and J.L. Downs. 1992. *Vascular Plants of the Hanford Site.* WHC-EP-0554, Westinghouse Hanford Operations, Richland, Washington.

Sanders, C.L. 1976. "Pulmonary Macrophage and Epithelial Cells." *Proceedings of the 16th Annual Hanford Biology Symposium,* Energy Research and Development Agency and Pacific Northwest Laboratory, Oak Ridge, Tennessee.

Sanders, C.L. 1979. "Pulmonary Toxicology of Respirable Particles." *Proceedings of the 19th Annual Hanford Life Sciences Symposium,* U.S. Department of Energy and Pacific Northwest Laboratory, Oak Ridge, Tennessee.

Sanger, S.L. 1989. *Hanford and the Bomb: An Oral History of World War II.* Living History Press, Seattle, Washington.

Sanger, S.L. 1995. *Working on the Bomb: An Oral History of WW II Hanford.* Continuing Education Press, Portland State University, Portland, Oregon.

Schalliol, W.L., and O.J. Wick. 1951. *Technical Activities Report: Research and Development & 234-5 Metallurgy Groups, Metallurgy - Pile Technology Unit.* HW-22572, General Electric Company, Richland, Washington.

Schmidt, B.J. 1957. *Production Test: 300 Series—A Bibliography.* HW-47000, General Electric Company, Richland, Washington.

Schuette, J.B. 1998. Telephone call from Joy Keating (Pacific Northwest National Laboratory) to John Schuette (Emergency Preparedness Program Manager at the Pacific Northwest National Laboratory), November 20, 1998. Copy of notes on file at the Library of the U.S. Department of Energy's Hanford Cultural Resources Laboratory, Richland, Washington.

Seaborg, G.T. 1994. *The Plutonium Story: The Journals of Professor Glenn T. Seaborg 1939-1946.* R.L. Kathren, J.B. Gough, and G.T. Benefiel (eds.), Battelle Press, Columbus, Ohio.

Selby, J.M., and J.K. Soldat. 1958. *Summary of Environmental Incidents at Hanford 1952-1957.* HW-54636, General Electric Company, Richland, Washington.

Sharpe, J.J., 1999. *Pre-Hanford Agricultural History: 1900-1943.* BHI-01326, Rev 0, Bechtel Hanford, Inc., Richland, Washington.

Shaw, D.F. 1953a. *Expansion of 234-5 Facilities (Project No. CG-551).* Directive No. HW-305, U.S. Atomic Energy Commission, Richland, Washington.

Shaw, D.F. 1953b. *Expansion of 234-5 Facilities (Project No. CG-551).* Directive No. HW-305, Modification No. 1, U.S. Atomic Energy Commission, Richland, Washington

Shaw, D.F. 1953c. *Expansion of 234-5 Facilities (Project No. CG-551).* Directive No. HW-305, Modification No. 2, U.S. Atomic Energy Commission, Richland, Washington.

Shaw, D.F. 1954. *Expansion of 234-5 Facilities (Project No. CG-551).* Directive No. HW-305, Modification No. 3, U.S. Atomic Energy Commission, Richland, Washington.

Shaw, D.F. 1955. *Additional Air-Drying Facilities, Building 234-5 (Project No. CG-617).* Directive No. HW-350, U.S. Atomic Energy Commission, Richland, Washington.

Shaw, H.P. 1960. *Physical Completion Notice: RMC Button Line, Project CG-734.* General Electric Company, Richland, Washington.

Shaw, H.P. 1966. *Physical Completion Notice: Plutonium Buy-Back Facility, Project CGC-180.* Isochem, Inc., Richland, Washington.

Simes, R.L. 1996. *Lise Meitner: a Life in Physics.* University of California Press, Berkeley, California.

Sloat, R.J. 1963a. *Critical Mass Control Specification, Waste Hutment.* HW-65616 Rev, General Electric Company, Richland, Washington.

Sloat, R.J. 1963b. *Critical Mass Control Specification: Transporting Plutonium-Bearing Materials within the 234-5 Building.* HW-79829, General Electric Company, Richland, Washington.

Sloat, R.J. 1963c. *Critical Mass Control Specification: Transporting of Plutonium-Bearing Materials By Motor Truck.* HW-74716, General Electric Company, Richland, Washington.

Sloat, R.J. 1967. *Hanford Low Level Waste Management Reevaluation Study.* ARH-231 DEL, Atlantic Richfield Hanford Company, Richland, Washington.

Smetana, R.J. 1976. *Z Plant Liquid Waste Disposal Through the 241-Z Vault.* ARH-CD-323, Atlantic Richfield Hanford Company, Richland, Washington.

Smith, B.L.R. 1990. *American Science Policy Since World War II.* The Brookings Institute, Washington, D.C.

Smith, R.E. 1958. "Continuous Task I-II." Letter to W.N. Mobley, January 9, 1958, General Electric Company, Richland, Washington.

Smyth, H.D. 1989. *Atomic Energy for Military Purposes.* Stanford University Press, Stanford, California.

Soldat, J. K. 1960. "Monitoring for Air-Borne Radioactive Materials at Hanford Atomic Products Operation." *Journal of the Air Pollution Control Association,* 10, no. 4: 265-270.

Soldat, J. K. 1999. Interview of Joe Soldat (Health Physicist at the Hanford Site, 1948-1995) by Brian Freer (CH2M HILL Hanford, Inc.), April 9, 1999 with changes in review comments received from Mary Jarvis (DOE-RL), August 2, 2001. Copy of notes and review comments on file at the Library of the U.S. Department of Energy's Hanford Cultural Resources Laboratory, Richland, Washington.

Sorenson, D. 1996. "Pioneers Found Opportunity and Homes." *Hanford Reach,* February 26, 1996, Richland, Washington.

Sorenson, D. 1999. "Veterans from Priest Rapids Valley Served Proudly." *Hanford Reach,* November 8, 1999, Richland, Washington.

Souza, P.A., G.D. DuVall, and N. Tinker. 1996. *Department of Energy Oak Ridge Operations Office Cultural Resource Management Plan, Anderson and Roane Counties, Tennessee.* ORNL/M-580, Oak Ridge National Laboratory, Oak Ridge, Tennessee.

Stannard, J.N. 1988. *Radioactivity and Health: A History.* DOE/RL/01830-T59, Pacific Northwest Laboratory, Richland, Washington.

Stapp, D.C., and T.E. Marceau. 1996. *The Hanford Site N Reactor Buildings Task, Identification and Evaluation of Historic Properties.* BHI-00627, Rev. 0, Bechtel Hanford, Inc., Richland, Washington. [Online] Available URL: http://www.bhi-erc.com/library/bhi.htm

Stark, B.M. 1952. *Separations P.E.S. Report No. 11: Task I, Activation Study, RMA Line, 234-5 Building.* HW-24046, General Electric Company, Richland, Washington.

Stewart, A.M., and G.K. Kneale. 1991. "An Overview of the Hanford Controversy." *Occupational Medicine (State of the Art Reviews),* 6(4):641-664.

Stone, R.S. 1951. "General Introduction to Reports on Medicine, Health Physics, and Biology." *Industrial Medicine on the Plutonium Project,* R.S. Stone (ed.), McGraw-Hill, New York.

Streid, D.D. 1949. *234-5 Building Program Review.* HM-253, General Electric Company, Richland, Washington.

Strom, D.J. 1997. "A Comparison of Fatal Accident Rates at Hanford with National Rates." Conference Proceedings, *The Health of the Hanford Site,* First Annual Conference: 333-339. December 3-4, Richland, Washington. University of Washington, Seattle, Washington.

Sublette, C. 1997. "Nuclear Weapons Frequently Asked Question, Section 10.0 Chronology for The Origin of Atomic Weapons." High Energy Weapons Archive. [Online] Available URL: http://sun00781.dn.net/nuke/hew/Nwfaq/Nfaq10.html

Sublette, C. 1999. "The Manhattan Project (and Before)." High Energy Weapons Archive. [Online] Available URL: http://www.fas.org/nuke/hew/Usa/Med/Med.html

Sudmann, R.H. 1992. *Etiology of Contaminated Wounds.* WHC-SA-1484-FP, U.S. Department of Energy, Richland, Washington.

Sullivan, L.H. 1896. "The Tall Office Building Artistically Considered." *Lippincotts Magazine.* March, pp. 403-409.

Sullivan, M.F. 1968. *Gastrointestinal Radiation Injury.* Excerpta Medica Foundation, Amsterdam, Monographs on Nuclear Medicine and Biology No. 2.

Swain, E.O. 1959. *Review of Radiation Shielding Requirements and Exposure to Operators, RMC Button Line, Project CG-734.* HW-61140, General Electric Company, Richland, Washington.

Szempruch, R.W., and M.J. Szulinski. 1977. *Alternatives for Near-Term Production of Plutonium at Hanford.* ARH-CD-909, Atlantic Richfield Hanford Company, Richland, Washington.

Szilard, L. 1945. "A Petition to the President of the United States." [Online] Available URL: http://www.atomicarchive.com/Docs/SzilardPetution.shtml

Tabasinske, R.C. 1959. *26" Vacuum System Improvements, 234-5 Building.* HW-58897, General Electric Company, Richland, Washington.

Takaro, T. 1998. "Worker Safety and Hanford's Integrated Environment, Safety and Health Management System (ISMS)." *The Health of the Hanford Site,* Second Annual Conference, November 3-4, Richland, Washington.

Taylor, K. 1998a. "Hanford Steam Plants Closed." Press Release No. 98-030, March 19, 1998, Richland, Washington.

Taylor, W. 1998b. Interview of William Taylor (Industrial Safety Engineer, 1948- 1994) by Brian Freer (CH2M HILL Hanford, Inc.), December 10, 1998. Copy of notes on file at the Library of the U.S. Department of Energy's Hanford Cultural Resources Laboratory, Richland, Washington.

Teague, W.D. 1940. *Design this Day: The Technique of Order in the Machine Age.* Harcourt, Brace & Company, New York.

Teeple, B.E., and R.C. Tabasinske. 1970. *Design Criteria for Fire Protection Improvements, Z Plant.* ARH-1886, Atlantic Richfield Hanford Company, Richland, Washington.

Thayer, H. 1996. *Management of the Hanford Engineer Works in World War II.* American Society of Civil Engineers Press, New York.

Thielman, J.W. 1995. *The Changing Face of Hanford Security 1990-1994.* RL-D95-002, U.S. Department of Energy. Copy available in the U.S. Department of Energy Richland Operations Office Public Reading Room, Richland, Washington.

Thomas, E. 2000. "Bobby at the Brink." *Newsweek,* Vol. CXXXVI, No. 7, pp. 49-59.

Thomas, J.I. 1956. *Improved Task I and Task II Facilities - 234-5 Building (Project No. CG-691).* Directive No. HW-394, Modification #1, U.S. Atomic Energy Commission, Richland Washington.

Thompson, R.C. 1962. "The Biology of the Transuranic Elements." *Health Physics* 8(6):561-780.

Thompson, R.C., and W.J. Bair. 1972. "The Biological Implications of the Transuranium Elements." *Proceedings of the 11th Hanford Biology Symposium,* Pergamon Press, London.

Thompson, R.C., and J.A. Mahaffey. 1983. "Life-Span Radiation Effects Studies in Animals: What They Can Tell Us." *Proceedings of the 22nd Hanford Life Sciences Symposium,* U.S. Department of Energy and Pacific Northwest Laboratory, Oak Ridge, Tennessee.

Toffer, H. 1989. *Evolution of the Hanford Graphite Reactor Technology in 50 Years with Nuclear Fission.* American Nuclear Society, La Grange, Illinois.

Tokarz, R.D. 1984. *Fuel Accountability Enhancement Feasibility Study.* Pacific Northwest Laboratory, Richland, Washington.

Towne, G. 1998. Interview of Gordon Towne (Assistant Fire Chief, Hanford Fire Dpeartment, 1947-1978) by Brian Freer (CH2M HILL Hanford, Inc.), December 10, 1998. Copy of notes on file at the Library of the U.S. Department of Energy's Hanford Cultural Resources Laboratory, Richland, Washington.

Travis, J.E. 1956. *Permanent Mold Casting, Building 234-5 (Project No. CG-727).* Directive No. HW-421, U.S. Atomic Energy Commission, Richland, Washington.

Travis, J.E. 1958. *RMC Button Line - 234-5 Building Project No. CG-734.* Directive No. HW-426, Modification #1, U.S. Atomic Energy Commission, Richland, Washington.

Travis, J.E. 1960. *Waste Treatment Facility - Z-Plant Project No. CGC-912.* Directive No. HW-522, U.S. Atomic Energy Commission, Richland, Washington.

Travis, J.E. 1962. *Plutonium Storage Ingoting and Auxiliaries Facility - RMA Line - 234-5 Building (Project No. CGC-983).* Directive No. HW-546, U.S. Atomic Energy Commission, Richland Washington.

Travis, J.E. 1963. *Metal Stabilization Facility - 234-5 Building (Project No. CAC-987).* Directive No. AEC-219, U.S. Atomic Energy Commission, Richland, Washington.

Tri-City Herald. 1961. "Camp Hanford in North Richland Is Deserted." March 31, 1961, Kennewick, Washington.

Tri-City Herald. 1995. "Small Town Recalls WWII Blast." May 5, 1995, p. A4, Kennewick, Washington.

Uebelacker, D.L. 1975. *Plutonium Processing Manual.* ARH-779, Atlantic Richfield Hanford Company, Richland, Washington.

UNI - United Nuclear Industries, Inc. 1984. *Fuel Operations Process Work Request No. B-377, Richland, Washington.* Document in the Seattle Federal Records Center Box 105032.

UNI - United Nuclear Industries, Inc. n.d. "Hanford's N Reactor." Brochure published in the early 1980s by the operating contractor for the N Reactor, Richland, Washington.

Upington, G.W. 1968. *Safety Analysis Report and Hazards Review, Hood No. 3 — Miscellaneous Treatment Facility.* ARH-421, Atlantic Richfield Hanford Company, Richland, Washington.

UPOA - University Publications of America. 1977a. *A Guide to Manhattan Project: Official History and Documents.* Book 4, Instrument Department; Book 10, Production Department, Part I – II; Washington, D.C

UPOA - University Publications of America. 1977b. A *Guide to Manhattan Project: Official History and Documents.* Books 1 and 4, Washington, D.C.

USACE - U.S. Army Corps of Engineers. 1947. *Manhattan District History, Book IV - Pile Project.* In Manhattan Project, Official History and Documents (1977 microfilm), University Publications of America, Washington, D.C.

UW - University of Washington. 1977. *Hanford - Yesterday, Today, and Tomorrow.* For the Hanford Project, Richland, Washington.

Van Arsdol, T. 1980. *The Wahluke Slope and Hanford.* Document on file with the Washington State Library, Olympia, Washington.

Van Arsdol, T. 1992. *Hanford: The Big Secret.* Privately published, Vancouver, Washington. Copy available in the U.S. Department of Energy Richland Operations Office Public Reading Room, Richland, Washington.

Vaughan, B.E., and W.H. Rickard. 1977. *Hanford National Environmental Research Park (NERP). A Descriptive Summary of the Site and Site-Related Research Programs, 1952-1977.* PNL-2299, Pacific Northwest Laboratory, Richland, Washington.

VEC - Vitro Engineering Corp. 1980. *Title I Design Report - Z Plant Ventilation and Services Upgrade.* RHO-CD-826, Richland, Washington.

Veneziano, T.B. 1983. *Operating Specifications for 234-5Z, 236-Z, Room 227, Load-In and Load-Out Gloveboxes.* OSD-Z-184-00012, Rockwell Hanford Operations, Richland, Washington.

Vinther, P.A. 1996. Telephone call from Ian Lindsay (Pacific Northwest National Laboratory) to Paul A.Vinther (retired Vice President and Director, Fuels Operation), July 11, 1996, Richland, Washington. Copy of notes on file at the Library of the U.S. Department of Energy's Hanford Cultural Resources Laboratory, Richland, Washington.

Voelz, G.L. 1991. "Health Considerations for Workers Exposed to Plutonium." *Occupational Medicine (State of the Art Reviews)*, 6(4):681-694.

Vogt, E.C. 1996. "Plutonium Finishing Plant Operational Activities During Remainder of Fiscal Year 1996." Letter to J.E. Mecca, Correspondence #9651517, May 17, 1996, Westinghouse Hanford Company, Richland, Washington.

Wahlen, R.K. 1989. *History of 100-B Area.* WHC-EP-0273, Westinghouse Hanford Company, Richland, Washington.

Walker, A. 1999. Telephone call from Joy Keating (Pacific Northwest National Laboratory) to Alan Walker (Safeguards and Security Division at the U.S. Department of Energy), May 3, 1999. Copy of notes on file at the Library of the U.S. Department of Energy's Hanford Cultural Resources Laboratory, Richland, Washington.

Walker, J.S. 1992. *Containing the Atom: Nuclear Regulation in a Changing Environment 1963-1971.* University of California Press, Berkeley, California.

Walker, M. 1995. *The Cold War. A History.* First Owl Book Edition, Henry Holt and Company, Inc., New York.

Walser, R.W., and R.Y. Lyon. 1965. *Z-Plant Weekly Report: Task I-II and Recovery Operations, January 1, 1964 through December 31, 1964.* HW-80356 RD, General Electric Company, Richland, Washington.

Walser, R.W., and R.E. Van Der Cook. 1979a. *Z Plant RMA Oxide Line Campaign Process Control Plan.* RHO-CD-644, Rockwell Hanford Operations, Richland, Washington.

Walser, R.W., and R.E. Van Der Cook. 1979b. *Z Plant RMA Oxide Line Flowsheet Addendum - Filtrate Concentration.* RHO-CD-616 Add., Rockwell Hanford Operations, Richland, Washington.

Walton, C.W. 1999. "Physical Protection." *Future Safeguards at the Hanford Site,* Symposium by the Institute of Nuclear Materials Management, Pacific Northwest Chapter, April 29, 1999, Richland, Washington.

Warner, G.T. 1993. "Hanford Historical Accident/Incident Experience." Letter from G.T. Warner (Westinghouse Hanford Company) to R.A. Holten (U.S. Department of Energy), November 2, 1993, Richland, Washington. Copy on file at the Library of the U.S. Department of Energy's Hanford Cultural Resources Laboratory, Richland, Washington.

Warner, M. 2000. "Hanford History at www.hanford.gov/doe/culres/mpd." Electronic mail message from Mark Warner (son of 1st Lieutenant John B. Warner who served with Company D, 744th Military Police Battalion assigned from Fort Lewis to the Hanford Site) to Dave Harvey and Darby Stapp (Pacific Northwest National Laboratory), March 26, 2000. Copy on file at the Library of the U.S. Department of Energy's Hanford Cultural Resources Laboratory, Richland, Washington.

Warren, J.H. 1952. *Production Loss Due to Ruptured Uranium Slugs.* HW-228133, General Electric Company, Richland, Washington.

Warren, J.H. 1963. *Chemical Form of Plutonium Buy Back.* HW-76496, General Electric Company, Richland, Washington.

Wasserman, H., and N. Soloman. 1982. *Killing our Own: The Disaster of America's Experience with Atomic Radiation.* Delacorte Press, New York.

Watson, E.C., I.C. Nelson, D.H. Wood, R.O. McClellan, and L.K. Bustad. 1964. "Effect of Varying Stable Iodine in Diets of Cows Fed I131 on Uptake of I131 in Man Drinking the Milk – An Abstract." In *Biology of Radioiodine: Proceedings of the Hanford Symposium on the Biology of Radioiodine, Richland, Washington, July 17-19, 1964.* Pergamon Press, Oxford.

Weakley, E.A. 1958. *Technology of the HAPO Lead Dip Fuel Element Canning Process.* HW-58115, General Electric Company, Richland, Washington.

Weakley, E.A. 1976. *History and Status of Environmental Improvements for Fuels Production Division.* UNI-652, United Nuclear Industries, Inc., Richland, Washington.

Weakley, E.A. 1980. *Operation of N Reactor and Fuels Fabrication Facilities.* DOE/EA-0030, Rev. 1, U.S. Department of Energy, Richland, Washington.

Weakley, E.A. 1999. Interview of Everett A. Weakley (Chemical Engineer with General Electric Company and Westinghouse Hanford Company, 1951-1994) by Dave Harvey (Pacific Northwest National Laboratory), July 1999. Copy of notes on file at the Library of the U.S. Department of Energy's Hanford Cultural Resources Laboratory, Richland, Washington.

Welander, A.D., L.R. Donaldson, R.F. Foster, K. Bonham, and A.H Seymour. 1948. "The Effects of Roentgen Rays on the Embryos and Larvae of Chinook Salmon." *Growth* 12:201-242.

Wellington, D. 1998. Telephone call from David Harvey (Pacific Northwest National Laboratory) to Denny Wellington (Camp Hanford Veteran), August 20 and August 24, 1998, Richland, Washington.

Wellock, T. 1998. "Radical Initiatives and Moderate Alternatives: California's 1976 Nuclear Safeguards Initiative." In *The Atomic West,* B. Hevly and J.M. Findlay (eds.), University of Washington Press, Seattle.

Welsome, E. 1999. *The Plutonium Files: America's Secret Medical Experiments in the Cold War.* The Dial Press, Random House, Inc., New York.

WHC - Westinghouse Hanford Company. 1989. *N-Reactor Updated Safety Analysis Report (NUSAR).* WHC-SP-0297, Vol. 5, Amendment 21, Richland, Washington.

WHC - Westinghouse Hanford Company. 1994. *Fast Flux Test Reactor.* Richland, Washington.

WHC - Westinghouse Hanford Company. 1996. *DNFSB Recommendation 94-1, Hanford Site Integrated Stabilization Management Plan.* WHC-EP-0853 Rev 4, Richland, Washington.

WIDS - *Waste Information Data System.* 1998. Bechtel Hanford Inc., Richland, Washington. [Online] Available URL: http://www.bhi-erc.com/projects/p_m/eis/wids/wids.htm

Wildung, R.E., and E.A. Jenne. 1981. "Biological Availability of Trace Metals." *Proceedings of the 21st Annual Hanford Life Sciences Symposium,* Elsevier, Amsterdam.

Wilk, J.L. 1993. *Plutonium Finishing Plant Instrument Upgrade, B-610-CENRTC.* WHC-SD-610-FDC-001 Rev. 3, Westinghouse Hanford Company, Richland, Washington.

Wilkinson, G.S. 1991. "Epidemiologic Studies of Nuclear and Radiation Workers: An Overview of What Is Known about Health Risks Posed by the Nuclear Industry." *Occupational Medicine (State of the Art Reviews),* 6(4):715-724.

Williams, D.G. 1966. Letter from D.G. Williams to General Electric Company, Hanford Atomic Products Operation, March 25, 1966. Document in CGN 184 File in Federal Records Center Box 84671, Seattle, Washington.

Williams, G.R. 1976. *Exterior Door Control System of Buildings 234-5 and 236-Z.* ARH-CD-711, Atlantic Richfield Hanford Company, Richland, Washington.

Wilson, G.R., and J.K. McClusky. 1983. "RMA Line Backup Support to PUREX." RHO Internal Letter NMM04000-83-041, March 31, 1983, Rockwell Hanford Operations, Richland, Washington.

Wilson, R.H. 1987. *Historical Review of Personnel Dosimetry Development and Its Use in Radiation Protection Programs at Hanford, 1944 to the 1980s.* PNL-6125, Pacific Northwest Laboratory, Richland, Washington.

Wirta, R.W. 1954. *Coil Design for Task III Replacement Equipment.* HW-32810, General Electric Company, Richland, Washington.

Wirth, J.E. 1951. "Medical Services of the Plutonium Project." *Industrial Medicine on the Plutonium Project,* Robert S. Stone (ed.), McGraw-Hill, New York.

Work, J.B. 1947a. *234-5 Building Design Philosophy.* HW-8140, General Electric Company, Richland, Washington.

Work, J.B. 1947b. *Analytical Requirements of the 234-5 Project Chemical Development Program.* HW-8252, General Electric Company, Richland, Washington.

Work, J.B. 1948. *Additional Design Forms for 234-5 Project.* HW-11112, General Electric Company, Richland, Washington.

Yoo, S. 1985. *Remote Mechanical C Line Startup Criteria and Requirements.* SD-CP-CR-009, Rockwell Hanford Operations, Richland, Washington.

Young, J.S., and J.S. Furchter. 1991. *Addendum to Data Compilation Task Report for the Source Investigation of the 300-FF-1 Operable Unit Phase 1 Remedial Investigations.* EMO-1026, Pacific Northwest Laboratory, Richland, Washington.

Zangar, C.N. 1962. *Summary Report of Accidental Nuclear Excursion, RECUPLEX Operation, 234-5 Facility.* TID-18431, U.S. Atomic Energy Commission, Richland, Washington.

Bibliography for Further Information

BIBLIOGRAPHY FOR FURTHER INFORMATION

59 CFR 28848. 1994. "Defense Nuclear Facilities Safety Board Recommendation 94-1, Improved Schedule for Remediation in the Defense Nuclear Facilities Complex." U.S. Code of Federal Regulations.

Allen, C.R. 1980. *Unusual Occurrence Report, Fire in Digester Glove Box 200.* No. 80-23, Hanford Engineering Development Laboratory, Richland, Washington.

Allen, C.R., J.R. Divine, and R.G. Cowan. 1978. *Safety Analysis Report for the Radioactive Acid Digestion Test Unit RADTU.* TC-893, Hanford Engineering Development Laboratory, Richland, Washington.

American Society of Mechanical Engineers. 1952. *Uranium, Plutonium and Industry: A Summary of the U.S. Atomic Energy Program.* New York.

Anderson, M.D. 1979. *1979 Z Plant RMA Oxide Line Campaign Evaluation.* RHO-CD-878, Rockwell Hanford Operations, Richland, Washington.

Atlantic Richfield Hanford Company. 1967. *Monthly Report, 200 Areas Operation.* Monthly issues for October through December 1967. ARH-60 DEL, ARH-61 DEL, ARH-62 DEL, Richland, Washington.

Atlantic Richfield Hanford Company. 1968. *Monthly Report, 200 Areas Operation.* Monthly issues for January through August, October through December. ARH-300 DEL, ARH-301 DEL, ARH-302, ARH-303 DEL, ARH-304 DEL, ARH-305 DEL, ARH-306 DEL, ARH-307 DEL, ARH-309 DEL, ARH-310 DEL, ARH-311 DEL, Richland, Washington.

Atlantic Richfield Hanford Company. 1969. *Monthly Report, 200 Areas Operation.* Monthly issues for January, February, April through September, December 1969. ARH-1100 DEL, ARH-1101 DEL, ARH-1103 DEL, ARH-1104 DEL, ARH-1105 DEL, ARH-1106 DEL, ARH-1107 DEL, ARH-1108 DEL, ARH-1109 DEL, ARH-1110 DEL, Richland, Washington.

Atlantic Richfield Hanford Company. 1970. *200 Areas Operation Monthly Report.* Monthly issues for January through December 1970. ARH-1111 DEL, ARH-1501 DEL, ARH-1502 DEL, ARH-1503 DEL, ARH-1504 DEL, ARH-1505 DEL, ARH-1506 DEL, ARH-1507 DEL, ARH-1508 DEL, ARH-1509 DEL, ARH-1510 DEL, ARH-1511 DEL, ARH-1512 DEL, Richland, Washington.

Atlantic Richfield Hanford Company. 1971. *200 Areas Operation Monthly Report.* Monthly issues for January through December 1971. ARH-1901 DEL, ARH-1902 DEL, ARH-1903 DEL, ARH-1904 DEL, ARH-1905 DEL, ARH-1906 DEL, ARH-1907 DEL, ARH-1908 DEL, ARH-1909 DEL, ARH-1910 DEL, ARH-1911 DEL, ARH-1912 DEL Richland, Washington.

Atlantic Richfield Hanford Company. 1972. *Atlantic Richfield Hanford Company Monthly Report.* Monthly issues for January-August, October-December 1972, ARH-2301-A DEL, ARH-2302-A DEL, ARH-2303-A DEL, ARH-2304-A DEL, ARH-2305-A DEL, ARH-2306-A DEL, ARH-2307-A DEL, ARH-2308-A DEL, ARH-2310-A DEL, ARH-2311-A DEL, ARH-2312-A DEL, Richland, Washington.

Atlantic Richfield Hanford Company. 1972. *Contaminated Puncture Wound Requiring Excising.* RL Occurrence Report 72-34, Richland, Washington.

Atlantic Richfield Hanford Company. 1973. *Atlantic Richfield Hanford Company Monthly Report.* Monthly issues for January, April, August through October 1973 ARH-2701-B DEL, ARH-2707-B DEL, ARH-2708-B DEL, ARH-2709-B DEL, ARH-2710-B DEL, Richland, Washington.

Atlantic Richfield Hanford Company. 1973. *Loss of Contamination Control, Plastic Bag Operation.* RL Occurrence Report 73-4, Richland, Washington.

Atlantic Richfield Hanford Company. 1973. *Potential Lung Deposition.* RL Occurrence Report 73-51, Richland, Washington.

Atlantic Richfield Hanford Company. 1974. *Atlantic Richfield Hanford Company Monthly Report.* Monthly issues for December 1973, January, June, December 1974. ARH-2712-B DEL, ARH-3001-B DEL, ARH-3006-B DEL, ARH-3011-B DEL, Richland, Washington.

Atlantic Richfield Hanford Company. 1975. *Atlantic Richfield Hanford Company Monthly Report.* Monthly issues for January, April and September 1975. ARH-LD-201-B, ARH-LD-204-B, ARH-LD-209-B, Richland, Washington.

Atlantic Richfield Hanford Company. 1975. *Decommissioning of Division of Military Application Equipment at Richland Operations Office.* ARH-R-215-1, ARH-R-215-2, Richland, Washington.

Atlantic Richfield Hanford Company. 1975. *Leakage of Plutonium Nitrate from Hood 179-9.* RL Occurrence Report 75-81, Richland, Washington.

Atlantic Richfield Hanford Company. 1975. *Loss of Contamination Control.* RL Occurrence Report 75-8, Richland, Washington.

Atlantic Richfield Hanford Company. 1975. *Potential Plutonium Inhalation.* RL Occurrence Report 75-129, Richland, Washington.

Atlantic Richfield Hanford Company. 1975. *Ruptured Container and Gross Contamination, Room 173.* RL Occurrence Report 75-126, Richland, Washington.

Atlantic Richfield Hanford Company. 1976. *Atlantic Richfield Hanford Company Production Report, July 1976.* ARH-LD-219-A, Richland, Washington.

Atlantic Richfield Hanford Company. 1976. *Decommissioning of Division of Military Application Equipment at Richland Operations Office.* ARH-R-216-3, Richland, Washington.

Atlantic Richfield Hanford Company. n.d. *Hanford Patrol Handbook.* Richland, Washington. Copy available in the U.S. Department of Energy Richland Operations Office Public Reading Room.

Bair, W.J. 1970. *Plutonium Inhalation Studies.* BNWL-1221, Pacific Northwest Laboratory, Richland, Washington.

Bard, J.C., and J.B. Cox. 1997. "Euro-American Resettlement of the Hanford Site (Lewis and Clarke 1805 – Hanford Engineer Works 1943), Washington." DOE/RL-97-02, Rev. 0, in *National Register of Historic Places Multiple Property Documentation Form – Historic, Archaeological and Traditional Cultural Properties of the Hanford Site, Washington,* U.S. Department of Energy, Richland, Washington.

Beard, S.J. 1970. *Production of Ceramic Grade Plutonium Dioxide.* ARH-1851, Atlantic Richfield Hanford Company, Richland, Washington.

Beary, M.M. 1976. *Functional Design Criteria, Plutonium Processing Plant Radiation Shielding.* ARH-CD-626, Atlantic Richfield Hanford Company, Richland, Washington.

Beary, M.M., A.H. Hinkson, L.L. Lundgren, and M.G. Wheeler. 1972. *Hazards Evaluation of Prototype Waste Compactor for 234-5 Building.* ARH-2547, Atlantic Richfield Hanford Company, Richland, Washington.

Bell, R.S. 1951. *S Division Monthly Report, 234-5 Operations, February 1951.* HW-20506 DEL, General Electric Company, Richland, Washington.

Bensen, D.W. 1960. *Review of Soil Chemistry Research at Hanford.* HW-67201, General Electric Company, Richland, Washington.

Bergquist, G.G. 1994. *System Design Description, Plutonium Finishing Plant Low Level Waste Treatment Facility.* WHC-SD-CP-SDD-018, Westinghouse Hanford Company, Richland, Washington.

Berry, E.C., and J.F. Cline. 1950. *First Annual Report of the Botany Field Station.* HW-16056, General Electric Company, Richland, Washington.

Bond, R.H. 1964. *Corrosion of Plutonium Metal When Stored in Containers Having Plastic Components.* HW-81240, General Electric Company, Richland, Washington.

Bonham, K., L.R. Donaldson, R.F. Foster, A.D. Welander, and A.H. Seymour. 1948. "The Effect of X-Ray on Mortality, Weight, Length, and Counts of Erythrocytes and Hematopoietic Cells in Fingerling Chinook Salmon, *Oncorhynchus tshawytscha* Walbaum." *Growth* 12:101-121.

Borgeson, M.E. 1964. "Furnace Conveyor Belt." Memo to J.W. Fillmore, December 31, 1964, General Electric Company, Richland, Washington.

Braden, D.E. 1972. "Plutonium Recovery From 216-Z-9 Cavern Soil - Criteria for Scrap and Waste." Memo to L.M. Knights, August 17, 1972, Atlantic Richfield Hanford Company, Richland, Washington.

Brecke, L.I. 1958. *Investigation of Rapid Oxidation of Compact Alpha Plutonium Chips.* HW-57240, General Electric Company, Richland, Washington.

Brown, A.C. 1977. *The Secret History of the Atomic Bomb.* The Dial Press/James Wade, New York.

Brown, C.L. 1963. *Nuclear Safety Review, New Plutonium Reclamation Facility, Z Plant - 200 West Area (CAC-880).* HW-78880, G.E. Hanford Company, Richland, Washington.

Bruns, L.E. 1971. *Prototype Solids Handling Systems for Plutonium Finishing.* ARH-2078, Atlantic Richfield Hanford Company, Richland, Washington.

Bruns, L.E. 1972. *Utilization of Z Plant for Plowshare Program.* ARH-2585 DEL, Atlantic Richfield Hanford Company, Richland, Washington.

Brunstad, A. 1956. *Corrosion of Stainless Steel, Titanium, and Tantalum in Plutonium-Nitric Acid Solution.* HW-46779, General Electric Company, Richland, Washington.

Buchanan, J.A., E.S. Gilbert, and E. Omohundro. 1993. "Mortality of Workers at the Hanford Site 1945-1986." *Health Physics* 64(6):577-590.

Burke, J.E. 1981. "Recollections of Processing Uranium Hydride and Plutonium at Wartime Los Alamos." *Journal of Nuclear Materials Management,* 100:11-16.

Burton, G. 1975. "Americium 241 Recovery." Letter to O.J. Elgert, May 20, 1975, Atlantic Richfield Hanford Company, Richland, Washington.

Bush, V. 1945. *Science the Endless Frontier.* U.S. Government Printing Office, Washington, D.C.

Bush, V. 1970. *Pieces of the Action.* William Morrow and Company, Inc., New York.

Cameron, B. 1998. "Hanford Patrol Pistol Team best in the West." *Hanford Reach*, September 21, 1998, U.S. Department of Energy, Richland, Washington.

Cammann, J.W., E.H. Carbaugh, W.G. Jasen, H.E. McGuire, and J.E. Pierce. 1979. *Analysis of 200 Area Effluent Sampler-Monitor Systems.* RHO-CD-667, Rockwell Hanford Operations, Richland, Washington.

Cantril, S.T. 1951. "Biological Bases for Maximum Permissible Exposures." In *Industrial Medicine on the Plutonium Project*, R. S. Stone (ed.), McGraw-Hill, New York.

Cantril, S.T., and H.M. Parker. 1945. *The Tolerance Dose.* CH-2812, Health Division, Metallurgical Laboratory, Chicago, Illinois.

Cantril, S.T., E.O. Wollan, and K.S. Cole. 1942. *Health and Safety.* CH-376, Health Division, Metallurgical Laboratory, Chicago, Illinois.

Carpenter, E.R. 1977. *Facilities and Equipment Recommendations for Z Plant Rag Incinerator.* RHO-CD-211, Rockwell Hanford Operations, Richland, Washington.

Chapman, V.R. 1952. *Separations Section - Operations, 234-5 Building Monthly Report.* Monthly issues for December 1951, January, February 1952. HW-23125 DEL, HW-23423 DEL, HW-23692 DEL, General Electric Company, Richland, Washington.

Chapman, V.R. 1953. *Separations Section - Operations Sub-Section, 234-5 Building Monthly Report, September 1953.* HW-29499 DEL, General Electric Company, Richland, Washington.

Chapman, V.R. 1954. *Separation Section - Operations Sub-Section, 234-5 Building Monthly Report, February 1954.* HW-30969 DEL, General Electric Company, Richland, Washington.

Chew, M.H., and Associates, Inc. 1992. *A Facility History of Building 771 at Rocky Flats Plant.* DE-AC-001-90-EW-48063, U.S. Department of Energy, Washington, D.C.

Cockeram, D.J. 1978. *Rockwell Hanford Operations Monthly Report - July 1978.* RHO-LD-10-B Jul 78, Rockwell Hanford Operations, Richland, Washington.

Compton, J.A., and C.E. White. 1978. *Laboratory Development of the 236-Z Canyon and Process Hood Floor Cleanup System.* RHO-CD-76, Rockwell Hanford Operations, Richland, Washington.

Cooper, V.R. 1951. *Recommendations - Recovery of Product from 234-5 Building Miscellaneous Liquid Wastes.* HW-21277, General Electric Company, Richland, Washington.

Corbell, D.W. 1975. "Ventilation Flow Test, 236-Z Building." Memo to D.L. Uebelacker December 2, 1975, with attachment "Ventilation Duct Replacement 236-Z Building, Unique Procedure 75-12," D.W. Corbell. Atlantic Richfield Hanford Company, Richland, Washington.

Corlew, R.P. 1972. "Solids Removal Prototype." Letter to O.J. Elgert, December 28, 1972, Atlantic Richfield Hanford Company, Richland, Washington.

Crawford, M. 1997. *Toxic Waste Sites: An Encyclopedia of Endangered America.* ABC-Clio, Santa Barbara, California.

Crocker, H.W. 1961. *Continuous Dissolution of Plutonium-Bearing Slag and Crucible Residues.* HW-69415, General Electric Company, Richland, Washington.

Crocker, R.L. 1984. *RMA History VI Sample Schedules and FFTF Oxide Specifications.* SD-CP-TI-041, Rockwell Hanford Operations, Richland, Washington.

Curren, E.F. 1966. *Hazards Evaluation - Plutonium Nitrate Receiving and Blending Hood HC-4.* ISO-59, Isochem, Inc., Richland, Washington.

Curtis, M.H. 1962. *Continuous Electrowinning of Plutonium Metal, Status Report.* HW-75363, General Electric Company, Richland, Washington.

Curtis, M.H. 1965. *Continuous Electrowinning of Plutonium Metal.* RL-SEP-30, General Electric Company, Richland, Washington.

Curtis, M.H. 1973. *Major Program Issue for Fiscal Year 1975, Improved Control of Aqueous Releases, Z Plant.* ARH-2828, Atlantic Richfield Hanford Company, Richland, Washington.

Curtis, M.H. 1974. *Liquid Waste Management - Z Plant.* ARH-2963, Atlantic Richfield Hanford Company, Richland, Washington.

Curtis, M.H. 1983. *Criteria for Interim Metal Production.* SD-CP-CR-003, Rockwell Hanford Operations, Richland, Washington.

Davis, M.V., and H.A. Fowler. 1956. *Hanford Test Pile.* General Electric Company, Richland, Washington.

DeFord, D.H. 1991. *PUREX Aggregate Area Management Study, Technical Baseline Report.* WHC-IP-0795, Westinghouse Hanford Company, Richland, Washington.

DeFord, D.H. 1991. *T Plant Aggregate Area Baseline Study, Technical Baseline Report.* WHC-IP-787, Westinghouse Hanford Company, Richland, Washington.

DeFord, D.H. 1991. *U Plant Aggregate Area Baseline Study, Technical Baseline Report.* WHC-IP-0775, Westinghouse Hanford Company, Richland, Washington.

Demiter, J.A., D.R. Duncan, W.E. Meeuwsen, R.W. Reddinger, and M.S. Zvonar. 1991. *Historical Report on Retrievably Stored Division of Military Application Equipment at the Hanford Site.* WHC-IP-0791, Westinghouse Hanford Company, Richland, Washington.

Dittner, G.R. 1993. "1952 Reflected Full Year as Protectors for Camp Hanford Soldiers." *The Courier,* East Benton County Historical Society, Kennewick, Washington.

Dixon, D.R. 1977. *236-Z Canyon Utilization Study.* ARH-CD-908, Atlantic Richfield Hanford Company, Richland, Washington.

E.I. Du Pont de Nemours and Company. 1943. "Memorandum covering technical basis of contract." HAN-43508, [Online]. Available URL: http://www2.hanford.gov/declass/d30pyfindpage.py?AKey=D197294669&Pages=22

E.I. Du Pont de Nemours and Company. 1945. *Design and Procurement History of Hanford Engineer Works and Clinton Semi-Works.* IN-6263, Vols. 1 and 2, Wilmington, Delaware.

E.I. Du Pont de Nemours and Company. 1946. *Book 6.* HAN-73214, Richland, Washington.

E.I. Du Pont de Nemours and Company. 1946. *Book 10.* HAN-73214-DEL, Richland, Washington.

Elderkin, C.E., and J.J Fuquay. 1965. *Recent Hanford Diffusion Results Significant for Estimating Pollution Potential.* BNWL-SA-429, Pacific Northwest Laboratory, Richland, Washington.

Fecht, J.B. 1972. "Process Prototype - Z-9 Enclosed Trench Mining Equipment." Letter to O.J. Elgert, February-18, 1972, Atlantic Richfield Hanford Company, Richland, Washington.

Fecht, J.B., R.D. Anderson, H.A. Lee, L.L. Lundgren, W.E. Matheison, M.T. Slind, and F.R. Vollert. 1975. *Hazards Review - Contaminated Soil Removal Facility - 216-Z-9 Enclosed Trench, Project HCP-687.* ARH-CD-268, Atlantic Richfield Hanford Company, Richland, Washington.

Felt, R.E. 1972. *Plutonium Scrap Management Plan.* ARH-2618 DEL, Atlantic Richfield Hanford Company, Richland, Washington.

Fix, J.J., R.H. Wilson, and W.V. Baumgartner. 1997. *Retrospective Assessment of Personnel Neutron Dosimetry for Workers at the Hanford Site.* PNNL-11196, Pacific Northwest National Laboratory, Richland, Washington.

Fleckenstein, D.C. 1956. *Radiological Sciences Department Annual Report 1955.* HW-41315, General Electric Company, Richland, Washington.

Foster, R.F., and R.W. Coopey. 1948. *The Aquatic Biology Laboratory (Rough Draft).* HW-23791, General Electric Company, Richland, Washington.

Freer, B. 1994. "Atomic Pioneers and Environmental Legacy at the Hanford Site." *The Canadian Review of Sociology and Anthropology,* 31(3): 305-324.

Fremling, A.G. 1976. *Statement RE: Investigation of the Chemical Explosion of an Ion Exchange Resin Column in the 242-Z Building on August 30, 1976.* U.S. Energy Research and Development Administration, Richland, Washington.

Fritz, J.R. 1957. "Basis for Cost Estimate, Project CG-734 - RMC Button Line." Memo to J.S. Crowder, General Electric Company, Richland, Washington.

Fuquay, J.J. 1964. *Progress in Atmospheric Physics: A Summary of Hanford Laboratories Work on This Program Under General Electric 1959-1964.* HW-83602, General Electric Company, Richland, Washington.

Galison, P., and B. Hevly. 1992. *Big Science, the Growth of Large-Scale Research.* Stanford University Press, Stanford, California.

Gamertsfelder, C.C. 1963. *Plans and Hazards Analysis for the First Hanford I^{131} Field Release Test.* HW-78312, General Electric Company, Richland, Washington.

Gast, P.F. 1951. *Analysis of Supernatant, 234-5 Building.* HW-22004, General Electric Company, Richland, Washington.

General Electric Company. 1947. *Hanford Engineer Works Monthly Report.* Monthly issues for March, April through June, September, October 1947. HW-7-6048 DEL, HW-7-6184 DEL, HW-7-6391 DEL, HEW-7096 DEL, HW-7979 DEL, Richland, Washington.

General Electric Company. 1948. *Hanford Works Monthly Report.* Monthly issues for December 1947, January, February, June though November 1948. HW-8438 DEL, HW-8931 DEL, HW-9191 DEL, HW-10378 DEL, HW-10714 DEL, HW-10993 DEL, HW-11226 DEL, HW-11499 DEL, HW-11835 DEL, Richland, Washington.

General Electric Company. 1949. *Hanford Works Monthly Report.* Monthly issues for December 1948, January through April, June 1949. HW-12086 DEL, HW-12391 DEL, HW-12666 DEL, HW-12937 DEL, HW-13190 DEL, HW-13793 DEL, Richland, Washington.

General Electric Company. 1950. *Hanford Works Monthly Report.* Monthly issues for January through November 1950. HW-15843 DEL, HW-17056 DEL, HW-17410 DEL, HW-17660 DEL, HW-17971 DEL, HW-18221 DEL, HW-18473 DEL, HW-18740 DEL, HW-19021 DEL, HW-19301 DEL, HW-19622 DEL, Richland, Washington.

General Electric Company. 1950. *Project Proposal: Skull Recovery Facilities for Building 234-5.* HW-19577, Richland, Washington.

General Electric Company. 1951. *Hanford Works Monthly Report.* Monthly issues for January, April, June through November. HW-20161 DEL, HW-20991 DEL, HW-21506 DEL, HW-21802 DEL, HW-22304 DEL, HW-22610 DEL, HW-22875 DEL, Richland, Washington.

General Electric Company. 1951. *The Second Annual Report of the Botany Field Station.* HW-20999, Richland, Washington.

General Electric Company. 1952. *Hanford Works Monthly Report.* Monthly issues for December 1951, July 1941. HW-23140 DEL, HW-25227 DEL, Richland, Washington.

General Electric Company. 1953. *Hanford Works Monthly Report.* Monthly issues for June, July, November 1953. HW-28576 DEL, HW-28906 DEL, HW-30130 DEL, Richland, Washington.

General Electric Company. 1954. *Annual Report Research and Development Fiscal Year 1954.* HW-33128, Richland, Washington.

General Electric Company. 1954. *Hanford Works Monthly Report.* Monthly issues for June, September 1954. HW-32317 DEL, HW-33200 DEL, Richland, Washington.

General Electric Company. 1955. *1954 at Hanford.* HW-34000-DEL, Richland, Washington.

General Electric Company. 1955. *Hanford Works Monthly Report for December 1954.* HW-34147 DEL, Richland, Washington.

General Electric Company. 1955. *Hanford Works Monthly Report.* Monthly issues for February, May, August, September 1955. HW-35530 DEL, HW-36928 DEL, HW-38828 DEL, HW-39260 DEL, Richland, Washington.

General Electric Company. 1955. *Radiological Sciences Department Annual Report 1954.* HW-35905, Richland, Washington.

General Electric Company. 1956. *Hanford Works Monthly Report for December 1955.* HW-40692 DEL, Richland, Washington.

General Electric Company. 1957. *1956 Annual Report, Hanford Atomic Products Operation.* HW-50000-DEL, Richland, Washington.

General Electric Company. 1957. *Quarterly Progress Report Research and Development in the Field of Radiological Sciences, July-September, 1957.* HW-52866, Richland, Washington.

General Electric Company. 1958. *1957 Annual Report.* HW-54000-DEL, Richland, Washington.

General Electric Company. 1958. *Chemical Processing Department Monthly Report.* Monthly issues for March and November 1958. HW-55571 DEL, HW-58305 DEL, Richland, Washington.

General Electric Company. 1958. *Quarterly Progress Report Research and Development Activities in the Field of Radiological Sciences, October-December, 1957.* HW-54938, Richland, Washington.

General Electric Company. 1959. *Annual Report 1958.* HW-60000-DEL, Richland, Washington.

General Electric Company. 1959. *Chemical Processing Department Monthly Report for January 1959.* HW-59079 DEL, Richland, Washington.

General Electric Company. 1960. *Chemical Processing Department Monthly Report for August 1960.* HW-66646 DEL, Richland, Washington.

General Electric Company. 1961. *Annual Report 1960 Hanford Atomic Products Operation.* HW-67700-DEL, Richland, Washington.

General Electric Company. 1961. *Chemical Processing Department Monthly Report.* Monthly issues for May through November 1961. HW-69803 DEL, HW-70182 DEL, HW-70588 DEL, HW-70918 DEL, HW-71187 DEL, HW-71577 DEL, HW-71895 DEL, Richland, Washington.

General Electric Company. 1962. *1961 Annual Report Hanford Atomic Products Operation.* HW-72000-DEL, Richland, Washington.

General Electric Company. 1962. *Chemical Processing Department Monthly Report.* Monthly issues for December 1961, January, March, April, June, September 1962. HW-72154 DEL, HW-72551 DEL, HW-73193 DEL, HW-73525 DEL, HW-74151 DEL, HW-75145 DEL, Richland, Washington.

General Electric Company. 1962. *Project Proposal; Personnel Protection, Storage Hood Shielding - 234-5 Building.* HW-72643, Richland, Washington.

General Electric Company. 1963. *Chemical Processing Department Monthly Report.* Monthly issues for December 1962, April, June 1963. HW-76054 DEL, HW-77504 DEL, HW-78076 DEL, Richland, Washington.

General Electric Company. 1963. *Description of the 100-B, 100-D, 100-DR, 100-F, and 100-H Production Reactor Plants.* HW-74094, Vol. 3, Richland, Washington.

General Electric Company. 1963. *Description of the 100-KE and the 100-KW Production Reactor Plants.* HW-74095, Vol. 3, Richland, Washington.

General Electric Company. 1963. *Project Proposal, Process Waste Disposal Facility - Plutonium Reclamation Operations - Z Plant Project CAC-118.* HW-78967, Richland, Washington.

General Electric Company. 1964. *Chemical Processing Department Monthly Report.* Monthly issues for March through June 1964. HW-81620 DEL, HW-82089 DEL, HW-82526 DEL, HW-83102 DEL, Richland, Washington.

General Electric Company. 1965. *Chemical Processing Department Monthly Report for December 1964.* RL-SEP-197 DEL, Richland, Washington.

Gerber, M.S. 1996. *"Pac-Men" of the Hanford Site Await Decommissioning.* WHC-SA-2454-FP, U.S. Department of Energy, Richland, Washington.

Gibbs, L.M. 1998. *Love Canal: The Story Continues.* New Society Publications, Branford, Connecticut.

Golden Gate National Park Association. n.d. *NIKE Missile Site SF88L, Fort Barry.* Sausalito, California.

Grimm, K.G. 1961. *The Processing of Pu Scrap at Hanford.* HW-70361, General Electric Company, Richland, Washington.

Grossman, D. 1994. *A Policy History of Hanford's Atmospheric Releases.* Massachusetts Institute of Technology, Cambridge, Massachusetts.

Haddad, T.R. 1973. *Design Criteria, Mining Phase - Mining, Packaging, and Storage, Contaminated Soil Removal Facility.* ARH-2651, Atlantic Richfield Hanford Company, Richland, Washington.

Hales, P.B. 1992. "Topographies of Power: The Forced Spaces of the Manhattan Project." In *Mapping American Culture*, W. Franklin and M. Steiner (eds.), University of Iowa Press, Iowa City, Iowa.

Hall, J.C, 1972. *Engineering Study: Liquid Organic Waste Collection and Disposal.* ARH-2424, Atlantic Richfield Hanford Company, Richland, Washington.

Hanford Engineer Works. n.d. *History of Operations (1 January 1944 – 20 March 1945).* OUT-1462. Richland, Washington.

Hanford Project. 1977. *Hanford: Yesterday, Today, and Tomorrow.* Richland, Washington.

Hanlon, B.M. 1998. *Waste Tank Summary Report for Month Ending March 31, 1998.* HNF-EP-0182-120, Fluor Daniel Hanford, Inc., Richland, Washington.

Harden, B. 1996. *A River Lost: The Life and Death of the Columbia.* W.W. Norton & Company, New York.

Harlow, D.G. 1975. *Waste Handling and Plutonium Finishing Process Engineering Monthly Reports, January 1, 1975 Through December 31, 1975.* ARH-CD-236 RD, Books 1 and 2, Atlantic Richfield Hanford Company, Richland, Washington.

Harlow, D.G. 1976. *Waste Handling and Plutonium Finishing Process Engineering Monthly Reports, January 1, 1976 Through December 31, 1976.* ARH-CD-533 RD, Books 1 and 2, Atlantic Richfield Hanford Company, Richland, Washington.

Harmon, M.K. 1969. *Non-Defense Plutonium Status Report as of December 31, 1968.* ARH-1057 DEL, Atlantic Richfield Hanford Company, Richland, Washington.

Harrington, P.M. 1995. "Everyone Tied Into TEDF." *Hanford Reach,* June 5, 1995, p. 3, Westinghouse Hanford Company, Richland, Washington.

Hayes, D.F. 1957. *A Summary of Incidents Involving Radioactive Materials in Atomic Energy Activities, January-December 1956.* TID-5360, Suppl., U.S. Atomic Energy Commission, Washington, D.C.

Healy, J.W. 1950. *Health Instruments: Methods and History.* HW-89078, U.S. Department of Energy, Richland, Washington.

Healy, J.W., B.V. Anderson, H.V. Clukey, and J.K. Soldat. 1963. "Radiation Exposure to People in the Environs of a Major Production Atomic Energy Plant." *Proceeding from the Second UN Geneva Conference on the Peaceful Uses of Atomic Energy,* 18:309-318.

Heubach, E.C. 1979. *Functional Design Criteria, Project B-252.* RHO-CD-459 FDC, Rockwell Hanford Operations, Richland, Washington.

Hewlett, R.G., and F. Duncan. 1990. *Atomic Shield: A History of the United States Atomic Energy Commission, Volume II, 1947-1952.* University of California Press, Berkeley, California.

Hogan, J.J. 1980. *Rockwell Hanford Operations Production Report, July 1980.* RHO-LD-10-A-July 80, Rockwell Hanford Operations, Richland, Washington.

Hogan, J.J. 1981. *Rockwell Hanford Operations Production Report, January 1981.* RHO-LD-10-A-Jan 81, Rockwell Hanford Operations, Richland, Washington.

Hogan, J.J. 1981. *Rockwell Hanford Operations Production Report, July 1981.* RHO-LD-10-A-July 81, Rockwell Hanford Operations, Richland, Washington.

Hopkins, H.H. 1965. "Recovery of Plutonium From Polystyrene Cubes." Memo to L.M. Knights, December 1, 1965, General Electric Company, Richland, Washington.

Huginnie, A.Y. 1998. "Containment and Emancipation: Race, Class, and Gender in the Cold War West." In *The Cold War American West, 1945-1989,* K.J. Fernlund (ed.), University of New Mexico Press, Albuquerque, New Mexico.

Illinois Historic Preservation Agency. 1992. "Cold War Relics." *Historic Illinois,* Vol. 15, No. 4, December 1992.

Ingalls, W.P. 1973. "Contaminated Soil Removal Facility." Memo to L.M. Knights, January 15, 1973, Atlantic Richfield Hanford Company, Richland, Washington.

Irish, J.R. 1974. "Estimation of Costs - Waste Treatment Column Replacement." Memo to N.C. Dresser, June 7, 1974, Atlantic Richfield Hanford Company, Richland, Washington.

Irwin, C.K. 1956. *Research, Development, and Experience Resources in Radiological Sciences at Hanford Atomic Products Operation.* HW-43781, General Electric Company, Richland, Washington.

Isochem, Inc. 1966. *Chemical Processing Division Monthly Report.* Monthly issues for January through November 1966. ISO-89 DEL, ISO-143 DEL, ISO-210 DEL, ISO-276 DEL, ISO-315 DEL, ISO-365 DEL, ISO-428 DEL, ISO-476 DEL, ISO-512 DEL, ISO-563 DEL, ISO-610 DEL, Richland, Washington.

Isochem, Inc. 1967. *Chemical Processing Division Monthly Report for December 1966.* ISO-642 DEL, Richland, Washington.

Isochem, Inc. 1967. *Chemical Processing Division Monthly Report.* Monthly issues for January through July 1967," ISO-707 DEL, ISO-708 DEL, ISO-709 DEL, ISO-710 DEL, ISO-711 DEL, ISO-712 DEL, ISO-713 DEL, Richland, Washington.

Isochem, Inc. 1967. *Fabrication of Fuel Elements for the Zero Power Plutonium Reactor.* ISO-1099, Richland, Washington.

Isochem, Inc. 1967. *Z Plant Report, Task I-II & Recovery Operations, January 1, 1966 through December 31, 1966, January 1, 1967 through December 31, 1967.* ISO-57 RD, ISO-659 RD, Richland, Washington.

Joint Committee on Atomic Energy. 1960. *The Future Role of the Atomic Energy Commission Laboratories.* Congress of the United States, Washington, D.C.

Jones, T.E. 1993. *Hanford Technical Exchange Program, Process Chemistry at Hanford (Genesis of Hanford Wastes).* PNL-SA-23121 S, Pacific Northwest Laboratory, Richland, Washington.

Kaplan, L. 1992. *A Slow Motion Emergency: Radiation Health Effects and Hanford.* Brandeis University, Waltham, Massachusetts.

Kelly, J.R. 1996. *Final Report, Hanford Environmental Compliance Project 89-D-172.* WHC-SD-WO24H-PRS-001, Westinghouse Hanford Company, Richland, Washington.

Kerr, W.B., and J.M. Hay. 1952. *Production Test - 234-5: Plant Process Evaluation, Precipitation of Plutonium (IV) Oxalate.* HW-23203, General Electric Company, Richland, Washington.

Kesel, G.P., and R.B. Laws. 1966. *The Semi-Remotely Maintained Plutonium Reclamation Facility.* ISO-SA-14, Isochem, Inc., Richland, Washington.

Kingsley, R.S. 1964. *Americium Recovery, Interim Startup Flow Sheet.* HW-81236, General Electric Company, Richland, Washington.

Kinzer, J.E. 1979. *Radioactive Acid Digestion Test Unit RADTU Program Plan.* RHO-CD-697, Rockwell Hanford Operations, Richland, Washington.

Kneale, G.W., A.M. Stewart, and T.F. Mancuso. 1978. "A Re-Analysis of Data Relating to the Hanford Study of the Cancer Risks of Radiation Workers." *Late Biological Effects of Ionizing Radiation,* Vol. 1, pp. 387-412, International Atomic Energy Agency, Vienna, Austria.

Knights, L.M. 1957. *RECUPLEX Operation.* HW-48073, General Electric Company, Richland, Washington.

Knights, L.M. 1961. *Solids in RECUPLEX Waste.* HW-71997, General Electric Company, Richland, Washington.

Knights, L.M. 1966. *RECUPLEX Clean-Out, Final Status Report.* ISO-320, Isochem, Inc., Richland, Washington.

Knights, L.M. 1966. *Safety Analysis Report for Hood No. 4, Miscellaneous Treatment Facility.* ISO-371, Isochem, Inc., Richland, Washington.

Knights, L.M. 1968. "Resumption of Polystyrene Burning." Memo to R.E. Olson. Atlantic Richfield Hanford Company, Richland, Washington.

Knights, L.M. 1969. "Plutonium Recovery From 216-Z-9 Cavern." Memo to L.W. Finch, February 24, 1969, Atlantic Richfield Hanford Company, Richland, Washington.

Krich, F.A. 1977. "236-Z Process Control Plan." Memo to R.B. Gelman, November 14, 1977, Rockwell Hanford Operations, Richland, Washington.

Kuletz, V. 1998. *The Tainted Desert: Environmental and Social Ruin in the American West.* Routledge, London.

Larson, M.P. 1982. *The Plutonium Finishing Plant Program Plan.* RHO-CD-488, Rev. 5, Rockwell Hanford Operations, Richland, Washington.

Laws, R.B. 1979. *Z Ventilation and Upgrading Study.* RHO-CD-687, Rockwell Hanford Operations, Richland, Washington.

Laws, R.B., T.J. Mueller, and M.G. Wheeler. 1974. *Engineering Study, Z Plant Environs Protection Facilities.* ARH-CD-202, Atlantic Richfield Hanford Company, Richland, Washington.

Low, A.J. 1972. "Incinerator Operations Study." Memo to R.E. Olson, April 7, 1972, Atlantic Richfield Hanford Company, Richland, Washington.

Lyneis, C.A. 1969. *Chemical Processing Division, Major Program Issue: Storage of Plutonium Products and Scrap.* ARH-1490 DEL, Atlantic Richfield Hanford Company, Richland, Washington.

Manry, C.W., T.R. McKenzie, and R.E. Smith. 1974. *Hanford Production and Waste Management Master Plan.* ARH-2956, Atlantic Richfield Hanford Company, Richland, Washington.

Marshall, L.D. 1976. *Conceptual Design Report, Plutonium Finishing Plant Environs Protection Facilities.* ARH-CD-473 Rev, Atlantic Richfield Hanford Company, Richland, Washington.

Merrick, D.L. 1971. *Liquid Waste Effluents from Plutonium Finishing Facility.* ARH-2292, Atlantic Richfield Hanford Company, Richland, Washington.

Mickelson, M.L. 1950. *Annual Report of the Health Instruments Divisions.* HW-21699, General Electric Company, Richland, Washington.

Miller, R.L., and J.M. Steffes. 1986. *Fuel Storage Basins Cleanup and Stabilization Project Report.* UNI-3958, UNC Nuclear Services, Richland, Washington.

Motyka, A. 1959. "Z-12 Underground Crib." Memo to J.J. Courtney, March 11, 1959, General Electric Company, Richland, Washington.

Murphy, L.T. 1980. *Project Quality Requirements Plan for Project No. B-2185, Z Plant Laboratory Upgrade.* RHO-CD-36-B-185, Rockwell Hanford Operations, Richland, Washington.

Nassar, N.G. 1977. *236-Z Liner Replacement.* VITRO-R-488, Vitro Engineering Corp., Richland, Washington.

National Park Service. n.d. *Site SF-88.* San Francisco, California. [Online] Available URL: http://www.nikemissile.org/nike.html

National Research Council. 1987. *Safety Issues at the Defense Production Reactors: A Report to the U.S. Department of Energy.* National Academy Press, Washington, D.C.

Nelson, D.W. 1996. *PRF Deactivation Summary Plan.* WHC-SD-CP-SAR-001, Westinghouse Hanford Company, Richland, Washington.

Newell, D.M. 1955. *Z Plant Process Improvements, Tasks I, II and III.* HW-40321, General Electric Company, Richland, Washington.

Olympic Engineering Corp. 1977. *Engineering Study and Cost Estimate for Liner Repair of 236-Z Hot Cell.* Richland, Washington.

Organization for Economic Co-Operation and Development. 1996. *Trends in Nuclear Research Institutes.* Nuclear Energy Agency, Paris, France.

Oscarson, E.E., P.B. Fisk, and M.E. Muller. 1972. *Hazards Review - Plutonium Waste Concentrator, Plutonium Finishing Analytical Laboratory.* ARH-2070 Sup 1, Atlantic Richfield Hanford Company, Richland, Washington.

Pacific Northwest Laboratory. 1966. *Pacific Northwest Laboratory Annual Report for 1965 in the Physical Sciences, Volume 4: Instrumentation.* BNWL-235 v.4, Richland, Washington.

Pacific Northwest Laboratory. 1970. *Pacific Northwest Laboratory Annual Report for 1969 to the USAEC Division of Biology and Medicine, Volume 2: Physical Sciences, Part 3: Instrumentation.* BNWL-1307 Part 3, Richland, Washington.

Pacific Northwest Laboratory. 1979. *Pacific Northwest Laboratory Annual Report for 1978 to the DOE Assistant Secretary for Environment, Part 5: Environmental Assessment, Control, Health and Safety.* PNL-2850 Part 5, Richland, Washington.

Pacific Northwest Laboratory. 1981. *Pacific Northwest Laboratory Annual Report for 1980 to the DOE Assistant Secretary for Environment, Part 5: Environmental Assessment, Control, Health and Safety.* PNL-3700 Part 5, Richland, Washington.

Pacific Northwest Laboratory. 1983. *Pacific Northwest Laboratory Annual Report for 1982 to the DOE Office of Energy Research, Part 5: Environmental and Occupational Protection, Assessment, and Engineering.* PNL-4600 Part 5, Richland, Washington.

Pacific Northwest Laboratory. 1984. *Pacific Northwest Laboratory Annual Report for 1983 to the DOE Office of Energy Research, Part 5: Overview and Assessment.* PNL-5000 Part 5, Richland, Washington.

Pacific Northwest Laboratory. 1985. *Pacific Northwest Laboratory Annual Report for 1984 to the DOE Office of Energy Research, Part 5: Overview and Assessment.* PNL-5500 Part 5, Richland, Washington.

Pacific Northwest Laboratory. 1986. *Pacific Northwest Laboratory Annual Report for 1985 to the DOE Office of Energy Research, Part 4: Physical Sciences.* PNL-5750 Part 4, Richland, Washington.

Pacific Northwest Laboratory. 1986. *Pacific Northwest Laboratory Annual Report for 1985 to the DOE Office of Energy Research, Part 5: Overview and Assessment.* PNL-5750 Part 5, Richland, Washington.

Pacific Northwest Laboratory. 1987. *Pacific Northwest Laboratory Annual Report for 1986 to the DOE Office of Energy Research, Part 5: Nuclear and Operational Safety.* PNL-6100 Part 5, Richland, Washington.

Pacific Northwest Laboratory. 1988. *Pacific Northwest Laboratory Annual Report for 1987 to the DOE Office of Energy Research, Part 5: Environment, Safety, Health, and Quality Assurance.* PNL-6500 Part 5, Richland, Washington.

Pacific Northwest Laboratory. 1989. *Pacific Northwest Laboratory Annual Report for 1988 to the DOE Office of Energy Research, Part 5: Environment, Safety, Health, and Quality Assurance.* PNL-6800 Part 5, Richland, Washington.

Panesko, J.V. 1984. *Evaluation of Plutonium Finishing Plant Issues.* SD-CP-DIC-001, Rockwell Hanford Operations, Richland, Washington.

Parker, H.M. 1949. *Health Instrument Divisions: Report for Month of November 1949.* HW-15267E, General Electric Company, Richland, Washington.

Parker, H.M. 1950. *Health Instrument Divisions: Report for Month of March 1950.* HW-17410E, General Electric Company, Richland, Washington.

Parker, H.M. 1977. *The Squares of the Natural Numbers in Radiation Protection: Presented March 17, 1977.* Lauriston S. Taylor Lectures in Radiation Protection and Measurements, National Council on Radiation Protection and Measurements, Washington, D.C.

Parker, H.M., R.L. Kathren, R.W. Baalman, and W.J. Bair. 1986. *Herbert M. Parker: Publications and Other Contributions to Radiological and Health Physics.* Battelle Press, Columbus, Ohio.

Parsegian, V.L. 1965. *Industrial Management in the Atomic Age: Patterns, Restraints, Opportunities, and Trends in Government-Sponsored Technology.* Addison-EC, Washington, D.C.

Patterson, W.C. 1976. *Nuclear Power.* Penguin Books, New York.

Patterson, W.C. 1984. *The Plutonium Business and the Spread of the Bomb.* Sierra Club Books, San Francisco.

Pierick, E.G. 1954. *Final Report: The Evaluation of Carbon Dioxide as Replacement for Helium in the Machining, Briquetting and Recovery Operations, 234-5 Building.* HW-33952, General Electric Company, Richland, Washington.

Pierick, E.G. 1954. *Plutonium Metal Turnings Fire.* HW-33125, General Electric Company, Richland, Washington.

Pinto, A.P. 1977. *Functional Design Criteria, Project B-183, Z-Plant Liquid Effluent Improvements.* RHO-CD-200, Rockwell Hanford Operations, Richland, Washington.

Pinto, A.P. 1978. *Conceptual Design Report, Z-Plant Liquid Effluent Improvements, B-183a, 26 Vacuum System Modifications.* RHO-CD-246, Rockwell Hanford Operations, Richland, Washington.

Present, T. 1985. *Dear Margaret: Letters from Oak Ridge to Margaret Mead.* East Tennessee Historical Society, Knoxville, Tennessee.

Price, K.R. 1971. "The Arid Land Ecology Reserve: An Outdoor Laboratory." Presented at the Oregon Museum of Science and Industry, Portland, Oregon, BNWL-SA-4087, Pacific Northwest Laboratory, Richland, Washington.

Raile, M.N. 1972. "Status Report - Prototype Solids Handling Facilities, 236-Z & 242-Z." Memo to L.M. Knights, February 7, 1972, Atlantic Richfield Hanford Company, Richland, Washington.

Raile, M.N. 1974. P-11 *Facility Cleanup: Summary Report.* ARH-ST-106, U.S. Atomic Energy Commission, Richland, Washington.

Raile, M.N. 1975. *Demolition and Removal of Plutonium-Contaminated Facilities at Hanford.* ARH-SA-223, U.S. Energy Research and Development Administration, Richland Washington.

Rasmussen, M.J., and R.L. Crocker. 1962. *Recovery Of Plutonium From Incinerator Ashes.* HW-72285, General Electric Company, Richland, Washington.

Rees, D.M., and L.H. Taylor. 1976. *Evaluation of Nondestructive Assay System Performance on Z-9 Crib Soil Measurements, Phase I, Section II.* ARH-CD-811 Pt 2, Atlantic Richfield Hanford Company, Richland, Washington.

Richardson, B. 1998/1999. "Integrated Safety Management Policy." December 1998/January 1999, Vol. 1, *Risk Excellence Notes,* U.S. Department of Energy Center for Risk Excellence, Argonne, Illinois. [Online] Available URL: http://riskcenter.doe.gov/docs/riskxnotes/ren_v1n4/index.htm

Robkin, M.A. 1992. "Experimental Release of ^{131}I: The Green Run." *Health Physics* 62(6):487-495.

Rockwell Hanford Operations. 1979. *Contamination Resulting From a L-3 Shipping Container.* Unusual Occurrence Report 79-92, Richland, Washington.

Rockwell Hanford Operations. 1979. *Rockwell Hanford Operations Monthly Report, July 1979.* RHO-PB-SR-10-B-CP-Jul 79, Richland, Washington.

Rockwell Hanford Operations. 1979. *Ruptured Sealout Polyethylene Bag - Hood 9E - RMA Oxide Line.* Unusual Occurrence Report 79-111, Richland, Washington.

Rockwell Hanford Operations. 1979. *Skin Contamination.* Unusual Occurrence Report 79-102, Richland, Washington.

Rockwell Hanford Operations. 1980. *Rockwell Hanford Operations Monthly Report.* Monthly issues for January, July. RHO-PB-SR-10-B-CP-Jan 80, RHO-PB-SR-10-B-CP-Jul 80, Richland, Washington.

Rockwell Hanford Operations. 1980. *Z Plant Plutonium Processing Operations Training Manual.* RHO-MA-234, Richland, Washington.

Rockwell Hanford Operations. 1981. *Rockwell Hanford Operations Monthly Report.* Monthly issues for January, July. RHO-PB-SR-10-B-CP-Jan 81, RHO-PB-SR-10-B-CP-Jul 81, Richland, Washington.

Rockwell Hanford Operations. 1982. *Rockwell Hanford Operations Monthly Report.* Monthly issues for January, July. RHO-PB-SR-10-B-CP-Jan 82, RHO-PB-SR-10-B-CP-Jul 82, Richland, Washington.

Rockwell Hanford Operations. 1983. *Rockwell Hanford Operations Chemical Processing Monthly Report.* Monthly issues for January, July. RHO-PB-SR-10-B-CP-Jan 83, RHO-PB-SR-10-B-CP-Jul 83, Richland, Washington.

Rockwell Hanford Operations. 1985. *7% Overbatched Powder Pan Resulting in a Criticality Prevention Specification Violation.* Unusual Occurrence Report 85-81, Richland, Washington.

Rockwell Hanford Operations. 1985. *Cooling Coil Leak Resulting in a Criticality Specification Violation.* Unusual Occurrence Report 85-73, Richland, Washington.

Rockwell Hanford Operations. 1985. *Emergency Actions Taken Resulting in a Glove Box Criticality Posting Violation.* Unusual Occurrence Report 85-5, Richland, Washington.

Rockwell Hanford Operations. 1985. *Minor Fire in Remote Mechanical C (RMC) Line Glove Box HC-9B.* Unusual Occurrence Report 85-75, Richland, Washington.

Rockwell Hanford Operations. 1985. *Misrouting of Plutonium Feed Solution Resulting in a Criticality Specification Violation.* Unusual Occurrence Report 85-70, Richland, Washington.

Rockwell Hanford Operations. 1985. *Momentary Interruption of Fresh Air Supply Resulting in Personnel Contamination.* Unusual Occurrence Report 85-97, Richland, Washington.

Rockwell Hanford Operations. 1985. *Skin Contamination Due to a Faulty Glass Vial.* Unusual Occurrence Report 85-19, Richland, Washington.

Rockwell Hanford Operations. 1986. *Accidental Discharging of Sprinkler System in RMC Line Room 228B.* Unusual Occurrence Report 86-14, Richland, Washington.

Rockwell Hanford Operations. 1986. *Contaminated of an Instrument Technician and Rooms 263 and 228A.* Unusual Occurrence Report 86-26, Richland, Washington.

Rockwell Hanford Operations. 1986. *Contaminated Puncture Wounds to Two Rockwell Craftsmen.* Unusual Occurrence Report 86-19, Richland, Washington.

Rockwell Hanford Operations. 1986. *High Hydrogen Fluoride (HF) Levels in 9B Glove Box Exhaust System Causing Failure of 9B and 9 AB HEPA Filters.* Unusual Occurrence Report 86-15, Richland, Washington.

Rockwell Hanford Operations. 1986. *Solution in the 26" Vacuum System Resulting in a Violation.* Unusual Occurrence Report 86-40, Richland, Washington.

Rockwell Hanford Operations. 1986. *Water Routed to the Vessel Ventilation System and Subsequent Failure of a Criticality Drain to Operate as Designed.* Unusual Occurrence Report 86-44, Richland, Washington.

Roesch, W.C., C.C. Gamertsfelder, H.V. Larson, E.C. Watson, and J.M. Nielsen. 1962. *Dosimetry Investigation of the RECUPLEX Accident.* HW-SA-2730, General Electric Company, Richland, Washington.

Roos, L.C. 1958. *Construction Operation Investigation Report, Radioactive Contamination Spread.* No. 58-1R, General Electric Company, Richland, Washington.

Sawicki, J.A. n.d. *Antiaircraft Artillery Battalions of the Army.* Vols. 1-2, U.S. Army Military History Institute, Carlisle Barracks, Pennsylvania.

Schilling, A.E. 1989. *Plutonium Finishing Plant PFP Remediation Project Management Plan.* SD-CP-PMP-003, Westinghouse Hanford Company, Richland, Washington.

Schulz, W.W., E.T. Abramowski, and E.P. Horwitz. 1985. *Solvent Extraction Processing of Pu Residues and Waste Solutions, Overview of U.S. Experience, Solvent Extraction Processing at the PRF: The TRUEX Process.* RHO-RE-SA-105, Rockwell Hanford Operations, Richland, Washington.

Schwartz, S.I., ed. 1998. *Atomic Audit: The Costs and Consequences of U.S. Nuclear Weapons Since 1940.* Brookings Institution Press, Washington, D.C.

Seaborg, G.T., and D.M. Wilkes. 1964. *Education and the Atom: An Evaluation of Government's Role in Science Education and Information, Especially as Applied to Nuclear Energy.* McGraw-Hill, New York.

Seese, W.S., and G.H. Daub. 1985. *Basic Chemistry.* Prentice-Hall, Englewood Cliffs, New Jersey.

Sehmel, G.A. 1977. *Radioactive Particle Resuspension Research Experiments on the Hanford Reservation.* BNWL-2081, Pacific Northwest Laboratory, Richland, Washington.

Shaw, D.F. 1952. *RECUPLEX Installation, Building 234-5 (Project No. C-496).* Directive No. HW-279, U.S. Atomic Energy Commission, Richland, Washington.

Shaw, D.F. 1952. *RECUPLEX Installation, Building 234-5 (Project No. C-496).* Directive No. HW-279, Modification No. 1, Directive No. HW-279, Modification No. 2, Directive No. HW-279, Modification No. 4, U.S. Atomic Energy Commission, Richland, Washington.

Shaw, D.F. 1952. *Skull Recovery Facilities for Building 234-5 (Project No. C-422).* Directive No. HW-223, Modification #2, U.S. Atomic Energy Commission, Richland, Washington.

Shaw, D.F. 1953. *Activate Task I, RMA Line, Building 234-5 (Project No. CG-549).* Directive No. HW-304, Directive No. HW-304, Modification No. 1, Directive No. HW-304, Modification No. 2, U.S. Atomic Energy Commission, Richland, Washington.

Shaw, D.F. 1954. *RECUPLEX Installation, Building 234-5 (Project No. C-496).* Directive No. HW-279, Modification No. 5, Directive No. HW-279, Modification No. 6, Directive No. HW-279, Modification No. 8, U.S. Atomic Energy Commission, Richland, Washington.

Skiba, W.F. 1981. *Process Specifications for the Operation of 241-Z.* PSD-Z-184-00010, Rockwell Hanford Operations, Richland, Washington.

Sloat, R.J. 1963. "Z-9 Crib Replacement." Memo to W.J. Gartin, General Electric Company, Richland, Washington.

Smith, A.E. 1973. "Program Status - 216-Z-9 and 216-Z-1Aa Facilities." Letter to O.J. Elgert, May 14, August 13, August 21, Atlantic Richfield Hanford Company, Richland, Washington.

Smith, A.E. 1973. "Program Status - 216-Z-9 Facilities." Letter to O.J. Elgert, August 6, 1973, Atlantic Richfield Hanford Company, Richland, Washington.

Smith, A.E. 1973. *Nuclear Reactivity Evaluations of 216-Z-9 Enclosed Trench.* ARH-2915, Atlantic Richfield Hanford Company, Richland, Washington.

Smith, H.D. 1973. *Contaminated Soil Removal Facility, 216-Z-9 Enclosed Trench (Project No. HCP-687).* Directive No. HW-643, U.S. Atomic Energy Commission, Richland, Washington.

Smith, H.D. 1973. *Plutonium Effluent Cleanup, Plutonium Finishing Plant (Project No. HCP-669).* Directive No. HW-611, Modification #2, U.S. Atomic Energy Commission, Richland, Washington.

Smith, R.E., and R.T. Wilde. 1981. *Hanford Monitored Retrievable Storage Study.* RSD-BWI-ES-009, Rockwell Hanford Operations, Richland, Washington.

Smyth, H.D. 1945. *Atomic Energy for Military Purposes: The Official Report on the Development of the Atomic Bomb under the Auspices of the United States Government 1940-1945.* Princeton University, Princeton, New Jersey.

Snow, C.P. 1961. *Science and Government.* Harvard University Press, Cambridge, Massachusetts.

Snyder, D.A. 1958. *Preliminary Project Proposal: Plutonium Recovery from Contaminated Materials, 234-5 Building.* HW-56562, General Electric Company, Richland, Washington.

Snyder, D.A. 1958. *Project Proposal, Crib and Test Wells for 234-5 Building Wastes Project CAC-798.* HW-55497, General Electric Company, Richland, Washington.

Stafford, C.A. 1996. "It's Fluor." *Hanford Reach,* August 12, 1996, p. 1, Fluor Hanford, Richland, Washington.

Stannard, J.N., and R.L. Kathren. 1995. "Radiation Protection and Medical Practice with Special Reference to Health Physicists and the Health Physics Society." *Health Physics,* 69(5):837-844.

Steele, K.D. 1988. "Hanford's Bitter Legacy." *Bulletin of the Atomic Scientists* 44(1):17-23.

Steele, K.D. 1989. "Hanford: America's Nuclear Graveyard." *Bulletin of the Atomic Scientists* 45(8):15-23.

Stenner, R.D., K.H. Cramer, K.A. Higley, S.J. Jette, D.A. Lamar, A.J. McLaughlin, D.R. Sherwood, and N.C. Van Houten. 1988. *Hazard Ranking System Evaluation of CERCLA Inactive Waste Sites at Hanford.* PNL-6456, Pacific National Laboratory, Richland, Washington.

Stiffler, G.L. 1961. *The Removal of Plutonium From 234-5 Building Swamp Water.* HW-70406, General Electric Company, Richland, Washington.

Stocking, G.T. 1975. "Solidification of Z Plant Aqueous Salt Waste." Letter to F.R. Standerfer, March 13, 1975, Atlantic Richfield Hanford Company, Richland, Washington.

Stone, W.A. 1964. *Meteorological Instrumentation of the Hanford Area.* HW-62455, General Electric Company, Richland, Washington.

Theis, J.A. 1981. *Construction Design Criteria, Replacement Disposal Facilities for 216-Z-19 Ditch Influents.* RHO-CD-1405, Rockwell Hanford Operations, Richland, Washington.

Thomas, R.J. 1970. *Design Criteria: Plutonium Oxide Handling Facilities, 234-5 Building.* ARH-1848, Atlantic Richfield Hanford Company, Richland, Washington.

Travis, J.E. 1956. *Improved Task I and Task II Facilities - 234-5 Building (Project No. CG-691).* Directive No. HW-394, U.S. Atomic Energy Commission, Richland, Washington.

Travis, J.E. 1958. *Crib and Test Wells for 234-5 Building Wastes Project No. CAC-798.* Directive No. AEC-134, U.S. Atomic Energy Commission, Richland, Washington.

Travis, J.E. 1958. *Reduction of Radiation Exposure, RMA Line - 234-5 (Project No. CG-800).* Directive No. HW-458, Directive No. HW-458, Modification #1, U.S. Atomic Energy Commission, Richland, Washington.

Travis, J.E. 1958. *RMC Button Line - 234-5 Building (Project No. CG-734).* Directive No. HW-426, Modification #2, U.S. Atomic Energy Commission, Richland, Washington.

Travis, J.E. 1958. *RMC Fabrication Line - 234-5 Building (Project No. CG-745).* Directive No. HW-427, Modification #3, U.S. Atomic Energy Commission, Richland, Washington.

Travis, J.E. 1960. *Plutonium Reclamation Facility - Z Plant Project No. CAC-880.* Directive No. AEC-172, Directive No. AEC-172, Modification #1, U.S. Atomic Energy Commission, Richland, Washington.

Travis, J.E. 1961. *Waste Treatment Facility - Z-Plant (Project No. CGC-912).* Directive No. HW-522, Modification #1, Directive No. HW-522, Modification #2, U.S. Atomic Energy Commission, Richland, Washington.

Travis, J.E. 1962. *Additional Plutonium Storage Facilities - 234-5 Building (Project No. CGC-968).* Directive No. HW-545, U.S. Atomic Energy Commission, Richland, Washington.

Travis, J.E. 1962. *RMC Button Line Filtrate Handling Facility - Z-Plant (Project No. CGC-978).* Directive No. HW-542, U.S. Atomic Energy Commission, Richland, Washington.

Travis, J.E. 1963. *Plutonium Reclamation Facility - Z Plant (Project No. CAC-880).* Directive No. AEC-172, Modification #4, Directive No. AEC-172, Modification #5, Directive No. AEC-172, Modification #6, U.S. Atomic Energy Commission, Richland, Washington.

Travis, J.E. 1963. *Process Waste Disposal Facility - Plutonium Reclamation Operations - Z Plant (Project No. CAC-118).* Directive No. AEC-226, U.S. Atomic Energy Commission, Richland, Washington.

Travis, J.E. 1963. *Waste Treatment Facility - Z-Plant (Project No. CGC-912).* Directive No. HW-522, Modification #4, U.S. Atomic Energy Commission, Richland, Washington.

Tri-City Herald. 1964. *Hanford to Cut Back in 1965.* From the January 8, 1964, issue, Kennewick, Washington.

Unzicker, W.F., and R.Y. Lyon. 1964. *Z Plant Weekly Report: Task I-II and Incinerator, From January 1, 1963, to December 1, 1963.* HW-76162 RD, General Electric Company, Richland, Washington.

U.S. Atomic Energy Commission. 1959. *A Summary of Industrial Accidents in USAEC Facilities.* TID-5360, Suppl. 2, Washington, D.C.

U.S. Atomic Energy Commission. 1960. *The Future Role of the Atomic Energy Commission Laboratories, Volume II: Supplementary Materials.* Washington, D.C.

U.S. Atomic Energy Commission. 1961. *A Summary of Industrial Accidents in USAEC Facilities.* Suppl. 3, Revised, Washington, D.C.

U.S. Atomic Energy Commission. 1962. *Final Report of Accidental Nuclear Excursion, RECUPLEX Operation, 234-5 Facility.* HW-74723, Richland, Washington.

U.S. Atomic Energy Commission. 1963. *A Summary of Industrial Accidents in USAEC Facilities 1961-1962.* Suppl. 4, Washington, D.C.

U.S. Atomic Energy Commission. 1965. *A Summary of Industrial Accidents in USAEC Facilities, 1963-1964.* TID-5360, Suppl. 5, Washington, D.C.

U.S. Atomic Energy Commission. 1967. *A Summary of Industrial Accidents in USAEC Facilities 1965-1966.* TID-5360, Suppl. 6, Washington, D.C.

U.S. Atomic Energy Commission. 1967. *A Summary of Industrial Accidents in USAEC Facilities 1956.* TID-5360, Suppl. 1, Revised, Washington, D.C.

U.S. Atomic Energy Commission. 1968. *Major Activities in the Atomic Energy Programs, January-December 1967.* U.S. Government Printing Office, Washington, D.C.

U.S. Atomic Energy Commission. 1968. *Report to the Advisory Committee on Reactor Safeguards on the Matter of the Hanford N Reactor Effluent Control Program.* Division of Reactor Licensing, Washington, D.C.

U.S. Atomic Energy Commission. n.d. Index and Synopses. TID-5360, Washington, D.C.

Air Defenses of Hanford - Camp Hanford: The Forward Positions 1950 -1964. On file at the Library of the U.S. Department of Energy's Hanford Cultural Resources Laboratory, Richland, Washington.

U.S. Committee on Science and Technology. 1986. *Research Funding as an Investment: Can we Measure the Returns?* Science Policy Background Report No. 12., Washington, D.C.

U.S. Committee on Science and Technology. 1986. *The Regulatory Environment for Science.* Science Policy Background Report No. 10., Washington, D.C.

U.S. Department of Energy. 1984. *GUIDAR Security System.* Video Tape No. 1820, March 28, 1984, Lockheed Martin Services, Inc., Video Productions, Richland, Washington.

U.S. Department of Energy. 1984. *User's Guide to DOE Facilities.* DOE/ER-0174, Washington, D.C.

U.S. Department of Energy. 1986. *Capsule Review of DOE Research and Development and Field Activities.* DOE/ER-0305, Washington, D.C.

U.S. Department of Energy. 1988. *A Review of the Exploratory Research and Development Programs at the Five Multiprogram Energy Laboratories.* DOE/ER-0361, Washington, D.C.

U.S. Department of Energy. 1991. *Sanitary Waste Program.* DOE/EM-0022P, Washington, D.C.

U.S. Department of Energy. 1994. *Plutonium Working Group Report.* DOE/EH-0415, Washington, D.C.

Walmsley, M.B. 1981. *Environmental Evaluation EE, Replacement Disposal Facilities for 216-Z-19 Ditch Influents.* RHO-CD-1453, Rockwell Hanford Operations, Richland, Washington.

Walser, R.W., and R.Y. Lyon. 1966. *Z-Plant Weekly Report: Task I-II and Recovery Operations, January 1, 1965 through December 31, 1965.* RL-SEP-249 RD, General Electric Company, Richland, Washington.

Walters, F.F., and J.A. Teal. 1988. *Review of Plutonium Accumulations in Service Piping and Ductwork, Monitoring Enhancement Plan, Plutonium Finishing Plant.* SD-CP-ER-019, Westinghouse Hanford Company, Richland, Washington.

Warren, J.H. 1966. *Recovery of Pu-238 Scrap.* ISO-468, Isochem, Inc., Richland, Washington.

Washington General Support Services Contractor. 1994. *Phase 1 Environmental Site Assessment Report, Proposed Environmental and Molecular Science Laboratory Site, Richland, Washington.* Prepared for the U.S. Department of Energy, Richland Office.

Weinberg, A.M. 1994. *The First Nuclear Era: The Life and Times of a Technological Fixer.* American Institute of Physics Press, New York.

Westinghouse Hanford Company. 1989. *Trend of Degraded Performance, Inoperability of Alpha CAMs.* Unusual Occurrence Report 88-45-PFP-10, Richland, Washington.

Westinghouse Hanford Company. 1993. *300 Area Building Catalog.* Richland, Washington.

Westinghouse Hanford Company. 1995. *Plutonium Finishing Plant Final Safety Analysis Report.* WHC-SD-CP-SAR-021, Richland, Washington.

Wheeler, M.G. 1975. *Conceptual Design Report, Z Plant Environs Protection Facilities.* ARH-CD-914, Atlantic Richfield Hanford Company, Richland, Washington.

Wiggin, E.A., ed. 1961. "A Forum Report, Nuclear Frontiers 1960." *Proceedings of the Annual Conference for Members and Guests,* Atomic Industrial Forum, Inc., New York.

Wilging, J.F. 1970. "ZPPR Residue Resume." Letter to F.E. Abbott, February 12, 1970, Dow Chemical Company, Golden, Colorado.

Williams, D.G. 1967. *231-Z Building Waste Disposal Crib 216-Z-16 (Project No. AEC-290).* Directive No. AEC-290, U.S. Atomic Energy Commission, Richland, Washington.

Williams, L.C. 1994. *Project 90L-EWB-680H, Plutonium Finishing Plant PFP Liquid Low Level Waste System Modification, Official Acceptance of Construction Section III.* Correspondence #EN-94-019, U.S. Department of Energy, Richland, Washington.

Work, J.B. 1947. *234-5 Project Specification Letter 234-10, 235-10, Ventilation.* HW-08323, General Electric Company, Richland, Washington.

Work, J.B. 1947. *234-5 Project Specification Letter 234-6, 235-6 Revision 1, Waste Disposal.* HW-8254, General Electric Company, Richland, Washington.

Work, J.B. 1947. *234-5 Project Specification Letter 234-7, 235-7, Essential Material Storage & Process Solution Preparation.* HW-08272, General Electric Company, Richland, Washington.

Work, J.B. 1948. *234-5 Project Ventilation Specifications.* HW-10482, General Electric Company, Richland, Washington.

Work, J.B. 1948. *234-5 Project, Specification Letter 234-41. 235-41 (Supplement D), Chemical Development Laboratory.* HW-11321, General Electric Company, Richland, Washington.

Zachary, G.P. 1997. *Endless Frontier: Vannevar Bush, Engineer of the American Century.* The Free Press, New York.

Zorpette, G. 1996. "Hanford's Nuclear Wasteland." *Scientific American,* May 1996, pp. 88-97.

GLOSSARY

GLOSSARY

100 Areas	site of the Hanford production reactors, which include B, C, D, DR (replacement reactor), F, H, KE (K East), KW (K West), and N Reactors
105 buildings	buildings housing the production reactors
181 buildings	buildings housing the pumps that drew in water from the Columbia River
182 buildings	buildings where water drawn from the Columbia River was stored and then pumped when needed to the treatment facilities
183 buildings	buildings where water was treated prior to cooling the reactors
190 buildings	buildings where the treated water was stored and then pumped to the reactors
200 Areas	sites of the Hanford chemical separations plants, which include the bismuth phosphate process plants (B and T Plants), plutonium uranium extraction plant (PUREX/A Plant), and reduction and oxidation plant (REDOX/S Plant)
300 Area	site of the research, development and fuel fabrication operations; the 300 Area had most of Hanford's research and development activities, chemical process laboratories, and test reactors
400 Area	site of the Fast Flux Test Facility
500 Area	a non-geographic area designating the electrical lines, distribution substations, fence and road lighting, fire alarm systems, and telephones and telephone cables
600 Area	in 1990 all land within the Hanford Site not occupied by the 100, 200, 300, 400, 1100, or 3000 Areas
700 Area	enclosed (fenced) area that housed the administrative functions during the Manhattan Project and early Cold War years until the Atomic Energy Commission sold it in 1958-1959
800 Area	a non-geographic area designating the steam power plants and overhead pipe line facilities, specifically all pipe supports, steam lines, air lines, and process lines
900 Area	a non-geographic area designating the underground pipe line facilities, specifically the export water lines and valve houses, raw water and fire protection lines, sanitary sewers, process sewers, wells, and pumps
1100 Area	in 1990, site of the warehouse, vehicle maintenance, and transportation operations center; commercial district/ downtown Richland was part of the 1100 Area, except for the 700 Area, which was the administrative district for the site
3000 Area	in1990, site of engineering, construction, and research and development activities
ACHP	Advisory Council on Historic Preservation
AEA	*Atomic Energy Act* (42 USC 2011 et seq.) federal law that administers and regulates the production and use of atomic power; passed in 1946 and amended substantially in 1954 and several times since then
AEC	Atomic Energy Commission, created by the *Atomic Energy Act* in 1946 as the civilian agency responsible for the production of nuclear weapons and research and regulation of atomic energy; federal overseer of the Hanford Site from January 1, 1947 to January 19, 1975; successor to the U.S. Army Corps of Engineers and predecessor to the Energy Research and Development Agency (ERDA), now the U.S. Department of Energy
ALARA	as low as reasonably achievable
ALE	Fitzner/Eberhardt Arid Lands Ecology Reserve
background level	measured level at which the concentration of a hazardous substance is consistently present in the environment that has not been influenced by local human activities
billet - uranium	semi-finished, short, thick bar of uranium in the form of a cylinder; produced offsite from ingots and began arriving at the Hanford Site November 1944 in preparation for Hanford's capability to extrude the billets into rods which started early 1945

breeder reactor nuclear reactor that produces more fissionable material than it consumes

bus set of two or more electric conductors that serve as common connections between load circuits and each of the polarities (in direct-current systems) or phases (in alternating-current systems) of the source of electric power

button metal chunks of plutonium (resembling a hockey puck), which were the result of converting plutonium nitrate into plutonium metal

calcine process that uses heat to convert liquid, high level waste into a dry, powdery form; also the powdered waste that results from this process

canning process of sealing the fuel core in either aluminum or zirconium cladding (also called a jacket) to create the fuel element; jacketing is another term used for canning

canyon vernacular term used at the Hanford Site for the chemical separations plants, inspired by their long, high, narrow structure

CERCLA *Comprehensive Environmental Response, Compensation, and Liability Act of 1980* (42 USC 9601 et seq. as amended); federal law, enacted in 1980 and amended in 1986, that governs the cleanup of hazardous, toxic, and radioactive substances; the Act and its amendments created a trust fund, commonly known as Superfund, to finance the investigation and cleanup of releases of hazardous substances; the 1986 amendments included provisions that require the U.S. Department of Energy and other federal agencies to clean up their facilities under Federal Facility agreements with the U.S. Environmental Protection Agency

cesium element chemically similar to sodium and potassium; isotope, cesium-137, is one of the most important fission products with a half-life of about 30 years

cfs cubic feet per second

chain reaction succession of nuclear divisions; the neutrons set free in the nuclear divisions split the fissile nuclei (uranium-233, uranium-235, plutonium-239)

charging process of loading the uranium fuel elements into the process channels located in the front face of the reactor block

chemical separation process for extracting uranium, plutonium, and other radionuclides from dissolved spent nuclear fuel and irradiated targets; the fission products left behind are high level waste

cladding outer layer of metal over the fissile material of a nuclear fuel element; usually aluminum or zirconium; jacket is another term used for cladding

Clinton Engineer Works original name of the Oak Ridge Reservation

Clinton Semiworks pilot plant to produce plutonium at the Oak Ridge Reservation in Tennessee

co-extrusion process used to clad nuclear fuel elements for the N Reactor; a press extrudes uranium billets welded inside aluminum or zirconium cladding material into tubes, bonding the uranium to the cladding materials

Cold War state of tension and military rivalry that existed between the United States and the former Soviet Union and their allies from 1945 to 1990

Corps U.S. Army Corps of Engineers, federal overseer of the Hanford Site from 1943 to January 1, 1947; predecessor to the Atomic Energy Commission

council Advisory Council on Historic Preservation

CREHST The Columbia River Exhibition of History, Science, and Technology

crib underground chambers used to dispose of large volumes of low level, radioactive liquid waste, usually constructed of loosely spaced timbers several feet below ground level, creating a chamber of more than 1000 cubic feet; the liquid would percolate through the chamber to the underlying soil; numerous cribs exist in the 100, 200, and 300 Areas

critical condition in which a material undergoes nuclear fission at a self-sustaining rate

criticality event of a material undergoing nuclear fission at a self-sustaining rate

curie unit which describes the intensity (strength) of radioactivity in a material; a curie is a measure of the rate at which a radioactive material gives off particles and disintegrates; one curie gives off 37 billion disintegrations per second; a typical home smoke detector contains about 1 millionth of a curie of radioactivity

deactivation activities that ensure surplus facilities are secure in a safe and stable condition pending their ultimate disposition; includes eliminating immediate safety and environmental hazards as well as removing most contaminants within the facility

decommissioning retirement of nuclear facilities, including decontamination and/or dismantlement

decontamination removal of unwanted radioactive or hazardous contamination by a chemical or mechanical process

discharging process of irradiated fuel elements being pushed out from the rear face of the reactor block as operators loaded the reactor with new fuel elements

DOE U.S. Department of Energy; cabinet-level federal government agency responsible for nuclear weapons production, energy research, and the cleanup of hazardous and radioactive waste at its sites; federal overseer of the Hanford Site from 1977 to present; successor to the Energy Research and Development Agency

DOE-RL U.S. Department of Energy - Richland Operations Office

dose specific amount of ionizing radiation or of a toxic substance absorbed by a living being; measured in rads or rem

dose – external whole body measure of radiological impact for a situation wherein all of the organs and tissues of the body are essentially uniformly irradiated by radiation originating from a source outside of the body

dose - reference estimate established by the U.S. Environmental Protection Agency for specific chemicals (with uncertainty spanning perhaps an order of magnitude) of the daily exposure of the human population to a potential hazard that is likely to be without risk of deleterious effects during a lifetime

dose - tolerance dose which can be tolerated for a prolonged period of time without ultimately suffering injury

dosimetry measurement and calculation of radiation doses

downcomer vertical pipe that delivers fluid from above to below

Du Pont E. I. Du Pont de Nemours and Company, Inc.; prime contractor from 1942-1946 of the Hanford Site for the U. S. Army Corps of Engineers

Ecology Washington State Department of Ecology

EIS environmental impact statement

environmental contamination release into the environment of radioactive, hazardous, and toxic materials

EPA U.S. Environmental Protection Agency; established in 1970; responsible for enforcing environmental laws, including the *Resource Conservation and Recovery Act* (RCRA); the *Comprehensive Environmental Response, Compensation and Liability Act* (CERCLA); and the *Toxic Substances Control Act* (TSCA)

ERA *Energy Reorganization Act*; act signed October 11, 1974 by President Ford abolishing the Atomic Energy Commission and establishing the Energy Research and Development Administration, which was activated January 19, 1975

ERDA Energy Research and Development Agency; created in 1974 to take over the weapons production and research responsibilities of the Atomic Energy Commission; overseer of the Hanford Site from 19751977; successor to the Atomic Energy Commission and predecessor to the U.S. Department of Energy

ExHPIF Expanded Historic Property Inventory Form

Fat Man second atomic bomb used in combat by the United Stated; Fat Man was dropped on Nagasaki, Japan on August 9, 1945

FFTF Fast Flux Test Facility

fissile capable of being split by a low energy neutron; most common fissile isotopes are uranium-235 and plutonium-239

fission nuclear reaction in which the nucleus of an atom breaks up into two or more nuclei and releases energy (radiation)

French drain — shallow shaft used to dispose of low volumes of low-level, radioactive liquid waste with low activity, typically a 3-6 foot concrete pipe about 36 inches in diameter buried vertically and filled with rock or gravel; the liquid usually entered the shaft from a waste line at the top (surface) of the pipe and flowed through the pipe to the soil below; hundred of French drains are scattered throughout the Hanford production areas

fuel core — uranium part of the fuel element

fuel element — pipe cylinder that contained the nuclear fuel for the plutonium production reactors at the Hanford Site; commonly referred to as a "slug" at the Hanford Site; the process of creating a fuel element consisted of

- ingot — initial form of uranium
- billet — produced offsite from ingots and began arriving at the Hanford Site November 1944 in preparation for Hanford's capability to extrude the billets into rods which started January 1945
- rod — initially produced offsite from billets and began arriving at the Hanford Site October 1943; produced at the Hanford Site beginning January 1945
- fuel core — uranium part of the fuel element produced from rods

fuel – nuclear/reactor — natural or enriched uranium that sustains the fission chain reaction in a nuclear reactor; also known as reactor fuel

fuel- spent nuclear — irradiated fuel discharged from a reactor; the constituent elements of spent fuel have not been separated out

fusion — process whereby the nuclei of lighter elements, especially the isotopes of hydrogen (deuterium and tritium) combine to form the nucleus of a heavier element with the release of substantial amounts of energy

GE — General Electric Company

Geographic Information System — computerized system designed to efficiently capture, store, update, manipulate, analyze, and display all forms of geographically referenced information

geometrically favorable equipment — equipment built in sizes and configurations that would avoid an accumulation of plutonium large enough to cause a criticality

Green Run — secret military experiment conducted at the Hanford Site in December 1949 at the 221-T chemical separations facility; this experiment became known as the Green Run because irradiated uranium fuel was still "green" when the chemical separations process began. "Green" meant the fuel had not been aged in a holding basin for the length of time recommended by Health Instruments; the nuclear arms race fueled this experiment because the United States wanted to know if the Soviet Union was neglecting to age their irradiated fuel; if the Soviet Union was not taking the same amount of time as the United States to age the fuel, this would mean revising estimates of nuclear weapon stockpiles

HABS — Historic American Buildings Survey

HAER — Historic American Engineering Record

half-life — time required for an initial number of radioactive atoms to be reduced to half that number by radioactive decay; each isotope has its own characteristic half-life; they range from fractions of a second to billions of years

Hanford Site — from 1978-present, name of the site where the U.S. Government produced plutonium in south-central Washington State
- **Hanford Engineer Works** — name of the Hanford Site from 1943-1946
- **Hanford Works** — name of the Hanford Site from 1947-1974
- **Hanford Reservation** — name of the Hanford Site from 1975-1977

Hanford Reach — segment of the Columbia River that extends 85 kilometers (51 miles) downstream from Priest Rapids Dam to the head of the McNary Pool near the city of Richland, Washington

HAPO — Hanford Atomic Products Operation; term used by the General Electric Company for its internal corporate division that managed the Hanford Site; not to be confused with Hanford Works, which was the name of the site when the General Electric Company managed it

HCRMP — Hanford Cultural Resources Management Plan

heavy metals — metallic elements with high atomic weights (for example, arsenic, lead, mercury, plutonium, and uranium) that can damage living organisms at low concentrations and tend to accumulate in the food chain

HEDR — Hanford Environmental Dose Reconstruction - scientific study to estimate radiation doses the public may have received as a result of these releases

HEIS Hanford Environmental Information System; an electronic database that consolidates the data gathered during environmental monitoring and restoration of the Hanford Site

heritage tourism visitation of sites of historic or cultural importance

HEW Hanford Engineer Works - name of the Hanford Site from 1943-1946

HGIS Hanford Geographic Information System database

historic district significant concentration, linkage, or continuity of sites, buildings, structures, or objects united historically or aesthetically by plan or physical development

historic property any pre-historic or historic building, structure, object, site, or district included in or eligible for inclusion in the National Register of Historic Places

holdup time length of time a parent radionuclide spends in the reactor core, usually expressed in seconds; also length of time between harvest and consumption of food products

hoods term workers used for glove boxes

HPIF Historic Property Inventory Form

HW Hanford Works; name of the Hanford Site from 1947-1975

ICRP International Commission on Radiological Protection

industrial hygiene science and art devoted to recognition, evaluation, and control of those environmental factors or stresses, arising in or from the work place, which may cause sickness, impaired health and well being, or significant discomfort and inefficiency among workers or among citizens of the community

ingot initial form of mined uranium

initiator device that produces a timed burst of neutrons to initiate a fission chain reaction in a nuclear weapon; initiators made of polonium-210 and beryllium were located at the center of the fissile cores of early atomic weapons

ionization process by which a neutral atom or molecule loses or gains electrons, thereby acquiring a net charge and becoming an ion

irradiation exposure of an object to radiation usually in a nuclear reactor; targets are irradiated to produce isotopes

isotope form of the same chemical element that differs only by the number of neutrons in its nucleus; most elements have more than one naturally occurring isotope; many more isotopes have been produced in nuclear reactors and accelerators

jacket outer layer of metal over the fissile material of a nuclear fuel element; usually aluminum or zirconium (also called cladding)

jacketing process of sealing the fuel core in either aluminum or zirconium cladding (also called a jacket) to create the fuel element; canning is another term used for jacketing

Little Boy first atomic bomb used in combat by the United States; Little Boy was dropped on Hiroshima, Japan on August 6, 1945

lockdown activity where gates were closed and all personnel searched before exiting facility

Manhattan Project U.S. Government project, named for the Manhattan Engineer District, that produced the first atomic bombs during World War II; started in 1942, the Manhattan Project formally ended in 1946; the Hanford Site, Oak Ridge Reservation, and the Los Alamos National Laboratory were created for this effort

MED Manhattan Engineer District established August 1942; name given by the U.S. Army Corps of Engineers to their organization that oversaw the design, production, and testing of the first nuclear weapons; January 1, 1947, the district transferred authority over nuclear weapons stewardship to the civilian authority of the newly established Atomic Energy Commission

meltdown accident in a nuclear reactor in which melting of the fuel core occurs

mill tailings sand-like materials left over from separating uranium from its ore; more than 99 percent of the ore becomes tailings, which typically contain about 85 percent of the radioactivity present in unprocessed ore

mitigation actions that limit or compensate for the damage an undertaking does to historic properties

M.R.E.P. unit of measurement of dose equivalence; REP is an obsolete unit of absorbed dose in tissue; originally derived from roentgen equivalent physical and is approximately equivalent to the rem

NARA National Archives and Records Administration

NCRP National Council on Radiation Protection and Measurement

NEPA *National Environmental Policy Act of 1969* (42 USC 4321 et seq. as amended); federal law requiring the federal government to consider the environmental impacts of and alternatives to major proposed actions in its decision making process

neutron massive, uncharged particle that comprises part of an atomic nucleus; uranium and plutonium atoms fission when they absorb neutrons; the chain reactions that make nuclear reactors and weapons work thus depend on neutrons; man-made elements can be manufactured by bombarding other elements with neutrons in production reactors

NHPA *National Historic Preservation Act* (16 USC 470 et seq. as amended); passed in 1966 directing federal agencies to assume responsibility for all cultural resources under their jurisdiction; Section 110 requires agencies to survey the lands under their control and evaluate all historic properties for eligibility for the National Register of Historic Places; Section 106 requires agencies to consider the effects of their actions on properties listed, or eligible for listing, in the National Register

NPDES National Pollution Discharge Elimination System

NRC Nuclear Regulatory Commission; an independent agency of the federal government created by the *Energy Reorganization Act* of 1974, which abolished the Atomic Energy Commission and transferred its regulatory function to the NRC

nuclear materials All materials so designated by the Department of Energy. Presently, these materials are : depleted uranium, enriched uranium, americium-241, americium-243, curium, berkelium, californium, plutonium 238-242, lithium-6, uranium-233, normal uranium, neptunium-237, deuterium, tritium, thorium.

nuclear weapons complex chain of foundries, uranium enrichment plants, nuclear reactors, chemical separation plants, factories, laboratories, assembly plants, and test sites that produce nuclear weapons

nucleus cluster of protons and neutrons at the center of an atom that determines its identify and chemical and nuclear properties

NWPA *Nuclear Waste Policy Act of 1982* (Public Law 97-425); federal law that provides for the development of geologic repositories for disposal of high level waste and spent nuclear fuel and establishes a program of research, development, and demonstration regarding disposal of high level waste and spent nuclear fuel

OSRD Office of Scientific Research and Development

PA *Programmatic Agreement Among the U.S. Department of Energy-Richland Operations Office, the Advisory Council on Historic Preservation, and the Washington State Historic Preservation Office for the Maintenance, Deactivation, Alteration, and Demolition of the Built Environment on the Hanford Site, Washington*; document (DOE/RL-96-77) published by the U.S. Department of Energy in 1996

PFP plutonium finishing plant

PFPP Plutonium Fabrication Pilot Plant

PRCF Plutonium Recycle Critical Facility

PRTR Plutonium Recycle Test Reactor

pH term used to indicate the acidity and alkalinity of a solution; pH 7, the value for pure water, is regarded as neutral; pH values from 0-7 indicate acidity and pH values from 7-14 indicate alkalinity

PIC Pocket Ionization Chambers

pile early term for plutonium production reactors

pit central core of the primary stage of a nuclear weapon fabricated from plutonium metal buttons by casting, grinding, and shaping the buttons into forms needed for weapons

plume volume of air, soil, or water containing contaminants released from a contaminant source

plutonium fabricated fissile element; pure plutonium is a silvery metal heavier than lead; material rich in plutonium-239 isotope is preferred for manufacturing nuclear weapons; the half-life of plutonium-239 is 24,000 years

plutonium – fuel-grade plutonium containing more than 7 percent plutonium-240 isotope by mass

plutonium – weapons-grade plutonium containing 93 percent plutonium-239 isotope by mass

PNNL Pacific Northwest National Laboratory

postum early samples of plutonium made in B Reactor

production operations activities connected with the production reactors in the 100 Areas (B, C, D, DR, F, H, KE, KW, or N reactors) in which uranium or other fuel was irradiated with neutrons to produce radioactive materials; used primarily at the Hanford Site to produce plutonium for weapons; used also for research

proton positively charged particle which, in conjunction with the neutron, forms all atomic nuclei

PUREX plutonium-uranium extraction; chemical separations plant (A Plant/202-A Building at the Hanford Site), which operated from 1955-1972 and 1983-1988

rad radiation absorbed dose, unit of measurement used to describe absorbed dose

radiation energy transferred through space or other media in the form of particles or waves; certain radiation types are capable of breaking up atoms or molecules; splitting or decay of unstable atoms emits ionizing radiation

radiation – gamma high energy, highly penetrating electromagnetic radiation emitted in the radioactive decay of many radionuclides; gamma rays are similar to x-rays

radiation – ionizing particles or photons that have sufficient energy to produce ionization directly in their passage through a substance; also particles that are capable of nuclear interactions in which sufficient energy is released to produce ionization

radioactivity spontaneous emission of radiation (alpha, beta, gamma rays, and/or neutrons) by some nuclides as they transform into other nuclides

radionuclide radioactive isotope of an element; for example, strontium-90 and tritium are radionuclides of elements of strontium and hydrogen, respectively

RCRA *Resource Conservation and Recovery Act of 1976* (42 USC 6901 et seq. as amended, Public Law 94-580); federal law enacted in 1976 to address the treatment, storage, and disposal of hazardous waste

reactor – graphite nuclear reactor using graphite blocks surrounding the nuclear fuel to slow the neutrons to low energy so that self-sustaining chain reaction is achieved; the first nuclear reactors built at the Hanford Site were graphite reactors

reactor – nuclear device that sustains a controlled nuclear fission chain reaction

reactor - power nuclear reactor designed to produce heat for conversion into electrical energy or mechanical propulsion

reactor – production nuclear reactor designed to produce fabricated isotopes; plutonium and tritium are made in production reactors; the United States has 14 such reactors, 9 at the Hanford Site and 5 at the Savannah River Site; all have been closed

reactor – research class of nuclear reactors used for research on nuclear physics, reactor materials and design, and nuclear medicine; some research reactors also produce isotopes for industrial and medical use

RECUPLEX RECovery of Uranium and PLutonium by EXtraction process

REDOX reduction oxidation; chemical separations plant (S Plant/202-S Building at the Hanford Site), which operated from 1952-1967

release discharge of a substance into the environment

release factor ratio of amount released to the amount processed

rem roentgen equivalent man; unit of measurement used to describe dose equivalence; the amount of any type of radiation that produces the same biological effect as is obtained from an absorbed dose of 1 rad of 200 kilovolt potential x-rays

retention basin	excavated and lined area used to hold contaminated fluids until radioactive decay reduced activities to levels permissible for release
retention time	length of time effluent water was held in a retention basin before discharge to the Columbia River; normally expressed in hours
reverse wells	deep, well-like shafts with perforated casings, which were drilled to a depth above groundwater; use of reverse wells was discontinued after about a year and no new ones were constructed
RG Line	Rubber Glove Line; initial processing line in the Plutonium Finishing Plant where rubber gloves were used inside a protective box to manually manipulate the plutonium
RHA	Hanford Site Records Holding Area
Richland Village	planned community of Richland known as the Hanford Engineer Works Village or Richland Village; the village included all of the original town of Richland and the new town designed and built by Du Pont and its architect G. Albin Pehrson
RMA Line	Remote Mechanical A Line; second processing line in the Plutonium Finishing Plant, which used a remote process to mechanically manipulate the plutonium
RMC Line	Remote Mechanical C Line; third processing line in the Plutonium Finishing Plant, which used a remote process to mechanically manipulate the plutonium
rod	extruded billet; initially produced offsite, arriving at the Hanford Site October 1943; produced at the Hanford Site beginning early 1945
roentgen	unit of exposure of ionizing radiation that produces a charge of 1 coulomb of electric charge per kilogram of dry air
S&C	slag and crucible
salt cake	cake of dry crystals of radionuclides found in high level waste tanks
SCRAM	safety control rod ax man; initially a manual process, later an automatic insertion of control and/or safety rods to shut down production reactors
semiworks	pilot plant or small scale plant sufficient to test the actual operating capabilities of a new technique or process
SHPO	State Historic Preservation Officer
single-pass reactors	water cooled nuclear reactors which discharge their cooling water after a single use rather than recirculating it; the first eight production reactors at the Hanford Site were single-pass reactors; N Reactor, the ninth reactor, had a water recirculation system.
slug	commonly used term at the Hanford Site for fuel element
SNM	special nuclear material; plutonium, uranium-233, or uranium enriched in the isotope 235, and any other material which, pursuant to the provisions of Section 51 of the Atomic Energy Act of 1954, as amended, has been determined to be special nuclear material, but which does not include source material; or it also includes any material artificially enriched by any of the foregoing, not including source material
stack	tall chimney that was the primary release point of exhaust air from a reactor or separations plant building
storage basin	water-filled facility for holding irradiated reactor fuels with the water acting as a shield
strike	addition of hydrogen peroxide to further separate plutonium from its carrier solutions
supernatant liquid	liquid above settled solids, as in a gravity separator
Supply System	Washington Public Power Supply System
tank farm	complex of underground vessels used to store waste materials
target	material placed in a nuclear reactor to be bombarded with neutrons to produce radioactive materials; uranium-238 targets are used to make plutonium; lithium targets are used to make tritium; thoria targets are used to make fissionable uranium-233 for nuclear weapons

TLD	thermoluminescent dosimeters
traditional cultural site	any site included in or eligible for inclusion in the National Register because of its association with cultural practices or beliefs of a living community that is rooted in that community's history and is important in maintaining the continuing cultural identity of the community
traditional religious leader	anyone who is recognized by members of that Indian tribe as being responsible for performing cultural duties relating to the ceremonial or religious traditions of that group or exercising a leadership role in an Indian tribe based on the group's culture, ceremonial, or religious practices
transmutation	process whereby one nuclide changes or is changed into another, usually by the addition of nuclear particles
transuranic elements	all elements beyond uranium on the periodic table, including neptunium, plutonium, americium, and curium; all transuranic elements are fabricated
Trinity Test	test of first atomic bomb July 16, 1945 at what was then called the Alamogordo Bombing Range and today is called the White Sands Missile Range
Tri-Party agencies	three government agencies (U.S. Department of Energy, U.S. Environmental Protection Agency, and the Washington State Department of Ecology) that are signatories to the Tri-Party Agreement
Tri-Party Agreement	Federal Facility Compliance Agreement signed in 1989 by the U.S. Department of Energy, the State of Washington, and the U.S. Environmental Protection Agency; commonly known as the Tri-Party Agreement
triple dip	first process used to clad reactor fuel at the Hanford Site; process involves successive baths of molten bronze, tin, and aluminum-silicon mixture
tritium	heaviest isotope of the element hydrogen; tritium is produced in nuclear reactors and is three times heavier than ordinary hydrogen; tritium gas is used to boost the explosive power of most modern nuclear weapons; tritium has a half-life of approximately 12 years
TSCA	*Toxic Substances Control Act* (Public Law 94-469); federal law enacted in 1976 to protect human health and the environment from unreasonable risk caused by exposure to or the manufacturing, distribution, use, or disposal of substances containing toxic chemicals
uranium – natural	naturally occurring uranium (0.7 percent uranium-235 and 99.3 percent uranium-238), which is naturally slightly radioactive and can be refined to a heavy metal more dense than lead
uranium - highly enriched	uranium with more than 20 percent of the uranium-235 isotope, used for making nuclear weapons and also as fuel for some isotope production, research, and power reactors; weapons-grade uranium is a subset of this group
uranium – low enriched	uranium enriched until it consists of about 3 percent uranium-235 and 97 percent uranium-238; used as nuclear reactor fuel
uranium – mining, milling, and refining	process where uranium ore is extracted (mined) from the earth's crust and chemically processed (milled) to prepare uranium concentrate (U_3O_8), sometimes called uranium octaoxide or yellow cake; uranium concentrate is chemically converted (refined) to purify it into the form needed as feed
uranium – weapon-grade	uranium made up of over 90 percent of the fissile uranium-235 isotope
USACE	U.S. Army Corps of Engineers
USGS	United States Geological Survey
waste – hazardous	non-radioactive materials, such as lead and other heavy metals, solvents, paints, oils, other hazardous organic materials, or materials that exhibit certain characteristics of ignitability, corrosivity, or reactivity; defined under RCRA and its implementing regulations in title 40 of the Code of Federal Regulations, Parts 260-279, and corresponding state regulations
waste – high level	highly radioactive material resulting from the reprocessing of spent nuclear fuel, including liquid waste produced directly in reprocessing and any solid material derived from such liquid waste that contains fission products in sufficient concentrations; the term is used to indicate the source of the waste not the level of radioactivity; low level waste can be more radioactive than high level waste

waste – low level any radioactive material that is not spent fuel, high level, transuranic waste, or the tailings or waste produced by the extraction or concentration of uranium or thorium from any ore processed primarily for its source material

waste – mixed material containing both chemically hazardous waste, as defined under RCRA, and source, special nuclear, or byproduct materials as defined under the *Atomic Energy Act*

waste – sanitary materials that does not contain radioactive or hazardous constituents sufficient to require special management; sanitary waste includes municipal solid waste, construction/demolition debris, and some waste water

waste – transuranic materials contaminated with uranium-233 or transuranic elements having half-lives of over 20 years in concentrations more than 1 ten-millionth of a curie per gram of waste

WHC Westinghouse Hanford Company

WIDS Waste Information Data Systems

yellow cake common uranium compound (U_3O_8) named for its typical color; uranium is sent from the uranium mill to the refinery in this form

INDEX

INDEX

f = figure, s = sidebar box, t = table

A

f = figure, s = sidebar box, t = table

B

f = figure, s = sidebar box, t = table

C

f = figure, s = sidebar box, t = table

D

f = figure, s = sidebar box, t = table

E

F

f = figure, s = sidebar box, t = table

G

f = figure, s = sidebar box, t = table

H

f = figure, s = sidebar box, t = table

I

J

K

L

f = figure, s = sidebar box, t = table

M

f = figure, s = sidebar box, t = table

N

O

P

f = figure, s = sidebar box, t = table

f = figure, s = sidebar box, t = table

Q

R

f = figure, s = sidebar box, t = table

S

f = figure, s = sidebar box, t = table

f = figure, s = sidebar box, t = table

T

U

f = figure, s = sidebar box, t = table

V

W

f = figure, s = sidebar box, t = table

X

Y

Z

f = figure, s = sidebar box, t = table